Introduction to Robust Estimation and Hypothesis Testing

Introduction to Robust Estimation and Hypothesis Testing

3rd Edition

Rand Wilcox

AMSTERDAM • BOSTON • HEIDELBERG • LONDON
NEW YORK • OXFORD • PARIS • SAN DIEGO
SAN FRANCISCO • SINGAPORE • SYDNEY • TOKYO

Academic Press is an imprint of Elsevier

Academic Press is an imprint of Elsevier
225 Wyman Street, Waltham, MA 02451, USA
525 B Street, Suite 1900, San Diego, CA 92101-4495, USA
The Boulevard, Langford Lane, Kidlington, Oxford OX5 1GB, UK
Radarweg 29, PO Box 211, 1000 AE Amsterdam, The Netherlands

First edition 2012

Notice

No responsibility is assumed by the publisher for any injury and/or damage to persons or property as a matter of products liability, negligence or otherwise, or from any use or operation of any methods, products, instructions or ideas contained in the material herein. Because of rapid advances in the medical sciences, in particular, independent verification of diagnoses and drug dosages should be made.

Library of Congress Cataloging-in-Publication Data
A catalog record for this book is available from the Library of Congress.
Application submitted.

British Library Cataloguing-in-Publication Data
A catalogue record for this book is available from the British Library.

ISBN: 978-0-12-386983-8

For information on all Academic Press publications
visit our web site at *www.elsevierdirect.com*

Typeset by: diacriTech, India

Printed and bound in USA
12 13 14 15   10 9 8 7 6 5 4 3 2 1

Contents

Preface

This book focuses on the practical aspects of modern and robust statistical methods. The increased accuracy and power of modern methods, versus conventional approaches to the analysis of variance (ANOVA) and regression, is remarkable. Through a combination of theoretical developments, improved and more flexible statistical methods, and the power of the computer, it is now possible to address problems with standard methods that seemed insurmountable only a few years ago.

The most common approach when comparing two or more groups is to compare means, assuming that observations have normal distributions. When comparing independent groups, it is further assumed that distributions have a common variance. Conventional wisdom is that these standard ANOVA methods are robust to violations of assumptions. This view is based in large part on studies, published before the year 1960, showing that if groups do not differ (meaning that they have identical distributions), then good control over the probability of a type I error is achieved. However, if groups differ, hundreds of more recent journal articles have described serious practical problems with standard techniques and how these problems might be addressed. One concern is that the sample mean can have a relatively large standard error under slight departures from normality. This in turn can mean low power. Another problem is that probability coverage, based on conventional methods for constructing confidence intervals, can be substantially different from the nominal level, and undesirable power properties arise as well. In particular, power can go down as the difference between the means gets large. The result is that important differences between groups are often missed, and the magnitude of the difference is poorly characterized. Put another way, groups probably differ when null hypotheses are rejected with standard methods, but in many situations, standard methods are the least likely to find a difference, and they offer a poor summary of how groups differ and the magnitude of the difference. Yet another fundamental concern is that the *population* mean and variance are not robust, roughly mean that under arbitrarily small shifts from normality, their values can be substantially altered and potentially mislead. Thus, even with arbitrarily large sample sizes, the sample mean and variance might provide an unsatisfactory summary of the data.

When dealing with regression, the situation is even worse. That is, there are even more ways in which analyses, based on conventional assumptions, can be misleading. The very foundation of standard regression methods, namely estimation via the least squares principle, leads to practical problems, as do violations of other standard assumptions. For example, if the error term in the standard linear model has a normal distribution, but is heteroscedastic, the least squares estimator can be highly inefficient, and the conventional confidence interval for the regression parameters can be extremely inaccurate.

In 1960, it was unclear how to formally develop solutions to the many problems that have been identified. It was the theory of robustness developed by P. Huber and F. Hampel that paved the road for finding practical solutions. Today, there are many asymptotically correct ways of substantially improving on standard ANOVA and regression methods. That is, they converge to the correct answer as the sample sizes get large, but simulation studies have shown that when sample sizes are small, not all methods should be used. Moreover, for many methods, it remains unclear how large the sample sizes must be before reasonably accurate results are obtained. One of the goals of this book is to identify those methods that perform well in simulation studies, as well as those that do not.

This book does not provide an encyclopedic description of all the robust methods that might be used. Although some methods are excluded because they perform poorly relative to others, many methods have not been examined in simulation studies, so their practical value remains unknown. Indeed, there are so many methods, a massive effort is needed to evaluate them. Moreover, some methods are difficult to study with current computer technology. That is, they require so much execution time that simulations remain impractical. Of course, this might change in the near future, but what is needed now is a description of modern robust methods that have practical value in applied work.

Although the goal is to focus on the applied aspects of robust methods, it is important to discuss the foundations of modern methods, so this is done in Chapters 2 and 3, and to some extent in Chapter 4. One general point is that modern methods have a solid mathematical foundation. Another goal is to impart the general flavor and aims of robust methods. This is important because misconceptions are rampant. For example, some individuals firmly believe that one of the goals of modern robust methods is to find better ways of estimating μ, the population mean. From a robust point of view, this goal is not remotely relevant, and it is important to understand why. Another misconception is that robust methods only perform well when distributions are symmetric. In fact, both theory and simulations indicate that robust methods offer an advantage over standard methods when distributions are skewed.

A practical concern is applying the methods described in this book. Many of the recommended methods have been developed in only the last few years and are not available in standard statistical packages for the computer. To deal with this problem, easy-to-use R functions are supplied. (With this third edition, S-PLUS functions are no longer supported.) They can be obtained as indicated in Section 1.8 of Chapter 1. With one R command, all of the functions described in this book become a part of your version of R. Illustrations, using these functions, are included.

The book assumes that the reader has had an introductory statistics course. That is, all that is required is some knowledge about the basics of ANOVA, hypothesis testing, and regression. The foundations of robust methods, described in Chapter 2, are written at a relatively nontechnical level, but the exposition is much more technical than the rest of the book, and it might be too technical for some readers. It is recommended that Chapter 2 be read or at least skimmed, but those who are willing to accept certain results can skip to Chapter 3. One of the main points in Chapter 2 is that the robust measures of location and scale that are used are not arbitrary, but were chosen to satisfy specific criteria. Moreover, these criteria eliminate from consideration the population mean, variance, and the usual correlation coefficient.

From an applied point of view, Chapters 4–11, which include methods for addressing common problems in ANOVA and regression, form the heart of the book. Technical details are kept to a minimum. The goal is to provide a simple description of the best methods available, based on theoretical and simulation studies, and to provide advice on which methods to use. Usually, no single method dominates all others, one reason being that there are multiple criteria for judging a particular technique. Accordingly, the relative merits of the various methods are discussed. Although no single method dominates, standard methods are typically the least satisfactory, and many alternative methods can be eliminated.

CHAPTER 1

Introduction

Introductory statistics courses describe methods for computing confidence intervals and testing hypotheses about means and regression parameters based on the assumption that observations are randomly sampled from normal distributions. When comparing independent groups, standard methods also assume that groups have a common variance, even when the means are unequal, and a similar homogeneity of variance assumption is made when testing hypotheses about regression parameters. Currently, these methods form the backbone of most applied research. There is, however, a serious practical problem: Many journal articles have illustrated that these standard methods can be highly unsatisfactory. Often the result is a poor understanding of how groups differ and the magnitude of the difference. Power can be relatively low compared to recently developed methods, least squares regression can yield a highly misleading summary of how two or more random variables are related as can the usual correlation coefficient, the probability coverage of standard methods for computing confidence intervals can differ substantially from the nominal value, and the usual sample variance can give a distorted view of the amount of dispersion among a population of participants. Even the population mean, if it could be determined exactly, can give a distorted view of what the typical participant is like.

Although the problems just described are well known in the statistics literature, many textbooks written for nonstatisticians still claim that standard techniques are completely satisfactory. Consequently, it is important to review the problems that can arise and why these problems were missed for so many years. As will become evident, several pieces of misinformation have become part of statistical folklore resulting in a false sense of security when using standard statistical techniques.

1.1 Problems with Assuming Normality

To begin, distributions are never normal. For some this seems obvious, hardly worth mentioning, but an aphorism given by Cramér (1946) and attributed to the mathematician Poincaré remains relevant: "Everyone believes in the [normal] law of errors, the experimenters because they think it is a mathematical theorem, the mathematicians because they think it is an experimental fact." Granted, the normal distribution is the most important distribution in all aspects of statistics. But in terms of approximating the distribution of any continuous distribution, it can fail to the point that practical problems arise, as will become evident at numerous points in this book. To believe in the normal distribution implies that only two numbers are required to tell us everything about the probabilities associated with a random variable: the population mean μ and population variance σ^2. Moreover, assuming normality implies that distributions must be symmetric.

Of course, nonnormality is not, by itself, a disaster. Perhaps a normal distribution provides a good approximation of most distributions that arise in practice, and there is the central limit theorem, which tells us that under random sampling, as the sample size gets large, the limiting distribution of the sample mean is normal. Unfortunately, even when a normal distribution provides a good approximation to the actual distribution being studied (as measured by the Kolmogorov distance function described later) practical problems arise. Also, empirical investigations indicate that departures from normality, that have practical importance, are rather common in applied work (e.g., Hill & Dixon, 1982; Micceri, 1989; Wilcox, 2009a). Even over a century ago, Karl Pearson and other researchers were concerned about the assumption that observations follow a normal distribution (e.g., Hand, 1998, p. 649). In particular, distributions can be highly skewed, they can have heavy tails (tails that are thicker than a normal distribution), and random samples often have outliers (unusually large or small values among a sample of observations). Outliers and heavy-tailed distributions are serious practical problems because they inflate the standard error of the sample mean, so power can be relatively low when comparing groups. Modern robust methods provide an effective way of dealing with this problem. Fisher (1922), for example, was aware that the sample mean could be inefficient under slight departures from normality.

A classic way of illustrating the effects of slight departures from normality is with the *contaminated* or *mixed normal* distribution (Tukey, 1960). Let X be a standard normal random variable having distribution $\Phi(x) = P(X \leq x)$. Then for any constant $K > 0$, $\Phi(x/K)$ is a normal distribution with standard deviation K. Let ϵ be any constant, $0 \leq \epsilon \leq 1$. The *contaminated normal* distribution is

$$H(x) = (1 - \epsilon)\Phi(x) + \epsilon\Phi(x/K), \tag{1.1}$$

which has mean 0 and variance $1 - \epsilon + \epsilon K^2$. (Stigler, 1973, finds that the use of the contaminated normal dates back at least to Newcomb, 1896.) In other words, the contaminated normal arises by sampling from a standard normal distribution with probability $1 - \epsilon$; otherwise, sampling is from a normal distribution with mean 0 and standard deviation K.

To provide a more concrete example, consider the population of all adults, and suppose that 10% of all adults are at least 70 years old. Of course, individuals at least 70 years old might have a different distribution from the rest of the population. For instance, individuals under the age of 70 might have a standard normal distribution, but individuals at least 70 years old might have a normal distribution with mean 0 and standard deviation 10. Then, the entire population of adults has a contaminated normal distribution with $\epsilon = .1$ and $K = 10$. In symbols, the resulting distribution is

$$H(x) = 0.9\Phi(x) + 0.1\Phi(x/10), \qquad (1.2)$$

which has mean 0 and variance 10.9. Moreover, Eq. (1.2) is not a normal distribution, verification of which is left as an exercise.

To illustrate problems that arise under slight departures from normality, we first examine Eq. (1.2) more closely. Figure 1.1 shows the standard normal and the contaminated normal probability density function corresponding to Eq. (1.2). Notice that the tails of the contaminated normal are above the tails of the normal, so the contaminated normal is said to have heavy tails. It might seem that the normal distribution provides a good approximation of the contaminated normal, but there is an important difference. The standard normal has variance 1, but the contaminated normal has variance 10.9. The reason for the seemingly large difference between the variances is that σ^2 is very sensitive to the tails of a distribution. In essence, a small proportion of the population of participants can have an inordinately large effect on its value. Put another way, even when the variance is

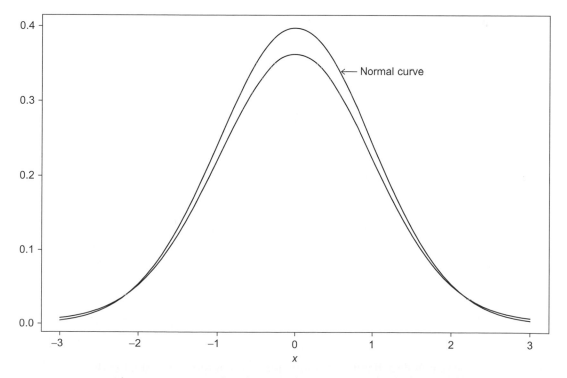

Figure 1.1: Normal and contaminated normal distributions.

known, if sampling is from the contaminated normal, the length of the standard confidence interval for the population mean, μ, will be over three times longer than it would be when sampling from the standard normal distribution instead. What is important from a practical point of view is that there are location estimators other than the sample mean that have standard errors that are substantially less affected by heavy tailed distributions. By "measure of location," it is meant that some measure intended to represent the typical participant or object, the two best-known examples being the mean and the median. (A more formal definition is given in Chapter 2.) Some of these measures have relatively short confidence intervals when distributions have a heavy tail, yet the length of the confidence interval remains reasonably short when sampling from a normal distribution instead. Put another way, there are methods for testing hypotheses that have good power under normality, but that continue to have good power when distributions are nonnormal, in contrast to methods based on means. For example, when sampling from the contaminated normal given by Eq. (1.2), both Welch's and Student's method for comparing the means of two independent groups have power approximately 0.278 when testing at the 0.05 level with equal sample sizes of 25 and when the difference between the means is 1. In contrast, several other methods, described in Chapter 5, have power exceeding 0.7.

In an attempt to salvage the sample mean, it might be argued that in some sense the contaminated normal represents an extreme departure from normality. The extreme quantiles of the two distributions do differ substantially, but based on various measures of the difference between two distributions, they are very similar as suggested by Figure 1.1. For example, the *Kolmogorov distance* between any two distributions, F and G, is the maximum value of

$$\Delta(x) = |F(x) - G(x)|,$$

the maximum being taken over all possible values of x. (If the maximum does not exist, the supremum or least upper bound is used.) If distributions are identical, the Kolmogorov distance is 0, and its maximum possible value is 1, as is evident. Now consider the Kolmogorov distance between the contaminated normal distribution, $H(x)$, given by (1.2), and the standard normal distribution, $\Phi(x)$. It can be seen that $\Delta(x)$ does not exceed 0.04 for any x. That is, based on a Kolmogorov distance function, the two distributions are similar. Several alternative methods are often used to measure the difference between distributions. (Some of these are discussed by Huber and Ronchetti, 2009.) The choice among these measures is of interest when dealing with theoretical issues, but these issues go beyond the scope of this book. Suffice it to say that the difference between the normal and contaminated normal is again small. Gleason (1993) discusses the difference between the normal and contaminated normal from a different perspective and also concludes that the difference is small.

Even if it could be concluded that the contaminated normal represents a large departure from normality, concerns over the sample mean would persist, for reasons already given.

In particular, there are measures of location having standard errors similar in magnitude to the standard error of the sample mean when sampling from normal distributions, but that have relatively small standard errors when sampling from a heavy-tailed distribution instead. Moreover, experience with actual data indicates that the sample mean does indeed have a relatively large standard error in some situations. In terms of testing hypotheses, there are methods for comparing measures of location that continue to have high power in situations where there are outliers or sampling from a heavy-tailed distribution. Other problems that plague inferential methods based on means are also reduced when using these alternative measures of location. For example, the more skewed a distribution happens to be, the more difficult it is to get an accurate confidence interval for the mean, and problems arise when testing hypotheses. Theoretical and simulation studies indicate that problems are reduced substantially when using certain measures of location discussed in this book.

When testing hypotheses, a tempting method for reducing the effects of outliers or sampling from a heavy-tailed distribution is to check for outliers, and if any are found, they are thrown out and standard techniques are applied to the remaining data. This strategy cannot be recommended, however, because it yields incorrect estimates of the standard errors, for reasons given in Chapter 3.

Yet another problem needs to be considered. If distributions are skewed enough, doubts begin to rise about whether the population mean is a satisfactory reflection of the typical participant under study. Figure 1.2 shows a graph of the probability density function corresponding to a mixture of two chi-squared distributions. The first has four degrees of freedom and the second is again chi-squared with four degrees of freedom, only the observations are multiplied by 10. This is similar to the mixed normal already described, only chi-squared distributions are used instead. Observations are sampled from the first distribution with probability .9, otherwise sampling is from the second. As indicated in Figure 1.2, the population mean is 7.6, a value that is relatively far into the right tail. In contrast, the population median is 3.75, and this would seem to be a better representation of the typical participant under study.

1.2 Transformations

Transforming data has practical value in a variety of situations. Emerson and Stoto (1983) provide a fairly elementary discussion of the various reasons one might transform data and how it can be done. The only important point here is that simple transformations can fail to deal effectively with outliers and heavy-tailed distributions. For example, the popular strategy of taking logarithms of all the observations does not necessarily reduce problems due to outliers, and the same is true when using Box–Cox transformations instead (e.g., Doksum & Wong, 1983; Rasmussen, 1989). Other concerns were expressed by Thompson and Amman (1990). Better strategies are described in subsequent chapters.

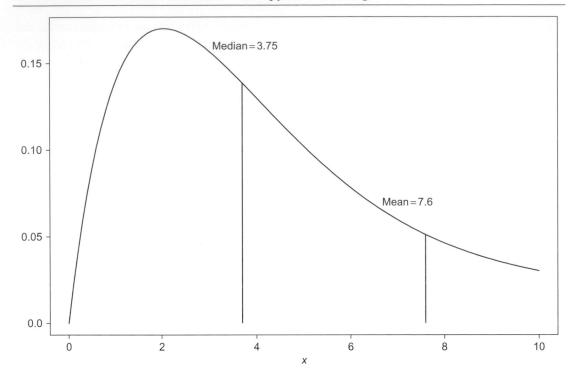

Figure 1.2: Mixed chi-square distribution.

Skewness can be a source of concern when using methods based on means, as will be illustrated in subsequent chapters. Transforming data is often suggested as a way of dealing with skewness. More precisely, the goal is to transform the data so that the resulting distribution is approximately symmetric about some central value. There are situations where this strategy is reasonably successful. But even after transforming data, a distribution can remain severely skewed. In practical terms, this approach can be highly unsatisfactory, and assuming that it performs well can result in erroneous and misleading conclusions. When comparing two independent groups, with say a Student's t test, the assumption is that the same transformation applied to group 1 is satisfactory when transforming the data associated with group 2. A seemingly better way to proceed is to use a method that deals well with skewed distributions even when data are not transformed and when the distributions being compared differ in the amount of skewness.

Perhaps it should be noted that when using simple transformations on skewed data, if inferences are based on the mean of the transformed data, then attempts at making inferences about the mean of the original data, μ, have been abandoned. That is, if the mean of the transformed data is computed and we transform back to the original data, in general we do not get an estimate of μ.

1.3 The Influence Curve

This section gives one more indication of why robust methods are of interest by introducing the influence curve as described by Mosteller and Tukey (1977). It bears a close resemblance to the *influence function*, which plays an important role in subsequent chapters, but the influence curve is easier to understand. In general, the *influence curve* indicates how any statistic is affected by an additional observation having the value x. In particular it graphs the value of a statistic versus x.

As an illustration, let $\bar{X}$ be the sample mean corresponding to the random sample $X_1, \ldots, X_n$. Suppose we add an additional value, x, to the n values already available, so now there are $n+1$ observations. Of course this additional value will in general affect the sample mean, which is now $(x + \sum X_i)/(n+1)$. It is evident that as x gets large, the sample mean of all $n+1$ observations increases. The influence curve plots x versus

$$\frac{1}{n+1}\left(x + \sum X_i\right), \tag{1.3}$$

the idea being to illustrate how a single value can influence the value of the sample mean. Note that for the sample mean, the graph is a straight line with slope $1/(n+1)$, the point being that the curve increases without bound. Of course, as n gets large, the slope decreases, but in practice there might be two or more unusual values that dominate the value of $\bar{X}$.

Now consider the usual sample median, M. Let $X_{(1)} \leq \cdots \leq X_{(n)}$ be the observations written in ascending order. If n is odd, let $m = (n+1)/2$, in which case $M = X_{(m)}$, the mth largest order statistic. If n is even, let $m = n/2$ in which case $M = (X_{(m)} + X_{(m+1)})/2$. To be more concrete, consider the values

2 4 6 7 8 10 14 19 21 28.

Then $n = 10$ and $M = (8+10)/2 = 9$. Suppose an additional value, x, is added, so that now $n = 11$. If $x > 10$, then $M = 10$, regardless of how large x might be. If $x < 8$, $M = 8$ regardless of how small x might be. As x increases from 8 to 10, M increases from 8 to 10 as well. The main point is that in contrast to the sample mean, the median has a bounded influence curve. In general, if the goal is to minimize the influence of a relatively small number of observations on a measure of location, attention might be restricted to those measures having a bounded influence curve. A concern with the median, however, is that its standard error is large relative to the standard error of the mean when sampling from a normal distribution, so there is interest in searching for other measures of location having a bounded influence curve, but that have reasonably small standard errors when distributions are normal.

Also notice that the sample variance, s^2, has an unbounded influence curve, so a single unusual value can inflate s^2. This is of practical concern because the standard error of $\bar{X}$ is estimated with $s/\sqrt{n}$. Consequently, conventional methods for comparing means can have low power and relatively long confidence intervals due to a single unusual value. This problem does indeed arise in practice, as illustrated in subsequent chapters. For now the only point is that it is desirable to search for measures of location for which the estimated standard error has a bounded influence curve. Such measures are available that have other desirable properties as well.

1.4 The Central Limit Theorem

When working with means or least squares regression, certainly the best-known method for dealing with nonnormality is to appeal to the central limit theorem. Put simply, under random sampling, if the sample size is sufficiently large, the distribution of the sample mean is approximately normal under fairly weak assumptions. A practical concern is the description sufficiently large. Just how large must n be to justify the assumption that $\bar{X}$ has a normal distribution? Early studies suggested that $n = 40$ is more than sufficient, and there was a time when even $n = 25$ seemed to suffice. These claims were not based on wild speculations, but more recent studies have found that these early investigations overlooked two crucial aspects of the problem.

The first is that early studies looking into how quickly the sampling distribution of $\bar{X}$ approaches a normal distribution focused on very light-tailed distributions where the expected proportion of outliers is relatively low. In particular, a popular way of illustrating the central limit theorem was to consider the distribution of $\bar{X}$ when sampling from a uniform or exponential distribution. These distributions look nothing like a normal curve, the distribution of $\bar{X}$ based on $n = 40$ is approximately normal, so a natural speculation is that this will continue to be the case when sampling from other nonnormal distributions. But more recently it has become clear that as we move toward more heavy-tailed distributions, a larger sample size is required.

The second aspect being overlooked is that when making inferences based on Student's t, the distribution of T can be influenced more by nonnormality than the distribution of $\bar{X}$. In particular, even if the distribution of $\bar{X}$ is approximately normal based on a sample of n observations, the actual distribution of T can differ substantially from a Student's t-distribution with $n - 1$ degrees of freedom. *Even when sampling from a relatively light-tailed distribution*, practical problems arise when using Student's t as will be illustrated in Section 4.1. When sampling from heavy-tailed distributions, even $n = 300$ might not suffice when computing a 0.95 confidence interval via Student's t.

1.5 Is the ANOVA F Robust?

Practical problems with comparing means have already been described, but some additional comments are in order. For many years, conventional wisdom held that standard analysis of variance (ANOVA) methods are robust, and this point of view continues to dominate applied research. In what sense is this view correct? What many early studies found was that if two groups are *identical*, meaning that they have identical distributions, Student's *t* test and more generally the ANOVA *F*-test are robust to nonnormality in the sense that the actual probability of a type I error would be close to the nominal level. Tan (1982) reviews the relevant literature. Many took this to mean that the *F*-test is robust when groups differ. In terms of power, some studies seemed to confirm this by focusing on standardized differences among the means. To be more precise, consider two independent groups with means μ_1 and μ_2 and variances σ_1^2 and σ_2^2. Many studies have investigated the power of Student's *t* test by examining power as a function of

$$\delta = \frac{\mu_1 - \mu_2}{\sigma},$$

where $\sigma = \sigma_1 = \sigma_2$ is the assumed common variance. What these studies failed to take into account is that small shifts away from normality, toward a heavy-tailed distribution, lowers δ, and this can mask power problems associated with Student's *t* test. The important point is that for a given difference between the means, $\mu_1 - \mu_2$, modern methods can have substantially more power.

To underscore concerns about power when using Student's *t*, consider the two normal distributions in the left panel of Figure 1.3. The difference between the means is 0.8 and both distributions have variance 1. With a random sample of size 40 from both the groups, and when testing at the 0.05 level, Student's *t* has power approximately equal to 0.94. Now look at the right panel. The difference between the means is again 0.8, but now power is 0.25, despite the obvious similarity to the right panel. The reason is that the distributions are contaminated normals, each having variance 10.9.

More recently it has been illustrated that standard confidence intervals for the difference between means can be unsatisfactory and that the *F*-test has undesirable power properties. One concern is that there are situations where, as the difference between the means increases, power goes down, although eventually it goes up. That is, the *F*-test can be *biased*. For example, Wilcox (1996a) describes a situation involving lognormal distributions where the probability of rejecting is .18, when testing at the $\alpha = 0.05$ level, even though the means are equal. When the first mean is increased by 0.4 standard deviations, power *drops* to 0.096, but increasing the mean by 1 standard deviation, power increases to 0.306. Cressie and Whitford (1986) show that for unequal sample sizes, and when distributions differ in skewness, Student's *t* test is not even asymptotically correct. More specifically, the variance of the test

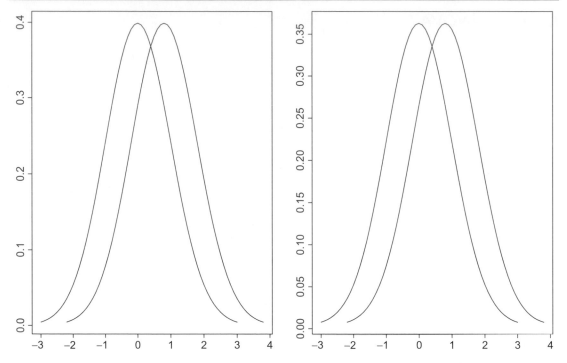

Figure 1.3: Small changes in the tails of distributions can substantially lower power when using means. In the left panel, Student's t **has power approximately equal to 0.94. But in the right panel, power is 0.25.**

statistic does not converge to one as is typically assumed, and there is the additional problem that the null distribution is skewed. The situation improves by switching to heteroscedastic methods, but problems remain (e.g., Algina, Oshima, & Lin, 1994). The modern methods described in this book address these problems.

1.6 Regression

Outliers, as well skewed or heavy-tailed distributions, also affect the ordinary least squares regression estimator. In some ways the practical problems that arise are even more serious than those associated with the ANOVA F-test.

Consider two random variables, X and Y, and suppose

$$Y = \beta_1 X + \beta_0 + \lambda(X)\epsilon,$$

where ϵ is a random variable having variance σ^2, X and ϵ are independent, and $\lambda(X)$ is any function of X. If ϵ is normal and $\lambda(X) \equiv 1$, standard methods can be used to compute confidence intervals for β_1 and β_0. However, even when ϵ is normal but $\lambda(X)$ varies with X, probability coverage can be poor, and problems get worse under nonnormality. There is the additional problem that under nonnormality, the usual least squares estimate of the parameters can have relatively low efficiency, and this can result in relatively low power. In fact, low efficiency occurs even under normality when λ varies with X. There is also the concern that a single unusual Y value, or an usual X value, can greatly distort the least squares estimate of the slope and intercept. Illustrations of these problems and how they can be addressed are given in subsequent chapters.

1.7 More Remarks

Problems with means and the influence of outliers have been known since at least the 19th century. Prior to the year 1960, methods for dealing with these problems were ad hoc compared to the formal mathematical developments related to the analysis of variance and least squares regression. What marked the beginning of modern robust methods, resulting in mathematical methods for dealing with robustness issues, was a paper by Tukey (1960) discussing the contaminated normal distribution. A few years later, a mathematical foundation for addressing technical issues was developed by a small group of statisticians. Of particular importance is the theory of robustness developed by Huber (1964) and Hampel (1968). These results, plus other statistical tools developed in recent years, and the power of the computer, provide important new methods for comparing groups and studying the association between two or more variables.

1.8 Using the Computer: R

Most of the methods described in this book are not yet available in standard statistical packages for the computer. Consequently, to help make these methods accessible, a library of over 950 easy-to-use R functions has been supplied for applying them to data. The (open source) software R (R Development Core Team, 2010) is free and can be downloaded from www.R-project.org. Many books are now available that cover the basics of R (e.g., Crawley, 2007; Venables & Smith, 2002; Verzani, 2004; Zuur, 2009). The book by Verzani is available on the web at http://cran.r-project.org/doc/contrib/Verzani-SimpleR.pdf. R has a built-in manual as well.

The R functions written for this book are available in an R package, or they can be downloaded from the author's web page. To install the R package, created by Felix Schönbrodt, use the R command

        install.packages("'WRS", repos="http://R-Forge.R-project.org").

Access to the functions is gained via the R command

        library(WRS)

Alternatively, go to the web page http://college.usc.edu/labs/rwilcox/home, or the web page www-rcf.usc.edu/~rwilcox/, and download the file Rallfun. (Currently, the most recent version is Rallfun-v15.) Then use the R command

source("Rallfun-v15")

Now all of the functions written for this book are part of your version of R until you remove them. An advantage of the R package is that it contains help files. An advantage of downloading the functions from the author's web page is that updates are made more frequently. (Information about updates are available on the author's web page; see the file update_info.) The author's web page also contains some of the data sets used in this book.

In case it helps, here is a list of the R packages that are utilized in this book:

- akima
- cobs
- MASS
- mgcv
- multicore
- plotrix
- pwr
- quantreg
- robust
- robustbase
- rrcov
- scatterplot3d
- stats

All of these packages can be installed with the install.packages command (assuming you are connected to the web). For example, the R command

install.packages("akima")

will install the R package akima, which is used when creating three-dimensional plots.

Nearly all of the R functions written for this book have fairly low execution time. But when the sample size is large and a bootstrap method is used in conjunction with certain multivariate methods, execution time can be relatively high. To reduce this problem, some of the R functions include the ability of taking advantage of a multicore processor if one is available. More information is supplied when the need arises.

It is noted that there are books that focus on S-PLUS (e.g., Becker, Chambers, & Wilks, 1988; Chambers, 1998; Chambers & Hastie, 1992; Fox, 2002; Krause & Olson, 2002; Venables & Ripley, 2000), which can be useful when using R. However, many of the R functions written for this book now rely on R packages that are not readily accessible via S-PLUS. And because R is free, S-PLUS versions of the functions in this book are no longer described or updated.

1.9 Some Data Management Issues

Some of the R functions written for this book are aimed at manipulating and managing data in a manner that might be helpful, some of which are summarized in this section. Subsequent chapters provide more details about when and how the functions summarized here might be used.

A common situation is where data are stored in columns with one of the columns indicating the group to which a participant belongs and one or more other columns contain the measures of interest. For example, the data for eight participants might be stored as

```
10 2 64
 4 2 47
 8 3 59
12 3 61
 6 2 73
 7 1 56
 8 1 78
15 2 63
```

where the second column indicates to which group a participant belongs. There are three groups because the numbers in column 2 have one of three distinct values. For illustrative purposes, suppose that for each participant, two measures of reduced stress are recorded in columns 1 and 3. Then two of the participants belong to group 1, on the first measure of reduced stress their scores are 7 and 8, and on the second their scores are 56 and 78. Some of the R functions written for this book require storing data associated with different groups either in a matrix (with columns corresponding to groups) or in list mode. What is needed is a simple method of sorting the observations just described into groups based on the values in column 2. By storing the data in list mode, various R functions (to be described) can now be used. The R function

$$fac2list(x,g)$$

is supplied for accomplishing this goal, where x is an R variable, typically the column of some matrix or a data frame, containing the data to be analyzed, and g is an R variable indicating the levels of the groups to be compared. For a one-way ANOVA, g is assumed to be a single column of values. For a two-way ANOVA, g would have two columns, and for a three-way ANOVA it would have three columns, each column corresponding to a factor. A maximum of four columns is allowed.

■ Example

R has a built-in data set, stored in the R variable ChickWeight, which is a matrix containing four columns of data. The first column contains the weight of chicks, column 4 indicates which of four diets was used, and the second column gives the number of days since birth when the measurement was made, which were 0, 2, 4, 6, 8, 10, 12, 14, 16, 18, 20, and 21. So for each chick, measurements were taken on 12 different days. Imagine that the goal is to sort data on weight into four groups based on the four groups indicated in column 4 and that the results are to be stored in list mode. This is accomplished with the R command

z=fac2list(ChickWeight[,1],ChickWeight[,4]).

The data for group 1 are stored in z[[1]], the data for group 2 are stored in z[[2]], and so on. If the levels of the groups are indicated by numeric values, fac2list puts the levels in ascending order. If the levels are indicated by a character string, the levels are put in alphabetical order.

The R function

fac2Mlist(x,grp.col,lev.col,pr=T)

is like the R function fac2list; it can be useful when dealing with a multivariate analysis of variance (MANOVA) design using the methods in Section 7.10. Roughly, it sorts data into groups based on the data in column of x indicated by the argument grp.col. See Sections 7.10.2 and 7.10.3 for more details. When dealing with a between-by-between MANOVA design, the function

fac2BBMlist(x,grp.col,lev.col,pr=T)

can be used.

Now consider between-by-between or a between-by-within ANOVA design. Some of the functions written for this book assume that the data are stored in list mode, or a matrix with columns corresponding to groups, and that the data are arranged in a particular order: the first K groups belong to the first level of the first factor, the next K group belong to the second level of the second factor, and so on.

■ Example

For a 2-by-4 design, with the data stored in the R variable x, having list mode, the data are assumed to be arranged as follows:

	Factor B			
Factor	$x[[1]]$	$x[[2]]$	$x[[3]]$	$x[[4]]$
A	$x[[5]]$	$x[[6]]$	$x[[7]]$	$x[[8]]$

■

■ Example

Consider again the previous example dealing with the R variable ChickWeight, only now the goal is to store the data in list mode in the order just described. The R command

$$z=fac2list(ChickWeight[,1],ChickWeight[,c(4,2)]).$$

accomplishes this goal.

■

Look closely at the argument ChickWeight[,c(4,2)] and note the use of c(4,2). The 2 comes after the 4 because column 2 corresponds to the within group factor, which in this book always corresponds to the second factor. If ChickWeight[,c(2,4)] had been used, functions in this book aimed at a between-by-within design would assume that column 4 corresponds to the within group factor, which is incorrect.

Earlier editions of this book provided another way of sorting the data into groups via the R function selby, which is still available and has the form

$$selby(m,grpc,coln),$$

where m is any matrix having n rows and at least two columns. The argument grpc is used to indicate which column contains the group identification numbers. The argument coln indicates which column of data is to be analyzed.

■ Example

Consider again the data

```
10 2 64
 4 2 47
 8 3 59
```

```
12 3 61
 6 2 73
 7 1 56
 8 1 78
15 2 63
```

If the data are stored in the matrix mat, the command

$$tdat=selby(mat,2,3)$$

sorts the data into three groups and stores the values in the third column of mat into the R variable tdat$x which will have list mode. In particular, the variable tdat$x[[1]] contains the data for the first group, namely the values 7 and 8. Similarly, tdat$x[[2]] contains the values 64, 47, 73, and 63, and tdat$x[[3]] contains 59 and 61.

∎

The function selby also returns the values of the group numbers that are stored in column grpc. The values are stored in selby$grpn. In the illustration, the command tdat=selby(mat,2,3) causes these values to be stored in the R vector tdat$grpn.

In the last example, tdat$grpn[1] contains 1 meaning that tdat$x[[1]] contains all of the data corresponding to group 1. If the only group numbers had been 3, 6, and 8, then tdat$grpn[1] would have the value 3, and all of the corresponding data would be stored in tdat$x[[1]]. Similarly, tdat$grpn[2] would have the value 6, and the data for this group would be stored in tdat$x[[2]]. Finally, the data for the third group, numbered 8, would be stored in tdat$x[[3]].

An extension of the function selby, called selby2, deals with situations where there is more than one factor. It has the form

$$selby2(m,grpn,coln)$$

where grpn is a vector of length 2 indicating the column numbers of m where the group numbers are stored. The third argument, coln, indicates which column contains the data to be analyzed. It accomplishes the same goal as the function fac2list. Although fac2list is more flexible and seems a bit easier to use, selby2 is illustrated here in case some readers prefer to use it.

Suppose the following data are stored in the R matrix m having 13 rows and 4 columns.

```
10 2 64 1
 4 2 47 1
 8 3 59 1
12 3 61 2
 6 2 73 2
```

```
 7 1 56 2
 8 1 78 2
15 2 63 2
 9 3 71 1
 2 3 81 1
 4 1 68 1
 5 1 53 1
21 3 49 2
```

The goal is to perform a 3-by-2 ANOVA, where the numbers in column 2 indicate the levels of the first factor, and the numbers in column 4 indicate the levels of the second. Further assume that the values to be analyzed are stored in column 1. For example, the first row of data indicates that the value 10 belongs to level 2 of the first factor and level 1 of the second. Similarly, the third row indicates that the value 8 belongs to the third level of the first factor and the first level of the second. Chapter 7 describes R functions for comparing the groups. Using these functions requires storing the data in list mode or a matrix, and the function selby2 is supplied to help accomplish this goal with the R command

$$\text{dat}=\text{selby2}(m,c(2,4),1),$$

The output stored in dat is

```
$x:
$x[[1]]:
[1] 4 5

$x[[2]]:
[1] 7 8

$x[[3]]:
[1] 10  4

$x[[4]]:
[1]  6 15

$x[[5]]:
[1] 8 9 2

$x[[6]]:
[1] 12 21


$grpn:
     [,1]  [,2]
[1,]   1     1
[2,]   1     2
```

```
[3,]    2    1
[4,]    2    2
[5,]    3    1
[6,]    3    2
```

The R variable dat$x[[1]] contains the data for level 1 of both factors. The R variable dat$x[[2]] contains the data for level 1 of the first factor and level 2 of the second. The R variable dat$grpn contains the group numbers found in columns 2 and 4, and the ith row indicates which group is stored in $x[[i]]. For example, the third row of $grpn has 2 in the first column and 1 in the second meaning that for level 2 of the first factor and level 1 of the second, the data are stored in $x[[3]]. It is note that the data are stored in the form expected by the ANOVA functions covered in Chapter 7. One of these functions is called t2way. In the illustration, the command

$$t2way(3,2,dat$x,tr=0)$$

would compare means using a heteroscedastic method appropriate for a 3-by-2 ANOVA design, where the outcome measure corresponds to the data in column 1 of the R variable m. To perform a 3-by-2 ANOVA for the data in column 3, first enter the command

$$dat=selby2(m,c(2,4),3)$$

and then

$$t2way(3,2,dat$x,tr=0).$$

However, for the situation just described, it seems easier to use the function fac2list. And fac2list allows the data to be stored in a data frame. In contrast, selby only accepts data stored in a matrix. The R commands

$$z=fac2list(m[,3],m[,c(2,4)]) \ t2way(3,2,z,tr=0).$$

perform the same operations just illustrated. Recently, variations of some of the R functions written for this book have been added that make it possible to avoid using both the R function fac2list as well as selby2. They will be described in subsequent chapters.

Another goal that is sometimes encountered is splitting a matrix of data into groups based on the values in one of the columns. For example, column 6 might indicate whether participants are male or female, denoted by the values 0 and 1, and it is desired to store the data for females and males in separate R variables. This can be done with the R function

$$matsplit(m,coln=NULL),$$

which sorts the data in the matrix m into separate R variables corresponding to the values indicated by the argument coln. The function is similar to fac2list, only now two or more columns of a matrix can be sorted into groups rather than a single column of data, as is the case when using fac2list. Also, matsplit returns the data stored in a matrix rather than list mode.

The R function

$$mat2grp(m,coln)$$

also splits the data in a matrix into groups based on the values in column coln of the matrix m. Unlike matsplit, mat2grp can handle more than two values. That is, the column of m indicated by the argument coln can have more than two unique values. The results are stored in list mode.

The R function

$$qsplit(x,y,split.val=NULL)$$

splits the data in x into three groups based on a range of values stored in y. The length of y is assumed to be equal to the number of rows in the matrix x. (The argument x can be a vector rather than a matrix.) If split.val=NULL, the function computes the lower and upper quartiles based on the values in y. Then the corresponding rows of data in x that correspond to y values less than or equal to the lower quartile are returned in qsplit$lower. The rows of data for which y has a value between the lower and upper quartiles are returned in qsplit$middle, and the rows for which y has a value greater than or equal to the upper quartile are returned in qsplit$upper. If two values are stored in the argument split.val, they will be used in place of the quartiles.

■ Example

R has a built-in data set stored in the R variable ChickWeight (a matrix with 4 columns) that deals with weight gain over time and based on different diets. The amount of weight gained is stored in column 1. For illustrative purposes, imagine the goal is to separate the data in column 1 into three groups. The first group is to contain those values that are less than or equal to the lower quartile, the next is to contain the values between the lower and upper quartiles, and the third group is to contain the values greater than or equal to the upper quartile. The command

$$qsplit(ChickWeight[,1],ChickWeight[,1])$$

accomplishes this goal.

■

Two other functions are provided for manipulating data stored in a matrix:

- bw2list
- bbw2list.

These two functions are useful when dealing with a between-by-within design and a between-between-by-within design and will be described and illustrated in Chapter 8.

To illustrate the next R function, consider data reported by Potthoff and Roy (1964) dealing with an orthodontic growth study where for each of 27 children, the distance between the pituitary and pterygomaxillary fissure was measured at ages 8, 10, 12, and 14 years of age. The data can be accessed via the R package nlme and are stored in the R variable Orthodont. The first 10 rows of the data are:

```
    Distance  Age  Subject   Sex
1      26.0    8      M01    Male
2      25.0   10      M01    Male
3      29.0   12      M01    Male
4      31.0   14      M01    Male
5      21.5    8      M02    Male
6      22.5   10      M02    Male
7      23.0   12      M02    Male
8      26.5   14      M02    Male
9      23.0    8      M03    Male
10     22.5   10      M03    Male
```

It might be useful to store the data in a matrix where each row contains the outcome measure of interest, which is distance in the example. For the orthodontic growth study, this means storing the data in a matrix having 27 rows corresponding to the 27 participants, where each row has four columns corresponding to the four times that measures were taken. The R function

$$long2mat(x,Sid.col,dep.col)$$

accomplishes this goal. The argument x is assumed to be a matrix or a data frame. The argument dep.col is assumed to have a single value that indicates which column of x contains the data to be analyzed. The argument Sid.col indicates the column containing a participant's identification. So for the orthodontic growth study, the command m=long2mat(Orthodont,3,1) would create a 27×4 matrix with the first row containing the values 26, 25, 29, and 31, the measures associated with the first participant.

The R function

$$longcov2mat(x,Sid.col,dep.col)$$

is like the function long2mat, only the argument dep.col can have more than one value and a matrix of covariates is stored in list mode for each of the n participants. Continuing the last example, the command m=long2mat(Orthodont,3,1) would result in m having list mode, m[[1]] would be a 4×1 matrix containing the values for the first participant, m[[2]] would be the values for the second participant, and so on.

A few other R functions that might be useful. One is

$$listm(x),$$

which stores data in list mode (having length J, say) in the J columns of a matrix. That is, x[[1]] becomes column 1, x[[2]] becomes column 2, and so on. The R function

$$matl(x),$$

stores the data in the J columns of a matrix in list mode having length J, and

$$l2v(x)$$

converts data in list mode into a single vector of values.

Consider the following data:

```
1   1   1   Easy    6
1   1   2   Easy    3
1   1   3   Easy    2
1   1   4   Hard    7
1   1   5   Hard    4
1   1   6   Hard    1
1   2   1   Easy    2
1   2   2   Easy    2
1   2   3   Easy    7
1   2   4   Hard    7
1   2   5   Hard    3
1   2   6   Hard    2
2   1   1   Easy    1
2   1   2   Easy    4
2   1   3   Easy    4
2   1   4   Hard    7
2   1   5   Hard    7
2   1   6   Hard    6
2   2   1   Easy    2
2   2   2   Easy    3
2   2   3   Easy    1
2   2   4   Hard    7
2   2   5   Hard    5
2   2   6   Hard    5
```

Imagine that column 2 indicates a participants identification number, columns 1, 3, and 4 indicate categories, and column 5 is some outcome of interest. Further imagine it is desired to compute some measure of location for each category indicated by the values in columns 1 and 4. This can be accomplished with the R function

$$M2m.loc(m, grpc, col.dat, locfun = tmean, \ldots),$$

where the argument locfun indicates the measure of location that will be used, which defaults to a 20% trimmed mean, grpc indicates the columns of m that indicate the category (or levels of a factor), and col.dat indicates the column containing the outcome measure of interest. For the situation at hand, assuming the data are stored in the data frame x, the command M2m.loc(x,c(1,4),5,locfun=mean) returns

```
  V1    V4         loc
1 Easy  3.666667
1 Hard  4.000000
2 Easy  2.500000
2 Hard  6.166667
```

So, for example, participants who are in both category 1 and category easy, the mean is 3.67.

1.9.1 Eliminating Missing Values

From a statistical point of view, a simple strategy for handling missing values is to simply eliminate them. There are other methods for dealing with missing values (e.g., Little and Rubin, 2002), a few of which are covered in subsequent chapters. Here it is merely noted that when data are stored in a matrix or a data frame, say m, the R function

na.omit(m)

will eliminate any row having missing values. (The R function elimna accomplishes the same goal.)

CHAPTER 2

A Foundation for Robust Methods

Measures that characterize a distribution, such as measures of location and scale, are said to be *robust* if slight changes in a distribution have a relatively small effect on their value. As indicated in Chapter 1, the population mean and standard deviation, μ and σ, as well as the sample mean and sample standard deviation, $\bar{X}$ and s^2, are not robust. This chapter elaborates on this problem by providing a relatively nontechnical description of some of the tools used to judge the robustness of parameters and estimators. Included are some strategies for identifying measures of location and scale that are robust. The emphasis in this chapter is on finding robust analogs of μ and σ, but the results and criteria described here are directly relevant to judging estimators as well, as will become evident. This chapter also introduces some technical tools that are of use in various situations.

This chapter is more technical than the remainder of the book. When analyzing data, it helps to have some understanding of how robustness issues are addressed, and providing a reasonably good explanation requires some theory. Also, many applied researchers, who do not religiously follow developments in mathematical statistics, might still have the impression that robust methods are ad hoc procedures. Accordingly, although the main goal is to make robust methods accessible to applied researchers, it needs to be emphasized that modern robust methods have a solid mathematical foundation. It is stressed, however, that many mathematical details arise that are not discussed here. The goal is to provide an indication of how technical issues are addressed without worrying about the many relevant details. Readers interested in mathematical issues can refer to the excellent books by Huber and Ronchetti (2009) as well as Hampel, Ronchetti, Rousseeuw, and Stahel (1986). The monograph by Reider (1994) is also of interest. For a book written at an intermediate level of difficulty, see Staudte and Sheather (1990).

2.1 Basic Tools for Judging Robustness

There are three basic tools that are used to establish whether quantities such as measures of location and scale have good properties: qualitative robustness, quantitative robustness, and infinitesimal robustness. This section describes these tools in the context of location measures, but they are relevant to measures of scale as will become evident. These tools not only provide formal methods for judging a particular measure, they can be used to help derive measures that are robust.

Before continuing, it helps to be more formal about what is meant by a measure of location. A quantity that characterizes a distribution, such as the population mean, is said to be a measure of location if it satisfies four conditions, and a fifth is sometimes added. To describe them, let X be a random variable with distribution F, and let $\theta(X)$ be some descriptive measure of F. Then $\theta(X)$ is said to be a measure of location if for any constants a and b,

$$\theta(X+b) = \theta(X)+b \tag{2.1}$$

$$\theta(-X) = -\theta(X) \tag{2.2}$$

$$X \geq 0 \text{ implies } \theta(X) \geq 0 \tag{2.3}$$

$$\theta(aX) = a\theta(X). \tag{2.4}$$

The first condition is called *location equivariance*. It simply requires that if a constant b is added to every possible value of X, a measure of location should be increased by the same amount. Let $E(X)$ denote the expected value of X. From basic principles, the population mean is location equivariant. That is, if $\theta(X) = E(X) = \mu$, then $\theta(X+b) = E(X+b) = \mu+b$. The first three conditions, taken together, imply that a measure of location should have a value within the range of possible values of X. The fourth condition is called *scale equivariance*. If the scale by which something is measured is altered by multiplying all possible values of X by a, a measure of location should be altered by the same amount. In essence, results should be independent of the scale of measurement. As a simple example, if the typical height of a man is to be compared to the typical height of a woman, it should not matter whether the comparisons are made in inches or feet.

The fifth condition that is sometimes added was suggested by Bickel and Lehmann (1975). Let $F_x(x) = P(X \leq x)$ and $F_y(x) = P(Y \leq x)$ be the distributions corresponding to the random variables X and Y. Then X is said to be stochastically larger than Y if for any x, $F_x(x) \leq F_y(x)$ with strict inequality for some x. If all the quantiles of X are greater than the corresponding quantiles of Y, then X is stochastically larger than Y. Bickel and Lehmann argue that if X is stochastically larger than Y, then it should be the case that $\theta(X) \geq \theta(Y)$ if θ is to qualify as a measure of location. The population mean has this property.

2.1.1 Qualitative Robustness

To understand qualitative robustness, it helps to begin by considering any function $f(x)$, not necessarily a probability density function. Suppose it is desired to impose a restriction on this function so that it does not change drastically with small changes in x. One way of doing this is to insist that $f(x)$ be continuous. If, for example, $f(x) = 0$ for $x \leq 1$, but $f(x) = 10,000$

for any $x > 1$, the function is not continuous, and if $x = 1$, an arbitrarily small increase in x results in a large increase in $f(x)$.

A similar idea can be used when judging a measure of location. This is accomplished by viewing parameters as functionals. In the present context, a functional is just a rule that maps every distribution into a real number. For example, the population mean can be written as

$$T(F) = E(X),$$

where the expected value of X depends on F. The role of F becomes more explicit if expectation is written in integral form, in which case this last equation becomes

$$T(F) = \int x \, dF(x).$$

If X is discrete and the probability function corresponding to $F(x)$ is $f(x)$,

$$T(F) = \sum x f(x),$$

where the summation is over all possible values x of X.

One advantage of viewing parameters as functionals is that the notion of continuity can be extended to them. Thus, if the goal is to have measures of location that are relatively unaffected by small shifts in F, a requirement that can be imposed is that when viewed as a functional, it is continuous. Parameters with this property are said to have *qualitative robustness*.

Let $\hat{F}$ be the usual empirical distribution. That is, for the random sample $X_1, \ldots, X_n$, $\hat{F}(x)$ is just the proportion of X_i values less than or equal to x. An estimate of the functional $T(F)$ is obtained by replacing F with $\hat{F}$. For example, when $T(F) = E(X) = \mu$, replacing F with $\hat{F}$ yields the sample mean, $\bar{X}$. An important point is that qualitative robustness includes the idea that if $\hat{F}$ is close to F, in a sense to be made precise, then $T(\hat{F})$ should be close to $T(F)$. For example, if the empirical distribution represents a close approximation of F, then $\bar{X}$ should be a good approximation of μ, but this is not always the case.

One more introductory remark should be made. From the technical point of view, continuity leads to the issue of how the difference between distributions should be measured. Here, the Kolmogorov distance is used. Other metrics play a role when addressing theoretical issues, but they go beyond the scope of this book. Readers interested in pursuing continuity, as it relates to robustness, can refer to Hampel (1968).

To provide at least the flavor of continuity, let F and G be any two distributions and let $D(F, G)$ be the Kolmogorov distance between them, which is the maximum value of $|F(x) - G(x)|$, the maximum being taken over all possible values of x. If the maximum does not exist, the supremum or least upper bound is used instead. That is, the Kolmogorov distance is the least upper bound on $|F(x) - G(x)|$ over all possible values of x. More succinctly, $D(F, G) = \sup |F(x) - G(x)|$, where the notation *sup* indicates *supremum*. For readers unfamiliar with the notion of a *least upper bound*, the Kolmogorov distance is the smallest value of A such that $|F(x) - G(x)| \leq A$. Any A satisfying $|F(x) - G(x)| \leq A$ is called an upper bound on $|F(x) - G(x)|$ and the smallest (least) upper bound is the Kolmogorov distance. Note that $|F(x) - G(x)| \leq 1$ for any x, so for any two distributions, the maximum possible value for the Kolmogorov distance is 1. If the distributions are identical, $D(F, G) = 0$.

Now consider any sequence of distributions, G_n, $n = 1, 2, \ldots$. For example, G_n might be the empirical distribution based n observations randomly sampled from some distribution F. Another sequence of distributions is the contaminated normal with $\epsilon = 1/n$. The functional T is said to be continuous at F if for *any* sequence G_n, such that $D(G_n, F)$ approaches 0 as n gets large, $|T(G_n) - T(F)|$ approaches 0. In particular, the functional evaluated at the empirical distribution, $T(\hat{F})$, should approach $T(F)$, the functional evaluated at the distribution from which observations are being sampled. Under random sampling, the empirical distribution approaches the true distribution as n gets large, and from standard results the sample mean approaches the population mean as well. However, there are sequences of distributions for which $D(G_n, F)$ approaches 0 for any F, but the mean of the empirical distribution, the sample mean, does not approach the mean of the true distribution, $T(F) = \mu$, as n gets large. Details are given by Staudte and Sheather (1990, p. 66). Thus, for the Kolmogorov metric, $T(F) = E(X)$ is not continuous. That is, if we require a measure of location that has a continuous functional, the population mean, μ, is ruled out.

An example of a continuous functional, that plays a central role in this book, is the γ-trimmed mean, $0 < \gamma \leq 0.5$. A γ-trimmed mean is the mean of a distribution after the distribution has been transformed in a particular way. More specifically, it is trimmed by truncating the distribution at the γ and $1 - \gamma$ quantiles. Note that if a probability density function is trimmed, it no longer qualifies as a probability density function because the area under the curve is no longer equal to 1, it is equal to $1 - 2\gamma$. Consequently, dividing the trimmed probability density function by $1 - 2\gamma$, the resulting function is again a probability density function. Here, two-sided trimming is assumed unless stated otherwise. (Some authors, when referring to a γ-trimmed mean, assume one-sided trimming, but others assume two-sided trimming instead.) In general, when referring to a trimmed distribution, this means that the probability density function, $f(x)$, is transformed to

$$\frac{1}{1 - 2\gamma} f(x), \quad x_\gamma \leq x \leq x_{1-\gamma},$$

where x_γ and $x_{1-\gamma}$ are the γ and $1 - \gamma$ quantiles. In essence, trimming results in focusing on the middle portion of a distribution.

As a simple example, consider a standard normal distribution after it has been trimmed 20% ($\gamma = 0.2$) and rescaled so that the area under the curve is equal to one. The 0.2 and 0.8 quantiles of the standard normal distribution are -0.84 and 0.84, respectively. Thus, the 20% trimmed analog of the standard normal distribution is defined for $-.84 \leq x \leq .84$. The standard normal probability density function is

$$\frac{1}{\sqrt{2\pi}} \exp(-x^2/2), \quad -\infty \leq x \leq \infty,$$

so the 20% trimmed analog of the standard normal probability density function is

$$f(x) = \frac{1}{0.6} \frac{1}{\sqrt{2\pi}} \exp(-x^2/2), \quad -0.84 \leq x \leq 0.84. \tag{2.5}$$

2.1.2 Infinitesimal Robustness

To provide a relatively simple explanation of infinitesimal robustness, it helps to again consider the situation where $f(x)$ is any function, not necessarily a probability density function. Once more consider what restrictions might be imposed so that small changes in x do not result in large changes in $f(x)$. One such condition is that it be differentiable and that the derivative be bounded. In symbols, if $f'(x)$ is the derivative, it is required that $f'(x) < B$ for some constant B. The function $f(x) = x^2$, for example, does not satisfy this condition because its derivative, $2x$, increases without bound as x gets large.

Analogs of derivatives of functionals exist and so a natural way of searching for robust measures of location is to focus on functionals that have a bounded derivative. In the statistics literature, the derivative of a functional, $T(F)$, is called the *influence function* of T at F, which was introduced by Hampel (1968, 1974). Roughly, the influence function measures the relative extent a small perturbation in F has on $T(F)$. Put another way, it reflects the (normed) limiting influence of adding one more observation, x, to a very large sample.

To provide a more precise description of the influence function, let Δ_x be a distribution where the value x occurs with probability one. As is fairly evident, if Y has distribution Δ_x, then $P(Y \leq y) = 0$ if $y < x$, and the mean of Y is $E(Y) = x$.

Next, consider a mixture of two distributions where an observation is randomly sampled from distribution F with probability $1 - \epsilon$, otherwise sampling is from the distribution Δ_x. That is, with probability ϵ, the observed value is x. The resulting distribution is

$$F_{x,\epsilon} = (1 - \epsilon)F + \epsilon \Delta_x. \tag{2.6}$$

It might help to notice the similarity between $F_{x,\epsilon}$ and the contaminated or mixed normal described in Chapter 1. In the present situation, F is any distribution, including normal distributions as a special case. Also notice the similarity with the influence curve in Chapter 1.

Here, interest is in how the value x affects the value of some functional when x occurs with probability ϵ. For example, if F has mean μ, then $F_{x,\epsilon}$ has mean $(1-\epsilon)\mu + \epsilon x$, and the difference between the mean of $F_{x,\epsilon}$ and the mean of F is $\epsilon(x-\mu)$.

Notice that when ϵ is small, $F_{x,\epsilon}$ is similar to F, as measured by the Kolmogorov distance function. To see this, first note that if the distributions are evaluated at any value, say y,

$$|F_{x,\epsilon}(y) - F(y)| = |-\epsilon[F(y) - \Delta_x(y)]|.$$

But F and Δ_x are distributions, so $|F(y) - \Delta_x(y)| \leq 1$. Consequently, the Kolmogorov distance between $F_{x,\epsilon}$ and F is at most ϵ. Moreover, $F_{x,\epsilon}$ and F can be made arbitrarily close by choosing ϵ sufficiently small.

The relative influence on $T(F)$ of having the value x occur with probability ϵ is

$$\frac{T(F_{x,\epsilon}) - T(F)}{\epsilon},$$

and the *influence function* of T at F is

$$IF(x) = \lim \frac{T(F_{x,\epsilon}) - T(F)}{\epsilon}, \tag{2.7}$$

where the limit is taken as ϵ approaches 0 from above. Roughly, $IF(x)$ is the relative influence of x on some measure that characterizes a distribution, $T(F)$, when the probability of observing the value x is arbitrarily close to 0. $T(F)$ is said to be *B robust*, or to have *infinitesimal robustness*, if $IF(x)$ is bounded. The *gross error sensitivity* of $T(F)$ is $\sup_x |IF(x)|$.

As already indicated, if $T(F) = E(X)$, $T(F_{x,\epsilon}) - T(F) = \epsilon(x-\mu)$, so $(T(F_{x,\epsilon}) - T(F))/\epsilon = x - \mu$. Thus, the influence function of the population mean is

$$IF(x) = x - \mu,$$

which *does not* depend on F. Especially important is that the influence function is unbounded in x. That is, μ does not have infinitesimal robustness. And its gross error sensitivity is ∞.

2.1.3 Quantitative Robustness

The third approach to judging some quantity that characterizes a distribution is the *breakdown point*, which addresses the notion of quantitative robustness. The general idea is to describe quantitatively the effect a small change in F has on some functional $T(F)$.

Again consider $F_{x,\epsilon} = (1-\epsilon)F + \epsilon\Delta_x$, which has mean $(1-\epsilon)\mu + \epsilon x$. Thus, for any $\epsilon > 0$, the mean goes to infinity as x gets large. In particular, even when ϵ is arbitrarily close to 0, in which case the Kolmogorov distance between $F_{x,\epsilon}$ and F is small, the mean of $F_{x,\epsilon}$ can be made arbitrarily large by increasing x. The minimum value of ϵ, for which a functional goes to infinity as x gets large, is called the *breakdown point*. When necessary, the minimum value is replaced by the infimum or greatest lower bound. (This definition oversimplifies technical issues, but it suffices for present purposes. See Huber, 1981, Section 1.4 for more details.) In the illustration, any $\epsilon > 0$ causes the mean to go to infinity, so the breakdown point is 0. In contrast, the median of a distribution has a breakdown point of 0.5, and more generally the γ-trimmed mean, μ_t, has a breakdown point of γ.

When searching for measures of dispersion, the breakdown point turns out to have considerable practical importance. In some cases the breakdown point is more important than the efficiency of any corresponding estimator. For the moment, it is merely noted that the standard deviation, σ, has a breakdown point of 0, and this renders it unsatisfactory in various situations.

2.2 Some Measures of Location and Their Influence Function

There are many measures of location. (See Andrews et al., 1972.) This section describes some measures that are particularly important based on what is currently known. (And a few additional measures of location are introduced in Chapters 3 and 6.)

2.2.1 Quantiles

It is convenient to begin with quantiles. For any random variable X with distribution F, the qth quantile, say x_q, satisfies $F(x) = P(X \leq x_q) = q$, where $0 < q < 1$. For example, if X is standard normal, the 0.8 quantile is $x_{0.8} = 0.84$ and $P(X \leq 0.84) = 0.8$.

In the event that there are multiple x values such that $F(x) = q$, the standard convention is to define the qth quantile as the smallest value x such that $F(x) \geq q$. For completeness, it is sometimes necessary to define the qth quantile as $x_q = \inf\{x : F(x) \geq q\}$, where inf indicates infimum or greatest lower bound, but this is a detail that is not important here.

The qth quantile has location and scale equivariance and it satisfies the other conditions for a measure of location given by Eqs (2.1) through (2.4), plus the Bickel–Lehmann condition. In so far as it is desired to have a measure of location that reflects the typical subject under study, the median, $x_{0.5}$, is a natural choice. The breakdown point of the median is 0.5, and more generally the breakdown point of the qth quantile is $1 - q$ (e.g., Staudte and Sheather, 1990, p. 56).

For some distributions x_q has qualitative robustness, but for others, including discrete distributions, it does not. In fact, even if x_q has qualitative robustness at F, it is not qualitatively robust at $F_{x,\epsilon}$. That is, there are distributions that are arbitrarily close to F for which x_q is not qualitatively robust.

Letting $f(x)$ represent the probability density function, and assuming $f(x_q) > 0$ and that $f(x_q)$ is continuous at x_q, the influence function of x_q is

$$IF_q(x) = \begin{cases} \frac{q-1}{f(x_q)}, & \text{if } x < x_q \\ 0, & \text{if } x = x_q \\ \frac{q}{f(x_q)}, & \text{if } x > x_q. \end{cases} \quad (2.8)$$

This influence function is bounded, so x_q has infinitesimal robustness.

2.2.2 The Winsorized Mean

One problem with the mean is that the tails of a distribution can dominate its value, and this is reflected by an unbounded influence function, a breakdown point of 0, and a lack of qualitative robustness. Put in more practical terms, if a measure of location is intended to reflect what the typical subject is like, the mean can fail because its value can be inordinately influenced by a very small proportion of the subjects who fall in the tails of a distribution. One strategy for dealing with this problem is to give less weight to values in the tails and pay more attention to those near the center. One specific strategy for implementing this idea is to *Winsorize* the distribution.

Let F be any distribution, and let x_γ and $x_{1-\gamma}$ be the γ and $1 - \gamma$ quantiles. Then a γ-Winsorized analog of F is the distribution

$$F_w(x) = \begin{cases} 0, & \text{if } x < x_\gamma \\ \gamma, & \text{if } x = x_\gamma \\ F(x), & \text{if } x_\gamma < x < x_{1-\gamma} \\ 1, & \text{if } x \geq x_{1-\gamma}. \end{cases}$$

In other words, the left tail is pulled in so that the probability of observing the value x_γ is γ, and the probability of observing any value less than x_γ, after Winsorization, is 0. Similarly, the right tail is pulled in so that, after Winsorization, the probability of observing a value greater than $x_{1-\gamma}$ is 0. The mean of the Winsorized distribution is

$$\mu_w = \int_{x_\gamma}^{x_{1-\gamma}} x \, dF(x) + \gamma(x_\gamma + x_{1-\gamma}).$$

In essence, the Winsorized mean pays more attention to the central portion of a distribution by transforming the tails. The result is that μ_w can be closer to the central portion of a

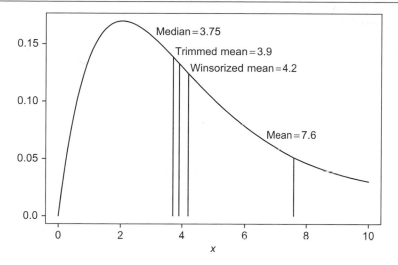

Figure 2.1: Mixed chi-square Distribution.

distribution. It can be shown that μ_w satisfies Eqs (2.1) through (2.4), so it qualifies as a measure of location and it also satisfies the Bickel–Lehmann condition.

For the mixed chi-square distribution described in Chapter 1, the 20% Winsorized mean is approximately $\mu_w = 4.2$, based on simulations with 10,000 replications. Figure 2.1 shows the position of the Winsorized mean relative to the median, $x_{0.5} = 3.75$, and the mean, $\mu = 7.6$. As is evident, Winsorization results in a measure of location that is closer to the bulk of the distribution. For symmetric distributions, $\mu_w = \mu$.

Like quantiles, there are distributions arbitrarily close to any distribution F for which the Winsorized mean is not qualitatively robust. On the positive side, its breakdown point is γ. This suggests choosing $\gamma = 0.5$ to achieve the highest possible breakdown point, but there are some negative consequences if γ is too far from 0, as will become evident in Chapter 3.

Let

$$C = \mu_w - \frac{\gamma^2}{f(x_\gamma)} - \frac{\gamma^2}{f(x_{1-\gamma})},$$

where again f is the probability density function corresponding to F. The influence function of the Winsorized mean is

$$IF_w(x) = \begin{cases} x_\gamma - \frac{\gamma}{f(x_\gamma)} - C, & \text{if } x < x_\gamma \\ x - C, & \text{if } x_\gamma < x < x_{1-\gamma} \\ x_{1-\gamma} + \frac{\gamma}{f(x_{1-\gamma})} - C, & \text{if } x > x_{1-\gamma}. \end{cases}$$

Notice that the influence function is bounded but not smooth; it has jumps at x_γ and $x_{1-\gamma}$.

2.2.3 The Trimmed Mean

Rather than Winsorize, another strategy for reducing the effects of the tails of a distribution is to simply remove them, and this is the strategy employed by the trimmed mean. The γ-trimmed mean is

$$\mu_t = \frac{1}{1-2\gamma} \int\limits_{x_\gamma}^{x_{1-\gamma}} x \, dF(x).$$

In words, μ_t is the mean of a distribution after it has been trimmed as described in Section 2.1.1. The trimmed mean is both location and scale equivariant, more generally it satisfies Eqs. (2.1) through (2.4), and it also satisfies the Bickel–Lehmann condition for a measure of location.

The influence function of the trimmed mean is

$$IF_t(x) = \begin{cases} \frac{1}{1-2\gamma}(x_\gamma - \mu_w), & \text{if } x < x_\gamma \\ \frac{1}{1-2\gamma}(x - \mu_w), & \text{if } x_\gamma \le x \le x_{1-\gamma} \\ \frac{1}{1-2\gamma}(x_{1-\gamma} - \mu_w), & \text{if } x > x_{1-\gamma}. \end{cases}$$

The influence function is bounded, but there are jumps at x_γ and $x_{1-\gamma}$. As already indicated, μ_t is qualitatively robust when $\gamma > 0$, and its breakdown point is γ. It can be seen that $E[IF(X)] = 0$.

For the mixed chi-square distribution described in Chapter 1, the 20% trimmed mean is $\mu_t = 3.9$, and its position relative to the mean, median, and 20% Winsorized mean is shown in Figure 2.1.

It should be noted that the influence function of the trimmed mean can be derived under very general conditions which include both symmetric and asymmetric distributions (Huber, 1981). Staudte and Sheather (1990) derive the influence function assuming distributions are symmetric, but their results are easily extended to the asymmetric case.

2.2.4 M-Measures of Location

M-measures of location form a large class of location measures that include the population mean, μ, as a special case. Typically, μ is viewed as $E(X)$, the expected value of the random variable X. However, to gain some insight into the motivation for M-measures of location, it helps to first view μ in a different way.

When searching for a measure of location, one strategy is to use some value, say c, that is in some sense close, on average, to all the possible values of the random variable X. One way of quantifying how close a value c is from all possible values of X is in terms of its expected squared distance. In symbols, $E(X - c)^2$ represents the expected squared distance from c. If c is intended to characterize the typical subject or thing under study, a natural approach is to use the value c that minimizes $E(X - c)^2$. Viewing $E(X - c)^2$ as a function of c, the value of c minimizing this function is obtained by differentiating, setting the result equal to 0, and solving for c. That is, c is given by the equation

$$E(X - c) = 0, \tag{2.9}$$

so $c = \mu$. In other words, μ is the closest point to all possible values of X in terms of expected squared distance. But μ is not robust and in the present context the problem is that $E(X - c)^2$ gives an inordinate amount of weight to values of X that are far from c. Put another way, the function $(x - c)^2$ increases too rapidly as x moves away from c.

The approach just described for deriving a measure of location can be improved by considering a class of functions for measuring the distance from a point and then searching for a function within this class that has desirable properties. To this end, let $\xi(X - \mu_m)$ be some function that measures the distance from μ_m, and let Ψ be its derivative with respect to μ_m. Attention is restricted to those functions for which $E[\xi(X - \mu_m)]$, viewed as a function of μ_m, has a derivative. As mentioned earlier, $\xi(X - \mu_m) = (X - \mu_m)^2$ and $\Psi(X - \mu_m) = -2(X - \mu_m)$, then a measure of location that is closest to all possible values of X, as measured by its expected distance, is the value μ_m that minimizes $E[\xi(X - \mu_m)]$. This means that μ_m is determined by the equation

$$E[\Psi(X - \mu_m)] = 0. \tag{2.10}$$

Typically the function Ψ is assumed to be odd, meaning that $\Psi(-x) = -\Psi(x)$ for any x. (The reason for this will become clear in Chapter 3.) The value μ_m that satisfies Eq. (2.10) is called an *M-measure of location*. Obviously the class of odd functions is too large for practical purposes, but this problem can be corrected, as will be seen. Huber (1981) describes general conditions under which M-measures of location have both quantitative and qualitative robustness.

M-measures of location are estimated with M-estimators obtained by replacing F in Eq. (2.10) with the empirical distribution $\hat{F}$. (Details are given in Chapter 3.) It should be remarked that many books and journal articles do not make a distinction between M-measures and M-estimators. Ordinarily, the term M-estimator is used to cover both situations.

Of course, to make progress, criteria are needed for choosing ξ or Ψ. One criterion for a robust measure of location is that its influence function be bounded. It turns out that when μ_m is determined with Eq. (2.10), its influence function has a relatively simple form:

$$IF_m(x) = \frac{\Psi(x - \mu_m)}{E[\Psi'(X - \mu_m)]},$$

Table 2.1: Some Choices for ξ and Ψ.

Criterion	$\xi(x)$	$\Psi(x)$	Range
Huber	$\frac{1}{2}x^2$	x	$\|x\| \leq K$
	$\|x\|K - \frac{1}{2}K^2$	$K\,\text{sign}(x)$	$\|x\| > K$
Andrews	$a[1 - \cos(x/a)]$	$\sin(x/a)$	$\|x\| \leq a\pi$
	$2a$	0	$\|x\| > a\pi$
Hampel	$\frac{1}{2}x^2$	x	$\|x\| \leq a$
	$a\|x\| - \frac{1}{2}a^2$	$a\,\text{sign}(x)$	$a < \|x\| \leq b$
	$\dfrac{a(c\|x\| - \frac{1}{2}x^2)}{c-b} - \frac{7}{6}a^2$	$\dfrac{a\,\text{sign}(x)(c - \|x\|)}{c-b}$	$b < \|x\| \leq c$
	$a(b+c-a)$	0	$\|x\| > c$
Biweight		$x(1 - x^2)^2$	$\|x\| < 1$
		0	$\|x\| \geq 1$

where $\Psi'(X - \mu_m)$ is the derivative of Ψ. That is, the influence function is Ψ rescaled by the expected value of its derivative, $E[\Psi'(X - \mu_m)]$. Thus, to obtain a bounded influence function, attention can be restricted to those Ψ that are bounded. From results already given, this rules out the choice $\Psi(X - \mu_m) = X - \mu_m$, which yields $\mu_m = \mu$.

Table 2.1 lists some choices for ξ and Ψ that have been proposed. The function $\text{sign}(x)$ in Table 2.1 is equal to -1, 0, or 1 according to whether x is less than, equal to, or greater than 0. The constants a, b, c, and K can be chosen so that the resulting measure of location has desirable properties. A common strategy is to choose these constants so that when estimating μ_m, the estimator has reasonably high efficiency when sampling from a normal distribution, but continues to have high efficiency when sampling from a heavy-tailed distribution instead. For now, these constants are left unspecified. As will be seen, further refinements can be made that make it a relatively simple matter to choose Ψ in applied work.

Figure 2.2 shows a graph of Ψ for the Huber, Andrews, Hampel, and biweight given in Table 2.1. (The biweight also goes by the name of Tukey's bisquare.) Notice that all four graphs are linear, or approximately so, for an interval around 0. This turns out to be desirable when properties of estimators are considered, but the details are postponed for now. Also notice that the biweight and Andrews' Ψ redescend to 0. That is, extreme values are given less weight in determining μ_m, and x values that are extreme enough are ignored.

As a measure of location, μ_m, given with by Eq. (2.10), satisfies Eqs. (2.1) through (2.3), but it does not satisfy Eq. (2.4), scale equivariance, for the more interesting choices for Ψ, including those shown in Table 2.1. This problem can be corrected by incorporating a measure of scale into Eq. (2.10), but not just any measure of scale will do. In particular, a measure of scale with a high breakdown point is needed if the M-measure of location is to

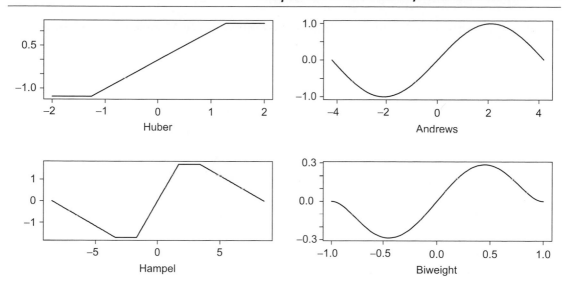

Figure 2.2: Possible Choices for Influence Functions.

have a reasonably high breakdown point as well. The standard deviation, σ, has a breakdown point of 0, so some other measure must be used. A method of dealing with scale equivariance is described in Section 2.4.

2.2.5 R-Measures of Location

R-measures of location do not play a role in this book, but for completeness they are briefly described here. Generally, R-measures of location are derived by inverting tests of hypotheses based on ranks. Let J be some specified function. In functional form, an R-measure of location, μ_r, satisfies

$$\int J \left\{ \frac{1}{2} \left[q + 1 - F\left(2\mu_r - x_q \right) \right] \right\} dq = 0.$$

A common choice for J is $J(x) = x - \frac{1}{2}$. This leads to the Hodges–Lehmann estimator, but no details are given here. For symmetric distributions, the Hodges–Lehmann estimator has a well behaved influence function, it is bounded and smooth (Staudte and Sheather, 1990). However, for asymmetric distributions, the denominator of the influence function can be very small (Huber, 1981, p. 65) suggesting that practical problems might arise.

Another concern was pointed by Bickel and Lehmann (1975). For any R-estimator, which estimates an R-measure of location, there are distributions such that the asymptotic efficiency of the R-estimator relative to $\bar{X}$ is zero. Again it is skewed distributions that create problems. For more details on R-estimators, including the Hodges–Lehmann estimator, see Hettmansperger (1984). For a description of situations where R-estimators exhibit practical concerns even when sampling from a symmetric distribution, see Morgenthaler and Tukey (1991).

2.3 Measures of Scale

This section briefly describes some measures of scale that play an important role in robust methods, plus some popular measures of scale that are not robust. (Some additional measures of scale are described in Chapter 3.) As with measures of location, it helps to start with a precise definition of what constitutes a measure of scale.

Any nonnegative functional, $\tau(X)$, is said to be a *measure of scale* if for any constants $a > 0$ and b,

$$\tau(aX) = a\tau(X) \tag{2.11}$$

$$\tau(X + b) = \tau(X) \tag{2.12}$$

$$\tau(X) = \tau(-X) \tag{2.13}$$

The first of these conditions is called *scale equivariance*, the second is called *location invariance*, and the third is *sign invariance*. From basic principles, σ qualifies as a measure of scale.

Suppose X and Y have a symmetric distribution and that the distribution of $|X|$ is stochastically larger than the distribution of $|Y|$. Bickel and Lehmann (1976) call a measure of scale a *measure of dispersion* if $\tau(X) \geq \tau(Y)$. It can be seen that σ is a measure of dispersion, but as already mentioned, σ has a breakdown point of 0, and its influence function is unbounded.

Currently, there are two general approaches to measuring scale that are of particular importance: L-measures and M-measures. L-measures are estimated with linear combinations of the order statistics, and M-measures are similar to M-measures of location in the sense that τ is defined by the equation

$$E[\chi\left(\frac{x}{\tau}\right)] = 0,$$

where χ is some specified function. Typically χ is an even function, meaning that $\chi(-x) = \chi(x)$.

Mean Deviation from the Mean. A reasonable choice for a measure of scale is

$$\tau(F) = E|X - \mu|.$$

However, its breakdown point is 0 and its influence function is unbounded. On the positive side, the natural estimate of this measure of scale is relatively efficient when sampling from heavy-tailed distributions.

Mean Deviation from the Median. Another popular choice for a measure of scale is

$$\tau(F) = E|X - x_{0.5}|.$$

It might appear that this measure of scale is robust because it uses the median, $x_{0.5}$, but its breakdown point is 0 and its influence function is unbounded.

Median Absolute Deviation. The median absolute deviation, ω, is defined by

$$P(|X - x_{0.5}| \leq \omega) = 0.5.$$

In other words, ω is the median of the distribution associated with $|X - x_{0.5}|$, the distance between X and its median. This measure of scale is an M-measure of scale with $\chi(x) = \mathrm{sign}(|x| - 1)$. The breakdown point is 0.5, and this makes it attractive for certain purposes, as will be seen. Its influence function is

$$IF_\omega(x) = \frac{\mathrm{sign}(|x - x_{0.5}| - \omega) - \frac{f(x_{0.5}+\omega) - f(x_{0.5}-\omega)}{f(x_{0.5})} \mathrm{sign}(x - x_{0.5})}{2[f(x_{0.5} + \omega) + f(x_{0.5} - \omega)]}, \tag{2.14}$$

where $f(x)$ is the probability density function associated with X. Assuming $f(x_{0.5})$ and $2[f(x_{0.5} + \omega) + f(x_{0.5} - \omega)]$ are not equal to 0, IF_ω is defined and bounded. (Alternatives to the median absolute deviation measure of variation were studied by Rousseeuw and Croux, 1993.)

The q-Quantile Range. The q-quantile range is an L-measure of scale given by

$$\tau(F) = x_{1-q} - x_q, \quad 0 < q < 0.5.$$

A special case in common use is the interquartile range where $q = 0.25$, so τ is the difference between the 0.75 and 0.25 quantiles. Its breakdown point is 0.25. Recalling that the influence functions of $x_{0.75}$ and $x_{0.25}$ are given by Eq. (2.8), the influence function of the interquartile range is $IF_{0.75} - IF_{0.25}$. Letting

$$C = q \left\{ \frac{1}{f(x_q)} + \frac{1}{f(x_{1-q})} \right\},$$

a little algebra shows that the influence function of the q-quantile range is

$$IF_{\mathrm{range}} = \begin{cases} \frac{1}{f(x_q)} - C, & \text{if } x < x_q \\ -C, & \text{if } x_q \leq x \leq x_{1-q} \\ \frac{1}{f(x_q)} - C, & \text{if } x > x_{1-q}. \end{cases}$$

The Winsorized Variance. The γ-Winsorized variance is

$$\sigma_w^2 = \int_{x_\gamma}^{x_{1-\gamma}} (x - \mu_w)^2 dF(x) + \gamma [(x_\gamma - \mu_w)^2 + (x_{1-\gamma} - \mu_w)^2].$$

In other words, σ_w^2 is the variance of F after it has been Winsorized. (For a standard normal distribution, $\sigma_w^2 = 0.4129$. It can be shown that σ_w^2 is a measure of scale, it is also a measure of dispersion, and it has a bounded influence function. Welsh & Morrison (1990) report the influence function of a large class of L-measures of scale that contains the Winsorized variance as a special case.

2.4 Scale Equivariant M-Measures of Location

M-measures of location can be made scale equivariant by incorporating a measure of scale in the general approach described in Section 2.2.4. That is, rather than determine μ_m with Eq. (2.10), use

$$E\left[\Psi\left(\frac{X-\mu_m}{\tau}\right)\right] = 0, \qquad (2.15)$$

where τ is some appropriate measure of scale.

When considering which measure of scale should be used in Eq. (2.15), it helps to notice that τ plays a role in determining whether a value for X is unusually large or small. To illustrate this, consider Huber's Ψ which, in the present context, is given by

$$\Psi\left(\frac{x-\mu_m}{\tau}\right) = \begin{cases} -K, & \text{if } (x-\mu_m)/\tau < -K \\ \frac{x-\mu_m}{\tau}, & \text{if } -K \leq (x-\mu_m)/\tau \leq K \\ K, & \text{if } (x-\mu_m)/\tau > K \end{cases}$$

Then according to Ψ, the distance between x and μ_m, $|x-\mu_m|$, is not unusually large or small if $-K \leq (x-\mu_m)/\tau \leq K$. In this case, the same Ψ used to define the population mean, μ, is being used. If $x - \mu_m > K\tau$, Ψ considers the distance to be relatively large, and the influence of x on μ_m is reduced. Similarly, if $x - \mu_m < -K\tau$, x is considered to be unusually far from μ_m.

For the special case where X is normal, and τ is taken to be the standard deviation, σ, x is considered to be unusually large or small if it is more than K standard deviations from μ. A problem with σ is that its value is inflated by heavy-tailed distributions, and this can mask unusually large or small x values. For example, suppose $K = 1.28$, the 0.9 quantile of the standard normal distribution. Then, if X has a standard normal distribution, $K\sigma = 1.28$, so $x = 3$ is considered to be unusually large by Ψ. Now suppose X has the contaminated normal distribution shown in Figure 1.1 of Chapter 1. Then $x = 3$ is still fairly far into the right tail, it should be considered unusually large, but now $K\sigma = 1.28 \times 3.3 = 4.224$, so $x = 3$ is not labeled as being unusually large. What is required is a measure of scale that is relatively insensitive to heavy-tailed distributions so that unusual values are not masked. In particular, a measure of scale with a high breakdown point is needed. Among the measures of scale described in Section 2.5, the median absolute deviation, ω, has a breakdown point of 0.5. This is the highest possible breakdown point, and it is higher than any other measure of scale described in Section 2.5. This suggests using ω in Eq. (2.15) and this choice is typically made. There are other considerations when choosing τ in Eq. (2.15), such as efficiency, but ω remains a good choice. M-measures of location, defined by Eq. (2.15), satisfy the four requirements for measures of location given by Eqs. (2.1) through (2.4).

The influence function of M-measures of location, defined by Eq. (2.15), takes on a more complicated form versus the influence function associated with Eq. (2.10). Moreover, it depends on the choice of scale, τ. As an illustration, suppose $\tau = \omega$ is used, where ω is the median absolute deviation measure of scale introduced in Section 2.5. Let $y = (x - \mu_m)/\omega$. Then, if Eq. (2.15) is used to define a measure of location, the influence function is

$$I F_m(x) = \frac{\omega \Psi(y) - I F_\omega(x)\{E[\Psi'(y)y]\}}{E[\Psi'(y)]},$$

where $I F_\omega$ is given by Eq. (2.14). Note that because the influence function of ω, $I F_\omega$, is bounded, $I F_m$ is bounded as well.

The breakdown point depends on the choice for K and the measure of scale, τ. For the common choice $\tau = \omega$, the breakdown point does not depend on K and is equal to 0.5. Despite this, μ_m can have a value that is further from the median than the 20% trimmed mean. For example, with Huber's Ψ and the common choice of $K = 1.28$, $\mu_m = 4.2$, approximately, for the mixed chi-square distribution in Figure 2.1. In contrast, $\mu_t = 3.9$ and the median is $x_{0.5} = 3.75$. Lowering K to 1, μ_m drops to 4.0.

2.5 Winsorized Expected Values

One final tool is introduced that has practical value in various situations: *Winsorized expected values*. What will be needed is a generalization of $E(X)$ that maintains standard properties of expected values.

Let $g(X)$ be any function of the continuous random variable X. When working with a single random variable, the γ-*Winsorized expected value* of $g(X)$ is defined to be

$$E_w[g(X)] = \int_{x_\gamma}^{x_{1-\gamma}} g(x)dF(x) + \gamma[g(x_\gamma) + g(x_{1-\gamma})].$$

That is, the expected value of $g(X)$ is defined in the usual way, only with respect to the Winsorized distribution corresponding to F. However, a generalization of E_w is needed which provides Winsorized expected values of linear combinations of random variables.

Let X and Y be any two continuous random variables with joint distribution F and probability density function $f(x, y)$. What is needed is an analog of Winsorization for any bivariate

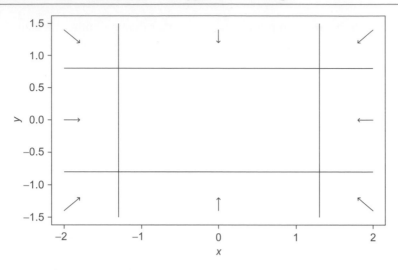

Figure 2.3: Winsorization of a Bivariate Distribution.

distribution. Note that any point (x, y) falls in one of nine regions shown in Figure 2.3, where the corners of the rectangle are determined by the γ and $1 - \gamma$ quantiles of X and Y. That is, the rectangle is given by the four points (x_γ, y_γ), $(x_\gamma, y_{1-\gamma})$, $(x_{1-\gamma}, y_\gamma)$, and $(x_{1-\gamma}, y_{1-\gamma})$. Winsorization of any bivariate distribution consists of pulling in any point outside the rectangle formed by these four points, as indicated by the arrows in Figure 2.3. For any point inside this rectangle, the Winsorized distribution has probability density function $f(x, y)$. The corners of the rectangle become discrete distributions, even when working with continuous random variables. For example, the point (x_γ, y_γ) has probability $P(X \leq x_\gamma, Y \leq y_\gamma)$. Similarly, the point $(x_\gamma, y_{1-\gamma})$ has probability equal to the probability that $X \leq x_\gamma$ and $Y \geq y_{1-\gamma}$, simultaneously. However, the sides of the rectangle, excluding the four corners, have a continuous distribution when X and Y are continuous. Taking expected values with respect to this Winsorized distribution provides the generalization of E_w that will be needed.

More formally, let X and Y be any two random variables with joint distribution F, and let $g(X, Y)$ be any function of X and Y. Following Wilcox (1993b, 1994b), the *Winsorized expected value* of $g(X, Y)$ is defined to be

$$E_w[g(X, Y)] = \int_{x_\gamma}^{x_{1-\gamma}} \int_{y_\gamma}^{y_{1-\gamma}} g(x, y)dF(x, y)$$

$$+ \int_{-\infty}^{x_\gamma} \int_{y_\gamma}^{y_{1-\gamma}} g(x_\gamma, y)dF(x, y) + \int_{-\infty}^{x_\gamma} \int_{-\infty}^{y_\gamma} g(x_\gamma, y_\gamma)dF(x, y)$$

$$+ \int_{-\infty}^{x_\gamma} \int_{y_{1-\gamma}}^{\infty} g(x_\gamma, y_{1-\gamma}) dF(x, y) + \int_{x_{1-\gamma}}^{\infty} \int_{y_\gamma}^{y_{1-\gamma}} g(x_{1-\gamma}, y) dF(x, y)$$

$$+ \int_{x_{1-\gamma}}^{\infty} \int_{-\infty}^{y_\gamma} g(x_{1-\gamma}, y_\gamma) dF(x, y) + \int_{x_{1-\gamma}}^{\infty} \int_{y_{1-\gamma}}^{\infty} g(x_{1-\gamma}, y_{1-\gamma}) dF(x, y)$$

$$+ \int_{x_\gamma}^{x_{1-\gamma}} \int_{-\infty}^{y_\gamma} g(x, y_\gamma) dF(x, y) + \int_{x_\gamma}^{x_{1-\gamma}} \int_{y_{1-\gamma}}^{\infty} g(x, y_{1-\gamma}) dF(x, y).$$

Figure 2.4 illustrates the first step when Winsorizing a bivariate distribution. The bivariate distribution of X and Y is trimmed by removing any points outside the rectangle formed by the four points (x_γ, y_γ), $(x_\gamma, y_{1-\gamma})$, $(x_{1-\gamma}, y_\gamma)$, and $(x_{1-\gamma}, y_{1-\gamma})$.

It can be seen that $E_w(X + Y) = E_w(X) + E_w(Y) = \mu_{wx} + \mu_{wy}$, the sum of the Winsorized means. More generally, for n random variables $X_1, \ldots, X_n$, and constants $c_1, \ldots, c_n$,

$$E_w\left(\sum c_i X_i\right) = \sum c_i E_w(X_i).$$

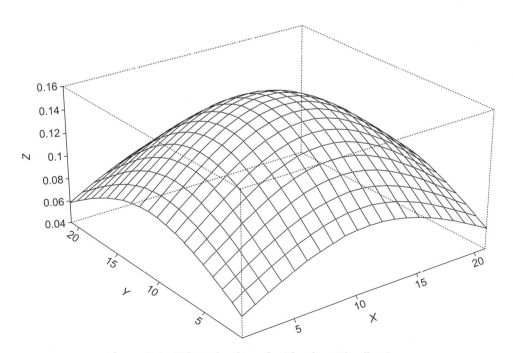

Figure 2.4: Winsorization of a Bivariate Distribution.

Also, if X and Y are independent, $E_w(XY) = E_w(X)E_w(Y)$, the Winsorized variance of $X + Y$ is $\sigma_{wx}^2 + \sigma_{wy}^2$, the sum of the Winsorized variances. These results do not always generalize to other measures of scale. A Winsorized covariance is $\text{COV}_w(X, Y) = E_w[(X - \mu_{wx})(Y - \mu_{wy})]$, and for the situation where the X_i are dependent with possibly different distributions, the Winsorized variance of $E_w(\sum c_i X_i)$ is

$$\sum \sum c_i c_j \text{COV}_w(X_i, X_j).$$

Finally, the definition of E_w makes it a simple matter to find estimates of Winsorized parameters. For a random sample, $X_1, \ldots, X_n$, suppose

$$E_w[g(X_1, \ldots, X_n)] = \xi. \tag{2.16}$$

This indicates that ξ be estimated with $\hat{\xi}_w = g(W_1, \ldots, W_n)$, where

$$W_i = \begin{cases} X_{(k+1)}, & \text{if } X_i \leq X_{(k+1)} \\ X_i, & \text{if } X_{(k+1)} < X_i < X_{(n-k)} \\ X_{(n-k)}, & \text{if } X_i \geq X_{(n-k)}, \end{cases}$$

$X_{(1)} \leq \cdots \leq X_{(n)}$ are the order statistics, and $k = [\gamma n]$, the greatest integer less than or equal to γn. When Eq. (2.16) holds, $\hat{\xi}_w$ is said to be a Winsorized unbiased estimate of ξ. For example, $\bar{W} = \sum W_i / n$ is a Winsorized unbiased estimate of μ_w, and $s_w^2 = \sum (W_i - \bar{W})^2 / (n - 1)$ is a Winsorized unbiased estimate of σ_w^2. Numerical illustrations are given in Chapter 3.

Estimating Measures of Location and Scale

This chapter describes methods of estimating the measures of location and scale introduced in Chapter 2, and it introduces some additional measures of location and scale that have practical importance. Also, two general approaches to estimating standard errors are described and illustrated. One is based on estimating expressions for the standard errors of estimators, which is perhaps the more common strategy to use, and the other is based on a so-called bootstrap method. As will be seen, estimating standard errors is often done in a way that is neither intuitive nor obvious based on standard statistical training. Another goal is to introduce some outlier detection methods plus some graphical methods for summarizing data that will be used in later chapters.

This chapter is less technical than Chapter 2, but it is important at least to touch on theory so that readers understand why common strategies in applied research turn out to be inappropriate. For example, why is it incorrect to discard outliers and apply standard techniques? Although this chapter gives the reader some indication of how theoretical problems are addressed, mathematical details are kept to a minimum. Readers interested in a more rigorous description of mathematical issues can refer to Huber (1981) as well as Hampel et al. (1986). For a book written at an intermediate level of difficulty, see Staudte and Sheather (1990).

3.1 A Bootstrap Estimate of a Standard Error

It is convenient to begin with a description of the most basic bootstrap method for estimating a standard error. Let $\hat{\theta}$ be any estimator based on a random sample of observations, $X_1, \ldots, X_n$. The goal is to estimate $\text{VAR}(\hat{\theta})$, the squared standard error of $\hat{\theta}$. The strategy used by the bootstrap method is based on a very simple idea. Temporarily assume that observations are randomly sampled from some *known* distribution, F. Then for a given sample size, n, the sampling distribution of $\hat{\theta}$ could be determined by randomly generating n observations from F, computing $\hat{\theta}$, randomly generating another set of n observations, computing $\hat{\theta}$, and repeating this process many times. Suppose this is done B times and the resulting values for $\hat{\theta}$ are labeled $\hat{\theta}_1, \ldots, \hat{\theta}_B$. If B is large enough, the values $\hat{\theta}_1, \ldots, \hat{\theta}_B$

provide a good approximation of the distribution of $\hat{\theta}$. In particular, they provide an estimate of the squared standard error of $\hat{\theta}$, namely,

$$\frac{1}{B-1}\sum_{b=1}^{B}(\hat{\theta}_b - \bar{\theta})^2,$$

where

$$\bar{\theta} = \frac{1}{B}\sum_{b=1}^{B}\hat{\theta}_b.$$

That is, VAR($\hat{\theta}$) is estimated with the sample variance of the values $\hat{\theta}_1, \ldots, \hat{\theta}_B$. If, for example, $\hat{\theta}$ is taken to be the sample mean, $\bar{X}$, the squared standard error would be found to be σ^2/n, approximately, provided B is reasonably large. Of course when working with the mean, it is known that its squared standard error is σ^2/n, so the method just described is unnecessary. The only point is that a reasonable method for estimating the squared standard error of $\hat{\theta}$ has been described.

In practice F is not known, but it can be estimated with

$$\hat{F}(x) = \frac{\#\{X_i \le x\}}{n},$$

the proportion of observations less than or equal to x, which provides a nonparametric maximum likelihood estimate of F. The empirical distribution assigns probability $1/n$ to each X_i, so the estimated probability of observing the value X_i is f_i/n, where f_i is the number of times the value X_i occurred among the n observations. All other possible values (values not observed) have an estimated probability of zero. The bootstrap estimate of the standard error is obtained as described in the previous paragraph, except that $\hat{F}$ replaces F. In practical terms, a *bootstrap sample* is obtained by resampling with replacement n observations from $X_1, \ldots, X_n$. This is easily done with the R command

sample(x,size=length(x),replace=T)

To estimate the sampling distribution of $\hat{\theta}$, generate a bootstrap sample from the observations $X_1, \ldots, X_n$ and compute $\hat{\theta}$ based on the obtained bootstrap sample. The result will be labeled $\hat{\theta}^*$ to distinguish it from $\hat{\theta}$, which is based on the observed values $X_1, \ldots, X_n$. Repeat this process B times yielding $\hat{\theta}_1^*, \ldots, \hat{\theta}_B^*$. These B values provide an estimate of the sampling distribution of $\hat{\theta}$ and in particular an estimate of its squared standard error given by

$$S^2 = \frac{1}{B-1}\sum_{b=1}^{B}(\hat{\theta}_b^* - \bar{\theta}^*)^2,$$

where $\bar{\theta}^* = \sum \hat{\theta}_b^*/B$.

How many bootstrap samples should be used? That is, how should B be chosen? This depends, of course, on the goals and criteria that are deemed important. Suppose, for example, estimated standard errors are used to compute a confidence interval. One perspective is to choose B so that the actual probability coverage is reasonably close to the nominal level. Many of the methods in this book are based on this view. However, another approach is to choose B so that if a different collection of bootstrap samples were used, the results would change by a negligible amount. That is, choose B to be sufficiently large so that if the seed in the random number generator is altered, essentially the same conclusions would be obtained. Booth and Sarkar (1998) derived results on choosing B from this latter point of view, and with the increased speed of computers in recent years, some of the newer methods in this book take this latter view into account.

Although the bootstrap estimate of the sampling distribution of a statistic can be argued to be reasonable, it is not immediately clear the extent to which it has practical value. The basic bootstrap methods covered in this book are not a panacea for the many problems that confront the applied researcher, as will become evident in subsequent chapters. But with over 1000 journal articles on the bootstrap, including both theoretical and simulation studies, all indications are that it has great practical value, particularly when working with robust measures of location and scale, as will be seen. Also, there are many proposed ways of possibly improving upon the basic bootstrap methods used in this book, summaries of which are given by Efron and Tibshirani (1993). Some of these look very promising, but the extent to which they have practical value for the problems considered here has not been determined. When testing hypotheses or computing confidence intervals, for some problems, a bootstrap method is the only known method that provides reasonably accurate results.

3.1.1 R Function bootse

As explained in Section 1.7 of Chapter 1, R functions have been written for applying the methods described in this book. The software written for this book is free, and a single command incorporates them into your version of R. Included is the function

$$\text{bootse(x,nboot=1000,est=median)},$$

which can be used to compute a bootstrap estimate of the standard error of virtually any estimator covered in this book. Here, x is any R variable containing the data. The argument nboot represents B, the number of bootstrap samples, and defaults to 1000 if not specified. (As is done with all R functions, optional arguments are indicated by an $=$ and they default to the value shown. Here, for example, the value of nboot is taken to be 1000 if no value is specified by the user.) The argument est indicates the estimator for which the standard error is to be computed. If not specified, est defaults to the median. That is, the standard error of the usual sample median will be estimated. So, for example, if data are stored in the R variable blob, the command bootse(blob) will return the estimated standard error of the usual sample median.

3.2 Density Estimators

Before continuing with the main issues covered in this chapter, it helps to first touch on a related problem that plays a role here as well as in subsequent chapters. The problem is estimating $f(x)$, the probability density function, based on a random sample of observations. Such estimators play an explicit role when trying to estimate the standard error of certain location estimators to be described. More generally, density estimators provide a useful perspective when trying to assess how groups differ and by how much.

Generally, kernel density estimators take the form

$$\hat{f}(x) = \frac{1}{nh} \sum_{i=1}^{n} K\left(\frac{x - X_i}{h}\right),$$

where K is some probability density function and h is a constant to be determined. The constant h has been given several names including the span, the window width, the smoothing parameter, and the bandwidth. Some explicit choices for h are discussed later in this section. Often K is taken to be a distribution symmetric about zero, but there are exceptions. There is a vast literature on kernel density estimators (Silverman, 1986; Scott, 1992; Wand & Jones, 1995; Simonoff, 1996) and research in this area remains active. (For some recent results, see for example, Clements, Hurn, & Lindsay, 2003; Devroye & Lugosi, 2001; Messer & Goldstein, 1993; Yang & Marron, 1999; cf. Liu & Brown, 1993.) Here, four types of kernel density estimators are summarized for later reference.

3.2.1 Normal Kernel

The first of the four methods covered here simply takes K to be the standard normal density. For reasons to be illustrated, the method can be unsatisfactory, but it is the default method used by some software packages, so it is included merely to illustrate potential problems. Following Silverman (1986), as well as the recommendation made by Venables and Ripley (2002, p. 127), the span is taken to be

$$h = 1.06 \min(s, \text{IQR}/1.34)n^{-1/5},$$

where s is the usual sample standard deviation and IQR is some estimate of the interquartile range. That is, IQR estimates the difference between the 0.75 and 0.25 quantiles.

3.2.2 Rosenblatt's Shifted Histogram

The second method, Rosenblatt's shifted histogram estimator, uses results derived by Scott (1979) and Freedman and Diaconis (1981). The computational details are as follows. Set

$$h = \frac{1.2(\text{IQR})}{n^{1/5}}.$$

(Here, IQR will be based on the ideal fourths described in Section 3.12.) Let A be the number of observations less than or equal $x + h$. In symbols,

$$A = \#\{X_i \leq x + h\},$$

where the notation $\#\{X_i \leq x + h\}$ indicates the cardinality of the set of observations satisfying $X_i \leq x + h$. Similarly, let

$$B = \#\{X_i < x - h\},$$

the number of observations less than $x - h$. Then the estimate of $f(x)$ is

$$\hat{f}(x) = \frac{A - B}{2nh}.$$

3.2.3 The Expected Frequency Curve

The next estimator, *the expected frequency curve*, is basically a variation of what is called the naive density estimator, and it is related to certain regression smoothers discussed later in this book. It also has similarities to the nearest neighbor method for estimating densities as described in Silverman (1986). The basic idea when estimating $f(x)$, for a given value x, is to use the proportion of observed values among $X_1, \ldots, X_n$ that are "close" to x.

The method begins by computing the median absolute deviation (MAD) statistic, which is just the sample median of the n values $|X_1 - M|, \ldots, |X_n - M|$, where M is the usual sample median. (For relevant asymptotic results on MAD, see Hall & Welsh, 1985.) Let MADN=MAD/$z_{0.75}$, where $z_{0.75}$ is the 0.75 quantile of a standard normal distribution. Then x is said to be close to X_i if $|X_i - x|/\text{MADN} \leq h$, where h again plays the role of a span. Typically, $h = 0.8$ gives good results. (As is evident, there is no particular reason here to use MADN rather than MAD; it is done merely to follow certain conventions covered in Section 3.6 where MAD is introduced in a more formal manner.) Let N_x be the number of observations close to x in which case $\hat{p}_i = N_i/n$ estimates p_i, the probability that a randomly sampled value is close to X_i. An estimate of the density at x is

$$\hat{f}(x) = \frac{N_x}{2hn\text{MADN}}.$$

In contrast is the naive density estimator discussed by Silverman (1986, Section 2.3) where essentially MADN is replaced by the value 1. That is, the width of the interval around each point when determining N_x depends in no way on the data, but only on the investigators choice for h.

On rare occasions, data are encountered where MAD is zero. In the event this occurs when computing an expected frequency curve, here MAD is replaced by IQR (the interquartile range) estimated via the ideal fourths, as described in Section 3.12.5, and MADN is replaced by IQRN, which is IQR divided by $z_{0.75} - z_{0.25}$, where again $z_{0.75}$ and $z_{0.25}$ are the 0.75 and 0.25 quantiles, respectively, of a standard normal distribution. Now an estimate of the density at x is taken to be

$$\hat{f}(x) = \frac{N_x}{2hn\text{IQRN}}.$$

A criticism of the expected frequency curve is that it can miss bimodality when the span is set to $h = 0.8$. This can be corrected by lowering h to say 0.2, but a criticism of routinely using $h = 0.2$ is that it often yields a rather ragged approximation of the true probability density function. With $h = 0.8$ and n small, again a rather ragged plot can result, but an appealing feature of the method is that often it improves upon the normal kernel in terms of capturing the overall shape of the true distribution.

3.2.4 An Adaptive Kernel Estimator

With large sample sizes, the expected frequency curve typically gives a smooth approximation of the true density, but with small sample sizes a rather ragged approximation can be had. A possible method for smoothing the estimate is to use an adaptive kernel estimate that is known to compete well with other estimators that have been proposed (Silverman, 1986; cf. Politis & Romano, 1997). There are, in fact, many variations of the adaptive kernel estimator, but only one is described here. Following Silverman (1986), let $\tilde{f}(X_i)$ be an initial estimate of $f(X_i)$. Here, $\tilde{f}(X_i)$ is based on the expected frequency curve. Let

$$\log g = \frac{1}{n} \sum \log \tilde{f}(X_i)$$

and

$$\lambda_i = (\tilde{f}(X_i)/g)^{-a},$$

where a is a *sensitivity parameter* satisfying $0 \leq a \leq 1$. Based on comments by Silverman (1986), $a = 0.5$ is used unless stated otherwise. Then the adaptive kernel estimate of f is taken to be

$$\hat{f}(t) = \frac{1}{n} \sum \frac{1}{h\lambda_i} K\{h^{-1}\lambda_i^{-1}(t - X_i)\},$$

where

$$K(t) = \frac{3}{4}\left(1 - \frac{1}{5}t^2\right)\bigg/ \sqrt{5}, \quad |t| < \sqrt{5}$$
$$= 0, \qquad\qquad\qquad \text{otherwise,}$$

is the Epanechnikov kernel, and following Silverman (1986, pp. 47–48), the span is

$$h = 1.06\frac{A}{n^{1/5}},$$

where

$$A = \min(s, \ \text{IQR}/1.34),$$

s is the standard deviation, and IQR is the interquartile range. Again the interquartile range is estimated as described in Section 3.12.5 (using what are called the ideal fourths).

When using an adaptive kernel estimator, perhaps there are advantages to using some initial estimator other than the expected frequency curve. The relative merits of this possibility have not been explored. One reason for using the expected frequency curve as the preliminary estimate is that it reduces problems due to a restriction in range that are known to be a concern when using the normal kernel as described in Section 3.2.1.

3.2.5 R Functions skerd, kerden, kdplot, rdplot, akerd, and splot

It is noted that R has a built-in function called density that computes a kernel density estimate based on various choices for K. (This function also contains various options not covered here.) By default, K is taken to be the standard normal density. Here, the R function

$$\text{skerd(x,op=T,kernel=``gaussian'')}$$

is supplied in the event there is a desire to plot the data based on this collection of estimators. When op=T, the function uses the default density estimator employed by R; otherwise it uses the method recommended by Venables and Ripley (2002, p. 127). (When using R, the default density estimator differs from the one used by S-PLUS, but with op=F, R and S-PLUS use the same method.) To use the Epanechnikov kernel, set the argument kernel=epanechnikov.

The function

$$\text{kerden(x,q=0.5,xval=0)},$$

written for this book, computes the kernel density estimate of $f(x_q)$ for the data stored in the R vector x using the Rosenblatt shifted histogram method, described in Section 3.2.2. (Again, see Section 1.7 on how to obtain the functions written for this book.) If unspecified, q defaults to 0.5. The argument xval is ignored unless q=0, in which case the function estimates f when x is equal to value specified by the argument xval. The function

$$kdplot(x,rval=15)$$

plots the estimate of $f(x)$ based on the function kerden, where the argument rval indicates how many quantiles will be used. The default value, 15, means that $f(x)$ is estimated for 15 quantiles evenly spaced between 0.01 and 0.99, and then the function plots the estimates to form an estimate of $f(x)$.

The R function

$$rdplot(x,fr=NA,plotit=T,pts=NA,pyhat=F)$$

computes the expected frequency curve. The argument fr is the span, h. If not specified, fr=0.8 is used in the univariate case, otherwise fr=0.6 is used. By default, pts=NA (for not available) in which case a plot of the estimated density is based on the points $(X_i, \hat{f}(X_i))$, $i = 1, \ldots, n$. If values are stored in pts, the plot is created based on these points. For example, the command rdplot(mydat,pts=c(0,mydat)) will create a plot based on all of the points in mydat plus the point $(0, \hat{f}(0))$. If pyhat=T, for true, the function returns the $\hat{f}$ values that were computed. So rdplot(mydat,pts=0,pyhat=T) returns $\hat{f}(0)$, and rdplot(mydat,pts=c(1,2),pyhat=T) returns $\hat{f}(1)$ and $\hat{f}(2)$. Setting plotit=F (for false) suppresses the plot. (The function can handle multivariate data and produces a plot in the bivariate case; the computational details are outlined in Chapter 6.)

The R function

$$akerd(x,hval=NA,aval=0.5,op=1,fr=0.8,pts=NA,pyhat=F)$$

applies the adaptive kernel estimate as described in Section 3.2.4, where the argument hval is the span, h, aval is a, the sensitivity parameter, and fr is the span used by the initial estimate based on the expected frequency curve. If the argument op is set to 2, the Epanechnikov kernel is replaced by the normal kernel. The argument hval defaults to NA meaning that if not specified, the span h is determined as described in Section 3.2.4, otherwise h is taken to be the value given by hval. Setting pyhat=T, the function returns the $\hat{f}(X_i)$ values. If pts contains values, the function returns $\hat{f}$ for values in pts instead. (The function can be used with multivariate data and produces a plot in the bivariate case.)

For convenience, when working with discrete data, the function

$$splot(x,op=T,xlab="X",ylab="Rel. Freq.")$$

is supplied which plots the relative frequencies of all distinct values found in the R variable x. With op=T, a line connecting points marking the relative frequencies is added to the plot.

■ Example

Table 3.1 shows data from a study dealing with hangover symptoms for two groups of individuals: sons of alcoholics and a control group. Note that for both groups, zero is the most common value, and it is fairly evident that the data do not have a bell-shaped distribution. The top two panels of Figure 3.1 show an estimate of the distributions using the R function skerd. (The plots are based on the data for Group 1.) This illustrates a well-known problem with certain kernel density estimators: a restriction in

Table 3.1: The Effect of Alcohol.

Group 1	0	0	0	0	0	0	0	0	2	2
	3	3	6	9	11	11	11	18	32	41
Group 2	0	0	0	0	0	0	0	0	0	0
	0	0	0	0	1	2	3	8	12	32

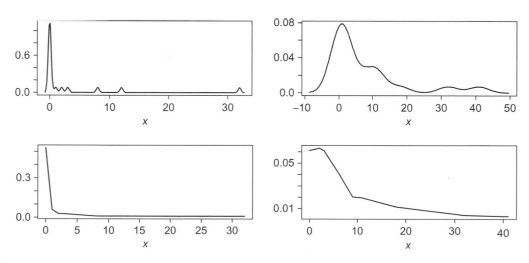

Figure 3.1: An example comparing four plots of data. The upper left panel shows a kernel density estimate using a normal kernel based on the Group 1 data in Table 3.1. The upper right panel is the estimate using the Group 2 data. The bottom left panel used the same data as in the upper left panel, only the adaptive kernel density estimator was used. The lower right panel used the adaptive kernel density estimate with the Group 2 data.

the range of possible values can lead to highly unsatisfactory results. This is clearly the case here because values less than zero are impossible, in contrast to what is suggested, particularly in the top right panel. The bottom two panels are plots of the data using the adaptive kernel density estimator in Section 3.2.4. The method handles the restriction in range reasonably well and provides what seems like a much more satisfactory summary of the data.

■

■ Example

Figure 3.2 shows the plots created by the R functions just described for $n = 30$ observations randomly sampled from a standard normal distribution. The upper left panel was produced by the function skerd, the upper right panel shows the curve produced by kdplot (which uses the method in Section 3.2.2), the lower left panel is the expected frequency curve (using the function rdplot), and the lower right panel (created by the function akerd) is based on the adaptive kernel estimator in Section 3.2.4. Plots based on akerd are typically smoother than the plot returned by rdplot, particularly when using small sample sizes. Generally, the expected frequency curve and the adaptive kernel density estimator seem more likely to give a better overall sense of what the data

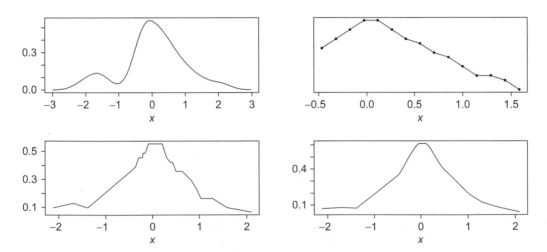

Figure 3.2: Another example comparing four plots of data. The upper left panel shows a kernel density estimate using a normal kernel based on $n = 30$ observations sampled from a standard normal distribution. The upper right panel is the plot (based on the same data) using Rosenblatt's shifted historgram. The lower left panel is the expected frequency curve and lower right panel is based on the adaptive kernel estimator.

are like when sampling from a heavy-tailed distribution, but this issue has not been studied extensively.

■

■ Example

To add perspective, 500 observations were generated from the lognormal distribution shown in Figure 3.3. This particular distribution is defined only for $X > 0$. That is, $P(X < 0) = 0$. The upper left panel of Figure 3.4 shows the plot created by R using the normal kernel as described in Section 3.2.1. The upper right panel is based on Rosenblatt's shifted histogram (described in method in Section 3.2.2), the lower left panel is based on the expected frequency curve (using the function kdplot), and the lower right panel is the plot based on the adaptive kernel estimator described in Section 3.2.4. Notice that both rdplot and kdplot do a better job of capturing the shape of the true density. The output from skerd is too Gaussian on the left (for $x \leq 5$), meaning that it resembles a normal curve when it should not, and it performs rather poorly for $X < 0$. (A similar problem arises when sampling from an exponential distribution.) Setting op=F when using skerd, the restriction in range associated with the lognormal distribution is less of a problem, but the plot becomes rather ragged. Increasing the sample size to $n = 1000$ and changing the seed in the R random number generator produces results very similar to those in Figure 3.4. The R function density has optional arguments that replace the normal kernel with other functions, several of these

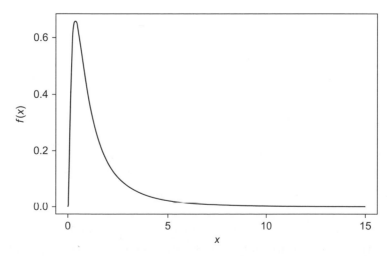

Figure 3.3: A lognormal distribution.

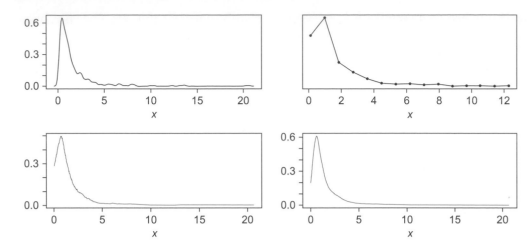

Figure 3.4: Some kernel density estimators can perform poorly when the variable under study is bounded, even with large sample sizes. The upper left panel is a plot based on the function skerd and $n = 500$ randomly sampled observations from the lognormal distribution in Figure 3.3. The upper right panel is Rosenblatt's shifted histogram (using on the function rdplot), the lower left panel shows the plot created by kdplot, and the lower right panel is based on the adaptive kernel estimator.

were considered for the situation at hand, but similar results were obtained. So although situations are encountered where the R function skerd produces a smoother, more visually appealing plot versus rdplot, kdplot, and akerd, blind use of this function can be misleading.

3.3 The Sample Trimmed Mean

As already indicated, the standard error of the sample mean can be relatively large when sampling from a heavy-tailed distribution, and the sample mean estimates a nonrobust measure of location, μ. The sample trimmed mean addresses these problems.

The sample trimmed mean, which estimates the population trimmed μ_t (described in Section 2.2.3), is computed as follows. Let $X_1, \ldots, X_n$ be a random sample and let $X_{(1)} \le X_{(2)} \le \cdots \le X_{(n)}$ be the observations written in ascending order. The value $X_{(i)}$ is called the ith *order statistic*. Suppose the desired amount of trimming has been chosen to be $\gamma, 0 \le \gamma < 0.5$. Let $g = [\gamma n]$, where $[\gamma n]$ is the value of γn rounded down to the nearest integer. For example, $[10.9] = 10$. The sample trimmed mean is computed by removing the g largest and g smallest observations and averaging the values that remain. In symbols, the sample trimmed mean is

$$\bar{X}_t = \frac{X_{(g+1)} + \cdots + X_{(n-g)}}{n - 2g}.$$ (3.1)

In essence, the empirical distribution is trimmed in a manner consistent with how the probability density function was trimmed when defining μ_t. As indicated in Chapter 2, two-sided trimming is assumed unless stated otherwise.

The definition of the sample trimmed mean given by Eq. (3.1) is the one most commonly used. However, for completeness, it is noted that the term trimmed mean sometimes refers to a slightly different estimator (e.g., Reed, 1998; cf. Hogg, 1974), namely,

$$\frac{1}{n(1 - 2\gamma)} \left(\sum_{i=g+1}^{n-g} X_{(i)} + (g - \gamma n)(X_{(g)} + X_{(n-g+1)}) \right).$$

Also see Patel, Mudholkar, and Fernando (1988) as well as Kim (1992a). Here, however, the definition given by (3.1) is used exclusively.

For the trimmed mean to have any practical importance, a value for γ must be chosen. One approach is to choose γ so that $\bar{X}_t$ tends to have a relatively small standard error among commonly occurring situations, and a related restriction might be that little accuracy is lost when sampling from a normal distribution. Based on this view, and other criteria to be described, a good choice for general use is $\gamma = 0.2$. If γ is too small, the standard error of the trimmed mean, $\sqrt{\mathrm{VAR}(\bar{X}_t)}$, can be drastically inflated by outliers or sampling from a heavy-tailed distribution. If γ is too large, the standard error can be relatively large compared with the standard error of the sample mean when sampling from a normal distribution. (Some illustrations are given in Section 3.11.) Empirical investigations based on data from actual studies suggest that the optimal amount of trimming, in terms of minimizing the standard error, is usually between 0 and 0.25 (e.g., Hill & Dixon, 1982; Wu, 2002). Although $\gamma = 0.1$, for example, might be more optimal than $\gamma = 0.2$ in certain situations, one argument for using $\gamma = 0.2$ is that it can result in a standard error that is much smaller than the standard error associated with $\gamma = 0.1$ or $\gamma = 0$, but the reverse is generally untrue. That is, $\gamma = 0.2$ guards against complete disaster but sacrifices relatively little in situations where $\gamma = 0.1$ and $\gamma = 0$ are more optimal. In some cases, however, more than 20% trimming might be desirable.

Another approach is to determine γ empirically according to some criterion such as the standard error. That is, estimate the standard error of $\bar{X}_t$ when, for example, $\gamma = 0, 0.1$, and 0.2 and then use the value of γ corresponding to the smallest estimate. These so-called adaptive trimmed means have been studied by Leger and Romano (1990a,b) and Leger, Politis, and Romano (1992). Or one could determine the amount of trimming based on some measure of skewness and heavy-tailedness. For a recent summary and comparison of such methods, see Reed (1998) as well as Reed and Stark (1996). The properties of this approach, in the context of testing hypotheses and computing confidence intervals, have not been studied to the extent where γ is chosen to be a prespecified constant. In particular, the practical utility of adaptive trimmed means needs further investigation, so they are not discussed here, but further investigation seems warranted. It should be remarked, however, that empirically determining how much trimming to do is fraught with difficulties that are not always obvious. Interested readers can read the discussions of a paper by Hogg (1974), especially the comments by P. Huber. For some results on so-called hinge estimators, regarding control over the probability of a type I error, see Keselman, Wilcox, Lix, Algina, and Fradette (2003).

■ Example

Dana (1990) conducted a study dealing with self-awareness and self-evaluation. One segment of his study measured the time subjects could keep a portion of an apparatus in contact with a specified target. Table 3.2 shows some data for one of the groups. The sample mean and the sample trimmed means with $\gamma = 0.1$ and 0.2 are 448, 343, and 283, respectively. In this particular case, there is an obvious difference between the three measures of location. However, even if they had been nearly equal, this is not necessarily an indication that the sample mean is satisfactory because the standard error of the trimmed mean can be substantially smaller than the standard error of the mean.

Table 3.2: Self-Awareness Data.

77	87	88	114	151	210	219	246	253	262
296	299	306	376	428	515	666	1310	2611	

It might seem that $\gamma = 0.2$ is equivalent to randomly throwing away 40% of the data, but this is not the case. To see why, notice that the order statistics are dependent even though the observations $X_1, \ldots, X_n$ are independent. This result is covered in basic texts on mathematical statistics. For readers unfamiliar with this result, a brief explanation will help shed some light on other practical problems covered in this chapter.

If the random variables X and Y are independent, then the probability function of X is not altered given the value of Y. This means in particular that the range of possible values of X cannot depend on the value of Y. Suppose $X_1, \ldots, X_n$ is a random sample, and for the sake of illustration, suppose the value of each random variable can be any of the integers between 1 and 10 inclusive, each value occurring with some positive probability. Then knowing that

$X_2 = 3$, say, tells us nothing about the probability of observing a particular value for X_1. However, suppose $X_{(2)} = 3$. Then, the smallest value, $X_{(1)}$, cannot be 4. More generally, $X_{(1)}$ cannot be any number greater than 3. In contrast, if we do not know the value of $X_{(2)}$, or any of the other order statistics, $X_{(1)}$ could have any of the values $1, 2, \ldots, 10$, and these values occur with some positive probability. Thus, knowing the value of $X_{(2)}$ alters the probabilities associated with $X_{(1)}$. That is, $X_{(1)}$ and $X_{(2)}$ are dependent and dependence occurs because knowing the value of $X_{(2)}$ restricts the range of possible values for $X_{(1)}$. More generally, any two order statistics, say $X_{(i)}$ and $X_{(j)}$, $i \neq j$, are dependent.

3.3.1 R Functions mean, tmean, and lloc

R has a built-in function that evaluates the trimmed mean. If observations are stored in the vector x, the R command

$$\text{mean(x,trim=0)}$$

computes the γ-trimmed mean where the argument trim determines the amount of trimming. By default, the amount of trimming is 0. For example, mean(x,0.2) returns the 20% trimmed mean. The value 283 is returned for the data in Table 3.2, assuming the data are stored in the R variable x. Because it is common to use 20% trimming, for convenience the R function

$$\text{tmean(x,tr=0.2)}$$

has been supplied, which computes a 20% trimmed mean by default using the data stored in the R variable x. The amount of trimming can be altered using the argument tr. So tmean(blob) will compute a 20% trimmed mean for the data stored in blob, and tmean(blob,tr=0.3) will use 30% trimming instead. For convenience, the function

$$\text{lloc(x,est=tmean,}\ldots\text{)}$$

is supplied for computing a trimmed mean when data are stored in list mode, in a data frame, or a matrix. If x is a matrix or data frame, lloc computes the trimmed mean for each column. Other measures of location can be used via the argument est. (For example, est=median will compute the median.) The argument $\ldots$ means that an optional argument associated with est can be used.

3.3.2 Estimating the Standard Error of the Trimmed Mean

To have practical value when making inferences about μ_t, properties of the sampling distribution of $\bar{X}_t$ need to be determined. This subsection takes up the problem of estimating $\sqrt{\text{VAR}(\bar{X}_t)}$, the standard error of the sample trimmed mean.

At first glance the problem might appear to be trivial. The standard error of the sample mean is $\sigma/\sqrt{n}$, which is estimated with $s/\sqrt{n}$, where

$$s^2 = \frac{1}{n-1}\sum(X_i - \bar{X})^2$$

is the usual sample variance. A common mistake in applied work is to estimate the standard error of the trimmed mean by simply computing the sample standard deviation of the untrimmed observations, and then dividing by $\sqrt{n-2g}$, the square root of the number of observations left after trimming. That is, apply the usual estimate of the standard error using the untrimmed values. To see why this simple idea fails, let $X_1, \ldots, X_n$ be any random variables, possibly dependent with unequal variances, and let $a_1, \ldots, a_n$ be any n constants. Then the variance of $\sum a_i X_i$ is

$$\text{VAR}\left(\sum a_i X_i\right) = \sum_{i=1}^{n}\sum_{i=1}^{n} a_i a_j \text{COV}(X_i, X_j), \tag{3.2}$$

where $\text{COV}(X_i, X_j)$ is the covariance between X_i and X_j. That is,

$$\text{COV}(X_i, X_j) = E\{(X_i - \mu_i)(X_j - \mu_j)\},$$

where $\mu_i = E(X_i)$. When $i = j$, $\text{COV}(X_i, X_j) = \sigma_i^2$, the variance of X_i. When the random variables are independent, Eq. (3.2) reduces to

$$\text{VAR}\left(\sum a_i X_i\right) = \sum_{i=1}^{n} a_i^2 \sigma_i^2. \tag{3.3}$$

Under random sampling, in which case the variance of each of the n random variables has a common value σ^2, the variance of $\bar{X}$ can be seen to be σ^2/n by taking $a_i = 1/n, i = 1, \ldots, n$, in (3.3). The problem with the sample trimmed mean is that it is a linear combination of dependent random variables, namely a linear combination of the order statistics, so Eq. (3.3) does not apply, Eq. (3.2) must be used instead. For $i \neq j$, there are asymptotic results that can be used to estimate $\text{COV}(X_{(i)}, X_{(j)})$, the covariance between the ith and jth order statistics, this suggests a method for estimating the standard error of a trimmed mean, but a simpler method for estimating $\text{VAR}(\bar{X}_t)$ is typically used and found to give good results.

The influence function of the trimmed mean, $IF_t(x)$, introduced in Chapter 2, provides a convenient and useful way of dealing with the dependence among the order statistics. It can be shown that

$$\bar{X}_t = \mu_t + \frac{1}{n}\sum_{i=1}^{n} IF_t(X_i), \tag{3.4}$$

plus a remainder term that goes to zero as n gets large. Moreover, $E(IF_t(X_i)) = 0$. In words, the sample trimmed mean can be written as μ_t plus a sum of independent, identically distributed random variables (assuming random sampling) having mean 0, plus a term that can be ignored provided n is not too small. The central limit theorem, applied to (3.4), shows that the distribution of $\bar{X}_t$ approaches a normal distribution as $n \to \infty$. Fortunately, all indications are that the error term can be ignored even when n is as small as 10. Because $IF_t(X)$ has mean 0, Eq. (3.3) can be used to show that

$$\text{VAR}(\bar{X}_t) = \frac{1}{n^2} \sum E\{(IF_t(X_i))^2\}, \tag{3.5}$$

ignoring the error term. From Chapter 2,

$$(1 - 2\gamma)IF_t(X) = \begin{cases} x_\gamma - \mu_w, & \text{if } x < x_\gamma \\ X - \mu_w, & \text{if } x_\gamma \leq X \leq x_{1-\gamma} \\ x_{1-\gamma} - \mu_w, & \text{if } x > x_{1-\gamma}, \end{cases}$$

where μ_w is the Winsorized population mean, and x_γ is the γ quantile. The main point here is that an estimate of $E\{(IF_t(X_i))^2\}$ yields an estimate of $\text{VAR}(\bar{X}_t)$ via Eq. (3.5). Note that

$$P\left(IF_t(X) = \frac{x_\gamma - \mu_w}{1 - 2\gamma}\right) = \gamma$$

$$P\left(IF_t(X) = \frac{x_{1-\gamma} - \mu_w}{1 - 2\gamma}\right) = \gamma.$$

The first step in estimating $E\{(IF_t(X_i))^2\}$ is estimating the population Winsorized mean, μ_w. This is done by Winsorizing the empirical distribution and computing the sample mean of what results. *Winsorization of a random sample* consists of setting

$$W_i = \begin{cases} X_{(g+1)}, & \text{if } X_i \leq X_{(g+1)} \\ X_i, & \text{if } X_{(g+1)} < X_i < X_{(n-g)} \\ X_{(n-g)}, & \text{if } X_i \geq X_{(n-g)}. \end{cases} \tag{3.6}$$

The *Winsorized sample mean* is

$$\bar{X}_w = \frac{1}{n} \sum W_i,$$

which estimates μ_w, the population Winsorized mean introduced in Chapter 2. In words, Winsorization means that the g smallest values are pulled in and set equal to $X_{(g+1)}$, and the g largest values are pulled in and set equal to $X_{(n-g)}$. The sample mean of the resulting values is the Winsorized sample mean. (For a detailed study of the sample Winsorized mean when sampling from a skewed distribution, see Rivest, 1994.) Put another way, Winsorization

Table 3.3: Winsorized Values for the Self-Awareness Data.

114	114	114	114	151	210	219	246	253	262
296	299	306	376	428	515	515	515	515	

consists of estimating the γ and $1 - \gamma$ quantiles with $X_{(g+1)}$ and $X_{(n-g)}$, respectively, and estimating the population Winsorized distribution, described in Chapter 2, with the resulting W_i values. Also, results on the Winsorized expected value provide a more formal way of justifying $\bar{X}_w$ as an estimate of μ_w. In particular, it is readily verified that $E_w(\bar{X}) = \mu_w$, so $\bar{X}_w$ is a Winsorized unbiased estimate of μ_w.

Table 3.3 shows the Winsorized values for the data in Table 3.2 when $\gamma = 0.2$, in which case $g = [0.2(19)] = 3$. Thus, Winsorizing the observations in Table 3.2 consists of replacing the three smallest observations with $X_{(4)} = 114$. Similarly, because $n - g = 19 - 3 = 16$, the three largest observations are replaced by $X_{(16)} = 515$. The sample mean of the values in Table 3.3 is 293, and this is equal to the 20% Winsorized sample mean for the data in Table 3.2.

The expression for the influence function of the trimmed mean involves three unknown quantities: x_γ, $x_{1-\gamma}$, and μ_w. As already indicated, these three unknown quantities are estimated with $X_{(g+1)}$, $X_{(n-g)}$, and $\bar{X}_w$, respectively. A little algebra shows that an estimate of $E[IF_t(X_i)]$ is $(W_i - \bar{W})/(1 - 2\gamma)$, so an estimate of $E\{(IF(X_i))^2\}$ is $(W_i - \bar{W})^2/(1 - 2\gamma)^2$. Referring to Eq. (3.5), the resulting estimate of VAR($\bar{X}_t$) is

$$\frac{1}{n^2(1-2\gamma)^2}\sum (W_i - \bar{W})^2.$$

When there is no trimming, this last equation becomes

$$\frac{n-1}{n} \times \frac{s^2}{n},$$

but typically s^2/n is used instead. Accordingly, to be consistent with how the standard error of the sample mean is usually estimated,

$$\frac{1}{n(n-1)(1-2\gamma)^2}\sum (W_i - \bar{W})^2 \tag{3.7}$$

will be used to estimate VAR($\bar{X}_t$).

The quantity

$$s_w^2 = \frac{1}{n-1}\sum (W_i - \bar{W})^2 \tag{3.8}$$

is called the *sample Winsorized variance*. A common way of writing (3.7) is in terms of the Winsorized variance:

$$\frac{s_w^2}{(1-2\gamma)^2 n}. \tag{3.9}$$

In other words, to estimate the squared standard error of the trimmed mean, compute the Winsorized observations W_i using Eq. (3.6), compute the sample variance using the resulting values, then divide by $(1-2\gamma)^2 n$. Using the notion of Winsorized expected values as described in Chapter 2, $E_w(s^2) = \sigma_w^2$, and this helps to justify s_w^2 as an estimate σ_w^2, the population Winsorized variance. That is, s_w^2 is a Winsorized unbiased estimate of the population Winsorized variance. It can be seen that $E_w\{(IF(X_i))^2\} = \sigma_w^2/(1-2\gamma)^2$, and this provides another way of justifying (3.9) as an estimate of $\mathrm{VAR}(\bar{X}_t)$. Consequently, the standard error of the sample trimmed mean is estimated with

$$\sqrt{\frac{s_w^2}{(1-2\gamma)^2 n}} = \frac{s_w}{(1-2\gamma)\sqrt{n}}.$$

Table 3.4 summarizes the calculations used to estimate the standard error of the trimmed mean.

■ Example

For the data in Table 3.3, the sample variance is 21,551.4 and this is the Winsorized sample variance for the data in Table 3.2. Because 20% trimming was used, $\gamma = 0.2$, and the estimated standard error of the sample trimmed mean is

$$\frac{\sqrt{21,551.4}}{[1-2(0.2)]\sqrt{19}} = 56.1$$

In contrast, the standard error of the sample mean is $s/\sqrt{n} = 136$, a value that is approximately 2.4 times larger than the standard error of the trimmed mean.

■

Table 3.4: Summary of How to Estimate the Standard Error of the Trimmed Mean.

To estimate the standard error of the trimmed mean based on a random sample of n observations, first Winsorize the observations by transforming the ith observation, X_i, to W_i using Eq. (3.6). Compute the sample variance of the W_i values yielding s_w^2, the Winsorized sample variance. The standard error of the trimmed mean is estimated to be

$$\frac{s_w}{(1-2\gamma)\sqrt{n}},$$

where γ is the amount of trimming chosen by the investigator.

3.3.3 Estimating the Standard Error of the Sample Winsorized Mean

An estimate of the standard error of the sample Winsorized mean, $\bar{X}_w$, can be derived from the influence function of the population Winsorized mean given in Section 2.2.2. Dixon and Tukey (1968) suggest a simpler estimate:

$$\frac{n-1}{n-2g-1} \times \frac{s_w}{\sqrt{n}},$$

where $g = [\gamma n]$ is the number of observations Winsorized in each tail, so $n - 2g$ is the number of observations that are not Winsorized.

3.3.4 R Functions winmean, winvar, trimse, and winse

Included in the R functions written for this book is a function called winmean that computes the Winsorized mean. If the data are stored in the R variable x, it has the form

$$\text{winmean}(x, tr=0.2).$$

The optional argument tr is the amount of Winsorizing, which defaults to 0.2 if unspecified. (The R function win also computes the Winsorized mean.) For example, the command winmean(dat) computes the 20% Winsorized mean for the data in the R vector dat. The command winmean(x,0.1) computes the 10% Winsorized mean. If there are any missing values (stored as NA in R), the function automatically removes them.

The function winvar computes the Winsorized sample variance, s_w^2. It has the form

$$\text{winvar}(x, tr=0.2).$$

Again, tr is the amount of Winsorization which defaults to 0.2 if unspecified. The function

$$\text{trimse}(x, tr=0.2)$$

estimates the standard error of the trimmed mean and

$$\text{winse}(x, tr=.2)$$

estimates the standard error of the Winsorized mean. For example, the R command trimse(x,0.1) estimates the standard error of the 10% trimmed mean for the data stored in the vector x, and winvar(x,0.1) computes the Winsorized sample variance using 10% Winsorization. The R command winvar(x) computes s_w^2 using 20% Winsorization.

3.3.5 Estimating the Standard Error of the Sample Median, M

Trimmed means contain the usual sample median, M, as a special case where the maximum amount of trimming is used. When using M and the goal is to estimate its standard error, alternatives to Eq. (3.9) should be used. Many methods have been proposed, comparisons of which were made by Price and Bonett (2001). In terms of hypothesis testing, an effective and fairly simple estimate appears to be one derived by McKean and Schrader (1984). To apply it, compute

$$k = \frac{n+1}{2} - z_{0.995}\sqrt{\frac{n}{4}},$$

where k is rounded to the nearest integer and $z_{0.995}$ is the 0.995 quantile of a standard normal distribution. Put the observed values in ascending order yielding $X_{(1)} \leq \cdots \leq X_{(n)}$. Then the McKean–Schrader estimate of the squared standard error of M is

$$\left(\frac{X_{(n-k+1)} - X_{(k)}}{2z_{0.995}}\right)^2.$$

(Price & Bonett, 2001 recommend a slightly more complicated estimator, but when computing a confidence interval for the median, currently it seems that their method offers little or no advantage.)

3.3.6 R Function msmedse

The R function

$$\text{msmedse(x)}$$

estimates the standard error of M the square root of the last equation.

3.4 The Finite Sample Breakdown Point

Before describing additional measures of location, it helps to introduce a technical device for judging any estimator that is being considered. This is the *finite sample breakdown point* of a statistic, which refers to the smallest proportion of observations that, when altered sufficiently, can render the statistic meaningless. More precisely, the finite sample breakdown point of an estimator refers to the smallest proportion of observations that when altered can cause the value of the statistic to be arbitrarily large or small. The finite sample breakdown point of an estimator is a measure of its *resistance* to contamination. For example, if the ith observation among the observations $X_1, \ldots, X_n$ goes to infinity, the sample mean $\bar{X}$ goes to infinity as well. This means that the finite sample breakdown point of the sample mean is only $1/n$. In contrast, the finite sample breakdown point of the γ-trimmed mean is γ. For example, if $\gamma = 0.2$, about 20% of the observations can be made arbitrarily large without driving the sample trimmed mean to infinity, but it is possible to alter 21% of the observations so that $\bar{X}_t$ becomes arbitrarily large. Typically, the limiting value of the finite sample breakdown point is equal to the breakdown point, as defined in Chapter 2, of the parameter being estimated. For example, the breakdown point of the population mean, μ, is 0, which equals $1/n$ as n goes to infinity. Similarly, the breakdown point of the trimmed mean is γ.

Two points should be stressed. First, having a high finite-sample breakdown point is certainly a step in the right direction when trying to deal with unusual values that have an inordinate influence, but it is no guarantee that an estimator will not be unduly influenced by even a small number of outliers. (Examples will be given when dealing with robust regression estimators.) Second, various refinements regarding the definition of a breakdown point have been proposed (e.g., Genton & Lucas, 2003), but no details are given here.

3.5 Estimating Quantiles

When comparing two or more groups, the most common strategy is to use a single measure of location, and the median or 0.5 quantile is an obvious choice. It can be highly advantageous to compare other quantiles as well, but the motivation for doing this is best explained in Chapter 5. For now, attention is focused on estimating quantiles and the associated standard error.

There are many ways of estimating quantiles, comparisons of which are reported by Parrish (1990), Sheather and Marron (1990), as well as Dielman, Lowry, and Pfaffenberger (1994). Here, two are described and their relative merits are discussed.

For any q, $0 < q < 1$, let x_q be the qth quantile. For a continuous random variable, or a distribution with no flat spots, x_q is defined by the equation $P(X \leq x_q) = q$. This definition is satisfactory in the sense that there is only one value that qualifies as the qth quantile, so there is no ambiguity when referring to x_q. However, for discrete random variables or distributions with flat spots, special methods must be used to avoid having multiple values that qualify as the qth quantile. There are methods for accomplishing this goal, but they are not directly relevant to the topics of central interest in this book, at least based on current technology, so this issue is not discussed.[1]

Setting $m = [qn + 0.5]$, where $[qn + 0.5]$ is the greatest integer less than or equal to $qn + 0.5$, the simplest estimate of x_q is

$$\hat{x}_q = X_{(m)},$$

the mth observation after the data are put in ascending order. For example, if the goal is to estimate the median, then $q = 1/2$, and if $n = 11$, then $m = [11/2 + 0.5] = 6$, and the estimate

---

[1] The usual method for defining quantiles is as follows. If F is the distribution of the random variable X, then the qth quantile is the greatest lower bound, or infimum, for the set of values $\{x : F(x) \geq q\}$. Usually this is written as $x_q = \inf\{x : F(x) \geq q\}$.

of $x_{.5}$ is the usual sample median, M. Of course, if n is even, this estimator does not yield the usual sample median, it is equal to what is sometimes called the *upper empirical cumulative distribution function estimator*.

3.5.1 Estimating the Standard Error of the Sample Quantile

Assuming that observations are randomly sampled from a continuous distribution, and that $f(x_q) > 0$, the influence function of the qth quantile is

$$IF_q(x) = \begin{cases} \frac{q-1}{f(x_q)}, & \text{if } x < x_q \\ 0, & \text{if } x = x_q \\ \frac{q}{f(x_q)}, & \text{if } x > x_q, \end{cases} \tag{3.10}$$

and

$$\hat{x}_q = x_q + \frac{1}{n}\sum IF_q(X_i)$$

plus a remainder term that goes to zero as n gets large. That is, the situation is similar to the trimmed mean in the sense that the estimate of the qth quantile can be written as x_q, the population parameter being estimated, plus a sum of independent identically distributed random variables having a mean of zero, plus a term that can be ignored as the sample size gets large. Consequently, the influence function of the qth quantile can be used to determine the (asymptotic) standard error of $\hat{x}_q$. The result is

$$VAR(\hat{x}_q) = \frac{q(1-q)}{n[f(x_q)]^2}. \tag{3.11}$$

For example, when estimating the median, $q = 0.5$, and the variance of $\hat{x}_{.5}$ is

$$\frac{1}{4n[f(x_{.5})]^2},$$

so the standard error of $\hat{x}_{0.5}$ is

$$\frac{1}{2\sqrt{n}f(x_{.5})}.$$

Moreover, for any q between 0 and 1,

$$2\sqrt{n}f(x_q)(\hat{x}_q - x_q)$$

approaches a standard normal distribution as n goes to infinity.

Using Eq. (3.11) to estimate the standard error of $\hat{x}_q$ requires an estimate of $f(x_q)$, the probability density function of X evaluated at x_q, and this can be done using one of the methods described in Section 3.2. It is suggested that the adaptive kernel estimator be used in most cases, but all four kernel density estimators can be used with the software provided in case there are known reasons for preferring one kernel density estimator over another. The advantages of using a boostrap estimate of the standard error, over the method outlined here, have not been investigated.

■ Example

The data in Table 3.2 are used to illustrate how the standard error of $\hat{x}_{.5}$ can be estimated when using Rosenblatt's shifted histogram estimate of $f(x)$. There are 19 observations, so $[0.25n + 0.5] = 5$, $[0.75n + 0.5] = 14$, and an estimate of the interquartile range is $X_{(14)} - X_{(5)} = 376 - 151 = 225$, so

$$h = \frac{1.2(225)}{19^{1/5}} = 149.8.$$

The sample median is $M = \hat{x}_{0.5} = X_{(10)} = 262$, so $x_{0.5} + h = 411.8$, and the number of observations less than or equal to 411.8 is $A = 14$. The number of observations less than $\hat{x}_{0.5} - h = 112.2$ is $B = 3$, so

$$\hat{f}(\hat{x}_{0.5}) = \frac{14 - 3}{2(19)(149.8)} = 0.00193.$$

Consequently, an estimate of the standard error of the sample median is

$$\frac{1}{2\sqrt{19}(0.00193)} = 59.4.$$

■

3.5.2 R Function qse

The R function

$$qse(x,q=0.5,op=3)$$

estimates the standard error of $\hat{x}_q$ using Eq. (3.11). As indicated, the default value for q is 0.5. The argument op determines which density estimator is used to estimate $f(x_q)$. The choices are:

- op=1, Rosenblatt's shifted histograms
- op=2, expected frequence curve
- op=3, adaptive kernel method.

For example, storing the data in Table 3.2 in the R vector x, the command qse(x,op=1) returns the value 64.3. In contrast, using op=2 and op=3, the estimates are 58.94 and 47.95, respectively. So the choice of density estimator can make a practical difference.

3.5.3 The Maritz–Jarrett Estimate of the Standard Error of $\hat{x}_q$

Maritz and Jarrett (1978) derived an estimate of the standard error of sample median, which is easily extended to the more general case involving $\hat{x}_q$. That is, when using a single order statistic, its standard error can be estimated using the method outlined here. It is based on the fact that $E(\hat{x}_q)$ and $E(\hat{x}_q^2)$ can be related to a beta distribution. The beta probability density function, when a and b are positive integers, is

$$f(x) = \frac{(a+b+1)!}{a!b!} x^a (1-x)^b, \quad 0 \le x \le 1. \tag{3.12}$$

Details about the beta distribution are not important here. Interested readers can refer to Johnson and Kotz (1970, Chapter 24).

As before, let $m = [qn + 0.5]$. Let Y be a random variable having a beta distribution with $a = m - 1$ and $b = n - m$, and let

$$W_i = P\left(\frac{i-1}{n} \le Y \le \frac{i}{n}\right).$$

Many statistical computing packages have functions that evaluate the beta distribution, so evaluating the W_i values is relatively easy to do. In R, there is the function pbeta(x,a,b) that computes $P(Y \le x)$. Thus, W_i can be computed by setting $x = i/n$, $y = (i-1)/n$, in which case W_i is pbeta(x,m-1,n-m) minus pbeta(y,m-1,n-m).

Let

$$C_k = \sum_{i=1}^{n} W_i X_{(i)}^k.$$

When $k = 1$, C_k is a linear combination of the order statistics. Linear sums of order statistics are called *L-estimators*. Other examples of L-estimators are the trimmed and Winsorized means already discussed. The point here is that C_k can be shown to estimate $E(X_{(m)}^k)$, the kth moment of the mth order statistic. Consequently, the standard error of the mth order statistic, $X_{(m)} = \hat{x}_q$, is estimated with

$$\sqrt{C_2 - C_1^2}.$$

Note that when n is odd, this last equation provides an alternative to the McKean–Schrader estimate of the standard error of M described in Section 3.3.4. Based on limited studies, it seems that when computing confidence intervals or testing hypotheses based on M, the McKean–Schrader estimator is preferable.

3.5.4 R Function mjse

The R function

$$\text{mjse(x,q=0.5)}$$

computes the Maritz–Jarrett estimate of the standard error of $\hat{x}_q = X_{(m)}$, the mth order statistic, where $m = [qn + 0.5]$. If unspecified, q defaults to 0.5. The command mjse(x,0.4), for example, estimates the standard error of $\hat{x}_{.4} = X_{(m)}$. If the the data in Table 3.2 are stored in the R variable xv, and if the median is estimated with $X_{(10)}$, the command mjse(xv) reports that the Maritz–Jarrett estimate of the standard error is 45.8. Using instead the method in Section 3.5.1, based on the adaptive kernel density estimator, the estimate is 43.95. Note that both estimates are substantially less than the estimated standard error of the sample mean, which is 136.

All indications are that the Maritz–Jarrett estimator is more accurate than the method based on Eq. (3.11) used in conjunction with Rosenblatt's shifted histogram described in Section 3.2.2. There are some weak indications that the Maritz–Jarrett estimator remains more accurate when Rosenblatt's shifted histogram is replaced by the adaptive kernel estimator, but an extensive study of this issue has not been conducted. Regardless, the kernel density estimator plays a useful role when dealing with M-estimators of location or when summarizing data. (Also, there are no simple methods for computing the beta distribution with some programming languages.)

3.5.5 The Harrell–Davis Estimator

A concern when estimating the qth quantile with $\hat{x}_q = X_{(m)}$, $m = [qn + .5]$, is that its standard error can be relatively high. The problem is of particular concern when sampling from a light-tailed or normal distribution. A natural strategy for addressing this problem is to use all of the order statistics to estimate x_q, as opposed to a single order statistic, and several methods have been proposed. One such estimator was derived by Harrell and Davis (1982). To compute it, let Y be a random variable having a beta distribution with parameters $a = (n + 1)q$ and $b = (n + 1)(1 - q)$. That is, the probability density function of Y is

$$\frac{\Gamma(a + b)}{\Gamma(a)\Gamma(b)} y^{a-1}(1 - y)^{b-1}.$$

(Γ is the gamma function, the details of which are not important for present purposes.) Let

$$W_i = P\left(\frac{i-1}{n} \leq Y \leq \frac{i}{n}\right).$$

Then the Harrell–Davis estimate of the qth quantile is

$$\hat{\theta}_q = \sum_{i=1}^{n} W_i X_{(i)}. \tag{3.13}$$

This is another example of an L-estimator. Asymptotic normality of $\hat{\theta}_q$ was established by Yoshizawa, Sen, and Davis (1985) for $q = 0.5$, only.

In some cases, the Harrell–Davis estimator is much more efficient than $\hat{x}_q$ and this can translate into substantial gains in power when testing hypotheses, as illustrated in Chapter 5. This is not to say, however, that the Harrell–Davis estimator always dominates $\hat{x}_q$ in terms of its standard error. In fact, if the tails of a distribution are heavy enough, the standard error of $\hat{x}_q$ can be substantially smaller than the standard error of $\hat{\theta}_q$, as is illustrated later in this chapter. The main advantage of $\hat{\theta}_q$ is that it guards against extremely poor efficiency under normality, but as the sample size gets large, it seems that this becomes less of an issue (Sheather & Marron, 1990). There are kernel density estimators of quantiles, but they are not discussed because they seem to behave in a manner very similar to the Harrell–Davis estimator used here. (For comparisons of various quantile estimators, see Parrish, 1990; as well as Dielman, Lowry, and Pfaffenberger (1994). For a possible improvement on the Harrel–Davis estimator, see Sfakianakis & Verginis, 2008.)

3.5.6 R Function hd

The R function

$$hd(x, q = 0.5)$$

computes $\hat{\theta}_q$, the Harrell–Davis estimate of the qth quantile. If any missing values (stored as NA) are detected, they are automatically removed. The default value for q is 0.5. Storing the data in Table 3.2 in the R vector x, the command hd(x) returns the value $\hat{\theta}_{.5} = 271.7$ as the estimate of the median. Similarly, the estimate of the 0.4 quantile is computed with the command hd(x,0.4), and for the data in Table 3.2 it returns the value 236.

3.5.7 A Bootstrap Estimate of the Standard Error of $\hat{\theta}_q$

The influence function of the Harrell–Davis estimator has not been derived, and there is no simple equation giving its standard error. However, its standard error can be obtained using the bootstrap method in Section 3.1. That is, in Section 3.1, simply replace $\hat{\theta}$ with $\hat{\theta}_q$.

3.5.8 R Function hdseb

The R function

$$\text{hdseb}(x, q = 0.5, \text{nboot} = 100)$$

computes the bootstrap estimate of the standard error of the Harrell–Davis estimator for the data stored in the vector x. Of course, the R function bootse in Section 3.1.1 could be used as well; the function hdseb is provided merely for convenience. If, for example, the R command bootse(x,nboot$=100$,est$=$hd) is used, this yields the same estimate of the standard error returned by hdseb. When using hdseb, the default value for q is 0.5 and the default value for nboot, which represents B, the number of bootstrap samples, is 100. (But the default estimate when using bootse is nboot$=1000$.) For example, hdseb(x) uses $B = 100$ bootstrap samples to estimate the standard error when estimating the median. For the data in Table 3.2, this function returns the value 50.8. This is a bit smaller than the estimated standard error of the 20% trimmed mean, which is 56.1, it is a bit larger than the Maritz–Jarrett estimate of the standard error of $\hat{x}_{0.5}$, 45.8, and it is substantially smaller than the estimated standard error of the sample mean, 136. With $B = 25$, the estimated standard error of $\hat{\theta}_{.5}$ drops from 50.8 to 49.4. When using the Harrell–Davis estimator to estimate the qth quantile, $q \neq 0.5$, an estimate of the standard error is obtained with the command hdseb(x,q), and $B = 100$ will be used. The command hdseb(x,0.3,25) uses $B = 25$ bootstrap samples to estimate the standard error when estimating the 0.3 quantile.

3.6 An M-Estimator of Location

The trimmed mean is based on a predetermined amount of trimming. That is, you first specify the amount of trimming that is desired, after which the sample trimmed mean, $\bar{X}_t$, can be computed. Another approach is to empirically determine the amount of trimming. For example, if sampling is from a light-tailed distribution, or even a normal distribution, it might be desirable to trim very few observations or none at all. If a distribution is skewed to the right, a natural reaction is to trim more observations from the right versus the left tail of the empirical distribution. In essence, this is what the M-estimator of location does. There are, however, some practical difficulties that arise when using M-estimators of location, and in some cases, trimmed means have important advantages. But there are also important advantages to using M-estimators, especially in the context of regression.

Before describing how an M-estimator is computed, it helps to elaborate on the line of reasoning leading to M-estimators (beyond what was covered in Chapter 2) and to comment on some technical issues. Chapter 2 put μ in the context of minimizing the expected squared difference between X and some constant c, and this was used to provide some motivation for the general approach used to define M-measures of location. In particular, setting $c = \mu$

minimizes $E(X - c)^2$. One practical concern was that if a measure of location is defined as the value of c minimizing $E(X - c)^2$, extreme X values can have an inordinately large effect on the resulting value for c. For a skewed distribution, values of X that are extreme and relatively rare can "pull" the value of μ into the tail of the distribution. A method of addressing this concern is to replace $(X - c)^2$ with some other function that gives less weight to extreme values. In terms of estimators of location, a similar problem arises. The sample mean is the value of c minimizing $\sum(X_i - c)^2$. From basic calculus, minimizing this sum turns out to be equivalent to choosing c such that $\sum(X_i - c) = 0$, and the solution is $c = \bar{X}$. The data in Table 3.2 illustrate that the sample mean can be quite far into the tail of a distribution. The sample mean is 448, yet 15 of the 19 observations have values less than 448. In fact, the data suggest that 448 is somewhere near the 0.8 quantile. M-estimators of location address this problem by replacing $(X_i - c)^2$ with some function that gives less weight to extreme X_i values (cf. Martin & Zamar, 1993).

From Chapter 2, an M-measure of location is the value μ_m such that

$$E\left\{\Psi\left(\frac{X - \mu_m}{\tau}\right)\right\} = 0, \tag{3.14}$$

where τ is some measure of scale and Ψ is an odd function meaning that $\Psi(-x) = -\Psi(x)$. Some choices for Ψ are listed and described in Table 2.1. Once a random sample of observations is available, the M-measure of location is estimated by replacing expected value with summation in Eq. (3.14). That is, an M-estimator of location is the value $\hat{\mu}_m$ such that

$$\sum \Psi\left(\frac{X_i - \hat{\mu}_m}{\tau}\right) = 0. \tag{3.15}$$

If $\Psi\{(X_i - \hat{\mu}_m)/\tau)\} = (X_i - \hat{\mu}_m)/\tau$, $\hat{\mu}_m = \bar{X}$.

There are three immediate problems that must be addressed if M-measures of location are to have any practical value: Choosing an appropriate Ψ, choosing an appropriate measure of scale, τ, and finding a method for estimating μ_m once a choice for Ψ and τ has been made.

First consider the problem of choosing Ψ. There are many possible choices, so criteria are needed for deciding whether a particular choice has any practical value. Depending on the choice for Ψ, there can be 0, 1, or multiple solutions to Eq. (3.15), and this helps to limit the range of functions one might use. If there are zero solutions, this approach to estimation has little value, as is evident. If there are multiple solutions, there is the problem of choosing which solution to use in practice. A reasonable suggestion is to use the solution closest to the median, but estimation problems can persist. One of the more important examples of this (see Freedman & Diaconis, 1982) arises when Ψ is taken to be the so-called biweight:

$$\Psi(x) = \begin{cases} x(1 - x^2)^2, & \text{if } |x| < 1 \\ 0, & \text{if } |x| \geq 1. \end{cases} \tag{3.16}$$

All indications are that it is best to limit attention to those Ψ that yield a single solution to Eq. (3.14). This can be done by limiting attention to Ψ that are monotonic increasing.

Insisting on a single solution to Eq. (3.15) provides a criterion for choosing Ψ, but obviously more is needed. To make progress, it helps to replace Eq. (3.15) with an equivalent approach to defining an estimator of location. First note that from basic calculus, defining a measure of location with Eq. (3.15) is equivalent to defining $\hat{\mu}_m$ as the value minimizing

$$\sum \xi \left(\frac{X_i - \hat{\mu}_m}{\tau} \right), \tag{3.17}$$

where Ψ is the derivative of ξ. Now, if sampling is from a normal distribution, the optimal estimator, in terms of minimum variance, is the sample mean $\bar{X}$, and the sample mean can be viewed as the value minimizing $\sum (X_i - \hat{\mu}_m)^2$. That is, using

$$\xi \left(\frac{X_i - \hat{\mu}_m}{\tau} \right) = (X_i - \hat{\mu}_m)^2 \tag{3.18}$$

yields $\hat{\mu}_m = \bar{X}$, which is optimal under normality. As already indicated, the problem with this function is that it increases too rapidly as the value of X_i moves away from $\hat{\mu}_m$, and this can cause practical problems when sampling from nonnormal distributions for which extreme values can occur. But because this choice of ξ is optimal under normality, a natural strategy is to search for some approximation of Eq. (3.18) that gives nearly the same results when sampling from a normal distribution. In particular, consider functions that are identical to Eq. (3.18) provided X_i is not too extreme.

To simplify matters, temporarily consider a standard normal distribution, and take τ to be σ, the standard deviation, which in this case is 1. Then the optimal choice for ξ is $(x - \hat{\mu}_m)^2$, as already explained. Suppose instead that ξ is taken to be

$$\xi(x - \hat{\mu}_m) = \begin{cases} -2K(x - \hat{\mu}_m), & \text{if } x < -K \\ (x - \hat{\mu}_m)^2, & \text{if } -K \leq x \leq K \\ 2K(x - \hat{\mu}_m), & \text{if } x > K, \end{cases} \tag{3.19}$$

where K is some constant to be determined. Thus, when sampling from a normal distribution, the optimal choice for ξ is being used provided an observation is not too extreme, meaning that its value does not exceed K or is not less than $-K$. If it is extreme, ξ becomes a linear function, rather than a quadratic function, this linear function increases less rapidly than Eq. (3.18), so extreme values are having less of an influence on $\hat{\mu}_m$.

The strategy for choosing ξ, outlined earlier, is illustrated in the left panel of Figure 3.5 which shows a graph of $\xi(x - \hat{\mu}_m) = (x - \hat{\mu}_m)^2$ when $\hat{\mu}_m = 0$, and this is the optimal choice for ξ when sampling from a standard normal distribution. Also shown is the approximation of the optimal ξ, given by Eq. (3.19), when $K = 1.28$. When $-1.28 \leq x \leq 1.28$, the approximation

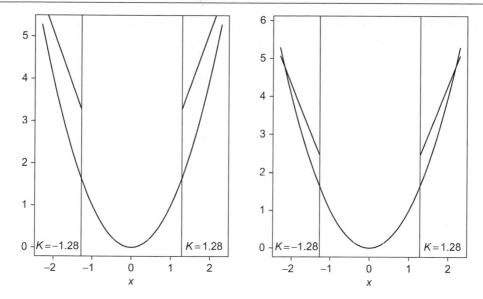

Figure 3.5: An approximation of the optimal function.

is exact. When $x < -1.28$ or $x > 1.28$, the straight line above the curve is used to approximate ξ. Because $K = 1.28$ is the 0.9 quantile of a standard normal distribution, there is a 0.8 probability that a randomly sampled observation will have a value between $-K$ and K. Note how Figure 3.5 suggests that Eq. (3.19) with $K = 1.28$ is a reasonable approximation of $\xi (x - \hat{\mu}_m)^2 = (x - \hat{\mu}_m)^2$.

The left panel of Figure 3.5 suggests lowering the straight lines to get a better approximation of ξ. The right panel shows what happens when the lines are lowered by $K^2/2$. That is, Eq. (3.19) is replaced by

$$\xi(x - \hat{\mu}_m) = \begin{cases} -2K(x - \hat{\mu}_m) - \frac{K^2}{2}, & \text{if } x < -K \\ (x - \hat{\mu}_m)^2, & \text{if } -K \leq x \leq K \\ 2K(x - \hat{\mu}_m) - \frac{K^2}{2}, & \text{if } x > K. \end{cases}$$

However, this modification yields the same equation for determining $\hat{\mu}$, as given by Eq. (3.22) in the next paragraph.

Now, $\hat{\mu}_m$ is the value minimizing Eq. (3.17). Taking the derivative of this equation, with ξ given by Eq. (3.19), and setting the result equal to zero, $\hat{\mu}_m$ is determined by

$$2 \sum \Psi(X_i - \hat{\mu}_m) = 0, \tag{3.20}$$

where

$$\Psi(x) = \max[-K, \min(K, x)] \tag{3.21}$$

is Huber's Ψ. (For a graph of Huber's Ψ, see Chapter 2.) Of course, the constant 2 in Eq. (3.20) is not relevant to solving for $\hat{\mu}_m$, and typically (3.20) is simplified to

$$\sum \Psi(X_i - \hat{\mu}_m) = 0. \tag{3.22}$$

There remains the problem of choosing K. One strategy is to choose K so that the large sample (asymptotic) standard error of $\hat{\mu}_m$ is reasonably close to the standard error of the sample mean when sampling from a normal distribution, yet the standard error of $\hat{\mu}_m$ is relatively unaffected when sampling from a heavy-tailed distribution. A common choice is $K = 1.28$, the 0.9 quantile of the standard normal distribution, and this will be used unless stated otherwise. For a more detailed discussion about choosing K, see Huber (1981). In a given situation, some other choice might be more optimal, but $K = 1.28$ guards against relatively large standard errors while sacrificing very little when sampling from a normal distribution. A more efficacious choice might be made based on knowledge about the distribution being sampled, but the extent to which this strategy can be recommended is unclear.

One more technical issue must be addressed. From Chapter 2, a requirement of a measure of location is that it be scale equivariant. In the present context, this means that if μ_m is the M-measure of location associated with the random variable X, aX should have $a\mu_m$ as a measure of location for any constant a. If $\hat{\mu}_m$ is estimated with Eq. (3.22), this requirement is not met, Eq. (3.15) must be used instead. Using Eq. (3.15) means in particular that a measure of scale, τ, must be chosen. It turns out that the measure of scale need not be efficient in order for $\hat{\mu}_m$ to be efficient. The main concern is that it be reasonably resistant. In particular, it should have a finite-sample breakdown point that is reasonably high. A common choice for a measure of scale is the value of ω determined by

$$P(|X - x_{.5}| < \omega) = \frac{1}{2},$$

where $x_{.5}$ is the population median. That is, ω is the 0.5 quantile of the distribution of $|X - x_{.5}|$. If, for example, sampling is from a standard normal distribution, in which case $x_{.5} = 0$, ω is determined by

$$P(-\omega \le Z \le \omega) = 0.5,$$

where Z has a standard normal distribution. That is, ω is the 0.75 quantile of the standard normal distribution, which is approximately equal to 0.6745.

The standard estimate of ω is the median absolute deviation statistic given by

$$\text{MAD} = \text{MED}\{|X_1 - M|, \ldots, |X_n - M|\},$$

where M is the usual sample median, which is computed as described in Chapter 1. That is, MAD is the sample median of the n values $|X_1 - M|, \ldots, |X_n - M|$, and its finite sample breakdown point is approximately 0.5. (For more details about the finite sample breakdown point of MAD, see Gather & Hilker, 1997.)

If observations are randomly sampled from a normal distribution, MAD does not estimate σ, the standard deviation, it estimates $z_{0.75}\sigma$, where $z_{0.75}$ is the 0.75 quantile of the standard normal distribution. To put MAD in a more familiar context, it is typically rescaled so that it estimates σ when sampling from a normal distribution. In particular,

$$\text{MADN} = \frac{\text{MAD}}{z_{0.75}} \approx \frac{\text{MAD}}{0.6745}$$

is used, and this convention will be followed here. Then for a random sample, Eq. (3.14) says that an M-estimator of location is the value $\hat{\mu}_m$ satisfying

$$\sum \Psi \left(\frac{X_i - \hat{\mu}_m}{\text{MADN}} \right) = 0 \tag{3.23}$$

3.6.1 R Function mad

The R function

$$\text{mad}(x)$$

computes MADN. That is, R assumes that MAD is to be re-scaled to estimate σ when sampling from a normal distribution. To use R to compute MAD, simply use the command qnorm(0.75)*mad(x). The command qnorm(0.75) returns the 0.75 quantile of a standard normal random variable.

3.6.2 Computing an M-estimator of Location

Solving Eq. (3.23) for $\hat{\mu}_m$ is usually accomplished with an iterative estimation procedure known as the Newton–Raphson method. It involves the derivative of Ψ, which is given by

$$\Psi'(x) = \begin{cases} 1, & \text{if } -K \le x \le K \\ 0, & \text{otherwise.} \end{cases} \tag{3.24}$$

The steps used to determine $\hat{\mu}_m$ are shown in Table 3.5. Typically, $K = 1.28$ is used, and this choice is assumed henceforth unless stated otherwise.

Table 3.5: How to Compute the M-Estimator of Location $\hat{\mu}_m$.

Set $k = 0$, $\hat{\mu}_k = M$, the sample median, and choose a value for K. A common choice is $K = 1.28$.

Step 1. Let

$$A = \sum \Psi\left(\frac{X_i - \hat{\mu}_k}{\text{MADN}}\right).$$

Here, Ψ given by Eq. (3.21) is used.

Step 2. Let

$$B = \sum \Psi'\left(\frac{X_i - \hat{\mu}_k}{\text{MADN}}\right),$$

where Ψ' is the derivative of Ψ given by Eq. (3.24). B is just the number of observations X_i satisfying $-K \leq (X_i - \hat{\mu}_k)/\text{MADN} \leq K$.

Step 3. Set

$$\hat{\mu}_{k+1} = \hat{\mu}_k + \frac{\text{MADN} \times A}{B}.$$

Step 4. If $|\hat{\mu}_{k+1} - \hat{\mu}_k| < 0.0001$, stop and set $\hat{\mu}_m = \hat{\mu}_{k+1}$. Otherwise, increment k by one and repeat steps 1–4.

■ Example

For the data in Table 3.2, $\hat{\mu}_0 = M = 262$ and $\text{MADN} = 169$. Table 3.6 shows the resulting values of $\Psi\{(X_i - \mu_0)/\text{MADN}\}$ corresponding to each of the 19 values. The sum of the values in Table 3.6 is $A = 2.05$. The number of Ψ values between -1.28 and 1.28 is $B = 15$, so the first iteration using the steps in Table 3.5 yields

$$\hat{\mu}_1 = 262 + \frac{169 \times 2.05}{15} = 285.1.$$

The iterative estimation process consists of using $\hat{\mu}_1$ to recompute the Ψ values, yielding a new value for A and B, which in turn yields $\hat{\mu}_2$. For the data at hand, it turns out that there is no difference between $\hat{\mu}_2$ and $\hat{\mu}_1$, so the iterative process stops and $\hat{\mu}_m = 285.1$.

Table 3.6: Values of Huber's Ψ for the Self-Awareness Data.

−1.09	−1.04	−1.04	−0.88	−0.66	−0.31	−0.25	−0.09	−0.05	0.0
0.20	0.22	0.26	0.67	0.98	1.28	1.28	1.28	1.28	

If there had been a difference, the Ψ values would be computed again using $\hat{\mu}_2$, and this would continue until $|\hat{\mu}_{k+1} - \hat{\mu}_k| < 0.0001$.

∎

When computing the M-estimator of location as described in Table 3.5, the measure of scale, MADN, does not change when iterating. There are also M-estimators where a measure of scale is updated. That is, a measure of scale is simultaneously determined in an iterative fashion. (See Huber, 1981, p. 136.) Currently, it seems that this alternative estimation procedure offers no practical advantage, so it is not discussed. In fact, if a measure of scale is estimated simultaneously with a measure of location, using Huber's Ψ, the Bickel–Lehmann condition for a measure of location is no longer satisfied (Bickel & Lehmann, 1975).

Notice that the M-estimator in Table 3.5 empirically determines whether an observation is unusually large or small. In the first step where $\hat{\mu}_0 = M$, the sample median, X_i is considered unusually small if $(X_i - M)/\text{MADN} < -1.28$, where the typical choice of $K = 1.28$ is being used, and it is unusually large if $(X_i - M)/\text{MADN} > 1.28$. This becomes clearer if the first step in the iterative process is written in a different form. Let i_1 be the number of observations X_i for which $(X_i - M)/\text{MADN} < -1.28$, and let i_2 be the number of observations such that $(X_i - M)/\text{MADN} > 1.28$. Some algebra shows that the value of $\hat{\mu}_1$ in Table 3.5 is

$$\frac{1.28(\text{MADN})(i_2 - i_1) + \sum_{i=i_1+1}^{n-i_2} X_{(i)}}{n - i_1 - i_2}, \tag{3.25}$$

the point being that the sum in this expression is over only those values that are not too large or too small.

To conclude this section, it is noted that there are more formal methods for motivating Huber's Ψ, but no details are given here. Readers interested in technical issues can refer to Huber (1981).

3.6.3 R Functions mest

The R function

$$\text{mest(x,bend=1.28)}$$

performs the calculations in Table 3.5. The argument bend corresponds to K in Huber's Ψ and defaults to 1.28 if unspecified. For example, the command mest(x) computes the M-estimator of location for the data in the vector x using $K = 1.28$. The command mestx(x,1.5) uses $K = 1.5$. Increasing K increases efficiency when sampling from a normal distribution, but it increases sensitivity to the tails of the distribution, and efficiency can be lower as well when sampling from a heavy-tailed distribution. To illustrate sensitivity to the tail of a distribution, the mean, median, 20% trimmed mean, M-estimator, with $K = 1.28$, and MOM (described in Section 3.10) are equal to 448, 262, 282.7, 258.1, and 245.4, respectively, for the data in Table 3.2. With $K = 1.2$, the M-estimator is 281.6, and with $K = 1$, the M-estimator is equal to 277.4. Note that MOM has a value less than the median, in contrast to the other location estimators that were used. This illustrates a curious property about the population value of MOM. Suppose a distribution is skewed to the right. Then the population value of MOM can lie between the median and the mode of this distribution, in contrast to any trimmed mean or M-estimator which typically lie to the right of the median.

3.6.4 Estimating the Standard Error of the M-estimator

This subsection describes the first of two methods for estimating the standard error of the M-estimator of location. The method here is based on the influence function of μ_m and the other uses a bootstrap.

As was the case with the trimmed mean and $\hat{x}_q$, the justification for the non-bootstrap estimate of the standard error follows from the result that

$$\hat{\mu}_m = \mu_m + \frac{1}{n}\sum IF_m(X_i),$$

plus a remainder term that goes to zero as n gets large. That is, an M-estimator can be written as its population value plus a sum of independent random variables having mean zero.

The influence function of the M-measure of location has a somewhat complicated form. It depends in part on the measure of scale that is used in Ψ, and here this is $\omega_N = \omega/0.6745$. The influence function of ω, which is estimated by MAD, is given in Chapter 2. The influence function of ω_N is just the influence function of ω divided by 0.6745. Let

$$A(x) = \text{sign}(|x - \theta| - \omega),$$

where θ is the population median and $\text{sign}(x)$ equals $-1, 0$, or 1 according to whether x is less than, equal to, or greater than 0. Let

$$B(x) = \text{sign}(x - \theta),$$

and

$$C(x) = A(x) - \frac{B(x)}{f(\theta)}\{f(\theta + \omega) - f(\theta - \omega)\}.$$

The influence function of ω_N is

$$IF_{\omega_N}(x) = \frac{C(x)}{2(0.6745)\{f(\theta + \omega) + f(\theta - \omega)\}}.$$

Estimating $IF_{\omega_N}(X_i)$, the value of the influence function of ω_N at X_i, requires an estimate of the probability density function, and this can be done as described in Section 3.2. Here, the adaptive kernel estimator in Section 3.2.4 will be used unless stated otherwise. Denoting the estimate of the probability density function $f(x)$ with $\hat{f}(x)$, and computing

$$\hat{A}(X_i) = \text{sign}(|X_i - M| - \text{MAD}),$$

$$\hat{B}(X_i) = \text{sign}(X_i - M),$$

$$\hat{C}(X_i) = \hat{A}(X_i) - \frac{\hat{B}(X_i)}{\hat{f}(M)}\{\hat{f}(M + \text{MAD}) - \hat{f}(M - \text{MAD})\},$$

an estimate of $IF_{\omega_N}(X_i)$ is

$$V_i = \frac{\hat{C}(X_i)}{2(.6745)\{\hat{f}(M + \text{MAD}) + \hat{f}(M - \text{MAD})\}}. \tag{3.26}$$

Letting $y = (x - \mu_m)/\omega_N$, the influence function of μ_m is

$$IF_m(x) = \frac{\omega_N \Psi(y) - IF_{\omega_N}(x)\{E(\Psi'(y)y)\}}{E[\Psi'(y)]}. \tag{3.27}$$

Having described how to estimate $IF_{\omega_N}(X_i)$, and because MADN estimates ω_N, all that remains when estimating $IF_m(X_i)$ is estimating $E[\Psi'(Y)Y]$ and $E[\Psi'(Y)]$, where $Y = (X - \mu_m)/\omega_N$. Set

$$Y_i = \frac{X_i - \hat{\mu}_m}{\text{MADN}},$$

and

$$D_i = \begin{cases} 1, & \text{if } |Y_i| \leq K \\ 0, & \text{otherwise.} \end{cases}$$

Then $E[\Psi'(y)]$ is estimated with

$$\bar{D} = \frac{1}{n}\sum D_i.$$

Finally, estimate $E[\Psi'(y)y]$ with

$$\bar{C} = \frac{1}{n}\sum D_i Y_i.$$

The sum in this last equation is just the sum of the Y_i values satisfying $|Y_i| \leq K$. The value of the influence function of μ_m, evaluated at X_i, is estimated with

$$U_i = \{(\text{MADN})\Psi(Y_i) - V_i\bar{C}\}/\bar{D}. \tag{3.28}$$

The squared standard error of $\hat{\mu}_m$ can now be estimated from the data. The estimate is

$$\hat{\sigma}_m^2 = \frac{1}{n(n-1)} \sum U_i^2. \tag{3.29}$$

Consequently, the standard error, $\sqrt{VAR(\hat{\mu}_m)}$, is estimated with $\hat{\sigma}_m$. Note that the sum in Eq. (3.29) is divided by $n(n-1)$, not n^2 as indicated by Eq. (3.5). This is done because if no observations are flagged as being unusually large or small by Ψ, $\hat{\mu}_m = \bar{X}$, and Eq. (3.29) reduces to s^2/n, the estimate that is typically used.

The computations just described are straightforward but tedious, so no detailed illustration is given. Interested readers can use the R function mestse, which is described in the next subsection.

3.6.5 R Function mestse

The R function

$$\text{mestse(x,bend}=1.28,\text{op}=2)$$

estimates the standard error of the M-estimator using the method just described. The argument bend corresponds to K in Huber's Ψ and defaults to 1.28 if not specified. The argument op indicates which density estimator is used to estimate the influence function. By default (op=2), the adaptive kernel estimator is used, otherwise Rosenblatt's shifted histogram is used. If the data in Table 3.2 are stored in the R variable x, the command mestse(x) returns the value 54.1, and this is reasonably close to 56.1, the estimated standard error of the trimmed mean. The 20% trimmed mean and M-estimator have similar influence functions, so reasonably close agreement was expected. A difference between the two estimators is that the M-estimator identifies the four largest values as being unusually large. That is, $(X_i - \hat{\mu}_m)/\text{MADN}$ exceeds 1.28 for the four largest values, whereas the trimmed mean trims the three largest values only, so the expectation is that $\hat{\mu}_m$ will have a smaller standard. Also, the M-estimator does not identify any of the lower values as being unusual, but the trimmed mean automatically trims three values.

3.6.6 A Bootstrap Estimate of the Standard Error of $\hat{\mu}_m$

The standard error can also be estimated using a bootstrap method. The computational details are essentially the same as those described in Section 3.1. Begin by drawing a bootstrap sample, $X_1^*, \ldots, X_n^*$ from the observed values $X_1, \ldots, X_n$. That is, randomly sample n observations with replacement from $X_1, \ldots, X_n$. Compute the value of $\hat{\mu}_m$ using the bootstrap sample and call the result $\hat{\mu}_m^*$. Repeat this process B times yielding $\hat{\mu}_{m1}^*, \ldots, \hat{\mu}_{mB}^*$. Let

$$\bar{\mu}^* = \frac{1}{B} \sum_{b=1}^{B} \hat{\mu}_{mb}^*,$$

in which case the bootstrap estimate of the squared standard error is

$$\hat{\sigma}_{m\text{boot}}^2 = \frac{1}{B-1} \sum_{b=1}^{B} (\hat{\mu}_{mb}^* - \bar{\mu}^*)^2. \tag{3.30}$$

Using $B = 25$ might suffice, whereas $B = 100$ appears to be more than adequate in most situations (Efron, 1987).

A negative feature of the bootstrap is that if n is large, execution time can be high, even on a mainframe computer, when working with various software packages designed specifically for doing statistics. The accuracy of the bootstrap method versus the kernel density estimator has not been examined when n is small. Early attempts at comparing the two estimators, via simulations, were complicated by the problem that the kernel density estimator that was used can be undefined because of division by zero. Yet another problem is that the bootstrap method can fail when n is small because a bootstrap sample can yield MAD=0, in which case $\hat{\mu}_m^*$ cannot be computed because of division by zero. A few checks were made with $n = 20$ and $B = 1000$ when sampling from a normal or lognormal distribution. Limited results suggest that the bootstrap is more accurate, but a more detailed study is needed to resolve this issue.

3.6.7 R Function mestseb

The R function

$$\text{mestseb(x,nboot=1000,bend=1.28)}$$

computes the bootstrap estimate of the standard error of $\hat{\mu}_m$ for the data stored in the R variable x. The argument nboot is B, the number of bootstrap samples to be used, which defaults to $B = 100$ if unspecified. The default value for bend, which corresponds to K in Huber's Ψ, is 1.28. For example, mestseb(x,50) will compute a bootstrap estimate of the standard error using $B = 50$ bootstrap replications, whereas mestseb(x) uses $B = 100$. For the data in Table 3.2, mestseb(x) returns the value 53.7, which is in reasonable agreement with 53.2, the estimated standard error using the influence function. The function mestseb sets the seed of the random number generator in R so that results will be duplicated if mestseb is executed a second time with the same data. Otherwise, if mestseb is invoked twice, slightly different results would be obtained because the bootstrap method would use a different sequence of random numbers.

3.7 One-Step M-estimator

Typically, when computing $\hat{\mu}_m$ with the iterative method in Table 3.5, convergence is obtained after only a few iterations. It turns out that if only a single iteration is used, the resulting estimator has good asymptotic properties (Serfling, 1980). In particular, for large sample sizes, it performs in a manner very similar to the fully iterated M-estimator. An expression for the first iteration was already described, but it is repeated here for convenience, assuming $K = 1.28$. Let i_1 be the number of observations X_i for which $(X_i - M)/\text{MADN} < -1.28$, and let i_2 be the number of observations such that $(X_i - M)/\text{MADN} > 1.28$. The one-step M-estimate of location is

$$\hat{\mu}_{os} = \frac{1.28(\text{MADN})(i_2 - i_1) + \sum_{i=i_1+1}^{n-i_2} X_{(i)}}{n - i_1 - i_2}. \tag{3.31}$$

Although the one-step M-estimator is slightly easier to compute than the fully iterated M-estimator, its influence function has a much more complicated form when distribution are skewed (Huber, 1981, p. 140). In terms of making inferences about the corresponding population parameter, it seems that there are no published results suggesting that the influence function plays a useful role when testing hypotheses or computing confidence intervals. Consequently, details about the influence function are not given here. As for estimating the standard error of the one-step M-estimator, only the bootstrap method will be used. The basic strategy is the same as it was when working with $\hat{\mu}_m$ or the Harrell–Davis estimator already discussed. In particular, draw a bootstrap sample by re-sampling n observations with replacement from the n observations available and compute $\hat{\mu}_{os}$. Consistent with previous notation, the result will be labeled $\hat{\mu}_{os}^*$ to distinguish it from $\hat{\mu}_{os}$ based on the original observations $X_1, \ldots, X_n$. Repeat this process B times yielding $\hat{\mu}_{os1}^*, \ldots, \hat{\mu}_{osB}^*$. Let

$$\bar{\mu}_{os}^* = \frac{1}{B} \sum_{b=1}^{B} \hat{\mu}_{osb}^*,$$

in which case the bootstrap estimate of the squared standard error is

$$\hat{\sigma}_{osboot}^2 = \frac{1}{B-1} \sum_{b=1}^{B} (\hat{\mu}_{osb}^* - \bar{\mu}_{os}^*)^2. \tag{3.32}$$

Again $B = 25$ might suffice, whereas $B = 100$ appears to be more than sufficient.

3.7.1 R Function onestep

The R function

$$\text{onestep(x,bend=1.28)}$$

computes the one-step M-estimator given by Eq. (3.31).

3.8 W-estimators

W-estimators of location are closely related to M-estimators and usually they give identical results. However, when extending M-estimators to regression, the computational method employed by W-estimators is typically used. This method is just another way of solving Eq. (3.23). For completeness, W-estimators are briefly introduced here.

Let

$$w(x) = \frac{\Psi(x)}{x}.$$

In this last equation, Ψ could be any of the functions associated with M-estimators, and the generic measure of scale τ, used to define the general class of M-estimators, could be used. If for example τ is estimated with MADN, then $\hat{\mu}_m$ is determined by solving Eq. (3.23), which becomes

$$\sum \left(\frac{X_i - \hat{\mu}_m}{\text{MADN}} \right) w \left(\frac{X_i - \hat{\mu}_m}{\text{MADN}} \right) = 0. \tag{3.33}$$

Rearranging terms in Eq. (3.33) yields

$$\hat{\mu}_m = \frac{\sum X_i w\{(X_i - \hat{\mu}_m)/\text{MADN}\}}{\sum w\{(X_i - \hat{\mu}_m)/\text{MADN}\}}.$$

This last equation does not yield an immediate value for $\hat{\mu}_m$ because $\hat{\mu}_m$ appears on both sides of the equation. However, it suggests an iterative method for obtaining $\hat{\mu}_m$ that has practical value.

Set $k = 0$ and let $\hat{\mu}_0$ be some initial estimate of $\hat{\mu}_m$. For example, $\hat{\mu}_0$ could be the sample mean. Let

$$U_{ik} = \frac{X_i - \hat{\mu}_k}{\text{MADN}}.$$

Then the iteration formula is

$$\hat{\mu}_{k+1} = \frac{\sum X_i w(U_{ik})}{\sum w(U_{ik})}.$$

That is, given $\hat{\mu}_k$, which is an approximate value for $\hat{\mu}_m$ that solves (3.33), an improved approximation is $\hat{\mu}_{k+1}$. One simply keeps iterating until $|\hat{\mu}_{k+1} - \hat{\mu}_k|$ is small, say less than 0.0001.

The iterative method just described is an example of what is called *iteratively re-weighted least squares*. To explain, let $\hat{\mu}$ be any estimate of a measure of location and recall that $\hat{\mu} = \bar{X}$ is the value that minimizes $\sum(X_i - \hat{\mu})^2$. In the context of regression, minimizing this sum is based on the least squares principle taught in every introductory statistics course. Put another way, the sample mean is the ordinary least squares (OLS) estimator. Weighted least squares, based on *fixed* weights, w_i, determines a measure of location by minimizing $\sum w_i(X_i - \hat{\mu}_m)^2$, and this is done by solving $\sum w_i(X_i - \hat{\mu}_m) = 0$, which yields

$$\hat{\mu} = \frac{\sum w_i X_i}{\sum w_i}.$$

The problem in the present context is that the weights in Eq. (3.33) are not fixed, they depend on the value of $\hat{\mu}_m$ which is not known but updated with each iteration, so this last equation for $\hat{\mu}$ does not apply. Instead, the weights are recomputed according to the value of $\hat{\mu}_k$.

3.8.1 Tau Measure of Location

A variation of the W-estimator just described plays a role in some settings. Called the *tau measure of location*, it is computed as follows: Let

$$W_c(x) = \left(1 - \left(\frac{x}{c}\right)^2\right)^2 I(|x| \le c)$$

where the indicator function $I(|x| \le c) = 1$ if $|x| \le c$; otherwise $I(|x| \le c) = 0$. The weights are

$$w_i = W_c\left(\frac{X_i - M}{\text{MAD}}\right),$$

and the resulting measure of location is denoted by

$$\hat{\mu}_\tau = \frac{\sum w_i X_i}{\sum w_i}.$$

Following Maronna and Zamar (2002), $c = 4.5$ is used unless stated otherwise.

3.8.2 R Function tauloc

The R function

$$\text{tauloc}(x, \text{cval}=4.5)$$

computes the tau measure of location.

3.8.3 Zuo's Weighted Estimator

Yet another approach to choosing the weights when computing a W-estimator was suggested by Zuo (2010). (It is related to a class multivariate W-estimators introduced in Section 6.3.7.) Let

$$D_i = 1/(1 + |X_i - M|/\text{MAD}).$$

The weights are taken to be

$$w_i = I_{D_i \geq c} + \frac{e^{-k(1 - D_i^2/c^2)^2} - e^{-k}}{(1 - e^{-k})I_{D_i < c}},$$

where the indicator function $I_{D_i \geq c} = 1$ if $D_i \geq c$, otherwise $I_{D_i \geq c} = 0$. The constant c satisfies $0 \leq c \leq 1$ and $k > 0$. Zuo suggests using $k = 3$ and $c = 0.2$. The practical advantages of this estimator, relative to the many other robust location estimators that have been studied extensively, are unclear.

3.9 The Hodges–Lehmann Estimator

Chapter 2 mentioned some practical concerns about R-measures of location in general and the Hodges and Lehmann (1963) estimator in particular. But the Hodges–Lehmann estimator plays a fundamental role when applying standard rank-based methods (in particular, the Wilcoxon signed-rank test), so for completeness the details of this estimator are given here.

The *Walsh averages* of n observations refers to all pairwise averages: $(X_i + X_j)/2$, for all $i \leq j$. The Hodges–Lehmann estimator is the median of all Walsh averages, namely,

$$\hat{\theta}_{\text{HL}} = \text{med}_{i \leq j} \frac{X_i + X_j}{2}.$$

3.10 Skipped Estimators

Skipped estimators of location refer to the natural strategy of checking the data for outliers, removing any that are found, and averaging the values that remain. The first skipped estimator appears to be one proposed by Tukey (see Andrews et al., 1972) where checks for outliers were based on a boxplot rule. (Boxplot methods for detecting outliers are described in Section 3.13.) The one-step M-estimator given by Eq. (3.32) is almost a skipped estimator. If we ignore the term $1.28(\text{MADN})(i_2 - i_1)$ in the numerator of Eq. (3.32), an M-estimator removes the value X_i if

$$\frac{|X_i - M|}{\text{MADN}} > 1.28$$

and averages the values that remain. In essence, X_i is declared an outlier if it satisfies this last equation. But based on how the M-estimator is defined, the term $1.28(\text{MADN})(i_2 - i_1)$ arises.

When testing hypotheses, a slight variation of the skipped estimator, just described, has practical value. This *modified one-step M-estimator* (MOM) simply averages values not declared outliers, but to get reasonably good efficiency under normality, the outlier detection rule used by the one-step M-estimator is modified. Now X_i is declared an outlier if

$$\frac{|X_i - M|}{\text{MADN}} > 2.24$$

(which is a special case of a multivariate outlier detection method derived by Rousseeuw & van Zomeren, 1990). This last equation is known as the *Hampel identifier*, only Hampel used 3.5 rather than 2.24. When using 3.5, this will be called the Hampel version of MOM (HMOM.)

3.10.1 R Functions mom and bmean

The R function

$$\text{mom}(x, \text{bend}=2.24)$$

computes the MOM estimate of location, where the argument bend is the constant used in the Hampel identifier. The function

$$\text{bmean}(x, \text{mbox}=T)$$

computes a skipped estimator where outliers are identified by a boxplot rule covered in Section 3.13. The default value for mbox is T, indicating that Carling's method (described in Section 3.13.3) is used, and mbox=F uses the boxplot rule based on the ideal fourths.

3.11 Some Comparisons of the Location Estimators

Illustrations given in the previous sections of this chapter, based on data from actual studies, demonstrate that the estimated standard errors associated with robust estimates of location can be substantially smaller than the standard error of the sample mean. It has been hinted that these robust estimators generally compete well with the sample mean when sampling from a normal distribution, but no details have been given. There is also the concern of how estimators compare under various nonnormal distributions, including skewed distributions as a special case. Consequently, this section briefly compares the standard error of the robust estimators to the standard error of the sample mean for a few distributions.

One of the nonnormal distributions considered here is the lognormal shown in Figure 3.3. The random variable X is said to have a lognormal distribution if the distribution of $Y = \ln(X)$ is normal. It is a skewed distribution for which standard methods for computing confidence intervals for the mean can be unsatisfactory, even with $n = 160$. (Details are given in Chapters 4 and 5.) Consequently, there is a general interest in how methods based on alternative measures of location perform when sampling from this particular distribution. The immediate concern is whether robust estimators have relatively small standard errors for this special case.

Table 3.7 shows the variance of several estimators for a few distributions when $n = 10$. (Results for MOM, HMOM, and the Harrell–Davis estimator are based on simulations with 10,000 replications.) The distribution *one-wild* refers to sampling from a normal distribution and multiplying one of the observations by 10. Observations are generated from the *slash* distribution by generating an observation from the standard normal distribution and dividing by an independent uniform random variable on the interval (0, 1). Both the one-wild and slash distributions are symmetric distributions with heavier than normal tails. The slash distribution has an extremely heavy tail. In fact, it has infinite variance. The motivation for considering these distributions, particularly the slash distribution, is to see how an estimator performs under extreme conditions. It is unclear how heavy the tails of a distribution might be in practice, so it is of interest to see how an estimator performs for a distribution that represents an extreme departure from normality that is surely unrealistic. If an estimator performs reasonably well under normality and continues to perform well when sampling from a slash

Table 3.7: Variances of Selected Estimators, $n = 10$.

Estimator	Distribution			
	Normal	Lognormal	One-Wild	Slash
Mean	0.1000	0.4658	1.0900	∞
$\bar{X}_t \ (\gamma = 0.1)$	0.1053	0.2238	0.1432	∞
$\bar{X}_t \ (\gamma = 0.2)$	0.1133	0.1775	0.1433	0.9649
Median	0.1383	0.1727	0.1679	0.7048
$\hat{\mu}_m$ (Huber)	0.1085	0.1976	0.1463	0.9544
$\hat{\theta}_5$	0.1176	0.1729	0.1482	1.4731
MOM	0.1243	0.2047	0.1409	0.7331
HMOM	0.1092	0.2405	0.1357	1.0272
$\hat{\mu}_\tau$	0.1342	0.3268	2.1610	

distribution, this suggests that it has practical value for any distribution that might arise in practice by providing protection against complete disaster, disaster meaning standard errors that are extremely large compared with some other estimator that might have been used.

The small-sample efficiency of the tau measure of location, which is not included in Table 3.7, does not compete well with a 20% trimmed, MOM, and the one-step M-estimator (Özdemir & Wilcox, 2010).

The ideal estimator would have a standard error as small or smaller than any other estimator. None of the estimators in Table 3.7 satisfies this criterion. When sampling from a normal distribution, the sample mean has the lowest standard error, but the improvement over the 10% trimmed mean ($\gamma = 0.1$), the 20% trimmed mean, the Harrell–Davis estimator, and the M-estimator using Huber's Ψ, $\hat{\mu}_m$, is relatively small. Using the sample median is relatively unsatisfactory. For heavy-tailed distributions, the sample mean performs poorly and its performance can be made as bad as desired by making the tails of the distribution sufficiently heavy. The two estimators that do reasonably well for all of the distributions considered are the 20% trimmed mean and $\hat{\mu}_m$. Note that even the standard error of the Harrell–Davis estimator, $\hat{\theta}_{.5}$, becomes relatively large when the tails of a distribution are sufficiently heavy. Again, there is the possibility that in practice, $\hat{\theta}_{.5}$ competes well with the 20% trimmed mean and the M-estimator, but an obvious concern is that exceptions might occur. In situations where interest is specifically directed at quantiles, and in particular the population median, the choice between the sample median and the Harrell–Davis estimator is unclear. The Harrell–Davis estimator has a relatively small standard error when sampling from a normal distribution, but as the tails of a distribution get heavier, eventually the sample median performs substantially better. Although MOM has a lower standard error than the median under normality, all other estimators have a lower standard error than MOM for this special case. Switching to the Hampel identifier when using MOM (HMOM), efficiency now competes well with a 20% trimmed mean and the M-estimator based on Huber's Ψ, but for the lognormal distribution, HMOM performs rather poorly, and it is substantially worse than MOM when sampling from the slash distribution. The tau measure of location does not perform all that well, particularly when dealing with the slash distribution. In exploratory studies one might consider two or more estimators, but in the context of testing hypotheses, particularly in a confirmatory study, some might object to using multiple estimators of location because this will inflate the probability of at least one type I error. There are methods for adjusting the individual tests so that the probability of at least one type I error does not exceed a specified value, but such an adjustment might lower power by a substantial amount.

If a skipped estimator is used where outliers are detected via a boxplot rule (described in Section 3.13), good efficiency can be obtained under normality, but situations arise where other estimators offer a distinct advantage. For the situations in Table 3.7, under normality, the variance of this skipped estimator is 0.109 when using Carling's modification of the boxplot method to detect outliers, and switching to the boxplot rule based on the ideal fourths, nearly the same result is obtained. For the lognormal distribution, however, the variances of the skipped estimators are 0.27 and 0.28, approximately, making them the least accurate estimators, on average, excluding the sample mean. They perform the best for the one-wild distribution, but for the slash, they are the least satisfactory excluding the 10% mean and mean.

It cannot be stressed too strongly that no single measure of location always has the lowest standard error. For the data in Table 3.2, the lowest estimated standard error was 45.8, obtained for $\hat{x}_{.5}$ using the Maritz–Jarrett method. (A bootstrap estimate of the standard error is 42.) The appeal of the 20% trimmed mean and M-estimator of location is that they guard against relatively large standard errors. Moreover, the potential reduction in the standard error using other estimators is relatively small compared with the possible reduction using the 20% trimmed mean or M-estimator instead.

Based purely on achieving a high breakdown point, the median and an M-estimator (based on Huber's Ψ) are preferable to a 20% trimmed or the Hodges–Lehmann estimator. (For an analysis when sample sizes are very small, see Rousseeuw & Verboven, 2002.) But in terms of achieving accurate probability coverage, methods based on a 20% trimmed mean often are more satisfactory.

3.12 More Measures of Scale

Although measures of location are often the focus of attention versus measures of scale, measures of scale are of interest in their own right. Some measures of scale have already been discussed, namely ω estimated by MAD and the Winsorized variance σ_w^2. Many additional measures of scale appear in the literature. Two additional measures are described here, which play a role in subsequent chapters.

To begin, it helps to be precise about what is meant by a *scale estimator*. It is any nonnegative function, $\hat{\zeta}$, such that for any constants a and b,

$$\hat{\zeta}(a+bX_1, \ldots, a+bX_n) = |b|\hat{\zeta}(X_1, \ldots, X_n). \tag{3.34}$$

From basic principles, the sample standard deviation, s, satisfies this definition. In words, a scale estimator ignores changes in location and it responds to uniform changes in scale in a manner consistent with what is expected based on standard results related to s. In the terminology of Chapter 2, $\hat{\zeta}$ should be location-invariant and scale-equivariant.

A general class of measures of scale, that has been found to have practical value, stems from the influence function of M-estimators of location when distributions are symmetric. For this special case, the influence function of μ_m takes on a rather simple form:

$$IF_m(X) = \frac{E(\Psi^2(Y))}{\{E(\Psi'(Y))\}^2},$$ (3.35)

$$Y = \frac{X - \mu_m}{K\tau},$$

where τ and Ψ are as in Section 3.6. The (asymptotic) variance of $\sqrt{n}\hat{\mu}_m$ is

$$\zeta^2 = \frac{K^2\tau^2 E(\Psi^2(Y))}{\{E(\Psi'(Y))\}^2}$$ (3.36)

and this defines a broad class of measures of scale. (In case it is not obvious, the reason for considering the variance of $\sqrt{n}\hat{\mu}_m$, rather than the variance of $\hat{\mu}_m$, is that the latter goes to 0 as n gets large, and the goal to define a measure of scale for the distribution under study, not the sampling distribution of the M-estimator that is being used.) Included as a special case among the possible choices for ζ is the usual population standard deviation, σ. To see this, take $\Psi(x) = x$, $K = 1$, and $\tau = \sigma$, in which case $\zeta = \sigma$.

3.12.1 The Biweight Midvariance

There is the issue of choosing Ψ when defining a measure of scale with Eq. 3.36. For reasons to be described, there is practical interest in choosing Ψ to be the biweight given by Eq. (3.16). The derivative of the biweight is $\Psi'(x) = (1 - x^2)(1 - 5x^2)$ for $|x| < 1$, otherwise it is equal to 0.

Let K be any positive constant. For reasons given later, $K = 9$ is a common choice. Also, let τ be ω, which is estimated by MAD. To estimate ζ, set

$$Y_i = \frac{X_i - M}{K \times \text{MAD}},$$

$$a_i = \begin{cases} 1, & \text{if } |Y_i| < 1 \\ 0, & \text{if } |Y_i| \geq 1, \end{cases}$$

in which case the estimate of ζ is

$$\hat{\zeta}_{bi} = \frac{\sqrt{n}\sqrt{\sum a_i(X_i - M)^2(1 - Y_i^2)^4}}{|\sum a_i(1 - Y_i^2)(1 - 5Y_i^2)|}.$$ (3.37)

The quantity $\hat{\zeta}_{bi}^2$ is called a *biweight midvariance*. It appears to have a finite sample breakdown point of approximately 0.5 (Goldberg & Iglewicz, 1992), but a formal proof has not been found.

Explaining the motivation for $\hat{\zeta}_{bi}$ requires some comments on methods for judging estimators of scale. A tempting approach is to compare the standard errors of any two estimators, but this can be unsatisfactory. The reason is that if $\hat{\zeta}$ is a measure of scale, so is $b\hat{\zeta}$, a result that follows from the definition of a measure of scale given by Eq. (3.34). But the variance of $\hat{\zeta}$ is larger than the variance of $b\hat{\zeta}$ if $0 < b < 1$, and in fact the variance of $b\hat{\zeta}$ can be made arbitrarily small by choosing b appropriately. What is needed is a measure for comparing scale estimators that is not affected by b. A common method for dealing with this problem (e.g., Lax, 1985; Iglewicz, 1983) is to compare two scale estimators with $\text{VAR}(\ln(\hat{\zeta}))$, the variance of the natural logarithm of the estimators being considered. Note that $\ln(b\hat{\zeta}) = \ln(b) + \ln(\hat{\zeta})$ for any $b > 0$ and scale estimator $\hat{\zeta}$, so $\text{VAR}(\ln(b\hat{\zeta})) = \text{VAR}(\ln(\hat{\zeta}))$. That is, the variance of the logarithm of $\hat{\zeta}$ is not affected by the choice of b.

Another method of comparing scale estimators is in terms of the variance of $\hat{\zeta}/\zeta$. Note that if $\hat{\zeta}$ and ζ are replaced by $b\hat{\zeta}$ and $b\zeta$ for any $b > 0$, the ratio $\hat{\zeta}/\zeta$ remains unchanged, so the problem mentioned in the previous paragraph has been addressed.

Lax (1985) compared over 150 methods of estimating measures of scale, several of which belong to the class of measures defined by Eq. (3.36). Comparisons were made in terms of what is called the *triefficiency* of an estimator. To explain, let $V_{\min}$ be the smallest known value of $\text{VAR}(\ln(\hat{\zeta}))$ among all possible choices for an estimator of scale, $\hat{\zeta}$. Then

$$E = 100 \times \frac{V_{\min}}{\text{VAR}(\ln(\hat{\zeta}))}$$

is a measure of *efficiency*. For some measures of scale, $V_{\min}$ can be determined exactly for a given distribution, while in other situations $V_{\min}$ is replaced by the smallest variance obtained, via simulations, among the many estimators of scale that are being considered. For example, when sampling from a normal distribution with $n = 20$, the smallest attainable value of $\text{VAR}(\ln(\hat{\zeta}))$ is 0.026 which is attained by s, the sample standard deviation. Thus, for normal distributions, s has efficiency $E = 100$, the best possible value. The main point is that the efficiency of many estimators has been determined for a variety of distributions. Moreover, three distributions have played a major role: normal, one-wild, and slash, already described. The smallest efficiency of an estimator, among these three distributions, is called its *triefficiency*. For example, if $n = 20$, s has efficiency 100, 0, and 0 for these three distributions, the smallest of these three efficiencies is 0, so its triefficiency is 0. Using $K = 9$ in Eq. (3.37) when defining $\hat{\zeta}_{bi}$, the efficiencies corresponding to these three distributions are 86.7, 85.8, and 86.1, so the triefficiency is 85.8, and this is the highest triefficiency among the measures of scale considered by Lax (cf. Croux, 1994).

Because the choice $K = 9$ in the scale estimator $\hat{\zeta}$, given by Eq. (3.37), yields the highest triefficiency of any of the estimators studied by Lax, the term biweight midvariance will

Table 3.8: How to Compute the Biweight Midvariance, $\hat{\zeta}^2_{\text{bimid}}$.

Set

$$Y_i = \frac{X_i - M}{9 \times MAD},$$

$$a_i = \begin{cases} 1, & \text{if } |Y_i| < 1 \\ 0, & \text{if } |Y_i| \geq 1, \end{cases}$$

in which case

$$\hat{\zeta}_{\text{bimid}} = \frac{\sqrt{n}\sqrt{\sum a_i(X_i - M)^2(1 - Y_i^2)^4}}{|\sum a_i(1 - Y_i^2)(1 - 5Y_i^2)|}, \tag{3.38}$$

and the biweight midvariance is $\hat{\zeta}^2_{\text{bimid}}$.

assume $K = 9$ unless stated otherwise. Table 3.8 summarizes how to compute this measure of scale, which will be labeled $\hat{\zeta}^2_{\text{bimid}}$.

3.12.2 R Function bivar

The R function

$$\text{bivar}(x)$$

computes the biweight midvariance as described in Table 3.8 using the data stored in the R variable x. For the data in Table 3.2, the function bivar returns the value 25,512 as the estimated biweight midvariance.

3.12.3 The Percentage Bend Midvariance and tau Measure of Variation

Two other measures of scale should be mentioned. The first, replaces Ψ, the biweight in Eq. (3.36), with Huber's Ψ. Following Shoemaker and Hettmansperger (1982), the particular form of Huber's Ψ used here is

$$\Psi(x) = \max[-1, \min(1, x)]. \tag{3.39}$$

Also, rather than use $\tau = \omega$ when defining Y in Eq. (3.36), a different measure is used instead. Again let θ be the population median. For any β, $0 < \beta < 0.5$ define ω_β to be the measure of scale determined by

$$P(|X - \theta| < \omega_\beta) = 1 - \beta.$$

Thus, ω_β is the $1 - \beta$ quantile of the distribution of $|X - \theta|$. If X has a standard normal distribution, then ω_β is the $1 - \beta/2$ quantile. For example, if $\beta = 0.1$, $\omega_{.1} = 1.645$, the 0.95

quantile. Note that when $\beta = 0.5$, ω_β is just the measure of scale ω already discussed and estimated by MAD.

The parameter ω could be rescaled so that it estimates the population standard deviation, σ, when sampling from a normal distribution. That is,

$$\omega_{N,\beta} = \frac{\omega_\beta}{z_{1-\frac{\beta}{2}}}$$

could be used where $z_{1-\beta/2}$ is the $1 - \beta/2$ quantile of the standard normal distribution. When $\beta = 0.5$, $\omega_{N,\beta}$ is just ω_N which is estimated by MADN. Shoemaker and Hettmansperger (1982) do not re-scale ω, and this convention is followed here. Shoemaker and Hettmansperger choose $\beta = 0.1$, but here $\beta = 0.2$ is used unless stated otherwise. The resulting measure of scale given by Eq. (3.36), with $K = 1$, is called the *percentage bend midvariance* and labeled ζ_{pb}^2. When $\beta = 0.1$ and sampling is from a standard normal distribution, $\zeta_{pb}^2 = 1.03$, while for $\beta = 0.2$, $\zeta_{pb}^2 = 1.05$. A method of estimating ζ_{pb}^2 is shown in Table 3.9

Table 3.9: How to Compute the Percentage Bend Midvariance, $\hat{\zeta}_{pb}^2$.

Set $m = [(1 - \beta)n + 0.5]$, the value of $(1 - \beta)n + 0.5$ rounded down to the nearest integer. For good efficiency, under normality, versus the usual sample variance, $\beta = 0.1$ is a good choice, in which case $m = [0.9n + 0.5]$. For example, if $n = 56$, $m = [0.9 \times 56 + 0.5] = [50.9] = 50$. Let $W_i = |X_i - M|$, $i = 1, \ldots, n$, and let $W_{(1)} \leq \ldots \leq W_{(n)}$ be the W_i values written in ascending order. But a concern with $\beta = 0.1$ is that the breakdown point is a bit low, in which case something like $\beta = 0.2$ might be preferable. The estimate of ω_β is

$$\hat{\omega}_\beta = W_{(m)},$$

the mth largest of the W_i values. Put another way, $W_{(m)}$ is the estimate of the $1 - \beta$ quantile of the distribution of W.

Next, set

$$Y_i = \frac{X_i - M}{\hat{\omega}_\beta}$$

$$a_i = \begin{cases} 1, & \text{if } |Y_i| < 1 \\ 0, & \text{if } |Y_i| \geq 1, \end{cases}$$

in which case the estimated percentage bend midvariance is

$$\hat{\zeta}_{pb}^2 = \frac{n\hat{\omega}_\beta^2 \sum \{\Psi(Y_i)\}^2}{(\sum a_i)^2}, \tag{3.40}$$

where

$$\Psi(x) = \max[-1, \min(1, x)].$$

There are two reasons for including the percentage bend midvariance in this book. First, Bickel and Lehmann (1976) argue that if both X and Y have symmetric distributions about zero, and if $|X|$ is stochastically larger than $|Y|$, then it should be the case that a measure of scale should be larger for X than it is for Y. That is, if ζ_x is some proposed measure of scale for the random variable X, it should be the case that $\zeta_x > \zeta_y$. Bickel and Lehmann define a measure of scale that satisfies this property to be a *measure of dispersion*. The biweight midvariance is not a measure of dispersion (Shoemaker & Hettmansperger, 1982). In contrast, if Huber's Ψ is used in Eq. (3.36) with $K = 1$, the resulting measure of scale, the percentage bend midvariance, is a measure of dispersion. However, if Huber's Ψ is used with $K > 1$, the resulting measure of scale is not a measure of dispersion (Shoemaker & Hettmansperger, 1982). (The Winsorized variance is also a measure of dispersion.) The second reason is that a slight modification of the percentage bend midvariance yields a useful measure of association (a robust analog of Pearson's correlation coefficient) when testing for independence.

The inclusion of the biweight and percentage bend midvariance is motivated by results in Lax (1985). Note that Lax refers to these measures of variation as *A-estimators*, but here, following Shoemaker and Hettmansperger (1982), the terms biweight and percentage bend midvariance are used. More recently, Randal (2008) compared these measures of scale to more recently proposed estimators and again concluded that the biweight and percentage bend midvariances perform relatively well. Two measures of scale not included in the study by Randal are Rocke's (1996) TBS estimator, which is introduced in Section 6.3.3 in the more general setting of multivariate data, and the tau measure of scale described in Yohai and Zamar (1988). Checks on the efficiency of these estimators indicate that under normality, the percentage bend midvariance and the tau measure of variation perform relatively well. For the one-wild distribution and the contaminated normal distribution in Section 1.1, the biweight and percentage bend midvariances are best. But for a sufficiently heavy-tailed distribution (the slash distribution), the tau measure of scale offers an advantage.

Finally, the other measure of variation that should be mentioned is the *tau measure of variation* given by

$$\zeta_\tau^2 = \frac{\text{MAD}^2}{n} \sum \rho_c \left(\frac{X_i - \mu_{tau}}{MAD} \right),$$

where $\rho_c = \min(x^2, c^2)$ and μ_{tau} is the tau measure of location introduced in Section 3.8.1. Following Maronna and Zamar, $c = 3$ is used unless stated otherwise.

3.12.4 R Functions pbvar, tauvar

The R function

```
pbvar(x,beta=0.2)
```

computes the percentage bend midvariance for the data stored in the vector x. The default value for the argument beta, which is β in Table 3.9, is 0.2. An argument for using beta=0.1 is that the resulting estimator is about 85% as efficient as the sample variance under normality. With beta=0.2, it is only about 67% as efficient, but a concern about beta=0.1 is that the breakdown point is only 0.1. For the data in Table 3.2, the function returns the value 54,422 with beta=0.1, while with beta=0.2 the estimate is 30,681, beta=0.5 the estimate is 35,568. In contrast, the biweight midvariance is estimated to be 25,512. In terms of resistance, beta=0.5 is preferable to beta=0.1 or .2, but for other goals discussed in subsequent chapters, beta=0.1 or 0.2 might be preferred for general use. The R function

$$tauvar(x, cval=3)$$

computes the tau measure of variation.

3.12.5 The Interquartile Range

The population interquartile range is the difference between the 0.75 and 0.25 quantiles, $x_{.75} - x_{.25}$; it plays a role when dealing with a variety problems to be described. As previously noted, many quantile estimators have been proposed, so there are many ways the interquartile range might be estimated. A simple quantile estimator, $\hat{x}_q$, was described in Section 3.3, this leads to a simple estimate of the interquartile range, but for various purposes alternative estimates of the interquartile range have been found to be useful. In particular, when checking data for outliers, results in Frigge, Hoaglin, and Iglewicz (1989) suggest using what are called the *ideal fourths* (cf. Carling, 2000; Cleveland, 1985; Hoaglin & Iglewicz, 1987; Hyndman & Fan, 1996).

The computations are as follows. Let $j = [(n/4) + (5/12)]$. That is, j is $(n/4) + (5/12)$ rounded down to the nearest integer. Let

$$h = \frac{n}{4} + \frac{5}{12} - j.$$

Then the estimate of the lower quartile (the 0.25 quantile) is given by

$$q_1 = (1 - h)X_{(j)} + hX_{(j+1)} \tag{3.41}$$

Letting $k = n - j + 1$, the estimate of the upper quartile, is

$$q_2 = (1 - h)X_{(k)} + hX_{(k-1)}. \tag{3.42}$$

So the estimate of the interquartile range is

$$\text{IQR} = q_2 - q_1.$$

3.12.6 R Function idealf

The R function

$$\text{idealf}(x),$$

written for this book, computes the ideal fourths for the data stored in the R variable x.

3.13 Some Outlier Detection Methods

This section summarizes some outlier detection methods, two of which are variations of so-called boxplot techniques. One of these methods has, in essence, already been described, but it is convenient to include it here along with a description of relevant software.

3.13.1 Rules Based on Means and Variances

We begin with a method for detecting outliers that is known to be unsatisfactory, but it is a natural strategy to consider, so its limitations should be made explicit. The rule is to declare X_i an outlier if

$$\frac{|X_i - \bar{X}|}{s} > K,$$

where K is some constant. Basic properties of normal distributions suggest appropriate choices for K. For illustrative purposes, consider $K = 2.24$. A concern about this rule, and indeed any rule based on the sample mean and standard deviation, is that if suffers from *masking*, meaning that the very presence of outliers masks their detection. Outliers affect the sample means, but in a certain sense they have a bigger impact on the standard deviation.

■ Example

Consider the values 2, 2, 3, 3, 3, 4, 4, 4, 100,000, 100,000. Obviously, the value 100,000 is an outlier and surely any reasonable outlier detection method would flag the value 100,000 as being unusual. But the method just described fails to do so. ■

3.13.2 A Method Based on the Interquartile Range

The standard boxplot approach to detecting outliers is based on the interquartile range. As previously noted, numerous quantile estimators have been proposed, and when checking for outliers, a good method for estimating the quartiles appears to be the ideal fourths, q_1 and q_2, described in Section 3.12. Then a commonly used rule is to declare X_i an outlier if

$$X_i < q_1 - k(q_2 - q_1) \text{ or } X_i > q_2 + k(q_2 - q_1), \tag{3.43}$$

where $k = 1.5$ is used unless stated otherwise.

3.13.3 Carling's Modification

One useful way of characterizing an outlier detection method is with its *outside rate per observation*, p_n, which refers to the expected proportion of observations declared outliers. So if m represents the number of points declared outliers based on a sample of size n, $p_n = E(m/n)$. A common goal is to have p_n reasonably small, say approximately equal to 0.05, when sampling from a normal distribution. A criticism of the boxplot rule given by Eq. (3.43) is that p_n is somewhat unstable as a function of n; p_n tends to be higher when sample sizes are small. To address this, Carling (2000) suggests declaring X_i an outlier if

$$X_i < M - k(q_2 - q_1) \text{ or } X_i > M + k(q_2 - q_1), \tag{3.44}$$

where M is the usual sample median, q_1 and q_2 are given by Eqs (3.41) and (3.42), respectively, and

$$k = \frac{17.63n - 23.64}{7.74n - 3.71}. \tag{3.45}$$

3.13.4 A MAD-Median Rule

Henceforth, the MAD-median rule for detecting outliers will refer to declaring X_i an outlier if

$$\frac{|X_i - M|}{\text{MAD}/0.6745} > K,$$

where K is taken to be $\sqrt{\chi^2_{0.975,1}}$, the square root of the 0.975 quantile of a chi-squared distribution with one degree of freedom (cf. Davies & Gather, 1993).[2] (So K is approximately 2.24.)

Detecting outliers based on MAD and the median has the appeal of being able to handle a large number of outliers because both MAD and the median have the highest possible breakdown, 0.5. We will see, however, that for certain purposes, the choice between a boxplot rule and a MAD-Median rule is not always straightforward.

---

[2] This rule is a special case of a multivariate outlier detection method proposed by Rousseeuw and van Zomeren (1990).

3.13.5 R Functions outbox, out, and boxplot

The R function

$$\text{outbox(x,mbox=F,gval=NA)}$$

checks for outliers using one of two boxplot methods just described. As usual, the argument x is any vector containing data. Using mbox=F (for false), results in using the method in Section 3.13.2. Setting mbox=T results in using Carling's modification in Section 3.13.3. The argument gval can be used to alter the constant k. If unspecified, $k = 1.5$ when using the method in Section 3.13.2, and k is given by Eq. (3.45) when using the method in Section 3.13.3.

The R function

$$\text{out(x)}$$

checks for outliers using the MAD-Median rule in Section 3.13.4. (This function contains additional arguments that are related to detecting outliers among multivariate data, but the details are postponed for now.)

The built-in R function

$$\text{boxplot(x)}$$

creates the usual graphical version of the boxplot, examples of which are shown in Figures 3.6 and 3.7. (But this function does not use the ideal fourths.) Several variations of this method for plotting data have been proposed that were recently summarized by Marmolejo-Ramos and Tian (2010).

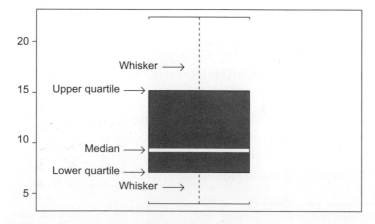

Figure 3.6: An example of a boxplot when there are no outliers.

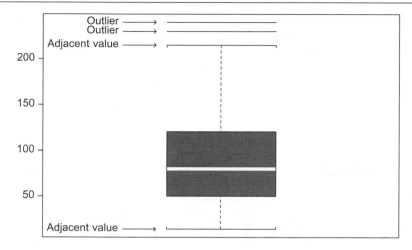

Figure 3.7: An example of a boxplot when there are outliers.

3.13.6 Skewness and the Boxplot Rule

It should be noted that the boxplot rule for detecting outliers has been criticized on the grounds that it might declare too many points outliers when there is skewness. More precisely, if a distribution is skewed to the right, among the larger values that are observed, too many might be declared outliers. Hubert and Vandervieren (2008) review the literature and suggest a modification of the boxplot rule that is based in part on a robust measure of skewness, called the *medcouple*, which was introduced by Brys, Hubert, and Struyf (2004) and is given by

$$MC = \text{med}(h(X_i, X_j)), \ X_i \le M \le X_j,$$

where for all $X_i \ne X_j$,

$$h(X_i, X_j) = \frac{(X_j - M) - (M - X_i)}{X_j - X_i}.$$

If $MC > 0$, declare X_i values outside the interval

$$[q_1 - 1.5e^{-4MC}(q_2 - q_1), \ q_2 + 1.5e^{3MC}(q_2 - q_1)]$$

as potential outliers. If $MC < 0$, declare X_i values outside the interval

$$[q_1 - 1.5e^{-3MC}(q_2 - q_1), \ q_2 + 1.5e^{4MC}(q_2 - q_1)]$$

as potential outliers.

There is, however, a feature of this adjusted boxplot rule that should mentioned. Imagine a distribution that is skewed to the right. Among the larger values, the adjusted boxplot rule might declare fewer points outliers, as intended, but among the lower values it might declare points outliers that are not flagged as outliers by the boxplot rule.

3.13.7 R Function adjboxout

The R function

$$adjboxout(x),$$

applies the adjusted boxplot rule just described. The function returns the values flagged as outliers. It also returns values stored in $cl and $cu, which are the lower and upper ends of the interval used to determine whether a value is an outlier.

■ Example

Consider the values

12, 33, 47, 55, 85, 87, 87, 96, 97, 99, 113, 118, 128, 138, 165, 202, 213, 218, 275, 653.

Both the boxplot rule and the adjusted boxplot rule declare the value 653 to be an outlier, which certainly seems reasonable based on a casual inspection of the data. But unlike the boxplot rule, the adjusted boxplot rule declares the values 12, 33, and 47 outliers as well.

■

3.14 Exercises

1. Included among the R functions written for this book is the function ghdist(n,g=0,h=0) which generates *n* observations from a so-called g-and-h distribution (which is described in more detail in Chapter 4). The command ghdist(30,0,.5) will generate 30 observations from a symmetric, heavy-tailed distribution. Generate 30 observations in this manner, create the density estimates using the functions skerd, kdplot, rdplot, and akerd. Repeat this 20 times and comment on the pattern of results.

2. In the study by Dana (1990) on self-awareness, described in this chapter (in connection with Table 3.2), a second group of subjects yielded the observations
59 106 174 207 219 237 313 365 458 497 515 529 557 615 625 645 973 1065 3215.
Compute the sample median, the Harrell–Davis estimate of the median, the M-estimate of location (based on Huber's Ψ), and the 10% and 20% trimmed means. Estimate the standard errors for each location estimator and compare the results.

3. For the data in Exercise 1, compute MADN, the biweight midvariance, and the percentage bend midvariance. Compare the results to those obtained for the data in Table 3.2. What do the results suggest about which group is more dispersed?

4. For the data in Exercise 1, estimate the deciles using the Harrell–Davis estimator. Do the same for the data in Table 3.2. Plot the difference between the deciles as a function of the estimated deciles for the data in Exercise 1. What do the results suggest? Estimate the standard errors associated with each decile estimator.

5. Comment on the strategy of applying the boxplot to the data in Exercise 2, removing any outliers, computing the sample mean for the data that remain, and then estimating the standard error of this sample mean based on the sample variance of the data that remain.

6. Cushny and Peebles (1904) conducted a study on the effects of optical isomers of hyoscyamine hydrobromide in producing sleep. For one of the drugs, the additional hours of sleep for 10 patients were

   0.7, −1.6, −0.2, −1.2, −0.1, 3.4, 3.7, .8, 0, and 2.

   Compute the Harrell–Davis estimate of the median, the mean, the 10% and 20% trimmed means, and the M-estimate of location. Compute the corresponding standard errors.

7. Use results on Winsorized expected values in Chapter 2 to show that if the error term in Eq. (3.4) is ignored, $\bar{X}_t$ is a Winsorized unbiased estimate of μ_t.

8. Use results on Winsorized expected values in Chapter 2 to show that $\bar{X}_w$ is a Winsorized unbiased estimate of μ_w.

9. Set $X_i = i$, $i = 1, \ldots, 20$, and compute the 20% trimmed mean and the M-estimate of location based on Huber's Ψ. Next, set $X_{20} = 200$ and compute both estimates of location. Replace X_{19} with 200 and again estimate the measures of location. Keep doing this until the upper half of the data is equal to 200. Comment on the resistance of the M-estimator versus 20% trimming.

10. Repeat the previous exercise, only this time compute the biweight midvariance, the 20% Winsorized variance, and the percentage bend midvariance. Comment on the resistance of these three measures of scale.

11. Set $X_i = i$, $i = 1, \ldots, 20$ and compute the Harrell–Davis estimate of the median. Repeat this, but with X_{20} equal to 1000 and then 100,000. When $X_{20} = 100,000$, would you expect $\hat{x}_{0.5}$ or the Harrell–Davis estimator to have the smaller standard error? Verify your answer.

12. Argue that if Ψ is taken to be the biweight, it approximates the optimal choice for Ψ under normality when observations are not too extreme.

13. Verify that Eq. (3.29) reduces to s^2/n if no observations are flagged as being unusually large or small by Ψ.

CHAPTER 4

Confidence Intervals in the One-Sample Case

A fundamental problem is testing hypotheses and computing confidence intervals for the measures of location described in Chapters 2 and 3. As will be seen, a method that provides accurate probability coverage for one measure of location can perform poorly with another. That is, the recommended method for computing a confidence interval depends in part on which the measure of location is of interest. An appeal of the methods in this chapter is that when computing confidence intervals for robust measures of location, it is possible to get reasonably accurate probability coverage in situations where no known method for the mean gives good results.

4.1 Problems when Working with Means

It helps to first describe problems associated with Student's t. When testing hypotheses or computing confidence intervals for μ, it is assumed that

$$T = \frac{\sqrt{n}(\bar{X} - \mu)}{s} \tag{4.1}$$

has a Student's t-distribution with $v = n - 1$ degrees of freedom. This implies that $E(T) = 0$, and that T has a symmetric distribution. From basic principles, this assumption is correct when observations are randomly sampled from a normal distribution. However, at least three practical problems can arise. First, there are problems with power and the length of the confidence interval. As indicated in Chapters 1 and 2, the standard error of the sample mean, $\sigma/\sqrt{n}$, becomes inflated when sampling from a heavy-tailed distribution, so power can be poor relative to methods based on other measures of location, and the length of confidence intervals, based on Eq. (4.1), become relatively long – even when σ is known. (For a detailed analysis of how heavy-tailed distributions affect the probability coverage of the t-test, see Benjamini, 1983.) Second, the actual probability of a type I error can be substantially higher or lower than the nominal α level. When sampling from a symmetric distribution, generally the actual level of Student's t-test will be less than the nominal level (Efron, 1969). When sampling from a symmetric, heavy-tailed distribution the actual probability of type I error can

Introduction to Robust Estimation and Hypothesis Testing

104 Introduction to Robust Estimation and Hypothesis Testing

be substantially lower than the nominal α level, and this further contributes to low power and relatively long confidence intervals. From theoretical results reported by Basu and DasGupta (1995), problems with low power can arise even when n is large. When sampling from a skewed distribution with relatively light tails, the actual probability coverage can be substantially less than the nominal $1 - \alpha$ level resulting in inaccurate conclusions and this problem becomes exacerbated as we move toward (skewed) heavy-tailed distributions. Third, when sampling from a skewed distribution, T also has a skewed distribution, it is no longer true that $E(T) = 0$, and the distribution of T can deviate enough from a Student's t-distribution so that practical problems arise. These problems can be ignored if the sample size is sufficiently large, but given data it is difficult knowing just how large n has to be. When sampling from a lognormal distribution, it is known that $n > 160$ is required (Westfall & Young, 1993). As we move away from the lognormal distribution toward skewed distributions where outliers are more common, $n > 300$ might be required. Problems with controlling the probability of a type I error are particularly serious when testing one-sided hypotheses. And this has practical implications when testing two-sided hypotheses because it means that a biased hypothesis testing method is being used, as will be illustrated.

Problems with low power were illustrated in Chapter 1, so further comments are omitted. The second problem, that probability coverage and type I error probabilities are affected by departures from normality, is illustrated with a class of distributions obtained by transforming a standard normal distribution in a particular way. Suppose Z has a standard normal distribution, and for some constant $h \geq 0$, let

$$X = Z \exp\left(\frac{hZ^2}{2}\right).$$

Then X has what is called an *h distribution*. When $h = 0$, $X = Z$, so X is standard normal. As h gets large, the tails of the distribution of X get heavier, and the distribution is symmetric about 0. (More details about the h distribution are described in Section 4.2)

Suppose sampling is from an h distribution with $h = 1$, which has very heavy tails. Then with $n = 20$ and $\alpha = 0.05$, the actual probability of a type I error, when using Student's t to test $H_0 : \mu = 0$, is approximately .018 (based on simulations with 10,000 replications). Increasing n to 100, the actual probability of a type I error is approximately .019. A reasonable suggestion for dealing with this problem is to inspect the empirical distribution to determine whether the tails are relatively light. This might be done in various ways, but there is no known empirical rule that reliably indicates whether the type I error probability will be substantially lower than the nominal level when attention is restricted to using Student's t-test.

To illustrate the third problem, and provide another illustration of the second, consider what happens when sampling from a skewed distribution with relatively light tails. In particular, suppose X has a lognormal distribution, meaning that for some normal random variable,

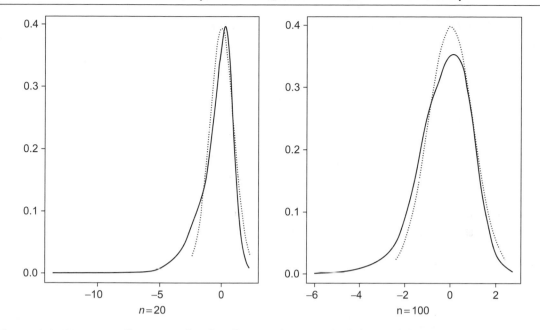

Figure 4.1: Nonnormality can seriously affect Students t. The left panel shows an approximation of the actual distribution of Students t when sampling from a lognormal distribution and $n = 20$ and the right panel is when $n = 100$.

Y, $X = \exp(Y)$. This distribution is light-tailed in the sense that the expected proportion of values declared an outlier, using the MAD-Median rule used to define the MOM estimator in Section 3.7, is relatively small.[1]

For convenience, assume Y is standard normal in which case $E(X) = \sqrt{e}$, where $e = \exp(1) \approx 2.71828$, and the standard deviation is approximately $\sigma = 2.16$. Then Eq. (4.1) assumes that $T = \sqrt{n}(\bar{X} - \sqrt{e})/s$ has a Student's t-distribution with $n - 1$ degrees of freedom. The left panel of Figure 4.1 shows a (kernel density) estimate of the actual distribution of T when $n = 20$; the symmetric distribution is the distribution of T under normality. As is evident, the actual distribution is skewed to the left, and its mean is not equal to 0. Simulations indicate that $E(T) = -0.54$, approximately. The right panel shows an estimate of the probability density function when $n = 100$. The distribution is more symmetric compared to $n = 20$, but it is clearly skewed to the left.

Let μ_0 be some specified constant. The standard approach to testing H_0: $\mu \leq \mu_0$ is to evaluate T with $\mu = \mu_0$ and reject H_0 if $T > t_{1-\alpha}$, where $t_{1-\alpha}$ is the $1 - \alpha$ quantile of Student's t-distribution with $\nu = n - 1$ degrees of freedom, and α is the desired probability of a

---

[1]   Some journal articles characterize the lognormal distribution as having heavy tails, but based on the expected proportion of points labeled outliers, it is relatively light-tailed.

type I error. If $H_0: \mu \le \sqrt{e}$ is tested when X has a lognormal distribution, H_0 should not be rejected, and the probability of a type I error should be as close as possible to the nominal level, α. If $\alpha = 0.05$ and $n = 20$, the actual probability of a type I error is approximately .008 (Westfall & Young, 1993, p. 40). As indicated in Figure 4.1, the reason is that T has a distribution that is skewed to the left. In particular, the right tail is much lighter than the assumed Student's t-distribution, and this results in a type I error probability that is substantially smaller than the nominal 0.05 level. Simultaneously, the left tail, below the point -1.73, the 0.95 quantile of Student's t-distribution with 19 degrees of freedom, is too thick. Consequently, when testing $H_0: \mu \ge \sqrt{e}$ at the 0.05 level, the actual probability of rejecting is .153. Increasing n to 160, the actual probability of a type I error is .022 and .109 for the one-sided hypotheses being considered. And when observations are sampled from a heavy-tailed distribution, control over the probability of a type I error deteriorates.

Generally, as we move toward a skewed distribution with heavy tails, the problems illustrated by Figure 4.1 become exacerbated. As an example, suppose sampling is from a squared lognormal distribution that has mean exp(2). (i.e., if X has a lognormal distribution, $E(X^2) = \exp(2)$.) Figure 4.2 shows plots of T values based on sample sizes of 20 and 100. (Again, the symmetric distributions are the distributions of T under normality.)

Of course, the seriousness of a type I error depends on the situation. Presumably there are instances where an investigator does not want the probability of a type I error to exceed .1, otherwise the common choice of $\alpha = 0.05$ would be replaced by $\alpha = 0.1$ in order to increase power. Thus, assuming Eq. (4.1) has a Student's t-distribution might be unsatisfactory when

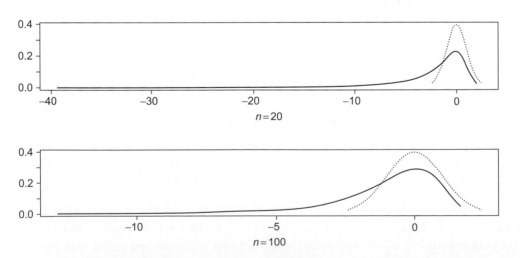

Figure 4.2: The same as Figure 4.1, only now sampling is from a squared lognormal distribution. This illustrates that as we move toward heavy-tailed distributions, problems with nonnormality are exacerbated.

testing hypotheses, and the probability coverage of the usual two-sided confidence interval, $\bar{X} \pm t_{1-\alpha/2} s/\sqrt{n}$ might be unsatisfactory as well. Bradley (1978) argues that if a researcher makes a distinction between $\alpha = 0.05$ and $\alpha = 0.1$, the actual probability of a type I error should not exceed .075, the idea being that otherwise it is closer to .1 than .05, and he argues that it should not drop below .025. He goes on to suggest that ideally, at least in many situations, the actual probability of a type I error should be between .045 and .055 when $\alpha = 0.05$.

It is noted that when testing H_0: $\mu < \mu_0$, and when a distribution is skewed to the right, improved control over the probability of a type I error can be achieved using a method derived by Chen (1995). However, even for this special case, problems with controlling the probability of a type I error remain in some situations, and power problems plague any method based on means. (A generalization of this method to some robust measure of location might have some practical value, but this has not been established as yet.) Banik and Kibria (2010) compared numerous methods for computing a (two-sided) confidence interval for the mean. In terms of probability coverage, none of the methods were completely satisfactory when the sample size is small. For $n \geq 50$, Chen's method performed reasonably well among the distributions considered, including situations where sampling is from a lognormal distribution. But the lognormal distribution is relatively light tailed. How well Chen's method performs when sampling from a skewed, heavy-tailed distribution, or even a symmetric, heavy-tailed distribution (such as the contaminated normal), appears to be unknown.

4.2 The g-and-h Distribution

One of the main goals in this chapter is to recommend certain procedures for computing confidence intervals and testing hypotheses, and to discourage the use of others. These recommendations are based in part on simulations, some of which generate observations from a so-called g-and-h distribution. This section is included for readers interested in the motivation and details of such studies. Readers primarily concerned with how methods are applied, or which methods are recommended, can skip or skim this section.

A basic problem is establishing whether a particular method for computing a confidence interval has probability coverage reasonably close to the nominal $1 - \alpha$ level when the sample size is small or even moderately large. When investigating the effect of nonnormality, there is the issue of deciding which nonnormal distributions to consider when checking the properties of a particular procedure via simulations. One approach, which provides a partial check on how a method performs, is to consider four types of distributions: normal, symmetric with a heavy tail, asymmetric with a light tail, and asymmetric with a heavy tail. But how heavy-tailed and asymmetric should they be? A natural approach is to use distributions that are similar to those found in applied settings. But coming to terms with what constitutes a reasonable range of values is difficult at best. Several papers have been published with the goal of characterizing the range of heavy-tailedness and skewness that a researcher is likely to encounter (e.g., Pearson & Please, 1975; Sawilowsky & Blair, 1992; Micceri, 1989; Hill & Dixon, 1982; Wilcox, 1990a). The most striking feature of these studies is the extent to which they differ. For example, some papers suggest that distributions are never extremely skewed, whereas others indicate the exact opposite. In a sexual attitude study by Pedersen, Miller, Putcha-Bhagavatula, and Yang (2002), the skewness and kurtosis, based on 105 participants, is 15.9 and 256.3, respectively. In a related study based on 16,288 participants, the 10 variables had estimated skewness that ranged between 52.1 and 115.5, and kurtosis that ranged between 3290 and 13,357. In a review of 440 large-sample psychological studies, Micceri (1989) reported that 97% (35 of 36 studies) "of those distributions exhibiting kurtosis beyond the double exponential (3.00) also showed extreme or exponential skewness." Moreover, 72% (36 of 50) distributions that exhibited skewness greater than two also had tail weights that were heavier than the double exponential.

One way of attempting to span the range of skewness and heavy-tailedness that one might encounter is to run simulations where observations are generated from a g-and-h distribution. An observation X is generated from a g-and-h distribution by first generating Z from a standard normal distribution and then setting

$$X = \frac{\exp(gZ) - 1}{g} \exp(hZ^2/2),$$

where g and h are nonnegative constants that can be chosen so that the distribution of X has some characteristic of interest. When $g = 0$, this last equation is taken to be

$$X = Z \exp(hZ^2/2).$$

When $g = h = 0$, $X = Z$, so X has a standard normal distribution. When $g = 0$, X has a symmetric distribution. As h increases, the tails of the distribution get heavier. As g increases, the distribution becomes more skewed. The case $g = 1$ and $h = 0$ corresponds to a lognormal distribution that has been shifted to have a median of zero. Note that within the class of g-and-h distributions, the lognormal is skewed with a relatively light tail. Hoaglin (1985) provides a detailed description of various properties of the g-and-h distribution, but only a few properties are listed here. Table 4.1 summarizes the skewness and kurtosis values for four selected situations that have been used in published studies and are considered at various points in this book. In Table 4.1, skewness and kurtosis are measured with $\kappa_1 = \mu_{[3]}/\mu_{[2]}^{1.5}$ and $\kappa_2 = \mu_{[4]}/\mu_{[2]}^2$, where $\mu_{[k]} = E(X - \mu)^k$. When $g > 0$ and $h \geq 1/k$, $\mu_{[k]}$ is not defined and the corresponding entry is left blank.

A possible criticism of simulations based on the g-and-h distribution is that observations generated on a computer have skewness and kurtosis that are not always the same as the

Table 4.1: Some Properties of the g-and-h Distribution.

g	h	κ_1	κ_2	$\hat{\kappa}_1$	$\hat{\kappa}_2$	μ	$\mu_t(20\%)$	μ_m
0.0	0.0	0.00	3.00	0.0	3.00	0.0000	0.0000	0.0000
0.0	0.5	0.00	—	0.00	11,986.2	0.0000	0.0000	0.0000
0.5	0.0	1.75	8.9	1.81	9.7	0.2653	0.0541	0.1047
0.5	0.5	—	—	120.10	18,393.6	0.8033	0.0600	0.0938

theoretical values listed in Table 4.1. The reason is that observations generated on a computer come from some bounded interval on the real line, so $\mu_{[k]}$ is finite even when in theory it is not. For this reason, Table 4.1 also reports $\hat{\kappa}_1$ and $\hat{\kappa}_2$, the estimated skewness and kurtosis based on 100,000 observations. (Skewness is not estimated when $g = 0$ because it is known that $\kappa_1 = 0$.) The last three columns of Table 4.1 show the value of μ, the 20% trimmed mean, μ_t, and μ_m, the M-measure of location, which were determined via numerical quadrature. For completeness, it is noted that for the lognormal distribution, $\kappa_1 = 6.2$, $\kappa_2 = 114$, the 20% trimmed mean is $\mu_t = 1.111$, and $\mu_m = 1.1857$.

Ideally, a method for computing a confidence interval will have accurate probability coverage when sampling from any of the four g-and-h distributions in Table 4.1. It might be argued that when g or h equals 0.5, the corresponding distribution is unrealistically nonnormal. The point is that if a method performs well under seemingly large departures from normality, this offers some reassurance that it will perform well for distributions encountered in practice. Of course, even if a method gives accurate results for the four distributions in Table 4.1, this does not guarantee accurate probability coverage for any distribution that might arise in practice. In most cases, there is no known method for proving that a particular technique always gives good results.

Another possible criticism of the four g-and-h distributions in Table 4.1 is that perhaps the skewed, light-tailed distribution ($g = 0.5$ and $h = 0$) does not represent a large enough departure from normality. In particular, Wilcox (1990a) found that many random variables he surveyed had estimated skewness greater than 3, but the skewness of this particular g-and-h distribution is only 1.8, approximately. For this reason, it might also be important to consider the lognormal distribution when studying the small-sample properties of a particular method.

Table 4.2 shows the estimated probability of a type I error (based on simulations with 10,000 replications) when using Student's t to test H_0: $\mu = 0$ with $n = 12$ and $\alpha = 0.05$. (The notation $t_{0.025}$ refers to the 0.025 quantile of Student's t-distribution.) For example, when sampling from a g-and-h distribution with $g = 0.5$ and $h = 0$, the estimated probability of a type I error is $.000 + .420 = .420$, which is about eight times as large as the nominal level. Put another way, if H_0: $\mu < 0$ is tested with $\alpha = 0.025$, the actual probability of rejecting when $\mu = 0$ is approximately 0.42, over 16 times larger than the nominal level. Note that for

Table 4.2: One-Sided Type I Error Probabilities when Using Student's t, $n = 12$, $\alpha = 0.025$.

g	h	$P(T > t_{0.975})$	$P(T < t_{0.025})$
0.0	0.0	.025	.025
0.0	0.5	.015	.016
0.5	0.0	.000	.420
0.5	0.5	.000	.295

fixed g, as the tails get heavier (h increases from 0 to 0.5), the probability of a type I error decreases. This is not surprising because sampling from a heavy-tailed distribution inflates s which in turn results in longer confidence intervals. A similar result, but to a lesser extent, is found when using robust measures of location.

Multivariate g-and-h Distributions

It is noted that multivariate distributions having some specified correlation matrix **R** can be generated as follows. Generate **X** where the marginal distributions are independent. Form the Cholesky decomposition $U'U = R$, where U is the matrix of factor loadings of the principal components of the square-root method of factoring a correlation matrix, and U' is the transpose of U. Then **XU** produces a matrix of data that has population correlation matrix **R**.

4.2.1 R Functions ghdist and rmul

The R function

$$\text{ghdist}(n, g=0, h=0)$$

generates n observations from a g-and-h distribution. By default, observations are generated from a standard normal distribution ($g = h = 0$). The R function

$$\text{rmul}(n, p = 2, \text{cmat} = \text{diag}(\text{rep}(1, p)), \text{rho} = \text{NA}, \text{mar.fun} = \text{rnorm}, ...)$$

generates n vectors of observations from a p-variate distribution having correlation matrix specified by the argument cmat and marginal distributions specified by the argument mar.fun. By default, data are generated from a bivariate normal distribution with Pearson's correlation equal to 0. If the argument rho is specified, all pairs of variables will have correlation rho. The command

$$\text{rmul}(30, p = 3, \text{rho} = 0.4, \text{mar.fun} = \text{ghdist}, g=1, h=0.2)$$

would first generate data from a trivariate distribution for which the marginal distributions are independent with marginal g-and-h distributions, where $g = 1$ and $h = 0.2$, after which the data are transformed so that all pairs of variables have correlation 0.4. It should be noted that changing the correlation via the argument rho can alter the marginal measures of location when $g > 0$, in which case the marginal distributions are skewed.

4.3 Inferences About the Trimmed and Winsorized Means

When working with the trimmed mean, μ_t, an analog of Eq. (4.1) is

$$T_t = \frac{(1-2\gamma)\sqrt{n}(\bar{X}_t - \mu_t)}{s_w}. \tag{4.2}$$

When $\gamma = 0$, $T_t = T$ given by Eq. (4.1). Tukey and McLaughlin (1963) suggest approximating the distribution of T_t with a Student's t-distribution having $n - 2g - 1$ degrees of freedom, where, as in Chapter 3, $g = [\gamma n]$ is the integer portion of γn. Then $n - 2g$ is the number of observations left after trimming. The resulting two-sided $1 - \alpha$ confidence interval for μ_t is

$$\bar{X}_t \pm t_{1-\alpha/2}\frac{s_w}{(1-2\gamma)\sqrt{n}}, \tag{4.3}$$

where $t_{1-\alpha/2}$ is the $1 - \alpha/2$ quantile of Student's t-distribution with $n - 2g - 1$ degrees of freedom. Let μ_0 be some specified constant. Then under the null hypothesis H_0: $\mu_t = \mu_0$, T_t becomes

$$T_t = \frac{(1-2\gamma)\sqrt{n}(\bar{X}_t - \mu_0)}{s_w}, \tag{4.4}$$

and H_0 is rejected if $|T_t| > t_{1-\alpha}$. One-sided tests can be performed in the usual way. In particular, reject H_0: $\mu_t \leq \mu_0$ if $T_t > t_{1-\alpha}$, the $1 - \alpha$ quantile of Student's t-distribution with $n - 2g - 1$ degrees of freedom. Similarly, reject H_0: $\mu_t \geq \mu_0$ if $T_t < t_\alpha$.

Based on various criteria, plus a slight variation of the sample trimmed mean used here, Patel, Mudholkar, and Fernando (1988) found the Tukey–McLaughlin approximation to be reasonably accurate when sampling from various distributions. They also report that for $\gamma = 0.25$, a better approximation is a Student's t-distribution with $n - 2.48g - 0.15$ degrees of freedom. For $n < 18$, they suggest a more refined approximation, but another method, to be described, gives more satisfactory results.

Additional support for using a Student's t-distribution with $n - 2g - 1$ degrees of freedom is reported by Wilcox (1994a). Using a Winsorized analog of a Cornish–Fisher expansion of T_t, a correction term for skewness was derived and compared to the correction term used when there is no trimming. As the amount of trimming increases, the magnitude of the correction term decreases indicating that the probability coverage using Eq. (4.3) should be closer to the nominal level than the probability coverage when $\gamma = 0$. Numerical results for the lognormal distribution indicate that as γ increases, the magnitude of the correction term decreases rapidly up to about $\gamma = 0.2$.

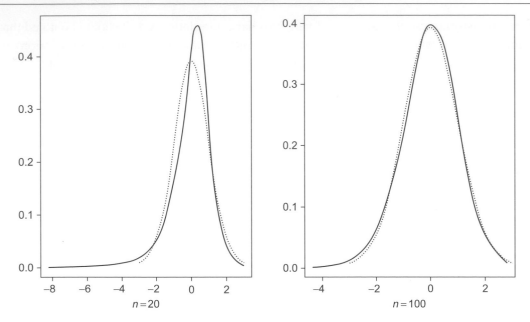

Figure 4.3: The distribution of T_t with 20% trimming when sampling from a lognormal distribution. Compare this to the distribution of T shown in Figure 4.1.

The left panel of Figure 4.3 shows the probability density function of T_t with 20% trimming when $n = 20$ and sampling is from a lognormal distribution. (The symmetric distribution is the assumed distribution of T_t when testing hypotheses.) The actual distribution is skewed to the left, but the tail of the distribution is not as heavy as the tail of the distribution shown in Figure 4.1. The result is that when testing H_0: $\mu_t > 0$, the probability of a type I error will be greater than the nominal level, but not as much versus no trimming. For example, if $\alpha = 0.025$, the actual probability of a type I error is approximately .062 with 20% trimming versus .132 when using the mean to test H_0: $\mu > \sqrt{e}$. The right panel of Figure 4.3 shows the distribution of T_t when n is increased to 100. Note that the distribution is reasonably symmetric, as is assumed when using T_t, versus the right panel of Figure 4.1 which is clearly skewed to the left. This illustrates the general expectation that when using the 20% trimmed mean, probability coverage will improve more rapidly as the sample size increases, versus confidence intervals based on means. If a distribution is both skewed and sufficiently heavy-tailed, problems with controlling the probability of a type I error can persist unless n is fairly large. That is, as the amount of trimming increases, problems with controlling the probability of a type I error decrease, but even with 20% trimming, not all practical problems are eliminated using the method in this section. Increasing the amount of trimming beyond 20%, such as using a median, could be used, but at the risk of low power if indeed a distribution is normal or relatively light-tailed. A better strategy seems to be to use a bootstrap method described in Section 4.4.

It was previously noted that when sampling from a symmetric, heavy-tailed distribution (an h distribution with $h = 1$), the actual probability of a type I error can be as low as .018 when testing H_0: $\mu = 0$ with Student's t-test, $n = 20$, and $\alpha = 0.05$. In contrast, with 20% trimming, the probability of a type I error is approximately .033. Generally, if a symmetric distribution is sufficiently heavy-tailed, roughly meaning that the expected proportion of values declared outliers is relatively high, actual type I error probabilities can drop below the nominal level. In some situations it currently seems that this problem can persist no matter which location estimator is used.

■ Example

Table 4.3 shows the average LSAT scores for the 1973 entering classes of 15 American law schools. (LSAT is a national test for prospective lawyers.) The sample mean is $\bar{X} = 600.3$ with an estimated standard error of 10.8. The 20% trimmed mean is $\bar{X}_t = 596.2$ with an estimated standard error of 14.92, and with $15 - 6 - 1 = 8$ degrees of freedom, the 0.95 confidence interval for μ_t is (561.8, 630.6). In contrast, the 0.95 confidence interval for μ is (577.1, 623.4), assuming T given by Eq. (4.1) does indeed have a Student's t-distribution with 14 degrees of freedom. Note that the length of the confidence interval for μ is smaller, and in fact is a subset of the confidence interval for μ_t. This might seem to suggest that the sample mean is preferable to the trimmed mean for this particular set of data, but closer examination suggests the opposite conclusion. As already illustrated, if sampling is from a light-tailed, skewed distribution, the actual probability coverage for the sample mean can be smaller than the nominal level. For the situation at hand, the claim that (577.1, 623.4) is a 0.95 confidence interval for the mean might be misleading and overly optimistic. Figure 4.4 shows a boxplot of the data. The sample median is 580 indicating that the central portion of the data is skewed to the right. Moreover, there are no outliers suggesting the possibility that sampling is from a relatively light-tailed distribution. Thus, the actual probability coverage of the confidence interval for the mean might be too low – a longer confidence interval might be needed to achieve .95 probability coverage. It is not being suggested, however, that if there had been outliers, there is reason to believe that probability coverage is not too low. For example, boxplots of data generated from a lognormal distribution frequently have values flagged as outliers, and as already noted, sampling from a lognormal distribution can result in a confidence interval for μ that is too short.

Table 4.3: Average LSAT Scores for 15 Law Schools.

545	555	558	572	575	576	578	580
594	605	635	651	653	661	666	

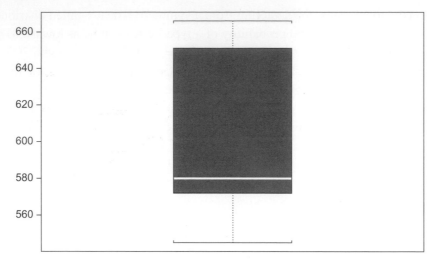

Figure 4.4: A boxplot of the LSAT scores.

As for computing a confidence interval for the population Winsorized mean, results in Dixon and Tukey (1968) suggest using

$$\bar{X}_w \pm t_{1-\alpha/2} \left(\frac{n-1}{n-2g-1} \right) \left(\frac{s_w}{\sqrt{n}} \right),$$

where again the degrees of freedom are $n - 2g - 1$. It appears that the accuracy of this confidence interval, when sampling from a skewed distribution, has not been studied.

4.3.1 R Functions trimci and winci

The R function

$$\text{trimci(x, tr=.2, alpha=0.05)},$$

written for this book, computes a $1 - \alpha$ confidence interval for μ_t using Eq. (4.3) based on the data stored in the R vector x, where x is any R variable containing data, tr is the desired amount of trimming (the value of γ), and alpha is α. The default amount of trimming is $\gamma = 0.2$ (20%), and the default value for α is 0.05. For example, the command trimci(w,0.1,0.025) returns two values: the lower and upper end of the 0.975 confidence interval for the 10% trimmed mean using the data stored in w. The command trimci(w) returns a 0.95 confidence interval for the 20% trimmed mean. The R function

$$\text{winci(x,tr=0.2,alpha=0.05)},$$

computes a confidence interval for the population Winsorized mean.

4.4 Basic Bootstrap Methods

The method used to compute a confidence interval for a trimmed mean, described in Section 4.3, is based on the fundamental strategy developed by Laplace about two centuries ago: When using $\hat{\theta}$ to estimate some parameter of interest, θ, estimate the standard error of $\hat{\theta}$ with say $\hat{\Upsilon}$, and try to approximate the distribution of

$$\frac{\hat{\theta} - \theta}{\hat{\Upsilon}}.$$

Laplace accomplished this by appealing to his central limit theorem which he publicly announced in 1810. That is, assume this last equation has a standard normal distribution.

An alternative approach is to use some type of bootstrap method. There are many variations; see Efron and Tibshirani (1993), Chernick (1999), Davison and Hinkley (1997), Hall and Hall (1995), Lunneborg (2000), Mooney and Duval (1993), and Shao and Tu (1995). Here attention is focused on two basic types (with some extensions described in subsequent chapters). Alternative methods are not considered because either they currently seem to have no practical advantage for the problems considered here, in terms of controlling the probability of a type I error or yielding accurate probability coverage, or the practical advantages of these alternative methods have not been adequately investigated when sample sizes are small or moderately large.

4.4.1 The Percentile Bootstrap Method

The first basic version is the so-called percentile bootstrap. It begins by obtaining a *bootstrap sample* of size n. That is, values are obtained by randomly sampling with replacement n values from $X_1, \ldots, X_n$ yielding $X_1^*, \ldots, X_n^*$.

Let $\hat{\theta}^*$ be an estimate of θ based on this bootstrap sample. Of course, a new bootstrap sample can be generated to yield a new bootstrap estimate of θ. Repeating this process B times yields B bootstrap estimates: $\hat{\theta}_1^*, \ldots, \hat{\theta}_B^*$. Let $\ell = \alpha B / 2$, rounded to the nearest integer, and let $u = B - \ell$. Letting $\hat{\theta}_{(1)}^* \leq \cdots \leq \hat{\theta}_{(B)}^*$ represent the B bootstrap estimates written in ascending order, an approximate $1 - \alpha$ confidence interval for θ is

$$\left(\hat{\theta}_{(\ell+1)}^*, \hat{\theta}_{(u)}^* \right).$$

An outline of the theoretical justification of the method is as follows. Imagine that the goal is to test

$$H_0 : \theta = \theta_0,$$

where θ_0 is some given constant. Let $p^* = P(\hat{\theta}^* < \theta_0)$. That is, p^* is the probability that a bootstrap estimate of θ is less than the hypothesized value, θ_0. The value of p^* is not known, but it is readily estimated with

$$\hat{p}^* = \frac{A}{B},$$

where A is the number of bootstrap estimates among $\hat{\theta}_{(1)}^* \leq \cdots \leq \hat{\theta}_{(B)}^*$ that are less than θ_0. Under fairly general conditions, if the null hypothesis is true, the distribution of $\hat{p}^*$ approaches a uniform distribution as n and B get large (e.g., Liu & Singh, 1997; Hall, 1988a, 1988b). This suggests rejecting H_0 when $\hat{p}^* \leq \alpha/2$ or $\hat{p}^* \geq 1 - \alpha/2$. A little algebra shows that this leads to the percentile bootstrap confidence interval described in the previous paragraph. A (generalized) p-value is $2\min(\hat{p}^*, 1 - \hat{p}^*)$.

A practical problem is choosing B. If the goal is to control the probability of a type I error, $B = 500$ suffices for some problems, even with n very small, but $B = 2000$ or larger might be needed for other situations. And in some instances – such as when making inferences about the population mean, the method performs poorly even when both B and n are fairly large. A rough characterization is that if a location estimator has a low finite sample breakdown point, the percentile method might be unsatisfactory, but with a relatively high finite sample breakdown, it performs reasonably well, even with small sample sizes, and in fact appears to be the method of choice in many situations. More details are provided as we consider various parameters of interest. Also, when dealing with regression, we will see situations where even with a low finite sample breakdown point, a percentile bootstrap method performs relatively well.

4.4.2 R Function onesampb

The R function

$$\text{onesampb(x,est=onestep,alpha=0.05,nboot=500,...)}$$

can be used to compute a percentile bootstrap $1 - \alpha$ confidence interval when using virtually any estimator available through R. The argument est indicates the estimator to be used, which defaults to the one-step M-estimator. The argument ... can be used to supply values for any additional parameters associated with the estimator indicated by the argument est. For example, to compute a 0.9 confidence interval based on 10% trimmed means, using 1000 bootstrap samples, use the command

$$\text{onesampb(x,est=mean,alpha=0.1,nboot=1000,tr=0.1)}.$$

The command

$$\text{onesampb(x,est=pbvar)}$$

computes a 0.95 confidence based on the percentage bend midvariance.

4.4.3 Bootstrap-t Method

The main alternative to the percentile bootstrap is the *bootstrap-t* method, which also has been called a *percentile-t* technique. When working with means, for example, the strategy is to use the observed data to approximate the distribution of

$$T = \frac{\sqrt{n}(\bar{X} - \mu)}{s}$$

by proceeding as follows:

1. Generate a bootstrap sample $X_1^*, \ldots, X_n^*$.
2. Compute $\bar{X}^*$, s^* and $T^* = \sqrt{n}(\bar{X}^* - \bar{X})/s^*$ based on the bootstrap sample generated in step 1.
3. Repeat steps 1 and 2 B times yielding T_b^*, $b = 1, \ldots, B$.

The T_b^* values provide an approximation of the distribution of T and in particular an estimate of the $\alpha/2$ and $1 - \alpha/2$ quantiles.

When testing $H_0: \mu - \mu_0$, there are two variations of the bootstrap-t method that deserve comment. The first is the *equal-tailed* method. Let $T_{(1)}^* \leq \cdots \leq T_{(B)}^*$ be the T_b^* values written in ascending order, let $\ell = \alpha B/2$, rounded to the nearest integer, and let $u = B - \ell$. Then H_0 is rejected if

$$T \leq T_{(\ell)}^* \text{ or } T \geq T_{(u)}^*.$$

Rearranging terms, a $1 - \alpha$ confidence interval for μ is

$$\left(\bar{X} - T_{(u)}^* \frac{s}{\sqrt{n}}, \bar{X} - T_{(\ell)}^* \frac{s}{\sqrt{n}} \right). \tag{4.5}$$

This last equation might appear to be incorrect because $T_{(u)}^*$, the estimate of the $1 - \alpha/2$ quantile of the distribution of T, is used to compute the lower end of the confidence interval. Simultaneously, $T_{(\ell)}^*$, an estimate of the $\alpha/2$ quantile is used to compute the upper end of the confidence interval. It can be seen, however, that this last equation follows from the decision rule that rejects $H_0: \mu = \mu_0$ if $T \leq T_{(\ell)}^*$ or $T \geq T_{(u)}^*$. Also, when computing the upper end of the confidence interval, $T_{(\ell)}^*$ will be negative, which is why the term $T_{(\ell)}^* \frac{s}{\sqrt{n}}$ is subtracted from $\bar{X}$.

The second variation of the bootstrap-t method, yielding a so-called *symmetric confidence interval*, uses

$$T^* = \frac{\sqrt{n}|\bar{X}^* - \bar{X}|}{s^*}.$$

Let $c = (1 - \alpha)B$, rounded to the nearest integer. Now a $1 - \alpha$ confidence interval for μ is

$$\bar{X} \pm T_{(c)}^* \frac{s}{\sqrt{n}}.$$

Asymptotic results (Hall, 1988a, 1988b) suggest that it tends to have more accurate probability coverage than the equal-tailed confidence interval, but some small-sample exceptions are noted later.

An interesting theoretical property of the bootstrap-t method is that it is second-order correct. Roughly, when using T, as the sample size increases, the discrepancy between the actual probability coverage and the nominal level goes to zero at the rate $1/\sqrt{n}$ as n gets large, meaning that the method is first-order correct. But when using the bootstrap-t method, the discrepancy goes to zero at the rate $1/n$. That is, the discrepancy goes to zero faster versus methods that rely on the central limit theorem.

Again there is the practical issue of choosing B, the number of bootstrap samples. The default choices for B used by the R functions in this book are based on the goal of achieving reasonably good control over the probability of a type I error. But arguments can be made that perhaps a larger value for B has practical value, the concern being that otherwise there might be some loss of power. Racine and MacKinnon (2007a) discuss this issue at length and proposed a method for choosing the number of bootstrap samples. (Also see Jöckel, 1986). Davidson and MacKinnon (2000) proposed a pretest procedure for choosing B. Theoretical results derived by Olive (2010) suggest using $B \geq [n \log(n)]$.

4.4.4 Bootstrap Methods when Using a Trimmed Mean

As previously indicated, the 20% trimmed mean can be expected to provide better control over the probability of a type I error and more accurate probability coverage, versus the mean, in various situations. In some cases, however, even better probability coverage and control of type I error probabilities might be desired, particularly when the sample size is small. Some type of bootstrap method can make a substantial difference, with the choice of method depending on how much trimming is done.

First it is noted that the bootstrap methods in Sections 4.4.1 and 4.4.2 are readily applied when using a trimmed mean. When using the percentile bootstrap method, generate a bootstrap sample and compute the sample trimmed mean yielding $\bar{X}_{t1}^*$. Repeat this process B times yielding $\bar{X}_{t1}^*, \ldots, \bar{X}_{tB}^*$. Then an approximate $1 - \alpha$ confidence interval for μ_t is given by

$$\left(\bar{X}_{t(\ell+1)}^*, \bar{X}_{t(u)}^* \right),$$

where again ℓ is $\alpha B/2$ rounded to the nearest integer, $u = B - \ell$, and $\bar{X}_{t(1)}^* \leq \cdots \leq \bar{X}_{t(B)}^*$ are the B bootstrap trimmed means written in ascending order.

The bootstrap-t extends to trimmed means in a straightforward manner as well, and to be sure the details are clear, they are summarized in Table 4.4. In the context of testing $H_0: \mu_t = \mu_0$

versus $H_1\!:\!\mu_t \neq \mu_0$, reject if $T_t < T^*_{t(\ell)}$ or $T_t > T^*_{t(u)}$, where

$$T^*_t = \frac{(1-2\gamma)\sqrt{n}(\bar{X}^*_t - \bar{X}_t)}{s^*_w},$$ (4.6)

As for the symmetric, two-sided confidence interval, now use

$$T^*_t = \frac{(1-2\gamma)\sqrt{n}|\bar{X}^*_t - \bar{X}_t|}{s^*_w},$$ (4.7)

in which case a two-sided confidence interval for μ_t is

$$\bar{X}_t \pm T^*_{t(c)} \frac{s_w}{(1-2\gamma)\sqrt{n}}.$$ (4.8)

The choice between the percentile bootstrap versus the bootstrap-t, based on the criterion of accurate probability coverage, depends on the amount of trimming. With no trimming, all indications are that the bootstrap-t is preferable (e.g., Westfall & Young, 1993). Consequently, early investigations based on means suggested using a bootstrap-t when making inferences about a population trimmed mean, but more recent studies indicate that as the amount of trimming increases, at some point the percentile bootstrap method offers an advantage. In particular, simulation studies indicate that when the amount of trimming is 20%, the percentile bootstrap confidence interval should be used rather then the bootstrap-t (e.g., Wilcox, 2001a). Perhaps with slightly less trimming the percentile bootstrap continues to give more accurate probability coverage in general, but this issues has not been studied extensively.

Table 4.4: Summary of the Bootstrap-t Method for a Trimmed Mean.

To apply the bootstrap-t (or percentile-t) method when working with a trimmed mean, proceed as follows:

1. Compute the sample trimmed mean, $\bar{X}_t$.
2. Generate a bootstrap sample by randomly sampling with replacement n observations from $X_1, \ldots, X_n$, yielding $X^*_1, \ldots, X^*_n$.
3. When computing an equal-tailed confidence interval, use the bootstrap sample to compute T^*_t given by Eq. (4.6). When computing a symmetric confidence interval, compute T^*_t using Eq. (4.7) instead.
4. Repeat steps 2 and 3 yielding $T^*_{t1}, \ldots, T^*_{tB}$. $B = 599$ appears to suffice in most situations when $n \geq 12$.
5. Put the $T^*_{t1}, \ldots, T^*_{tB}$ values in ascending order yielding $T^*_{t(1)}, \ldots, T^*_{t(B)}$.
6. Set $\ell = \alpha B/2$, $c = (1-\alpha)B$, round both ℓ and c to the nearest integer, and let $u = B - \ell$.

The equal-tailed $1-\alpha$ confidence interval for μ_t is

$$\left(\bar{X}_t - T^*_{t(u)} \frac{s_w}{\sqrt{n}}, \; \bar{X}_t - T^*_{t(\ell)} \frac{s_w}{\sqrt{n}} \right).$$ (4.9)

and the symmetric confidence interval is given by Eq. (4.9).

One issue is whether Eq. (4.6) yields a confidence interval with reasonably accurate probability coverage when sampling from a light-tailed, skewed distribution. To address this issue, attention is again turned to the lognormal distribution, which has $\mu_t = 1.111$. First consider what happens when the bootstrap-t is not used. With $n = 20$ and $\alpha = 0.025$, the probability of rejecting H_0: $\mu_t > 1.111$ when using Eq. (4.4) is approximately .065, about 2.6 times as large as the nominal level. In contrast, the probability of rejecting H_0: $\mu_t < 1.111$ is approximately .010. Thus, the probability of rejecting H_0: $\mu_t = 1.111$ when testing at the 0.05 level is approximately $.065 + .010 = .075$. If the bootstrap-t method is used instead, with $B = 599$, the one-sided type I error probabilities are now .035 and .020, so the probability of rejecting H_0: $\mu_t = 1.111$ is approximately .055 when testing at the 0.05 level. (The reason for using $B = 599$, rather than $B = 600$, stems from results in Hall, 1986, showing that B should be chosen so that α is a multiple of $(B + 1)^{-1}$. On rare occasions this small adjustment improves matters slightly, so it is used here.) As we move toward heavy-tailed distributions, generally the actual probability of a type I error tends to decrease.

For completeness, when testing a two-sided hypothesis or computing a two-sided confidence interval, asymptotic results reported by Hall (1988a, 1988b) suggest modifying the bootstrap-t method by replacing T_t^* with

$$T_t^* = \frac{(1 - 2\gamma)\sqrt{n}|\bar{X}_t^* - \bar{X}_t|}{s_w^*}. \tag{4.10}$$

Now the two-sided confidence interval for μ_t is

$$\bar{X}_t \pm T_{t(c)}^* \frac{s_w}{(1 - 2\gamma)\sqrt{n}}, \tag{4.11}$$

where $c = (1 - \alpha)B$, rounded to the nearest integer. This is an example of a *symmetric* two-sided confidence interval. That is, the confidence interval has the form $(\bar{X}_t - \hat{c}, \bar{X}_t + \hat{c})$, where $\hat{c}$ is determined with the goal that the probability coverage be as close as possible to $1 - \alpha$. In contrast, an *equal-tailed* two-sided confidence interval has the form $(\bar{X}_t - \hat{a}, \bar{X}_t + \hat{b})$, where $\hat{a}$ and $\hat{b}$ are determined with the goal that $P(\mu_t < \bar{X}_t - \hat{a}) \approx P(\mu_t > \bar{X}_t + \hat{b}) \approx \alpha/2$. The confidence interval given by Eq. (4.9) is equal-tailed. In terms of testing H_0: $\mu_t = \mu_0$ versus H_1: $\mu_t \neq \mu_0$, Eq. (4.11) is equivalent to rejecting if $T_t < -1 \times T_{t(c)}^*$, or if $T_t > T_{t(c)}^*$. When Eq. (4.11) is applied to the lognormal distribution with $n = 20$, a simulation estimate of the actual probability of a type I error is .0532 versus .0537 using (4.9). Thus, in terms of type I error probabilities, there is little separating between these two methods for this special case, but in practice, the choice between these two methods can be important, as will be seen.

Table 4.5: Values of $\hat{\alpha}$ Corresponding to Three Critical Values, $n = 12$, $\alpha = 0.025$.

g	h	$P(T_t < -t)$	$P(T_t > t)$	$P(T_t < T^*_{t(\ell)})$	$P(T_t > T^*_{t(u)})$	$P(T_t < -T^*_{t(c)})$	$P(T_t > T^*_{t(c)})$
0.0	0.0	.031	.028	.026	.030	.020	.025
0.0	0.5	.025	.022	.024	.037	.012	.024
0.5	0.0	.047	.016	.030	.023	.036	.017
0.5	0.5	.040	.012	.037	.028	.025	.011

Table 4.5 summarizes the values of $\hat{\alpha}$, an estimate of the probability of a type I error when performing one-sided tests with $\alpha = 0.025$, and when the critical value is estimated with one of the three methods described in this section. The first estimate of the critical value is t, the $1 - \alpha/2$ quantile of Student's t-distribution with $n - 2g - 1$ degrees of freedom. That is, reject if T_t is less than $-t$ or greater than t depending on the direction of the test. The second estimate of the critical value is $T^*_{t(\ell)}$ or $T^*_{t(u)}$ (again depending on the direction of the test), where $T^*_{t(\ell)}$ and $T^*_{t(u)}$ are determined with the equal-tailed bootstrap-t method. The final method uses $T^*_{t(c)}$ resulting from the symmetric bootstrap t as used in Eq. (4.11). Estimated type I error probabilities are reported for the four g-and-h distributions discussed in Section 4.2. For example, when sampling is from a normal distribution ($g = h = 0$), $\alpha = 0.025$, and when H_0 is rejected because $T_t < -t$ the actual probability of rejecting is approximately .031. In contrast, when $g = 0.5$ and $h = 0$, the probability of rejecting is estimated to be .047, about twice as large as the nominal level. (The estimates in Table 4.5 are based on simulations with 1000 replications when using one of the bootstrap methods, and 10,000 replications when using Student's t.) If sampling is from a lognormal distribution, not shown in Table 4.5, the estimate increases to 0.066, which is 2.64 times as large as the nominal 0.025 level. For $(g, h) = (0.5, 0.0)$ and $\alpha = 0.05$, the tail probabilities are .094 and .034.

Note that the choice between Eqs (4.9) and (4.11), the equal-tailed and symmetric bootstrap methods, is not completely clear based on the results in Table 4.5. An argument for Eq. (4.11) is that the largest estimated probability of a type I error in Table 4.5, when performing a two-sided test, is $.036 + .017 = .053$, whereas when using Eq. (4.9) the largest estimate is $.037 + .028 = .065$. A possible objection to (4.11) is that in some cases it is too conservative – the tail probability can be less than half the nominal .025 level. Also, if one can rule out the possibility that sampling is from a skewed distribution with very heavy tails, Table 4.5 suggests using Eq. (4.9) over Eq. (4.11), at least based on probability coverage.

There are other bootstrap techniques that might have a practical advantage over the bootstrap-t method, but at the moment this does not appear to be the case when γ is close to zero. However, extensive investigations have not been made, so future investigations might alter this view. One approach is to use a bootstrap estimate of the actual probability coverage when using T_t with Student's t-distribution and then adjust the α level so that the actual probability coverage is closer to the nominal level (Loh, 1987a, 1987b). When sampling from a lognormal distribution with $n = 20$, the one-sided tests considered above now have actual type I error probabilities approximately equal to .011 and .045, which is a bit worse than the results with the bootstrap-t. Westfall and Young (1993) advocate yet another method that estimates the p-value of T_t. For the situation considered here, simulations (based on 4000 replications and $B = 1000$) yield estimates of the type I error probabilities equal to .034 and .017. Thus, at least for the lognormal distribution, these two alternative methods appear to have no practical advantage when $\gamma = 0.2$, but of course a more definitive study is needed. Another interesting possibility is the ABC method discussed by Efron and Tibshirani (1993). The appeal of this method is that accurate confidence intervals might be possible with a substantially smaller choice for B, but there are no small-sample results on whether this is the case for the problem at hand. Additional calibration methods are summarized by Efron and Tibshirani (1993).

■ Example

Consider again the law data in Table 4.3 which has $\bar{X}_t = 596.2$ based on 20% trimming. The symmetric bootstrap-t confidence interval, based on Eq. (4.11), is (541.6, 650.9), which was computed with the R function trimcibt described in Section 4.4.6. As previously indicated, the confidence interval for μ_t, based on Student's t-distribution and given by Eq. (4.3), is (561.8, 630.6), which is a subset of the interval based on Eq. (4.11). In fact, the length of this confidence is 68.8 versus 109.3 using the bootstrap-t method. The main point here is that the choice of method can make a substantial difference in the length of the confidence interval, the ratio of the lengths being 68.8/109.3=0.63. This might seem to suggest that using Student's t-distribution is preferable, because the confidence interval is shorter. However, as previously noted, it appears that sampling is from a light-tailed, skewed distribution, and this is a situation where using Student's t-distribution can yield a confidence interval that does not have the nominal probability coverage – the interval can be too short. The 0.95 confidence interval for μ is (577.1, 623.4), which is even shorter and probably very inaccurate in terms of probability coverage. If instead the equal-tailed bootstrap-t method is used, given by (4.9), the resulting .95 confidence interval for the 20% trimmed mean is (523.0, 626.3), which is also substantially longer than the confidence interval based on Student's t-distribution. To reiterate, all indications are that trimming, versus no trimming, generally improves probability coverage when using Eq. (4.3) and sampling is from a skewed, light-tailed distribution, but the percentile bootstrap or bootstrap-t method can give even better results, at least when n is small.

■

4.4.5 Singh's Modification

Consider a random sample where say 15% of the observations are outliers. Of course, if a 20% trimmed mean is used, these outliers do not have an undue influence on the estimate as well as the standard error. Note, however, that when generating a bootstrap sample, by chance the number of outliers could exceed 20% which can result in a relatively long confidence interval. Singh (1998) derived theoretical results showing that this problem can be addressed by Winsorizing the data before taking a bootstrap sample, provided the amount of Winsorizing does not exceed the amount of trimming. So if inferences based on a 20% trimmed are to be made, theory allows taking bootstrap samples from the Winsorized data provided the amount of Winsorizing does not exceed 20%. When using a percentile bootstrap method, for example, confidence intervals are computed in the usual way. That is, the only difference from the basic percentile bootstrap method in Section 4.4.1 is that observations are resampled with replacement from the Winsorized data.

Although theory allows the amount of Winsorizing to be as large as the amount of trimming, if we Winsorize as much as we trim, probability coverage can be unsatisfactory, at least with small to moderate sample sizes (Wilcox, 2001a). However, if for example 10% Winsorizing is done when making inferences based on a 20% trimmed mean, good probability coverage is obtained.

Singh's results extend to the bootstrap-t method. But all indications are that achieving accurate probability coverage is difficult. Presumably this problem becomes negligible as the sample size increases, but just how large the sample must be to obtain reasonably accurate probability coverage is unknown.

4.4.6 R Functions trimpb and trimcibt

The R function trimpb (written for this book) computes a 0.95 confidence interval using the percentile bootstrap method. It has the general form

$$\text{trimpb}(x,tr=0.2,alpha=0.05,nboot=2000,WIN=F,plotit=F,win=0.1,pop=1),$$

where x is any R vector containing data, tr again indicates the amount of trimming, alpha is α, and nboot is B which defaults to 2000. The argument WIN controls whether Winsorizing is done. If plotit is set to T (for true), a plot of the bootstrap trimmed means is created, and the type of plot created is controlled by the argument pop. The choices are:

*   pop=1, expected frequency curve
*   pop=2, kernel density estimate (using a normal kernel)
*   pop=3, boxplot
*   pop=4, stem-and-leaf

- pop=5, histogram
- pop=6, adaptive kernel density estimate

The function

$$\text{trimcibt(x,tr=0.2,alpha=0.05,nboot=2000,WIN=F,plotit=F,win=0.1,op=1),}$$

performs the bootstrap-t method. Now if plotit=T, a plot of the $T_{t1}^*, \ldots, T_{tB}^*$ values is created based on the adaptive kernel estimator if op=1. If op=2, an expected frequency curve is used.

4.5 Inferences About M-Estimators

A natural way of computing a confidence interval for μ_m, an M-measue of location, is to estimate the standard error of $\hat{\mu}_m$ with $\hat{\sigma}_m$, as described in Chapter 3, and consider intervals having the form $(\hat{\mu}_m - c\hat{\sigma}_m, \hat{\mu}_m + c\hat{\sigma}_m)$ for some appropriate choice for c. This strategy seems to have merit when sampling from a symmetric distribution, but for asymmetric distributions it can be unsatisfactory (Wilcox, 1992a). If, for example, c is determined so that the probability coverage is exactly $1 - \alpha$ when sampling from a normal distribution, the same c can yield a confidence interval with probability coverage substantially different from $1 - \alpha$ when sampling from asymmetric distributions instead. Moreover, it is unknown how large n must be so that the resulting confidence interval has probability coverage reasonably close to the nominal level.

One alternative approach is to apply a bootstrap-t method, but simulations do not support this approach, at least when $n \le 40$. Could the bootstrap-t method be improved by using something like the adaptive kernel density estimator when estimating the standard of $\hat{\mu}_m$? All indications are that probability coverage remains unsatisfactory. Currently, the most effective method is the percentile bootstrap (but direct comparisons with the method studied by Kuonen, 2005, have not been made).

As before, generate a bootstrap sample by randomly sampling n observations, with replacement, from $X_1, \ldots, X_n$, yielding $X_1^*, \ldots, X_n^*$. Let $\hat{\mu}_m^*$ be the M-estimator of location based on the bootstrap sample just generated. Repeat this process B times yielding $\hat{\mu}_{m1}^*, \ldots, \hat{\mu}_{mB}^*$. The $1 - \alpha$ confidence interval for μ_m is

$$(\hat{\mu}_{m(\ell+1)}^*, \hat{\mu}_{m(u)}^*) \tag{4.12}$$

where $\ell = \alpha B / 2$, rounded to the nearest integer, and $u = B - \ell$, and $\hat{\mu}_{m(1)}^* \le \cdots \le \hat{\mu}_{m(B)}^*$ are the B bootstrap values written in ascending order.

The percentile bootstrap method appears to give fairly accurate probability coverage when $n \ge 20$ and $B = 399$, but for smaller sample sizes the actual probability coverage can be less than .925, with $\alpha = 0.05$. Increasing B to 599 does not appear to improve the situation very

**Table 4.6: Values of $\hat{\alpha}$ when Using
(4.12), $B = 399$, $\alpha = 0.05$, $n = 20$.**

g	h	$P(\hat{\mu}^*_{m(\ell)} > 0)$	$P(\hat{\mu}^*_{m(u)} < 0)$
0.0	0.0	.030	.034
0.0	0.5	.029	.036
0.5	0.0	.023	.044
0.5	0.5	.023	.042

much. Another problem is that the iterative method of computing $\hat{\mu}_m$ can break down when applying the bootstrap and n is small. The reason is that if more than half of the observations have a common value, MAD $= 0$ resulting in division by zero when computing $\hat{\mu}_m$. Because the bootstrap is based on sampling with replacement, as n gets small, the probability of getting MAD $= 0$, within the bootstrap, increases. Of course, problems might also arise in situations where some of the X_i have a common value. Similar problems arise when using $\hat{\mu}_{os}$ instead. This might suggest abandoning the M-estimator, but as noted in Chapter 3, there are situations where it might be preferred over the trimmed mean.

Table 4.6 shows the estimated probability of observing $\hat{\mu}^*_{m(\ell)} > 0$, and the probability of $\hat{\mu}^*_{m(u)} < 0$, when observations are generated from a g-and-h distribution that has been shifted so that $\mu_m = 0$. For example, when sampling from a normal distribution, the probability of a type I error, when testing H_0: $\mu_m = 0$, is $.030 + .034 = .064$. If sampling is from a lognormal distribution, the two tail probabilities are estimated to be .019 and .050, so the probability of a type I error when testing H_0: $\mu_m = 0$ is .069. Increasing B to 599, the estimated probability of a type I error is now .070. Thus, there is room for improvement, but probability coverage and control over the probability of a type I error might be deemed adequate in some situations.

Tingley and Field (1990) suggest yet another method for computing confidence intervals based on *exponential tilting* and a *saddlepoint approximation* of a distribution. (Also see Gatto & Ronchetti, 1996, as well as Robinson, Ronchetti, & Young, 2003, for related results.) They illustrate the method when dealing with M-estimators, but their results are quite general and might have practical interest when using other measures of location. While preparing this chapter, the authors ran a few simulations to determine how their method performs when working with M-estimators. When sampling from a standard normal distribution, with $n = 25$ and simulations based on 10,000 replications, the estimated type I error probability was $\hat{\alpha} = 0.078$ when testing at the $\alpha = 0.05$ level. In contrast, $\hat{\alpha} = 0.064$ when using the percentile bootstrap. Perhaps there are situations where the Tingley–Field method offers a practical advantage, but this has not been established as yet.

From an efficiency point of view, the one-step M-estimator (with Huber's Ψ) given by Eq. (3.25) can be a bit more satisfactory than the modified one-step M-estimator (MOM) in Section 3.10. (With sufficiently heavy tails, MOM can have better efficiency.) However, with very small sample sizes, it seems that reasonably accurate confidence intervals are easier to obtain when using MOM.

4.5.1 R Functions mestci and momci

The R function onesampb can be used to compute confidence intervals based on MOM or an M-estimator. For convenience, the R function mestci is supplied for the special case where the goal is to compute a $1 - \alpha$ confidence interval for μ_m (an M-measure of location based on Huber's Ψ) using the percentile bootstrap method. The function has the form

$$\text{mestci(x,alpha=0.05,nboot=399,bend=1.28,os=F)}.$$

The default value for alpha (α) is 0.05, nboot is the number of bootstrap samples to be used, which defaults to 399, and bend is the bending constant used in Huber's Ψ, which defaults to 1.28. (See Chapter 3.) The argument os is a logical variable that defaults to F, for false, meaning that the fully iterated M-estimator is to be used. Setting os=T causes the one-step M-estimator, $\hat{\mu}_{os}$, to be used. The R function

$$\text{momci(x,alpha=0.05,nboot=500)}.$$

is supplied for situations where there is specific interest in the modified one-step M-estimator.

■ Example

If the law data in Table 4.3 are stored in the R variable x, the command mestci(x) returns a 0.95 confidence interval for μ_m equal to (573.8, 629.1). For this data, the function also prints a warning that because the number of observations is less than 20, division by zero might occur when computing the bootstrap M-estimators, but in this particular case this problem does not arise. Note that the length of the confidence interval is shorter than the length of the confidence interval for the trimmed mean, based on Eq. (4.3), but with such a small sample size, and because sampling appears to be from a light-tailed distribution, the probability coverage of both confidence intervals might be less than 0.95. The command mestci(x,os=T) computes a 0.95 confidence interval using the one-step M-estimator. This yields (573.8, 629.5), which is nearly the same as the 0.95 confidence interval based on $\hat{\mu}_m$.

■

4.6 Confidence Intervals for Quantiles

This section addresses the problem of computing a confidence interval for x_q, the qth quantile. Many strategies are available, but only a few are listed here.

Consider the interval $(X_{(i)}, X_{(j)})$. As a confidence interval for the qth quantile, the exact probability coverage of this interval is

$$\sum_{k=i}^{j-1} \binom{n}{k} q^k (1-q)^{n-k}$$

(e.g., Arnold, Balakrishnan, & Nagaraja, 1992). An issue is whether alternative methods might give shorter confidence intervals, but it seems that among the alternatives listed here, little is known about this possibility. Imagine that a confidence interval is sought that has probability coverage at least .95. If n is small and fixed, as q goes to zero or one, it becomes impossible to achieve this goal. For example, if $n = 30$ and $q = 0.05$, the highest possible probability coverage is .785. So an issue is whether other methods can be found that perform reasonably well in this case.

Next consider techniques based on the Harrell–Davis estimator, $\hat{\theta}_q$. A simple method that seems to be reasonably effective, at least for $\alpha = 0.05$ and $n \geq 20$, is to use the percentile bootstrap. However, another approach is used here, one that appears to be about as effective as the percentile bootstrap, its main advantage being that it continues to give good results in situations covered in Chapter 5, whereas the percentile bootstrap does not. (For other methods that have been considered, see Wilcox, 1991b.)

Let $\hat{\sigma}_{\text{hd}}$ be the bootstrap estimate of the standard error of $\hat{\theta}_q$, which is described in Chapter 3. Here, $B = 100$ bootstrap samples are used to compute $\hat{\sigma}_{\text{hd}}$. Temporarily assume that sampling is from a normal distribution and suppose c is determined so that the interval

$$(\hat{\theta}_q - c\hat{\sigma}_{\text{hd}}, \hat{\theta}_q + c\hat{\sigma}_{\text{hd}}) \tag{4.13}$$

has probability coverage $1 - \alpha$. Then simply continue to use this interval when sampling from nonnormal distributions. There is the practical problem that c is not known, but it is easily estimated by running simulations on a computer. Suppose c is to be chosen with the goal of computing a 0.95 confidence interval. For normal distributions, simulations indicate that for n fixed, c does not vary much as a function of the quantile being estimated, provided $n \geq 11$ and attention is restricted to those quantiles between 0.3 and 0.7. For convenience, c was determined for $n = 11, 15, 21, 31, 41, 61, 81, 121$, and 181, and then a regression line was fitted to the resulting pairs of points yielding

$$\hat{c} = 0.5064n^{-0.25} + 1.96, \tag{4.14}$$

where the exponent, -0.25, was determined using the half-slope ratio of Tukey's resistant regression line. (See, e.g., Velleman & Hoaglin, 1981; Wilcox, 1996a.) When dealing with the 0.2 or 0.8 quantile, (4.14) gives reasonably good results for $n > 21$. For $11 \leq n \leq 21$, use

$$\hat{c} = \frac{-6.23}{n} + 5.01.$$

Critical values have not been determined for $n < 11$. For the 0.1 and 0.9 quantiles, use

$$\hat{c} = \frac{36.2}{n} + 1.31$$

when $11 \leq n \leq 41$, otherwise use Eq. (4.14).

As a partial check on the accuracy of the method, it is noted that when observations are generated from a lognormal distribution, the actual probability coverage when working with the median, when $n = 21$ and $\alpha = 0.05$, is approximately .959, based on a simulation with 10,000 replications. For the 0.1 and 0.9 quantiles it is .974 and .928, respectively. However, with $n = 30$ and $q = 0.05$, this method performs poorly.

Another approach is to use $\hat{x}_q$ to estimate the qth quantile as described in Section 3.5, estimate the standard error of $\hat{x}_q$ with $\hat{\sigma}_{mj}$, the Maritz–Jarrett estimator described in Section 3.5.3, and then assume that

$$Z = \frac{\hat{x}_q - x_q}{\hat{\sigma}_{mj}}$$

has a standard normal distribution. Then an approximate $1 - \alpha$ confidence interval for the qth quanitle is

$$(x_q - z_{1-\alpha/2}\hat{\sigma}_{mj}, \ x_q + z_{1-\alpha/2}\hat{\sigma}_{mj}), \tag{4.15}$$

where $z_{1-\alpha/2}$ is the $1 - \alpha/2$ quantile of the standard normal distribution. Table 4.7 shows $\hat{\alpha}$, the estimate of one minus the probability coverage, for the four g-and-h distributions discussed in Section 4.2 when $q = 0.5$, $\alpha = 0.05$, and $n = 13$. The estimates are based on simulations with 10,000 replications. When sampling from a lognormal distribution, $\hat{\alpha} = 0.067$.

A variation of this last method is to replace the Maritz–Jarrett estimate of the standard error with an estimate based on Eq. (3.11), which requires an estimate of $f(x_q)$, the probability density function evaluated at x_q. If $f(x_q)$ is estimated with the adaptive kernel method in Section 3.2.4, a relatively accurate 0.95 confidence interval can be had with $n = 30$ and

Table 4.7: Values of $\hat{\alpha}$ when Using (4.15), $n = 13$ and $\alpha = 0.05$.

g	h	$\hat{\alpha}$
0.0	0.0	0.067
0.0	0.5	0.036
0.5	0.0	0.062
0.5	0.5	0.024

$q = 0.05$. In fact, this is the only method known to perform reasonably well for this special case. As we move from normality toward heavy-tailed distributions, this method continues to perform tolerably well up to a point (a g-and-h distribution with $g = h = 0.2$), but eventually it will fail. For example, with $g = h = 0.5$, the probability coverage is approximately .92, but increasing n to 40, the probability coverage is approximately .95.

4.6.1 Beware of Tied Values when Using the Median

When making inferences about the median in particular (and more generally any quantile), tied values can create serious practical problems when computing confidence intervals and testing hypotheses. For the special case where the goal is to compute a confidence interval for the population median, the method in the next section can be used when tied values occur. But when comparing the median of two or more distributions, techniques based on estimates of the standard error of M, which simultaneously assume M has a normal distribution, can be highly unsatisfactory, even with large sample sizes. The first general problem is getting a reasonably accurate estimate of the standard error. As noted in Section 3.3.5, all known estimates of the standard error of the sample median can be extremely inaccurate. The second difficulty is that the sampling distribution of M can be poorly approximated by a normal distribution, even with a large sample size.

To underscore why tied values can cause problems when working with the median, and to illustrate a limitation of the central limit theorem, imagine a random sample $X_1, \ldots, X_n$, where each X_i has the binomial probability function

$$\binom{15}{x} 0.7^x 0.3^{15-x},$$

So, for example, the probability that a randomly sampled participant responds with the value 13 is 0.09156. As is evident, with a sample size of $n = 20$, tied values are guaranteed since there are only 16 possible responses. The left panel of Figure 4.5 shows a plot of the relative frequencies associated with 5000 sample medians, with each sample median based on $n = 20$ randomly sampled observations. The plot resembles somewhat a normal curve, but note that only five values for the sample median occur. Now look at the right panel, which was created in the same manner as the left panel, only with a sample size of $n = 100$ for each sample median. Blind reliance on the central limit theorem would suggest that the plot will look more like a normal distribution than the left panel, but clearly this is not the case. Now only three values for the sample median are observed. In practical terms, methods for making inferences about the median, which assume the sample median has a normal distribution, can be disastrous when tied values can occur.

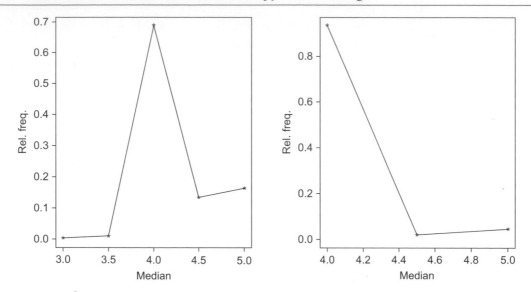

Figure 4.5: When tied values can occur, the sample median might not be asymptotically normal. The left panel shows the sampling distribution of the median with $n = 20$. The right panel is the sampling distribution with $n = 100$.

4.6.2 Alternative Method for the Median

When the goal is to compute a confidence interval for the population median, the following method can be used even when there are tied values. Suppose W is a binomial random variable with probability of success $p = .5$ and n trials. For any integer k between 0 and $[n/2]$, let $\gamma_k = P(k \leq W \leq n - k)$, the probability that the number of successes, W, is between k and $n - k$, inclusive. Then a distribution-free γ_k confidence interval for the median is

$$(X_{(k)}, X_{(n-k+1)}).$$

That is, the probability coverage is exactly γ_k under random sampling (e.g., Hettmansperger & McKean, 1998; also see Yohai & Zamar, 2004). This is just a special case of the first method described in the previous section.

Because the binomial distribution is discrete, it is not possible, in general, to choose k so that the probability coverage is exactly equal to $1 - \alpha$. For example, if $n = 10$, 0.891 and 0.978 confidence intervals can be computed, but not a 0.95 confidence interval as is often desired. However, linear interpolation can be used along the lines suggested by Hettmansperger and Sheather (1986) so that the probability coverage is approximately $1 - \alpha$. First determine k such that $\gamma_{k+1} < 1 - \alpha < \gamma_k$. Next, compute

$$I = \frac{\gamma_k - 1 - \alpha}{\gamma_k - \gamma_{k+1}}$$

and

$$\lambda = \frac{(n-k)I}{k + (n-2k)I}.$$

Then an approximate $1 - \alpha$ confidence interval is

$$(\lambda X_{(k+1)} + (1 - \lambda) X_{(k)}, \lambda X_{(n-k)} + (1 - \lambda) X_{(n-k+1)}). \tag{4.16}$$

Results reported by Sheather and McKean (1987), as well as Hall and Sheather (1988), support the use of this method.

4.6.3 R Functions qmjci, hdci, sint, sintv2, qci, and qint

The R function

$$\text{qmjci(x,q=0.5,alpha=0.05,op=1),}$$

computes a $1 - \alpha$ confidence interval for the qth quantile using Eq. (4.15) and the data stored in the R vector x. The function returns the lower and upper values of the confidence interval. The default value for q is 0.5 and the default value for alpha (α) is 0.05. (The accuracy of this confidence interval for $q \neq 0.5$ and n small has not been studied.) With op=1, the Maritz–Jarrett estimate of the standard error is used, and with op=2, the McKean–Schrader estimate (described in Section 3.3.5) is used instead. (With op=2, only $q=0.5$ is allowed.) With op=3, the function estimates the standard error via the adaptive kernel estimate of $f(x_q)$. The function

$$\text{qci(x,q=0.5,alpha=0.05)}$$

returns the same confidence interval as qmjci with op=3 and is provided in case it is convenient.

The R function

$$\text{hdci(x,q=0.5,nboot=100)}$$

computes a 0.95 confidence interval using the Harrell–Davis estimator. As indicated, the default value for q is 0.5, and the default number of bootstrap samples, nboot, is 100.

Finally, the function

$$\text{sint(x,alpha=0.05),}$$

computes a confidence interval for the median using Eq. (4.16), where α is taken to be 0.05 if not specified. (Version 6 of Minitab also has a command called sint that computes a confidence interval for the median using (4.16)). To get a p-value when testing the hypothesis that the population median is equal to some specified value, use the R function

$$\text{sintv2(x,alpha=0.05,nullval=0),}$$

where the null value is specified by the argument nullval, which defaults to 0. Confidence intervals for general q, that do not use interpolation, are computed by the function

$$\text{qint(x,q=0.5,alpha=0.05).}$$

The exact probability coverage, based on the first method in Section 4.6, is reported as well.

■ **Example**

Staudte and Sheather (1990) illustrate the use of Eq. (4.16) with data from a study on the lifetimes of EMT6 cells. The values are 10.4, 10.9, 8.8, 7.8, 9.5, 10.4, 8.4, 9.0, 22.2, 8.5, 9.1, 8.9, 10.5, 8.7, 10.4, 9.8, 7.7, 8.2, 10.3, and 9.1. Both the sample median, M, and $\hat{x}_{0.5}$ are equal to 9.1. The resulting 0.95 confidence interval reported by sint is (8.72, 10.38). In contrast, the confidence interval based on Eq. (4.15), as computed by the R function qmjci, is (8.3, 9.9). The length of the confidence intervals are about the same. The main difference is that the confidence interval based on (4.15) is centered about the sample median, 9.1, while the confidence interval based on Eq. (4.16) is not. The Harrell–Davis estimate of the median is 9.26, and a 0.95 confidence interval based on Eq. (4.13), computed with the R function hdci, is (8.45, 10.08).

■

4.7 Empirical Likelihood

Empirical likelihood methods (Owen, 2001) represent another nonparametric approach for computing a confidence interval for the population mean that should be noted. Asymptotic results suggest that a Bartlett corrected empirical likelihood approach is superior to using a bootstrap-t method (DiCiccio, Hall, & Romano, 1991).

The empirical likelihood method can be used to construct a confidence interval for μ, but for simplicity it is described in terms of testing H_0: $\mu = \mu_0$. Consider distributions F_p, $p = (p_1, \ldots, p_n)$ supported on the sample $X_1, \ldots, X_n$, where X_i is assigned mass p_i. For a specified value of μ, the empirical likelihood $L(\mu)$ is defined to be the maximum value of Πp_i over all such distributions that satisfy $\sum X_i p_i = \mu$. Because Πp_i attains its overall maximum when $p_i = n^{-1}$, it follows that the empirical likelihood is maximized when $\mu = \bar{X}$. The empirical likelihood ratio for testing H_0 is

$$W = -2\log\{L(\mu_0)/L(\bar{X})\}.$$

When the null hypothesis is true, W has, approximately a chi-squared distribution with 1 degree of freedom. In particular, reject H_0 at the α level if $W \geq c$, where c is the $1 - \alpha$ quantile of a chi-squared distribution with 1 degree of freedom.

4.7.1 Bartlett Corrected Empirical Likelihood

The Bartlett corrected empirical likelihood method is applied as follows. Let $\hat{\mu}_j = n^{-1} \sum (X_i - \bar{X})^j$ and

$$a = \frac{1}{2} \hat{\mu}_4 \hat{\mu}_2^{-2} - \frac{1}{3} \hat{\mu}_3^2 \hat{\mu}_2^{-3}.$$

Then the null hypothesis is rejected if $W(1 - an^{-1}) \geq c$.

Table 4.8 reports simulation estimates (based on 1000 replications) of the type I error probability for the empirical likelihood (EL) method, the Bartlett corrected empirical likelihood (BCEL), the equal-tailed bootstrap-t (BEQ) and the symmetric bootstrap-t (BSYM). The distributions considered are normal, chi-squared with 1 degree of freedom (χ_1^2), a Student's t with 5 degrees of freedom (t_5), a lognormal distribution (LogN), the contaminated normal (cnorm) shown in Figure 1.1, and some g-and-h distributions. Glenn and

Table 4.8: Estimated Type I Error Probabilities.

n	Distribution	EL	BCEL	BEQ	BSYM
20	Normal	.074	.064	.058	.045
	χ_1^2	.117	.103	.068	.080
	t_5	.075	.059	.067	.036
	LogN	.137	.120	.099	.104
	Cnorm	.169	.138	.116	.010
	(g,h)=(0.2,0.0)	.090	.072	.083	.035
	(g,h)=(0.2,0.2)	.094	.080	.083	.047
	(g,h)=(0.5,0.5)	.270	.241	.231	.186
50	Normal	.052	.050	.055	.049
	χ_1^2	.074	.069	.055	.059
	t_5	.062	.058	.072	.048
	LogN	.068	.062	.058	.054
	Cnorm	.137	.125	.145	.011
	(g,h)=(0,0.2)	.061	.057	.073	.037
	(g,h)=(0.2,0.2)	.074	.066	.080	.050
	(g,h)=(0.5,0.5)	.215	.203	.207	.194

EL=empirical likelihood.
BCEL=Bartlett corrected empirical likelihood.
BEQ=bootstrap-t, equal-tailed.
BSYM=bootstrap-t, symmetric.

Zhao (2007) derived theoretical results indicating that the empirical likelihood methods can be unsatisfactory when sampling from contaminated normal. And the results in Table 4.8 illustrate that they can indeed be highly unsatisfactory. (Also see Wilcox, 2010g.)

Although the seriousness of a type I error depends on the situation, Bradley (1978) has suggested that generally, at a minimum, the actual type I error probability should be between .025 and .075 when testing at the 0.05 level. Based on this criterion, none of the methods are satisfactory. However, for skewed distributions for which the median proportion of outliers does not exceed 0.05, the symmetric bootstrap method gives satisfactory results. The symmetric bootstrap method can be too conservative when sampling from a symmetric heavy-tailed distribution, but this might be judged to be less serious than having an actual Type I error greater than 0.075, as is the case when using the empirical likelihood methods. Note that with $n = 20$, the symmetric bootstrap method has a type I error probability of .080 when sampling from a chi-squared distribution with 1 degree of freedom. Increasing the sample size to $n = 25$, the estimate drops to .065, and for $n = 30$ it is .059.

Some additional simulations were run with $n = 100$ and it was found that the empirical likelihood methods continue to perform poorly when sampling from the heavy-tailed distributions considered here. With $n = 200$ they perform well when sampling from the contaminated normal but estimates exceed .15 when sampling from the g-and-h distribution when $g = h = 0.5$.

Recent results on how to improve the empirical likelihood method, when working with the mean, are reported by Vexler, Liu, Kang, and Hutson (2009), but control over the type I error probability remains rather poor when dealing with nonnormal distributions. Also see Glenn (2002) as well as Glenn and Zhao (2007). For a review of empirical likelihood methods when dealing with regression, see Chen and Van Keilegem (2009).

As for $n = 50$, the empirical likelihood methods compete better with the bootstrap-t methods, but the symmetric bootstrap-t performs well in situations where the empirical likelihood methods are unsatisfactory based on Bradley's criterion. Again a criticism of the symmetric bootstrap-t is that for a symmetric heavy-tailed distribution (the contaminated normal), the type I error probability drops below .025. But the other three methods have estimates greater than .12. So for general use, the symmetric bootstrap-t seems best.

Some additional simulations were run with $n = 100$ and it was found that the empirical likelihood methods continue to perform poorly when sampling from the heavy-tailed distributions considered here. With $n = 200$ they perform well when sampling from the contaminated normal but estimates exceed .15 when sampling from the g-and-h distribution when $g = h = 0.5$.

4.8 Concluding Remarks

To summarize a general result in this chapter, there is a plethora of methods one might use to compute confidence intervals and test hypotheses. Many methods can be eliminated based on published studies, but several possibilities remain. As noted in Chapter 3, there are arguments for preferring trimmed means over M-estimators, and there are arguments for preferring M-estimators instead, so the choice between the two is not particularly obvious. In terms of computing confidence intervals, all indications are that when working with the 20% trimmed mean, reasonably accurate probability coverage can be obtained over a broader range of situations versus an M-estimator or mean. As already stressed, the 20% trimmed mean can have a relatively small standard error when sampling from a heavy-tailed distribution, but other criteria can be used to argue for some other measure of location. If the sample size is at least 20, M-estimators appear to be a viable option based on the criterion of accurate probability coverage. An advantage of the modified one-step M-estimator (MOM) is that accurate confidence intervals can be computed with small sample sizes in situations where methods based on M-estimators are not quite satisfactory, and it is flexible about how many observations are trimmed, in contrast to a trimmed mean. From an efficiency point of view, M-estimators based on Huber's Ψ generally have a bit of an advantage over MOM, when sampling from a normal distribution or a distribution where the expected proportion of outliers is less an 0.1. However, as the expected proportion of outliers increases, MOM can have a smaller standard error than the one-step M-estimator (Özdemir & Wilcox, 2010). And in terms of controlling type I error probabilities, it seems that using MOM in conjunction with a percentile bootstrap method is a bit more satisfactory than using a one-step M-estimator, particularly when dealing with skewed, relatively light-tailed distributions. Inferences about quantiles might appear to be rather uninteresting at this point, but they can be used to address important issues that are ignored by other measures of location, as will be seen in Chapter 5. Put more generally, different methods for summarizing data can reveal important and interesting features that other methods miss.

4.9 Exercises

1. Describe situations where the confidence interval for the mean might be too long or too short. Contrast this with confidence intervals for the 20% trimmed mean and μ_m.
2. Compute a 0.95 confidence interval for the mean, 10% mean, and 20% mean using the data in Table 3.1 of Chapter 3. Examine a boxplot of the data and comment on the accuracy of the confidence interval for the mean. Use both Eq. (4.3) and the bootstrap-t method.

3. Compute a 0.95 confidence interval for the mean, 10% mean, and 20% mean using the lifetime data listed in the example of Section 4.6.3. Use both Eq. (4.3) and the bootstrap-t method.

4. Use the R functions qmjci, hdci, and sint to compute a 0.95 confidence interval for the median based on the LSAT data in Table 4.3. Comment on how these confidence interval compare to one another.

5. The R function rexp generates data from an exponential distribution. Use R to estimate the probability of getting at least one outlier, based on a boxplot, when sampling from this distribution. Discuss the implications for computing a confidence interval for μ.

6. If the exponential distribution has variance $\mu_{[2]} = \sigma^2$, then $\mu_{[3]} = 2\sigma^3$ and $\mu_{[4]} = 9\sigma^4$. Determine the skewness and kurtosis. What does this suggest about getting an accurate confidence interval for the mean?

7. Do the skewness and kurtosis of the exponential distribution suggest that the bootstrap-t method will provide a more accurate confidence interval for μ_t versus the confidence interval given by Eq. (4.3)?

8. For the exponential distribution, would the sample median be expected to have a relatively high or low standard error? Compare your answer to the estimated standard error obtained with data generated from the exponential distribution.

9. Discuss the relative merits of using the R function sint versus qmjci and hdci.

10. Verify Eq. (4.5) using the decision rule about whether to reject H_0 described in Section 4.4.3.

11. For the LSAT data in Table 4.3, compute a 0.95 bootstrap-t confidence interval for mean using the R function trimcibt with plotit=T. Note that a boxplot finds no outliers. Comment on the plot created by trimcibt in terms of achieving accurate probability coverage when using Student's t. What does this suggest about the strategy of using Student's t if no outliers are found by a boxplot?

12. Generate 20 observations from a g-and-h distribution with $g = h = 0.5$. (This can be done with the R function ghdist, written for this book.) Examine a boxplot of the data. Repeat this 10 times. Comment on the strategy of examining a boxplot to determine whether the confidence interval for the mean has probability coverage at least as high as the nominal level.

CHAPTER 5

Comparing Two Groups

A natural and reasonable approach to comparing two distributions is to compare robust measures of location and scale, but first attention is focused on global comparisons of two distributions. The motivation for global comparisons is that if two distributions differ, they might do so in complicated and interesting ways that are not revealed by differences between single measures of location or scale. For example, if one or both distributions are skewed, the difference between the means might be large compared with the difference between the trimmed means, or any other measure of location that might be used. As is evident, the reverse can happen when the difference between the trimmed means is large and the difference between the means is not. Of course, two or more measures of location might be compared, but this might miss interesting differences and trends among subpopulations of participants.

To elaborate, it helps first to consider a simple but unrealistic situation. Consider two normal distributions that have means $\mu_1 = \mu_2$. Then any test of the hypothesis $H_0: \mu_1 = \mu_2$ should not reject. But suppose the variances differ. To be concrete, suppose an experimental method is being compared with a control group, and that the control group has variance $\sigma_1^2 = 1$, whereas the experimental method has $\sigma_2^2 = 0.5$. Then the experimental group is effective in the sense that low-scoring participants in the experimental group have higher scores than low-scoring participants in the control group. Similarly, the experimental method is detrimental in the sense that high-scoring participants in the experimental group tend to score lower than high-scoring participants in the control group. That is, different subpopulations of participants respond in different ways to the experimental method. Of course, in this simple example, one could compare the variances of the two groups, but for various reasons to be explained and illustrated, it can be useful to compare the quantiles of the two groups instead.

As another example, consider the two distributions in Figure 5.1. The distributions differ, the effectiveness of one method over the other depends on which quantiles are compared, yet the distributions have identical means and variances. (The skewed distribution is chi-square with four degrees of freedom, so the mean and variance are 4 and 8, respectively, and the symmetric distribution is normal.)

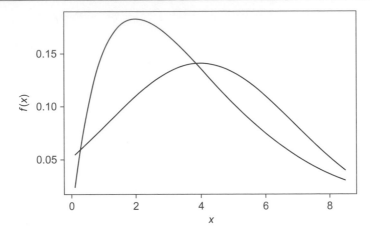

Figure 5.1: Two different distributions with equal means and variances.

In this book, an experimental method is defined to be *completely effective* compared with a control group if each quantile of the experimental group is greater than the corresponding quantile of the control group. In symbols, if x_q and y_q are the qth quantiles of the control and experimental group, respectively, the experimental method is said to be *completely effective* if $y_q > x_q$ for any q. This implies that the experimental method is stochastically larger than the distribution associated with the control. If $x_q > y_q$ for some q, but $x_q < y_q$ for others, the experimental method is defined to be *partially effective*. Both of the illustrations just described correspond to situations where an experimental method is only partially effective. There are situations where comparing measures of location and scale can be used to establish whether an experimental method is completely effective (e.g., Wilcox, 1990b), but this requires assumptions that are not always met and cannot always be tested in an effective manner, so this approach is not pursued here. Note that Student's t-test assumes that $\sigma_1 = \sigma_2$ even when $\mu_1 \neq \mu_2$. If this assumption is met, and distributions are normal, then $\mu_1 > \mu_2$ implies that the experimental method is completely effective. The practical concern is that, when these two assumptions are not met, as is commonly the case, such a conclusion can be highly misleading and inaccurate.

5.1 The Shift Function

There are various ways entire distributions might be compared. This section describes an approach based on the so-called shift function. The basic idea, which was developed by Doksum (1974, 1977) as well as Doksum and Sievers (1976), is to plot the quantiles of the control group versus the differences between the quantiles. That is, letting X be the random

Table 5.1: Weight Gain, in Grams, for Large Babies.

Group 1 (heartbeat)			
Subject	Gain	Subject	Gain
1	190	11	10
2	80	12	10
3	80	13	0
4	75	14	0
5	50	15	−10
6	40	16	−25
7	30	17	−30
8	20	18	−45
9	20	19	−60
10	10	20	−85

Group 2 (no heartbeat)							
Subject	Gain	Subject	Gain	Subject	Gain	Subject	Gain
1	140	11	−25	21	−50	31	−130
2	100	12	−25	22	−50	32	−155
3	100	13	−25	23	−60	33	−155
4	70	14	−30	24	−75	34	−180
5	25	15	−30	25	−75	35	−240
6	20	16	−30	26	−85	36	−290
7	10	17	−45	27	−85		
8	0	18	−45	28	−100		
9	−10	19	−45	29	−110		
10	−10	20	−50	30	−130		

variable associated with the control, plot x_q versus

$$\Delta(x_q) = y_q - x_q, \tag{5.1}$$

where y_q is the qth quantile of the experimental method. $\Delta(x_q)$ is called a *shift function*. It measures how much the control group must be shifted so that it is comparable with the experimental method at the qth quantile.

The shift function is illustrated with some data from a study by Salk (1973). The goal was to study weight gain in newborns. Table 5.1 shows the data for a portion of the study based on infants who weighed at least 3500 g at birth. The experimental group was continuously exposed to the sound of a mother's heartbeat. For the moment, attention is focused on comparing the deciles rather than all of the quantiles. For the control group (not exposed to the sound of a heartbeat), the Harrell-Davis estimates of the deciles (the 0.1, 0.2, 0.3, 0.4, 0.5, 0.6, 0.7, 0.8, and 0.9 quantiles) are −171.7, −117.4, −83.1, −59.7, −44.4, −32.1, −18.0, 7.5, and 64.9. For the experimental group, the estimates are −55.7, −28.9, −10.1, 2.8, 12.2, 22.8, 39.0, 61.9, and 102.7, and this yields an estimate of the shift function. For example, an estimate of $\Delta(x_{0.1})$ is $\hat{\Delta}(-171.7) = -55.7 - (-171.7) = 116$. That is, the weight gain among

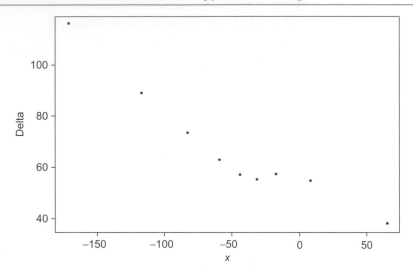

Figure 5.2: The *x*-axis indicates the deciles for the first group and the *y*-axis indicates the difference between the deciles of the second group versus the first.

infants at the 0.1 quantile of the experimental group is estimated to be 116 g higher than the infants corresponding to the 0.1 quantile of the control. Figure 5.2 shows a plot of the estimated deciles for the control group versus $\hat{\Delta}$. Notice the apparent monotonic decreasing relationship between x_q, weight gain in the control group, versus Δ. The plot suggests that exposure to the sound of a heartbeat is most effective for infants who gain the least amount of weight after birth. As is fairly evident, this type of detailed comparison can be important and useful. If, for example, an experimental method is expensive or invasive, knowing how different subpopulations compare might affect the policy or strategy one is willing to adopt when dealing with a particular problem.

Next, attention is turned to the more general setting where the goal is to compare all of the quantiles rather than just the deciles. Suppose the value x satisfies $P(X \leq x) = q$. As noted in Chapter 3, from a strictly technical point of view, x is not necessarily the qth quantile. If x is the qth quantile, the difference between the two distributions at x_q is measured with $\Delta(x_q)$ given by Eq. (5.1). If x is not the qth quantile, as might be when sampling from a discrete distribution, the difference between the two distributions is measured with

$$\Delta(x) = y_q - x.$$

Let $X_1, \ldots, X_n$ and $Y_1, \ldots, Y_m$ be random samples from a control and experimental group, respectively. As usual, let $X_{(1)} \leq \cdots \leq X_{(n)}$ be the order statistics. Following Doksum and Sievers (1976), $\Delta(x)$ can be estimated as follows. Let $\hat{q} = \hat{F}(x)$ be the proportion of observations in the control group that are less than or equal to x. In terms of the order statistics, $\hat{q} = i/n$, where i is the largest integer such that $X_{(i)} \leq x$. Note that x qualifies as a reasonable estimate of x_q, the qth quantile. Then to estimate $\Delta(x)$, all that is needed is an estimate of the qth quantile of Y. The simplest estimate is $Y_{(\ell)}$, where $\ell = [\hat{q}m + 0.5]$ and as usual the notation $[\hat{q}m + 0.5]$ means to round $\hat{q}m + 0.5$ down to the nearest integer. In other words, use the quantile estimator described in Section 3.5 of Chapter 3. Finally, estimate $\Delta(x)$ with

$$\hat{\Delta}(x) = Y_{(\ell)} - x. \tag{5.2}$$

One can then plot $\hat{\Delta}(x)$ versus x to get an overall sense of how the two groups compare.

■ Example

As a simple illustration, suppose it is desired to estimate $\Delta(-160)$ for the weight data in Table 5.1. There are $n = 36$ participants in the control group, three participants have values less than or equal to -160, so $\hat{q} = 3/36$. To estimate the $3/36$ quantile of the experimental group, note that there are $m = 20$ participants, so $\ell = [(3/36)(20) + 0.5] = 2$. Therefore, the estimate of $y_{3/36}$ is the second smallest value in the experimental group, which is -60. Hence, $\Delta(-160)$ is estimated to be $-60 - (-160) = 100$ suggesting that the typical infant who would lose 160 g at birth would lose 100 g less if exposed to the sound of a heartbeat. There are also three values in the control group less than or equal to -180, so $\Delta(-180)$ is estimated to be $\hat{\Delta}(-180) = -60 - (-180) = 120$. Note that the shift function is just a series of straight lines with jumps at the points $(X_i, \hat{\Delta}(X_i))$, $i = 1, \ldots, n$. (A graphical illustration is given in Figure 5.3 which is discussed at the end of Section 5.1.4.)

There remains the problem of how to make inferences about $\Delta(x)$. Three approaches are described for comparing independent groups plus an extension of one of the methods to the case of dependent groups. One approach is based on comparing the deciles only, when they are estimated with the Harrell–Davis estimator, and the other two compute a confidence band for all quantiles. The latter two methods are based on two versions of the Kolmogorov–Smirnov test which are summarized in the next section of this chapter, after which attention is returned to making inferences about $\Delta(x)$. ■

5.1.1 The Kolmogorov–Smirnov Test

The Kolmogorov–Smirnov test is designed to test

$$H_0 : F(x) = G(x), \text{ all x,} \tag{5.3}$$

versus H_1: $F(x) \neq G(x)$ for at least one x, where F and G are the distributions associated with two independent groups (cf. Li, Tiwari, & Wells, 1996). The Kolmogorov–Smirnov test is of interest in its own right, and it is of interest in the present situation because it yields a confidence band for the shift function (cf. Fan, 1996).

Let $X_1, \ldots, X_n$, and $Y_1, \ldots, Y_m$ be random samples from two independent groups. Let $\hat{F}(x)$ and $\hat{G}(x)$ be the usual empirical distribution functions. Thus, $\hat{F}(x)$ is just the proportion of X_i values less than or equal to x, and $\hat{G}(x)$ is the proportion of Y_i values less than or equal to x. The Kolmogorov–Smirnov statistic for testing Eq. (5.3) is based on $\max|\hat{F}(x) - \hat{G}(x)|$, the maximum being taken over all possible values of x. That is, the test statistic is based on an estimate of the Kolmogorov distance between the two distributions. Let Z_i be the $n+m$ pooled observations. In symbols, $Z_i = X_i$, $i = 1, \ldots, n$, and $Z_{n+i} = Y_i$, $i = 1, \ldots, m$. Then the Kolmogorov–Smirnov test statistic is

$$D = \max|\hat{F}(Z_i) - \hat{G}(Z_i)|, \tag{5.4}$$

the maximum being taken over all $n+m$ values of i. That is, for each i, $i = 1, \ldots, n+m$, compute $|\hat{F}(Z_i) - \hat{G}(Z_i)|$ and set D equal to the largest of these values.

When sampling from continuous distributions, in which case ties occur with probability zero, percentage points of the null distribution of D can be obtained using a recursive method (Kim & Jennrich, 1973). Table 5.2 outlines the calculations. The method is too tedious to do by hand, an R function is provided that computes the exact significance level, so no illustration is given on how to perform the calculations.

Suppose the null hypothesis of identical distributions is true. If H_0 is rejected when $D > c$, and the algorithm in Table 5.2 indicates that $P(D > c) = \alpha$, given n and m, then the probability of a type I error is exactly α when ties are impossible. Moreover, the Kolmogorov–Smirnov test is distribution free—the probability of a type I error is exactly α regardless of which distributions are involved. However, if there are ties, the probability of a type I error is less than α (Schroër & Trenkler, 1995), but an exact significance level can be computed. Given the pooled data, Z, and letting $Z_{(1)} \leq \cdots \leq Z_{(N)}$ be the order statistics, the significance level of D, given Z, can be determined by proceeding as described in Table 5.2, only $C(i, j)$ is also set equal to 1 if $i + j < N$ and $Z_{(i+j)} = Z_{(i+j+1)}$ (Schroër & Trenkler, 1995). (Also see Hilton, Mehta, & Patel, 1994.)

There is another version of the Kolmogorov–Smirnov test worth considering that is based on a weighted analog of the Kolmogorov distance between two distributions. Let $N = m+n$, $M = mn/N$, $\lambda = n/N$, and $\hat{H}(x) = \lambda\hat{F}(x) + (1-\lambda)\hat{G}(x)$. Now, the difference between any

Table 5.2: Computing the Percentage Points of the Kolmogorov–Smirnov Statistic.

To compute $P(D \le c)$, where D is the Kolmogorov–Smirnov test statistic given by (5.4), let $C(i,j) = 1$ if

$$\left| \frac{i}{n} - \frac{j}{m} \right| \le c, \tag{5.5}$$

otherwise $C(i,j) = 0$, where the possible values of i and j are $i = 0, \ldots, n$ and $j = 0, \ldots, m$. Note that there are $(m+1)(n+1)$ possible values of D based on sample sizes of m and n. Let $N(i,j)$ be the number of paths over the lattice

$$\{(i,j) : i = 0, \ldots, n; \ j = 0, \ldots, m\},$$

from $(0,0)$ to (i,j), satisfying (5.5). Because the path to (i,j) must pass through either the point $(i-1,j)$ or $(i, j-1)$, $N(i,j)$ is given by the recursion relation

$$N(i,j) = C(i,j)[N(i, j-1) + N(i-1, j)],$$

subject to the initial conditions $N(i,j) = C(i,j)$ when $ij = 0$. When ties occur with probability zero, and H_0: $F(x) = G(x)$ is true,

$$P(D \le c) = \frac{m! n! N(m,n)}{(n+m)!},$$

where the binomial coefficient, $(n+m)!/(m!n!)$, is the number of paths from $(0,0)$ to (n,m).

When working with the weighted version of the Kolmogorov–Smirnov test, D_w, proceed exactly as before only set $C(i,j) = 1$ if

$$\sqrt{\frac{mn}{n+m}} \left| \frac{i}{n} - \frac{j}{m} \right| \left[\frac{i+j}{n+m} \left(1 - \frac{i+j}{n+m} \right) \right]^{-1/2} \le c.$$

Then

$$P(D_w \le c) = \frac{m! n! N(m,n)}{(n+m)!}.$$

two distributions, at the value x, is estimated with

$$\frac{\sqrt{M} |\hat{F}(x) - \hat{G}(x)|}{\sqrt{\hat{H}(x)[1 - \hat{H}(x)]}}. \tag{5.6}$$

Then H_0: $F(x) = G(x)$ can be tested with an estimate of the largest weighted difference over all possible values of x. (Also see Büning, 2001.) The test statistic is

$$D_w = \max \frac{\sqrt{M} |\hat{F}(Z_i) - \hat{G}(Z_i)|}{\sqrt{\hat{H}(Z_i)(1 - \hat{H}(Z_i))}}, \tag{5.7}$$

where the maximum is taken over all values of i, $i = 1, \ldots, N$, subject to $\hat{H}(Z_i)[1 - \hat{H}(Z_i)] > 0$. An exact significance level can be determined as described in Table 5.2. An argument for D_w is that it gives equal weight to each x in the sense that the large sample (asymptotic) variance of Eq. (5.6) is independent of x. Put another way, $|\hat{F}(x) - \hat{G}(x)|$, the estimate of the Kolmogorov distance at x, tends to have a larger variance when x is in the tails of the distributions. Consequently, inferences based on the unweighted Kolmogorov–Smirnov test statistic, D, tend to be more sensitive to differences that occur in the middle portion of the distributions. In contrast, Eq. (5.5) is designed so that its variance remains fairly stable as a function of x. Consequently, D_w is more sensitive than D to differences that occur in the tails.

When using D_w, and both m and n are less than or equal to 100, an approximate 0.05 critical value is

$$\frac{1}{95}\{0.48[\max(n, m) - 5] + 0.44|n - m|\} + 2.58,$$

the approximation being derived from the percentage points of D_w reported by Wilcox (1989). When using D, an approximation of the α critical value, when performing a two-sided test, is

$$\sqrt{-\frac{n + m}{2nm}\log(\alpha/2)}$$

(e.g., Hollander & Wolfe, 1973). This approximate critical value is reported by the R function ks described in the next section of this chapter.

5.1.2 R Functions ks, kssig, kswsig, and kstiesig

The R function

$$ks(x,y,w=F,sig=T,alpha=0.05),$$

ks, written for this book, performs the Kolmogorov–Smirnov test, where x and y are any R vectors containing data. (Again, the R functions written for this book can be obtained as described in Section 1.8.) The default value for w is F for false, indicating that the unweighted test statistic, D, is to be used. Using w=T results in the weighted test statistic, D_w. The default value for sig is T, meaning that the exact p-value is computed using the method in Table 5.2. If sig=F, ks uses the approximate α critical value, where by default, $\alpha = 0.05$ is used. The function returns the value of D or D_w, the approximate α critical value if sig=F, and the exact p-value if sig=T.

■ Example

> If the weight-gain data in Table 5.1 are stored in the R vectors x and y, the command ks(x,y,sig=F) returns the value $D = 0.48$ and reports that the 0.05 critical value is approximately 0.38. (The value of D is stored in the R variable ks$test, and the approximate critical value is stored in ks$crit.) The command ks(x,y) reports the exact p-value, which is .0018 and stored in the R variable ks$siglevel. Thus, with $\alpha = 0.05$, one would reject the hypothesis of identical distributions. This leaves open the issue of where the distributions differ and by how much, but this can be addressed with the confidence bands and confidence intervals described in the remaining portion of this section. The command ks(x,y,T,F) reports that $D_w = 3.5$ and that an approximate 0.05 critical value is 2.81. The command ks(x,y,T) computes the p-value when using D_w, which in contrast to D assumes there are no ties. For the weight-gain data, there are ties and the function warns that the reported p-value is not exact.
>
> ■

For convenience, the functions kssig, kswsig, and kstiesig are also supplied, which compute exact probabilities for the Kolmogorov–Smirnov statistics. These functions are used by the function ks to determine significance levels, so in general they are not of direct interest when testing $H_0: F(x) = G(x)$, but they might be useful when dealing with other issues covered in this chapter. The function kssig has the form

$$kssig(n,m,c).$$

It returns the exact p-value level when using the critical value c, assuming there are no ties and the sample sizes are n and m. In symbols, it determines $P(D > c)$ when computing D with n and m observations randomly sampled from two independent groups having identical distributions. Continuing the illustration involving the weight-gain data, the R command

$$kssig(length(x),length(y),ks(x,y)\$test)$$

computes the p-value of the unweighted Kolmogorov–Smirnov statistic assuming there are no ties. The result is .021. Because there are ties in the pooled data, this is higher than the p-value reported by kstiesig which takes ties into account.

If there are ties among the pooled observations, the exact p-value can be computed with the R function kstiesig. (This is done automatically when using the function ks.) It has the general form

$$kstiesig(x,y,c)$$

and reports the value of $P(D > c|Z)$, where the vector $Z = (Z_1, \ldots, Z_N)$ is the pooled data. For the weight-gain data, $D = 0.48$, and kstiesig(x,y, and 0.48) returns the value 0.0018, the same value returned by the function ks. If there are no ties among the observations, kstiesig returns the same significance level as ks(x,y,sig=T), the significance level associated with the unweighted test statistic, D.

5.1.3 The S Band and W Band for the Shift Function

This section describes two methods for computing a simultaneous $1 - \alpha$ level confidence band for $\Delta(x)$. Suppose c is chosen so that $P(D \le c) = 1 - \alpha$. As usual, denote the order statistics by $X_{(1)} \le \cdots \le X_{(n)}$ and $Y_{(1)} \le \cdots \le Y_{(m)}$. For convenience, let $X_0 = -\infty$ and $X_{(n+1)} = \infty$. For any x satisfying $X_{(i)} \le x < X_{(i+1)}$, let

$$k_* = \left[m \left(\frac{i}{n} - \frac{c}{\sqrt{M}} \right) \right]^+ ,$$

where $M = mn/(m+n)$ and the notation $[x]^+$ means to round up to the nearest integer. For example, $[5.1]^+ = 6$. Let

$$k^* = \left[m \left(\frac{i}{n} + \frac{c}{\sqrt{M}} \right) \right],$$

where k^* is rounded down to the nearest integer. Then a level $1 - \alpha$ simultaneous, distribution-free confidence band for $\Delta(x)$ $(-\infty < x < \infty)$ is

$$[Y_{(k_*)} - x, Y_{(k^*+1)} - x), \tag{5.8}$$

where $Y_{(k_*)} = -\infty$ if $k_* < 0$ and $Y_{(k^*)} = \infty$ if $k^* \ge m+1$ (Doksum & Sievers, 1976). That is, with probability $1 - \alpha$, $Y_{(k_*)} - x \le \Delta(x) < Y_{(k^*+1)} - x$ for all x. The resulting confidence band is called an *S band*.

■ Example

Suppose a confidence band for Δ is to be computed for the data in Table 5.1. For the sake of illustration, consider computing the confidence band at $x = 77$. Because $n = 36$ and $m = 20$, $M = 12.86$. Note that the value 77 is between $X_{(33)} = 70$ and $X_{(34)} = 100$, so $i = 33$. From the previous subsection, the 0.05 critical value is approximately $c = 0.38$, so

$$k_* = \left[20 \left(\frac{33}{36} - \frac{0.38}{\sqrt{12.86}} \right) \right]^+ = 17.$$

Similarly, $k^* = 20$. From Table 5.1, the 17th value in the experimental group, after putting the values in ascending order, is $Y_{(17)} = 75$, $Y_{(20)} = 190$, so the interval around $\Delta(77)$ is

$$(75 - 77, 190 - 77) = (-2, 113).$$

An exact confidence band for Δ, called a *W band*, can also be computed with the weighted Kolmogorov–Smirnov statistic. Let c be chosen so that $P(D_w \le c) = 1 - \alpha$. This time, for any x satisfying $X_{(i)} \le x < X_{(i+1)}$, let $u = i/n$ and set

$$h_* = \frac{u + \{c(1-\lambda)(1-2\lambda u) - \sqrt{c^2(1-\lambda)^2 + 4cu(1-u)}\}/2}{1 + c(1-\lambda)^2},$$

and

$$h^* = \frac{u + \{c(1-\lambda)(1-2\lambda u) + \sqrt{c^2(1-\lambda)^2 + 4cu(1-u)}\}/2}{1 + c(1-\lambda)^2}.$$

Set $k_* = [h_* m]^+$ and $k^* = [h^* m]$. In words, k_* is the value of $h_* m$ rounded up to the nearest integer, where m is the number of observations in the second group (associated with Y). The value of k^* is computed in a similar manner, only its value is rounded down. Then the confidence band is again given by Eq. (5.8).

5.1.4 R Functions sband and wband

The R functions sband and wband are provided for determining confidence intervals for $\Delta(x)$ at each of the X_i values in the control group, and they can be used to compute confidence bands as well. The function sband has the general form

sband(x,y,crit=1.36((length(x)+length(y))/(length(x)*length(y))),
                 flag=F, plotit=T,sm=T, op=1).

As usual, x and y are any R vectors containing data. The optional argument crit is the critical value used to compute the simultaneous confidence band, which defaults to the approximate 0.05 critical value if unspecified. The default value for flag is F, for false, meaning that the exact probability of a type I error will not be computed. The command sband(x,y,flag=T) will report the actual probability of a type I error using the approximate .05 critical value, assuming there are no ties among the pooled data. The command sband(x,y,.2,T) computes confidence intervals using the critical value 0.2, and it reports the exact probability of a type I error when there are no ties. The argument plotit defaults to T for true meaning that a plot of the shift function will be created. If sm=T, the plot of the shift function is smoothed using a regression smoother called lowess when the argument op is equal to one. (Smoothers are described in Chapter 11.) If op is not equal to one, lowess is replaced by a running interval smoother (described in Chapter 11).

The function returns an n-by-3 matrix of numbers in the R variable sband\$m. The ith row of the matrix corresponds to the confidence band computed at $\hat{\Delta}(X_{(i)})$, the estimate of the shift function at the ith largest X value. For convenience, the first column of the n-by-3 matrix returned by sband contains $\hat{q} = 1/n$, $i = 1, \ldots, n$. The values in the second column are the lower ends of the confidence band, whereas the upper ends are reported in column 3. A value of NA in the middle column corresponds to $-\infty$, whereas NA in the last column means the upper end of the confidence interval is ∞. The function also returns the critical value being used in the R variable sband$crit, the number of significant differences in sband$numsig, and if flag=T, the exact probability coverage is indicated by sband$pc. If flag=F, the value of sband$pc will be NA for not available. The function wband, which is used exactly like sband, computes confidence intervals (W bands) using the weighted Kolmogorov–Smirnov statistic, D_w, instead.

■ Example

Doksum and Sievers (1976) report data from a study designed to assess the effects of ozone on weight gain in rats. The experimental group consisted of 22 rats of 70-day old kept in an ozone environment for 7 days. A control group of 23 rats, of the same age, were kept in an ozone-free environment. The weight gains, in grams, are listed in Table 5.3. Table 5.4 shows the 23-by-3 matrix reported by the R function sband. The ith row of the matrix reports the confidence interval for Δ at the ith largest value among the X values. If there had been 45 observations in the first group, a 45 by 3 matrix would be reported instead. The function reports that numsig=8 meaning there are eight confidence intervals not containing 0, and from Table 5.4 these are the intervals extending from the second smallest to the ninth smallest value in the control group. For example, the second smallest value in the control group is $X_{(2)} = 13.1$ which corresponds to an estimate of the $q = 2/23 \approx 0.09$ quantile of the control group, and the second row in Table 5.4 (labeled [2,]) indicates that the confidence interval for $\Delta(13.1)$ is (NA, -3.0), which means that the confidence interval is $(-\infty, -3.0)$. The interval does not contain 0 suggesting that rats at the 0.09 quantile of the control group tend to gain more weight compared with the rats at the 0.09 quantile of the experimental method. The next eight confidence intervals do not contain 0 either, but the remaining confidence intervals all contain 0. The function sband indicates that the default critical value corresponds to $\alpha = 0.035$. Thus, there is evidence that rats who ordinarily gain a relatively small amount of weight will gain even less weight in an ozone environment.

Table 5.3: Weight Gain of Rats in Ozone Experiment.

Control	41.0	38.4	24.4	25.9	21.9	18.3	13.1	27.3	28.5	−16.9
Ozone	10.1	6.1	20.4	7.3	14.3	15.5	−9.9	6.8	28.2	17.9
Control	26.0	17.4	21.8	15.4	27.4	19.2	22.4	17.7	26.0	29.4
Ozone	−9.0	−12.9	14.0	6.6	12.1	15.7	39.9	−15.9	54.6	−14.7
Control	21.4	26.6	22.7							
Ozone	44.1	−9.0								

Table 5.4: Confidence Intervals for Δ Using the Ozone Data.

```
            qhat lower upper
 [1,] 0.04347826    NA  24.2
 [2,] 0.08695652    NA  -3.0
 [3,] 0.13043478    NA  -3.3
 [4,] 0.17391304    NA  -3.4
 [5,] 0.21739130    NA  -3.4
 [6,] 0.26086957    NA  -2.8
 [7,] 0.30434783    NA  -3.5
 [8,] 0.34782609    NA  -3.5
 [9,] 0.39130435    NA  -1.4
[10,] 0.43478261 -37.8   6.3
[11,] 0.47826087 -37.1  17.5
[12,] 0.52173913 -35.6  21.4
[13,] 0.56521739 -34.3  30.2
[14,] 0.60869565 -34.9    NA
[15,] 0.65217391 -35.0    NA
[16,] 0.69565217 -19.9    NA
[17,] 0.73913043 -20.0    NA
[18,] 0.78260870 -20.5    NA
[19,] 0.82608696 -20.1    NA
[20,] 0.86956522 -18.4    NA
[21,] 0.91304348 -17.3    NA
[22,] 0.95652174 -24.4    NA
[23,] 1.00000000 -26.7    NA
```

The R command sband(x,y,flag=T) reports that when using the default critical value, which is reported to be 0.406, the actual probability coverage is $1 - \alpha = .9645$ assuming there are no ties. To find out what happens to $1 - \alpha$ when a critical value of 0.39 is used instead, type the command sband(x,y,.39,T). The function reports that now, $1 - \alpha = 0.9638$.

The S band suggests that there might be a more complicated relationship between weight gain and ozone than is suggested by a single measure of location. Figure 5.3 shows the plot of $\hat{\Delta}(x)$ versus x that is created by sband. (The + along the x-axis marks the position of the median in the first group, and the lower and upper quartiles are indicated by an o.) The solid line, between the two dotted lines, is a graph of $\hat{\Delta}(x)$ versus x. The dotted lines form the approximate 0.95 confidence band. As previously indicated, the actual probability coverage of the confidence band is .9645. Notice that the bottom dotted line starts at $X = 21.9$. This is because for $X \leq 21.9$ the lower end of the confidence band is $-\infty$. Also, the lower dotted line

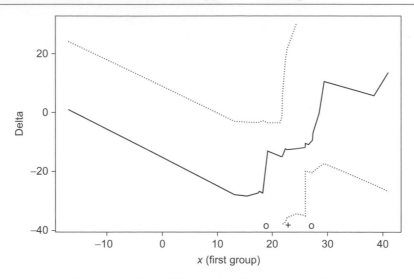

Figure 5.3: The shift function for the ozone data.

stops at $X = 41$ because $X = 41$ is the largest X value available. Similarly, for $X > 24.4$, the upper end of the confidence band is ∞, and this is why the upper dotted line in Figure 5.3 stops at $X = 24.4$. An interesting feature about Figure 5.3 is the suggestion that, as weight gain increases in the control group, ozone has less of an effect. In fact, rats in the control group who would ordinarily have a high weight gain might actually gain more weight when exposed to ozone. However, the confidence band makes it clear that more data are needed to resolve this issue.

5.1.5 Confidence Band for the Deciles Only

Confidence intervals for the difference between the deciles can be computed with the Harrell-Davis estimator such that the simultaneous probability coverage of all nine confidence intervals is approximately $1 - \alpha$. One advantage of this approach is that it might have more power than S or W bands when sampling from normal or light-tailed distributions.

Let $\hat{\theta}_{qx}$ be the Harrell-Davis estimate of the qth quantile of the distribution associated with X, $q = 0.1, \ldots, 0.9$. Let $\hat{\theta}_{qy}$ be the corresponding estimate for Y. The goal is to compute confidence intervals for $y_q - x_q$ such that the simultaneous probability coverage is $1 - \alpha$. A solution that appears to provide reasonably accurate probability coverage for a wide range of distributions begins by computing a bootstrap estimate of the standard errors for $\hat{\theta}_{qx}$ and $\hat{\theta}_{qy}$ as described in Section 3.5.7. Here, independent bootstrap samples are used for all 18 deciles being estimated. In particular, bootstrap samples are used to estimate the standard error of $\hat{\theta}_{0.1x}$, the Harrell-Davis estimate of the 0.1 quantile corresponding to X, and a different (independent) set of bootstrap samples is used to estimate the standard error of $\hat{\theta}_{0.2x}$. Let $\hat{\sigma}_{qx}^2$ be the bootstrap estimate of the squared standard error of $\hat{\theta}_{qx}$. Then a 0.95 confidence interval for $y_q - x_q$ is

$$(\hat{\theta}_{qy} - \hat{\theta}_{qx}) \pm c\sqrt{\hat{\sigma}_{qx}^2 + \hat{\sigma}_{qy}^2}, \tag{5.9}$$

where, when $n = m$,

$$c = \frac{80.1}{n^2} + 2.73. \tag{5.10}$$

The constant c was determined so that the simultaneous probability coverage of all nine differences is approximately .95 when sampling from normal distributions. Simulations suggest that when sampling form nonnormal distributions, the probability coverage remains fairly close to the nominal .95 level (Wilcox, 1995a). For unequal sample sizes, the current strategy for computing the critical value is to set n equal to the smaller of the two sample sizes and use c given by Eq. (5.10). This approach performs reasonably well provided the difference between the sample sizes is not too large, but if the difference is large enough, the actual probability of a type I error can be substantially smaller than the nominal level, especially when sampling from heavy-tailed distributions. Another approach is to use the nine-variate Studentized maximum modulus distribution to determine c, but Wilcox found this to be less satisfactory. Yet another approach is to use a percentile bootstrap method to determine an appropriate confidence interval, but this is less satisfactory as well.

5.1.6 R Function shifthd

The R function shifthd, that comes with this book, computes the 0.95 simultaneous confidence intervals for the difference between the deciles given by Eq. (5.9). The function has the form

shifthd(x,y,nboot=200,plotit=T,plotop=F).

The data corresponding to the two groups are stored in the R vectors x and y, and the default number of bootstrap samples used to compute the standard errors is 200. Thus, the command shifthd(x,y) will use 200 bootstrap samples when computing confidence intervals, while shifthd(x,y,100) will use 100 instead. The function returns a 9 by 3 matrix. The ith row corresponds to the results for the $i/10$ quantile. The first column contains the lower ends of the confidence intervals, the second column contains the upper ends, and the third column contains the estimated difference between the quantiles. With plotop=F and plotit=T, the function creates a plot where the x-axis contains the estimated quantiles of the first groups, as done by sband. With plotop=T, the function plots $q = 0.1, \ldots, 0.9$ versus $(\hat{\theta}_{qy} - \hat{\theta}_{qx})$.

■ Example

For the ozone data in Table 5.3, shifthd returns

```
             lower        upper    Delta.hat
[1,]   -47.75411     1.918352   -22.917880
[2,]   -43.63624    -6.382708   -25.009476
[3,]   -36.04607    -3.237478   -19.641772
[4,]   -29.70039    -0.620098   -15.160245
[5,]   -24.26883    -1.273594   -12.771210
[6,]   -20.71851    -1.740128   -11.229319
[7,]   -24.97728     7.280896    -8.848194
[8,]   -24.93361    19.790053    -2.571780
[9,]   -20.89520    33.838491     6.471643
```

The first row indicates that the confidence interval for $\Delta(x_{.1})$ is $(-47.75411, 1.918352)$, and that $\hat{\Delta}(\hat{\theta}_{x.1})$ is equal to -22.917880. The second row gives the confidence interval for Δ evaluated at the estimated 0.2 quantile of the control group, and so on. The confidence intervals indicate that the weight gain in the two groups differ at the 0.2, 0.3, 0.4, 0.5, and 0.6 quantiles of the control group. Note that in general, the third column, which reports $\hat{\Delta}(x)$, is increasing. That is, the differences between weight gain are getting smaller, and for the 0.9 quantile, there is the possibility that rats gain more weight in an ozone environment. However, the length of the confidence interval at the 0.9 quantile is too wide to be reasonably sure.

Figure 5.4 shows the plot created by shifthd for the ozone data. There are nine dots corresponding to the points $(\hat{\theta}_{xq}, \hat{\Delta}(\hat{\theta}_{xq}))$, $q = 0.1, \ldots, 0.9$. That is, the dots are the

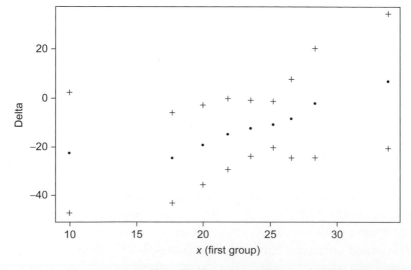

Figure 5.4: The plot created by the function shifthd using the ozone data.

estimated shift function plotted as a function of the estimated deciles corresponding to the data in the first argument, x. Note that in general, the dots are monotonic increasing, which is consistent with Figure 5.3. Above and below each dot is a + indicating the ends of the confidence interval.

5.1.7 R Functions g2plot and splotg2

To supplement the shift function, it might help to plot a density estimate for the two groups under study. The function

$$g2plot(x,y,op=4,rval=15,fr=0.8,aval=0.5)$$

is supplied to help accomplish this goal. (The density estimate for the second group is indicated by a dashed line.) The argument op controls the type of graph created. The choices are

- op=1, Rosenblatt shifted histogram
- op=2, kernel density estimate based on a normal kernel
- op=3, expected frequency curve
- op=4, adaptive kernel estimator

The other arguments are relevant to the various density estimators as described in Chapter 3.

When working with discrete data, the function

$$splotg2(x,y,op=T,xlab="X",ylab="Rel. Freq.")$$

is supplied in case it is desired to plot the relative frequencies for all distinct values found in each of two groups. With op=T, a line connecting the points corresponding to the relative frequencies is formed.

5.2 Student's t-test

This section reviews some practical concerns about comparing means with Student's t-test. From previous chapters, it is evident that Student's t-test can have low power under slight departures from normality toward a heavy-tailed distribution. There are some additional issues, however, that help motivate some of the heteroscedastic methods covered in this book.

It is a bit more convenient to switch notation slightly. For two independent groups, let X_{ij}, $i = 1, \ldots, n_j$; $j = 1, 2$ be a random sample of n_j observations from the jth group. Let μ_j and σ_j^2 be the mean and variance associated with the jth group. If the variances have a common value, say $\sigma_1^2 = \sigma_2^2 = \sigma^2$, and if sampling is from normal distributions, then from basic results,

$$T = \frac{\bar{X}_1 - \bar{X}_2 - (\mu_1 - \mu_2)}{\sqrt{\text{MSWG}\left(\frac{1}{n_1} + \frac{1}{n_2}\right)}} \tag{5.11}$$

has a Student's t-distribution with $\nu = n_1 + n_2 - 2$ degrees of freedom, where

$$\text{MSWG} = \frac{(n_1 - 1)s_1^2 + (n_2 - 1)s_2^2}{n_1 + n_2 - 2}$$

is the usual (means squares within groups) estimate of the assumed common variance, σ^2. If the assumptions of normality and equal variances are met, $E(T) = 0$ and the variance of T goes to one as the samples sizes get large. To quickly review, the hypothesis of equal means, $H_0: \mu_1 = \mu_2$, is rejected if $|T| > t$, the $1 - \alpha/2$ quantile of Student's t-distribution with $\nu = n_1 + n_2 - 2$ degrees of freedom, and a $1 - \alpha$ confidence interval for $\mu_1 - \mu_2$ is

$$(\bar{X}_1 - \bar{X}_2) \pm t \sqrt{\text{MSWG}\left(\frac{1}{n_1} + \frac{1}{n_2}\right)}. \tag{5.12}$$

Concerns about the ability of Student's t-test to control the probability of a type I error date back to at least Pratt (1964), who established that the level of the test is not preserved if distributions differ in dispersion or shape. If sampling is from normal distributions, the sample sizes are equal, but the variances are not equal, Eq. (5.12) provides reasonably accurate probability coverage no matter how unequal the variances might be, provided the common sample size is not too small (Ramsey, 1980). For example, if the common sample size is 15, and $\alpha = 0.05$, the actual probability coverage will not drop below .94. Put another way, in terms of testing H_0, the actual probability of a type I error will not exceed .06. However, if the sample sizes are equal, but sampling is from nonnormal distributions, probability coverage can be unsatisfactory, and if the sample sizes are unequal as well, probability coverage deteriorates even further. Even under normality with unequal sample sizes, there are problems. For example, under normality with $n_1 = 21$, $n_2 = 41$, $\sigma_1 = 4$, $\sigma_2 = 1$, and $\alpha = 0.05$, the actual probability of a type I error is approximately .15. Moreover, Fenstad (1983) argues that $\sigma_1/\sigma_2 = 4$ is not extreme, and various empirical studies support Fenstad's view (e.g., Grissom, 2000; Keselman et al., 1998; Wilcox, 1987a). The illustration just given might appear to conflict with results in Box (1954), but this is not the case. Box's numerical results indicate that under normality, and when $1/\sqrt{3} \leq \sigma_1/\sigma_2 \leq \sqrt{3}$, Student's t-test provides reasonably good control over the probability of a type I error, but more recent papers have shown that when $\sigma_1/\sigma_2 > \sqrt{3}$, Student's t-test becomes unsatisfactory (e.g., Brown & Forsythe, 1974; Tomarken & Serlin, 1986; Wilcox, Charlin, & Thompson, 1986).

To illustrate what can happen under nonnormality, suppose observations for the first group are sampled from a lognormal distribution that has been shifted to have a mean of zero, whereas the observations from second group have a normal distribution with mean 0 and standard deviation 0.25. With $n_1 = n_2 = 20$ and $\alpha = 0.025$, the probability of rejecting H_0: $\mu_1 < \mu_2$ is .136 (based on simulations with 10,000 replications), whereas the probability of rejecting H_0: $\mu_1 > \mu_2$ is .003. Moreover, Student's t-test assumes that $E(T) = 0$, but $E(T) = -0.52$, approximately, again based on a simulation with 10,000 replications. One implication is that, in addition to yielding a confidence interval with inaccurate probability coverage, the probability of rejecting H_0: $\mu_1 = \mu_2$ with Student's t-test has the undesirable property of not being minimized when H_0 is true. That is, Student's t-test is biased. If, for example, 0.5 is subtracted from each observation in the second group, the probability of rejecting H_0: $\mu_1 < \mu_2$ drops from .136 to .083. That is, the mean of the second group has been shifted by a half standard deviation away from the null hypothesis, yet power is less than the probability of rejecting when the null hypothesis is true.

When using Student's t-test, poor power properties and inaccurate confidence intervals are to be expected based on results in Chapter 4. To elaborate on why this is so, let $\mu_{[k]} = E(X - \mu)^k$ be the kth moment about the mean of the random variable X. The third moment, $\mu_{[3]}$, reflects skewness, the most common measure being $\kappa_1 = \mu_{[3]}/\mu_{[2]}^{1.5}$. For symmetric distributions, $\kappa_1 = 0$. It can be shown that for two independent random variables, X and Y, having third moments $\mu_{x[3]}$ and $\mu_{y[3]}$, the third moment of $X - Y$ is $\mu_{x[3]} - \mu_{y[3]}$. In other words, if X and Y have equal skewnesses, $X - Y$ has a symmetric distribution. If they have unequal skewnesses, $X - Y$ has a skewed distribution. From Chapter 4, it is known that when X has a skewed distribution, and when the tails of the distribution are relatively light, the standard confidence interval for μ can have probability coverage that is substantially lower than the nominal level. For symmetric distributions, this problem is much less severe, although probability coverage can be too high when sampling from a heavy-tailed distribution. Consequently, when X_{i1} and X_{i2} have identical distributions, and in particular have equal third moments, the third moment of $X_{i1} - X_{i2}$ is zero, suggesting that probability coverage of $\mu_1 - \mu_2$ will not be excessively smaller than the nominal level when using Eq. (5.12). Put another way, if two groups do not differ, $X_{i1} - X_{i2}$ has a symmetric distribution suggesting that the probability of a type I error will not exceed the nominal level by too much. (For results supporting this conclusion when dealing with highly discrete data, see Rasch, Teuscher, & Guiard, 2007.) However, when distributions differ, and in particular have different amounts of skewness, $X_{i1} - X_{i2}$ has a skewed distribution as well suggesting that probability coverage might be too low and that T given by Eq. (5.11) does not have a mean of zero as is commonly assumed. This in turn suggests that if groups differ, testing H_0: $\mu_1 = \mu_2$ with Student's t-test might result in an undesirable power property—the probability of rejecting might decrease as $\mu_1 - \mu_2$ increases, as was illustrated in the previous paragraph. In fact, if the groups differ, and have unequal variances and differ in skewness, and if the sample sizes differ as well, then confidence intervals based on Eq. (5.12) are not even asymptotically correct. In particular, the variance of T does not go to one as the sample sizes increase (Cressie & Whitford, 1986). In contrast, heteroscedastic methods are asymptotically correct, they give reasonably accurate probability coverage over a wider range of situations than Student's t, so only heterosecedastic methods are considered in the remainder of this chapter. (For a recent overview of heterosecedastic methods for means, in the two-sample case, see Sawilowsky, 2002.)

It should be noted, however, that even if two groups have the same amount of skewness, problems with probability coverage and control of type I error probabilities can arise when distributions differ in scale. This occurs, for example, when sampling from an exponential distribution. Figure 5.5 shows the probability density function of an exponential distribution, $f(x) = \exp(-x)$. The shape of this distribution is similar to the shape of empirical distributions found in various situations. (For an example based on psychometric data, see Sawilowsky & Blair, 1992.) The mean of this distribution is $\mu = 1$, the 20% trimmed mean is $\mu_t = 0.761$, and the M-measure of location (based on Huber's Ψ) is $\mu_m = 0.824$.

Consider two exponential distributions, shifted so that they both have a mean of 0, with the second distribution re-scaled so that its variance is four times as large as the first. With $n_1 = n_2 = 20$, the probability of a type I error is .133 when testing H_0: $\mu_1 = \mu_2$ at the $\alpha = 0.05$ level. Increasing n_1 to 40, the probability of a type I error is .165, while with $n_1 = n_2 = 40$ it is .08.

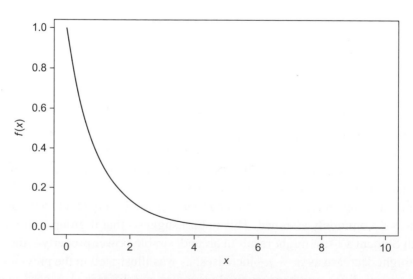

Figure 5.5: An exponential distribution.

A natural way of trying to salvage homoscedastic methods is to test for equal variances, and if not significant, assume the variances are equal. Even under normality, this strategy can fail because tests for equal variances might not have enough power to detect unequal variances in situations where the assumption should be abandoned, even when the test for equal variances is performed at the 0.25 level (e.g., Hayes & Cai, 2007; Markowski & Markowski, 1990; Moser, Stevens, & Watts, 1989; Wilcox, Charlin, & Thompson, 1986; Zimmerman, 2004).

5.3 Comparing Medians and Other Trimmed Means

This section considers the problem of testing

$$H_0 : \mu_{t1} = \mu_{t2},$$

the hypothesis that two independent groups have equal trimmed means, plus the related goal of computing a $1 - \alpha$ confidence interval for $\mu_{t1} - \mu_{t2}$. Included as a special case is a method for comparing medians, which requires specialized techniques.

Yuen's Method

Yuen (1974) derived a method for comparing trimmed means that is designed to allow unequal Winsorized variances. When there is no trimming ($\gamma = 0$), Yuen's method reduces to Welch's (1933) method for comparing means, which allows unequal variances.

Generalizing the notation of Chapters 3 and 4 in an obvious way, suppose the amount of trimming is γ. For the jth group, let $g_j = [\gamma n_j]$ be the number of observations trimmed from each tail, let $h_j = n_j - 2g_j$ be the number of observations left after trimming, and let s_{wj}^2 be the Winsorized sample variance. From Chapter 3, an estimate of the squared standard error of $\bar{X}_{tj}$ is $s_{wj}^2/\{(1 - 2\gamma)^2 n\}$. However, Yuen estimates the squared standard error with

$$d_j = \frac{(n_j - 1)s_{wj}^2}{h_j(h_j - 1)}. \tag{5.13}$$

It is left as an exercise to verify that both estimates give similar values. In terms of type I error probabilities and probability coverage, simulations indicate that Yuen's estimate gives slightly better results. Yuen's test statistic is

$$T_y = \frac{\bar{X}_{t1} - \bar{X}_{t2}}{\sqrt{d_1 + d_2}}. \tag{5.14}$$

The null distribution of T_y is approximated with a Student's t-distribution with estimated degrees of freedom

$$\hat{\nu}_y = \frac{(d_1 + d_2)^2}{\frac{d_1^2}{h_1 - 1} + \frac{d_2^2}{h_2 - 1}}.$$

The $1 - \alpha$ confidence interval for $\mu_{t1} - \mu_{t2}$ is

$$(\bar{X}_{t1} - \bar{X}_{t2}) \pm t\sqrt{d_1 + d_2}, \tag{5.15}$$

where t is the $1 - \alpha/2$ quantile of Student's t-distribution with $\hat{\nu}_y$ degrees of freedom. The hypothesis of equal trimmed means is rejected if

$$|T_y| \geq t.$$

As previously indicated, when two distributions differ, it can be difficult getting a confidence interval for the difference between the means that has probability coverage reasonably close to the nominal level. Theoretical results, supported by simulations, indicate that as the amount of trimming increases from 0 to 20%, Yuen's method yields confidence intervals for $\mu_{t1} - \mu_{t2}$ with probability coverage closer to the nominal level (Wilcox, 1994a). As an illustration, suppose the first group has a normal distribution, and the second group is skewed with $\kappa_1 = 2$ and $n_1 = n_2 = 12$. Wilcox (1994a) reports situations where $H_0: \mu_1 > \mu_2$ is tested with $\alpha = 0.025$, but the actual probability of a type I error is .054. (This result is based on simulations with 100,000 replications.) In contrast, with 20% trimming, the actual probability of a type I error is .022. With $n_1 = 80$ and $n_2 = 20$, the probability of a type I error can be as high as .093—nearly four times higher than the nominal level—when using Welch's test, while with 20% trimming the actual probability of type I error is approximately .042 for the same distributions. Of course, by implication, there are some situations where Welch's test will be unsatisfactory when dealing with a two-sided test and $\alpha = 0.05$.

■ Example

As another illustration that differences in skewness can make a practical difference, imagine that for the first group, 40 observations are generated from a normal, and for the second group, 20 observations are generated from a lognormal distribution that has been shifted so that it has a mean of zero. When testing at the 0.05 level, the actual level of Welch's test is approximately 0.11. And if instead observations for the second group are generated from a g-and-h distribution with $g = h = 0.5$, the actual level is approximately 0.20. Comparing 20% trimmed means instead, the actual levels for these two situations are 0.047 and 0.042, respectively. However, Section 5.3.2 notes that even Yuen's method can be unsatisfactory in terms of type I errors. And an alternative approach to comparing trimmed means is described that gives better results.

■

Table 5.5: Estimated Power, $n_1 = n_2 = 25$, $\alpha = .05$.

Distributions	δ	Welch	Yuen ($\gamma = 0.2$)	KS (exact)	KS ($\alpha = 0.052$)
Normal	0.6	0.536	0.464	0.384	0.464
Normal	0.8	0.780	0.721	0.608	0.700
Normal	1.0	0.931	0.890	0.814	0.872
CN1	1.0	0.278	0.784	0.688	0.780
CN2	1.0	0.133	0.771	0.698	0.772
Slash	1.0	0.054	0.274	0.235	0.308

From Randles and Wolfe (1979, p.384), the expectation is that the Kolmogorov–Smirnov test will have lower power than Welch's test when sampling from normal distributions with a common variance. More generally, it might seem that when distributions differ in location only, and are symmetric, the Kolmogorov–Smirnov test will have less power than the Yuen–Welch test. Table 5.5 shows the estimated power of these tests for four distributions, $n_1 = n_2 = 25$, and when δ is added to every observation in the first group. The notation KS (exact) means that the Kolmogorov–Smirnov critical value was chosen as small as possible with the property that the exact probability of a type I error will not exceed .05. The last column in Table 5.5 shows the power of the Kolmogorov–Smirnov test when the critical value is chosen so that the probability of a type I error is as close as possible to .05. For the situation at hand, the resulting probability of a type I error is .052. The notation CN1 refers to a contaminated normal where, in Eq. (1.1), $\epsilon = 0.1$ and $K = 10$. The notation CN2 refers to a contaminated normal with $K = 20$. As is seen, the exact test does have less power than Welch's test under normality, but the exact test has substantially more power when sampling from a heavy-tailed distribution. Moreover, with $\alpha = 0.052$, the Kolmogorov–Smirnov test has about the same amount of power as Yuen's test with 20% trimming. Another appealing feature of the Kolmogorov–Smirnov test, versus the Yuen–Welch test, is that the Kolmogorov–Smirnov test is sensitive to more features of the distributions. A negative feature of the Kolmogorov–Smirnov test is that when there are tied values among the pooled observations, its power can be relatively low.

Comparing Medians

As the amount of trimming approaches 0.5, Yuen's method breaks down; the method for estimating the standard error becomes highly inaccurate resulting inaccurate confidence intervals and poor control over the probability of a Type I error. If there are no tied values in either group, an approach that currently seems to have practical value is as follows. Let M_1 and M_2 be the sample medians corresponding to groups 1 and 2, respectively, and let S_1^2 and S_2^2 be the corresponding McKean–Schrader estimates of the squared standard errors. Then an approximate $1 - \alpha$ confidence interval for the difference between the population medians is

$$(M_1 - M_2) \pm c\sqrt{S_1^2 + S_2^2}$$

where c is the $1 - \alpha/2$ quantile of a standard normal distribution. Alternatively, reject the hypothesis of equal population medians if

$$\frac{|M_1 - M_2|}{\sqrt{S_1^2 + S_2^2}} \geq c.$$

But if there are tied values in either group, control over the probability of a Type I error can be very poor. There are two practical problems, which were noted in Chapter 4. First, with tied values, all known estimators of the standard of the sample median can be highly inaccurate. Second, the sampling distribution of the sample median does not necessarily approach a normal distribution as the sample size gets large. When tied values occur, the only known method for comparing medians that performs well in simulations, in terms of controlling the probability of a Type I error, is the percentile bootstrap method in Section 5.4.2.

5.3.1 R Function yuen

The R function

$$\text{yuen}(x,y,\text{tr}=0.2,\text{alpha}=0.05)$$

performs the Yuen–Welch method for comparing trimmed means. The default amount of trimming (tr) is 0.2, and the default value for α is 0.05. Thus, the command yuen(x,y) returns a 0.95 confidence interval for the difference between the 20% trimmed means using the data stored in the R vectors x and y. The confidence interval is returned in the R variable yuen$ci. The command yuen(x,y,0) returns a 0.95 confidence interval for the difference between the means based on Welch's method. The function also returns the value of the test statistic in yuen$teststat, a two-sided significance level in yuen$siglevel, a $1 - \alpha$ confidence interval in yuen$ci, the estimated degrees of freedom, the estimated difference between the trimmed means, and the estimated standard error.

■ **Example**

For the ozone data in Table 5.3 and 20% trimming, the R function yuen indicates that $T_y = 3.4$, the p-value is .0037, and the 0.95 confidence interval for $\mu_{t1} - \mu_{t2}$ is $(5.3, 22.85)$. In contrast, with zero trimming (Welch's method), $T_y = 2.46$, the p-value is .019, and the 0.95 confidence interval is $(1.96, 20.8)$. Both methods suggest that for the typical rat, weight gain is higher for rats living in an ozone-free environment, but they give a different picture of the extent to which this is true.

■

5.3.2 A Bootstrap-t Method for Comparing Trimmed Means

As previously indicated, when testing hypotheses with the Yuen–Welch method, control of type I error probabilities is generally better when using 20% trimming versus no trimming at all. However, problems might persist when using 20% trimming, especially when performing a one-sided test and the sample sizes are unequal. For example, if sampling is from exponential distributions with sample sizes of 15 and 30, and if the second group has a standard error four times as large as the first, the probability of a type I error can be twice as large as the nominal level. With $\alpha = 0.025$, $P(T_y < t_{0.025}) = 0.056$, whereas with $\alpha = 0.05$ the probability is .086. As in the one-sample case discussed in Chapter 4, a bootstrap-t method (sometimes called a percentile t method) can give better results. The bootstrap method advocated by Westfall and Young (1993) has been found to have a practical advantage over the Yuen–Welch method (Wilcox, 1996b), but it seems to have no practical advantage over the bootstrap-t, at least based on extant simulations, so it is not discussed here.

For the situation at hand, the general strategy of the bootstrap-t method is to estimate the upper and lower critical values of the test statistic, T_y, by running simulations on the available data. This is done by temporarily shifting the two empirical distributions so that they have identical trimmed means, and then generating bootstrap samples to estimate the upper and lower critical values for T_y that would result in a type I error probability equal to α. Once the critical values are available, a $1 - \alpha$ confidence interval can be computed, as is illustrated later.

One way of describing the bootstrap-t in a more precise manner is as follows. For fixed j, let $X_{1j}^*, \ldots, X_{n_j j}^*$ be a bootstrap sample from the jth group, and set $C_{ij}^* = X_{ij}^* - \bar{X}_{tj}$, $i = 1, \ldots, n_j$. Then C_{ij}^* represents a sample from a distribution that has a trimmed mean of zero. That is, the hypothesis of equal trimmed means is true for the distributions associated with the C_{ij}^* values. Consequently, applying the Yuen–Welch method to the C_{ij}^* values should not result in rejecting the hypothesis of equal trimmed means. Let T_y^* be the value of T_y based on the C_{ij}^* values. To estimate the distribution of T_y when the null hypothesis is true, repeat the process just described B times, each time computing T_y^* based on the resulting C_{ij}^* values. Label the resulting T_y^* values T_{yb}^*, $b = 1, \ldots, B$. Let $T_{y(1)}^* \leq \cdots \leq T_{y(B)}^*$ be the T_{yb}^* values written in ascending order. Set $\ell = \alpha B/2$, round ℓ to the nearest integer, and let $u = B - \ell$. Then an estimate of the lower and upper critical values is $T_{y(\ell+1)}^*$ and $T_{y(u)}^*$. That is, reject H_0: $\mu_{t1} = \mu_{t2}$ if $T_y < T_{y(\ell+1)}^*$ or $T_y > T_{y(u)}^*$. A little algebra shows that a $1 - \alpha$ confidence interval for $\mu_{t1} - \mu_{t2}$ is

$$\left(\bar{X}_{t1} - \bar{X}_{t2} - T_{y(u)}^* \sqrt{d_1 + d_2}, \; \bar{X}_{t1} - \bar{X}_{t2} - T_{y(\ell+1)}^* \sqrt{d_1 + d_2} \right), \tag{5.16}$$

where d_j, given by Eq. (5.13), is the estimate of the squared standard error of $\bar{X}_{tj}$ used by Yuen. (As in Chapter 4, it might appear that $T_{y(u)}^*$ should be used to compute the upper end of the confidence interval, but this is not the case. Details are relegated to the exercises.) When

Table 5.6: Summary of the Bootstrap-t Method for Trimmed Means.

1. Compute the sample trimmed means, $\bar{X}_{t1}$ and $\bar{X}_{t2}$, and Yuen's estimate of the squared standard errors, d_1 and d_2, given by Eq. (5.13).
2. For the jth group, generate a bootstrap sample by randomly sampling with replacement n_j observations from $X_{1j}, \ldots, X_{nj}$, yielding $X_{1j}^*, \ldots, X_{nj}^*$.
3. Using the bootstrap samples just obtained, compute the sample trimmed means plus Yuen's estimate of the squared standard error, and label the results $\bar{X}_{tj}^*$ and d_j^*, respectively, for the jth group.
4. Compute

$$T_y^* = \frac{(\bar{X}_{t1}^* - \bar{X}_{t2}^*) - (\bar{X}_{t1} - \bar{X}_{t2})}{\sqrt{d_1^* + d_2^*}}.$$

5. Repeat steps 2 through 4 B times yielding $T_{y1}^*, \ldots, T_{yB}^*$. $B = 599$ appears to suffice in most situations when $\alpha = 0.05$.
6. Put the $T_{y1}^*, \ldots, T_{yB}^*$ values in ascending order yielding $T_{y(1)}^* \leq \cdots \leq T_{y(B)}^*$. The T_{yb}^* values provide an estimate of the distribution of

$$\frac{(\bar{X}_{t1} - \bar{X}_{t2}) - (\mu_{t1} - \mu_{t2})}{\sqrt{d_1 + d_2}}.$$

7. Set $\ell = \alpha B/2$, rounding to the nearest integer, and let $u = B - \ell$.

The equal-tailed $1 - \alpha$ confidence interval for μ_t is

$$(\bar{X}_{t1} - \bar{X}_{t2} - T_{y(u)}^* \sqrt{d_1 + d_2}, \ \bar{X}_{t1} - \bar{X}_{t2} - T_{y(\ell+1)}^* \sqrt{d_1 + d_2}).$$

($T_{y(\ell)}^*$ will be negative, which is why it is subtracted from $\bar{X}_{t1} - \bar{X}_{t2}$.)

To get a symmetric two-sided confidence interval, replace step 4 with

$$T_y^* = \frac{|(\bar{X}_{t1}^* - \bar{X}_{t2}^*) - (\bar{X}_{t1} - \bar{X}_{t2})|}{\sqrt{d_1^* + d_2^*}}.$$

Set $a = (1 - \alpha)B$, rounding to the nearest integer. The confidence interval for $\mu_{t1} - \mu_{t2}$ is

$$(\bar{X}_{t1} - \bar{X}_{t2}) \pm T_{y(a)}^* \sqrt{d_1 + d_2}.$$

$\alpha = 0.05$, $B = 599$ appears to suffice in terms of probability coverage, and extant simulations suggest that little is gained using $B = 999$. However, in terms of power, $B = 999$ might make a practical difference. For $\alpha < 0.05$, no recommendations about B can be made for the goal of controlling the type I error probability.

In case it helps, Table 5.6 provides an equivalent way of describing how to apply the bootstrap-t to the two-sample case. The summary in Table 5.6 is very similar to the summary of the one-sample bootstrap-t method given in Table 4.4.

The confidence interval given by Eq. (5.16) is just an extension of the equal-tailed bootstrap-t method described in Chapter 4 to the two-sample case. Chapter 4 noted that there are theoretical results suggesting that when computing a two-sided confidence interval, a symmetric two-sided confidence interval should be used instead. A symmetric two-sided confidence interval can be obtained for the situation at hand by replacing T_y given by Eq. (5.14) with

$$T_y = \frac{|\bar{X}_{t1} - \bar{X}_{t2}|}{\sqrt{d_1 + d_2}}$$

and letting T_y^* represent the value of T_y based on the bootstrap sample denoted by C_{ij}^*. As before, repeatedly generate bootstrap samples yielding $T_{y1}^*, \ldots, T_{1B}^*$. Now, however, set $a = (1 - \alpha)B$, rounding to the nearest integer, in which case the critical value is $c = T_{y(a)}^*$, and the $1 - \alpha$ confidence interval is

$$(\bar{X}_{t1} - \bar{X}_{t1} - c\sqrt{d_1 + d_2}, \bar{X}_{t1} - \bar{X}_{t2} + c\sqrt{d_1 + d_2}). \tag{5.17}$$

A variation of this approach was derived by Guo and Luh (2000), which is based in part on a transformation stemming from Hall (1992). The basic idea is to transform Yuen's test statistic so that it is approximated reasonably well by a Student's t distribution. Results reported by Keselman, Othman, Wilcox, and Fradette (2004) indicate, however, that it is preferable to approximate the null distribution of the test statistic used by Guo and Luh using a bootstrap-t method. Among the situations considered by Keselman et al., a bootstrap-t method was found to perform relatively well when the amount of trimming is set at 10% or 15%. However, with small and unequal sample sizes, situations occur where the method is unsatisfactory when using 10% trimming. More details are given in Section 5.4.2. In practical terms, it currently seems that a percentile bootstrap method is a bit more satisfactory and that there can be an advantage in using 20% trimming in terms of controlling the probability of a type I error. Further evidence for preferring the use of a percentile bootstrap method is reported by Özdemir, Wilcox, and Yildiztepe (2010) who report simulation results when distributions differ in skewness. With no trimming, the bootstrap-t method studied by Keselman et al. can be highly unsatisfactory.

5.3.3 R Functions yuenbt and yhbt

The R function

$$\text{yuenbt(x,y,tr=0.2,alpha=0.05,nboot=599,side=F)},$$

computes a $1 - \alpha$ confidence interval for $\mu_{t1} - \mu_{t2}$ using the bootstrap-t method, where the default amount of trimming (tr) is 0.2, the default value for α is 0.05, and the default value for nboot (B) is 599. So far, simulations suggest that in terms of probability coverage, there is little or no advantage to using $B > 599$ when $\alpha = 0.05$. However, there is no recommended choice for B when $\alpha < 0.05$ simply because little is known about how the bootstrap-t performs for this special case. Finally, the default value for side is F, for false, indicating that the equal-tailed two-sided confidence interval is to be used. Using side=T results in the symmetric two-sided confidence interval.

■ Example

For the ozone data in Table 5.3, yuenbt reports that the 0.95 symmetric two-sided confidence interval for the difference between the trimmed means is $(4.75, 23.4)$. In contrast, the Yuen–Welch method yields a 0.95 confidence interval equal to $(5.3, 22.85)$. The equal-tailed bootstrap-t method yields a 0.95 confidence interval of $(3.78, 21.4)$. The symmetric two-sided confidence interval for the difference between the means is obtained with the command yuenbt(x,y,0.,side=T), assuming the data are stored in the R vectors x and y, and the result is $(1.64, 21.2)$. In contrast, yuenbt(x,y,0.) yields an equal-tailed confidence interval for the difference between the means of $(1.87, 21.6)$. Note that the lengths of the confidence intervals for the difference between the trimmed means are similar to each other and the length of the confidence interval for the difference between the means, but the next illustration demonstrates that this is not always the case.

■ Example

Table 5.7 shows data from a study dealing with the effects of consuming alcohol. (The data are from a portion of a study conducted by M. Earleywine.) Two groups of participants reported hangover symptoms the morning after consuming equal amounts of alcohol in a laboratory. Group 1 was a control and group 2 consisted of sons of alcoholic fathers. Figure 5.6 shows an adaptive kernel density estimate for the two groups. Note that the shapes are similar to an exponential distribution suggesting that confidence intervals for the difference between the means, with probability coverage close to the nominal level, might be difficult to obtain. In fact, even using 20% trimming, the Yuen–Welch method might yield inaccurate probability coverage, as already noted. The main point here is that the length of the confidence intervals based on the

Table 5.7: The Effect of Alcohol.

Group 1	0	32	9	0	2	0	41	0	0	0
	6	18	3	3	0	11	11	2	0	11
Group 2	0	0	0	0	0	0	0	0	1	8
	0	3	0	0	32	12	2	0	0	0

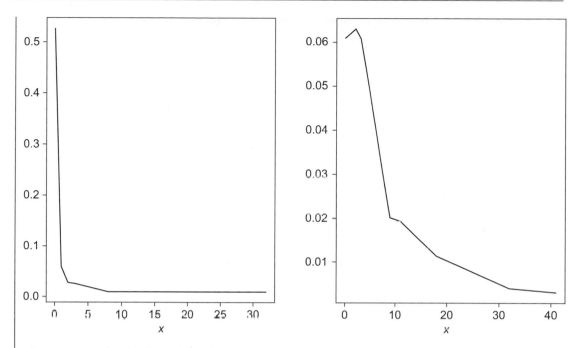

Figure 5.6: Adaptive kernel density estimates for the two groups in the study looking at sons of alcoholic fathers.

Yuen–Welch method can differ substantially from the length of the confidence interval using the bootstrap-t method. The Yuen–Welch method yields a 0.95 confidence interval equal to $(-0.455, 7.788)$ with a p-value of .076. In contrast, the equal-tailed bootstrap-t yields a 0.95 confidence interval of $(-4.897, 7.255)$. The ratio of the lengths of the confidence intervals is 0.678. The symmetric bootstrap-t confidence interval is $(-1.357, 8.691)$, and its length, divided by the length of the other bootstrap confidence interval, is .83

■

Although 20% trimming performs well under normality in terms of power and efficiency, situations might be encountered where it is desired to use 10% and 15% trimming instead. If this is the case, one strategy is to use the R function

$$\text{yhbt}(x, y, tr = 0.15, alpha = 0.05, nboot = 600, SEED = T, PV = F),$$

which uses the bootstrap-t version of the test statistic derived by Guo and Luh (2000) that was studied by Keselman et al. (2004). By default, 15% trimming is used. The function returns a confidence interval having probability coverage specified by the argument alpha. A p-value is returned if the argument PV=T, but on occasion this results in a numerical error causing the function to terminate. And even when not computing a confidence interval, situations are encountered where the function is unable to compute a confidence interval. The method is *not* recommended when the goal is to compare means. Another possibility is to use the percentile bootstrap method in Section 5.4.2. Limited studies suggest that even with 10% trimming, there is little or no advantage to using yhbt rather than a percentile bootstrap method. Moreover, the percentile bootstrap has faster execution time and computational problems do not arise when computing a *p*-value or a confidence interval.

5.3.4 Measuring Effect Size: Robust Analogs of Cohen's d

A common way of characterizing the extent two distributions differ is with the measure of effect size

$$\delta = \frac{\mu_1 - \mu_2}{\sigma},$$

where by assumption, $\sigma_1 = \sigma_2 = \sigma$. That is, homoscedasticity is assumed and the common variance is denoted by σ^2. Cohen (1988) suggests that as a general guide, $\delta = 0.2, 0.5$, and 0.8 correspond to small, medium, and large effect sizes, respectively, and often this suggestion is followed.

The usual estimate of δ, popularly knows as Cohen's d, is

$$d = \frac{\bar{X}_1 - \bar{X}_2}{s},$$

where $s^2 = [(n_1 - 1)s_1^2 + (n_2 - 1)s_2^2]/(n_1 + n_2 - 2)$ estimates the assumed common variance.

There are fundamental concerns regarding this measure of effect size. The first is that when dealing with heavy-tailed distributions, δ can be small even when from a graphical perspective the difference between the two distributions appears to be relatively large. The left panel of Figure 5.7 shows two normal distributions, both having variance 1, for which $\delta = 0.8$, which is often viewed as a large effect size. But now look at the right panel where again the difference between the means is 0.8. Despite the similarity with the left panel, $\delta = 0.24$, which is typically considered to be small. The reason δ is substantially smaller in the right panel is that the two distributions are contaminated normal distributions, which have variance 10.9. And outliers can result in d, the estimate of δ, being relatively small as well. A second general concern is that δ is based on the mean and variance, which are not robust. Yet another concern is that it assumes equal variances.

Algina, Keselman, and Penfield (2005) suggest using a generalization of δ based on 20% trimmed means and Winsorized variances. Their approach is homoscedastic in the sense that the groups are assumed to have a common (population) Winsorized variance. Moreover, the Winsorized variances are re-scaled so that under normality they estimate the variance. With 20% trimming, this means that the Winsorized variance is divided by 0.4121. That is, under

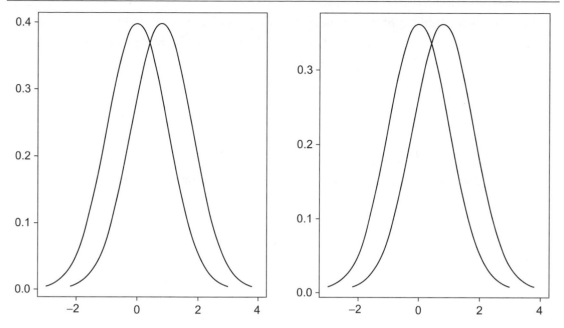

Figure 5.7: The left panel shows two normal distributions for which the measure of effect size δ is 0.8, which is often taken to be a large value. For the right panel, $\delta = 0.24$, despite the similarity to the left panel, illustrating that slight changes in the tails of the distributions can have a major impact on the magnitude of δ.

normality, $s_w^2/0.4142$ estimates σ^2. So now δ becomes

$$\delta_t = 0.642\frac{\bar{X}_{t1} - \bar{X}_{t2}}{S_w},$$

where

$$S_W^2 = \frac{(n_1 - 1)s_{w1}^2 + (n_2 - 1)s_{w2}^2}{n_1 + n_2 - 2}$$

is the pooled Winsorized variance. Under normality, and when the variances are equal, $\delta = \delta_t$. If the Winsorized variances are not equal, Algina et al. suggest using both

$$\delta_{t1} = 0.642\frac{\bar{X}_{t1} - \bar{X}_{t2}}{s_{w1}},$$

and

$$\delta_{t2} = 0.642\frac{\bar{X}_{t1} - \bar{X}_{t2}}{s_{w2}}.$$

A possible concern, however, is that δ_{t1} might suggest a large effect size, whereas δ_{t2} suggests the opposite.

A robust, heteroscedastic approach to measuring effect size was suggested by Wilcox and Tian (2011), which is based on a generalization of the notion of explanatory power (Doksum & Samarov, 1995). From a regression perspective, if $\hat{Y}$ is the predicted value of Y, given X, explanatory power is

$$\xi^2 = \frac{\sigma^2(\hat{Y})}{\sigma^2(Y)},$$

the variance of the predicted Y values divided by the variance of the observed Y values. (If $\hat{Y}$ is taken to be the usual least squares regression line, then $\xi^2 = \rho^2$, where ρ is Pearson's correlation.) Given that an observation is randomly sampled from the jth group, take $\hat{Y} = \mu_j$, in which case

$$\sigma^2(\hat{Y}) = \sum (\mu_j - \bar{\mu})^2,$$

where $\bar{\mu} = (\mu_1 + \mu_2)/2$. Momentarily assume that with probability 1, equal sample sizes are used. Let $\sigma^2(Y|j)$ be the variance of Y given that an observation is randomly sampled from the jth group, and let $\sigma^2(Y)$ be the unconditional variance of Y. Based on the random sample Y_{ij} ($i = 1, \ldots, n$; $j = 1, 2$), $\sigma^2(Y)$ is estimated with $\hat{\sigma}^2(Y)$, the usual sample variance based on these $2n$ (pooled) observations. So the estimate of ξ^2 is

$$\hat{\xi}^2 = \frac{\hat{\sigma}^2(\hat{Y})}{\hat{\sigma}^2(Y)}.$$

Now consider how to estimate ξ^2 when unequal sample sizes are used. First, it is stressed that a key component of the approach used here is defining $\sigma^2(Y)$ in terms of situations where equal sample sizes are used with probability 1. Put another way, $\sigma^2(Y)$ is the estimand associated with the sample variance of the pooled Y_{ij} values when $n_1 = n_2$. Given how $\sigma^2(Y)$ is defined, the problem is finding a reasonable estimate of $\sigma^2(Y)$ when dealing with unequal sample sizes. Kulinskaya & Staudte (2006, p. 101) conclude that a natural generalization of δ to the heteroscedastic case does not appear to be possible without taking into account the relative sample sizes. A simple strategy is to again estimate $\sigma^2(Y)$ with the sample variance based on all $n_1 + n_2$ Y_{ij} values, even when $n_1 \neq n_2$. But this estimation method can be shown to be unsatisfactory: the resulting estimate of ξ^2 can be severely biased. To deal with this, suppose the sample sizes are $n_1 < n_2$ for groups 1 and 2, respectively. If we randomly sample (without replacement) n_1 observations from the second group, we have equal sample sizes from both groups resulting in a satisfactory estimate of ξ^2. That is, use the estimation method for the equal sample case, where the both groups have sample size n_1. To use all of the data in the second group, repeat this process K times yielding a series of estimates for ξ^2, which are then averaged to get a final estimate, which we label $\hat{\xi}^2$. The estimate of ξ is just

$$\hat{\xi} = \sqrt{\hat{\xi}^2}$$

and is called the *explanatory measure of effect size*.

To get a robust version of ξ^2, simply replace the mean with some robust measure of location and replace $\sigma^2(Y)$ with some robust measure of variation. Here, unless stated otherwise, a 20% trimmed mean and a 20% Winsorized variance are used, where the Winsorized variance is re-scaled to estimate the usual variance, σ^2, when sampling from a normal distribution. For 20% Winsorization, this means that rather than compute the Winsorized variance of the pooled Y_{ij} values with say s_{wy}^2, use $s_{wy}^2/0.4121$. It is noted that under normality and homoscedasticity, $\delta = 0.2, 0.5$, and 0.8 roughly correspond to $\xi = 0.15, 0.35$, and 0.50, respectively. That is, if for example $\delta = 0.5$ is viewed as a medium effect size, as is often done, this corresponds to $\xi = 0.35$.

5.3.5 R Functions akp.effect, yuenv2, and ees.ci

The R function

$$\text{akp.effect(x,y,tr=0.2)}$$

estimates the effect size δ_t. The function automatically re-scales the Winsorized variance so that, based on the amount of trimming used, it estimates the usual variance under normality. The R function

$$\text{yuenv2(x,y,tr=0.2,alpha=0.05)}$$

is exactly like the R function yuen for comparing trimmed means, only the explanatory measure of effect size, $\hat{\xi}$, is reported. The R function

$$\text{ees.ci(x,y,SEED=T,nboot-400,tr=0.2,alpha=0.05)}$$

computes a $1 - \alpha$ confidence interval for $|\xi|$. A percentile bootstrap method is used, but modified so that if the p-value is greater than α when testing $H_0 : \mu_{t1} = \mu_{t2}$ with Yuen's method, the lower end of the $1 - \alpha$ confidence interval is set equal to zero. (If the goal is to compute a confidence interval for ξ rather than $|\xi|$, a percentile bootstrap method can be unsatisfactory.)

■ Example

A practical issue is the effect of ignoring heteroscedasticity when using δ rather than ξ to measure effect size. That is, can the choice of method alter the extent an effect size is deemed to be large? For illustrative purposes, we adopt the usual convention that $\delta = 0.2, 0.5$, and 0.8 correspond to small, medium, and large effect sizes, respectively. As already noted, under normality and homoscedasticity, these values roughly correspond to $\xi = 0.15, 0.3$, and 0.5. Note that if the group with the larger sample size also has the larger variance, this results in a relatively small value for d. To illustrate how d compares to $\hat{\xi}$, simulations were used to estimate both effect sizes with $n_1 = 80$ and

$n_2 = 20$, where the first group has a normal distribution with mean 0.8 and standard deviation 4, and the second group has a standard normal distribution. Based on 1000 replications, the median value of d was 0.22, which is typically considered to be a small effect size. (The mean value of d was nearly identical to the median.) The median value of $\hat{\xi}$ was 0.40, which suggests a medium effect size. So even under normality, a heteroscedastic measure of effect size can make a practical difference. If instead the first group has standard deviation 1 and the second has standard deviation 4, now the median estimates are 0.42 and 0.32. That is, in contrast to the first situation, the choice between homoscedastic and heteroscedastic measures of effect size makes little difference. If instead $n_1 = n_2 = 20$, now the median d value is 0.30, a somewhat small effect size, and the median $\hat{\xi}$ value is 0.34, which suggests a medium effect size instead. The effect of ignoring heteroscedasticity is less of an issue with equal sample sizes, compared with the first situation considered, but it has practical consequences. ∎

■ Example

In a study of sexual attitudes, 1327 men and 2282 women were asked how many sexual partners they desired over the next 30 years. (The data used in this example, supplied by Lynn Miller, are stored in the file miller.dat and can be downloaded from the author's web page given in Chapter 1.) Welch's test returns a p-value of .30, but Yuen's test has a p-value less than .001. Cohen's d is estimated to be less than 0.0001. In contrast, $\hat{\delta}_t = 0.48$, suggesting a medium effect size and $\hat{\xi} = 0.47$ suggesting a large effect size. ∎

5.3.6 Comments on Measuring Effect Size

It is not being suggested that ξ be used as the only measure of effect size. It would seem that in various situations, multiple perspectives are needed to get a good understanding of how groups compare. This might include graphical methods such as boxplots, plots of the distributions returned by the R function g2plot, and the shift function. And differences in variation might be used as well. Another potentially useful perspective is $P(X < Y)$, the probability that a randomly sampled observation from the first group is smaller than a randomly sampled observation from the second, which is discussed in Section 5.7.

5.4 Inferences Based on a Percentile Bootstrap Method

In recent years, inferences based on a percentile bootstrap method have been found to be particularly effective when working with a wide range of robust estimators. When comparing two independent groups, the method is applied as follows. First, generate bootstrap samples from each group as described in Table 5.6. Let $\hat{\theta}_j^*$ be the bootstrap estimate of θ_j, where θ_j is any parameter of interest associated with the jth group ($j = 1, 2$). Set

$$D^* = \hat{\theta}_1^* - \hat{\theta}_2^*.$$

Repeat this process B times yielding $D_1^*, \ldots, D_B^*$, let ℓ be $\alpha B/2$, rounded to the nearest integer, and let $u = B - \ell$, in which case an approximate $1 - \alpha$ confidence interval for $\theta_1 - \theta_2$ is

$$(D_{(\ell+1)}^*, \ D_{(u)}^*),$$

where $D_{(1)}^* \leq \cdots \leq D_{(B)}^*$.

The theoretical foundation for the method is similar to the theoretical foundation in the one-sample case described in Chapter 4. Imagine the goal is to test

$$H_0 : \theta_1 = \theta_2.$$

For the bootstrap estimates $\hat{\theta}_1^*$ and $\hat{\theta}_2^*$, let

$$p^* = P(\hat{\theta}_1^* > \hat{\theta}_2^*),$$

If the null hypothesis is true, then asymptotically (as both n and B get large), p^* has a uniform distribution. Consequently, reject H_0 if $p^* \leq \alpha/2$ or if $p^* \geq 1 - \alpha/2$. Although p^* is not known, it is readily estimated. Let A be the number of values among $D_1^*, \ldots, D_B^*$ that are greater than zero. Then an estimate of p^* is

$$\hat{p}^* = \frac{A}{B}.$$

For convenience, set

$$\hat{p}_m^* = \min(p^*, \ 1 - p^*).$$

Then $2\hat{p}_m^*$ is an estimate of what Liu and Singh (1997) call the generalized p-value, and H_0 is rejected if

$$2\hat{p}_m^* \leq \alpha.$$

This last equation leads to the confidence interval given in the previous paragraph.

5.4.1 Comparing M-Estimators

This section comments on the special case there the goal is to compare M-measures of location. Chapter 4 noted that, based on simulations conducted so far, the best approach to computing a confidence interval for μ_m, the M-measure of location, is to use a percentile bootstrap method. When comparing M-measures of location corresponding to two independent groups, the percentile bootstrap again is the best method based on current results. As in the one-sample case, there are bootstrap methods that have not been examined via simulations when comparing M-measures of location, so it is not being suggested that all other bootstrap techniques have no practical value for the problem at hand. A confidence interval based on an estimate of the standard error will provide good probability coverage when the sample sizes are sufficiently large, assuming the estimated difference is normally distributed, but it is unknown just how large the sample sizes should be before this approach can be recommended, particularly when distributions are skewed. If both distributions are symmetric, confidence intervals based on estimated standard errors seem to have merit when Student's *t*-distribution is used to determine an appropriate critical value, but there is no good decision rule, based on available empirical data, whether distributions are sufficiently symmetric. (One could test the assumption that distributions are symmetric, but how much power should such a test have to justify the use of a method that assumes symmetric distributions?) A bootstrap-t method might also be advantageous in certain situations, but it is unknown when, if ever, this approach should be used over the percentile bootstrap. When sample sizes are small, all indications are that the percentile bootstrap is best, so it is recommended until there is good evidence that some other method should be used instead.

5.4.2 Comparing Trimmed Means and Medians

When comparing trimmed means, and the amount of trimming is at least 20%, it currently seems that a percentile bootstrap method is preferable to the bootstrap-t method in Section 5.3.2. With a sufficiently small amount of trimming, a bootstrap-t method provides more accurate results, but there is uncertainty about when this is the case. (Comments on using 10% trimming are given at the end of this section.)

For the special case where the goal is to compare medians, a slight extension of the percentile bootstrap method is needed in case there are tied values. Let M_1^* and M_2^* be the bootstrap sample medians. Let

$$p^* = P(M_1^* > M_2^*) + 0.5P(M_1^* = M_2^*).$$

So among B bootstrap samples from each group, if A is the number of times $M_1^* > M_2^*$, and C is the number of times $M_1^* = M_2^*$, the estimate of p^* is

$$\hat{p}^* = \frac{A}{B} + 0.5\frac{C}{B}.$$

As usual, the *p*-value is

$$2\min(\hat{p}^*, 1 - \hat{p}^*).$$

In terms of controlling the Type I error probability, all indications are that this method performs very well regardless of whether tied values occur (Wilcox, 2006a). And in terms of handling tied values, this is the only known method that performs well in simulations.

Section 5.3.2 mentioned a bootstrap-t method (that is performed by the R function yhbt) that is based in part on a test statistic derived by Guo and Luh (2000). As previously noted, Keselman et al. (2004) found that it performs reasonably well in simulations when using 10% and 15% trimming. To extend slightly their results, consider a situation where $n_1 = 40$ observations are sampled from a standard normal distribution. And for the second group $n_2 = 20$ observations are sampled from a lognormal distribution shifted so that the trimmed mean is zero, after which the scale is increased by multiplying all observations by 4. When testing at the 0.05 level, and 10% trimming is used, the actual level of the bootstrap-t method is approximately 0.066 compared with 0.050 when using a percentile bootstrap method (based on a simulation with 1000 replications). Reducing the first sample size to $n_1 = 20$ and the second to $n_2 = 10$, the estimates are now 0.082 and 0.074, respectively. Increasing the amount of trimming to 0.2, again using sample sizes $n_1 = 20$ and $n_2 = 10$, the estimates are 0.081 and 0.063. So at least in some situations, the percentile bootstrap method has a bit of an advantage when using 10% trimming. And increasing the amount of trimming from 10% to 20% can improve control over the type I error probability.

5.4.3 R Functions trimpb2, pb2gen, m2ci, and medpb2

When comparing independent groups, the R function

$$pb2gen(x,y,alpha=0.05,nboot=2000,est=onestep,...)$$

can be used to compute a confidence interval for the difference between any two measures of location or scale using the percentile bootstrap method. As usual, x and y are any R vectors containing data. The default value for α is 0.05, the default for B (nboot) is 2000. The last argument, est, is any R function that is of interest. The default value for est is onestep, which is the R function described in Chapter 3 for computing a one-step M-estimator. The command pb2gen(dat1,dat2,est=mom), for example, will compare the modified one-step M-estimators based on the data stored in the R variables dat1 and dat2.

For convenience, a specific function for comparing robust M-estimators based on Huber's Ψ is provided. It has the general form

$$m2ci(x,y,nboot=1000,alpha=0.05,bend=1.28,os=F),$$

where the default value for B is nboot $= 1000$, the default value for α is 0.05, the default value for os is F, for false, meaning that the fully iterated M-estimator is used, and the default bending constant is 1.28.[1] Setting os=T means that the one-step M-estimator is used instead.

---

[1] An old version of this function used $B = 399$ which seems to suffice, in terms of probability coverage, when computing a 0.95 confidence interval.

If, for example, it is desired to compute a 0.95 confidence interval using the one-step M-estimator, type the command m2ci(x,y,os=T).

■ Example

For the ozone data in Table 5.3, m2ci returns a 0.95 confidence interval of $(3.67, 21.51)$ for the difference between the M-measures of location. Using the one-step M-estimator instead, the 0.95 confidence interval is $(3.64, 22.26)$.

■

Medians can be compared with the R function pb2gen by setting the argument est=median and trimmed means can be compared by setting est=tmean. But for convenience, the R function

$$\text{medpb2(x,y,alpha=0.05,nboot=2000)}$$

is supplied, which is designed specifically for comparing medians. And the R function

$$\text{trimpb2(x,y,alpha=0.05,nboot=2000)}$$

defaults to comparing 20% trimmed means.

5.5 Comparing Measures of Scale

In some situations, there is interest in comparing measures of scale. Based purely on efficiency, various robust estimators of scale have appeal. First, however, attention is focused on comparing the variances.

5.5.1 Comparing Variances

We begin with the goal of testing

$$H_0 : \sigma_1^2 = \sigma_2^2,$$

the hypothesis that two independent groups have equal variances. Numerous methods have been proposed. Virtually all have been found to be unsatisfactory with small to moderate sample sizes.

A variation of the percentile bootstrap method (Wilcox, 2002) that performs relatively well is performed as follows. Set $n_m = \min(n_1, n_2)$ and for the jth group ($j = 1, 2$), take a bootstrap sample of size n_m. Ordinarily, we take a bootstrap sample of size n_j from the jth group, but when sampling from heavy-tailed distributions, and when the sample sizes are unequal, control over the probability of a type I error can be extremely poor for the situation at hand.

Next, for each group, compute the sample variance based on the bootstrap sample and set D^* equal to the difference between these two values. Repeat this $B = 599$ times yielding 599 bootstrap values for D, which we label $D_1^*, \ldots, D_{599}^*$. As usual, when writing these values in ascending order, we denote this by $D_{(1)}^* \leq \cdots \leq D_{(B)}^*$. Then an approximate 0.95 confidence interval for the difference between the population variances is

$$(D_{(\ell+1)}^*, \ D_{(u)}^*), \tag{5.18}$$

where for $n_m < 40$, $\ell = 6$, and $u = 593$; for $40 \leq n_m < 80$, $\ell = 7$, and $u = 592$; for $80 \leq n_m < 180$, $\ell = 10$, and $u = 589$; for $180 \leq n_m < 250$, $\ell = 13$, and $u = 586$; and for $n_m \geq 250$, $\ell = 15$, and $u = 584$.

The method just described is based on a strategy similar to Gosset's derivation of Student's t: assume normality and then make adjustments so that for small sample sizes, accurate probability coverage is obtained. This method appears to perform reasonably well under nonnormality, but exceptions can occur when the distributions differ in skewness and the sample sizes are small. What appears to be more satisfactory is to use the method just described, only with $B = 1000$ and a corresponding adjustment to ℓ and u. Using $B = 999$, the actual level of the test can be substantially worse. A positive feature of this method is that in situations where the control over the type I error probability is not quite satisfactory due to small sample sizes, it appears to provide a reasonably good test of the hypothesis that the median value of s_1^2 is equal to the median value of s_2^2, but more research is needed to establish the extent this is the case.

When sampling from a distribution that is not too skewed and not very heavy-tailed, the method in Shoemaker (2003) might be used instead. Herbert et al. (2011) derived yet another method for comparing variances. How well it performs under nonnormality, including situations where distributions differ in skewness, needs more research. In terms of controlling the probability of a type I error, any practical advantages the method might have over the modified percentile bootstrap method have not been determined.

5.5.2 R Function comvar2

The R function

$$\text{comvar2(x,y,nboot=1000,SEED=T)}$$

compares variances using the bootstrap method just described. The method can only be applied with $\alpha = 0.05$; modifications based on other α values have not been derived. The function returns a 0.95 confidence interval for $\sigma_1^2 - \sigma_2^2$ plus an estimate of $\sigma_1^2 - \sigma_2^2$ based on the difference between the sample variances, $s_1^2 - s_2^2$, which is labeled vardif.

5.5.3 Comparing Biweight Midvariances

For some robust measures of scale, the percentile bootstrap method, described in Section 5.4, has been found to perform well. In particular, Wilcox (1993a) found that it gives good results when working with the biweight midvariance. (Other methods were considered but found to be unsatisfactory, so they are not discussed.) There is some indirect evidence that it will give good results when working with the percentage bend midvariance, but this needs to be checked before it can be recommended.

5.5.4 R Function b2ci

Robust measures of scale are easily compared with the R function pb2gen in Section 5.4.3. For convenience, the function

$$b2ci(x,y,alpha=0.05,nboot=2000,est=bivar)$$

has been supplied; it defaults to comparing the biweight midvariances. (When using pb2gen, setting est=bivar returns the same results when using the default settings of b2ci.)

■ **Example**

> For the ozone data in Table 5.3, 0.95 confidence interval returned by the R function b2ci is $(-538, -49)$ with a p-value of .012.

■

5.6 Permutation Tests

This section describes a permutation test for comparing the distributions corresponding to two independent groups, an idea introduced by R. A. Fisher in the 1930s. The method is somewhat similar to bootstrap techniques, but it accomplishes a different goal, as will become evident. There are many extensions and variations of the method about to be described, including a range of techniques aimed at multivariate data (e.g., Good, 2000; Pesarin, 2001; Rizzo & Székely, 2010), but only the basics are included here.

The permutation test in this section can be used with virtually any measure of location or scale, but regardless of which measure of location or scale is used, in essence the goal is to test the hypothesis that the groups under study have identical distributions. To illustrate the basics, the method is first described using means. The steps are as follows:

1. Compute $d = \bar{X}_1 - \bar{X}_2$, the difference between the sample means, where the sample sizes are n_1 and n_2.
2. Pool the data.

3. Consider any permutation of the pooled data, compute the sample mean of the first n_1 observations, compute the sample mean using the remaining n_2 observations, and compute the difference between these sample means.
4. Repeat the previous step for all possible permutations of the data yielding, say, L differences: $\hat{\delta}_1, \ldots, \hat{\delta}_L$.
5. Put these L differences in ascending order yielding $\hat{\delta}_{(1)} \leq \cdots \leq \hat{\delta}_{(L)}$.
6. Reject the hypothesis of identical distributions if $d < \hat{\delta}_{(\ell+1)}$ or if $d > \hat{\delta}_{(u)}$, where $\ell = \alpha L/2$, rounded to the nearest integer, and $u = L - \ell$.

Although this variation of the permutation test is based on the sample mean, it is known that it does not provide satisfactory inferences about the population means. In particular, it does not control the probability of a Type I error when testing H_0: $\mu_1 = \mu_2$, and it does not yield a satisfactory confidence interval for $\mu_1 - \mu_2$. For example, Boik (1987) established that when the goal is to compare means, unequal variances can affect the probability of a Type I error, even under normality, when testing H_0: $\mu_1 = \mu_2$. If the sample means are replaced by the sample variances, it can be seen that differences between the population means can affect the probability of a Type I error even when the population variances are equal. (The details are left as an exercise.) However, the method provides an exact distribution-free method for testing the hypothesis that the distributions are identical. For results on using a permutation test with the mean replaced by a robust estimator, see Lambert (1985). When the goal is to compare medians, again a permutation test can be unsatisfactory (Romano, 1990).

In practice, particularly with large sample sizes, generating all permutations of the pooled data can be impractical. A simple method for dealing with this problem is to simply use B random permutations instead. Now proceed as described above, only L is replaced by B.

5.6.1 R Function permg

The R function

$$\text{permg(x,y,alpha=0.05,est=mean,nboot=1000)}$$

performs the permutation test based on B random permutations of the pooled data. (The argument nboot corresponds to B.) By default, means are used, but any measures of location or scale can be used via the argument est.

5.7 Inferences About a Probabilistic Measure of Effect Size

There is another approach to comparing two independent groups that deserves consideration. Let

$$p = P(X_{i1} < X_{i2}),$$

be the probability that a randomly sampled observation from the first group is smaller than the a randomly sampled observation from the second. When there is no difference between the groups, and the distributions are identical, $p = 1/2$. The value of p has a natural interest, and some have argued that in many situations it is more interesting than the difference between any two measures of location (e.g., Cliff, 1993). Additional arguments for comparing groups based on p can be found in Acion, Peterson, Temple, and Arndt (2006), Kraemer and Kupfer (2006), and Vargha and Delaney (2000). For example, in clinical trials, of interest is the probability that method A is more effective than method B.

The best-known approach to comparing two independent groups, which is based on an estimate of p, is the Wilcoxon–Mann–Whitney test. The method might appear to provide a reasonable way of testing

$$H_0 : p = 0.5,$$

but a fundamental concern is that when distributions differ, there are general conditions under which the Wilcoxon–Mann–Whitney test uses the wrong standard error. More precisely, the standard error used by the Wilcoxon–Mann–Whitney is derived under the assumption that groups have identical distributions, and when the distributions differ, under general conditions the derivation no longer holds. Modern (heteroscedastic) methods are based on using an estimate of the correct standard error regardless of whether the distributions differ. Pratt (1964) established that the Wilcoxon–Mann–Whitney test is biased and documents its inability to control the probability of a type I error when testing $H_0 : p = 0.5$.

Often the Wilcoxon–Mann–Whitney test is described as a method for comparing the marginal medians, but it can be unsatisfactory in this regard (e.g., Fung, 1980; Hettmansperger, 1984). To elaborate, let $D = X - Y$ and let θ_D be the population median associated with D and let θ_X and θ_Y be the population medians associated with X and Y, respectively. It is left as an exercise to show that under general conditions, $\theta_D \neq \theta_X - \theta_Y$. Although the Wilcoxon–Mann–Whitney test does not provide a direct test of the hypothesis that X and Y have equal medians, it is based on an estimate of p, and when $p = 0.5$, $\theta_D = 0$.

Various attempts have been made to improve on the Wilcoxon–Mann–Whitney test, but not all of them are listed here. Interested readers can refer to Baumgartner, Weiss, and Schindler (1998), Ryu and Agresti (2008), Zhou (2008), Fligner and Policello (1981), Newcombe (2006a) plus the references they cite (cf. Neuhäuser, 2003).

Here, three methods are described, all of which appear to be viable options. The first method, which assumes that ties occur with probability zero, was derived by Mee (1990) and provides a confidence interval for p that compares well with many of the alternative methods that have been proposed. The computational steps are summarized in Table 5.8. (The quantity U in Table 5.8 is the Wilcoxon–Mann–Whitney statistic while $\hat{p}$ is an unbiased estimate of p. In terms of testing H_0: $p = 1/2$, Mee's method can maintain relatively high power when

Table 5.8: Mee's Confidence Interval for *p*.

Set $U_{ij} = 1$ if $X_{i1} < X_{j2}$ and $U_{ij} = 0$ if $X_{i1} \geq X_{j2}$. Let

$$U = \sum_{i=1}^{n_1} \sum_{j=1}^{n_2} U_{ij}$$

$$\hat{p} = \frac{U}{n_1 n_2}$$

$$\hat{p}_1 = \sum_{i=1}^{n_1} \sum_{j=1}^{n_2} \sum_{k \neq i}^{n_1} \frac{U_{ij} U_{kj}}{n_1 n_2 (n_1 - 1)}$$

$$\hat{p}_2 = \sum_{i=1}^{m} \sum_{j=1}^{n} \sum_{k \neq j}^{n} \frac{U_{ij} U_{ik}}{n_1 n_2 (n_2 - 1)}$$

$$b_1 = \frac{\hat{p}_1 - \hat{p}^2}{\hat{p} - \hat{p}^2}$$

$$b_2 = \frac{\hat{p}_2 - \hat{p}^2}{\hat{p} - \hat{p}^2}$$

$$A = \frac{(n_1 - 1)b_1 + 1}{1 - n_2^{-1}} + \frac{(n_2 - 1)b_2 + 1}{1 - n_1^{-1}}$$

$$\hat{N} = \frac{n_1 n_2}{A}$$

$$C = \frac{z_{1-\alpha/2}^2}{\hat{N}}$$

$$D = \sqrt{C[\hat{p}(1 - \hat{p}) + 0.25C]},$$

where $z_{1-\alpha/2}$ is the $1 - \alpha/2$ quantile of the standard normal distribution. The end points of the $1 - \alpha$ confidence interval for *p* are given by

$$\frac{\hat{p} + .5C \pm D}{1 + C}.$$

distributions are heavy-tailed. There are situations, however, where testing $H_0 : p = 1/2$ can result in low power compared with the Yuen–Welch method, the details of which are left as an exercise.

5.7.1 R Function mee

The R function

$$\text{mee}(x, y, \text{alpha} = 0.05).$$

performs the calculations in Table 5.8. As usual, the default value for α is 0.05, and x and y are any R vectors containing data. The function returns $\hat{p}$, an estimate of *p*, plus a $1 - \alpha$

confidence interval for p. (The function checks for tied values and prints a warning if any are found.)

■ Example

For the ozone data in Table 5.3, the R function Mee reports that the estimate of p is $\hat{p} = 0.239$, and the 0.95 confidence interval is $(0.116, 0.429)$. Thus, H_0: $p = 1/2$ would be rejected, and the data suggest that the weight gain for a typical rat in an ozone environment will be less than the weight gain for a typical rat in the control group.

■

5.7.2 The Cliff and Bruner–Munzel Methods: Handling Tied Values

This section describes two additional methods for testing H_0: $p = .5$. The first was derived by Cliff (1996), and the other was derived by Brunner and Munzel (2000). Both allow heteroscedasticity and are designed to perform well when tied values can occur. And when there are no tied values, it currently seems that they compete well with Mee's method.

Mee's method, which assumes tied values do not occur, is aimed at making inferences about $p = P(X_{i1} < X_{i2})$. With tied values, a different formulation is required. Let

$$p_1 = P(X_{i1} > X_{i2}),$$
$$p_2 = P(X_{i1} = X_{i2}),$$

and

$$p_3 = P(X_{i1} < X_{i2}).$$

For convenience, set $P = p_3 + .5p_2 = p + .5p_2$. The usual generalization to tied values replaces H_0: $p = .5$ with

$$H_0 : P = .5.$$

(So when tied values occur with probability zero, this hypothesis becomes H_0: $p = .5$.)

Cliff's Method

Cliff prefers a slightly different perspective, namely, testing

$$H_0 : \delta = p_1 - p_3 = 0.$$

It is readily verified that $\delta = 1 - 2P$.

For the ith observation in group 1 and the hth observation in group 2, let

$$d_{ih} = \begin{cases} -1 & \text{if } X_{i1} < X_{h2} \\ 0 & \text{if } X_{i1} = X_{h2} \\ 1 & \text{if } X_{i1} > X_{h2}. \end{cases}$$

An estimate of $\delta = P(X_{i1} > X_{i2}) - P(X_{i1} < X_{i2})$ is

$$\hat{\delta} = \frac{1}{n_1 n_2} \sum_{i=1}^{n_1} \sum_{h=1}^{n_2} d_{ih}, \tag{5.19}$$

the average of the d_{ih} values. Let

$$\bar{d}_{i.} = \frac{1}{n_2} \sum_h d_{ih},$$

$$\bar{d}_{.h} = \frac{1}{n_1} \sum_i d_{ih},$$

$$s_1^2 = \frac{1}{n_1 - 1} \sum_{i=1}^{n_1} (\bar{d}_{i.} - \hat{\delta})^2,$$

$$s_2^2 = \frac{1}{n_2 - 1} \sum_{h=1}^{n_2} (\bar{d}_{.h} - \hat{\delta})^2,$$

$$\tilde{\sigma}^2 = \frac{1}{n_1 n_2} \sum \sum (d_{ih} - \hat{\delta})^2.$$

Then

$$\hat{\sigma}^2 = \frac{(n_1 - 1)s_1^2 + (n_2 - 1)s_2^2 + \tilde{\sigma}^2}{n_1 n_2}$$

estimates the squared standard error of $\hat{\delta}$. Let z be the $1 - \alpha/2$ quantile of a standard normal distribution. Rather than use the more obvious confidence interval for δ, Cliff (1996, p. 140) recommends

$$\frac{\hat{\delta} - \hat{\delta}^3 \pm z\hat{\sigma} \sqrt{(1 - \hat{\delta}^2)^2 + z^2 \hat{\sigma}^2}}{1 - \hat{\delta}^2 + z^2 \hat{\sigma}^2}.$$

Cliff's confidence interval for δ is readily modified to give a confidence for P. Letting

$$C_\ell = \frac{\hat{\delta} - \hat{\delta}^3 - z\hat{\sigma} \sqrt{(1 - \hat{\delta}^2)^2 + z^2 \hat{\sigma}^2}}{1 - \hat{\delta}^2 + z^2 \hat{\sigma}^2}$$

and

$$C_u = \frac{\hat{\delta} - \hat{\delta}^3 + z\hat{\sigma} \sqrt{(1 - \hat{\delta}^2)^2 + z^2 \hat{\sigma}^2}}{1 - \hat{\delta}^2 + z^2 \hat{\sigma}^2},$$

a $1 - \alpha$ confidence interval for P is

$$\left(\frac{1 - C_u}{2}, \frac{1 - C_\ell}{2} \right). \tag{5.20}$$

Brunner–Munzel Method

To describe the Brunner–Munzel method, we begin by providing a formal definition of a midrank. Let

$$c^-(x) = \begin{cases} 0, & x \leq 0, \\ 1, & x > 0, \end{cases}$$

$$c^+(x) = \begin{cases} 0, & x < 0, \\ 1, & x \geq 0, \end{cases}$$

and

$$c(x) = \frac{1}{2}(c^+(x) + c^-(x)).$$

The *midrank* associated with X_i is

$$\frac{1}{2}\sum_{j=1}^{n} c(X_i - X_j).$$

In essence, midranks are the same as ranks when there are no tied values. If tied values occur, the ranks of tied values are averaged.

■ Example

Consider the values

$$7, 7.5, 7.5, 8, 8, 8.5, 9, 11, 11, 11.$$

If there were no tied values, their ranks would be 1, 2, 3, 4, 5, 6, 7, 8, 9, and 10. The midranks are easily determined as follows. Because there are two values equal to 7.5, their ranks are averaged yielding a rank of 2.5 for each. There are two values equal to 8, their original ranks were 4 and 5, so their midranks are both 4.5. There are three values equal to 11, their original ranks are 8, 9, and 10, the average of these ranks is 9, so their midranks are all equal to 9. The midranks corresponding to all ten of the original values are:

$$1, 2.5, 2.5, 4.5, 4.5, 6, 7, 9, 9, 9.$$

■

To apply the Brunner–Munzel method, first pool the data and compute midranks. Let $N = n_1 + n_2$ represent the total sample size (the number of observations among the pooled data), and let R_{ij} be the midrank associated with X_{ij} (the ith observation in the jth group) based on the pooled data. Let

$$\bar{R}_j = \frac{1}{n_j}\sum_{i=1}^{n_j} R_{ij}.$$

Compute the midranks for the data in group 1, ignoring group two, and label the results $V_{11}, \ldots V_{n_1 1}$. Do the same for group two (ignoring group one) and label the midranks $V_{12}, \ldots V_{n_2 2}$. The remaining calculations for testing H_0: $P = 0.5$, or for computing a confidence interval for P, are shown in Table 5.9.

Table 5.9: The Brunner–Munzel Method for Two Independent Groups.

Compute

$$S_j^2 = \frac{1}{n_j - 1} \sum_{i=1}^{n_j} \left(R_{ij} - V_{ij} - \bar{R}_j + \frac{n_j + 1}{2} \right)^2,$$

$$s_j^2 = \frac{S_j^2}{(N - n_j)^2}$$

$$s_e = \sqrt{N} \sqrt{\frac{s_1^2}{n_1} + \frac{s_2^2}{n_2}},$$

$$U_1 = \left(\frac{S_1^2}{N - n_1} + \frac{S_2^2}{N - n_2} \right)^2$$

and

$$U_2 = \frac{1}{n_1 - 1} \left(\frac{S_1^2}{N - n_1} \right)^2 + \frac{1}{n_2 - 1} \left(\frac{S_2^2}{N - n_2} \right)^2.$$

The test statistic is

$$W = \frac{\bar{R}_2 - \bar{R}_1}{\sqrt{N} s_e},$$

and the degrees of freedom are

$$\hat{v} = \frac{U_1}{U_2}.$$

Decision Rule: Reject H_0: $P = .5$ if $|W| \geq t$, where t is the $1 - \alpha/2$ quantile of a Student's t-distribution with $\hat{v}$ degrees of freedom. An estimate of P is

$$\hat{P} = \frac{1}{N} (\bar{R}_2 - \bar{R}_1) + \frac{1}{2}.$$

The estimate of $\delta = p_1 - p_3$ is

$$\hat{\delta} = 1 - 2\hat{P}.$$

An approximate $1 - \alpha$ confidence interval for P is

$$\hat{P} \pm t s_e.$$

In the second edition of this book, it was noted that situations can be constructed where, with many tied values, Cliff's method seems to be a bit better than the Brunner–Munzel method in terms of guaranteeing an actual Type I error probability less than the nominal α level. When testing at the 0.05 level, Cliff's method seems to do an excellent job of avoiding actual Type I error probabilities less than 0.04. In contrast, the Brunner–Munzel method can have an actual Type I error rate close to 0.07 when tied values are common and sample sizes are small. More recently, Neuhäuser, Lösch, and Jöckel (2007) provide a more comprehensive comparison of the Cliff and Brunner–Munzel methods in terms of their ability to control the probability of a Type I error. Again, with small sample sizes, it seems that Cliff's method has a bit of an advantage, and that generally there is little separating the two methods. Evidently, there are no published results on how Cliff's method compares to Mee's method when there are no tied values. With large sample sizes, the Brunner–Munzel method can have a lower execution time, especially when computing a p-value.

5.7.3 R Functions cid, cidv2, bmp, and wmwloc

The R function

$$\text{cid(x,y,alpha=0.05,plotit=F,pop=0,fr=0.8,rval=15,xlab=``",ylab=``")}$$

performs Cliff's method for making inferences about $\delta = P(X_{i1} > X_{i2}) - P(X_{i1} < X_{i2})$. The function also reports a confidence interval for $P = p_3 + 0.5 p_2$, which is labeled ci.p. The estimate of P is labeled phat. To get a *p*-value, use the function

$$\text{cidv2(x,y,plotit=F,xlab=``",ylab=``").}$$

The function

$$\text{bmp(x,y,alpha=0.05,plotit=T,pop=0,fr=0.8,rval=15,xlab=``",ylab=``")}$$

performs the Brunner–Munzel method. It returns the *p*-value when testing H_0: $P = .5$, plus an estimate of P labeled phat, and a confidence interval for P labeled ci.p (An estimate of $\delta = p_1 - p_3$, labeled d.hat, is returned as well.)

When plotit=T, these R functions create plots based on the $n_1 n_2$ differences, $D_{ih} = X_{i1} - X_{h2}$, $i = 1, \dots, n_1$ and $h = 1, \dots, n_2$. For reasons previously mentioned, these R functions can be viewed as methods aimed at testing the hypothesis that the distribution of $D = X - Y$ has a median of zero. With plotit=T, the function plots an estimate of the distribution of D, which should be symmetric about zero if the two distributions being compared are identical.

The argument pop determines the type of plot that will be created. The choices are:

* pop=0, adaptive kernel density estimate
* pop=1, expected frequency curve

- pop=2, Rosenblatt's shifted histogram
- pop=3, boxplot
- pop=4, stem-and-leaf
- pop=5, histogram
- pop=6, kernel density using a normal kernel.

The argument fr is the span when using a kernel density estimator, and rval indicates how many points are used by Rosenblatt's shifted histogram when creating the plot. (See Section 3.2.5.) Labels can be added to the x-axis and y-axis via the arguments xlab and ylab, respectively.

The R function

$$\text{wmwloc(x,y,alpha=0.05,na.rm=T)}$$

computes the median of the distribution of $D = X - Y$.

■ Example

For the data in Table 5.7, the Brunner–Munzel method has a *p*-value of .042, and its 0.95 confidence interval for P is (.167, .494), so H_0: $P = .5$ is rejected. Cliff's method also rejects at the 0.05 level, the 0.95 confidence interval for P being (0.198, 0.490). ■

■ Example

Measures of location provide some sense of how much groups differ, robust measures can provide more power versus methods based on means, and rank-based methods provide yet another perspective. But sometimes more might be needed to understand the nature and extent two groups differ, as illustrated here with data dealing with measures of self-regulation for children in grades in grades 6–7. The first group consisted of families with both parents, and the second group consisted of children from families with a single parent. (The sample sizes are 245 and 230, respectively.) Testing at the 0.05 level, no difference between the groups is found based on Student's *t*-test, Welch's heteroscedastic method for means, Yuen's method for trimmed means (in Section 5.3), the bootstrap methods for M-estimators and trimmed means covered in this chapter, and the rank-based methods as well. But is it possible that these methods are missing some true difference? That is, perhaps the distributions differ, but the hypothesis testing methods just listed are insensitive to this difference. The upper left panel of Figure 5.8 shows the shift function for these two groups. The function sband indicates that from about the 0.2–0.3 quantiles, the groups differ, and the 0.47

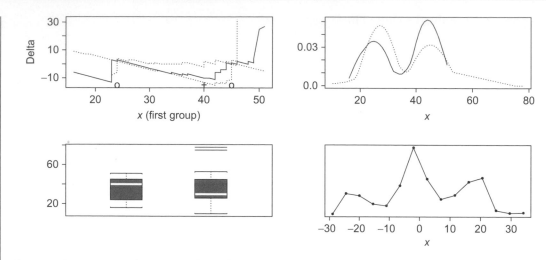

Figure 5.8: Four graphs summarizing how two groups differ based on a measure of self-regulation.

and 0.48 quantiles differ as well. To add perspective, the upper right plot shows the adaptive kernel density estimates created by the function g2plot. The lower left panel shows a boxplot of the data, and the lower right panel is a Rosenblatt shifted histogram created by the function cid (which performs Cliff's heteroscedastic analog of the Wilcoxon–Mann–Whitney test).

5.8 Comparing Two Independent Binomials

Many methods have been proposed for comparing two independent binomial distributions. Two methods described here were chosen based on results in Storer & Kim (1990) and Beal (1987) where comparisons of several methods were made. It is noted, however, that competing methods have been proposed that apparently have not been compared directly with the methods covered here (e.g., Berger, 1996; Coe & Tamhane, 1993). Results reported by Reed (2009) indicate that a method derived by Agresti and Caffo (2000) performs relatively well. More recently, Kulinskaya, Morgenthaler and Staudte (2010) derived yet another method, which will be denoted as the KMS method, which appears to be generally superior to the Agresti–Caffo method. It also competes well with a method derived by Newcombe (1998), which performed well among the methods compared by Brown and Li (2005). The best choice among the three methods covered here is not completely clear. It now appears that KMS is a bit preferable to Beal's method. An appeal of Beal's method and the KMS method is that they provide a confidence interval, whereas the Storer–Kim method does not. Situations arise in subsequent chapters where the Storer–Kim method has less power than Beal's method when comparing multiple groups of individuals. But when comparing two groups only, we find situations where the Storer–Kim method rejects and Beal's method does not. In terms of controlling the probability of a type I error when testing the hypothesis that the two binomial distributions have the same probability of success, all three methods appear to ensure that the actual level will be less than or equal to the nominal level. Limited comparisons indicate that typically the level of the Storer–Kim method is closest to the nominal level, when testing at the 0.05 level, suggesting that in general it will have the highest power.

Reiczigel, Abonyi-Tóth, and Singer (2008) generalized results derived by Sterne (1954) that yields a minimum volume confidence region for the two probabilities of success. Their method can be used, among other things, to compute a p-value when testing the hypothesis that two probabilities are equal. However, the Storer–Kim method appears to have a slight edge in terms of power, at least when testing at the 0.05 level. A more systematic study is needed to resolve this issue. The involved computational details are not described, but an R function for computing the confidence region derived by Reiczigel et al. is provided in Section 5.8.4.

5.8.1 Storer–Kim Method

Let p_j ($j = 1, 2$) be the probability of success associated with the jth group and let r_j be the number of successes among n_j trials. The goal is to test $H_0 : p_1 = p_2$. Note that the possible number of successes in the first group is any integer, x, between 0 and n_1, and for the second group it is any integer, y, between 0 and n_2. For any x and y, set

$$a_{xy} = 1$$

if

$$\left| \frac{x}{n_1} - \frac{y}{n_2} \right| \geq \left| \frac{r_1}{n_1} - \frac{r_2}{n_2} \right|;$$

otherwise

$$a_{xy} = 0.$$

Let

$$\hat{p} = \frac{r_1 + r_2}{n_1 + n_2}.$$

The test statistic is

$$T = \sum_{x=0}^{n_1} \sum_{y=0}^{n_2} a_{xy} b(x, \, n_1, \, \hat{p}) b(y, \, n_2, \, \hat{p}),$$

where

$$b(x, \, n_1, \, \hat{p}) = \binom{n_1}{x} \hat{p}^x (1 - \hat{p})^{n_1 - x},$$

and $b(y, n_2, \hat{p})$ is defined in an analogous fashion. You reject if

$$T \leq \alpha.$$

That is, T is the p-value. The Storer–Kim method does not provide a confidence interval, but it currently seems that typically it offers the most power among the methods that are available.

Note that the Storer–Kim method can be extended to comparing two multinomials by using a bootstrap method to determine a p-value. One possibility is to measure the overall difference between the two distributions using the sum of squared differences between the estimated cell probabilities as a test statistic. Next, estimate the assumed common probabilities as done here. Then generate bootstrap samples from a multinomial distribution having these cell probabilities, which can be used to compute a p-value. Put another way, one can perform a global test that for all possible x values $P(X = x) = P(Y = x)$, where X and Y are independent random variables. (More details can be found in Wilcox and Vigen (2011). Also see Alba-Fernández and Jiménez-Gamero, 2009.)

5.8.2 Beal's Method

Let $\hat{p}_1 = r_1/n_1$, $\hat{p}_2 = r_2/n_2$ and let $c = z_{1-\alpha/2}^2$ where $z_{1-\alpha/2}$ is the $1 - \alpha$ quantile of a standard normal distribution. (So c is the $1 - \alpha$ quantile of a chi-squared distribution with one degree of freedom.) Compute

$$a = \hat{p}_1 + \hat{p}_2$$

$$b = \hat{p}_1 - \hat{p}_2$$

$$u = \frac{1}{4}\left(\frac{1}{n_1} + \frac{1}{n_2}\right)$$

$$v = \frac{1}{4}\left(\frac{1}{n_1} - \frac{1}{n_2}\right)$$

$$V = u\{(2-a)a - b^2\} + 2v(1-a)b$$

$$A = \sqrt{c\{V + cu^2(2-a)a + cv^2(1-a)^2\}}$$

$$B = \frac{b + cv(1-a)}{1 + cu}.$$

The $1 - \alpha$ confidence interval for $p_1 - p_2$ is

$$B \pm \frac{A}{1 + cu}.$$

5.8.3 KMS Method

The confidence interval for $p_1 - p_2$, derived by Kulinskaya, Morgenthaler, and Staudte (2010) is

$$\frac{\hat{w}}{u} \sin\left(\arcsin\left[\frac{u\hat{\Delta} + \hat{v}}{\hat{w}} \right] \pm z_{1-\alpha/2} \sqrt{\frac{u}{2n_1 n_2 / N}} \right) - \frac{\hat{v}}{u},$$

where $z_{1-\alpha/2}$ is the $1 - \alpha/2$ quantile of a standard normal distribution, again r_1 and r_2 are the observed number of successes, $0 \leq A \leq 1$ is chosen by the user, $u = 2((1 - A)^2 \frac{n_2}{N} + A^2 \frac{n_1}{N})$, $\hat{\Delta} = (r_1 + 0.5)/(n_1 + 1) - (r_2 + 0.5)/(n_2 + 1)$, $\hat{\psi} = A(r_1 + 0.5)/(n_1 + 1) + (1 - A)(r_2 + 0.5)/(n_2 + 1)$, $\hat{v} = (1 - 2\hat{\psi})(A - \frac{n_2}{N})$, and $\hat{w} = \sqrt{2u\hat{\psi}(1 - \hat{\psi}) + \hat{v}^2}$. Here, following the suggestion made by Kulinskaya et al., $A = 0.5$ is used.

5.8.4 R Functions twobinom, twobici, bi2KMS, bi2KMSv2, and bi2CR

The R function

```
twobinom(r1 = sum(x), n1 = length(x), r2 = sum(y), n2 = length(y), x = NA, y = NA)
```

tests $H_0 : p_1 = p_2$ using the Storer–Kim method. The function can be used either by specifying the number of successes in each group (arguments r1 and r2) and the sample sizes (arguments n1 and n2), or the data can be in the form of two vectors containing 1s and 0s, in which case you use the arguments x and y.

Beal's method can be applied with the function

```
twobici(r1 = sum(x), n1 = length(x), r2 = sum(y), n2 = length(y), x = NA,
        y = NA, alpha = 0.05)
```

The R function

```
bi2KMS(r1 = sum(x), n1 = length(x), r2 = sum(y), n2 = length(y), x =NA, y = NA,
       alpha = 0.05)
```

performs method KMS. The R function

```
bi2KMSv2(r1=sum(x), n1=length(x), r2=sum(y), n2=length(y), x=NA, y=NA)
```

is the same as the R function bi2KMS, only it returns a p-value. The R function

```
bi2CR(r1, n1, r2, n2, alpha=0.05, xlab="p1", ylab="p2")
```

plots the $1 - \alpha$ confidence region for $(p1, p2)$ based on the method derived by Reiczigel et al. (2008).

■ Example

If for the first group we have 7 successes among 12 observations, for the second group we have 22 successes among 25 observations, the command twobinom(7,12,22,25) returns a p-value of .044, this is less than .05, so we would reject with $\alpha = .05$. The .95 confidence interval for $p_1 - p_2$ returned by the command twobici(7,12,22,25) is $(-0.61, 0.048)$, this interval contains zero, so in contrast to the Storer-Kim method we do not reject the hypothesis H_0: $p_1 = p_2$, the only point being that different conclusions might be reached depending on which method is used. The confidence interval returned by bi2KMS is $(-0.60, 0.025)$.

■

5.8.5 Comparing Discrete Distributions: R Functions binband and disc2com

When dealing with two discrete distributions, where the sample space is small, it might be desired to test

$$H_0 : P(X = x) = P(Y = x)$$

for each value x in the sample space. This can be done with the R function

$$\text{binband(x,y,KMS=F).}$$

By default, the individual probabilities are compared using the Storer–Kim method. Otherwise method KMS is used. The R function

$$\text{disc2com(x,y,alpha=0.05,nboot=500,SEED=TRUE)}$$

performs a global test using a bootstrap extension of the Storer–Kim method assuming the R package mc2d has been installed.

5.9 Comparing Dependent Groups

There are many ways two dependent groups might be compared, but as usual, no attempt is made to list all the possibilities. Rather, the goal is to describe methods similar in spirit to the methods for independent groups covered in this chapter.

5.9.1 A Shift Function for Dependent Groups

Lombard (2005) derived an extension of the shift function, described in Section 5.1, to dependent groups (cf. Wilcox, 2006f). Let $(X_1, Y_1), \ldots, (X_n, Y_n)$ be a random sample of

n pairs of observations from some bivariate distribution. Let $X_{(1)} \leq \cdots \leq X_{(n)}$ be the X_i values written in ascending order ($i = 1, \ldots, n$). Let

$$\hat{F}(x) = \frac{1}{n} \sum I(X_i \leq x)$$

be the estimate of $F(x)$, the marginal distribution of X, where the indicator function $I(X_i \leq x) = 1$ if $X_i \leq x$, otherwise $I(X_i \leq x) = 0$. The estimate of the marginal distribution Y, $\hat{G}(x)$, is defined in a similar manner. Denote the combined set $\{X_1, \ldots X_n, Y_1, \ldots Y_n\}$, written in ascending order, by $\{Z_{(1)} \leq \ldots \leq Z_{(2n)}\}$. Lombard's (2005) method for computing confidence intervals for the difference between each quantile stems from the test statistic

$$K = (n/2)^{1/2} \max |\hat{F}(Z_i) - \hat{G}(Z_i)|,$$

which can be used to test

$$H_0 : F(x) = G(x), \text{ for all } x,$$

versus

$$H_1 : F(x) \neq G(x), \text{ for at least one } x.$$

If $x > 0$, let

$$\psi_y(x) = (2\pi x^3)^{-1/2} y \exp(-y^2/2x),$$

otherwise, $\psi_y(x) = 0$. Let R_i be the rank of X_i values among the X values and let S_i be the rank of Y_i among the Y values. Let f_i be the frequency of occurrence of the value i among the values $\max\{R_1, S_1\}, \ldots, \max\{R_n, S_n\}$. Then the α level critical value used by Lombard is the value c solving

$$\frac{1}{n} \sum_{i=1}^{n} f_i \times \psi_c(i - f_1 - \cdots - f_i) = \alpha. \tag{5.21}$$

Here, the Nelder and Mead (1965) algorithm is used to determine c.

Let $[z]$ denote the integer portion of z, and for $k \geq 0$, let $Y_{(-k)} = X_{(-k)} = -\infty$ and $Y_{(n+1+k)} = X_{(n+1+k)} = \infty$. The quantile matching function q is given by $G(q(x)) = F(x)$. It specifies the functional relationship between the marginal distributions and reflects the difference between quantiles. Lombard's confidence interval for $q(X_{(j)})$, the quantile matching function evaluated at $X_{(j)}$, is

$$(Y_{(j-[(2n)^{1/2}c])}, \ Y_{(j+[(2n)^{1/2}c])}),$$

which is designed to have, approximately, simultaneous probability coverage $1 - \alpha$. So when the marginal distributions are identical, the interval

$$(Y_{(j-[(2n)^{1/2}c])} - X_{(j)}, \ Y_{(j+[(2n)^{1/2}c])} - X_{(j)}), \tag{5.22}$$

should contain zero for any j.

5.9.2 R Function lband

The R function

lband(x,y=NA,alpha=0.05,plotit=T,sm=T,ylab="delta", xlab="x (first group)")

computes Lombard's shift function for dependent groups. If the argument y=NA, the function assumes the argument x is a matrix with two columns or it has list mode. By default, the shift function is plotted. To avoid the plot, set the argument plotit=F. If the argument sm=T, a plot of shift function is smoothed using lowess.

5.9.3 Comparing Deciles

This section describes a method for comparing the deciles of two dependent groups, the idea being to capture the spirit of the shift function. The method is similar to the approach used to compare two independent groups based on the Harrell–Davis estimator, but certain details are different.

Let $(X_{11}, X_{12}), \ldots, (X_{n1}, X_{n2})$ be a random sample of n pairs of observations from some bivariate distribution. Let $\hat{\theta}_{jq}$ be the Harrell–Davis estimate of the qth quantile associated with the jth marginal distribution, $q = 0.1(0.1)0.9$. Then $\hat{d}_q = \hat{\theta}_{1q} - \hat{\theta}_{2q}$ estimates the difference between the qth quantiles. For the problem at hand, bootstrap samples are obtained by re-sampling with replacement n *pairs* of points. That is, n rows of data are sampled, with replacement, from the n by 2 matrix

$$\begin{pmatrix} X_{11}, X_{12} \\ \vdots \\ X_{n1}, X_{n2} \end{pmatrix}.$$

This is in contrast to the case of independent groups where bootstrap samples are obtained by re-sampling from $X_{11}, \ldots, X_{n1}$, and separate (independent) bootstrap samples are obtained by re-sampling from $X_{12}, \ldots, X_{n2}$.

Let $(X_{11}^*, X_{12}^*), \ldots, (X_{n1}^*, X_{n2}^*)$ be the bootstrap sample obtained by re-sampling n pairs of points, let $\hat{\theta}_{jq}^*$ be the Harrell–Davis estimate of x_{jq}, the qth quantile of the jth group, based on the values $X_{1j}^*, \ldots, X_{nj}^*$, and let $\hat{d}_q^* = \hat{\theta}_{1q}^* - \hat{\theta}_{2q}^*$. Repeat this bootstrap process B times yielding $\hat{d}_{q1}^*, \ldots, \hat{d}_{qB}^*$. Then an estimate of the squared standard error of $\hat{d}_q$ is

$$\hat{\sigma}_{dq}^2 = \frac{1}{B-1} \sum_{b=1}^{B} (\hat{d}_{qb}^* - \bar{d}_q)^2,$$

where

$$\bar{d}_q = \frac{1}{B}\sum_{b=1}^{B}\hat{d}^*_{qb}.$$

Setting

$$c = \frac{37}{n^{1.4}} + 2.75,$$

$$(\hat{\theta}_{1q} - \hat{\theta}_{2q}) \pm c\hat{\sigma}_{dq} \qquad (5.23)$$

yields a confidence interval for $x_{1q} - x_{2q}$, where c was determined so that the simultaneous probability coverage is approximately .95.

Notice that the method uses only one set of B bootstrap samples for all nine quantiles being compared. That is, the same bootstrap samples are used to compute $\hat{\sigma}^2_{dq}$ for each $q = 0.1, \dots, 0.9$. In contrast, when comparing independent groups, 18 sets of B bootstrap samples are used, one for each of the 18 quantiles being estimated, so execution time on a computer will be faster when working with dependent groups. The original motivation for using 18 sets of bootstrap values was to approximate the critical value using a nine-variate Studentized maximum modulus distribution. However, the approximation proved to be rather unsatisfactory when sample sizes are small. When working with independent groups, it might be possible to get accurate confidence intervals using only one set of B bootstrap samples, but this has not been investigated.

5.9.4 R Function shiftdhd

The R function

shiftdhd(x,y,nboot=200,plotit=T).

computes a confidence interval for the difference between the quantiles, when comparing two dependent random variables, using the method described in the previous subsection. The confidence intervals are designed so that the simultaneous probability coverage is approximately .95. As with shifthd, the default number of bootstrap samples (nboot) is $B = 200$ which appears to suffice in terms of controlling the probability of a Type I error. Simulations indicate that this has a practical advantage over $B = 100$ (Wilcox, 1995b). The last argument, plotit, defaults to T, for true, meaning that a plot of the shift function is created. The command shifthd(x,y,plotit=F) avoids the plot.

■ Example

An illustration in Section 5.3.3 is based on data from a study on the effects of consuming alcohol. Another portion of the study compared the effects of drinking alcohol for the same participants measured at three different times. Table 5.10 shows the data for the control group measured at times 1 and 3. No differences are found between the means or trimmed means, but this leaves open the possibility that one or more quantiles are significantly different. Comparing the deciles with shiftdhd, no significant differences are found with $\alpha = 0.05$. Figure 5.9 shows the results when using shiftdhd. The dots represent $\hat{\theta}_{1q}$, the Harrell–Davis estimate of the qth quantile based on the data stored in the R variable x, versus $\hat{d}_q = \hat{\theta}_{1q} - \hat{\theta}_{2q}$, the estimated difference between the qth quantiles, where $\hat{\theta}_{2q}$ is the estimate based on the data in the R variable y. Below and above each dot is a + marking the ends of the confidence interval for d_q.

Table 5.10: The Effect of Alcohol in the Control Group.

Time 1	0	32	9	0	2	0	41	0	0	0
	6	18	3	3	0	11	11	2	0	11
Time 3	0	25	10	11	2	0	17	0	3	6
	16	9	1	4	0	14	7	5	11	14

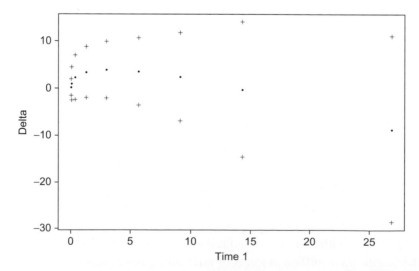

Figure 5.9: Plot created by the function shifthd using a portion of the alcohol data.

5.9.5 Comparing Trimmed Means

When comparing dependent groups, under general conditions, the difference between the marginal population medians does not equal the population median of the difference scores. This result generalizes to the γ-trimmed means, $\gamma > 0$. That is, if $D_i = X_{i1} - X_{i2}$, under general conditions $\mu_{td} \neq \mu_{t1} - \mu_{t2}$, where μ_{td} is the population trimmed mean corresponding to D. Note that inferences about μ_{td} are readily made using results in Chapter 4. For example, if pairs of observations are stored in the R variables time1 and time2, a confidence interval for μ_{td}, based on 20% trimming, can be computed with the R command trimci(time1–time2), and trimpb(time1–time2) will use a percentile bootstrap method instead. Alternatively, one can use the command onesampb(time1–time2,est=tmean). Of course, in some situations, there is little or no difference between comparing the trimmed means of the marginal distributions versus making inferences about a trimmed mean associated with difference scores. However, situations arise, particularly when comparing multiple groups, where the choice between the two strategies can make a considerable difference, as will be illustrated.

The remainder of this section focuses on a (nonbootstrap) method for comparing the trimmed means associated with the marginal distributions. Suppose $(X_{11}, X_{12}), \ldots, (X_{n1}, X_{n2})$ is a random sample of n pairs of observations from some bivariate distribution. The goal is to compute a confidence interval for $\mu_{t1} - \mu_{t2}$, the difference between the trimmed means. A simple approach is to estimate the squared standard error of $\bar{X}_{t1} - \bar{X}_{t2}$ and then use Student's t-distribution with appropriate degrees of freedom to get a confidence interval or test the hypothesis that the trimmed means are equal. This can be done using a simple generalization of Yuen's method.

Before continuing, it is remarked that for the special case where the goal is to compare the marginal medians, if there are tied values, only one method is known to perform well: a percentile bootstrap method. The R function dmedpb, described in Section 8.3.3 can be used to accomplish this goal. And even with no tied values, the method in this section should not be used because it is based on an unsatisfactory estimate of the standard error.

The process begins by Winsorizing the marginal distributions. In symbols, fix j and let $X_{(1)j} \leq X_{(2)j} \leq \cdots \leq X_{(n)j}$ be the n values in the jth group written in ascending order. Next, set

$$Y_{ij} = \begin{cases} X_{(g+1)j} & \text{if } X_{ij} \leq X_{(g+1)j} \\ X_{ij} & \text{if } X_{(g+1)j} < X_{ij} < X_{(n-g)j} \\ X_{(n-g)j} & \text{if } X_{ij} \geq X_{(n-g)j}, \end{cases}$$

where, as usual, g is the number of observations trimmed or Winsorized from each end of the distribution corresponding to the jth group. With 20% trimming, $g = [0.2n]$, where $[0.2n]$ means to round $0.2n$ down to the nearest integer. The expression for Y_{ij} says that $Y_{ij} = X_{ij}$ if X_{ij} has a value between $X_{(g+1)j}$ and $X_{(n-g)j}$. If X_{ij} is less than or equal to $X_{(g+1)j}$, set

$Y_{ij} = X_{(g+1)j}$, and if X_{ij} is greater than or equal to $X_{(n-g)j}$, set $Y_{ij} = X_{(n-g)j}$. Put another way, the observations are Winsorized with the dependent random variables remaining paired together, and this is consistent with the Winsorization of a bivariate distribution described in Chapter 2.

As an illustration, consider the eight pairs of observations

$$
\begin{array}{lcccccccc}
X_{i1}: & 18 & 6 & 2 & 12 & 14 & 12 & 8 & 9 \\
X_{i2}: & 11 & 15 & 9 & 12 & 9 & 6 & 7 & 10
\end{array}
$$

With 20% Winsorization, $g = 1$, so the smallest observation in each group is pulled up to the next smallest value. Thus, for the first row of data, the value 2 is Winsorized by replacing it with 6. Similarly, the largest value, 18, is replaced by the value 14. For the second row of data, 6 becomes 7 and 15 becomes 12. This yields

$$
\begin{array}{lcccccccc}
Y_{i1}: & 14 & 6 & 6 & 12 & 14 & 12 & 8 & 9 \\
Y_{i2}: & 11 & 12 & 9 & 12 & 9 & 7 & 7 & 10
\end{array}
$$

The population Winsorized covariance between X_{i1} and X_{i2} is, by definition, $\sigma_{w12} = E_w[(X_{i1} - \mu_w)(X_{i2} - \mu_{w2})]$, where E_w indicates the Winsorized expected value as defined in Chapter 2, and μ_{wj} is the population Winsorized mean of the jth group. It follows from the influence function of the trimmed mean that the squared standard error of $\bar{X}_{t1} - \bar{X}_{t2}$ is

$$
\frac{1}{(1-2\gamma)^2 n}\{\sigma_{w1}^2 + \sigma_{w2}^2 - 2\sigma_{w12}\},
$$

which reduces to a standard result when there is no trimming ($\gamma = 0$). The Winsorized covariance is estimated with the sample covariance between the Y_{i1} and Y_{i2} values:

$$
\frac{1}{n-1}\sum(Y_{i1} - \bar{Y}_1)(Y_{i2} - \bar{Y}_2),
$$

where $\bar{Y}_j$ is the Winsorized mean associated with the jth random variable. Generalizing Yuen's approach in an obvious way, the squared standard error of $\bar{X}_{t1} - \bar{X}_{t2}$ can be estimated with

$$
\frac{1}{h(h-1)}\left\{\sum(Y_{i1} - \bar{Y}_1)^2 + \sum(Y_{i2} - \bar{Y}_2)^2 - 2\sum(Y_{i1} - \bar{Y}_1)(Y_{i2} - \bar{Y}_2)\right\},
$$

where $h = n - 2g$ is the effective sample size. Letting

$$
d_j = \frac{1}{h(h-1)}\sum(Y_{ij} - \bar{Y}_j)^2,
$$

and

$$
d_{12} = \frac{1}{h(h-1)}\sum(Y_{i1} - \bar{Y}_1)(Y_{i2} - \bar{Y}_2),
$$

$H_0: \mu_{t1} = \mu_{t2}$ can be tested with

$$T_y = \frac{\bar{X}_{t1} - \bar{X}_{t2}}{\sqrt{d_1 + d_2 - 2d_{12}}}, \tag{5.24}$$

which is rejected if $|T_y| > t$, the $1 - \alpha$ quantile of Student's t-distribution with $h - 1$ degrees of freedom. A $1 - \alpha$ confidence interval for $\mu_{t1} - \mu_{t2}$ is

$$(\bar{X}_{t1} - \bar{X}_{t2}) \pm t\sqrt{d_1 + d_2 - 2d_{12}}.$$

5.9.6 R Functions yuend and yuendv2

The R function yuend computes the confidence interval for $\mu_{t1} - \mu_{t2}$, the difference between the trimmed means corresponding to two dependent groups, using the method described in the previous subsection of this chapter. The function has the form

yuend(x,y,tr=0.2,alpha=0.05).

As usual, the default amount of trimming is tr $= 0.2$, and alpha defaults to 0.05. The resulting confidence interval is returned in the R variable yuend$ci, the p-value level in yuend$siglevel, the estimated difference between the trimmed means in yuend$dif, the estimated standard error in yuend$se, the test statistic in yuend$teststat, and the degrees of freedom in yuend$df.

There are several ways the difference between two dependent groups might be characterized. One possibility is to use the explanatory measure of effect size as described in Section 5.3.4. The function

yuendv2(x,y,tr=.2,alpha=.05).

is exactly like the function yuen, only it also reports the explanatory measure of effect size.

■ Example

As a simple illustration, suppose the cholesterol levels of participants are measured before and after some treatment is administered yielding

Before: 190, 210, 300, 240, 280, 170, 280, 250, 240, 220
After: 210, 210, 340, 190, 260, 180, 200, 220, 230, 200.

Storing the before scores in the R vector x, and the after scores in y, the command yuend(x,y) returns

```
$ci: [1]  -8.29182  64.95849

$siglevel: [1] 0.1034335
```

```
$dif: [1] 28.33333

$se: [1] 14.24781

$teststat: [1] 1.98861

$df: [1] 5
```

Thus, the 0.95 confidence interval for $\mu_{t1} - \mu_{t2}$ is $(-8.3, 64.96)$, and the estimated difference between the trimmed means is 28.3. The test statistic is equal to 1.99, approximately, with a p-value of .103, the estimated standard error of the difference between the sample trimmed means is 14.2, and the degrees of freedom are 5. ∎

5.9.7 A Bootstrap-t Method for Marginal Trimmed Means

A bootstrap-t method can be used to compute a $1 - \alpha$ confidence interval for $\mu_{t1} - \mu_{t2}$. (Again, when using difference scores, simply proceed as in Chapter 4.) Begin by generating a bootstrap sample. That is, n pairs of observations are obtained by randomly sampling with replacement pairs of observations from the observed data. As usual, label the results (X_{i1}^*, X_{i2}^*), $i = 1, \ldots, n$. Now proceed along the lines in Section 5.3.2. More precisely, set $C_{ij}^* = X_{ij}^* - \bar{X}_{tj}$. Let T_y^* be the value of T_y, given by (5.24), based on the C_{ij}^* values just computed. Repeat this process B times yielding T_{yb}^*, $b = 1, \ldots, B$. Let $T_{y(1)}^* \leq \cdots \leq T_{y(B)}^*$ be the T_{yb}^* values written in ascending order. Set $\ell = \alpha B / 2$ and $u = (1 - \alpha/2)B$, rounding both to the nearest integer. Then an estimate of the lower and upper critical values is $T_{y(\ell+1)}^*$ and $T_{y(u)}^*$. An equal-tailed $1 - \alpha$ confidence interval for $\mu_{t1} - \mu_{t2}$ is

$$(\bar{X}_{t1} - \bar{X}_{t2} + T_{y(u)}^* \sqrt{d_1 + d_2 - 2d_{12}}, \ \bar{X}_{t1} - \bar{X}_{t2} + T_{y(\ell)}^* \sqrt{d_1 + d_2 - 2d_{12}}). \tag{5.25}$$

To get a symmetric confidence interval, replace T_{yb}^* by its absolute value, set $a = (1 - \alpha)B$, rounding to the nearest integer, in which case the $(1 - \alpha)$ confidence interval for $(\mu_{t1} - \mu_{t2})$ is

$$(\bar{X}_{t1} - \bar{X}_{t2}) \pm T_{y(a)}^* \sqrt{d_1 + d_2 - 2d_{12}}.$$

5.9.8 R Function ydbt

The R function ydbt is supplied for computing a bootstrap-t confidence interval for $\mu_{t1} - \mu_{t2}$ when dealing with paired data. It has the form

$$\text{ydbt(x,y,tr=0.2,alpha=0.05,nboot=599,side=F,plotit=F,op=1)}.$$

As usual, the default amount of trimming is tr=0.2, and α defaults to alpha=0.05. The number of bootstrap samples defaults to nboot = 599. Using side=F, for false, results in an equal-tailed confidence interval, while side $=$ T returns a symmetric confidence interval instead. Setting the argument plotit to T creates a plot of the bootstrap values, where the type of plot is controlled via the argument op. The possible values for op are 1, 2, and 3 which correspond to the an adaptive kernel estimator, the expected frequency curve, and a boxplot, respectively.

5.9.9 Inferences about the Distribution of Difference Scores

There is another perspective on comparing dependent groups that should be mentioned. For convenience, the focus is momentarily on trimmed means, but it is evident that some of the general remarks made here are relevant to any robust measure of location. As already noted, except for the special case of the sample mean, the trimmed mean of the pairwise differences can differ from the difference between the marginal trimmed means. That is, for n randomly sampled pairs of observations X_{ij} $(i = 1, \ldots, n; j = 1, 2)$ if

$$D_i = X_{i1} - X_{i2},$$

the trimmed mean based on the D_i values is not necessarily equal to $\bar{X}_{t1} - \bar{X}_{t2}$, the difference between the marginal trimmed means. Moreover, under general conditions, $\mu_{td} \neq \mu_{t1} - \mu_{t2}$, where μ_{td} is the population trimmed mean corresponding to D. Note that yet another way of characterizing how the two groups differ is in terms of the distribution of $D = X_1 - X_2$. That is, now compute the n^2 differences

$$\mathcal{D}_{ik} = X_{i1} - X_{k2}$$

for all i and k $(i = 1, \ldots, n; k = 1, \ldots, n)$. A way of comparing the two groups is to use some measure of location based on the $\mathcal{D}_{ik}$ values.

Let $M_{\mathcal{D}}$ denote the sample median based on all n^2 $\mathcal{D}_{ik}$ values. Of course, some other measure of location could be used, but in terms of efficiency, $M_{\mathcal{D}}$ compares well to smaller amounts of trimming, even under normality (Wilcox, 2006d). Moreover, letting $\theta_{\mathcal{D}}$ be the population median associated with $M_{\mathcal{D}}$, a basic percentile bootstrap method has been found to perform well, in terms of controlling the probability of a Type I error, when testing

$$H_0 : \theta_{\mathcal{D}} = 0.$$

That is, randomly sample with replacement n pairs of observation, compute $M_{\mathcal{D}}^*$ based on this bootstrap sample, and repeat this process B times. Note that when comparing independent groups, this hypothesis corresponds to H_0: $p = .5$, where p is the probabilistic measure of effect size discussed in Section 5.7.

To illustrate the different measures of location in more concrete terms, imagine a study based on randomly sample married couples where the goal is to compare cholesterol levels. One strategy is to compute the median of difference scores, which tells us something about the typical difference for a randomly sampled couple. A second approach is to compute the median of the marginal distributions, which provides information about how the typical cholesterol level of the males compares to the typical level for all females. A third approach is to compute the difference between cholesterol levels for each man and woman, not just men and women who are married. This provides information about the typical difference between any man and any woman.

5.9.10 R Functions loc2dif and l2drmci

The R function

loc2dif((x, y = NULL, est = median, na.rm = T, plotit = F, xlab = " ", ylab = " ", ...)

computes M_D for the data stored in x (time 1 for example) and y (time 2). If y is not specified, it is assumed x is a matrix with two columns. The argument na.rm=T means that the function will eliminate any pair where one or both values are missing. If it is desired to use all of the available data, set na.rm=F. If the argument plotit=T, the function plots an estimate of the distribution of $\mathcal{D}$. The R function

l2drmci(x,y=NA,est=median,alpha=0.05,na.rm=T)

tests H_0: $\theta_\mathcal{D} = 0$. The argument na.rm is used as was done with loc2dif.

■ Example

Rao (1948) reports data on cork boring weights taken from 28 trees, which are reproduced in Table 6.5. The borings were taken from the north, east, west, and south sides of each tree. Here the south and east sides of the trees are compared. The function loc2dif returns 1. That is, among all trees, the median difference between south and east sides is estimated to be 1. In contrast, the median difference for a randomly sampled tree, meaning the median of the difference scores associated with the 28 trees, is 3. So now we have information on how the two sides of the same tree compare, which is not the same as the difference among all the trees. Finally, the difference between the marginal medians is 1.5. This tells us something about how the typical weight for the south side of a tree compares to the typical weight of the east side. But it does not provide any direct information regarding the typical difference among all of the trees. The R function l2drmci returns a p-value of .336. So based on the differences among all pairs of trees, fail to reject the hypothesis H_0: $\theta_\mathcal{D} = 0$ at the 0.10 level. However, if the $n = 28$ difference scores are used, the R function sintv2 returns a p-value of .088. Now (at the 0.10 level), reject and conclude that for a randomly sampled tree, the median difference between the two weights differ from 0, the only point being that different perspectives can alter the *p*-value substantially.

■

5.9.11 Percentile Bootstrap: Comparing Medians, M-Estimators and Other Measures of Location and Scale

The percentile bootstrap method in Chapter 4 is readily extended to comparing various parameters associated with the marginal distributions of two dependent variables. When X_{i1} and X_{i2} are dependent, bootstrap samples are obtained by randomly sampling pairs of observations with replacement. That is, proceed as described in Section 5.9.7 yielding the pairs of bootstrap samples

$$(X_{11}^*, X_{12}^*)$$
$$\vdots$$
$$(X_{n1}^*, X_{n2}^*).$$

Let θ_j be any parameter of interest associated with the jth marginal distribution. Let $\hat{\theta}_j^*$ be the bootstrap estimate of θ_j based on $X_{1j}^*, \ldots, X_{nj}^*$ and let $d^* = \hat{\theta}_1^* - \hat{\theta}_2^*$. Repeat this process B times yielding $d_1^*, \ldots, d_B^*$, write these B values in ascending order yielding $d_{(1)}^* \leq \cdots \leq d_{(B)}^*$, in which case a $1 - \alpha$ confidence interval for $\theta_1 - \theta_2$ is

$$(d_{(\ell+1)}^*, d_{(u)}^*),$$

where $\ell = \alpha B/2$, rounded to the nearest integer, and $u = B - \ell$,

A (generalized) *p*-value can be computed as well. Let p^* be the probability that, based on a bootstrap sample, $\theta_1^* > \theta_2^*$. This probability will be estimated with $\hat{p}^*$, the proportion of bootstrap samples, among all B bootstrap samples, for which $\theta_1^* > \theta_2^*$. For a two-sided hypothesis, now reject if $\hat{p}^* \leq \alpha/2$ or if $\hat{p}^* \geq 1 - \alpha/2$. The estimate of the (generalized) *p*-value is

$$2\min(\hat{p}^*, 1 - \hat{p}^*).$$

The percentile bootstrap method just described can be unsatisfactory when the goal is to make inferences about means and variances, but it appears to be reasonably effective when working with M-estimators, the Harrell-Davis estimate of the median, as well as the biweight midvariance (Wilcox, 1996a). By this is meant that the actual probability of a type I error will not be much larger than the nominal level. However, a concern is that with small sample sizes, the actual probability of a type I error can drop well below the nominal level when working with M-estimators or the modified M-estimator described in Chapter 3. For example, there are situations where, when testing at the 0.05 level, the actual probability of a type I error can be less than .01. Wilcox and Keselman (2002) found a method that reduces this problem in simulations. Note that if based on the original data, $\hat{\theta}_1 = \hat{\theta}_2$, it should be the case that $\hat{p}^* = 0.5$, but situations arise where this is not case. The idea is to shift the data so that $\hat{\theta}_1 = \hat{\theta}_2$, compute $\hat{p}^*$ based on the shifted data, and then correct the bootstrap p-value given above. More precisely, let $\hat{q}^*$ be the value of $\hat{p}^*$ based on the shifted data. A so-called *bias-adjusted p-value* is

$$2\min(\hat{p}_a^*,\ 1 - \hat{p}_a^*),$$

where $\hat{p}_a^* = \hat{p}^* - 0.1(\hat{q}^* - 0.5)$. As the sample size increases, $\hat{q}^* - 0.5 \to 0$ and the adjustment becomes negligible.

Note that the pairs of bootstrap values $(\hat{\theta}_{1b}^*,\ \hat{\theta}_{2b}^*)$, $b = 1, \dots B$, provide an approximate $1 - \alpha$ confidence region for (θ_1, θ_2). The bootstrap method aimed at comparing the measures of location associated with the marginal distributions essentially checks to see how deeply a line through the origin, having slope one, is nested within the cloud of bootstrap values. Here, the depth of this line is measured by how many bootstrap points lie above it, which corresponds to how often $\hat{\theta}_{1b}^* < \hat{\theta}_{2b}^*$ among the B bootstrap pairs of points.

5.9.12 R Function bootdpci

The R function

$$\text{bootdpci(x,y,est=onestep,nboot=NA,alpha=0.05,plotit=T,dif=T,BA=F,\dots)}$$

performs a percentile bootstrap method using any estimator available through R. The argument est indicates the estimator that will be used and defaults to the one-step M-estimator (based on Huber's Ψ). The default value for nboot is NA, which in effect causes $B = 1000$ to be used. The argument dif controls whether inferences are made on difference scores; by default differences scores are used. To compare measures of location associated with the marginal distributions, set dif=F. If dif=F and BA=T, the bias adjusted p-value is computed.

■ Example

If the data in Table 5.10 are stored in the R variables t1 and t2, the command bootdpci(t1,t2,est=tmean) computes a 0.95 confidence interval for the 20% trimmed mean associated with the difference scores. The (generalized) p-value is .369. The left panel of Figure 5.10 shows the resulting plot of the bootstrap values. The p-value reflects the proportion of points below the horizontal line at zero. The command bootdpci(x,y,est=tmean,dif=F) compares the marginal distributions instead. Now the (generalized) p-value is .063. The right panel of Figure 5.10 shows the resulting plot. The p-value reflects the proportion of points below the line having slope one and

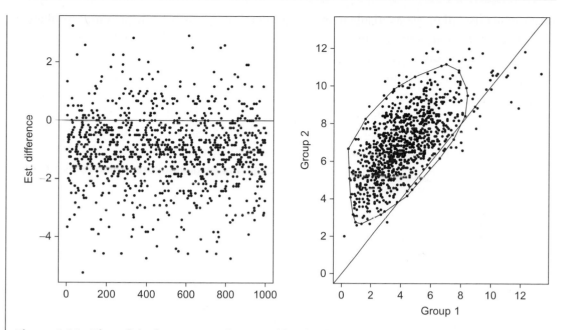

Figure 5.10: Plot of the bootstrap values used by the function bootdpci when analyzing the data in Table 5.10. The left panel is for dif=1, meaning that difference scores were analyzed. The right panel is for dif=F, meaning that marginal measures of location are compared.

intercept zero. The command bootdpci(x,y,est=tmean,dif=F,BA=T) again compares the marginal 20% trimmed means, but the bias-adjusted *p*-value is reported instead; it is .051, so now we nearly reject at the 0.05 level.

The command bootdpci(x,y,est=winvar,dif=F) would compute a 0.95 confidence interval for the difference between the 20% Winsorized variances. The command bootdpci(x,y) would attempt to compute a 0.95 confidence interval based on a one-step M-estimator, but for the data used here, eventually this results in an error in the R function hpsi. The reason, which was already discussed in Chapter 4, is that with small sample sizes, there is a good chance that an M-estimator based on a bootstrap sample cannot be computed due to division by zero.

5.9.13 *Handling Missing Values*

Numerous methods have been proposed for handling missing values when dealing with means, none of which are completely satisfactory. A simple approach is the so-called *complete case* method where any pair of observations is eliminated if one of the values is missing, after which methods previously described are applied to the data that remain. This section describes three methods for handling missing values when using a robust measure of location that use all of the available data, assuming missing values occur in a manner that does not alter the marginal measures of location. For example, it might be the case that values are missing completely at random (MCAR), meaning that the process resulting in missing values is independent of both the observed and the missing values. A weaker assumption, that allows the analysis to be performed without taking into account the mechanism that creates missing values, is called missing at random (MAR). Missing at random (MAR) is taken to mean that, given the observed data, the missingness mechanism does not depend on the unobserved data. (For a description of other mechanisms leading to missing values, see Little & Rubin, 2002.) At the end of this section, comments are made about the relative merits of the methods about to be described.

Method M1

The first method is based on a straightforward generalization of the method in Lin and Stivers (1974), assuming that the goal is to compare the marginal trimmed means, as opposed to the trimmed mean of the difference scores. It is assumed than n pairs of observations are randomly sampled where both values are available, which is denoted by $(X_1, Y_1), \ldots,$ (X_n, Y_n). The corresponding (marginal) γ-trimmed means are denoted by $\bar{X}_t$ and $\bar{Y}_t$. For the first marginal distribution, an additional n_1 observations are sampled for which the corresponding Y value is not observed. These observations are denoted by $X_{n+1}, \ldots, X_{n+n_1}$ and the trimmed mean of these n_1 observations is denoted by $\tilde{X}_t$. Similarly, n_2 observations are sampled for which the corresponding value for the first marginal distribution is not observed and the trimmed mean is denoted by $\tilde{Y}_t$. Let $h_j = [\gamma n_j] \, (j = 1, 2)$, and let $\lambda_j = h/(h + h_j)$, where $h = [\gamma n]$. Then an estimate of the difference between the marginal trimmed means, $\Delta_t = \mu_{t1} - \mu_{t2}$, is

$$\hat{\mu}_{tD} = \lambda_1 \bar{X}_{tD} - \lambda_2 \bar{Y}_{tD} + (1 - \lambda_1)\tilde{X}_{tD} - (1 - \lambda_2)\tilde{Y}_{tD},$$

a linear combination of three independent random variables. The squared standard of $\lambda_1 \bar{X} - \lambda_2 \bar{Y}$ is

$$\sigma_0^2 = \frac{1}{(1 - 2\gamma)^2 n}(\lambda_1^2 \sigma_{wx}^2 + \lambda_2 \sigma_{wy}^2 - 2\lambda_1 \lambda_2 \sigma_{wxy}), \tag{5.26}$$

where σ_{wxy} is the population Winsorized covariance between X and Y. The squared standard error of $(1 - \lambda_1)\tilde{X}$ is

$$\sigma_1^2 = \frac{(1 - \lambda_1)^2 \sigma_{wx}^2}{(1 - 2\gamma)^2 (n + n_1)} \tag{5.27}$$

and the squared standard error of $(1 - \lambda_2)\tilde{Y}$ is

$$\sigma_2^2 = \frac{(1-\lambda_2)^2\sigma_{wy}^2}{(1-2\gamma)^2(n+n_2)}. \tag{5.28}$$

So the squared standard error of $\hat{\mu}_{tD}$ is

$$\tau^2 = \sigma_0^2 + \sigma_1^2 + \sigma_2^2.$$

For convenience, let $N_1 = n + n_1$ and $g_1 = [\gamma N_1]$. The Winsorized values corresponding to $X_1, \ldots, X_{N_1}$ are

$$W_{xi} = \begin{cases} X_{(g_1+1)} & \text{if } X_i \le X_{(g_1+1)} \\ X_i & \text{if } X_{(g_1+1)} < X_i < X_{(N_1-g_1)} \\ X_{(N_1-g_1)} & \text{if } X_i \ge X_{(N_1-g_1)}. \end{cases}$$

The (sample) Winsorized mean is

$$\bar{W}_x = \frac{1}{N_1}\sum_{i=1}^{N_1} W_{xi},$$

an estimate of the Winsorized variance, σ_{wx}^2, is

$$s_{wx}^2 = \frac{1}{N_1-1}\sum(W_{xi} - \bar{W}_x)^2,$$

and an estimate of σ_{wy}^2 is obtained in a similar fashion. The Winsorized covariance between X and Y is estimated with

$$s_{wxy} = \frac{1}{n-1}\sum_{i-1}^{n}(W_{xi} - \tilde{W}_x)(W_{yi} - \tilde{W}_y),$$

where

$$\tilde{X}_w = \frac{1}{n}\sum_{i=1}^{n} W_{xi}$$

and $\tilde{Y}_w$ is defined in a similar manner.

The sample Winsorized variances yield estimates of σ_0^2, σ_1^2 and σ_2^2, say $\hat{\sigma}_0^2$, $\hat{\sigma}_1^2$ and $\hat{\sigma}_2^2$, in which case an estimate of the squared standard error of $\hat{\mu}_{tD}$ is

$$\hat{\tau}^2 = \hat{\sigma}_0^2 + \hat{\sigma}_1^2 + \hat{\sigma}_2^2.$$

So a reasonable test statistic for testing the hypothesis of equal (marginal) trimmed means is

$$T = \frac{\hat{\mu}_{tD}}{\hat{\tau}}. \tag{5.29}$$

There remains the problem of approximating the null distribution of T and here a basic bootstrap-t method is used. To make sure the details are clear, the method begins by randomly sampling with replacement $N = n + n_1 + n_2$ pairs of observations from $(X_1, Y_2), \ldots, (X_N, Y_N)$ yielding $(X_1^*, Y_2^*), \ldots, (X_N^*, Y_N^*)$. Based on this bootstrap sample, compute the absolute value of the test statistic as just described and label the result T^*. Repeat this process B times and put the resulting T^* values in ascending order yielding $T_{(1)}^* \leq \cdots \leq T_{(B)}^*$. Then an approximate $1 - \alpha$ confidence interval for Δ_t is

$$\hat{\Delta}_t \pm T_{(c)}^* \hat{\tau}$$

where $c = (1 - \alpha)B$ rounded to the nearest integer.

Method M2

Method M2 is based on the usual percentile bootstrap method. For the situation at hand, generate a bootstrap sample using all N pairs of observations and let $\tilde{D}_t^* = \tilde{X}_t^* - \tilde{Y}_t^*$, where $\tilde{X}_t^*$ is the trimmed mean based on all of the X_i^* values not missing and $\tilde{Y}_t^*$ is computed in a similar manner. Repeat this B times, put the resulting $\tilde{D}_t^*$ values in ascending order, and label the results $\tilde{D}_{t(1)}^* \leq \cdots \leq \tilde{D}_{t(B)}^*$. Then an approximate $1 - \alpha$ confidence interval for μ_{tD} is

$$(\tilde{D}_{t(\ell+1)}^*, \ \tilde{D}_{t(u)}^*),$$

where $\ell = \alpha B / 2$, rounded to the nearest integer, and $u = B - \ell$. A p-value is computed in the usual manner. That is, estimate $p = P(\hat{\mu}_{tD}^* > 0)$ with $\hat{p}$, the proportion of $\tilde{D}_t^*$ values greater than 0. Then a (generalized) p-value is

$$P = 2\min(\hat{p}, 1 - \hat{p}).$$

Method M3

Method M3 is based on θ_D, the median of the distribution of $D = X - Y$. The method begins by forming all pairwise differences among all of the observed X and Y values. That is, compute $D_{ij} = X_i - Y_j$ ($i = 1, \ldots, N_1$; $j = 1, \ldots, N_2$) resulting in $N_1 \times N_2$ D_{ij} values. Then an estimate of θ_D is obtained by computing the sample median of the D_{ij} values.

Again a basic percentile bootstrap method is used to make inferences about θ_D. Generate a bootstrap sample as done in Method M2 and let $\hat{\theta}_D^*$ be the resulting estimate of θ_D. Repeat this process B times yielding $\hat{\theta}_{Db}^*$, $b = 1, \ldots, B$. Next, put these B values in ascending order yielding $\hat{\theta}_{D(1)}^* \leq \cdots \leq \hat{\theta}_{D(B)}^*$ and let ℓ and u be defined as before. Then a $1 - \alpha$ confidence interval for θ_D is

$$(\hat{\theta}_{D(\ell+1)}^*, \ \hat{\theta}_{D(u)}^*).$$

This method can be applied with the R function l2drmci in Section 5.9.10 by setting the argument na.rm=F, meaning that any row of data that has a missing value is not removed.

Comments on Choosing a Method

Based on results in Wilcox (in press a), method M3, as well as method M2 coupled with a 20% trimmed mean, perform reasonably well in terms of controlling the probability of a type I error, with M2 having perhaps a slight advantage. Method M1 tends to have an actual type error probability less than the nominal level, again using a 20% trimmed mean, sometimes substantially so. However, as the amount of trimming decreases, at some point a percentile bootstrap method (method M2) will not perform well, suggesting that eventually, method M1 will be more satisfactory than method M2. But at what point this is the case has not been determined. Not surprisingly, method M1 can be highly unsatisfactory when working with means.

5.9.14 R Functions rm2miss and rmmismcp

The R function

$$\text{rmmismcp}(x, y = \text{NA}, \text{alpha} = 0.05, \text{con} = 0, \text{est} = \text{tmean}, \text{plotit} = \text{T}, \text{grp} = \text{NA}, \text{nboot} = 500, \text{SEED} = \text{T}, \text{xlab} = \text{"Group 1"}, \text{ylab} = \text{"Group 2"}, \text{pr} = \text{F}, ...),$$

has been supplied for dealing with missing values when the goal is to test the hypothesis H_0: $\mu_{t1} = \mu_{t2}$ using method M2 described in the previous section. In particular, rather than using the complete case analysis strategy, it uses all of the available data to compare the marginal trimmed means, assuming any missing values occur at random. With 20% trimming or more, it appears to be one of the better methods for general use when there are missing values. By default, a 20% trimmed mean is used, but other measures of location can be used via the argument est. For example, rmmismcp(x,y,est=onestep) would compare the groups with a one-step M-estimator. The function returns a confidence interval for the difference between the marginal measures of location. If the argument y=NA, it is assumed that the argument x is a matrix with columns corresponding to groups. (The function can handle more than two groups; see Section 8.1.5 for more details.) When there are two groups and the argument plotit=T, a plot of the bootstrap estimates is created.

The R function

$$\text{rm2miss}(x, y, \text{tr} = 0)$$

also tests H_0: $\mu_{t1} = \mu_{t2}$ using method M1.

5.9.15 Comparing Variances

For the special case where the goal is to compare the variances of dependent groups, a variation of the basic percentile bootstrap method is required that represents an analog of the so-called *Morgan–Pitman* test. Formally, the goal is to test

$$H_0 : \sigma_1^2 = \sigma_2^2. \tag{5.30}$$

Set

$$U_i = X_{i1} - X_{i2}$$

and

$$V_i = X_{i1} + X_{i2},$$

$(i = 1, \ldots, n)$, and let ρ_{uv} be the (population) value of Pearson's correlation between the U and V. It can be shown that if H_0 is true, then $\rho_{uv} = 0$, so a test the hypothesis of equal variances is obtained by testing

$$H_0 : \rho_{uv} = 0.$$

One approach to testing H_0: $\rho_{uv} = 0$, is to use a modified percentile bootstrap method, which allows heteroscedasticity. That is, take a bootstrap sample of n pairs of the U and V values, compute the correlation between these values, repeat this 599 times, label the results $r_1^*, \ldots, r_{599}^*$, in which case a 0.95 confidence interval for ρ_{uv} is

$$(r_{(\ell+1)}^*, \ r_{(u)}^*),$$

where for $n < 40$, $\ell = 6$ and $u = 593$; for $40 \leq n < 80$, $\ell = 7$ and $u = 592$; for $80 \leq n < 180$, $\ell = 10$ and $u = 589$; for $180 \leq n < 250$, $\ell = 13$ and $u = 586$; and for $n \geq 250$, $\ell = 15$ and $u = 584$. Another option is to use an adaptation of the method in Section 10.1.1, called the HC4 method, which also allows heteroscedasticity. Currently, it seems that there is little separating the two methods in terms of controlling the probability of a type I error. An appeal of the HC4 method is that it is not limited to $\alpha = 0.05$ and it provides a *p*-value.

The R functions pcorb and pcorhc4, described in a Section 9.3.14, have been supplied for testing the hypothesis that Pearson's correlation is zero. These function can be used to compare the variances of dependent groups using the method just described. If, for example, the data are stored in the R variables t1 and t2, the command pcorb(t1-t1,t1+t2) accomplishes this goal.

5.9.16 The Sign Test and Inferences about the Binomial Distribution

For completeness, one can also compare two dependent groups with the sign test. Let $p = P(X_{i1} < X_{i2})$. Then p is the probability that for a randomly sampled pair of observations, the first observation is less than the second. Letting $W_i = 1$ if $X_{i1} < X_{i2}$, $w = \sum W_i$ has a binomial distribution with probability of success p. Standard asymptotic theory can be used to compute a confidence interval for p, but various improvements have appeared in the literature, some of which are described here.

The goal is to determine c_L and c_U such that

$$P(c_L \le p \le c_U) = 1 - \alpha.$$

For the special cases where w is equal to $0, 1, n-1$, or n, results in Blyth (1986) can be used. In particular,

* If $w = 0$,

$$c_U = 1 - \alpha^{1/n}$$
$$c_L = 0.$$

* If $w - 1$,

$$c_L = 1 - \left(1 - \frac{\alpha}{2}\right)^{1/n}$$
$$c_U = 1 - \left(\frac{\alpha}{2}\right)^{1/n}.$$

* If $w = n - 1$,

$$c_L = \left(\frac{\alpha}{2}\right)^{1/n}$$
$$c_U = \left(1 - \frac{\alpha}{2}\right)^{1/n}.$$

* If $w = n$,

$$c_L = \alpha^{1/n},$$

    and

$$c_U = 1.$$

■ Example

If $n = 8$ and $w = 0$,
$$c_U = 1 - (0.05)^{1/8} = 0.312,$$
and a 0.95 one-sided confidence interval for p is $(0., 0.312)$. If $w = 1$ and $n = 8$, a two-sided 0.95 confidence interval would be $(0.003, 0.37)$.

Table 5.11: Pratt's Approximate Confidence Interval for p.

You observe w successes among n trials, and the goal is to compute a $1 - \alpha$ confidence interval for p.

Let c be the $1 - \alpha/2$ quantile of a standard normal distribution.

To determine c_U, the upper end of the confidence interval, compute

$$A = \left(\frac{w+1}{n-w}\right)^2$$

$$B = 81(w+1)(n-w) - 9n - 8$$

$$C = -3c\sqrt{9(w+1)(n-w)(9n+5-c^2)+n+1}$$

$$D = 81(w+1)^2 - 9(w+1)(2+c^2) + 1$$

$$E = 1 + A\left(\frac{B+C}{D}\right)^3$$

in which case

$$c_U = \frac{1}{E}.$$

To get the lower end of the confidence interval, compute

$$A = \left(\frac{w}{n-w-1}\right)^2$$

$$B = 81(w)(n-w-1) - 9n - 8$$

$$C = 3c\sqrt{9x(n-w-1)(9n+5-c^2)+n+1}$$

$$D = 81w^2 - 9w(2+c^2) + 1$$

$$E = 1 + A\left(\frac{B+C}{D}\right)^3$$

in which case

$$c_L = \frac{1}{E}.$$

For situations where the observed number of successes is not 0, 1, $n - 1$, or n, Pratt's (1968) approximation can be used, which is recommended by Blyth (1986) (cf. Chen, 1990). The computational details are given in Table 5.11.

For completeness, other methods for computing a confidence interval for the probability of success were compared by Brown, Cai, and DasGupta (2002), and they concluded that the Agresti–Coull method, which is a simple generalization of method derived by Agresti and Coull (1998), performs relatively well.

Let X represent the total number of successes among n observations, in which case

$$\hat{p} = \frac{X}{n},$$

the proportion of successes among the n observations. As before, let c be the $1 - \alpha/2$ quantile of a standard normal distribution. Compute

$$\tilde{n} = n + c^2,$$

$$\tilde{X} = X + \frac{c^2}{2},$$

and

$$\tilde{p} = \frac{\tilde{X}}{\tilde{n}}.$$

Then the Agresti–Coull $1 - \alpha$ confidence interval for the probability of success, p, is

$$\tilde{p} \pm c\sqrt{\frac{\tilde{p}(1 - \tilde{p})}{\tilde{n}}}.$$

5.9.17 R Functions binomci and acbinomci

The R function

$$\text{binomci(x = sum(y), nn = length(y), y = NA, n = NA, alpha = 0.05),}$$

which comes with this book, computes a $1 - \alpha$ confidence interval for p using Pratt's method described in the previous section. When the number of successes is 0, 1, $n - 1$, or n, Blyth's method is used instead. Here, x is the observed number of successes, nn is the number of observations, and alpha is α. The function can handle data stored as a vector of 1s and 0s via the argument y. If data are stored in y, and no values are specified by the arguments x and nn, the function takes x to be the number of successes in y and nn is taken to be the length of y. The R function

$$\text{acbinomci(x = sum(y), nn = length(y), y = NA, n = NA, alpha = 0.05),}$$

computes the Agresti–Coull confidence interval.

■ Example

Suppose there is 1 success in 80 trials. Then binomci(1,80) reports that the 0.95 confidence interval for p is $(0.00032, 0.0451)$. If mydat contains 0, 1, 1, 1, 0, 0, 1, 1, 0, 1, then the command binomci(y=mydat) returns an estimate (phat) of p equal to 0.6 and a 0.95 confidence interval equal to $(0.33, 0.88)$. The Agresti–Coull 0.95 confidence interval is $(0.31, 0.83)$.

5.10 Exercises

1. Compare the two groups of data in Table 5.1 using the weighted Kolmogorov–Smirnov test. Plot the shift function and its 0.95 confidence band. Compare the results with the unweighted test.

2. Compare the two groups of data in Table 5.3 using the weighted Kolmogorov–Smirnov test. Plot the shift function and its 0.95 confidence band. Compare the results with the unweighted test.

3. Summarize the relative merits of using the weighted versus unweighted Kolmogorov–Smirnov test. Also discuss the merits of the Kolmogorov–Smirnov test relative to comparing measures of location.

4. Consider two independent groups having identical distributions. Suppose four observations are randomly sampled from the first and three from the second. Determine $P(D = 1)$ and $P(D = 0.75)$, where D is given by Eq. (5.4). Verify your results with the R function kssig.

5. Compare the deciles only, using the Harrell–Davis estimator, using the data in Table 5.1.

6. Verify that if X and Y are independent, the third moment about the mean of $X - Y$ is $\mu_{x[3]} - \mu_{y[3]}$.

7. Apply the Yuen–Welch method to the data in Table 5.1 where the amount of trimming is 0, 0.05, 0.1, and 0.2. Compare the estimated standard errors of the difference between the trimmed means.

8. Describe a situation where testing H_0: $p = 1/2$ with Mee's method can have lower power than the Yuen–Welch procedure.

9. Comment on the relative merits of testing H_0: $p = 1/2$ with Mee's method versus comparing two independent groups with the Kolmogorov–Smirnov test.

10. Compute a confidence interval for p using the data in Table 5.1.

11. The example at the end of Section 5.3.3 examined some data from an experiment on the effects of drinking alcohol. Another portion of the study consisted of measuring the effects of alcohol over 3 days of drinking. The scores for the control group, for the first 2 days of drinking, are 4, 47, 35, 4, 4, 0, 58, 0, 12, 4, 26, 19, 6, 10, 1, 22, 54, 15, 4, and 22. The experimental group had scores 2, 0, 7, 0, 4, 2, 9, 0, 2, 22, 0, 3, 0, 0, 47, 26, 2, 0, 7, and 2. Verify that the hypothesis of equal 20% trimmed means is rejected with $\alpha = 0.05$. Next, verify that this hypothesis is not rejected when using the equal-tailed bootstrap-t method, but that it is rejected when using the symmetric percentile t procedure instead. Comment on these results.

12. Section 5.9.6 used some hypothetical data to illustrate the R function yuend with 20% trimming. Use the function to compare the means. Verify that the estimated standard error of the difference between the sample means is smaller than the standard error of the difference between the 20% trimmed means. Despite this, the p-value is smaller when comparing trimmed means versus means. Why? Make general comments on this

result. Next, compute the 20% trimmed mean of the difference scores. That is, set $D_i = X_{i1} - X_{i2}$ and compute the trimmed mean using the D_i values. Compare this to the difference between the trimmed means of the marginal distributions, and make additional comments about comparing dependent groups.

13. The file pyge.dat (see Section 1.8) contains pretest reasoning IQ scores for students in grades 1 and 2 who were assigned to one of three ability tracks. (The data are from Elashoff & Snow, 1970, and originally collected by R. Rosenthal.) The file pygc.dat contain data for a control group. The experimental group consisted of children for whom positive expectancies had been suggested to teachers. Compare the 20% trimmed means of the control group to the experimental group using the function yuen and verify that the 0.95 confidence interval is $(-7.12, 27.96)$. Thus, you would not reject the hypothesis of equal trimmed means. What reasons can be given for not concluding that the two groups have comparable IQ scores?

14. Continuing the last exercise, examine a boxplot of the data. What would you expect to happen if the 0.95 confidence interval is computed using a bootstrap-t method? Verify your answer using the R function yuenbt.

15. The file tumor.dat contains data on the number of days to occurrence of a mammary tumor in 48 rats injected with a carcinogen and subsequently randomized to receive either the treatment or the control. The data were collected by Gail, Santner, and Brown (1980) and represent only a portion of the results they reported. (Here, the data are the number of days to the first tumor. Most rats developed multiple tumors, but these results are not considered here.) Compare the means of the two groups with Welch's method for means and verify that you reject with $\alpha = 0.05$. Examine a boxplot and comment on what this suggests about the accuracy of the confidence interval for $\mu_1 - \mu_2$. Verify that you also reject when comparing M-measures of location. What happens when comparing 20% trimmed means or when using the Kolmogorov–Smirnov test?

16. Let $D = X - Y$, let θ_D be the population median associated with D, and let θ_X and θ_Y be the population medians associated with X and Y, respectively. Verify that under general conditions, $\theta_D \neq \theta_X - \theta_Y$.

17. Using R, generate 30 observations from a standard normal distribution and store the values in x. Generate 20 observations from a chi-squared distribution with one degree of freedom and store them in z. Compute y=4(z-1), so x and y contain data sampled from distributions having identical means. Apply the permutation test based on means with the function permg. Repeat this 200 times an determine how often the function rejects. What do the results indicate about controlling the probability of a type I error with the permutation test when testing the hypothesis of equal means? What does this suggest about computing a confidence interval for the difference between the means based on the permutation test?

CHAPTER 6

Some Multivariate Methods

The goal in this chapter is to discuss some basic problems and issues related to multivariate data and how they might be addressed. Then some inferential methods, based on the concepts introduced in this chapter, are described. This area has grown tremendously in recent years and, as usual, no attempt is made to provide an encyclopedic coverage of all techniques. Indeed, for some problems, many strategies are now available, for some purposes there are reasons for preferring certain ones over others, but the reality is that more needs to be done in terms of understanding the relative merits of these methods, and it seems that no single technique can be expected to be satisfactory among all situations encountered in practice.

6.1 Generalized Variance

It helps to begin with a brief review of a basic concept from standard multivariate statistical methods. Consider a random sample of n observations from some p-variate distribution and let

$$s_{jk} = \frac{1}{n-1} \sum_{i=1}^{n} (X_{ij} - \bar{X}_j)(X_{ik} - \bar{X}_k)$$

be the usual sample covariance between the jth and kth variables, where $\bar{X}_j = \sum_i X_{ij}/n$ is the sample mean of the jth variable. Letting **S** represent the corresponding sample covariance matrix, the generalized sample variance is

$$G = |\mathbf{S}|,$$

the determinant of the covariance matrix. The property of G that will be exploited here is that it is sensitive to outliers. Said another way, to get a value for G that is relatively small requires that all points be tightly clustered together. If even a single point is moved farther away from the others, G will increase.

Although not used directly, it is noted that an R function

$$\text{gvar(m)}$$

has been supplied to compute the generalized variance based on the data in m, where m can be any R matrix having *n* rows and *p* columns.

6.2 Depth

A general problem that has taken on increased relevance in applied work is measuring or characterizing how deeply a point is located within a cloud of data. Many strategies exist, but the focus here is on methods that have been found to have practical value when estimating location or testing hypotheses. (For a formal description and discussion of properties measures of depth should have, see Zuo & Serfling, 2000a; cf., Zuo & Serfling, 2000b. For a general theoretical perspective, see Mizera, 2002.) This is not to suggest that all alternative methods for measuring depth should be eliminated from consideration, but more research is needed to understand their relative merits when dealing with the problems covered in this book.

6.2.1 Mahalanobis Depth

Certainly the best-known approach to measuring depth is based on Mahalanobis distance. The squared *Mahalanobis distance* between a point $\mathbf{x}$ (a column vector having length p) and the sample mean, $\bar{\mathbf{X}} = (\bar{X}_1, \dots, \bar{X}_p)'$, is

$$d^2 = (\mathbf{x} - \bar{\mathbf{X}})' \mathbf{S}^{-1} (\mathbf{x} - \bar{\mathbf{X}}). \tag{6.1}$$

A convention is that the deepest points in a cloud of data should have the largest numerical depth. Following Liu and Singh (1997), *Mahalanobis depth* is taken to be

$$M_D(\mathbf{x}) = [1 + (\mathbf{x} - \bar{\mathbf{X}})' \mathbf{S}^{-1} (\mathbf{x} - \bar{\mathbf{X}})]^{-1}. \tag{6.2}$$

So the closer a point happens to be to the mean, as measured by Mahalanobis distance, the larger is its Mahalanobis depth.

Mahalanobis distance is not robust and is known to be unsatisfactory for certain purposes to be described. Despite this, it has been found to have value for a wide range of hypothesis testing problems and has the advantage of being fast and easy to compute with existing software.

6.2.2 Halfspace Depth

Another preliminary that is relevant to this chapter is the notion of halfspace depth, an idea originally developed by Tukey (1975); it reflects a generalization of the notion of ranks to multivariate data. In contrast to other strategies, halfspace depth does not use a covariance matrix.

The idea is, perhaps, best conveyed by first focusing on the univariate case, $p = 1$. Given some number x, consider any partitioning of all real numbers into two components: those values below x and those above. All values less than or equal to x form a *closed halfspace*, and all points strictly less than x is an open halfspace. In a similar manner, all points greater than or equal to x is a closed halfspace, and all points greater than x is an open halfspace. In statistical terms, $F(x) = P(X \leq x)$ is the probability associated with a closed halfspace formed by all points less than or equal to x. The notation

$$F(x^-) = P(X < x)$$

represents the probability associated with an open halfspace. Tukey's halfspace depth associated with the value x is intended to reflect how deeply x is nested within the distribution $F(x)$. Its formal definition is

$$T_D(x) = \min[F(x), \, 1 - F(x^-)]. \tag{6.3}$$

That is, the depth of x is the smallest probability associated with the two closed halfspaces formed by x. The estimate of $T_D(x)$ is obtained simply by replacing F with its usual estimate, $\hat{F}(x)$, the proportion of X_i values less than or equal to x, in which case $1 - \hat{F}(x^-)$ is the proportion of X_i values greater than or equal to x. So for $p = 1$, Tukey's halfspace depth is estimated to be the smaller of two proportions: the proportion of observed values less than or equal to x, and the proportion greater than or equal to x. When F is a continuous distribution, the maximum possible depth associated with any point is 0.5 and corresponds to the population median. However, the maximum possible depth in a sample of observations can exceed 0.5.

■ Example

Consider the values 2, 5, 9, 14, 19, 21, and 33. The proportion of values less than or equal to 2 is $1/7$ and the proportion greater than or equal to 2 is $7/7$, so the halfspace depth of 2 is $1/7$. The halfspace depth of 14 is $4/7$. Note that 14 is the usual sample median, which can be viewed as the average of the points having the largest halfspace depth.

■

■ Example

For the values 2, 5, 9, 14, 19, and 21, both the values 9 and 14 have the highest halfspace depth among the six observations, which is 0.5, and the average of these two points is the usual sample median. The value 1, relative to the samples 2, 5, 9, 14, 19, and 21, has a halfspace depth of zero.

■

Now we generalize to the bivariate case, $p = 2$. For any line, the points on or above this line form a closed halfspace, as do the points on or below the line. Note that for any bivariate distribution, there is a probability associated with the two closed halfspaces formed by any line. For $p = 3$, any plane forms two closed halfspaces: those points on or above the plane, as well as the points on or below the plane, and the notion of a halfspace generalizes in an obvious way for any p.

For the general p-variate case, consider any point $\mathbf{x}$, where again $\mathbf{x}$ is a column vector having length p, let $\mathcal{H}$ be any closed halfspace containing the point $\mathbf{x}$, and let $P(\mathcal{H})$ be the probability associated with $\mathcal{H}$. That is, $P(\mathcal{H})$ is the probability that an observation occurs in the halfspace $\mathcal{H}$. Then roughly, the halfspace depth of the point $\mathbf{x}$ is the smallest value of $P(\mathcal{H})$ among all halfspaces $\mathcal{H}$ containing $\mathbf{x}$. More formally, Tukey's halfspace depth is

$$T_D = \inf_{\mathcal{H}}[P(\mathcal{H}) : \mathcal{H} \text{ is a closed halfspace containing } \mathbf{x}]. \qquad (6.4)$$

For $p > 1$, halfspace depth can be defined instead as the least depth of any one-dimensional projection of the data (Donoho & Gasko, 1992). To elaborate, consider any point $\mathbf{x}$ and any p-dimensional (column) vector $\mathbf{u}$ having unit norm. That is, the *Euclidean norm* of $\mathbf{u}$ is $\|\mathbf{u}\| = \sqrt{u_1^2 + \cdots + u_p^2} = 1$. Then a one-dimensional projection of $\mathbf{x}$ is $\mathbf{u}'\mathbf{x}$ (where $\mathbf{u}'$ is the transpose of $\mathbf{u}$). For any projection, meaning any choice for $\mathbf{u}$, depth is defined by Eq. (6.3). In the p-variate case, the depth of a point is defined to be its minimum depth among all possible projections, $\mathbf{u}'\mathbf{X}$. Obviously, from a practical point of view, this does not immediately yield a viable algorithm for computing halfspace depth based on a sample of n observations, but it suggests an approximation that has been found to be relatively effective.

A data set is said to be in *general position* if there are no ties, no more than two points are on any line, no more than three points are in any plane, and so forth. It should be noted that the maximum halfspace depth varies from one situation to the next and in general does not equal 0.5. If the data are in general position, the maximum depth lies roughly between $1/(p + 1)$ and 0.5 (Donoho & Gasko, 1992).

Halfspace depth is metric free in the following sense. Let $\mathbf{A}$ be any nonsingular p-by-p matrix and let $\mathbf{X}$ be any n-by-p matrix of n points. Then the halfspace depths of these n points are unaltered under the transformation $\mathbf{X}\mathbf{A}$. More formally, halfspace depth is *affine invariant*.

6.2.3 Computing Halfspace Depth

For $p = 2$ and 3, halfspace depth, relative to $\mathbf{X}_1, \ldots, \mathbf{X}_n$, can be computed exactly (Rousseeuw & Ruts, 1996; Rousseeuw & Struyf, 1998). For $p > 3$, currently an approximation must be used and three are provided here. (For more details about the first two approximations, see Wilcox, 2003a. For yet another approximation of the point having the greatest depth, see Struyf & Rousseeuw, 2000.) Cuesta-Albertos and Nieto-Reyes (2008) suggest a method for determining Tukey's halfspace depth based on randomly chosen projections of the data, but it is unknown whether this has any practical advantage over the methods covered here, which are not based on random projections of the data.

Approximation A1 The first approximation is based on the one-dimensional projection definition of depth. First, an informal description is given after which the computational details are provided. The method begins by computing some multivariate measure of location, say $\hat{\theta}$. There are many reasonable choices, and for present purposes it seems desirable that it be robust. A simple choice is to use the marginal medians, but a possible objection is that they do not satisfy a criterion discussed in Section 6.3. (The marginal medians are not affine equivariant.) To satisfy this criterion, the MCD estimator in Section 6.3.2 is used with the understanding that the practical advantages of using some other estimator has received virtually no attention. Given an estimate of the center of the data, consider the line formed by the ith observation, $\mathbf{X}_i$, and the center. For convenience, call this line $\mathcal{L}$. Now, (orthogonally) project all points onto the line $\mathcal{L}$. That is, for every point $\mathbf{X}_j$, $j = 1, \ldots, n$, draw a line through it that is perpendicular to the line $\mathcal{L}$. Where this line intersects $\mathcal{L}$ is the projection of the point. Figure 6.1 illustrates the process where the point marked by a circle indicates the center of the data, the line going through the center is line $\mathcal{L}$, and the arrow indicates the projection of the point. That is, the arrow points to the point on the line $\mathcal{L}$ that corresponds to the projection. Next, repeat this process for each i, $i = 1, \ldots, n$. So for each projected point $\mathbf{X}_j$, we have a depth based on the ith line formed by $\mathbf{X}_i$ and $\hat{\theta}$. Call this depth d_{ij}. For fixed j, the halfspace depth of $\mathbf{X}_j$ is approximated by the minimum value among $d_{1j}, \ldots, d_{nj}$.

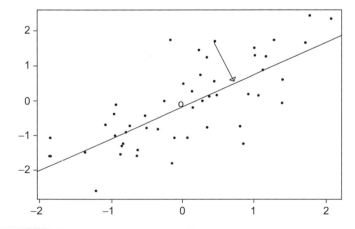

Figure 6.1: An illustration of projecting a point onto a line.

Now a more precise description of the calculations is given. For any i, $i = 1, \ldots, n$, let

$$\mathbf{U}_i = \mathbf{X}_i - \hat{\theta},$$

$$B_i = \mathbf{U}_i \mathbf{U}_i'$$

$$= \sum_{k=1}^{p} U_{ik}^2$$

and for any j ($j = 1, \ldots, n$) let

$$W_{ij} = \sum_{k=1}^{p} U_{ik} U_{jk},$$

and

$$T_{ij} = \frac{W_{ij}}{B_i}(U_{i1}, \ldots, U_{ip}). \tag{6.5}$$

The distance between $\hat{\theta}$ and the projection of $\mathbf{X}_j$ (when projecting onto the line connecting $\mathbf{X}_i$ and $\hat{\theta}$) is

$$D_{ij} = \text{sign}(W_{ij}) \, \| T_{ij} \|,$$

where $\| T_{ij} \|$ is the Euclidean norm associated with the vector T_{ij}. Let d_{ij} be the depth of $\mathbf{X}_j$ when projecting points onto the line connecting $\mathbf{X}_i$ and $\hat{\theta}$. That is, for fixed i and j, the depth of the projected value of $\mathbf{X}_j$ is

$$d_{ij} = \min[\#(D_{ij} \leq D_{ik}), \#(D_{ij} \geq D_{ik})],$$

where $\#\{D_{ij} \leq D_{ik}\}$ indicates how many D_{ik} ($k = 1, \ldots, n$) values satisfy $D_{ij} \leq D_{ik}$. Then the depth of $\mathbf{X}_j$ is taken to be

$$L_j = \min d_{ij},$$

the minimum being taken over all $i = 1, \ldots, n$.

Approximation A2 The second approximation of halfspace depth does not use a measure of location, rather, it uses all projections between any two points. That is, method A1 forms n lines, namely the lines passing through the center of the scatterplot and each of the n points. Method A2 uses $(n^2 - n)/2$ lines, namely all lines formed by any two (distinct) points. An advantage of method A2 is that, in situations where the exact depth can be determined, it has been found to be more accurate on average than method A1 or the method recommended by Rousseeuw and Struyf (1998); see Wilcox (2003a). Another disadvantage of method A1 is that in some situations to be described, the MCD estimate of location that it uses cannot be computed because the covariance matrix of the data is singular. (Switching to the MVE estimator described in Section 6.3.1 does not correct this problem.) A possible appeal of method A2 is that depth can still be computed in these situations. A negative feature of A2 is that with n large, execution time can be relatively high.

Although A2 is a more accurate approximation of halfspace depth than A1, perhaps in applied work this is not a concern. That is, we can think of A1 as a method of defining the depth of a point in a scatterplot, and for practical purposes, maybe any discrepancies between the exact halfspace depth and the depth of a point based on A1 have no negative consequences. This issue has received virtually no attention.

Approximation A3 A third approximation was derived by Rousseeuw and Struyf (1998), but no attempt is made to describe the involved details. A positive feature is that it has low execution time even for fairly large data sets, but it currently seems to be less accurate than the two previous approximations covered in this section. The practical consequences of using method A3 over A2 and A1, when dealing with inferential methods, have not been studied.

6.2.4 R Functions depth2, depth, fdepth, fdepthv2, and unidepth

The R function

$$unidepth(x,pts=NA)$$

is designed for univariate data only and computes the depth of the values in pts relative to the data stored in x. If pts is not specified, the function computes the depth of each value in x relative to all the values.

The function

$$depth(U,V,m)$$

handles bivariate data only and computes the exact halfspace depth of the point (U,V) within the data contained in the matrix m. For example, depth(0,0,m) returns the halfspace depth of $(0, 0)$. The function

$$depth2(m,pts=NA)$$

also handles bivariate data only, but unlike depth, it computes the depth of all the points stored in the matrix pts. If pts is not specified, the exact depths of all the points stored in m are determined. (The function depth is supplied mainly for convenience when computing the halfspace depth of a single point.)

For p-variate data, $p \geq 1$, the function

$$fdepth(m,pts=NA,plotit=T,cop=2)$$

computes the approximate halfspace depth of all the points in pts relative to the data in m using method A1. (For $p = 1$, the exact halfspace depth is computed by calling the function unidepth.) If pts is not specified, the function returns the halfspace depth of all points in m. The argument cop indicates the location estimator, $\hat{\theta}$, that will be used by method A1. By default, the MCD estimate is used. Setting cop=3 results in using the marginal medians, cop=4 uses the MVE estimator (discussed in Section 6.3.1). If simultaneously pts is not specified, m contains two columns of data (so bivariate data are being analyzed), and plotit=T, a scatterplot of the data is created that marks the center of the data corresponding to the location estimator specified by the argument cop, and it creates what Liu, Parelius, and Singh (1999, p. 789) call, the 0.5 depth contour. It is a polygon (a convex hull) containing the central half of the data as measured by the depth of the points. The function

$$\text{fdepthv2(m,pts=NA,plotit = T)}$$

uses method A2 to approximate halfspace depth.

6.2.5 Projection Depth

The approximation of halfspace depth, represented by method A1 in Section 6.2.3, suggests another measure of depth that has been found to have practical value when testing hypotheses. Roughly, for each of the n projections used in method A1, compute the distance between the estimated center ($\hat{\theta}$) and the ith point. The *projection distance* associated with the jth point, $\mathbf{X}_j$, is taken to be the largest distance among all n projections after the distances are standardized by dividing by some measure of scale. This can be converted to a measure of projection depth using the same strategy applied to the Mahalanobis distance.

More precisely, compute T_{ij} as given by Eq. (6.5) and let

$$D_{ij} = \| T_{ij} \| .$$

So for the projection of the data onto the line connecting $\mathbf{X}_i$ and $\hat{\theta}$, D_{ij} is the distance between $\mathbf{X}_j$ and $\hat{\theta}$. Now let

$$d_{ij} = \frac{D_{ij}}{q_2 - q_1}, \tag{6.6}$$

where for fixed i, q_2 and q_1 are the ideal fourths based on the values $D_{i1}, \ldots D_{in}$. The *projection distance* associated with $\mathbf{X}_j$, say $p_d(\mathbf{X}_j)$, is the maximum value of d_{ij}, the maximum being taken over $i = 1, \ldots, n$. To convert to a measure of depth, we simply use

$$P_D(\mathbf{X}_j) = \frac{1}{1 + p_d(\mathbf{X}_j)}. \tag{6.7}$$

A variation of the notion of projected distance is discussed by Donoho and Gasko (1992). The main difference is that they use MAD as a measure of scale for the D values in Eq. (6.6) rather than the interquartile range, $q_2 - q_1$. Here we take this to mean that Eq. (6.6) becomes

$$d_{ij} = \frac{0.6745 D_{ij}}{\text{MAD}}, \tag{6.8}$$

where for fixed i, MAD is based on the values $D_{i1}, \ldots D_{in}$. An obvious advantage of MAD over the interquartile range is that MAD has a higher breakdown point. However, based on other criteria (to be described), the use of the interquartile range has been found to have practical value.

6.2.6 R functions pdis and pdisMC

The R function

$$\text{pdis(m,MM=F,cop=3,dop=1,center=NA)}$$

computes projection distances, p_d, for the data stored in the matrix m. If MM=F is used, distances are scaled using the interquartile range, and MM=T uses MAD. The argument cop indicates which measure of location will be used. The choices, some of which are described in Section 6.3, are:

- cop=1, Donoho–Gasko median
- cop=2, MCD estimate of location
- cop=3, marginal medians
- cop=4, MVE estimate of location
- cop=5, OP (skipped) estimator

If a value is supplied for the argument center (a vector having length p), this value is used as a measure of location and the argument cop is ignored. (The argument dop is relevant when using the Donoho–Gasko analog of the trimmed mean. See Section 6.3.5.) If a multi-core processor is available, execution time can be reduced by using the R function

$$\text{pdisMC(m,MM=F,cop=3,dop=1,center=NA)}$$

instead of the function pdis.

6.2.7 Other Measures of Depth

For completeness, some additional measures of depth are mentioned. Two measures that are similar in spirit to halfspace depth are simplicial depth (Liu, 1990) and majority depth proposed by Singh. (See Liu & Singh, 1997, p. 268.) Both methods are geometric in nature and do not rely on some measure of multivariate scatter. (For some possible concerns about simplicial depth, see Zuo & Serfling, 2000a; cf., Zuo & Serfling, 2000b.) Both are nontrivial to compute and any practical advantages they might enjoy over halfspace depth have not been discovered as yet, so further details are omitted. Another approach is to use Mahalanobis distance but with the mean and usual covariance matrix replaced by robust estimators. Some possibilities are mentioned later in this chapter.

Zuo (2003) studied a notion of projection-based depth that is a broader generalization of other projection-based approaches. Again let **u** be any p-dimensional (column) vector having unit norm and let

$$O(\mathbf{x}; F) = \sup \frac{|\mathbf{u}'\mathbf{x} - \theta(F_u)|}{\sigma(F_u)},$$

be some measure of outlyingness of the point **x**, with respect to the distribution F, where F_u is the distribution of $\mathbf{u}'\mathbf{x}$, $\theta(F_u)$ is some (univariate) measure of location and $\sigma(F_u)$ is some measure of scale associated with F_u, and the supremum is taken over all possible choices for **u** such that $\|u\| = 1$. Then the projection depth of the point **x** is taken to be

$$P_D(\mathbf{x}) = \frac{1}{1 + O(\mathbf{x})}.$$

Inferential methods based on this notion of depth have not been studied as yet.

6.2.8 R Function zdepth

The R function

$$\text{zdepth(m,pts=m,zloc=median,zscale=mad)}$$

computes Zuo's notion of projection distance, $O(\mathbf{x}; F)$, for each point stored in the matrix pts, relative to the data stored in the matrix m. The arguments zloc and zscale correspond to $\theta(F_u)$ and $\sigma(F_u)$, respectively.

6.3 Some Affine Equivariant Estimators

One of the most difficult problems in robust statistics has been the estimation of multivariate shape and location. Many such estimators have been proposed (e.g., Davies, 1987; Donoho, 1982; Kent & Tyler, 1996; Lopuhaä , 1991; Maronna & Zamar, 2002; Rousseeuw's, 1984, 1985; Stahel, 1981; Tamura & Boos, 1986; Tyler, 1994; Wang & Raftery, 2002; Rocke & Woodruff, 1996.) A concern about early attempts, such as multivariate M-estimators as described in Huber (1981), is that when working with p-variate data, typically they have a breakdown point of at most $1/(p+1)$. So in high dimensions, a very small fraction of outliers can result in very bad estimates. Several estimators have been proposed that enjoy a high breakdown point. But simultaneously achieving relatively high accuracy, versus the vector of means when sampling from a multivariate normal distribution, has proven to be a very difficult problem.

In Section 2.1, a basic requirement for θ to qualify as a measure of location was that it be both scale and location equivariant. Moreover, a location estimator should satisfy this property as well. That is, if $T(X_1, \ldots, X_n)$ is to qualify as a location estimator, it should be the case that for constants a and b,

$$T(aX_1 + b, \ldots, aX_n + b) = aT(X_1, \ldots, X_n) + b.$$

So, for example, when transforming from feet to centimeters, the typical value in feet is transformed to the appropriate value in centimeters. In the multivariate case, a generalization of this requirement, called *affine equivariance*, is that for a p-by-p nonsingular matrix $\mathbf{A}$ and vector $\mathbf{b}$ having length p,

$$T(\mathbf{X}_1\mathbf{A} + \mathbf{b}, \ldots, \mathbf{X}_n\mathbf{A} + \mathbf{b}) = T(\mathbf{X}_1, \ldots, \mathbf{X}_n)\mathbf{A} + \mathbf{b}, \tag{6.9}$$

where now $\mathbf{X}_1, \ldots, \mathbf{X}_n$ is a sample from a p-variate distribution and each $\mathbf{X}_i$ is a (row) vector having length p. So in particular, the estimate is transformed properly under rotations of the data as well as changes in location and scale. The sample means of the marginal distributions are affine equivariant, but typically, when applying any of the univariate estimators in Chapter 3 to the marginal distributions, an affine equivariant estimator is not obtained.

A measure of scatter, say $\mathbf{V}(\mathbf{X})$, is said to be *affine equivariant* if

$$\mathbf{V}(\mathbf{A}\mathbf{X} + \mathbf{b}) = \mathbf{A}\mathbf{V}(\mathbf{X})\mathbf{A}'. \tag{6.10}$$

The usual sample covariance matrix is affine equivariant but not robust.

From Donoho and Gasko (1992, p. 1811), no affine equivariant estimator can have a breakdown point greater than

$$\frac{n - p + 1}{2n - p + 1}. \tag{6.11}$$

(Also see Lupuhaä & Rousseeuw, 1991.)

The rest of this section describes some of the estimators that have been proposed, and a particular variation of one of these methods is described in Section 6.5.

6.3.1 Minimum Volume Ellipsoid Estimator

One of the earliest affine equivariant estimators to achieve a breakdown point of approximately 0.5 is the so-called minimum volume ellipsoid (MVE) estimator, a detailed discussion of which can be found in Rousseeuw and Leroy (1987). Consider any ellipsoid containing half of the data. (An example in the bivariate case is shown in Figure 6.2.) The basic idea is to search among all such ellipsoids for the one having the smallest volume. Once this subset is found, the mean and covariance matrix of the corresponding points are taken as the estimated measure of location and scatter, respectively. Typically the covariance matrix is rescaled to obtain consistency at the multivariate normal model (e.g., Marazzi, 1993, p. 254). A practical problem is that it is generally difficult to find the smallest ellipse containing half of the data. That is, in general, the collection of all subsets containing half of the data is so large, determining the subset that has the minimum volume is impractical, so an approximation must be used. Let h be equal to $n/2 + 1$, rounded down to the nearest integer. An approach to computing the MVE estimator is to randomly select h points, without replacement, from the n points available, compute the volume of the ellipse containing these points, and then repeat this process many times. The set of points yielding the smallest volume is taken to be the minimum volume ellipsoid. (For relevant software, see Section 6.4.5.)

6.3.2 The Minimum Covariance Determinant Estimator

An alternative to the MVE estimator, which also has a breakdown point of approximately 0.5, is the so-called minimum covariance determinant (MCD) estimator. Rather than search for the subset of half the data that has the smallest volume, search for the half that has the smallest generalized variance. (For recent results on computing the MCD estimator, see Schnys, Haesbroeck, & Critchley, 2010.) Recall from Section 6.1 that for the determinant of the covariance (the generalized variance) to be relatively small, it must be the case that there are no outliers. That is, the data must be tightly clustered together. The MCD estimator searches for the half of the data that is most tightly clustered together among all subsets containing half of the data, as measured by the generalized variance. Like the MVE estimator, typically it is impractical considering all subsets of half the data, so an approximate method must be used. An algorithm for accomplishing this goal is described in Rousseeuw and van Driessen (1999); also see Atkinson (1994). For asymptotic results, see Butler, Davies, and Jhun (1993). Once an approximation of the subset of half of the data has been determined that minimizes the generalized variance, compute the usual mean and covariance matrix based on this subset. This yields the MCD estimate of location and scatter. Bernholt and Fischer (2004) indicate that this algorithm can provide a poor approximation of the MCD estimator. Results reported by Hawkins and Olive (2002) also raise concerns about this estimator. But as a diagnostic tool, MCD seems to have practical value when used in conjunction with other methods covered in this chapter. (For relevant software, see Section 6.4.5.)

Herwindiati, Djauhari, and Mashuri (2007) suggest a variation of the MCD estimator that searches for the subset of the data that minimizes the trace of the corresponding covariance matrix rather than the determinant, what they call the *minimum variance vector* (MVV) method. It has the same breakdown point as the MCD method and is simpler to compute.

Herwindiati et al. suggest that the method is applicable when dealing with large, high-dimensional data sets. In terms of identifying outliers, limited results suggest that it performs as well as the MCD estimator, but further study is needed.

6.3.3 S-Estimators and Constrained M-Estimators

One of the earliest treatments of S-estimators can be found in (Rousseeuw and Leroy, 1987, p. 263). A particular variation of this method that appears to be especially interesting is the translated biweight S-estimator (TBS) proposed by Rocke (1996). Generally, S-estimators of multivariate location and scatter are values for $\hat{\theta}$ and $\mathbf{S}$ that minimize $|\mathbf{S}|$, the determinant of $\mathbf{S}$, subject to

$$\frac{1}{n} \sum_{i=1}^{n} \xi\{[(\mathbf{X}_i - \hat{\theta})'\mathbf{S}^{-1}(\mathbf{X}_i - \hat{\theta})]^{1/2}\} = b_0, \tag{6.12}$$

where b_0 is some constant, and (as in Chapter 2) ξ is a nondecreasing function. Lopuhaä (1989) showed that S-estimators are in the class of M-estimators with standardizing constraints. Rocke (1996) showed that S-estimators can be sensitive to outliers even if the breakdown point is close to 0.5.

Rocke (1996) proposed a modified biweight estimator, which is essentially a constrained M-estimator, where for values of m and c to be determined, the function $\xi(d)$, when $m \leq d \leq m+c$, is

$$\xi(d) = \frac{m^2}{2} - \frac{m^2(m^4 - 5m^2c^2 + 15c^4)}{30c^4} + d^2\left(0.5 + \frac{m^4}{2c^4} - \frac{m^2}{c^2}\right)$$
$$+ d^3\left(\frac{4m}{3c^2} - \frac{4m^3}{3c^4}\right) + d^4\left(\frac{3m^2}{2c^4} - \frac{1}{2c^2}\right) - \frac{4md^5}{5c^4} + \frac{d^6}{6c^4},$$

for $0 \leq d < m$,

$$\xi(d) = \frac{d^2}{2},$$

and for $d > m+c$,

$$\xi(d) = \frac{m^2}{2} + \frac{c(5c + 16m)}{30}.$$

The values for m and c can be chosen to achieve the desired breakdown point and the *asymptotic rejection probability*, roughly referring to the probability that a point will get zero weight when the sample size is large. If the asymptotic rejection probability is to be γ say, then m and c are determined by

$$E_{\chi_p^2}(\xi(d)) = b_0,$$

and

$$m + c = \sqrt{\chi^2_{p,1-\gamma}},$$

where $\chi^2_{p,1-\gamma}$ is the $1 - \gamma$ quantile of a chi-squared distribution with p degrees of freedom.

6.3.4 R Function tbs

The R function

$$tbs(m)$$

computes the TBS measure of location and scatter just outlined using code supplied by David Rocke.[1]

6.3.5 Donoho–Gasko Generalization of a Trimmed Mean

Another approach to computing an affine equivariant measure of location was suggested and studied by Donoho and Gasko (1992). The basic strategy is to compute the halfspace depth for each of the n points, remove those that are not deeply nested within the cloud of data, and then average those points that remain. The *Donoho–Gasko* γ *trimmed mean* is the average of all points which are at least γ deep in the sample. That is, points having depth less than γ are trimmed and the mean of the remaining points is computed. An analog of the median, which has been called *Tukey's median*, is the average of all points having the largest depth (cf., Adrover & Yohai, 2002; Bai & He, 1999; Tyler, 1994). If the maximum depth of $\mathbf{X}_i$, $i = 1, \ldots, n$ is greater than or equal to γ, then the breakdown point of the Donoho–Gasko γ trimmed mean is $\gamma/(1 + \gamma)$. For symmetric distributions the breakdown point is approximately 0.5, but because the maximum depth among a sample of n points can be approximately $1/(1 + p)$, the breakdown point could be as low as $1/(p + 2)$. If the data are in general position, the breakdown point of Tukey's median is greater than or equal to $1/(p + 1)$. (The influence function, assuming a type of symmetry for the distribution, was derived by Chen & Tyler, 2002.)

■ **Example**

Table 6.1. shows results from Raine, Buchsbaum, and LaCasse (1997) who were interested in comparing EEG measures for murderers versus a control group. For the

---

[1] S-PLUS comes with a library called robust that contains the function covRob, which computes the TBS estimate of location and scatter. However, when checking for outliers, Rocke's code appears to be a bit more satisfactory (Wilcox, 2008a).

Table 6.1: EEG Measures for Murderers and a Control Group

Control				Murderers			
Site 1	Site 2	Site 3	Site 4	Site 1	Site 2	Site 3	Site 4
−0.15	−0.05	−0.33	−1.08	−0.26	−2.10	1.01	−0.49
−0.22	−1.68	0.20	−1.19	0.25	−0.47	0.77	−0.27
0.07	−0.44	0.58	−1.97	0.61	−0.91	−0.68	−1.00
−0.07	−1.15	1.08	1.01	0.38	−0.15	−0.20	−1.09
0.02	−0.16	0.64	0.02	0.87	0.23	−0.37	−0.83
0.24	−1.29	1.22	−1.01	−0.12	−0.51	0.27	−1.03
−0.60	−2.49	0.39	−0.69	0.15	−1.34	1.44	0.65
−0.17	−1.07	0.48	−0.56	0.93	−0.87	1.53	−0.10
−0.33	−0.84	−0.33	−1.86	0.26	−0.41	0.78	0.92
0.23	−0.37	0.50	−0.23	0.83	−0.02	−0.41	−1.01
−0.69	0.01	0.19	−0.22	0.35	−1.12	0.26	−1.81
0.70	−1.24	1.59	−0.68	1.33	−0.57	0.04	−1.12
1.13	−0.33	−0.28	−0.93	0.89	−0.78	−0.27	−0.32
0.38	0.78	−0.12	−0.61	0.58	−0.65	−0.60	−0.94

moment, consider the first two columns of data only. The exact halfspace depths, determined by the function depth2, are:

0.1428570 0.1428570 0.3571430 0.2142860 0.2142860 0.2142860 0.0714286
0.2142860 0.1428570 0.2142860 0.0714286 0.0714286 0.0714286 0.0714286

There are five points with depth less than 0.1. Eliminating these points and averaging the values that remain yields the Donoho–Gasko 0.1 trimmed mean, (−0.042, −0.783). The halfspace median corresponds to the deepest point, which is (0.07, −0.44). ∎

Another multivariate generalization of a trimmed mean was studied by Liu et al. (1999). Any practical advantages it might have over the Donoho–Gasko γ trimmed mean have not been discovered as yet, so for brevity, no details are given here.

6.3.6 R Functions dmean and dcov

The R function

$$\text{dmean(x,tr=.2,dop=1,cop=2)}$$

computes the Donoho–Gasko trimmed mean. When the argument tr is set equal to 0.5, it computes Tukey's median, namely, the average of the points having the largest halfspace depth. The argument dop controls how the halfspace depth is approximated. With dop=1, method A1 in halfspace depth is approximated. With dop=1, method A1 in Section 6.2.3 is used to approximate halfspace depth when $p > 2$, whereas whereas dop=2 uses method A2. If $p = 2$, halfspace depth is computed exactly. When using method A1, the center of the scatterplot is determined using the estimator indicated by the argument cop. The choices are:

- cop=2, MCD estimator
- cop=3, marginal medians
- cop=4, MVE estimato

When n is small relative to p, the MCD and MVE estimators cannot be computed, so in these cases, use dop=2. For small sample sizes, execution time is low.

Consider again Zuo's notion of projection depth described in Section 6.2.7. When $\theta(F_u)$ is taken to be the median and $\sigma(F_u)$ is MAD, and if the average of the deepest points are used as a measure of location, we get another affine equivariant generalization of the median. Comparisons with other affine equivariant median estimators are reported by Hwang et al. (2004) for the bivariate case. They conclude that this estimator and Tukey's median compare well to other estimators they considered.

6.3.7 The Stahel–Donoho W-Estimator

Stahel (1981) and Donoho (1982) proposed the first multivariate, equivariant estimator of location and scatter that has a high breakdown point. It is a weighted mean and covariance matrix where the weights are a function of how "outlying" a point happens to be. The more outlying a point, the less weight it is given. The notions of Mahalanobis depth, robust analogs of Mahalanobis depth based perhaps on the MVE or MCD estimators, and halfspace depth are examples of how to measure the outlyingness of a point. Attaching some weight w_i to $\mathbf{X}_i$, that is, a function of how outlying the point $\mathbf{X}_i$ happens to be, yields a generalization of W-estimators mentioned in Section 3.8 (cf., Hall & Presnell, 1999). Here, the estimate of location is

$$\hat{\theta} = \frac{\sum_{i=1}^{n} w_i \mathbf{X}_i}{\sum_{i=1}^{n} w_i} \tag{6.13}$$

and the measure of scatter is

$$\mathbf{V} = \frac{\sum_{i=1}^{n} w_i (\mathbf{X}_i - \hat{\theta})(\mathbf{X}_i - \hat{\theta})'}{\sum_{i=1}^{n} w_i}. \tag{6.14}$$

The Donoho–Gasko trimmed mean in Section 6.3.4 is a special case where the least deep points get a weight of zero; otherwise points get a weight of one. Other variations of this approach are based on the multivariate outlier detection methods covered in Section 6.4.

For general theoretical results on this approach to estimation, see Tyler (1994). Properties of certain variations were reported by Maronna and Yohai (1995). Also see Arcones, Chen, & Gine (1994), Bai and He (1999), He and Wang (1997), Donoho and Gasko (1992), Gather and Hilker (1997), Zuo (2003), Zuo, Cui, and He (2004), and Zuo, Cui, and Young (2004). Gervini (2002) derived the influence function assuming that sampling is from an elliptical distribution. (For an extension of M-estimators to the multivariate case that has a high breakdown point and deals with missing values, see Chen & Victoria-Feser, 2002.)

Zuo, Hengjian, and He (2004) and Zuo, Hengjian, and Young (2004) suggest a particular variation of the Donoho–Gasko W-estimator for general use. Let P_i be the projection depth of $\mathbf{x}_i$ described at the end of Section 6.2.7. Let C be the median of the P_i values. If $P_i < C$, set

$$w_i = \frac{\exp[-K(1 - P_i/C)^2] - \exp(-K)}{1 - \exp(-K)},$$

otherwise $w_i = 1$, and the measures of location and scatter are given by Eqs. (6.13) and (6.14), respectively. From Zuo et al., setting the constant $K = 3$ results in good asymptotic efficiency, relative to the sample mean, under normality.

6.3.8 R Function sdwe

The R function

$$\text{sdwe(x,K=3)}$$

computes the Stahel–Donoho W-estimator as suggested by Zuo, Hengjian, and He (2004) and Zuo, Hengjian, and Young (2004).

6.3.9 Median Ball Algorithm

This section describes a multivariate measure of location and scatter, introduced by Olive (2004), which is based on what he calls the reweighted median ball algorithm (RMBA). It is an iterative algorithm that begins with two initial estimates of location and scatter. The first, labeled $(T_{0,1}, \mathbf{C}_{0,1})$, is taken to be the usual mean and covariance matrix. The other starting value, $(T_{0,2}, \mathbf{C}_{0,2})$, is the usual mean and covariance based on the $c_n \approx n/2$ cases that are closest to the coordinate wise median in Euclidean distance. Compute all n Mahalanobis distances $D_i(T_{0,j}, \mathbf{C}_{0,j})$ based on the jth starting value. The next iteration consists of estimating the usual mean and covariance matrix based on the c_n cases corresponding to the smallest distances, yielding $(T_{1,j}, \mathbf{C}_{1,j})$. Repeating this process, based on $D_i(T_{1,j}, \mathbf{C}_{1,j})$, yields an updated measure of location and scatter, $(T_{2,j}, \mathbf{C}_{2,j})$. As done by Olive, unless stated otherwise, it is assumed five iterations are used yielding $(T_{5,j}, \mathbf{C}_{5,j})$. The RMBA estimator of location, labeled T_A, is taken to be $T_{5,i}$, where $i = 1$ if the determinant $|\mathbf{C}_{5,1}| \leq |\mathbf{C}_{5,2}|$, otherwise $i = 2$. And the measure of scatter is

$$\mathbf{C}_{\text{RMBA}} = \frac{\text{MED}[D_i^2(T_A, \mathbf{C}_A)]}{\chi_{p,0.5}^2}\mathbf{C}_A.$$

The RMBA estimator is $\sqrt{n}$ consistent. (Also see Olive & Hawkins, 2010.)

6.3.10 R Function rmba

The R function

$$\text{rmba(m,csteps=5)}$$

computes the RMBA measure of location and scatter, where the argument csteps controls the number of iterations. (The R code was graciously supplied by David Olive.)

6.3.11 OGK Estimator

Yet another estimator that is sometimes recommended is the orthogonal Gnanadesikan–Kettenring (OGK) estimator, derived by Maronna and Zamar (2002). In its general form, it is applied as follows. Let $\sigma(X)$ and $\mu(X)$ be any measure of dispersion and location, respectively. The method begins with the robust covariance between any two variables, say X and Y, which was proposed by Gnanadesikan and Kettenring (1972):

$$\text{cov}(X, Y) = \frac{1}{4}[\sigma(X+Y)^2 - \sigma(X-Y)^2]. \tag{6.15}$$

When $\sigma(X)$ and $\mu(X)$ are the usual standard deviation and mean, respectively, the usual covariance between X and Y results. Here, following Maronna and Zamar, $\sigma(X)$ is taken to be the tau scale of Yohai and Zamar (1988), which was introduced in Section 3.12.3. Using this measure of scale in Eq. (6.15), the resulting measure of covariance will be denoted by $v(X, Y)$.

Following the notation in Maronna and Zamar (2002), let $\mathbf{x}_i$ be the ith row of the n-by-p matrix $\mathbf{X}$. Then Maronna and Zamar define a scatter matrix $\mathbf{V}(X)$ and a location vector $\mathbf{t}(X)$ as follows:

1. Let $\mathbf{D} = \text{diag}[\sigma(X_1), \ldots, \sigma(X_p)]$ and $\mathbf{y}_i = \mathbf{D}^{-1}\mathbf{x}_i$, $i = 1, \ldots, n$.
2. Compute $\mathbf{U} = (U_{jk})$ by applying v to the columns of $\mathbf{Y}$. So $U_{jj} = 1$ and for $j \neq k$, $U_{jk} = v(Y_j, Y_k)$.
3. Compute the eigenvalues λ_j and eigenvectors $\mathbf{e}_j$ of $\mathbf{U}$ and let $\mathbf{E}$ be the matrix whose columns are the $\mathbf{e}_j$'s. (So $\mathbf{U} = \mathbf{E}\Lambda\mathbf{E}'$.)

4. Let $\mathbf{A} = \mathbf{DE}$, $\mathbf{z}_i = \mathbf{A}^{-1}\mathbf{x}_i$, in which case

$$\mathbf{V}(X) = \mathbf{A}\Gamma\mathbf{A}'$$

and

$$\mathbf{t}(X) = \mathbf{A}\nu,$$

where $\Gamma = \text{diag}(\sigma^2(Z_1), \ldots, \sigma^2(Z_p))$, $\nu = (\mu(Z_1), \ldots, \mu(Z_p))$ and μ is taken to be the tau measure of location in Section 3.8.1.

Maronna and Zamar (2002) note that the above procedure can be iterated and report results suggesting that a single iteration be used. More precisely, compute $\mathbf{V}$ and $\mathbf{t}$ for $\mathbf{Z}$ (the matrix corresponding to $\mathbf{z}_i$ computed in step 4) and then express them in the original coordinate system, namely, $\mathbf{V}_2 = \mathbf{AV}(\mathbf{Z})\mathbf{A}'$ and $\mathbf{t}_2(\mathbf{X}) = \mathbf{At}(\mathbf{Z})$. Maronna and Zamar show that the estimate can be improved by a reweighting step. Let

$$d_i = \sum_j \left[\frac{z_{ij} - \mu(Z_j)}{\sigma(Z_j)} \right]$$

and $w_i = I(d_i \leq d_0)$, where

$$d_0 = \frac{\chi^2_{p,\beta} \text{med}(d_1, \ldots, d_n)}{\chi^2_{p,.5}},$$

$\chi^2_{p,\beta}$ is the β quantile of the chi-squared distribution with p degrees of freedom and "med" denotes the sample median. The measure of location is now estimated to be

$$\mathbf{t}_w = \frac{\sum w_i \mathbf{x}_i}{\sum w_i},$$

and the measure of scatter is

$$\mathbf{V}_w = \frac{\sum w_i (\mathbf{x}_i - \mathbf{t}_w)(\mathbf{x}_i - \mathbf{t}_w)'}{\sum w_i}.$$

6.3.12 R Function ogk

The R function

```
ogk(x,sigmamu=taulc,v=gkcov,n.iter=1,beta=0.9, ...)
```

computes the OGK measure of location and scale.

6.3.13 An M-Estimator

As noted at the beginning of this section, a concern about (affine equivariant) M-estimators is that they have a breakdown point of at most $1/(p+1)$. Also, Devlin, Gnanadesikan, and Kettenring (1981, p. 361) found that M-estimators could tolerate even fewer outliers than indicated by this upper bound. Despite this, in situations where p is small, this approach might be deemed satisfactory. For example, Zu and Yuan (2010) suggest an approach to a mediation analysis that is based in part on an M-estimator with Huber weights, which was derived by Maronna (1976). A slight modification of the Zu and Yuan method has been found to perform relatively well in simulations, so for completeness, Maronna's M-estimator is outlined here. (Details of Zu and Yuan method for performing a mediation analysis are outlined in Section 11.7.2.)

The computation of this estimator is accomplished via an iterative scheme that corresponds to a multivariate version of the W-estimator in Section 3.8. Roughly, an initial estimate of the mean and covariance matrix is computed, which here is taken to be usual mean $\bar{\mathbf{X}}$ vector and covariance matrix $\mathbf{S}$. Based on this initial estimate, squared Mahalanobis distances are computed:

$$d_i^2 = (\mathbf{X}_i - \bar{\mathbf{X}})'\mathbf{S}^{-1}(\mathbf{X}_i - \bar{\mathbf{X}}).$$

Imagine that one wants to downweight a proportion κ of the observations. Let ϱ^2 be the $1 - \kappa$ quantile of a chi-squared distribution with p degrees of freedom. Let $w_i = 1$ if $d_i \leq \varrho$; otherwise $w_i = \varrho/d_i$. Then an updated estimate of the mean and covariance matrix is given by

$$\bar{\mathbf{X}} = \sum w_i \mathbf{X}_i / n$$

and

$$\mathbf{S} = \frac{1}{\tau n} \sum w_i^2 (\mathbf{X}_i - \bar{\mathbf{X}})(\mathbf{X}_i - \bar{\mathbf{X}})',$$

respectively, where τ is chosen so that $\mathbf{S}$ is an unbiased estimate of the covariance matrix under normality. These updated estimates are used to update the squared Mahalanobis distances, which in turn yields a new updated estimate of the mean and covariance matrix. This process is continued until convergence is achieved.

6.3.14 R Function MARest

The R function

MARest(x,kappa=0.1)

computes Maronna's M-estimator of location and scatter, where the argument kappa corresponds to κ in the previous section.

6.4 Multivariate Outlier Detection Methods

An approach to detecting outliers when working with multivariate data is to simply check for outliers among each of the marginal distributions using one of the methods described in Chapter 3. A concern about this approach, however, is that outliers can be missed because it does not take into account the overall structure of the data. In particular, any multivariate outlier detection method should be invariant under rotations of the data. Methods based on the marginal distributions do not satisfy this criterion.

To illustrate the problem, consider the observations in the upper left panel of Figure 6.2. The upper right panel shows a boxplot of both the X and Y values. As indicated, one Y value is flagged as an outlier. It is the point in the upper right corner of the scatterplot. If the points are rotated such that they maintain their relative positions, the outlier should remain an outlier, but for some rotations this is not the case. For example, if the points are rotated 45°, the scatterplot now appears as shown in the lower left panel with the outlier in the upper left corner. The lower right panel shows a boxplot of the X and Y values after the axes are rotated. Now the boxplots do not indicate any outliers because they fail to take into account the overall structure of the data. What is needed is a method that is invariant under rotations of

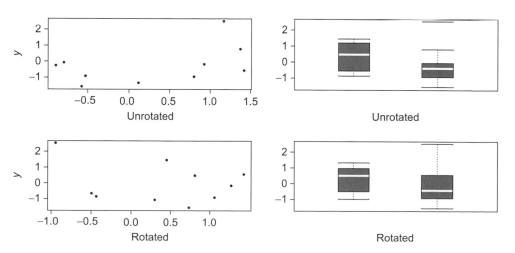

Figure 6.2: **The upper right panel shows a boxplot for the X and Y values shown in the upper left panel. The lower right panel shows a boxplot for the X and Y values after the points in the upper left panel are rotated to the position shown in the lower left panel. Now the boxplots find no outliers, in contrast to the unrotated case.**

the data. In addition, any outlier detection method should be invariant under changes in scale. All of the methods in this section satisfy these two criteria.

All but one of the multivariate outlier detection methods described in this section can be used with p-variate data for any $p > 1$. The one exception is a method called a relplot which is described first and is limited to bivariate data.

6.4.1 A Relplot

A relplot is a bivariate generalization of the boxplot derived by Goldberg and Iglewicz (1992). It is based in part on a bivariate generalization of M-estimators covered in Chapter 3 which belongs to the class of W-estimators described in Section 6.3.6. Let X_{ij} ($i = 1, \ldots, n$; $j = 1, 2$) be a random sample from some bivariate distribution. For fixed j, the method begins by computing M_j, MAD_j, and $\hat{\zeta}_j^2$ using the X_{ij} values, where M_j is the sample median, MAD_j is the median absolute deviation statistic, and $\hat{\zeta}_j^2$ is the biweight midvariance described in Section 3.12.1. Let

$$U_{ij} = \frac{X_{ij}}{9MAD_j},$$

and set $a_{ij} = 1$ if $|U_{ij}| < 1$, otherwise $a_{ij} = 0$. Let

$$T_j = M_j + \frac{\sum a_{ij}(X_{ij} - M_j)(1 - U_{ij}^2)^2}{\sum a_{ij}(1 - U_{ij}^2)^2}.$$

The remaining computational steps are given in Table 6.2 which yield a bivariate measure of location, (T_{b1}, T_{b2}), a robust measure of variance, s_{b1}^2 and s_{b2}^2, and a robust measure of correlation, R_b. These measures of location and scatter can be extended to $p > 2$ variates, but computational problems can arise (Huber, 1981).

The relplot consists of two ellipses. Once the computations in Table 6.2 are complete, the inner ellipse is constructed as follows. Let

$$Z_{ij} = \frac{X_{ij} - T_{bj}}{s_{bj}}$$

and

$$E_i = \sqrt{\frac{Z_{i1}^2 + Z_{i2}^2 - 2R_b Z_{i1} Z_{i2}}{1 - R_b^2}}.$$

Let E_m be the median of $E_1, \ldots, E_n$, and let $E_{\max}$ be the largest E_i value such that $E_i^2 < DE_m^2$, where D is some constant. Goldberg and Iglewicz recommend $D = 7$, and this value is used here. Let $R_1 = E_m\sqrt{(1 + R_b)/2}$ and $R_2 = E_m\sqrt{(1 - R_b)/2}$. For each υ between

Table 6.2: Computing Biweight M-estimators of Location, Scale, and Correlation.

Step 1. Compute $Z_{ij} = (X_{ij} - T_j)/\hat{\zeta}_j$.

Step 2. Recompute T_j, and $\hat{\zeta}_j^2$ by replacing the X_{ij} values with Z_{ij} yielding T_{zj}, and $\hat{\zeta}_{zj}^2$.

Step 3. Compute

$$E_i^2 = \left(\frac{Z_{i1} - T_{z1}}{\hat{\zeta}_{z1}}\right)^2 + \left(\frac{Z_{i2} - T_{z2}}{\hat{\zeta}_{z2}}\right)^2.$$

Step 4. For some constant C, let

$$W_i = \left(1 - \frac{E_i^2}{C}\right)^2$$

if $E_i^2 < C$, otherwise $W_i = 0$. Goldberg and Iglewicz (1992) recommend $C = 36$ unless more than half of the W_i values are equal to zero, in which case C can be increased until a minority of the W_i values is equal to zero.

Step 5. Compute

$$T_{bj} = \frac{\sum W_i X_{ij}}{\sum W_i}$$

$$S_{bj}^2 = \frac{\sum W_i (X_{ij} - T_{bj})^2}{\sum W_i}$$

$$R_b = \frac{\sum W_i (X_{i1} - T_{b1})(X_{i2} - T_{b2})}{S_{b1} S_{b2} \sum W_i}.$$

Step 6. Steps 4–8 are iterated. If step 4 has been performed only once, go to step 7; otherwise, let W_{oi} be the weights from the previous iteration, and stop if $\sum (W_i - W_{oi})^2 / (\sum W_i/n)^2 < \epsilon$.

Step 7. Store the current weight, W_i, into W_{oi}.

Step 8. Compute

$$Z_{i1} = \left(\frac{X_{i1} - T_{b1}}{S_{b1}} + \frac{X_{i2} - T_{b2}}{S_{b2}}\right) \frac{1}{\sqrt{2(1 + R_b)}}$$

$$Z_{i2} = \left(\frac{X_{i1} - T_{b1}}{S_{b1}} - \frac{X_{i2} - T_{b2}}{S_{b2}}\right) \frac{1}{\sqrt{2(1 - R_b)}}$$

$$E_i^2 = Z_{i1}^2 + z_{i2}^2.$$

Go back to step 4.

0 and 360, steps of 2 degrees, compute $\Upsilon_1 = R_1\cos(\upsilon)$, $\Upsilon_2 = R_2\sin(\upsilon)$, $A = T_{b1} + (\Upsilon_1 + \Upsilon_2)s_{b1}$, and $B = T_{b2} + (\Upsilon_1 - \Upsilon_2)s_{b2}$. The values for A and B form the inner ellipse. The outer ellipse is obtained by repeating this process with E_m replaced by $E_{\max}$.

6.4.2 R Function relplot

The R function

$$relplot(x,y,C=36,epsilon=0.0001,plotit=T)$$

performs the calculations in Table 6.2 and creates a relplot. Here, x and y are any R vectors containing data, C is a constant that defaults to 36 (see step 4 in Table 6.2), and epsilon is ϵ in step 6 of Table 6.2. (The argument epsilon is used to determine whether enough iterations have been performed. Its default value is 0.0001.) The function returns bivariate measures of location in relplot\$mest, measures of scale in relplot\$mvar, and a measure of correlation in relplot\$mrho. The last argument, plotit, defaults to T, for true, meaning that a graph of the bivariate boxplot (the relplot) will be created. To avoid the plot, simply set the last argument, plotit, to F for false. For example, the command relfun(x,y,plotit=F) will return the measures of location and correlation without creating the plot.

■ Example

Rousseeuw and Leroy (1987, p. 27) report the logarithm of the effective temperature at the surface of 47 stars versus the logarithm of its light intensity. Suppose the (Hertzsprung - Russell) star data are stored in the R variables u and v. Then the R command relplot(u,v) creates the plot shown in Figure 6.3. The smaller ellipse contains half of the data. Points outside the larger ellipse are considered to be outliers. The function reports a correlation of 0.7 which is in striking contrast to Pearson's correlation, $r = -0.21$.

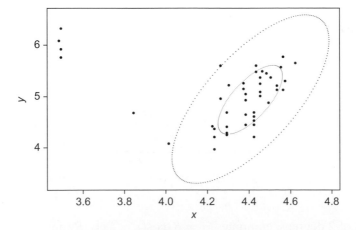

Figure 6.3: A relplot for the star data.

6.4.3 The MVE Method

A natural way of detecting outliers in p-variate data, $p \geq 2$, is to use Mahalanobis distance with the usual means and sample covariance matrix replaced by estimators that have a high breakdown point. One of the earliest such methods is based on the MVE estimators of location and scale (Rousseeuw & van Zomeren, 1990). Relevant theoretical results are reported by Lopuhaä (1999). Let the column vector $\mathbf{C}$, having length p, be the MVE estimate of location, and let the p-by-p matrix $\mathbf{M}$ be the corresponding measure of scatter. The distance of the point $\mathbf{x}_i' = (x_{i1}, \ldots, x_{ip})$ from $\mathbf{C}$ is given by

$$D_i = \sqrt{(\mathbf{x}_i - \mathbf{C})'\mathbf{M}^{-1}(\mathbf{x}_i - \mathbf{C})}. \qquad (6.16)$$

If $D_i > \sqrt{\chi^2_{.975,p}}$, the square root of the 0.975 quantile of a chi-square distribution with p degrees of freedom, then $\mathbf{x}_i$ is declared an outlier. Rousseeuw and van Zomeren recommend this method when there are at least five observations per dimension, meaning that $n/p > 5$. (Cook & Hawkins, 1990, illustrate that problems can arise when $n/p \leq 5$.) A criticism of this method is that it can declare too many points as being extreme (Fung, 1993).

6.4.4 The MCD Method

Rather than using the MVE measure of location and scatter to detect outliers, one could, of course, use the MCD estimators instead. That is, in Eq. (6.16), replace $\mathbf{M}$ and $\mathbf{C}$ with the MCD estimates of scatter and location. Cerioli (2010) derived a modification of this method with the goal that under multivariate normality, the probability of declaring one or more points an outlier is equal to some specified value.

6.4.5 R Functions covmve and covmcd

Both the MVE and MCD estimators can be computed with the R functions

cov.mve(m),

and

cov.mcd(m),

respectively, which are stored in the R library MASS. For convenience, the R functions

covmve(m),

and

covmcd(m)

are supplied, which eliminate the need to use the R command library(MASS) in order to access the R functions cov.mve and cov.mcd. Both of these functions return weights indicating whether a point is declared an outlier using the MVE and MCD methods. (It is noted that the S-PLUS version of the MVE and MCD outlier detection methods can give results that differ from the R versions used here.) However, these functions do not return the MVE and MCD estimate of location and scatter, but rather a W-estimate of location and scatter. (See Section 6.5.) In essence, points declared outliers are removed and the mean and covariance matrix are computed using the data that remain. R reports which subset of half of the data was used to compute the MVE and MCD estimates of location. So it is possible to determine the MVE and MCD estimates of location if desired.

In some situations it is convenient to have an R function that returns just the MVE measure of location. Accordingly, the R function

$$mvecen(m)$$

is supplied to accomplish this goal. The R function

$$mcdcen(m)$$

computes the MCD measure of location.

6.4.6 R function out

The R function

$$out(x, cov.fun = cov.mve, plotit = T, SEED = T, xlab = ``X", ylab = ``Y", qval = 0.975, \\ crit = NULL, \ldots)$$

identifies outliers using the MVE method for p-variate data using Eq. (6.15), where the argument x is an n-by-p matrix. The function returns a vector labeled out.id that identifies which rows of data are outliers. And another vector, labeled keep.id indicates the rows of data that are not declared outliers. In the bivariate case it creates a scatterplot of the data and marks outliers with an o. To avoid the plot, set the argument plotit to F, for false. Setting cov.fun=covmcd results in replacing the MVE estimator with the MCD estimator. (Other options for this argument are ogk, tbs, and rmba, which result in using the OGK, TBS, and median ball algorithm, respectively, but except for tbs, it seems that these options are relatively unsatisfactory.)

■ Example

If the star data in Figure 6.3 are stored in the matrix stardat, the R command out(stardat) returns the values 7, 9, 11, 14, 20, 30, and 34 in the R variable outmve$out.id. This means, for example, that row 7 of the data in the matrix stardat is an outlier. Six of these points correspond to the outliers in Figure 6.3 that are to the left of the outer ellipse. The seventh is the point near the upper middle portion of Figure 6.3 that lies on or slightly beyond the outer ellipse. (The point is at $x = 4.26$ and $y = 5.57$.) ∎

6.4.7 The MGV Method

An appeal of both the MVE and MCD outlier detection methods is that they are based on high breakdown estimators. That is, they provide an important step toward avoiding *masking*, roughly referring to an inability to detect outliers due to their very presence. (See Section 3.13.1.) But for certain purposes, two alternative methods for detecting outliers have been found to have practical value, one of which is the MGV method described here. (The other is a projection-type method described later in this chapter.)

As noted in Chapter 3, the *outside rate per observation* is the expected proportion of points declared outliers. That is, if among n points, A points are declared outliers, the outside rate per observation is $p_n = E(A/n)$. When sampling from multivariate normal distributions, for certain purposes it is desirable to have p_n reasonably close to zero; a common goal is to have p_n approximately equal to .05 (cf., Cerioli, 2010). When all variables are independent, it appears that both the MVE and MCD methods have an outside rate per observation approximately equal to .05. But under dependence, this is no longer true, it is higher when using the MCD method. Although the MVE method based on the R function cov.mve appears to have p_n approximately equal to 0.05 under normality, alternative outlier detection methods have been found to have practical advantages for situations to be described.

A multivariate outlier detection method for which p_n is reasonably close to 0.05 under normality, and which has practical value when dealing with problems to be addressed, is the so-called minimum generalized variance (MGV) method which is applied as follows:

1. Initially, all n points are described as belonging to set A.
2. Find the p points that are most centrally located. One possibility is as follows. Let

$$d_i = \sum_{j=1}^{n} \sqrt{\sum_{\ell=1}^{p} \left(\frac{X_{j\ell} - X_{i\ell}}{\mathrm{MAD}_\ell} \right)^2},$$
(6.17)

where MAD_ℓ is the value of MAD based on $X_{1\ell}, \ldots, X_{n\ell}$. The two most centrally located points are taken to be the p points having the smallest d_i values. Another possibility, in order to achieve affine equivariance, is to identify the p points having the largest halfspace depth or the largest depth based on the MVE or MCD methods.

3. Remove the p centrally located points from set A and put them into set B. At this step, the generalized variance of the points in set B is zero.

4. If among the points remaining in set A, the ith point is put in set B, the generalized variance of the points in set B will be changed to some value which is labeled s_{gi}^2. That is, associated with every point remaining in A is the value s_{gi}^2, which is the resulting generalized variance when it, and it only, is placed in set B. Compute s_{gi}^2 for every point in A.

5. Among the s_{gi}^2 values computed in the previous step, permanently remove the point associated with the smallest s_{gi}^2 value from set A and put it in set B. That is, find the point in set A which is most tightly clustered together with the points in set B. Once this point is identified, permanently remove it from A and leave it in B henceforth.

6. Repeat steps 4 and 5 until all points are now in set B.

The first p points removed from set A have a generalized variance of zero which is labeled $s_{g(1)}^2 = \cdots = s_{g(p)}^2 = 0$. When the next point is removed from A and put into B (using steps 4 and 5), the resulting generalized variance of the set B is labeled $s_{g(p+1)}^2$ and continuing this process, each point has associated with it some generalized variance when it is put into set B.

Based on the process just described, the ith point has associated with it one of the generalized variances just computed. For example, in the bivariate case, associated with the ith point (X_i, Y_i) is some value $s_{g(j)}^2$ indicating the generalized variance of the set B when the ith point is removed from set A and permanently put in set B. For convenience, this generalized variance associated with the ith point, $s_{g(j)}^2$, is labeled D_i. The p deepest points have D values of zero. Points located at the edges of a scatterplot have the highest D values meaning that they are relatively far from the center of the cloud of points. Moreover, we can detect outliers simply by applying one of the outlier detection rules in Chapter 3 to the D_i values. Note, however, that we would not declare a point an outlier if D_i is small, only if D_i is large.

In terms of maintaining an outside rate per observation that is stable as a function of n and p, and approximately equal to 0.05 under normality (and when dealing with certain regression problems to be described), a boxplot rule for detecting outliers seems best when $p = 2$, and for $p > 2$ a slight generalization of Carling's modification of the boxplot rule appears to perform well. In particular, if $p = 2$, then declare the ith point an outlier if

$$D_i > q_2 + 1.5(q_2 - q_1), \tag{6.18}$$

where q_1 and q_2 are the ideal fourths based on the D_i values. For $p > 2$ variables, replace Eq. (6.18) with

$$D_i > M_D + \sqrt{\chi_{.975,p}^2}(q_2 - q_1), \tag{6.19}$$

where $\sqrt{\chi^2_{.975,p}}$ is the square root of the 0.975 quantile of a chi-squared distribution with p degrees of freedom and M_D is the usual median of the D_i values.

A comment about detecting outliers among the D_i values, using a MAD-median rule, should be made. Using something like the Hampel identifier when detecting outliers has the appeal of using measures of location and scale that have the highest possible breakdown point. When $p = 2$, for example, this means that a point $\mathbf{X}_i$ is declared an outlier if

$$\frac{|D_i - M_D|}{\text{MAD}_D/0.6745} > 2.24, \tag{6.20}$$

where MAD_D is the value of MAD based on the D values. A concern about this approach is that the outside rate per observation is no longer stable as a function of n. This has some negative consequences when addressing problems in subsequent sections. Here, Eq. (6.19) is used because it has been found to avoid these problems and because it has received the most attention so far, but of course in situations where there are an unusually large number of outliers, using Eq. (6.19) might cause practical problems.

6.4.8 R Function outmgv

The R function

outmgv((x, y = NULL, plotit = T, outfun = outbox, se = T, op = 1, cov.fun = rmba, xlab = "X", ylab = "Y", SEED = T, …)

applies the MGV outlier detection method just described.[2] If the second argument is not specified, it is assumed that x is a matrix with p columns corresponding to the p variables under study and outmgv checks for outliers for the data stored in x. If the second argument, y, is specified, the function combines the data in x with the data in y and checks for outliers among these $p + 1$ variables. In particular, the data do not have to be stored in a matrix; they can be stored in two vectors (x and y) and the function combines them into a single matrix for you. If plotit=T is used and bivariate data are being studied, a plot of the data will be produced with outliers marked by a circle. The argument outfun can be used to change the outlier detection rule applied to the depths of the points (the D_i values in the previous section). By default, Eq. (6.19) is used. Setting outfun=out, Eq. (6.20) is used. The argument se=T ensures that the results do not change with changes in scale. (The marginal distributions are standardized when calling the R function apgdis.)

---

[2] If columns of the input matrix are reordered, this might affect the results due to rounding error when calling the built-in R function eigen.

6.4.9 A Projection Method

Consider a sample of n points from some p-variate distribution and consider any projection of the data (as described in Section 6.2.2). A projection-type method for detecting outliers among multivariate data is based on the idea that if a point is an outlier, then it should be an outlier for some projection of the n points. So if it were possible to consider all possible projections, and if for some projection a point is an outlier, then the point is declared an outlier. Not all projections can be considered, so the strategy here is to orthogonally project the data onto all n lines formed by the center of the data cloud, as represented by $\hat{\xi}$, and each $\mathbf{X}_i$. It seems natural that $\hat{\xi}$ should have a high breakdown point and that it should be affine equivariant. Two good choices appear to be the MVE and MCD estimators in Sections 6.3.1 and 6.3.2.

The computational details are as follows. Fix i, and for the point $\mathbf{X}_i$, orthogonally project all n points onto the line connecting $\hat{\xi}$ and $\mathbf{X}_i$, and let D_{ij} be the distance between $\hat{\xi}$ and $\mathbf{X}_j$ based on this projection. More formally, let

$$\mathbf{A}_i = \mathbf{X}_i - \hat{\xi},$$

$$\mathbf{B}_j = \mathbf{X}_j - \hat{\xi},$$

where both $\mathbf{A}_i$ and $\mathbf{B}_j$ are column vectors having length p, and let

$$\mathbf{C}_j = \frac{\mathbf{A}_i' \mathbf{B}_j}{\mathbf{B}_j' \mathbf{B}_j} \mathbf{B}_j,$$

$j = 1, \ldots, n$. Then when projecting the points onto the line between $\mathbf{X}_i$ and $\hat{\xi}$, the distance of the jth point from $\hat{\xi}$ is

$$D_{ij} = \|\mathbf{C}_j\|,$$

where

$$\|\mathbf{C}_j\| = \sqrt{C_{j1}^2 + \cdots + C_{jp}^2}.$$

Here, an extension of Carling's modification of the boxplot rule (similar to the modification used by the MGV method) is used to check for outliers among D_{ij} values. To be certain the computational details are clear, let $\ell = [n/4 + 5/12]$, where [.] is the greatest integer function, and let

$$h = \frac{n}{4} + \frac{5}{12} - \ell.$$

For fixed i, let $D_{i(1)} \leq \cdots \leq D_{i(n)}$ be the n distances written in ascending order. The ideal fourths associated with the D_{ij} values are

$$q_1 = (1-h)D_{i(h)} + hD_{i(h+1)}$$

and

$$q_2 = (1-h)D_{i(\ell)} + hD_{i(\ell-1)}.$$

Then the jth point is declared an outlier if

$$D_{ij} > M_D + \sqrt{\chi^2_{.95,p}}(q_2 - q_1), \tag{6.21}$$

where M_D is the usual sample median based on the $D_{i1}, \ldots, D_{in}$ values and $\chi^2_{.95,p}$ is the 0.95 quantile of a chi-squared distribution with p degrees of freedom.

The process just described is for a single projection; for fixed i, points are projected onto the line connecting $\mathbf{X}_i$ to $\hat{\xi}$. Repeating this process for each i, $i = 1, \ldots, n$, a point is declared an outlier if for any of these projections, it satisfies Eq. (6.21). That is, $\mathbf{X}_j$ is declared an outlier if for any i, D_{ij} satisfies Eq. (6.21). Note that this outlier detection method approximates an affine equivariant technique for detecting outliers, but it is not itself affine equivariant. However, it is invariant under rotations of the axes.

As was the case with the MGV method, a simple and seemingly desirable modification of the method just described is to replace the interquartile range $(q_2 - q_1)$ with the median absolute deviation (MAD) measure of scale based on the values $D_{i1}, \ldots, D_{in}$. So here, MAD is the median of the values

$$|D_{i1} - M_D|, \ldots, |D_{in} - M_D|,$$

which is denoted by MAD_i. Then the jth point is declared an outlier if for any i

$$D_{ij} > M_D + \sqrt{\chi^2_{.95,p}} \frac{\text{MAD}_i}{0.6745}. \tag{6.22}$$

Equation (6.22) represents an approximation of the method given by Eq. (1.3) in Donoho and Gasko (1992). Again, an appealing feature of MAD is that it has a higher finite sample breakdown point than the interquartile range. But a negative feature of Eq. (6.22) is that the outside rate per observation appears to be less stable as a function of n. In the bivariate case, for example, it is approximately 0.09 with $n = 10$ and drops below 0.02 as n increases. For the same situations, the outside rate per observation using Eq. (6.21) ranges, approximately, between 0.043 and 0.038.

A criticism of the projection method as just described is that changes in scale can alter decisions about whether a point is an outlier. That is, if the first variable only is multiplied by some constant $c \neq 0$, this might alter the decision about whether the ith point is an outlier. One possible way of dealing with this issue is to standardize the marginal distributions. Another approach is to use the MGV method instead, which seems to perform about as well as the projection method in terms of detecting true outliers (Wilcox, 2008a). But this comes at the cost of higher execution, which might be an issue with large sample sizes. An advantage of the projection method is that it can be applied even when n is small and p is large. This is not always the case when using the MGV method.

6.4.10 R functions outpro and out3d

The R function

$$\text{outpro(m,gval=NA,center=NA,plotit=T,op=T,MM=F,cop= 3,STAND=F)}$$

checks for outliers using the projection method just described. Here, m is any R variable containing data stored in a matrix (having n rows and p columns). The argument gval can be used to alter the values $\sqrt{\chi^2_{.95,p}}$ or $\sqrt{\chi^2_{.975,p}}$ in Eqs. (6.20) and (6.21). These values are replaced by the value stored in gval if gval is specified. Similarly, the argument center can be used to specify the center of the data cloud, $\hat{\xi}$, that will be used. If not specified, the center is determined by the argument cop. The choices are:

- cop=1, Donoho–Gasko median
- cop=2, MCD
- cop=3, median of the marginal distributions
- cop=4, MVE

When working with bivariate data, outpro creates a scatterplot of the data, marks outliers with a circle, and the plot includes a contour indicating the location of the deepest half of the data as measured by projection depth. More precisely, the depth of all points is computed, and among the points not declared outliers, all points having a depth less than or equal to the median depth are indicated. If op=T, the plot creates a 0.5 depth contour based on the data excluding points declared outliers. Setting op=F, the 0.5 depth contour is based on all of the data. If MM=T is used, the interquartile range is replaced by MAD. That is, Eq. (6.22) is used in place of (6.21). Setting the argument STAND=T, the marginal distributions are standardized, before checking for outliers, using the median and MAD. (The version of outpro in Rallfun-v13 and later contains this argument. But earlier versions of the R functions written for this book do not.)

When working with trivariate data, the R function

$$\text{out3d(x, outfun = outpro, xlab = "Var 1", ylab = "Var 2", zlab = "Var 3", reg.plane = F,}$$
$$\text{regfun = tsreg, COLOR = F)}$$

creates a three-dimensional scatterplot and marks the outliers, identified by the R function outpro, with *. (Setting the argument COLOR=T, outliers are marked with a red circle.) An alternative outlier detection method can be used via the argument outfun. This function also shows a regression plane when reg.plane=T, assuming the goal is to predict the third variable given value for the first two. (That is, column 3 of x is assumed to be the outcome variable, typically labeled y, and columns 1 and 2 contain the predictor variables.) The regression method used is controlled by the argument regfun, which defaults to the Theil–Sen estimator described in Chapter 10.

6.4.11 Outlier Identification in High Dimensions

Filzmoser, Maronna, and Werner (2008) noted that under normality, if the number of variables is large, the proportion of points declared outliers by the better-known outlier detection methods can be relatively high. This concern applies to all the methods covered here with the projection method seemingly best at avoiding this problem. But with more than nine variables ($p > 9$), it breaks down as well. Currently, it seems that one of the better ways of dealing with this problem is to use the projection method but with Eq. (6.22) replaced by

$$D_{ij} > M_D + c \frac{\text{MAD}_i}{0.6745},$$

where c is chosen so that the outside rate per observation is approximately equal to some specified value under normality, which is usually taken to be 0.05. Here, the constant c is determined via simulations. That is, n points are generated from a p-variate normal distribution, where all p variables are independent. This process is repeated say B times, and a value c is determined so that the expected proportion of points declared outliers is equal to the desired rate. A refinement of this strategy would be to generate data from a multivariate normal distribution that has the same covariance matrix as the data under study. Currently, this does not seem necessary or even desirable, but this issue is in need of further study.

A similar adjustment can be made when using the MGV method to detect outliers, which might be preferred because the MGV method is scale invariant. Direct comparisons of the performance of the adjusted MGV method and the adjusted projection method have not been made.

6.4.12 R Function outproad and outmgvad

The R function

    outproad(m, center = NA, plotit = T, op = T, MM = F, cop = 3, xlab = "VAR 1", ylab =
        "VAR 2", rate = 0.05, iter = 100, ip = 6, pr = T, SEED = T, STAND = F)

is like the R function outpro, only it uses simulations to adjust the decision rule for declaring a point an outlier as described in the previous section. The argument rate indicates the desired proportion of points declared outliers under normality. The R function

$$\text{outmgvad(m, center = NA, plotit = T, op = 1, xlab = "VAR 1", ylab = "VAR 2", rate =}$$
$$\text{0.05, iter = 100, ip = 6, pr = T)}$$

is like the R function outproad, only it is based on the MGV outlier detection technique.

6.4.13 Approaches Based on Geometric Quantiles

It is briefly noted that Chaudhuri (1996) derived a multivariate outlier detection technique that stems from the notion of geometric quantiles. Chaouch and Goga (2010) extended Chaudhuri's method to survey sampling situations. Direct comparisons with the projection method and the MGV method, based on the outside rate per observation, have not been made.

6.4.14 Comments on Choosing a Method

Choosing an outlier detection method is a nontrivial problem with no single method dominating all others; it seems that several methods deserve serious consideration. In addition to controlling the outside rate per observation, surely a desirable property of any outlier detection method is that it identifies points that are truly unusual based on a model that generated the data. Wilcox (2008a) compared several methods, and while no single method was always best, it was found the MGV and projection methods (applied with the functions outmgv and outpro, respectively) performed relatively well when the number of variables is not too large, meaning that $p \leq 9$. But as previously noted, with $p > 9$ variables, these two methods break down, in which case the projection method in Section 6.4.11 should be used.

It is worth noting that, *given some data*, the choice of method can matter. To illustrate this point, suppose both X and ϵ are independent standard normal variables and let $Y = X + \epsilon$. Table 6.3 shows 20 points generated in this manner. Suppose two additional points are added at $(X, Y) = (2.1, -2.4)$. These two points are clearly unusual compared to the model that generated the data in Table 6.3.

Using the projection method or the MGV method, the two points at $(2.1, -2.4)$ are flagged as outliers, and no other outliers are reported. These two points are declared outliers using the MVE method, but it flags two additional points as outliers. The R version of the MCD method finds no outliers, but the S-PLUS version finds five. So when using MVE or MCD, the choice of software can alter the results because the approximations of the MVE and MCD measures of location and scatter differ. This problem is avoided when using the projection-type method or the MGV method.

Table 6.3: Data Used to Illustrate Outlier Detection Methods.

0.49601344	0.2293483
−1.57153483	−2.1847545
0.55633893	1.4329751
1.17870964	2.0139325
−0.51404243	0.5547278
0.63358060	1.4263721
0.76638318	0.4863436
−0.65799729	−1.7676117
−0.78952475	−1.0790300
1.30434022	1.7647849
2.66008714	2.2311363
1.16480412	0.1011259
−1.24400427	−1.8162151
0.53216346	0.1244360
−0.21301222	−1.4762834
−0.08860754	1.8208403
−0.86177016	−0.9034465
0.53223243	0.3335626
−1.64469892	−1.5643544
0.09441703	1.9093186

Gleason (1993) argues that a lognormal distribution is light-tailed. In the univariate case, with $n = 20$, the MAD-median rule given by Eq. (3.45) has an outside rate per observation of approximately 0.13, and a boxplot rule has an outside rate per observation of approximately 0.067. As another illustration that the choice of method can make a difference, consider the case where X and Y are independent, each having a lognormal distribution. For this bivariate case, with $n = 20$, all of the methods considered here have an outside rate above 0.1. The MVE method seems to be highest, with an estimated rate of 0.17 when using S-PLUS, whereas for the outlier projection method and the MGV method the rates are approximately 0.15 and 0.13, respectively. When using the R version of MVE, the rate exceeds 0.2.

Figure 6.4 shows the plots created by the MGV, MVE, MCD and the projection method based on a sample of $n = 50$ pairs of points generated from two independent lognormal distributions. The upper left and lower right panels are based on the projection method and the MGV method, respectively. In this particular instance they give identical results and flag the fewest points as outliers, relative to the other methods used. The upper right panel is the output based on the MVE (S-PLUS) method, and the lower left panel is based on the MCD (S-PLUS) method. Although an argument can be made that in this particular instance, the MVE and MCD methods are less satisfactory, this must be weighed against the ability of the MVE and MCD methods to handle a larger number of outliers. (But if a large number of

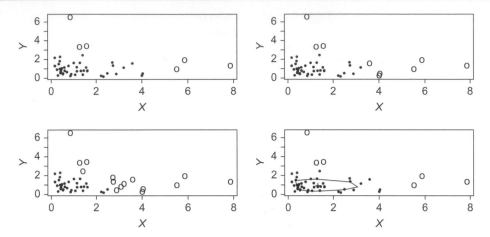

Figure 6.4: Output from four outlier detection methods. The upper left panel used the projection method in Section 6.4.9, the upper right used the MVE method, the lower left is based on the MCD method, and the lower right used the MGV method.

outliers is suspected, the versions of the projection-type method and the MGV method represented by Eqs. (6.20) and (6.19), respectively, might be used.) For more on detecting multivariate outliers, see Kosinski (1999), Liu, Parelius, and Singh (1999), Rocke and Woodruff (1996), Peña and Prieto (2001), Poon, Lew, and Poon (2000), Rousseeuw and Leroy (1987), Davies (1987), Fung (1993), and Rousseeuw and van Zomeren (1990).

6.5 A Skipped Estimator of Location and Scatter

Skipped estimators of location and scatter are estimators that search for outliers, discard any that are found, and then compute the mean and usual covariance matrix based on the data that remain. Such estimators are special cases of the W-estimator in Section 6.3.6, where points get a weight of 1 or 0 depending on whether they are declared outliers. Maronna and Yohai (1995) refer to such weight functions as hard rejection weights as opposed to soft rejection where the weights gradually descend toward zero as a function of how outlying a point happens to be. When using the outlier projection method in Section 6.4.8, with outliers getting a weight of zero, otherwise points get a weight of one, the corresponding W-estimator will be called the *OP-estimator*. When using the MVE outlier detection method, the skipped estimator will be called the *WMVE estimator*. And when using the MCD outlier detection method, the skipped estimator will be called the **WMCD estimator**.

Note that the methods just describe also yield robust analogs of the usual covariance matrix. If outliers are removed via the projection method, and the usual covariance matrix is computed based on the remaining data, this will be called the OP-estimate of scatter.

Table 6.4: Values of R (Accuracy), $n = 40$.

h	ρ	$\gamma = 0.10$	$\gamma = 0.15$	$\gamma = 0.20$	DGM	OP	M
0.0	0.0	0.73	0.62	0.50	0.45	0.92	0.81
0.5	0.0	5.99	5.92	5.40	4.11	6.25	8.48
1.0	0.0	4660.21	5764.79	5911.29	4643.16	5452.35	10820.14
0.0	0.7	0.80	0.71	0.61	0.48	0.95	0.44
0.5	0.7	4.74	4.76	4.50	3.20	4.64	5.44
1.0	0.7	1082.56	1300.44	1336.63	1005.24	1091.68	1760.98

To provide at least some sense of how the various locations estimators compare, some results on the expected squared standard error are provided when sampling from distributions that are symmetric about zero. More precisely, the measure of accuracy used is

$$R = \frac{\sqrt{E(\sum \bar{X}_j^2)}}{\sqrt{E(\sum \hat{\theta}_j^2)}},$$

where $\hat{\theta}_j^2$ is some competing estimator associated with the jth variable, $j = 1, \ldots, p$. Table 6.4 reports some results for four variations of the Donoho–Gasko trimmed mean, followed by the OP-estimator and the marginal medians. (In Table 6.4, h refers to the type of g-and-h distribution used, as described in Section 4.2, and ρ is the common Pearson correlation among the generated data.) Note that under normality, all four variations of the Donoho–Gasko trimmed mean are the least satisfactory, and method OP performs best among the robust estimators considered. As for the TBS estimator in Section 6.3.3, it performs in a manner similar to the Donoho–Gasko trimmed mean with $\gamma = 0.15$ when sampling from a normal distribution. That is, it is less satisfactory than other estimators that might be used. For $h = 0.5$ it performs nearly as well as the skipped estimator (method OP), and for $h = 1$ it is a bit more accurate. As for the WMVE skipped estimator, among the situations considered, it seems to have about the same accuracy as the OP-estimator, with OP offering a slight advantage.

Masse and Plante (2003) report more extensive results on the Donoho–Gasko trimmed mean, plus other estimators not described here. Their results further support the notion that the Donoho–Gasko trimmed mean is relatively inaccurate when sampling from light-tailed distributions. Among the 10 estimators they considered, Masse and Plante (2003) found the spatial median, studied by Haldane (1948) and Brown (1983), to be best. (They did not consider the OP-estimator in their study.) The *spatial median* is the value $\hat{\theta}$ that minimizes

$$\frac{1}{n} \sum \| \hat{\theta} - \mathbf{X}_i \|.$$

It is not affine equivariant, but it is translation equivariant and orthogonally equivariant. One way of computing the spatial median is via the Nelder and Mead (1965) algorithm for minimizing a function. (See Olsson and Nelson, 1975, for a discussion of the relative merits of the Nelder–Mead algorithm.)[3] An alternative algorithm for computing the spatial median can be found in Bedall and Zimmermann (1979) as well as Hössjer & Croux (1995). Ng and Wilcox (2010b) compared eight robust estimators for a wide range of situations and concluded that the OP-estimator generally performs best in terms of efficiency, as measured by the generalized variance of the sampling distribution.

6.5.1 R Functions smean, wmcd, wmve, mgvmean, L1medcen, spat, mgvcov, skip, skipcov, and dcov

The R function

$$\text{smean(m,cop=3,MM=F,op=1,outfun=outogk,cov.fun=rmba,MC=F, ...)}$$

computes the OP-estimator of location just described using the data stored in the n-by-p matrix m. The remaining arguments determine which outlier detection method is used. Setting op=1 results in using the projection-type method, and op=2 uses the MGV method. The initial measure of location used by the outlier detection method is determined by cop, the choices being

- cop=1, Tukey (halfspace) median
- cop=2, MCD
- cop=3, marginal medians

To take advantage of a multi-core processor, with the goal of reducing execution time, set the argument MC=T.

The R function

$$\text{skipcov(m,cop=6,MM=F,op=1,mgv.op=0,outpro.cop=3)}$$

computes the covariance matrix for the data stored in the argument m after outliers are removed. Like the R function smean, op=1 means that a projection method is used to identify outliers. When MM=F, Carling's modification of the boxplot rule is applied to each projection when checking for outliers. When MM=T, a MAD-median rule is used. Setting op=2, the MGV method is used to detect outliers. The argument outpro.cop controls which

---

[3] The Nelder–Mead algorithm is applied with the R function Nelder, written for this book. This code was modeled after the FORTRAN code in Olsson (1974).

measure of location is used to compute the projections; see the R function outpro for more details. The R function

$$\text{skip(m,cop=6,MM=F,op=1,mgv.op=0,outpro.cop=3)}$$

returns both the skipped measure of covariance and measure of location. The R function

$$\text{mgvcov(m,MM=F,op=1,cov.fun=rmba)}$$

computes the MGV covariance matrix. For an explanation of the remaining arguments, see the R function outmgv.

The R function

$$\text{spat(m)}$$

computes the spatial median as does

$$\text{L1medcen(X, tol = 1e-08, maxit = 200, m.init = apply(X, 2, median) trace = FALSE).}$$

The function spat uses the Nelder–Mead algorithm, whereas L1medcen uses the method described in Hössjer and Croux (1995). These two functions can give slightly different results. Currently it is unknown why one method might be preferred over the other.

A skipped estimator, with outliers detected via the MGV method, is called the MGV estimator of location and can be computed with the function smean. For convenience, the R function

$$\text{mgvmean(m,op=0,MM=F,outfun=outbox)}$$

is supplied for computing this measure of location. Setting op=0 results in the MGV outlier detection method using pairwise differences when searching for the centrally located points, op=1 uses the MVE method, and op=2 uses MCD.

The built-in R function cov.mve is designed to compute the WMVE estimate of location and scatter. (S-PLUS also comes with the function cov.mve that supposedly reports the WMVE estimate of location and scatter, but checks found that in some cases, it returns a measure of location that does not correspond to the weights it reports. The reason for this is unclear.) It should be noted that repeatedly applying the function cov.mve to the same data, there are instances where it gives a different estimate for both location and scatter. To get a measure of location and scatter that gives the same result when repeatedly applied to the same data, use the R function

$$\text{wmve(m,SEED=T).}$$

That is, cov.mve returns weights w_i ($i = 1, \ldots, n$) that are equal to zero or one according to whether a point is declared an outlier, and wmve uses these weights to compute measures of location and scatter based on Eqs. (6.13) and (6.14).

■ Example

For the data shown in Figure 6.3, the author got the following results based on the S-PLUS function cov.mve:

```
$cov:
              [,1]           [,2]
[1,] 0.009856233   0.03159224
[2,] 0.031592244   0.21197126

$center:
[1] 4.422632 4.973947
```

Entering the command cov.mve again yielded

```
$cov:
              [,1]           [,2]
[1,] 0.01121994   0.03755037
[2,] 0.03755037   0.23498775

$center:
[1] 4.41275 4.93350
```

A similar problem occurs when using cov.mve using R and when using the S-PLUS function cov.mcd. The R functions wmve always returns the same values by setting the seed of the random number generator used by R when the argument SEED=T.

■

6.6 Robust Generalized Variance

It is noted that one approach to measuring the overall variation of a cloud of points is with the generalized variance, where the usual covariance matrix is replaced by some robust analog. Based on the criterion of achieving good efficiency, a particular choice for the covariance matrix has been found to be relatively effective when distributions are normal or have moderately heavy tails: the OP-estimator of scatter where Carling's modification of the boxplot rule is applied to each projection of the data. For heavy-tailed distributions, use instead a MAD-median rule (Wilcox, 2006e).

6.6.1 R Function gvarg

The R function

$$gvarg(m, var.fun=cov.mba, \ldots)$$

a robust generalized variance for the data stored in the argument m. By default, the RMBA covariance matrix is used because other methods to be described appear to perform reasonably well based on this covariance matrix. The command

$$gvarg(x, skipcov, MM-F)$$

would compute the generalized variance based on the OP-estimate of scatter in conjunction with Carling's modification of the boxplot rule. The command

$$gvarg(x, skipcov, MM=T)$$

would use the MAD-median rule.

6.7 Inference in the One-Sample Case

This section describes two methods for making inferences about multivariate measures of location. The first is aimed at the population analog of the OP-estimator. The second is based on an extension of Hotelling's T^2 method to the marginal trimmed means.

6.7.1 Inferences Based on the OP Measure of Location

The immediate goal is to compute a $1 - \alpha$ confidence region for the population measure of location corresponding to the OP-estimator described in the previous section. Alternatively, the method in this section can be used to test the hypothesis that the population measure of location is equal to some specified value.

The basic strategy is to use a general percentile bootstrap method studied by Liu and Singh (1997). Roughly, generate bootstrap estimates and use the central $1 - \alpha$ bootstrap values as an approximate confidence region. A simple method for determining the central $1 - \alpha$ bootstrap values is to use Mahalanobis distance. Despite being nonrobust, this strategy performs well for a range of situations to be covered. Indeed, for many problems, there is no known reason to prefer another measure of depth, in terms of probability coverage. But for the problem at hand, Mahalanobis depth is unsatisfactory, at least with small to moderate sample sizes; the actual probability coverage can be rather unstable among various distributions, particularly as p gets large. That is, what is needed is a method for which the probability coverage is reasonably close to the nominal level regardless of the distribution associated with the data.

The method begins by generating a bootstrap sample by sampling with replacement n vectors of observations from $\mathbf{X}_1, \ldots, \mathbf{X}_n$, where again $\mathbf{X}_i$ is a vector having length p. Label the results $\mathbf{X}_1^*, \ldots, \mathbf{X}_n^*$. Compute the OP-estimate of location yielding $\hat{\theta}^*$. Repeat this B times yielding $\hat{\theta}_1^*, \ldots, \hat{\theta}_B^*$. Proceed as in Section 6.2.5, compute the projection distance of each bootstrap estimate, $\hat{\theta}_b^*$, relative to all B bootstrap values and label the result d_b^*, $b = 1, \ldots, B$. Put these B distances in ascending order yielding $d_{(1)}^* \leq \cdots \leq d_{(B)}^*$. Set $u = (1-\alpha)B$, rounding to the nearest integer. A direct application of results in Liu and Singh (1997) indicates that an approximate $1 - \alpha$ confidence region corresponds to the u bootstrap values having the smallest projection distances. As for testing

$$H_0 : \theta = \theta_0,$$

θ_0 given, let D_0 be the projection distance of θ_0. Set $I_b = 1$ if $D_0 \leq D_b^*$; otherwise $I_b = 0$. Then the (generalized) p-value is estimated to be

$$\hat{p} = \frac{1}{B} \sum_{b=1}^{B} I_b,$$

and a direct application of results in Liu and Singh (1997) indicates that H_0 be rejected if $\hat{p} \leq \alpha$.

However, when testing at the 0.05 level, this method can be unsatisfactory for $n \leq 120$ (and switching to Mahalanobis distance makes matters worse). A better approach is to adjust the decision rule when n is small. In particular, reject if $\hat{p} \leq \alpha_a$, where for $n \leq 20$, $\alpha_a = 0.02$; for $20 < n \leq 30$, $\alpha_a = 0.025$; for $30 < n \leq 40$, $\alpha_a = 0.03$; for $40 < n \leq 60$, $\alpha_a = 0.035$; for $60 < n \leq 80$, $\alpha_a = 0.04$; for $80 < n \leq 120$, $\alpha_a = 0.045$; and for $n > 120$, use $\alpha_a = 0.05$. Simulations (Wilcox, 2003b) suggest that for $p = 2, \ldots, 8$, reasonably good control over the probability of a type I error is obtained regardless of the correlations among the p variables under study. That is, the actual probability of a type I error will not be much larger than the nominal level. However, for $n = 20$, and when sampling from a heavy-tailed distribution, the actual probability of a type I error can drop below .01 when testing at the 0.05 level. So there is room for improvement, but currently, only the method just described has been found to be remotely successful for the problem at hand.

6.7.2 Extension of Hotelling's T^2 to Trimmed Means

Hotelling's T^2 test is a classic method for testing

$$H_0 : \mu = \mu_0,$$

where μ represents a vector of p population means and μ_0 is a vector of specified constants. The method is readily generalized to making inferences about the marginal trimmed means via the test statistic

$$T^2 = \frac{h(h-p)}{(n-1)p}(\bar{\mathbf{X}}_t - \mu_0)\mathbf{S}^{-1}(\bar{\mathbf{X}}_t - \mu_0)',$$

where $\mathbf{S}$ is the Winsorized variance–covariance matrix corresponding to the p measures under study and $\bar{\mathbf{X}}_t$ is the vector of marginal trimmed means and h is the number of observations left after trimming. When the null hypothesis is true, T^2 has, approximately, an F distribution with degrees of freedom $v_1 = p$ and $v_2 = h - p$. That is, reject at the α level if

$$T^2 \geq f,$$

where f is the $1 - \alpha$ quantile of an F distribution with $v_1 = p$ and $v_2 = h - p$ degrees of freedom.

6.7.3 R Functions smeancrv2 and hotel1.tr

The R function

$$\text{smeancrv2(m, nullv=rep(0, ncol(m)), nboot=500, plotit=T, MC=F, xlab="VAR 1",}$$
$$\text{ylab="VAR 2",STAND=F)}$$

tests the hypothesis $H_0: \theta = \theta_0$, where θ is the population value of the OP-estimator. The null value, θ_0, is specified by the argument nullvec and defaults to a vector of zeros. The argument cop determines the measure of location used by the projection outlier detection method and MM determines the measure of scale that is used; see Section 6.4.10. If m is a matrix having two columns and plotit=T, the function plots the bootstrap values and indicates the approximate 0.95 confidence region. Setting the argument MC=T, a multi-core processor can be used to compute the measure of location and the projection distances, which will help reduce execution time. (The function smeancrv2 is the same as the function smeancr, only smeancr does not have an option for using a multi-core processor.)

■ Example

Table 6.5 shows the cork boring weights for the north, east, south and west sides of 28 trees. (The data are from Rao, 1948.) For illustrative purposes, suppose the difference scores between the west and north sides of the trees are stored in column one of the R matrix m, and in column two are the difference scores between the west and east sides. Figure 6.5 shows the approximate 0.95 confidence region for the typical difference scores based on the OP-estimator and reported by the function smeancr. The (generalized) p-value, when testing $H_0: \theta = (0, 0)$ is .004, and the 0.05 critical p-value is $\alpha_a = 0.025$, so reject at the 0.05 level.

Table 6.5: Cork Boring Weights for the North, East, South and West Sides of Trees.

N	E	S	W	N	E	S	W
72	66	76	77	91	79	100	75
60	53	66	63	56	68	47	50
56	57	64	58	79	65	70	61
41	29	36	38	81	80	68	58
32	32	35	36	78	55	67	60
30	35	34	26	46	38	37	38
39	39	31	27	39	35	34	37
42	43	31	25	32	30	30	32
37	40	31	25	60	50	67	54
33	29	27	36	35	37	48	39
32	30	34	28	39	36	39	31
63	45	74	63	50	34	37	40
54	46	60	52	43	37	39	50
47	51	52	53	48	54	57	43

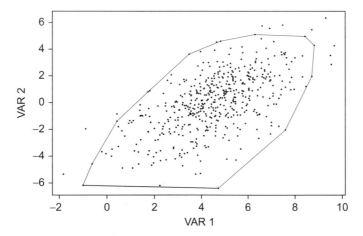

Figure 6.5: The 0.95 confidence region based on the OP-estimate of location using the cork data, where VAR 1 is the difference between the west and north sides and VAR 2 is the difference between the west and east sides.

The R function

$$\text{hotel1.tr(x,null.value=0,tr=0.2)}$$

performs the generalization of Hotelling's T^2 method to trimmed means. The argument null.value can contain a single value, which is taken to mean that all p hypothesized values are equal to the specified value, or the argument null.value can contain p values. The argument x is assumed to be a matrix or data frame. And as usual, tr=0.2 indicates that by default, 20% trimming is used.

6.7.4 Inferences Based on the MGV Estimator

A natural guess is that the inferential method based on the OP-estimator can be used with the MGV estimator as well. It appears, however, that some alternative modification of the bootstrap method is required. For example, if $n - 20$ and $p = 4$, the modified bootstrap method designed for the OP-estimator rejects at the 0.05 level if $\hat{p}^* \leq .02$ and appears to control the probability of a type I error for a wide range of distributions. However, if this method is applied with the OP-estimator replaced by the MGV estimator, the actual probability of a type I error exceeds .1 when sampling from a normal distribution; to achieve an actual probability of a type I error approximately equal to .05, reject if $\hat{p}^* \leq .006$. Switching to Mahalanobis distance makes matters worse. A better approach is to proceed exactly as was done with the OP-estimator, only use MGV distances when computing the (generalized) p-value.

6.7.5 R Function smgvcr

The R function

$$smgvcr(m,nullvec=rep(0,ncol(m)),SEED=T,op=0,nboot=500,plotit=T)$$

tests the hypothesis $H_0: \theta = \theta_0$, where θ is the population value of the MGV estimator. The null value, θ_0, is specified by the argument nullvec and defaults to a vector of zeros. The argument op determines how the central values are determined when using the MGV outlier detection method; see the function mgvmean.

6.8 Two-Sample Case

The method in Section 6.6 is readily extended to the two-sample case. That is, for two independent groups, θ_j represents the value of θ (the population OP measure of location) associated with the jth group ($j = 1, 2$), and the goal is to test

$$H_0 : \theta_1 = \theta_2.$$

Now, simply generate bootstrap samples from each group, compute the OP-estimator for each, label the results θ_1^* and θ_2^*, and set $d^* = \theta_1^* - \theta_2^*$. Repeat this process B times yielding $d_1^*, \ldots, d_B^*$. Then H_0 is tested by determining how deeply the vector $(0, \ldots, 0)$ is nested within the cloud of d_b^* values, $b = 1, \ldots, B$, again using the projection depth. If its depth, relative to all B bootstrap estimates, is low, meaning that it is relatively far from the center, then reject. More precisely, let D_b be the OP distance associated with the bth bootstrap sample and let D_0 be the distance associated with $(0, \ldots, 0)$. Set $I_b = 1$ if $D_b > D_0$, otherwise $I_b = 0$, in which case the estimated generalized p-value is

$$\hat{p} = \frac{1}{B} \sum_{b=1}^{B} I_b.$$

Currently, when $\alpha = 0.05$, it is recommended to set $n = \min(n_1, n_2)$ and use α_a as defined in the one sample case in Section 6.6. (Checks on this method, when the OP-estimator is replaced by the MGV estimator, have not been made.)

6.8.1 R Functions smean2, smean2v2, matsplit, and mat2grp

The R function

smean2v2(m1,m2,nullv=rep(0,ncol(m1)),cop=3,MM=F,SEED=NA,
nboot=500,plotit=T,MC=F)

tests the hypothesis that two multivariate distributions have the same measure of location using the method just described. Here, the data are assumed to be stored in the matrices m1 and m2, each having p columns. The argument nullv indicates the null vector and defaults to a vector of zeros. The arguments cop and MM control how outliers are detected when using the projection method; see Section 6.4.10. As usual, to avoid the plot, set plotit=F. To use a multi-core processor, set the argument MC=T. The function

smean2(m1,m2,nullv=rep(0,ncol(m1)),cop=3,MM=F,SEED=NA, nboot=500,plotit=T)

is exactly the same as smean2v2, only it does not have an option for using a multi-core processor.

Data Management

The R function

matsplit(m,coln)

is supplied in case it helps with data management. It splits the matrix m, into two matrices based on the values in the column of m indicated by the argument coln. This column is assumed to have two values only. Results are returned in $m1 and $m2.

The R function

mat2grp(m,coln)

also splits the data in a matrix into groups based on the values in column coln of the matrix m. Unlike matsplit, mat2grp can handle more than two values (i.e., more than two groups), and it stores the results in list mode.

■ Example

Thomson and Randall-Maciver (1905) report four measurements for male Egyptian skulls from five different time periods: 4000 B.C., 3300 B.C., 1850 B.C., 200 B.C., and 150 A.D. There are 30 skulls from each time period and four measurements: maximal breadth, basibregmatic height, basialveolar length, and nasal height. For illustrative purposes, assume the data are stored in the R variable skull, the four measurements are stored in columns 1–4, and the time period is stored in column 5. Here, the first and last time periods are compared, based on the OP measure of location. First, split the data into five groups based on the time period using, for example, the R command

$$z=mat2grp(skull,5).$$

So z[[1]] is a matrix containing the data corresponding to the first time period and z[[5]] is a matrix containing the data for the final (fifth) time period, 150 A.D. The R command

$$smean2v2(z[[1]][,1:4],z[[5]][,1:4])$$

compares the two groups based on all four measures. Figure 6.6 shows the plot created by smean2v2 when using the first two variables only. The polygon is an approximate 0.95 confidence region for the difference between the measures of location. The *p*-value is .002. (Using all four measures, the *p*-value is 0.)

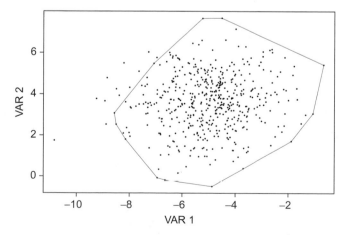

Figure 6.6: Using the first two skull measures for the first and last time periods, the plot shows the 0.95 confidence region for the difference between the OP measures of location. ■

6.8.2 Comparing Robust Generalized Variances

Robust generalized variances can be compared as well. A percentile bootstrap appears to avoid type I errors above the nominal level. But situations are encountered where the actual level can be substantially smaller than the nominal level. Corrections are available in some situations (Wilcox, 2006e), which are used by the R function described in the next section, but no details are given here.

6.8.3 R function gvar2g

The R function

$$\text{gvar2g(x, y, nboot} = 100, \text{DF} = \text{T, eop} = 1, \text{est} = \text{skipcov, alpha} = 0.05, \text{cop} = 3, \text{op} = 1,$$
$$\text{MM} = \text{F, SEED} = \text{T)}$$

compares two independent groups based on a robust version of the generalized variance. By default, the OP covariance matrix is used in conjunction with Carling's modification of the boxplot rule. Setting MM=T, a MAD-median rule is used. If DF=T, and if the sample sizes are equal, the function reports an adjusted critical p-value, assuming that the goal is to have a type I error probability equal to .05, the argument est=skipcov, and that other conditions are met. Otherwise, no adjusted critical value is reported. For information about the arguments op, cop, and eop, see the R function skipcov.

6.9 Multivariate Density Estimators

This section outlines two multivariate density estimators that will be used when plotting data. The first is based on a simple extension of the expected frequency curve described in Chapter 3 and the other is a multivariate analog of the adaptive kernel density estimator. An extensive discussion of multivariate density estimation goes beyond the scope of this book, but some indication of the method used here, when plotting data, seems warranted.

The strategy behind the expected frequency curve is to determine the proportion of points that are close to $\mathbf{X}_i$. There are various ways this might be done and here a method based on the MVE covariance matrix is used. Extant results suggest this gives a reasonable first approximation of the shape of a distribution in the bivariate case, but there are many alternative methods for determining which points are close to $\mathbf{X}_i$, and virtually nothing is known about their relative merits for the problem at hand.

Here, the point $\mathbf{X}_{i'}$ is said to be close to $\mathbf{X}_i$ if

$$\sqrt{(\mathbf{X}_{i'} - \mathbf{X}_i)' M^{-1} (\mathbf{X}_{i'} - \mathbf{X}_i)} \leq h,$$

where M is the MVE covariance matrix described in Section 6.3.1, and h is the span. Currently, $h = 0.8$ seems to be a good choice for most situations. Letting N_i represent the number of points close to $\mathbf{X}_i$, $f_i = N_i/n$ estimates the proportion of points close to $\mathbf{X}_i$. In the bivariate case, a plot of the data is created simply by plotting the points $(\mathbf{X}_i, f_i)$.

The expected frequency curve can be used as a first approximation when using an adaptive kernel density estimate. Here, once the expected frequency curve has been computed, the method described by Silverman (1986) is used based on the multivariate Epanechnikov kernel.

An outline of the method is as follows. First, rescale the p marginal distributions. More precisely, let $x_{i\ell} = X_{i\ell}/\min(s_\ell, \text{IQR}_\ell/1.34)$, where s_ℓ and IQR_ℓ are, respectively, the standard deviation and interquartile range based on $X_{1\ell}, \ldots, X_{n\ell}$, $\ell = 1 \ldots, p$. (Here, IQR is computed via the ideal fourths.) If $\mathbf{x}'\mathbf{x} < 1$, the multivariate Epanechnikov kernel is

$$K_e(\mathbf{x}) = \frac{(p+2)(1-\mathbf{x}'\mathbf{x})}{2c_p};$$

otherwise $K_e(\mathbf{x}) = 0$. The quantity c_p is the volume of the unit p-sphere: $c_1 = 2$, $c_2 = \pi$, and for $p > 2$ $c_p = 2\pi c_{p-2}/p$. Similar to Section 3.2, the estimate of the density function is

$$\hat{f}(t) = \frac{1}{n} \sum \frac{1}{h\lambda_i} K[h^{-1}\lambda_i^{-1}(t - X_i)],$$

where, following Silverman (1986, p. 86), the span is taken to be

$$h = A(p)n^{-1/(p+4)},$$

$A(1) = 1.77$, $A(2) = 2.78$ and for $p > 2$,

$$A(p) = \left[\frac{8p(p+2)(p+4)(2\sqrt{\pi})^p}{(2p+1)c_p}\right]^{1/(p+4)}.$$

The quantity λ_i is computed as described in Section 3.2.4, only now the initial estimate of f is based on the multivariate version of the expected frequency curve. The R functions rdplot and akerd, described in Section 3.2.5, perform the calculations.

6.10 A Two-Sample, Projection-Type Extension of the Wilcoxon–Mann–Whitney Test

There are various ways to generalize the Wilcoxon–Mann–Whitney test to the multivariate case, some of which are discussed in Chapter 7. Here, a projection-type extension is described that is based, in part, on the multivariate measures of location covered in this chapter. Consider two independent groups with p measures associated with each. Let θ_j be any measure of location associated with the jth group ($j = 1, 2$). The basic strategy is to (orthogonally) project the data onto the line connecting the points θ_1 and θ_2, and then consider the proportion of projected points associated with the first group that are "less than" the projected points associated with the second.

To elaborate, let $\mathcal{L}$ represent the line connecting the two measures of location and let d_j be the Euclidean distance of θ_j from the origin. For the moment, assume $\theta_1 \neq \theta_2$. Roughly, as we move along $\mathcal{L}$, the positive direction is taken to be the direction from θ_1 toward θ_2 if $d_1 \leq d_2$; otherwise the direction is taken to be negative. So if the projection of the point X onto $\mathcal{L}$ corresponds to the point U, and the projection of the point Y corresponds to the point V, and if moving from U to V corresponds to moving in the positive direction along $\mathcal{L}$, then it is said that X is "less than" Y.

For convenience, distances along the projected line are measured relative to the point midway between θ_1 and θ_2, namely, $(\theta_1 + \theta_2)/2$. That is, the distance of a projected point refers to how far it is from $(\theta_1 + \theta_2)/2$, where the distance is taken to be negative if a projected point lies in the negative direction from $(\theta_1 + \theta_2)/2$. If D_x and D_y are the projected distances associated with two randomly sampled observations, $\mathbf{X}$ and $\mathbf{Y}$, then it is said that $\mathbf{X}$ is "less than", "equal to", or "greater than" $\mathbf{Y}$ according to whether D_x is less than, equal to, or greater than D_y, respectively. In symbols, it is said that $\mathbf{X} \prec \mathbf{Y}$ if $D_x < D_y$, $\mathbf{X} \simeq \mathbf{Y}$ if $D_x = D_y$, and $\mathbf{X} \succ \mathbf{Y}$ if $D_x > D_y$. Extending a standard convention in rank-based methods in an obvious way, to deal with situations where $D_x = D_y$ can occur, let

$$\eta = P(\mathbf{X} \prec \mathbf{Y}) + 0.5 P(\mathbf{X} \simeq \mathbf{Y}).$$

The goal is to estimate η and test

$$H_0 : \eta = 0.5. \tag{6.23}$$

First consider estimation. Given an estimated measure of location $\hat{\theta}_j$ for the jth group ($j = 1$, 2), the projected distances are computed as follows. Let $\|\hat{\theta}_j\|$ be the Euclidean norm associated with $\hat{\theta}_j$, let $S = 1$ if $\|\hat{\theta}_1\| \geq \|\hat{\theta}_2\|$, otherwise $S = -1$. Let

$$\mathbf{C} = (\hat{\theta}_1 + \hat{\theta}_2)/2,$$

$$\mathbf{B} = S(\hat{\theta}_1 - \hat{\theta}_2),$$

$$A = \|\mathbf{B}\|^2,$$

$$\mathbf{U}_i = \mathbf{X}_i - \mathbf{C},$$

and for any i and $k = 1, \ldots, p$, let

$$W_i = \sum_{k=1}^{p} U_{ik} B_k,$$

$$T_{ik} = \frac{W_i}{A} B_{ik}$$

in which case the distance associated with the projection of $\mathbf{X}_i$ is

$$D_{xi} = \text{sign}(W_i) \sqrt{\sum_{k=1}^{p} T_{ik}^2},$$

$i = 1, \ldots, m$. The distances associated with the $\mathbf{Y}_i$ values are computed simply by replacing $\mathbf{X}_i$ with $\mathbf{Y}_i$ in the definition of $\mathbf{U}_i$. The resulting distances are denoted by D_{yi}, $i = 1, \ldots, n$.

To estimate η, let

$$V_{ii'} = \text{sign}(D_{xi} - D_{yi'}),$$

and

$$\bar{V} = \frac{1}{mn} \sum_{i=1}^{m} \sum_{i'=1}^{n} V_{ii'}.$$

Then extending results in Cliff (1996) in an obvious way,

$$\hat{\eta} = \frac{1 - \bar{V}}{2}$$

is an unbiased estimate of η and takes into account tied values.

When testing Eq. (6.23), Wilcox (2005a) found that a basic percentile bootstrap method is unsatisfactory in terms of controlling the probability of a type I error but that a slight modification of the method performs reasonably well in simulations. The method begins by subtracting θ_j from every observation in the jth group. In effect, shift the data so that null hypothesis is true. Now, for each group, generate bootstrap samples from the shifted data and estimate η based on the two bootstrap samples just generated. Label the result $\hat{\eta}^*$. Repeat this B times yielding $\hat{\eta}_1^*, \ldots, \hat{\eta}_B^*$ and put these B values in ascending order yielding $\hat{\eta}_{(1)}^* \leq \cdots \leq \hat{\eta}_{(B)}^*$. Then reject H_0 if $\hat{\eta}_{(\ell+1)}^* > \hat{\eta}$ or if $\hat{\eta}_{(u)}^* < \hat{\eta}$, where $\ell = \alpha B/2$, rounded to the nearest integer, and $u = B - \ell$. Here, $B = 1000$ is assumed unless stated otherwise.

6.10.1 R functions mulwmw and mulwmwv2

The R function

```
mulwmw(m1,m2,plotit=T,cop=3,alpha=0.05,nboot=1000,pop=4,fr=0.8,pr=F)
```

performs the multivariate extension of the Wilcoxon–Mann–Whitney test just described, where the arguments m1 and m2 are any matrices (having p columns) containing the data for the two groups. The argument pr can be used to track the progress of the bootstrap method used to compute a critical value. If plotit=T, a plot of the projected distances is created, the type of plot being controlled by the argument pop. The choices are

*   pop=1, dotplots
*   pop=2, boxplots
*   pop=3, expected frequency curve
*   pop=4, adaptive kernel density estimate

The argument cop controls which measure of location is used. The choices are:

*   cop=1, Donoho–Gasko Median
*   cop=2, MCD estimator
*   cop=3, marginal medians
*   cop=4, OP-estimator

The R function

mulwmwv2(m1,m2,plotit=T,cop=3,alpha=0.05,nboot=1000,pop=4,fr=0.8,pr=F)

is the same as mulwmw, only it also reports a robust explanatory measure of effect size, described in Section 5.3.4, based on the projected points.

■ Example

Figure 6.7 shows four plots corresponding to the various choices for the argument pop using the skull data used in Figure 6.6. The upper left panel used pop=1, the upper right panel used pop=2, the lower left used pop=3, and the lower right used pop=4.

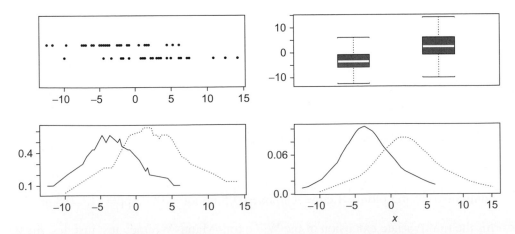

Figure 6.7: An example of the four types of plots created by the function mulwmw.

6.11 A Relative Depth Analog of the Wilcoxon–Mann–Whitney Test

This section describes another approach to generalizing the Wilcoxon–Mann–Whitney test to the multivariate case. To explain the strategy, first consider the univariate case and let $D = X - Y$, where X and Y are independent random variables. As explained in Section 5.7, heteroscedastic analogs of the Wilcoxon–Mann–Whitney test are concerned with how deeply zero is nested within the distribution of D. When tied values occur with probability zero, the usual null hypothesis is, in essence, that the depth of zero is equal to the highest possible depth. (That is, the hypothesis is that the median of the distribution of D is zero.) A slightly different formulation, which is useful for present purposes, is to say that the Wilcoxon–Mann–Whitney test is aimed at determining whether the depth of the value zero differs from the maximum possible depth associated with a distribution. A simple way of quantifying this difference is with Q, say, the depth of zero divided by the maximum possible depth, in which case the goal is to test

$$H_0 : Q = 1. \tag{6.24}$$

Put a bit more formally, imagine that $\mathbf{X}$ and $\mathbf{Y}$ are independent p-variate random variables and let $D_j = X_j - Y_j$ be the difference between the jth marginal distributions, $j = 1, \ldots, p$. Let A denote the depth of $\mathbf{0}$ (a vector having length p), relative to the joint distribution of $\mathbf{D} = \mathbf{X} - \mathbf{Y}$, and let B be the maximum possible depth for any point, again relative to the the joint distribution of $\mathbf{D}$. Then

$$Q = \frac{A}{B}.$$

To estimate Q, let X_{ij} ($i = 1, \ldots, n_1; j = 1, \ldots, p$) and $Y_{i'j}$ ($i' = 1, \ldots, n_2; j = 1, \ldots, p$) be random samples and for fixed i and i', consider the vector $\mathbf{D}$ formed by the p differences $X_{ij} - Y_{i'j}$, $j = 1, \ldots, p$. There are $L = n_1 n_2$ such vectors, one for each i and i', which are labeled $\mathbf{D}_\ell$, $\ell = 1, \ldots, L$. Let P_0 denote the depth of $\mathbf{0}$ relative to the these L vectors, and let P_ℓ be the depth of the ℓth vector, again relative to the L vectors $\mathbf{D}_\ell$, $\ell = 1, \ldots, L$. Let $P_m = \max P_\ell$, the maximum taken over $\ell = 1, \ldots, L$. Then an estimate of Q is

$$\hat{Q} = \frac{P_0}{P_m}.$$

Evidently, $\hat{Q}$ is not asymptotically normal when the null hypothesis is true. Note that in this case, Q lies on the boundary of the parameter space. Bootstrap methods have been considered for testing H_0, but their small-sample properties have proven to be difficult to study via simulations because of the high execution time required to compute the necessary depths. Let $N = \min(n_1, n_2)$ and suppose $\alpha = 0.05$. Currently, the only method that has performed well in simulations is to reject if

$$\hat{Q} \leq c,$$

where for $p = 2$ or 3,

$$c = \max(0.0057N + 0.466, \; 1),$$

for $p = 4$ or 5,

$$c = \max(0.00925N + 0.430, \; 1),$$

for $p = 6$ or 7,

$$c = \max(0.0264N + 0.208, \; 1),$$

for $p = 8$,

$$c = \max(0.0149N + 0.533, \; 1),$$

and for $p > 8$,

$$c = \max(0.04655p + 0.463, \; 1).$$

(See Wilcox, 2003f, for more details. Critical values for other choices of α have not been determined.)

6.11.1 R function mwmw

The R function

$$mwmw(m1,m2,cop=5,pr=T,plotit=T,pop=1,fr=0.8,dop=1,op=1)$$

performs the multivariate extension of the Wilcoxon–Mann–Whitney test just described, where the arguments m1 and m2 are any matrices (having p columns) containing the data for the two groups. The argument cop determines the center of the data that will be used when computing halfspace depth. The choices are:

- cop=1, Donoho–Gasko Median
- cop=2, MCD estimator
- cop=3, marginal medians
- cop=4, MVE estimator
- cop=5, OP-estimator

Setting the argument dop=2 causes halfspace depth to be approximated using method A2 in Section 6.2.3; by default, method A1 is used. For bivariate data a plot is created based on the value of the argument pop, the possible values being 1, 2, and 3, which correspond to a scatterplot, an expected frequency curve, and an adaptive kernel density estimate. The argument fr is the span used by the expected frequency curve. As usual, setting plotit=F avoids the plot. The function returns an estimate of η in the variable phat.

■ Example

The first two skull measures used in the example of Section 6.7.1 are used to illustrate the plot created by the R function mwmw. The plot is shown in Figure 6.8. The center of the data is marked by an o and based on the OP-estimator, and the null vector is indicated by a +. The function reports that phat is 0.33 indicating that the estimate of $\hat{Q}$ is 0.33. This is less than the critical value 0.62, so reject.

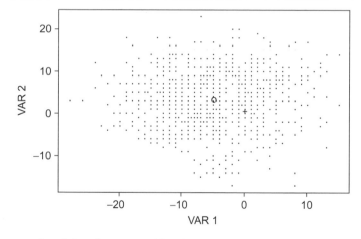

Figure 6.8: An example of the plot created by the function mwmw.for. The estimated center is marked by an o, and the null center is marked with a +.

6.12 Comparisons Based on Depth

This section describes yet another approach to comparing two independent groups based on multivariate data. The basic idea is that if groups do not differ, the typical depth of the points of the first group, relative to the second, should be the same as the typical depth of the second group, relative to the first. Roughly, the issue is the extent to which the groups are separated as measured by some notion of depth. Here, halfspace depth is used exclusively, simply because this special case has received the most attention from an inferential point of view.

Here we let $T_D(\mathbf{x}; F)$ represent Tukey's halfspace depth of $\mathbf{x}$ relative to the multivariate distribution F. As usual, let $\mathbf{X}$ and $\mathbf{Y}$ represent independent, p-variate random variables. The corresponding distributions are denoted by F and G. Let

$$R(\mathbf{y}; F) = P_F[T_D(\mathbf{X}; F) \leq T_D(\mathbf{y}; F)].$$

That is, $R(\mathbf{y}; F)$ is the probability that the depth of a randomly sampled $\mathbf{X}$, relative to F, is less than or equal to the depth of some particular point, $\mathbf{y}$, again relative to F. Said another way, $R(\mathbf{y}; F)$ is the fraction of the F population that is less central than the value $\mathbf{y}$. A *quality index* proposed by Liu and Singh (1993) is

$$Q(F, G) = E_G[R(\mathbf{Y}; F)],$$

the average of all $R(\mathbf{y}; F)$ values with respect to the distribution G. Put another way, for a randomly sampled $\mathbf{X}$ and $\mathbf{Y}$,

$$Q(F, G) = P[D(\mathbf{X}; F) \leq D(\mathbf{Y}; F)]$$

is the probability that the depth of Y is greater than or equal to depth of X. Liu and Singh show that the range of Q is [0, 1] and when $F = G$, $Q(F, G) = 1/2$. Moreover, when $Q < 1/2$, this reflects a location shift and/or scale increase from F to G. They also develop inferential methods based on Q where it is assumed that F is some reference distribution. Here a variation of their method is considered where the goal is to be sensitive to shifts in location. (For relevant asymptotic results, see Zuo & He, 2006.)

Suppose the sample sizes are m and n for the distributions F and G, respectively. Let

$$\bar{D}_{12} = \frac{1}{m} \sum T_D(\mathbf{X}_i; G_n)$$

be the average depth of the m vectors of observations sampled from F relative to the empirical distribution G_n associated with the second group. If $\bar{D}_{12}$ is relatively small, this can be due to a shift in location or differences in scale. But if

$$\bar{D}_{21} = \frac{1}{n} \sum T_D(\mathbf{Y}_i; F_m)$$

is relatively small as well, this reflects a separation of the two empirical distributions which is roughly associated with a difference in location. (Of course, groups can differ in scale as well when both $\bar{D}_{12}$ and $\bar{D}_{21}$ are small.) So a test of H_0: $F = G$ that is sensitive to shifts in location is one that rejects if $\bar{D}_M = \max(\bar{D}_{12}, \bar{D}_{21})$ is sufficiently small.

Assuming $m < n$, let $N = (3m + n)/4$. (If $m > n$, $N = [3n + m]/4$.) The only known method that performs well in simulations when testing at the 0.05 level, based on avoiding a type I error probability greater than the nominal level, is to reject if $\bar{D}_M \leq d_N$, where for $p = 1$,

$$d_N = \frac{-0.4578}{\sqrt{N}} + 0.2536;$$

for $p = 2$,

$$d_N = \frac{-0.3}{\sqrt{N}} + 0.1569;$$

for $p = 3$,

$$d_N = \frac{-0.269}{\sqrt{N}} + 0.0861;$$

for $p = 4$,

$$d_N = \frac{-0.1568}{\sqrt{N}} + 0.0540;$$

for $p = 5$

$$d_N = \frac{-0.0968}{\sqrt{N}} + 0.0367;$$

for $p = 6$,

$$d_N = \frac{-0.0565}{\sqrt{N}} + 0.0262;$$

for $p = 7$

$$d_N = \frac{-0.0916}{\sqrt{N}} + 0.0174;$$

and for $p > 8$ $d_N = .013$. In terms of type I errors, the main difficulty is that when sampling from heavy-tailed distributions, the actual type I error probability can drop well below .05 when testing at the 0.05 level (Wilcox, 2003c). (For $p > 8$, as p increases, the actual probability of a type I error decreases.) Determining a p-value via simulations when sampling from a normal distribution is, perhaps, more satisfactory, at the expense of higher execution time.

As for a method that is relatively sensitive to differences in scatter, which can be used for p-variate data, first estimate $Q(F, G)$ with

$$\hat{Q}(F, G) = \frac{1}{n} \sum_{i=1}^{n} R(Y_i; F_m). \tag{6.25}$$

(Properties of this estimator are reported by Liu & Singh, 1993.) Similarly, the estimate of $Q(G, F)$ is

$$\hat{Q}(G, F) = \frac{1}{m} \sum_{i=1}^{m} R(X_i; G_n). \tag{6.26}$$

The goal is to test

$$H_0 : Q(F, G) = Q(G, F). \tag{6.27}$$

Unlike the method based on $\bar{D}_{12}$ and $\bar{D}_{21}$, a basic percentile bootstrap method performs well in simulations. To begin, generate bootstrap samples from both groups in the usual way and let $\hat{Q}^*(F, G)$ and $\hat{Q}^*(G, F)$ be the resulting bootstrap estimates of $Q(F, G)$ and $Q(G, F)$. Set $D^* = \hat{Q}^*(F, G) - Q^*(G, F)$. Repeat this process B times yielding $D_b^*, b = 1, \ldots, B$. Put these B values in ascending order yielding $D_{(1)}^* \leq \cdots \leq D_{(B)}^*$. Then a $1 - \alpha$ confidence interval for $Q(F, G) - Q(G, F)$ is simply $(D_{(\ell+1)}^*, D_{(u)}^*)$, where $\ell = \alpha B/2$, rounded to the nearest integer, and $u = B - \ell$. Of course, reject H_0 if this interval does not contain zero.

6.12.1 R Functions lsqs3 and depthg2

The R function

> lsqs3(x,y,plotit=T,cop=2, cop=2, ap.dep=F, v2=F, pv=F, SEED=T, nboot=1000),

compares two independent groups based on the statistic $\hat{D}_M$ described in the previous section. For bivariate data, if plotit=T, a scatterplot of the data is produced with the points associated with the second group indicated by a circle. Setting the argument pv=T, a *p*-value is computed. The function

> depthg2(x,y,alpha=0.05,nboot=500,plotit=T,op=T)

tests (6.27). If the argument op is set to T, the function prints a message when each bootstrap step is complete.

■ Example

The left panel of Figure 6.9 shows the plot created by lsqs3 based on the skull data described in Section 6.7.1. (The same plot is created by depthg2.) The function lsqs3 rejects at the 0.05 level suggesting a shift in location, but depthg2, which is designed to be sensitive to differences in the amount of scatter, does not reject. The right panel shows a scatterplot where the both groups have bivariate normal distributions that

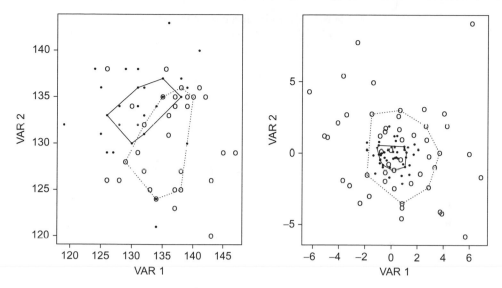

Figure 6.9: The left panel is the plot created by the function lsqs3 using the skull data in Section 6.8.1. The right panel is the plot based on data generated from a bivariate normal distribution where the marginal distributions have a common mean, but their standard deviations differ. One has a standard deviation of 1 and the other has a standard deviation of 3.

differ in scale only; the marginal distributions of the first group have standard deviation one, and for the other group the marginal distributions have standard deviation three. (Here, $m = n = 50$.) Now the function lsqs3 finds no difference between the groups, but depthg2 does (at the 0.05 level).

■

■ Example

Table 6.6 shows the data for 24 schizophrenia patients and 18 demographically matched controls. (The data are stored in the files schiz1.data and schiz2.data; see Chapter 1.) PP120 is a prepulse inhibition measure taken 120 ms following the onset of an attended stimulus, and PM120 is the prepulse inhibition measure taken 120 ms following the onset of an ignored stimulus. Figure 6.10 shows the scatterplot created by lsqs3; the test statistic is 0.049, the critical value is 0.089, so reject at the 0.05 level. (In Figure 6.10, VAR 1 is PP120 and VAR 2 is PM120.) It is left as an exercise to verify that comparing PM120 using means, trimmed means or an M-estimator, no difference is found at the 0.05 level, but for PP120, the reverse is true. Note, however, that in Figure 6.10, there is a sense in which PM120 (labeled VAR 2) for the control group lies

Table 6.6: Prepulse Inhibition Measures for Schizophrenia Patients and Controls.

Schizophrenia Patients		Control	
PP120	PM120	PP120	PM120
−88.82	−92.40	−100.00	−31.47
13.54	−36.70	−71.48	−15.38
−37.22	0.08	−87.01	−64.87
−43.26	−42.40	−100.00	−95.94
−43.35	−42.25	−100.00	−81.24
−31.64	−41.30	93.95	109.44
−98.73	−96.56	−59.89	−35.97
−37.35	−33.82	−79.88	−79.24
−8.48	50.59	−53.33	−38.19
−63.87	−8.80	−90.08	−42.69
26.55	32.13	−40.25	10.57
−91.05	−95.85	−33.78	−7.60
−8.07	9.82	−89.18	−61.35
−97.65	−95.80	−84.78	−58.98
−60.80	−52.63	−64.74	−39.57
−33.58	−56.36	−91.10	−82.26
−15.80	−38.51	−74.82	−66.60
−12.92	1.50	−86.52	−48.74
−77.35	−86.07		
−85.09	−84.71		
−33.53	−43.66		
−8.67	−9.40		
−89.21	−86.80		
−77.76	−75.83		

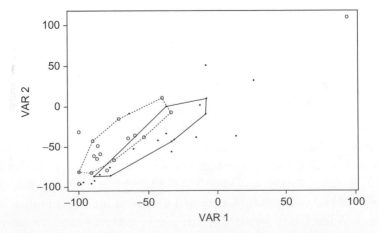

Figure 6.10: The plot created by the function lsqs3 using the schizophrenia data in Table 6.6.

above and to the left the points corresponding to the schizophrenia patients. Given PP120, PM120 tends to be greater for the control group.

∎

6.13 Comparing Dependent Groups Based on All Pairwise Differences

This section describes an affine invariant method for comparing J dependent groups that is based on a simple extension of the method in Section 6.11. For any $j < m$, let $D_{ijm} = X_{ij} - X_{im}$. Let F be the joint distribution of $\mathbf{D}_{jm}$, and let P be the depth of $\mathbf{0}$ relative to F, divided by the maximum possible depth. Then $0 \le P \le 1$ and the goal is to test

$$H_0 : P = 1. \tag{6.28}$$

A simple estimate of P is

$$\hat{P} = \frac{A}{C}, \tag{6.29}$$

where A is the halfspace depth of $\mathbf{0}$ among these n vectors and C is the maximum depth among these n points.

An alternative approach is to determine the halfspace median, which is just the average of the deepest points, and then use the depth of the halfspace median as an estimate of the maximum possible depth. Provided n is not too small, this alternative approach seems to have no practical value, but it can be useful when n is very small. For instance, if $n = 5$ and sampling is from a bivariate normal distribution with a correlation of zero, it is common to have all five depths equal to 0.05, but the depth of the halfspace median is typically close to 0.4.

It should be stressed that affine invariance refers to the D_{ijm} values because it is the depth of these difference scores that are used. It can be seen that the method is not affine invariant in terms of the X_{ij} values.

A technical problem is that generally, the MCD estimate of location cannot be computed when working with the D_{ijm} values because the corresponding covariance matrix is singular. Consequently, the approximation of halfspace depth with method A1 A1 in Section 6.2.3 is not immediately applicable. To deal with this problem, compute the MCD estimate of location based on the original X_{ij} values, yielding say $(\hat{\xi}_1, \ldots, \hat{\xi}_J)$, and then approximate the halfspace depth of the D_{ijm} values with method A1 by taking the center of location of the D_{ijm} to be $\hat{\theta}_{jm} = \hat{\xi}_j - \hat{\xi}_m$.

An alternative strategy is to use an approximation of halfspace depth that does not require that the D_{ijm} values have a nonsingular covariance matrix. This can be accomplished with method A2 in Section 6.2.3.

As was the case in Section 6.10, when the null hypothesis is true, the distribution of $\hat{P}$ is not asymptotically normal. The reason is that for this special case, P lies on the boundary of the parameter space and so for any general situation where $\hat{P}$ is a consistent estimate of P, it must be that $\hat{P} \leq 1$ with the probability of $\hat{P} = 1$ increasing as the sample sizes get large. For similar reasons, based on theoretical results in Liu and Singh (1997), the expectation is that when H_0 is true, a basic percentile bootstrap method for computing a confidence interval for P will fail, and this has been found to be the case in simulations.

Consider rejecting H_0 if $\hat{P} \leq c$. Based on results in Wilcox (2005b), when testing at the 0.05 level, the following approximations of c appear to perform well. For $J = 2$,

$$\hat{c} = -1.46n^{-0.5} + 0.95.$$

for $J = 3$,

$$\hat{c} = -1.71n^{-0.5} + 1.00,$$

for $J = 4$,

$$\hat{c} = -1.77n^{-0.5} + 1.057,$$

for $J = 5$,

$$\hat{c} = -1.76n^{-0.5} + 1.11,$$

for $J = 6$,

$$\hat{c} = -1.62n^{-0.3} + 1.41,$$

for $J = 7$,

$$\hat{c} = -1.71n^{-0.3} + 1.49,$$

and for $J = 8$,

$$\hat{c} = -1.38n^{-0.3} + 1.39.$$

Note that as $n \to \infty$, $c \to 1$. Moreover, as J increases, c converges to 1 more quickly. So in effect, reject if $\hat{P} < \min(\hat{c}, 1)$.

The method just described is affine invariant, roughly meaning that it is metric free. That is, if the n-by-p matrix of data is post multiplied by a nonsingular matrix $\mathbf{A}$, $\hat{P}$ is not altered.

6.13.1 R Function dfried

The R function

$$\text{dfried(m,plotit=T,pop=0,fr=0.8,v2=F,op=F)}$$

tests the hypothesis $H_0 : P = 1$ as just described. Here, m is any R variable having matrix mode with n rows and p columns. If $p = 2$ and plotit=T, a plot of the difference scores is created with the type of plot controlled by the argument pop. The choices are:

1. pop=0, adaptive kernel density
2. pop=1, expected frequency curve
3. pop=2, kernel density estimate using normal kernel
4. pop=3, R built-in kernel density estimate
5. pop=4, boxplot

The argument fr controls the span when using the expected frequency curve. Setting v2=T causes method A2 to be used to approximate halfspace depth, and op=T results in using the depth of Tukey's median as an estimate of the maximum possible halfspace depth.

6.14 Robust Principal Components Analysis

Roughly, principal components analysis (PCA) is aimed at finding p linear combinations of m ($p < m$) observed variables that explains most of the variability in the data. To quickly review the strategy underlying the classic approach, momentarily consider the situation where $p = 1$. Denoting the data for the jth variable by X_{ij} ($i = 1, \ldots, n$; $j = 1, \ldots m$), the goal is to reduce the m variables to a single variable via some linear combination of the m variables, denoted by

$$U_i = \sum_{j=1}^{m} h_j X_{ij},$$

with the constants $h_1, \ldots, h_m$ chosen so as to maximize the variance of the U_i values subject to $\sum h_j^2 = 1$. Now consider the problem of reducing the p variables down to two variables rather than just one. So for the ith participant, the goal is to compute two linear combinations of the p variables based on two sets of weights:

$$U_{i1} = h_{11} X_{i1} + \cdots + h_{1p} p X_{ip}$$

and

$$U_{i2} = h_{21} X_{i1} + \cdots + h_{2p} X_{ip}$$

($i = 1, \ldots, n$), where for fixed k, $\sum h_{jk}^2 = 1$ and the variance of the U_{ik} values is maximized subject to the condition that U_k and U_ℓ have correlation zero, $k \neq \ell$. More generally, m linear combinations are sought that maximize the variance of the marginal distributions with the property that any two linear combinations have zero correlation. This goal is accomplished by taking $\mathbf{h}_1, \ldots \mathbf{h}_m$ to be the eigenvectors of the usual covariance matrix. The columns $U_1, \ldots, U_m$ of the matrix $\mathbf{U}$ are called the *principal components* of $\mathbf{X}$. (For a recent discussion regarding the interpretation of principal components, see Anaya-Izquierdo, Critchley, & Vines, 2011.) Moreover, the variance of U_k is λ_k, where $\lambda_1 \geq \cdots \geq \lambda_m$, and λ_k is the eigenvalue corresponding to the eigenvector $\mathbf{h}_k$. The U_{ij} are called the *principal component scores*.

But because the usual covariance matrix is not robust, situations are encountered where upon closer scrutiny the resulting components explain a structure that has been created by a mere one or two outliers (e.g., Huber, 1981, p. 199). This has led to numerous suggestions regarding how the classic PCA method might be made more robust. A simple approach is to replace the covariance matrix with a robust scatter matrix or a robust correlation matrix. Devlin et al. (1981) and Campbell (1980) used an M estimator with a low breakdown point, so a relatively small number of outliers can cause practical problems. The minimum volume ellipsoid (MVE) estimator, as well as the (fast) minimum covariance determinant (MCD) estimator, might be used, but concerns about these estimators have already been noted. A method based on an S-estimator was studied by Croux and Haesbroeck (2000), and a fast and simple method was proposed by Locantore, Marron, Simpson, Tripoli, and Zhang (1999). Li and Chen (1985) suggest a projection pursuit approach meaning that directions are sought that maximize or minimize some robust measure of dispersion. (One appealing feature of projection-type methods is that they can be used when the number of variables exceeds the sample size.) Croux and Ruiz-Gazen (2005, section 5.1) describe an algorithm for implementing the Li and Chen method. (Also see Hubert, Rousseeuw, & Verboven, 2002; Salibián-Barrera, Van Aelst, & Willems, 2006.) One negative feature of the Li and Chen method is its computational complexity. Maronna (2005) extended this projection pursuit technique in a manner that improves computational efficiency and statistical performance. Yet another recent suggestion was made by Hubert, Rousseeuw, and Vanden Branden (2005) that was later refined by Engelen, Hubert, and Vanden Branden (2005), which is used here. Roughly, the first step is to compute a measure of outlyingness for each of the n points, where n is the sample size. Then for h chosen by the investigator, the h least outlying data points are used to compute a measure of location and scatter, which in turn are used to determine how many components will be retained, as well as the projected data points. Following Engelen et al. (2005), a reweighting step is added based on the orthogonal distances of the observations with respect to the first estimated PCA subspace. It is only at the first stage of the algorithm that the number of points eliminated must be specified via the choice for h. (For some additional results on robust approaches to PCA, see Serneels & Verdonck, 2008; Chen, Martin, & Montague, 2009.)

All of the methods just listed are based in part on maximizing some measure of variation associated with the marginal distributions of the m principal components. Another approach is to choose linear combinations (principal components) aimed at maximizing some robust generalized variance associated with the principal component scores (Wilcox, 2008c). That is, take into account the overall structure of the data when measuring variation, in contrast to maximizing the variance of the individual principal component scores. (Details are given in Section 6.14.6.)

There is yet another generalization of PCA that should be mentioned: kernel PCA (Schlölkopf, Smola, & Müller, 1998). Roughly, the method first maps the data into a higher-dimensional feature space. (It generalizes regular PCA by replacing the usual inner product with a broader class of functions.) A robust version of kernel PCA has been studied by Debruyne, Hubert, and van Horebeek (2010).

6.14.1 R Functions prcomp and regpca

The built-in R function

$$\text{prcomp(x,cor=F)},$$

performs the classic principal component analysis. By default it uses the covariance matrix rather than the correlation matrix. In case it is useful, the R function

$$\text{regpca(x, cor} = \text{T, loadings} = \text{T, SCORES} = \text{F, scree} = \text{T, xlab} = \text{"Principal Component"},$$
$$\text{ylab} = \text{"Proportion of Variance"}),$$

is provided, which performs the classic principal component analysis after first removing any rows of data for which one or more columns having missing values. Unlike prcomp, the function regpca uses the correlation matrix by default. And it creates a scree plot when the argument scree=T, which is a line segment that shows the fraction of the total variance among all m components as a function of the number of components. (The scree plot is illustrated in Section 6.14.8.)

6.14.2 Maronna's Method

This section provides a brief outline of the method proposed by Maronna (2005), which is based in part on an iterative algorithm. Let $\mathbf{x}_i$, $i = 1, \dots, n$, be an m-dimensional dataset, let $q = m - p$ and let $\mathbf{C}$ be an orthonormal $q \times m$ matrix. That is, $\mathbf{C}\mathbf{C}' = \mathbf{I}_q$. For some q-vector $\mathbf{a}$, let

$$r_i(\mathbf{C}, \mathbf{a}) = \|\mathbf{C}\mathbf{x}_i - \mathbf{a}\|^2,$$

and let $\sigma(\mathbf{r})$ be a scale statistic, where $\mathbf{r} = (r_1, \dots, r_n)$. The goal is to determine $\mathbf{C}$ and $\mathbf{a}$ so as to minimize $\sigma(\mathbf{r})$. Maronna considers two choices for $\sigma(\mathbf{r})$: an M-scale and an L-scale. Here the focus is on the L-scale

$$\sigma(\mathbf{r}) = \sum_{i=1}^{h} r_{(i)},$$

where $r_{(1)} \leq \cdots \leq r_{(h)}$, $h < n$, primarily because it is faster and easier to compute. Following Maronna (2005), h is taken to be the largest integer less than or equal to $(n + m - q + 2)/2$, where $q = m - p$.

6.14.3 The SPCA Method

The *spherical PCA*, called *method SPCA* procedure was derived by Locantore et al. (1999). Let μ be the L_1 median, which is computed by the R function spat, or the R function L1medcen. Let $\mathbf{y}_i = (\mathbf{x}_i - \mu)/\|\mathbf{x}_i - \mu\|$. The procedure consists of using the eigenvectors $\mathbf{b}_1, \ldots, \mathbf{b}_m$ of the covariance matrix of the $\mathbf{y}_i$. But the eigenvalues are in general not consistent, in which case they are replaced by

$$\lambda_j = S(\mathbf{b}_j'\mathbf{x}_1, \ldots, \mathbf{b}_j'\mathbf{x}_n)^2,$$

where S is any robust measure of scale. Following Maronna (2005), S is taken to be the median absolute deviation (MAD) statistic. The R package rrcov contains the function PcaLocantore that performs SPCA.

6.14.4 Method HRVB

Hubert et al. (2005) suggest a method that combines projection pursuit ideas with robust scatter matrix estimation. An adaptation of this method, called *method HRVB* was derived by Engelen et al. (2005) and is used here. The computational details are quite involved, and so only a brief outline of the method is provided.

The method begins by finding the h least outlying data points. The choice for h is made by the investigator and Hubert et al. consider choices of the form $h = \max\{[\alpha n], [(n + k_{\max} + 1)/2]\}$, where α is some value between 0.5 and 1 and $k_{\max}$ is the maximum number of components that will be computed; they use $\alpha = 0.75$ and $k_{\max} = 10$ and the same is done here. Next, outlyingness is measured using a maximum standardized distance among the class of all possible projections of the data onto a unidimensional space. Not all projections can be considered, so for n small they focus on all directions through two points, and for $\binom{n}{2} > 250$ they take at random 250 projections. They then focus on the mean and covariance matrix of the h points that have the smallest distances just computed. The next step computes fast MCD for the projected data resulting from the previous step, which is used to compute a reweighted mean and covariance matrix that increases statistical efficiency. A consistency factor is used to make the estimator unbiased at normal distributions.

6.14.5 Method OP

Method OP simply removes any outliers detected by the projection approach described in Section 6.4.9. Then the classic PCA is applied to the data that remain and the p-dimensional representation of the data is computed in the usual way.

Croux and Ruiz-Gazen (2005) suggest an algorithm that begins with projections based in part on the L_1 median, but it is evident that their approach differs from method OP. Method OP attempts to eliminate outliers in a manner that takes into account the overall structure of the data. The algorithm used by Croux and Ruiz-Gazen does not do this, but rather searches for projections that maximize a robust measure of scatter applied to the marginal distributions of the scores. Also, Croux, Filzmoser, and Oliveira (2007, pp. 6–7) note that the Croux and Ruiz-Gazen (2005) and Hubert et al. (2002) projection algorithms suffer from severe downward bias. It is unknown whether method OP suffers from the same problem.

6.14.6 Method PPCA

Method PPCA is aimed at finding principal components that maximize a robust generalized variance. Let **B** be any $p \times m$ matrix having the property that for any j $(1 \leq j \leq p)$,

$$\sum_{k=1}^{m} b_{jk}^2 = 1$$

and for any $j \neq \ell$

$$\sum_{k=1}^{m} b_{jk} b_{\ell j} = 0.$$

Given **B**, the resulting p-dimensional representation of the data is

$$\mathbf{z}_i = \mathbf{B}(\mathbf{x}_i - \theta), \tag{6.30}$$

where θ is some measure of location. (All of the methods outlined in this section use Eq. (6.30) and differ in how they determine **B** and θ.) The $\mathbf{z}_i$ $(i = 1, \ldots, n)$ are the *scores*. (Scores based on the other robust methods in this section are computed in a similar manner.) Let $\hat{\Xi}$ be an estimate of some robust generalized variance based on the $\mathbf{z}_i$ values. Here the covariance matrix based on the median ball algorithm is used unless stated otherwise and $\hat{\Xi}$ is taken to be the determinant of this covariance matrix that is computed with the $\mathbf{z}_i$ values.

The goal is to determine the matrix **B** that maximizes $\hat{\Xi}$. The method used here begins with an initial estimate of **B**, say $\mathbf{B}_0$, based on the Hubert et al. (2005) estimator (method HVRB). Then use the Nelder and Mead (1965) algorithm to search for the matrix **B** that maximizes $\hat{\Xi}$. (The Nelder–Mead algorithm is applied with the R function nelderv2, which improves on the random search method used by Wilcox, 2008c.)

Regarding the estimation of θ, Wilcox (2008c) considered the (fast) MCD estimator, the L_1 median, Olive (2004) estimator based on the median ball algorithm, and the mean of the data after points flagged as outliers by the projection method are removed. Simulation results indicate that the choice of location estimator makes little difference when using a random search for the matrix **B** that maximizes the generalized variance. However, when using the Nelder–Mead algorithm, Wilcox (2010c) found that the L_1 median, which is computed by the R function spat, performed relatively well, and so it is used here.

6.14.7 R Functions outpca, robpca, robpcaS, SPCA, Ppca, and Ppca.summary

The R function

$$\text{outpca(x,cor=F,SCORES=F,ADJ=F,scree=T, xlab="Principal Component",}$$
$$\text{ylab="Proportion of Variance")}$$

eliminates outliers via the projection method and applies the classic principal component analysis to the remaining data. Following the convention used by R, the covariance matrix is used by default. To use the correlation matrix, set the argument cor=T. Setting SCORES=T, the principal component scores are returned. If the argument ADJ=T, the R function outproad is used to check for outliers rather than the R function outpro, which is recommended if the number of variables is greater than 9. By default, the argument scree=T, meaning that a scree plot will be created. Another rule that is sometimes used is to retain those components for which the proportion of variance is greater than 0.1. When the proportion is less than 0.1, it has been suggested that the corresponding principal component rarely has much interpretive value.

The function

$$\text{robpcaS(x, SCORES=F)}$$

provides a summary of the results based on the method derived by Hubert et al. (2005), including a scree plot based on a robust measure of variation. A more detailed analysis is performed by the function

$$\text{robpca(x, scree=T, xlab = "Principal Component", ylab = "Proportion of Variance"),}$$

which returns the eigenvalues and other results discussed by Hubert et al. (2005), but these details are not discussed here.

For convenience, the R function

$$\text{SPCA(x, k} = 0, \text{kmax} = \text{ncol(x)}, \text{delta} = 0.001, \text{na.action} = \text{na.fail}, \text{scale} = \text{FALSE},$$
$$\text{signflip} = \text{TRUE}, \text{trace=FALSE}, \dots)$$

is provided for applying the spherical principal components method in Section 6.14.3. This function merely eliminates the need to issue the command library(rrcov) when calling the R function PcaLocantore. The argument x is assumed to be an n-by-p matrix. Information about the other arguments can be obtained via the R command ?PcaLocantor, assuming that the R command library(rrcov) has already been issued. The R command screeplot(SPCA(x)) would create a screeplot and summary(SPCA(x)) would return the standard deviations, the proportion of variance and the cumulative proportions.

The R function

$$\text{Ppca(x, p} = \text{ncol(x) - 1, locfun} = \text{L1medcen, loc.val} = \text{NULL, SCORES} = \text{F, gvar.fun} =$$
$$\text{cov.mba, pr} = \text{T, SEED} = \text{T, gcov} = \text{rmba, SCALE} = \text{T}, \dots)$$

applies the method aimed at maximizing a robust generalized variance. This particular function requires the number of principal components to be specified via the argument p, which defaults to $p - 1$. The argument SCALE=T means that the marginal distributions will be standardized based on the measure of location and scale corresponding to the argument gcv, which defaults to the median ball algorithm.

The R function

$$\text{Ppca.summary(x, MC=F, SCALE=T)}$$

is designed to deal with the issue of how many components should be used. It calls Ppca using all possible choices for the number of components, computes the resulting generalized standard deviations, and reports their relative size. If access to a multi-core processor is available, setting the argument MC=T will reduce execution time. Illustrations in the next section deal with the issue of how many components to use based on the output from the R function Ppca.summary.

6.14.8 Comments on Choosing the Number of Components

First focus on classic PCA. Regarding the choice for p, the number of components to use, a rule that is sometimes used is to retain those components for which the proportion of variance is greater than 0.1. When the proportion is less than 0.1, it has been suggested that the corresponding principal component rarely has much interpretive value. Another way of trying to judge how many principal components to use is by visual inspection of a scree plot, the strategy being to determine where the "elbow" of the curve occurs. This well-known strategy

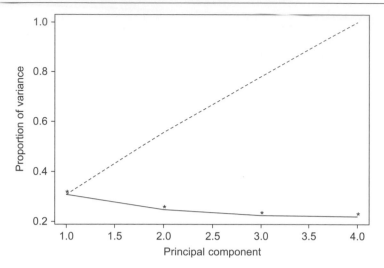

Figure 6.11: The scree plot returned by the R function regpca, where data are multivariate normal with all Pearson correlations equal to zero.

is illustrated with data generated from a multivariate normal distribution with all correlations equal to 0.0 and $n = 200$. The output from the R function regpca is

```
Importance of Components:
                          PC1     PC2    PC3    PC4
Standard Deviation      1.113   0.963  0.959  0.914
Proportion of Variance  0.316   0.236  0.235  0.213
Cumulative Proportion   0.316   0.552  0.787  1.000
```

Figure 6.11 shows the resulting scree plot. The bottom (solid) line shows the the variance associated with the principal components. The upper (dashed) line is the cumulative proportion. Note that the lower line is nearly horizontal with no steep declines, suggesting that all four components be used to capture the variability in the data. Also, for each component, the proportion of variance is greater than 0.1.

The output from the function Ppca.summary differs in crucial ways from the other functions described here. To illustrate it, multivariate normal data were generated with all correlations equal to 0.0. The output from Ppca.summary is

```
                    [,1]        [,2]       [,3]       [,4]
Num. of Comp.  1.0000000    2.000000   3.0000000  4.0000000
Gen.Stand.Dev  1.1735029    1.210405   1.0293564  1.0110513
Relative Size  0.9695129    1.000000   0.8504234  0.8353002
```

The second line indicates the (robust) generalized standard deviation given the number of components indicated by the first line. So when using two components, the generalized standard deviation is 1.210405. Note that the generalized standard deviations are not in descending order. Using two components results in the largest generalized standard deviation. But observe that all four generalized standard deviations are approximately equal, which is what we would expect for the situation at hand. The third line of the output is obtained by dividing each value in the second line by the maximum generalized standard deviation. Here, reducing the number of components from four to two does not increase the generalized standard deviation by very much, suggesting that four or maybe three components should be used. Also observe that there is no proportion of variance used here, in contrast to classic PCA. In classic PCA, an issue is how many components must be included to capture a reasonably large proportion of the variance. When using the robust generalized variance, it seems more appropriate to first look at the relative size of the generalized standard deviations using all of the components. If the relative size is small, reduce the number of components. In the example, the relative size using all four components is 0.835 suggesting that perhaps all four components should be used.

Now consider data that were generated from a multivariate normal distribution where all of the correlations are 0.9. Now the output from regpca is

```
Importance of Components:
                       PC1     PC2     PC3     PC4
Standard Deviation     1.869  0.3444  0.3044  0.2915
Proportion of Variance 0.922  0.0313  0.0244  0.0224
Cumulative Proportion  0.922  0.9531  0.9776  1.0000
```

Note that the first principal component has a much larger standard deviation than the other three principal components. The proportion of variance accounted for by PC1 is 0.922, suggesting that it is sufficient to use the first principal component only to capture the variability in the data. Figure 6.12 shows the scree plot.

Excluding method PPCA, the robust methods summarized in this section report results similar to the function regpca, only a robust measure of variation, associated with each component, is used. Scree plots can be created as well. However, when using method PPCA, the output from the R function Ppca is interpreted in a different manner. Generally, it is suggested that one first look at the sizes of the generalized standard deviations, relative to the largest generalized standard deviation, starting with $p = m$ components. If the relative size is close to 1, use all m components. If not, consider $p = m - 1$. If the relative size is close to 1, use $p - 1$ components. If not, continue in this manner.

Consider again the multivariate normal data with all correlations equal to 0, which were used to create the scree plot in Figure 6.11. The output from Ppca.summary is

```
                    [,1]        [,2]        [,3]        [,4]
Num. of Comp.  1.0000000   2.000000   3.0000000   4.0000000
Gen.Stand.Dev  1.1735029   1.210405   1.0293564   1.0110513
Relative Size  0.9695129   1.000000   0.8504234   0.8353002
```

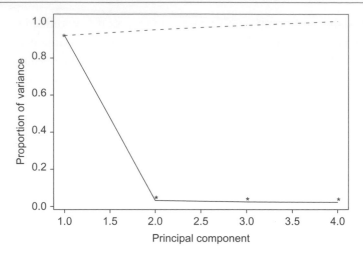

Figure 6.12: The scree plot for multivariate normal data with all Pearson correlations equal to 0.9.

The second line indicates the (robust) generalized standard deviation given the number of components indicated by the first line. So when using two components, the generalized standard deviation is 1.210405. Note that the generalized standard deviations are not in descending order. Using two components results in the largest generalized standard deviation. But observe that all four generalized standard deviations are approximately equal, which is what we would expect for the situation at hand. The third line of the output is obtained by dividing each value in the second line by the maximum generalized standard deviation. Here, reducing the number of components from four to two does not increase the generalized standard deviation by very much, suggesting that four or maybe three components should be used. Also observe that there is no proportion of variance used here, in contrast to classic PCA. In classic PCA, an issue is how many components must be included to capture a reasonably large proportion of the variance. Here, the relative size using all four components is 0.835 suggesting that perhaps all four components should be used.

Consider again the data used to create the scree plot in Figure 6.12. (The data have a multivariate normal distribution with all correlations equal to 0.9) The output from Ppca.summary is

```
                 [,1]        [,2]        [,3]        [,4]
Num. of Comp. 1.000000   2.0000000   3.0000000   4.00000000
Gen.Stand.Dev 2.017774   0.6632588   0.2167982   0.05615346
Relative Size 1.000000   0.3287082   0.1074442   0.02782942
```

As indicated, a single component results in a relatively large generalized standard deviation suggesting that a single component suffices. The relative sizes corresponding to three and four components are fairly small suggesting that using three or four components be ruled out. Even with two components the relative size is fairly small.

■ Example

In an unpublished study by L. Doi, a general goal was to study predictors of reading ability. Here, the focus is on five predictors: two measures of phonological awareness, a measure of speeded naming for digits, a measure of speeded naming for letters, and a measure of the accuracy of identifying lower case letters. Using the classic principal component analysis based on the correlation matrix, the R function regpca returns

```
Importance of Components:
                        Comp.1  Comp.2  Comp.3  Comp.4  Comp.5
Standard Deviation      1.4342  1.0360  0.9791  0.7651  0.57036
Proportion of Variance  0.4114  0.2146  0.1917  0.1170  0.06506
Cumulative Proportion   0.4114  0.6260  0.8178  0.9349  1.00000
```

Note that the proportion of variance exceeds 0.1 with four components or less, which some would take to suggest that four components be used. The R function robpcaS returns

```
                      [,1]     [,2]     [,3]     [,4]     [,5]
Number of Comp.       1.00000  2.00000  3.00000  4.00000  5.000000
Robust Stand. Dev.    2.23900  1.26512  1.21967  0.97752  0.606995
Proportion Robust var 0.53188  0.16981  0.15783  0.10138  0.039090
Cum. Proportion       0.53188  0.70169  0.85952  0.96090  1.000000
```

which is somewhat similar to the results based on the classic PCA.

However, Ppca.summary returns

```
                  [,1]      [,2]       [,3]       [,4]       [,5]
Num. of Comp.  1.000000  2.0000000  3.0000000  4.0000000  5.0000000
Gen.Stand.Dev  1.712513  1.5155318  0.7229315  0.4761138  0.3112773
Relative Size  1.000000  0.8849754  0.4221466  0.2780205  0.1817664
```

The second line shows the robust generalized standard deviations based on the number of components used. Because the relative sizes using three, four, or five components are rather small, the results suggest that two components suffice. (In fairness, it might be argued that a scree plot stemming from classic PCA also suggests that two components be used.)

6.15 Cluster Analysis

Cluster analysis is an exploratory data analysis tool aimed at sorting different objects into groups in a way that the degree of association between two objects is maximal if they belong to the same group and minimal otherwise. There are many relevant methods that go well beyond the scope of this book (e.g., Everitt, Landau, Leese, & Stahl, 2011). Here the goal is merely to mention a few R functions that might be useful.

6.15.1 R Functions Kmeans, kmeans.grp, TKmeans, and TKmeans.grp

R has a built-in function for performing cluster analysis called kmeans. Here the R function

$$Kmeans(x,k,xout=F,outfun=out)$$

is provided in case it helps. The argument x is a matrix or data frame containing the data and k indicates the number of clusters to be used. The function calls the built-in R function kmeans, but it automatically removes any rows of data that contain missing values. The R function kmeans uses the k-means method, which partitions the points into k groups such that the sum of squares from points to the assigned cluster centers is minimized. If the argument xout=T, the R function Kmeans removes any points declared outliers via the function specified by the argument out. The R function

$$Kmeans.grp(x,k,y,xout=F,outfun=out)$$

creates k groups based on the data stored in x and sorts the data in y into groups and stores the results in an R variable having list mode. For example, if k=2 and z=Kmeans.grp(x,k), z[[1]] will contain the data associated with the first cluster and z[[2]] will contain the data associated with the second cluster.

The R function

$$TKmeans(x,k)$$

applies the trimmed k means method derived by Cuesta-Albertos, Gordaliza, & Matran (1997). It removes any vectors of observations having missing values and then uses the R function trimkmeans in the R package trimcluster. The R function

$$TKmeans.grp(x,k,xout=F,outfun=out)$$

is like the function kmeans.grp, only it uses the R function TKmeans to determine the clusters.

6.16 Exercises

1. For the EEG data in Table 6.1, compute the MVE, MCD, OP, and the Donoho–Gasko .2 trimmed mean for group 1.
2. Repeat the last exercise using the data for group 2.
3. For the data in Table 6.1, check for outliers among the first group using the methods in Section 6.4. Comment on why the number of outliers found differs among the methods.

4. Repeat the last exercise using the data for group 2.
5. Repeat the last two exercises, but now use the data in Table 6.2.
6. Suppose that for each row of an n-by-p matrix, its depth is computed relative to all n points in the matrix. What are the possible values that the depths might be?
7. Give a general description of a situation where for $n = 20$, the minimum depth among all points is 3/20.
8. The average LSAT scores (X) for the 1973 entering classes of 15 American law schools, and the corresponding grade point averages (Y), are as follows.

X:	576 635 558 578 666 580 555 661 651 605 653 575 545 572 594
Y:	3.39 3.30 2.81 3.03 3.44 3.07 3.00 3.43 3.36 3.13 3.12 2.74 2.76 2.88 2.96

    Use a boxplot to determine whether any of the X values are outliers. Do the same for the Y values. Comment on whether this is convincing evidence that there are no outliers. Check for outliers using the MVE, MCD, and projection-type methods described in Section 6.4. Comment on the different results.
9. The MVE method of detecting outliers, described in Section 6.4.3, could be modified by replacing the MVE estimator of location with the Winsorized mean, and replacing the covariances with the Winsorized covariances described in Section 5.9.3. Discuss how this would be done and its relative merits.
10. The file read.dat contains data from a reading study conducted by L. Doi. Columns 4 and 5 contain measures of digit naming speed and letter naming speed. Use both the relplot and the MVE method to identify any outliers. Compare the results and comment on any discrepancies.
11. For the cork boring data in Table 6.5, imagine that the goal is to compare the north, east and south sides to the west side How might this be done with the software in Section 6.6.1? Perform the analysis and comment on the results. (The data are stored in the file corkall.dat; see Chapter 1.)
12. For the data in Table 6.1, compare the two groups with the method in Section 6.8.
13. For the data in Table 6.1, compare the two groups with the method in Section 6.10.
14. For the data in Table 6.1, compare the two groups with the method in Section 6.11.
15. For the data in Table 6.1, compare the two groups with the method in Section 6.12.
16. Argue that when testing Eq. (6.27), this provides a metric-free method for comparing groups based on scatter.
17. For the data in Table 6.6, compare the groups using the method in Section 6.8.

CHAPTER 7

One-Way and Higher Designs for Independent Groups

This chapters describes techniques for testing hypotheses in one-way and higher designs involving independent groups. Included are random effects models plus methods for performing multiple comparisons. This chapter makes no attempt at covering all the designs that are encountered in practice, but it does cover many of the more common designs that are used.

In this chapter, only heteroscedastic methods are considered. It might be hoped that as the number of groups increases, problems associated with homoscedastic methods, described in Chapter 5, might be reduced. In one-way designs, the exact opposite seems to be true. For example, even under normality with equal sample sizes but unequal variances, problems controlling the probability of a type I error can arise. With four independent groups, each having 50 observations, the usual analysis of variance F test of

$$H_0 : \mu_1 = \mu_2 = \mu_3 = \mu_4$$

can have a type I error probability approximately equal to .09 when testing at the 0.05 level (Wilcox, Charlin, & Thompson, 1986). With unequal sample sizes, the actual probability of a type I error can exceed .3. Under nonnormality, control over the probability of a type I error is even worse. Practical problems with more complicated designs have been found (e.g., Keselman, Keselman, & Lix, 1995).

It is sometimes suggested that one test the homoscedasticity assumption, and if an appropriate test fails to reject, use a method that assumes equal variances among the groups. However, published papers do not support this strategy (e.g., Hayes & Cai, 2007; Markowski & Markowski, 1990; Moser, Stevens, & Watts, 1989; Wilcox, Charlin, & Thompson, 1986; Zimmerman, 2004). As noted at the end of Section 5.2 when dealing with the two-sample case, the problem is that tests of the homoscedasticity assumption may not have enough power to detect situations where a violation of the homoscedasticity assumption creates practical problems.

One might try to salvage homoscedastic methods by arguing that if the variances are unequal, the means are unequal as well, in which case a type I error is not a concern. However, an inability to control the probability of a type I error often reflects an undesirable power property, the probability of rejecting the null hypothesis is not minimized when the null hypothesis is true. (The hypothesis testing method is biased.) This problem was already pointed out and illustrated in Chapter 5 where shifting one group by a half standard deviation results in a situation where the probability of rejecting is less compared with the situation where H_0 is true. Here, it is merely noted that this problem persists when comparing more than two groups (e.g., Wilcox, 1996a).

Another argument in support of the F-test is that it is reasonably good at controlling the probability of a type I error when distributions are identical. (Tan, 1982, reviews the relevant literature.) That is, it provides a test of the hypothesis that J groups have identical distributions. If the F-test is significant, a reasonable argument is that the means differ. However, if the goal is to derive a test that is exclusively sensitive to some measure of location, the F-test is unsatisfactory. Even if this problem can be ignored, concerns about low power, due to the low efficiency of the sample mean when distributions have heavy tails, remains a concern.

7.1 Trimmed Means and a One-Way Design

From a technical point of view, it is a simple matter to extend the methods in Chapter 5 to situations where the goal is to compare the trimmed means of more than two groups: simply select a heteroscedastic method for means and then proceed along the lines used to derive the Yuen–Welch test. In essence, replace the sample means by trimmed means, replace estimates of the standard errors with appropriate estimates based on the amount of trimming used, and adjust the degrees of freedom based in part on the number of observations left after trimming. For a one-way design, however, there are many heteroscedastic methods for comparing means, so it is not immediately obvious which to use. Two possibilities are described here, both of which have been examined in simulation studies and found to give relatively good control over the probability of a type I error. Other methods have been considered by Lix and Keselman (1998) as well as Luh and Guo (1999). For a method based on trimmed means that assumes equal variances, see Lee and Fung (1985). For the special case where the goal is to compare means, Krishnamoorthy, Lu, and Mathew (2007) review approaches for handling unequal variances, which are known to be unsatisfactory, and they suggest using instead a parametric bootstrap method that assumes normality. Cribbie, Fiksenbaum, Keselman, and Wilcox (in press) compared this parametric bootstrap method to several other techniques and found it to be unsatisfactory. The actual type I error probability, when testing at the 0.05 level, can exceed .25. A method that performed reasonably well was the Welch-type method for comparing trimmed means, which is described in the next section.

7.1.1 A Welch-Type Procedure and a Robust Measure of Effect Size

The goal is to test

$$H_0 : \mu_{t1} = \cdots = \mu_{tJ}, \tag{7.1}$$

where μ_{tj}, $j = 1, \ldots, J$, are the trimmed means corresponding to J independent groups. Table 7.1 describes a method for testing this hypothesis that reduces to Welch's (1951) adjusted degrees of freedom method for means when there is no trimming.

Table 7.1: Computations for Comparing Trimmed Means.

The goal is to test

$$H_0 : \mu_{t1} = \cdots = \mu_{tJ}.$$

For the jth group, let

$$d_j = \frac{(n_j - 1)s_{wj}^2}{h_j \times (h_j - 1)},$$

where h_j is the effective sample size of the jth group (the number of observations left after trimming) and s_{wj}^2 is the Winsorized variance. To test H_0, compute

$$w_j = \frac{1}{d_j}$$

$$U = \sum w_j$$

$$\tilde{X} = \frac{1}{U} \sum w_j \bar{X}_{tj}$$

$$A = \frac{1}{J - 1} \sum w_j (\bar{X}_{tj} - \tilde{X})^2$$

$$B = \frac{2(J - 2)}{J^2 - 1} \sum \frac{\left(1 - \frac{w_j}{U}\right)^2}{h_j - 1}$$

$$F_t = \frac{A}{1 + B}.$$

When the null hypothesis is true, F_t has, approximately, an F distribution with degrees of freedom

$$\nu_1 = J - 1$$

$$\nu_2 = \left[\frac{3}{J^2 - 1} \sum \frac{(1 - w_j/U)^2}{h_j - 1} \right]^{-1}.$$

A Robust, Heteroscedastic Measure of Effect Size

The robust explanatory measure of effect size, described in Section 5.3.4, is readily extended to more than two groups. For simplicity, first consider the situation where means are compared. Again we define the measure of effect size based on a situation where equal sample sizes are used with probability one, with the random sample being denoted by Y_{ij} ($i = 1, \ldots, n; j = 1, \ldots, J$). And then we use an appropriate estimation procedure when the sample sizes are not equal.

For equal sample sizes, let $\sigma^2(Y)$ be the estimand corresponding to

$$\hat{\sigma}^2(Y) = \frac{1}{N-1} \sum_{j=1}^{J} \sum_{i=1}^{n} (Y_{ij} - \bar{Y})^2,$$

where $\bar{Y} = \sum \sum Y_{ij}/N$ and $N = nJ$ is the total sample size. Adopting a regression perspective, and given that an observation is randomly sampled from the jth group, the predicted value is μ_j. Let

$$\sigma^2(\hat{Y}) = \frac{1}{J-1} \sum_{j=1}^{J} (\mu_j - \bar{\mu})^2,$$

where $\bar{\mu} = \sum \mu_j/J$ is the grand mean. The *explanatory measure of effect size* is

$$\xi = \sqrt{\frac{\sigma^2(\hat{Y})}{\sigma^2(Y)}}.$$

Letting

$$\hat{\sigma}^2(\hat{Y}) = \frac{1}{J-1} \sum_{j=1}^{J} (\bar{Y}_j - \bar{Y})^2,$$

an estimate of ξ is

$$\hat{\xi} = \sqrt{\frac{\hat{\sigma}^2(\hat{Y})}{\hat{\sigma}^2(Y)}}.$$

If the mean and variance are replaced by a trimmed mean and Winsorized variance (scaled to estimate the variance under normality), the resulting estimate of $\hat{\xi}$ can exceed 1 when there are $J > 2$ groups and the amount of trimming is greater than 0.

For unequal sample sizes, let m denote the smallest sample size among the J groups. Randomly sample (without replacement) m observations from each of the groups for which

$m < n_j$. Based on the resulting sample sizes of m observations from each group, compute $\hat{\xi}^2$ as just described. Repeat this process K times yielding a series of estimates for ξ^2, which are then averaged to get a final estimate, which we label $\hat{\bar{\xi}}^2$. The estimate of ξ is taken to be $\sqrt{\hat{\bar{\xi}}^2}$.

7.1.2 R Functions t1way, t1wayv2, esmcp, fac2list, and t1wayF

The R function

$$t1way(x,tr=0.2,grp=NA)$$

performs the calculations in Table 7.1. The data can be stored in any R variable having list mode, or it can be stored in a matrix, or a data frame. If x is a matrix or data frame, it is assumed that the columns correspond to groups. So the data for group 1 are stored in column one, and so on. The R function

$$t1wayv2(x,tr=0.2,grp=NA)$$

is the same as t1way, only the measure of effect size ξ is reported as well. The R function

$$esmcp(x,tr=0.2,grp=NA)$$

computes the robust explanatory effect size ξ for all pairs of groups.

Although familiarity with R is assumed, a brief description of *list mode* is provided for readers who are not be familiar with it. List mode is a convenient way of storing data corresponding to several groups under one variable name. For example, suppose two groups are to be compared, and the data for the two groups are stored in the R vectors x and y. The command

$$w=list()$$

creates a variable, called w, that has list mode. To get the first group of data (stored in x) into w, enter the command

$$w[[1]]=x.$$

Notice the use of the double brackets. In particular, w[[1]] is a vector of observations corresponding to the first group of subjects. This is one way R refers to a subset of data stored in list mode. To get the second group of data stored in w, type

$$w[[2]]=y.$$

More generally, in terms of a one-way design, w[[j]] would be where the data for the jth group are stored. (For more details about list mode, see Becker, Chambers, & Wilks, 1988, or consult the manual built into R.)

The second argument in t1way, tr, indicates the amount of trimming, which defaults to 0.2 (20% trimming). Thus, the command t1way(w) results in a test of the hypothesis that

the 20% trimmed means are equal. To compare 10% trimmed means, use the command
t1way(w,tr=0.1).

The third argument in t1way, grp, can be used to specify some subset of the populations to be compared. If not specified, all of the groups are used. If, for example, there are four groups, but the goal is to compare groups 1, 2, and 4, ignoring group 3, the command t1way(w,grp=c(1,2,4)) will test the hypothesis $H_0: \mu_{t1} = \mu_{t2} = \mu_{t4}$ using 20% trimmed means. The command t1way(w,0.1,grp=c(1,4)) would compare the 10% trimmed means of groups 1 and 4. Being able to compare some groups, while ignoring others, is useful when using a step-down multiple comparison procedure. (See Hochberg & Tamhane, 1987; or Wilcox, 1996a, 2003d, for a description of step-down methods plus comments on their relative merits.)

■ Example

Suppose that for three independent groups, the observations are

Group 1	1,2,3,4,5,6,7,8,9,10
Group 2	2,3,4,5,6,7,8,9,10,11
Group 3	5,6,7,8,9,10,11,12,13,14.

If the data are stored in the R variable w, in list mode, the command t1way(w,tr=0) tests the hypothesis of equal means. The function returns

```
$TEST:
[1] 4.558442

$nu1:
[1] 2

$nu2:
[1] 18

$siglevel:
[1] 0.02502042
```

In particular, $F_t = 4.56$ with a *p*-value of .025. The command t1way(w,tr=0,c(1,3)) compares the means of groups one and three and reports a *p*-value of .008. The command t1way(w,grp=c(1,3)) compares the 20% trimmed means of groups 1 and 3 and reports a *p*-value of .039.

■

Note that the third group has the largest sample mean, which is equal to 9.5. Increasing the largest observation in the third group to 40, the sample mean increases to 12.1 suggesting that there is now more evidence that the groups differ, but the *p*-value *increases* to .17. The reason is that the standard error of the sample mean also increases.

Data Management

It is common to have data stored in a matrix or data frame where one of the columns contains the outcome variable of interest and another column indicates the level (group identification) of the factor being studied. For example, data dealing with plasma retinol is available from a website maintained by Carnegie Mellon University; it can be downloaded from

```
http://lib.stat.cmu.edu/datasets/Plasma_Retinol.
```

For illustrative purposes, it is assumed that the data have been stored in the R variable plasma as a data frame. The variable names are:

```
1      AGE: Age (years)
2      SEX: Sex (1=Male, 2=Female).
3      SMOKSTAT: Smoking status (1=Never, 2=Former, 3=Current Smoker)
4      QUETELET: Quetelet (weight/(height^2))
5      VITUSE: Vitamin Use (1=Yes, fairly often, 2=Yes, not often, 3=No)
6      CALORIES. Number of calories consumed per day.
7      FAT: Grams of fat consumed per day.
8      FIBER: Grams of fiber consumed per day.
9      ALCOHOL: Number of alcoholic drinks consumed per week.
10     CHOLESTEROL: Cholesterol consumed (mg per day).
11     BETADIET: Dietary beta-carotene consumed (mcg per day).
12     RETDIET: Dietary retinol consumed (mcg per day)
13     BETAPLASMA: Plasma beta-carotene (ng/ml)
14     RETPLASMA: Plasma Retinol (ng/ml)
```

The first few lines of the data set look like this:

```
64  2  2 21.48380  1 1298.8 57.0  6.3  0.0 170.3 1945 890 200 915
76  2  1 23.87631  1 1032.5 50.1 15.8  0.0  75.8 2653 451 124 727
38  2  2 20.01080  2 2372.3 83.6 19.1 14.1 257.9 6321 660 328 721
40  2  2 25.14062  3 2449.5 97.5 26.5  0.5 332.6 1061 864 153 615
```

Now, imagine that the goal is to compare the three groups based on smoking status, which is indicated in column 3, in terms of of plasma beta-carotene, which is stored in column 13. To use the R function t1way, it is necessary to sort the data in column 13 into three groups based on the values stored in column 3. This can be done with R function

$$\text{fac2list(x,g),}$$

where the argument x is an R variable, usually some column of a matrix or column of a data frame, containing the data to be analyzed (the dependent variable) and g is a column of data indicating the group to which a corresponding value, stored in x, belongs. (When working with a data frame, this latter column of data can be a factor variable.) The output from fac2list is an R variable having list mode. If g contains numeric data, the groups are put in ascending order based on the values in g. If g contains character data, then the data are sorted into groups in alphabetical order.

■ Example

For the plasma retinol data, imagine the goal is to compare the trimmed means corresponding to the three smoking-status groups. The outcome measure of interest is plasma beta-carotene. The groups can be compared using the R commands

$$z=fac2list(plasma[,13],plasma[,3])$$
$$t1way(z)$$

The first command sorts the data stored in plasma[,13] into groups based on the values stored in plasma[,3], and it stores the data in the R variable z having list mode. The data stored in plasma[,3] has one of three values: 1, 2, and 3. So z[[1]] contains the data for the first group, z[[2]] contains the data for second, and z[[3]] the data for the third. If instead plasma[,3] contained one of three character strings, say "N", "Q", and "S", the data in z would be sorted alphabetically. So now z[[1]] would contain plasma retinol measures for participants designated by "N", z[[2]] would contain plasma retinol measures for participants designated by "Q", and z[[3]] would contain plasma retinol measures for participants designated by "S". ■

The R function

$$t1wayF(x,fac,tr=0.2,nboot=100,SEED=T)$$

is like the R function t1way, only x is assumed to be a column of data and fac is a factor variable. That is, this function eliminates the need to use the function fac2list.

■ Example

For the last example, the analysis could be done with the single command

$$t1wayF(plasma[,13],plasma[,3]).$$ ■

7.1.3 A Generalization of Box's Method

Again let h_j be the effective sample size associated with the jth group. Motivated by results in Box (1954) and Rubin (1983), Lix and Keselman (1998) considered testing (7.1), the hypothesis of equal trimmed means, with

$$F_b = \frac{\sum h_j (\bar{X}_{tj} - \bar{X}_t)^2}{\sum 1 - (h_j/H) S_j^2},$$

where $H = \sum h_j$, $\bar{X}_t = \sum h_j \bar{X}_{tj}/H$ and

$$S_j^2 = \frac{(n_j - 1)s_{wj}^2}{h_j - 1}.$$

When the null hypothesis is true, F_b has, approximately, an F distribution with

$$\hat{v}_1 = \frac{\left[\sum(1 - f_j)S_j^2\right]^2}{\left(\sum S_j^2 f_j\right)^2 + \sum_{j=1}^{J} S_j^4 (1 - 2f_j)}$$

and

$$\hat{v}_2 = \frac{\left[\sum_{j=1}^{J}(1 - f_j)S_j^2\right]^2}{\sum_{j=1}^{J} S_j^4 (1 - f_j)^2/(h_j - 1)}$$

degrees of freedom, where $f_j = h_j/H$. Currently, it seems that both F_t and F_b give similar protection against type I errors, with F_b being perhaps slightly better. When there are two groups ($J = 2$), these two methods give exactly the same results. In some situations, F_b has a type I error probability that exceeds α and is higher than the type I error probability associated with F_t, but there are situations where the reverse is true. Among the situations considered by Lix and Keselman, F_b is less likely to result in a type I error probability exceeding .075 when testing at the .05 level. However, F_b generally has less power.

7.1.4 R Function box1way

The R function

$$\text{box1way(x,tr=0.2,grp=NA)},$$

written for this book, performs the calculations described in the previous subsection. It is used in exactly the same manner as t1way. Thus, the command box1way(w,.1,c(1,3)) will test the hypothesis that the 10% trimmed means, associated with the first and third groups, are equal. When comparing only two groups, the R functions box1way, t1way, and yuen all give identical results.

■ Example

Suppose the data in Section 7.1.2, used to illustrate the function t1way, are stored in w. Then the command box1way(w) tests the hypothesis of equal 20% trimmed means, and the p-value is reported to be .077. Using t1way, the p-value is .025.

■

7.1.5 Comparing Medians

For the special case where the goal is to test the hypothesis of equal medians, the Yuen–Welch and Box methods for trimmed means are not recommended; an alternative estimate of the standard error is required. Here, the McKean–Schrader estimate of the standard error will be used in conjunction with a Welch-type test. It is stressed, however, that all known estimates of the standard error of the sample median can be highly inaccurate when tied values occur, even with large sample sizes. Consequently, the method in this section is not recommended when there are tied values. (Use instead the percentile bootstrap method in Section 5.4.2. The R function medpb, described in Section 7.4.8, is designed to compare medians in a manner that controls the probability of making one or more type I errors.)

Let M_j be the sample median for the jth group and let S_j^2 be the McKean–Schrader estimate of the squared standard error of M_j ($j = 1, \ldots, J$). Let

$$w_j = \frac{1}{S_j^2},$$

$$U = \sum w_j,$$

$$\tilde{M} = \frac{1}{U} \sum w_j M_j,$$

$$A = \frac{1}{J-1} \sum w_j (M_j - \tilde{M})^2,$$

$$B = \frac{2(J-2)}{J^2-1} \sum \frac{(1 - \frac{w_j}{U})^2}{n_j - 1},$$

$$F_m = \frac{A}{1+B}. \tag{7.2}$$

The hypothesis of equal population medians is rejected if $F_m \geq f$, the $1 - \alpha$ quantile of an F distribution with $\nu_1 = J - 1$ and $\nu_2 = \infty$ degrees of freedom.

7.1.6 R Function med1way

The R function

$$\text{med1way}(x,\text{grp}=\text{NA})$$

test the hypothesis of equal population medians. The argument x can be any variable having matrix mode or list mode. If a matrix, columns correspond to groups. The argument grp can be used to analyze a subset of the groups if desired. By default, all J groups are compared. So the command med1way(disdat,grp=c(1,3,5)) will compare groups 1, 3, and 5 using the data stored in the variable disdat. The function returns the value of the test statistic, F_m, and the p-value.

7.1.7 A Bootstrap-t method

Lix and Keselman (1998) found that two other methods for comparing trimmed means perform relatively well in terms of controlling the probability of a type I error, but with small sample sizes and with trimming less than or equal to 20%, none of the methods they considered, including the methods described in this chapter, always guaranteed that the actual probability of a type I error would be less than .075 when testing at the .05 level. In some situations, the probability of a type I error exceeds .08. It might be argued that this is satisfactory in some situations, but they did not consider situations where distributions have unequal skewnesses; among the nonnormal distributions in their study, they only considered situations where groups have unequal variances. From Chapter 5, if distributions have unequal skewnesses, the expectation is that control over the probability of a type I error will be worse. Again, one might try to salvage the situation by arguing that if groups have unequal variances or skewnesses, surely the trimmed means differ, so the probability of a type I error is not an issue. But as noted in Chapter 5, problems with controlling type I error probabilities often reflect an unsatisfactory characteristic: power can go down as the difference between the trimmed means increases, although eventually it will go up. Put another way, the probability of rejecting is not always minimized when the null hypothesis is true.

Chapter 5 described bootstrap methods for dealing with this problem. Provided the amount of trimming is relatively low (say less than 20%), it currently seems that a bootstrap-t method is relatively effective based on the criterion of controlling the probability of a type I error. In the present context, a bootstrap-t method refers to any bootstrap technique that is based in part on a test statistic that is a function of estimates of the standard errors of the location estimators being used. With sufficiently large sample sizes, a bootstrap method is not required, but it remains unclear how large the sample sizes must be. This subsection notes that a simple extension of the two-sample bootstrap-t method can be applied to the problem at hand. The strategy is to use the available data to estimate an appropriate critical value when using the Yuen–Welch method to compare trimmed means. Perhaps there is some practical advantage to replacing the Welch-type method with some other procedure, but this remains to be seen. (In the two-sample case, with the amount of trimming less than 20%, a generalization of the Yuen–Welch test seems to have merit; see Othman, Keselman, Wilcox, Fradette, & Padmanabhan, 2002.)

As was done in Section 5.3.2, the method begins by obtaining a bootstrap sample from each of the J groups: $X_{1j}^*, \ldots, X_{n_j j}^*$. Next, set $C_{ij}^* = X_{ij}^* - \bar{X}_{tj}$, $i = 1, \ldots, n_j$. Then $C_{1j}^*, \ldots, C_{n_j j}^*$ represents a sample from a distribution that has a trimmed mean of zero, so the hypothesis of equal trimmed means among these J distributions is true. Let F_t^* be the value of F_t (described in Table 7.1), when applied to the C_{ij}^* values. Repeat this process B times, each time obtaining bootstrap samples and computing F_t using the C_{ij}^* values that result. Label the resulting test statistics $F_{t1}^*, \ldots, F_{tB}^*$. Each time this process is applied, the null hypothesis is true, by construction, so the values $F_{t1}^*, \ldots, F_{tB}^*$ provide an estimate of an appropriate critical value. Letting $F_{t(1)}^* \leq \cdots \leq F_{t(B)}^*$ be the $F_{t1}^*, \ldots, F_{tB}^*$ values written in ascending order, an estimate of the α critical value is $F_{t(m)}^*$, where $u = (1 - \alpha)B$, rounded to the nearest integer. That is, reject the null hypothesis of equal trimmed means if F_t, computed as described in Table 7.1, is greater than or equal to $F_{t(u)}^*$.

7.1.8 R Functions t1waybt and btrim

The R function

$$\text{t1waybt(x,tr=0.2,alpha=0.05,grp=NA,nboot=599)}.$$

tests the hypothesis of equal trimmed means using the bootstrap-t method. As with t1way and box1way, the argument x can be any R variable that is a matrix or has list mode. If unspecified, the amount of trimming defaults to tr=0.2, and the argument alpha, corresponding to α, defaults to 0.05. Again the argument grp can be used to test the hypothesis of equal trimmed means for some subgroup of interest. If unspecified, all J groups are used. The default value for B is nboot=599 which appears to give good results, in terms of controlling the probability of a type I error, when $\alpha = 0.05$ and $n_j \geq 10$, $j = 1, \ldots, J$. Little is known about how the method performs when $\alpha < 0.05$. Cribbie et al. (in press) found that with 20% trimming, a parametric bootstrap technique performs a bit better than the method in Section 7.1.1. Limited checks indicate that the bootstrap-t method used here is better than the parametric bootstrap method in terms of avoiding type I error probabilities larger than the nominal level. But extensive comparisons have not been made.

■ Example

Again consider the data in Section 7.1.2. Assuming the data are stored in the variable w, the command t1waybt(w) reports that the 0.05 critical value is 4.97. The value of the test statistic is 2.87, which is the same value reported by the function t1way. The 0.05 critical value used by t1way is 4.1. That is, the bootstrap-t method estimates that t1way is using a critical value that is too small.

■

To add perspective, suppose the data in Section 7.1.2 are shifted so that the trimmed mean for each group is zero. This yields

Group 1	−4.5 −3.5 −2.5 −1.5 −0.5 0.5 1.5 2.5 3.5 4.5
Group 2	−4.5 −3.5 −2.5 −1.5 −0.5 0.5 1.5 2.5 3.5 4.5
Group 3	−4.5 −3.5 −2.5 −1.5 −0.5 0.5 1.5 2.5 3.5 4.5

Now suppose that these values represent the actual distributions associated with the three groups, each value within a group having the same probability of occurring. This is the process used by the bootstrap method. In group 1, for example, there are 10 possible values, every value occurs with equal probability, so the first value, -4.5, occurs with probability .1, the second value, -3.5, also occurs with probability .1, and so on. As is evident, all three distributions happen to be identical, but in general this will not be the case. By construction, each of the three distributions has a population trimmed mean equal to zero. Consequently, when F_t is computed using these observations, the probability of rejecting should be α. But if $n_1 = 10$ observations are randomly sampled from the first group, $n_2 = 10$ are randomly sampled from the second, and $n_3 = 10$ from the third, the actual probability of a type I error is .073 when $\alpha = 0.05$, based on a simulation with 1000 replications. Even without running a simulation, the expectation is that the type I error probability will be higher than .05. The reason is that the bootstrap-t method, when applied to the data in Section 7.1.2, simply performs simulations on the distributions being considered here, and it estimates that the .05 critical value is 4.97. But F_t uses a critical value of 4.1 (the 0.95 quantile of an F distribution with 2 and 10 degrees of freedom), which is too small. Put another way, if the bootstrap-t method estimates that the critical value is higher than the critical value used by F_t, in essence, a discrete distribution has been found for which F_t can be expected to have a type I error probability greater than the nominal level. For the situation at hand, the bootstrap-t estimates the critical value to be 4.97, which corresponds to the 0.968 quantile of an F distribution with 2 and 10 degrees of freedom. Moreover, the discrete distributions being used are estimates of the distributions under study, only shifted so that the null hypothesis of equal trimmed means is true.

The R function

$$\text{btrim(x,tr=0.2,grp=NA,g=NULL,dp=NULL,nboot=599)}.$$

is an updated version of t1waybt. In addition to the results reported by t1waybt, the function btrim reports the explanatory measure of effect size. And it has the ability to sort data into groups based on group identification values stored in column g of x, assuming x is a matrix or a data frame. The outcome (dependent variable) of interest is stored in the column indicated by the argument dp. In effect, this eliminates the need to call the function fac2list. For example, btrim(plasma,g=2,dp=4) would sort the data in column 4 of the R variable plasma into groups based on the values stored in column 2.

7.1.9 Percentile Bootstrap Methods

Chapter 5 noted that when comparing trimmed means, and when the amount of trimming is small, a bootstrap-t method generally performs better than a percentile bootstrap method, in terms of controlling the probability of a type I error, but with about 20% trimming or more, a percentile bootstrap method performs better than a bootstrap-t in simulations. When comparing multiple independent groups, all indications are that this continues to be the case. For a percentile bootstrap method that can be used to compare trimmed means, see Section 7.6. For a method designed specifically for a 20% trimmed mean, see Section 7.4.8.

7.2 Two-Way Designs and Trimmed Means

This section describes methods for testing hypotheses in a two-way design when working with trimmed means. It is assumed that the reader is familiar with the basic features and terminology of two-way designs, which are covered in numerous books on statistics. To briefly review, the basic goal is to compare groups, taking into account two main factors plus interactions. For example, Steele and Aronson (1995) conducted a study on how stereotype might affect performance on an aptitude test. They compared test scores of Black and White subjects taking into account how the purpose of the test was presented. The test was presented either as a diagnostic of intellectual ability, as a laboratory tool for studying problem solving, or as both a problem-solving tool and a challenge. This is a 2-by-3 design. The first factor, race, has two levels, whereas the second factor, type of presentation, has three. As is commonly done, the first factor is generically called factor A, and the second is called factor B. The term J-by-K ANOVA refers to a two-way design with factor A having J levels and factor B having K.

The groups are assumed to be arranged as shown in Table 7.2. Thus, μ_{tjk} is the population trimmed mean associated with the jth level of the first factor and the kth level of the second. Extending standard notation in an obvious way, the grand trimmed mean is the average of all

Table 7.2: Trimmed Means Corresponding to a *J*-by-*K* Design.

	Factor B				
	μ_{t11}	μ_{t12}	$\cdots$	μ_{t1K}	$\mu_{t1.}$
Factor	μ_{21}	μ_{22}	$\cdots$	μ_{2K}	$\mu_{2.}$
A	$\vdots$	$\vdots$	$\cdots$	$\vdots$	$\vdots$
	μ_{tJ1}	μ_{tJ2}	$\cdots$	μ_{tJK}	$\mu_{tJ.}$
	$\mu_{t.1}$	$\mu_{t.2}$	$\cdots$	$\mu_{t.K}$	

JK trimmed means. In symbols, the grand trimmed mean is

$$\bar{\mu}_t = \frac{1}{JK} \sum_{j=1}^{J} \sum_{k=1}^{K} \mu_{tjk}.$$

The main effects for factor A are defined to be

$$\alpha_1 = \mu_{t1.} - \bar{\mu}_t, \quad \alpha_J = \mu_{tJ.} - \bar{\mu}_t,$$

where

$$\mu_{tj.} = \frac{1}{K} \sum_{k=1}^{K} \mu_{tjk}.$$

The hypothesis of no main effects for factor A is

$$H_0 : \mu_{t1.} = \cdots = \mu_{tJ.}.$$

When the null hypothesis is true,

$$\alpha_{t1} = \cdots = \alpha_{tJ} = 0,$$

so another common way of writing the null hypothesis is

$$H_0 : \sum \alpha_j^2 = 0.$$

Similarly, the levels of factor B can be compared, ignoring factor A, by testing

$$H_0 : \mu_{t.1} = \mu_{t.2} = \cdots = \mu_{t.K},$$

where

$$\mu_{t.k} = \frac{1}{J} \sum_{j=1}^{J} \mu_{tjk}.$$

The effect size associated with the kth group is written as

$$\beta_k = \mu_{t.k} - \bar{\mu}_t,$$

and often the null hypothesis is written as

$$H_0 : \sum \beta_k^2 = 0.$$

The computational steps associated with a two-way design, when testing the hypotheses just listed, are much easier to describe in terms of matrices. Here, a generalization of results in Johansen (1980) is used. The implementation of the method is based on a generalization of the results in Algina and Olejnik (1984).

There are $p = JK$ independent groups with trimmed means $\mu_t = (\mu_{t1}, \ldots, \mu_{tJK})'$. The general strategy for testing main effects and interactions is to test

$$H_0 : \mathbf{C}\mu_t = 0, \tag{7.3}$$

where $\mathbf{C}$, which is constructed in a manner to be described, is a k-by-p contrast matrix of rank k, chosen to reflect the hypothesis of interest. For convenience, it is assumed that the sample trimmed means are are arranged in a 1×9 matrix

$$\mathbf{X}' = (\bar{X}_{t11} \ldots \bar{X}_{t1K} \ldots, \bar{X}_{tK1} \ldots \bar{X}_{t1K}),$$

where $\mathbf{X}'$ is the transpose of $\mathbf{X}$.

The construction of the contrast matrix $\mathbf{C}$ is accomplished as follows. For any integer $m \geq 2$, let $\mathbf{C}_m$ be an $(m-1)$-by-m matrix having the form

$$\begin{pmatrix} 1 & -1 & 0 & 0 & \ldots & 0 \\ 0 & 1 & -1 & 0 & \ldots & 0 \\ & & & \ldots & & \\ 0 & 0 & \ldots & 0 & 1 & -1 \end{pmatrix}.$$

That is, $c_{ii} = 1$ and $c_{i,i+1} = -1$, $i = 1, \ldots, m-1$. Let $\mathbf{j}_m'$ be a 1-by-m vector of 1s. For example, $\mathbf{j}_3' = (1, 1, 1)$. The matrix $\mathbf{C}$ for testing main effects and interactions can be constructed with what is called the (right) Kronecker product of matrices, applied to appropriate choices of $\mathbf{C}_m$ and $\mathbf{j}_m$. If $\mathbf{A}$ is any r-by-s matrix, and $\mathbf{B}$ is any t-by-u matrix, the Kronecker product of $\mathbf{A}$ and $\mathbf{B}$, written as $\mathbf{A} \otimes \mathbf{B}$, is

$$\begin{pmatrix} a_{11}B & a_{12}B & \ldots & a_{1s}B \\ & \vdots & & \\ a_{r1}B & a_{r2}B & \ldots & a_{rs}B \end{pmatrix}.$$

Table 7.3: How to Construct the Contrast Matrix, C, for a Two-Way Design.

Effect	C
A	$\mathbf{C}_J \otimes \mathbf{j}'_K$
B	$\mathbf{j}'_J \otimes \mathbf{C}_K$
A × B	$\mathbf{C}_J \otimes \mathbf{C}_K$

Table 7.3 shows how to construct the contrast matrix $\mathbf{C}$ for the main effects and interactions in a two-way design. For example, when testing for main effects for factor A, use $\mathbf{C} = \mathbf{C}_J \otimes \mathbf{j}'_K$.

Remembering that $p = JK$, the total number of groups, let $\mathbf{V}$ be a p-by-p diagonal matrix with

$$v_{jj} = \frac{(n_j - 1)s_{wj}^2}{h_j(h_j - 1)},$$

$j = 1, \ldots, p$. That is, v_{jj} is Yuen's estimate of the squared standard error of the sample trimmed mean corresponding to the jth group. The test statistic is

$$Q = \bar{\mathbf{X}}'\mathbf{C}'(\mathbf{C}\mathbf{V}\mathbf{C}')^{-1}\mathbf{C}\bar{\mathbf{X}}. \tag{7.4}$$

Let

$$\mathbf{R} = \mathbf{V}\mathbf{C}'(\mathbf{C}\mathbf{V}\mathbf{C}')^{-1}\mathbf{C},$$

and

$$A = \sum_{j=1}^{p} \frac{r_{jj}^2}{h_j - 1},$$

where r_{jj} is the jth diagonal element of $\mathbf{R}$. Asymptotically, a critical value for Q is c, the $1 - \alpha$ quantile of a chi-square distribution with k degrees of freedom. However, for small sample sizes, an adjusted critical value is needed, which is given by

$$c_{\text{ad}} = c + \frac{c}{2k}\left[A\left(1 + \frac{3c}{k+2}\right)\right].$$

If $Q \geq c_{\text{ad}}$, reject H_0.

7.2.1 R Functions t2way

The R function

$$t2way(J,K,x,grp=c(1:p),tr=0.2,alpha=0.05),$$

performs the tests on trimmed means described in the previous section, where J and K denote the number of levels associated with factors A and B. When the data are stored in list mode, the first K groups are assumed to be the data for the first level of factor A, the next K groups are assumed to be data for the second level of factor A, and so on. In R notation, x[[1]] is assumed to contain the data for level 1 of factors A and B, x[[2]] is assumed to contain the data for level 1 of factor A and level 2 of factor B, and so forth. If, for example, a 2-by-4 design is being used, the data are stored as follows:

Factor	Factor B			
A	x[[1]]	x[[2]]	x[[3]]	x[[4]]
	x[[5]]	x[[6]]	x[[7]]	x[[8]]

For instance, x[[5]] contains the data for the second level of factor A and the first level of factor B.

If the data are not stored in the assumed order, grp can be used to correct this problem. Suppose, for example, the data are stored as follows:

Factor	Factor B			
A	x[[2]]	x[[3]]	x[[5]]	x[[8]]
	x[[4]]	x[[1]]	x[[6]]	x[[7]]

That is, the data for level 1 of factors A and B are stored in the R variable x[[2]], the data for level 1 of A and level 2 of B is stored in x[[3]], and so forth. To use t2way, first enter the R command

$$grp<-c(2,3,5,8,4,1,6,7).$$

Then the command t2way(2,4,x,grp=grp) tells the function how the data are ordered. In the example, the first value stored in grp is 2, indicating that x[[2]] contains the data for level 1 of both factors A and B, the next value is 3, indicating that x[[3]] contains the data for level 1 of A and level 2 of B, while fifth value is 4, meaning that x[[4]] contains the data for level 2 of factor A and level 1 of B. As usual, tr indicates the amount of trimming, which defaults to 0.2, and alpha is α, which defaults to 0.05. The function returns the test statistic for factor A, V_a, in the variable t2way$test.A, and the significance level is returned in t2way$sig.A. Similarly, the test statistics for factor B, V_b, and interaction, V_{ab}, are stored in t2way$test.B and t2way$test.AB, with the corresponding significance levels stored in t2way$sig.B and t2way$sig.AB.

As a more general example, the command

$$t2way(2,3,z,tr=0.1,grp=c(1,3,4,2,5,6),alpha=0.1)$$

would perform the tests for no main effects and no interactions for a 2-by-3 design for the data stored in the R variable z, assuming the data for level 1 of factors A and B are stored in z[[1]], the data for level 1 of A and level 2 of B are stored in z[[3]], and so on. The analysis would be based on 10% trimmed means and $\alpha = 0.1$.

Note that the general form for t2way contains an argument p. It is used by t2way to check whether the total number of groups being passed to the function is equal to JK. If JK is not equal to the number of groups in x, the function prints a warning message. If, however, you want to perform an analysis using some subset of the groups stored in x, this can be done simply by ignoring the warning message. For example, suppose x contains data for 10 groups, but you want to use groups 3, 5, 1, and 9 in a 2-by-2 design. That is, groups 3 and 5 correspond to level 1 of the first factor and levels 1 and 2 of the second. The command

$$t2way(2,2,x,grp=c(3,5,1,9))$$

accomplishes this goal. Note that a value for p is not passed to the function. The only reason p is included in the list of arguments is to satisfy certain requirements of R. The details are not important here and therefore not discussed.

■ Example

Suppose participants are randomly assigned to one of two groups. The first group watches a violent film, and the other watches a nonviolent film. Afterwards, suppose aggressive affect is measured, and it is desired to compare both groups, taking gender into account as well. Some hypothetical data are shown in Table 7.4 to illustrate how t2way is used.

Suppose the data are stored in the R variable film, having list mode. In particular, assume film[[1]] contains the values for males watching a violent film (the values 8, 7,

Table 7.4: Hypothetical Data on the Effect of Watching a Violent Film.

	Film	
	Violent	Nonviolent
Male	8, 7, 5, 6, 10, 14, 2, 3, 16	2, 4, 6, 7, 11, 12, 12, 3, 4
Female	5, 6, 8, 2, 3, 4, 5, 2	12, 40, 23, 2, 2, 2, 2, 4, 8, 10

5, 6, 10, 14, 2, 3, and 16). The data for males watching a nonviolent film are stored in film[[2]], the data for females watching a violent film are stored in film[[3]], and the data for females watching a nonviolent film are stored in film[[4]]. Then the command t2way(2,2,film) would perform the appropriate tests for main effects and interactions using 20% trimmed means. If instead the data for males watching a violent film are stored in film[[2]], and the data for males watching a nonviolent film are stored in film[[1]], use the command t2way(2,2,film,grp=c(2,1,3,4)) to compare 20% means, while t2way(2,2,film,tr=0,grp=c(2,1,3,4)) compares means instead. ∎

7.2.2 Comparing Medians

For the special case where the goal is to compare medians, the method in Section 7.2.1 is not recommended. One way to proceed is to test global hypotheses as described here. (Another approach is to use the multiple comparison procedure for medians described in Section 7.4.7, which has the advantage of handling tied values in an effective manner.) Let M_{jk} be the sample median for the jth level of factor A and the kth level of B, and let n_{jk} and S_{jk}^2 be the corresponding sample size and estimate of the squared standard error of M_{jk}. Here S_{jk}^2 is the McKean–Schrader estimate. To perform hypotheses dealing with main effects, compute

$$R_j = \sum_{k=1}^{K} M_{jk}, \quad W_k = \sum_{j=1}^{J} M_{jk},$$

$$d_{jk} = S_{jk}^2,$$

$$\hat{v}_j = \frac{\left(\sum_k d_{jk}\right)^2}{\sum_k d_{jk}^2/(n_{jk}-1)}, \quad \hat{\omega}_k = \frac{\left(\sum_j d_{jk}\right)^2}{\sum_j d_{jk}^2/(n_{jk}-1)}$$

$$r_j = \frac{1}{\sum_k d_{jk}}, \quad w_k = \frac{1}{\sum_j d_{jk}}$$

$$r_s = \sum_{j=1}^{J} r_j, \quad w_s = \sum_{k=1}^{K} w_k,$$

$$\hat{R} = \frac{\sum_j r_j R_j}{r_s}, \quad \hat{W} = \frac{\sum_k w_k W_k}{w_s}$$

$$B_a = \sum_{j=1}^{J} \frac{1}{\hat{v}_j}\left(1 - \frac{r_j}{\sum r_j}\right)^2, \quad B_b = \sum_{k=1}^{K} \frac{1}{\hat{\omega}_k}\left(1 - \frac{w_k}{\sum w_k}\right)^2$$

$$V_a = \frac{\sum_j r_j(R_j - \hat{R})^2}{(J-1)\left(1 + \frac{2(J-2)B_a}{J^2-1}\right)}, \quad V_b = \frac{\sum_k w_k(W_k - \hat{W})^2}{(K-1)\left(1 + \frac{2(K-2)B_b}{K^2-1}\right)}.$$

The degrees of freedom for Factor A are $v_1 = J - 1$ and $v_2 = \infty$. For Factor B, the degrees of freedom are $v_1 = K - 1$ and $v_2 = \infty$. The hypothesis of no main effect for factor A is rejected if $V_a \geq f_{1-\alpha}$, the $1 - \alpha$ quantile of an F distribution with the degrees of freedom for factor A. Similarly, reject for factor B if $V_b \geq f_{1-\alpha}$, with the degrees of freedom for factor B.

A heteroscedastic test of the hypothesis of no interactions can be performed as follows. Again let $d_{jk} = S_{jk}^2$ be the McKean–Schrader estimate of the squared standard error of M_{jk}. Let

$$D_{jk} = \frac{1}{d_{jk}}$$

$$D_{.k} = \sum_{j=1}^{J} D_{jk}, \ D_{j.} = \sum_{k=1}^{K} D_{jk}$$

$$D_{..} = \sum_{j=1}^{J} \sum_{k=1}^{K} D_{jk}$$

$$\ddot{M}_{jk} = \sum_{\ell=1}^{J} \frac{D_{\ell k} M_{\ell k}}{D_{.k}} + \sum_{m=1}^{K} \frac{D_{jm} M_{jm}}{D_{j.}} - \sum_{\ell=1}^{J} \sum_{m=1}^{K} \frac{D_{\ell m} M_{\ell m}}{D_{..}}.$$

The test statistic is

$$V_{ab} = \sum_{j=1}^{J} \sum_{k=1}^{K} D_{jk} (M_{jk} - \tilde{M}_{jk})^2.$$

Let c be the $1 - \alpha$ quantile of a chi-square distribution with $v = (J - 1)(K - 1)$ degrees of freedom. Reject the null hypothesis if $V_{ab} \geq c$.

7.2.3 R Function med2way

The computations for comparing medians, just described, are performed by the R function

$$\text{med2way(J,K,x,alpha=0.05)}.$$

7.3 Three-Way Designs and Trimmed Means

This section extends the hypothesis testing technique in Section 7.2, based on trimmed means, to a three-way design. Again, a generalization of results in Johansen (1980) is used to test global hypotheses. It is assumed that a J-by-K-by-L design is being used, so there are a total of $p = JKL$ independent groups with trimmed means $\mu_t = (\mu_{t1}, \ldots, \mu_{tJKL})'$. The general strategy for testing main effects and interactions is to test

$$H_0 : \mathbf{C}\mu_t = 0, \tag{7.5}$$

where $\mathbf{C}$ is constructed in a manner to be described. For convenience, it is assumed that the sample trimmed means are

$$\bar{\mathbf{X}}' = (\bar{X}_{t111}, \ldots \bar{X}_{11L}, \bar{X}_{t121}, \ldots, \bar{X}_{t12L}, \ldots, \bar{X}_{t1KL}, \ldots, \bar{X}_{tJKL}).$$

That is, the third subscript, which corresponds to the third factor, is incrementing the fastest. Thus, for the first level of the first factor ($J = 1$), the data are arranged as

$$
\begin{array}{ccc}
\bar{X}_{t111} & \cdots & \bar{X}_{t11L} \\
\vdots & \vdots & \vdots \\
\bar{X}_{t1K1} & \cdots & \bar{X}_{t1KL}.
\end{array}
$$

For the second level of the first factor ($J = 2$), the data are arranged as

$$
\begin{array}{ccc}
\bar{X}_{t211} & \cdots & \bar{X}_{t21L} \\
\vdots & \vdots & \vdots \\
\bar{X}_{t2K1} & \cdots & \bar{X}_{t2KL}
\end{array}
$$

and so on. (The R function fac2list, illustrated in the next section, might help when dealing with data management.)

For any integer $m \geq 2$, again let $\mathbf{C}_m$ be an $(m-1)$-by-m matrix having the form

$$
\begin{pmatrix}
1 & -1 & 0 & 0 & \cdots & 0 \\
0 & 1 & -1 & 0 & \cdots & 0 \\
 & & & \cdots & & \\
0 & 0 & \cdots & 0 & 1 & -1
\end{pmatrix}.
$$

And as in Section 7.2, $\mathbf{j}'_m$ is a 1-by-m vector of ones. Table 7.5 shows how to construct the contrast matrix $\mathbf{C}$ for the main effects and interactions in a three-way design. For example, when testing for main effects for factor A, use $\mathbf{C} = \mathbf{C}_J \otimes \mathbf{j}'_K \otimes \mathbf{j}'_L$.

Table 7.5: How to Construct the Contrast Matrix, C, for a Three-Way Design.

Effect	C
A	$\mathbf{C}_J \otimes \mathbf{j}'_K \otimes \mathbf{j}'_L$
B	$\mathbf{j}'_J \otimes \mathbf{C}_K \otimes \mathbf{j}'_L$
C	$\mathbf{j}'_J \otimes \mathbf{j}'_K \otimes \mathbf{C}_L$
A × B	$\mathbf{C}_J \otimes \mathbf{C}_K \otimes \mathbf{j}'_L$
A × C	$\mathbf{C}_J \otimes \mathbf{j}'_K \otimes \mathbf{C}_L$
B × C	$\mathbf{j}'_J \otimes \mathbf{C}_K \otimes \mathbf{C}_L$
A × B × C	$\mathbf{C}_J \otimes \mathbf{C}_K \otimes \mathbf{C}_L$

Remembering that $p = JKL$, the total number of groups, let $\mathbf{V}$ be a p-by-p diagonal matrix with

$$v_{jj} = \frac{(n_j - 1)s^2_{wj}}{h_j(h_j - 1)},$$

$j = 1, \ldots, p$. That is, v_{jj} is Yuen's estimate of the squared standard error of the sample trimmed mean corresponding to the jth group. Then

$$Q = \bar{\mathbf{X}}'\mathbf{C}'(\mathbf{C}\mathbf{V}\mathbf{C}')^{-1}\mathbf{C}\bar{\mathbf{X}}. \tag{7.6}$$

can be used to test Eq. (7.5). Let

$$\mathbf{R} = \mathbf{V}\mathbf{C}'(\mathbf{C}\mathbf{V}\mathbf{C}')^{-1}\mathbf{C},$$

and

$$A - \sum_{j=1}^{p} \frac{r^2_{jj}}{h_j - 1},$$

where r_{jj} is the jth diagonal element of $\mathbf{R}$. Asymptotically, a critical value for Q is c, the $1 - \alpha$ quantile of a chi-square distribution with k degrees of freedom. However, for small sample sizes, an adjusted critical value is needed which is given by

$$c_{\mathrm{ad}} = c + \frac{c}{2k}\left[A\left(1 + \frac{3c}{k+2}\right)\right].$$

If $Q > c_{\mathrm{ad}}$, reject H_0.

7.3.1 R Functions t3way and fac2list

The R function

$$\text{t3way(J,K,L,x,tr=0.2,grp=c(1:p),alpha=0.05,p=J*K*L).}$$

performs tests the hypotheses of no main effects and no interactions in a three-way (J-by-K-by-L) design using the method described in the previous section. Again, x is any R variable containing the data, which is assumed to be stored in list mode. The default amount of trimming is tr=0.2, and the default value for alpha is $\alpha = 0.05$. The data are assumed to be arranged such that the first L groups correspond to level 1 of factors A and B ($J = 1$ and $K = 1$) and the L levels of factor C. The next L groups correspond to the first level of factor A, the second level of factor B, and the L levels of factor C. If, for example, a 3-by-2-by-4 design is being used, it is assumed that for $J = 1$ (the first level of the first factor), the data are stored in the R variables x[[1]],...,x[[8]] as follows:

	Factor C			
Factor	x[[1]]	x[[2]]	x[[3]]	x[[4]]
B	x[[5]]	x[[6]]	x[[7]]	x[[8]]

For the second level of the first factor, $J = 2$, it is assumed that the data are stored as

	Factor C			
Factor	x[[9]]	x[[10]]	x[[11]]	x[[12]]
B	x[[13]]	x[[14]]	x[[15]]	x[[16]]

If the data are not stored as assumed by t3way, grp can be used to indicate the proper ordering. As an illustration, consider a 2-by-2-by-4 design and suppose that for $J = 1$, the data are stored as follows:

	Factor C			
Factor	x[[15]]	x[[8]]	x[[3]]	x[[4]]
B	x[[6]]	x[[5]]	x[[7]]	x[[8]]

while for $J = 2$

	Factor C			
Factor	x[[10]]	x[[9]]	x[[11]]	x[[12]]
B	x[[1]]	x[[2]]	x[[13]]	x[[16]]

Then type the R command

$$grp< -c(15,8,3,4,6,5,7,8,10,9,11,12,1,2,13,16)$$

and the command t3way(2,2,3,x,grp=grp) will test all of the relevant hypotheses at the 0.05 level using 20% trimmed means.

The general form for t3way contains an argument p that is used to check whether $p = JKL$ is equal to the total number of groups contained in x, and it is also used to generate the default value for grp. As far as applications are concerned, this argument can be ignored. (It is necessary only to satisfy certain R requirements that are not relevant here.) If JKL is not equal to the number of groups passed to t3way, the function prints a warning message. If, however, you want to use some subset of the groups in a three-way design, you can do this simply by ignoring the error message and taking care that the proper groups are used in the analysis. In other words, proceed along the lines described in conjunction with t2way. As a simple illustration, if there are 10 groups, but it is desired to use only the first 8 groups in a 2-by-2-by-2 design, the command

$$t3way(2,2,2,x)$$

can be used, assuming the first two groups belong to level 1 of the first two factors and levels 1 and 2 of the third, and so on. If the groups are not in the proper order, grp can be used as already described and illustrated.

■ Example

The example in Section 7.2.1 involved a 2-by-2 design dealing with the effect of watching a violent versus nonviolent film. Extending the illustration, suppose that education is taken into account with one group having a college degree, and the other does not. Some hypothetical data for this 2-by-2-by-2 design are shown in Table 7.6. Suppose the data are stored in the assumed order in the R variable film. Thus, film[[1]] contains the data for level 1 of all three factors (the values 8, 7, 5, 6, 10, 14, 2, 3, and 16), film[[2]] contains the data for no degree, male subjects watching a nonviolent film, film[[4]] contains the data for no degree, female subjects watching a nonviolent film, and film[[6]] contains the data for male subjects, with a degree, who watch a nonviolent film. Then the command t3way(2,2,2,film) will test all relevant hypotheses using 20% trimmed means. If it had been the case that the data for no degree, male subjects watching a violent film were stored in film[[2]], and the data for no degree, males subjects watching a nonviolent film were in film[[1]], but otherwise the assumed order is correct, the R command t3way(2,2,2,film,grp=c(2,1,3,4,5,6,7,8)) would perform the correct computations.

Table 7.6: Hypothetical Data on the Effect of Watching a Violent Film.

	No Degree	
	Violent	Nonviolent
Male	8, 7, 5, 6, 10, 14, 2, 3, 16	2, 4, 6, 7, 11, 12, 12, 3, 4
Female	5, 6, 8, 2, 3, 4, 5, 2	12, 40, 23, 2, 2, 2, 2, 4
	Degree	
	Violent	Nonviolent
Male	8, 10, 12, 14, 2, 18, 20	2, 3, 2, 4, 5, 6, 7, 3, 4
Female	4, 5, 6,7, 6, 5, 4, 7,8	12, 1, 4, 19, 20, 22, 23, 24, 30

The function returns the various test statistics and corresponding critical values. The value of the test statistic, Q, for main effects for factor A, is returned in t3way\$Qa, for factor B it is returned in t3way\$Qb, and for factor C it is in t3way\$Qc. The corresponding critical values are returned in t3way\$Qa.crit, t3way\$Qb.crit, and t3way\$Qc.crit. The tests for two-way interactions are stored in t3way\$Qab, t3way\$Qac, and t3way\$Qbc; the critical values are in t3way\$Qab.crit, t3way\$Qac.crit, and t3way\$Qbc.crit; and the test for a three-way interaction is in t3way\$Qabc, with the critical value in t3way\$Qabc.crit.

If data are stored in a matrix, with some of the columns indicating the levels of the factors, it is noted that the function fac2list, described in Section 7.1.2, can be used to store the data in the manner required here. Suppose the data are stored in a matrix, say m, with group numbers for the three factors stored in columns 2, 4, and 6. If, for example, a 2-by-4-by-5 design is being examined, column 2 would contain the group identification numbers for the two levels of the first factor. The values in column 2 might be 1 or 2, or they might be a 10 and 16. That is, there are two distinct values only, but they can be any two numbers. If the outcome measures are stored in column 5, the R command

$$\text{dat=fac2list(m[,5],m[,c(2,4,6)])}$$

will store the data in dat, in list mode. If, for example, it is desired to compare 20% trimmed means, this is accomplished with the command

$$\text{t3way(2,4,6,dat).}$$

7.4 Multiple Comparisons Based on Medians and Other Trimmed Means

This section summarizes several methods for performing multiple comparisons based on trimmed means, including medians as a special case. Included are methods for testing hypotheses about linear contrasts associated with two-way and three-way designs, which are described and illustrated in Section 7.4.3. The role of linear contrasts is described in many books dealing with the analysis of variance, so for brevity, details are kept to a minimum.

A general goal is to control the probability of at least one type I error. And a related goal is computing confidence intervals that have some specified simultaneous probability coverage. But another goal that has received increased attention in recent years is to control the *false discovery rate*. To elaborate, when testing C hypotheses, let Q be the proportion of hypotheses that are true and rejected. That is, Q is the proportion of type I errors among the null hypotheses that are correct. If all hypotheses are false, then $Q = 0$, but otherwise Q can vary from one experiment to the next. The *false discovery rate* is the expected value of Q.

A common practice is to use multiple comparison procedures, such as those described in this section, only if a global test, such as those described in Section 7.1–7.3, reject. In terms of controlling the probability of one or more type I errors, the methods in this section do not require that a global test first be performed and rejected. Indeed, based on results reported by Bernhardson (1975), the expectation is that using the multiple comparisons procedures in this chapter, contingent on first rejecting a global hypothesis, would alter their ability to control the probability of at least one type I error in an unintended way. More precisely, the methods in this section are designed so that the probability of one more type I errors is α. If these methods are used contingent on a global test rejecting at the α level, the expectation is that the actual probability of one more type I errors will be less than α. In practical terms, a loss in power might result if the multiple comparison procedures in this chapter are used only if a global test rejects.

7.4.1 An Extension of Yuen's Method to Trimmed Means

A relatively simple strategy for performing multiple comparisons and tests about linear contrasts, when comparing trimmed means, is to use an extension of Yuen's method for two groups in conjunction with a simple generalization of Dunnett's (1980) heteroscedastic T3 procedure for means.

Letting $\mu_{t1}, \ldots, \mu_{tJ}$ be the trimmed means corresponding to J independent groups, a linear contrast is

$$\Psi = \sum_{j=1}^{J} c_j \mu_{tj},$$

where $c_1, \ldots, c_J$ are specified constants satisfying $\sum c_j = 0$. As a simple illustration, if $c_1 = 1$, $c_2 = -1$, and $c_3 = \cdots = c_J = 0$, $\Psi = \mu_{t1} - \mu_{t2}$, the difference between the first two trimmed means. Typically, C linear contrasts are of interest, a common goal being to compare all pairs of means. Linear contrasts also play an important role when dealing with two-way and higher designs.

Consider testing

$$H_0 : \Psi = 0. \tag{7.7}$$

An extension of the Yuen–Welch method accomplishes this goal. The estimate of Ψ is

$$\hat{\Psi} = \sum_{j=1}^{J} c_j \bar{X}_{tj}.$$

An estimate of the squared standard error of $\hat{\Psi}$ is

$$A = \sum d_j,$$

where

$$d_j = \frac{c_j^2(n_j - 1)s_{wj}^2}{h_j(h_j - 1)},$$

h_j is the effective sample size of the jth group, and s_{wj}^2 is the Winsorized variance. In other words, estimate the squared standard error of $\bar{X}_t$ as is done in Yuen's method, in which case an estimate of the squared standard error of $\hat{\Psi}$ is given by A. Let

$$D = \sum \frac{d_j^2}{h_j - 1},$$

set

$$\hat{v} = \frac{A^2}{D},$$

and let t be the $1 - \alpha/2$ quantile of Student's t-distribution with $\hat{v}$ degrees of freedom. Then an approximate $1 - \alpha$ confidence interval for Ψ is

$$\hat{\Psi} \pm t\sqrt{A}.$$

Let $\Psi_1, \ldots, \Psi_C$ be C linear contrasts of interest, where

$$\Psi_k = \sum_{j=1}^{J} c_{jk}\mu_{tj},$$

and let $\hat{v}_k$ be the estimated degrees of freedom associated with the kth linear contrast, which is computed as described in the previous paragraph. As previously noted, a common goal is to compute a confidence interval for each Ψ_k such that the simultaneous probability coverage is $1 - \alpha$. A related goal is to test $H_0 : \Psi_k = 0$, $k = 1, \ldots, C$, such that the *familywise error rate* (FWE), meaning the probability of at least one type I error among all C tests to be performed, is α, and the practical problem is finding a method that adjusts the critical value to achieve this goal. One strategy is to compute confidence intervals having the form

$$\hat{\Psi}_k \pm t_k\sqrt{A_k},$$

where A_k is the estimated squared standard error of $\hat{\Psi}_k$, computed as described when testing Eq. (7.7), and t_k is the $1 - \alpha$ percentage point of the C-variate Studentized maximum modulus distribution with estimated degrees of freedom $\hat{v}_k$. In terms of testing $H_0 : \Psi_k = 0$, $k = 1, \ldots, C$, reject $H_0 : \Psi_k = 0$ if $|T_k| > t_k$, where

$$T_k = \frac{\hat{\Psi}_k}{\sqrt{A_k}}.$$

(The R software written for this book determines t_k when $\alpha = 0.05$ or 0.01, and $C \leq 28$ using values computed and reported in Wilcox, 1986. For other values of α or $C > 28$, the function determines t_k via simulations with 10,000 replications. Bechhofer & Dunnett, 1982, report values up to $C = 32$.) When there is no trimming, the method just described reduces to Dunnett's (1980) T3 procedure for means when all pairwise comparisons are performed.

7.4.2 R Function lincon

The R function

$$\text{lincon(x,con=0,tr=0.2,alpha=0.05)}$$

is provided for testing linear contrasts involving trimmed means. The argument x is an R variable having list mode, tr indicates the amount of trimming, and con is a J-by-C matrix, the kth column containing the contrast coefficients for the kth linear contrast of interest, $k = 1, \ldots, C$. The argument alpha is α which defaults to 0.05. Any other value for the argument alpha results in $\alpha = 0.01$. As usual, x[[1]] contains the data for group 1, x[[2]] the data for group 2, and so on, and the default amount of trimming is tr=0.2, 20%. (The functions fac2list, selby, and selby2, described in Sections 1.9, can be used to store the data in list mode when initially the data are stored in a matrix, say m, with 1 or more columns of m containing group identification numbers.) If con is not specified, all pairwise comparisons are performed. The function returns two matrices called test and psihat. If all pairwise comparisons are to be performed, the first two columns of both matrices indicate which groups are being compared. The remaining columns of the first matrix, test, report the test statistic, the 0.95 critical value, the estimated standard error, and the degrees of freedom. If there is interest in using $\alpha = 0.01$, rather than 0.05, these results can be used to determine an appropriate critical value by referring to the table of Studentized maximum modulus distribution previously cited in this section. Columns 3–5 of psihat report $\hat{\Psi}$, and the lower and upper ends of the 0.95 confidence interval. These quantities are found in the columns labeled psihat, ci.lower, and ci.upper, respectively. If specific contrasts are of interest (meaning that a value for con is passed to lincon), the output is the same as just described, only the first two columns of the matrices returned by lincon are replaced by the number of the contrast being examined. That is, the first row of each matrix returned by lincon is numbered 1, meaning that it contains the results for Ψ_1, and so on.

The critical value used by lincon is determined with the goal that the probability of at least one type I error is less than or equal to 0.05 or 0.01, depending on the argument alpha, and that the simultaneous probability coverage of all the confidence intervals is greater than or equal to 0.95 or 0.99. This is accomplished by storing exact percentage points of the Studentized maximum modulus distribution in the R functions called smmcrit and smmcrit01 for $C = 2, \ldots, 28$ and selected degrees of freedom. These exact values were determined with the FORTRAN program in Wilcox (1986). For other degrees of freedom, linear interpolation on inverse degrees of freedom is used to determine the 0.95 and 0.99 quantiles. The function assumes that for $\nu \geq 200$, $\nu = \infty$. For $C > 28$, or values for α other than 0.05 and 0.01, the R function smmvalv2 is used to compute the required percentage point.

The optional argument con is a J-by-C matrix that contains the contrast coefficients to be used. The kth column is assumed to contain the contrast coefficients $c_{1k}, \ldots, c_{Jk}$, which correspond to the kth linear contrast, Ψ_k. As previously indicated, if con is not specified, then all pairwise comparisons are performed.

■ Example

Suppose four independent groups are being compared using the data in Table 7.7. If the data are stored in the R variable x, the command lincon(x) returns

```
$test
        Group  Group       test       crit         se          df
[1,]       1      2 0.4151210   3.120264   66.07368   11.374139
[2,]       1      3 0.2590833   3.094621   60.65340   11.900028
[3,]       1      4 0.6099785   3.372408   43.97761    7.892383
[4,]       2      3 0.1708173   3.101733   68.57785   11.749354
[5,]       2      4 0.9975252   3.464224   54.38857    7.177902
[6,]       3      4 0.8926159   3.410628   47.65732    7.578376


$psihat
        Group  Group     psihat   ci.lower   ci.upper    p.value
[1,]       1      2  -27.42857  -233.5959   178.7387  0.6857763
[2,]       1      3  -15.71429  -203.4136   171.9850  0.7999983
[3,]       1      4   26.82540  -121.4851   175.1358  0.5590247
[4,]       2      3   11.71429  -200.9959   224.4245  0.8672737
[5,]       2      4   54.25397  -134.1602   242.6682  0.3509425
[6,]       3      4   42.53968  -120.0017   205.0811  0.3995184
```

For example, when comparing groups 1 and 2 with 20% trimmed means, the third and fourth columns stored in $test indicate that the test statistic is 0.415, and the $\alpha = 0.05$ critical value is 3.12. The remaining two columns indicate the estimated standard error and degrees of freedom. The results in $psihat indicate that when comparing groups 2 and 3, $\hat{\Psi} = 11.7$, and the 0.95 confidence interval is $(-201, 224)$. The command lincon(x,alpha=0.01) would use $\alpha = 0.01$ instead. (For convenience, the R function mmean(x,est=tmean,...) has been provided, which computes measures of location for all groups stored in x.)

Table 7.7: Data Used to Illustrate the R Function lincon.

Group 1	119., −53., −77., 32., 194., −34., 48., −73., −69., −95., 175.
Group 2	−25., −22., 158., 208., 245., −70., −95., −68., 161., 28., −73.
Group 3	−95., 438., −72., 290., 3., −86., 136., 43., −27., 76., −79.
Group 4	−37., −88., −23., −50., 45., −36., −79., −86., −66., −73., −11., 16., 0., 47., 218.

■ Example

Again consider the data in Table 7.7, only now suppose that the fourth group is a control and it is desired to compare each of the first three groups to the control. That is, the goal is to compare group 1 to group 4, group 2 to group 4, and group 3 to group 4. Then the contrast coefficients for the first linear contrast are $c_{11} = 1$, $c_{21} = c_{31} = 0$, and $c_{41} = -1$, in which case

$$\Psi_1 = 1\mu_{t1} + 0\mu_{t2} + 0\mu_{t3} + (-1)\mu_{t4}$$
$$= \mu_{t1} - \mu_{t4}.$$

In a similar fashion, the contrast coefficients for $\Psi_2 = \mu_{t2} - \mu_{t4}$ are $c_{12} = c_{32} = 0$, $c_{22} = 1$, and $c_{42} = -1$. For $\Psi_3 = \mu_{t3} - \mu_{t4}$, they are $c_{13} = c_{23} = 0$, $c_{33} = 1$ and $c_{43} = -1$.

To use lincon, first store the matrix

$$\begin{pmatrix} 1 & 0 & 0 \\ 0 & 1 & 0 \\ 0 & 0 & 1 \\ -1 & -1 & -1 \end{pmatrix}$$

in any R variable. The first column contains the contrast coefficients for the first linear contrast, $(1, 0, 0, -1)$. The second columns contains the contrast coefficients for the second linear contrast, and the third column contains the contrast coefficients associated with Ψ_3. For example, the command

$$\text{MAT} <- \text{matrix}(c(1,0,0,-1,0,1,0,-1,0,0,1,-1),4,3)$$

stores the contrast coefficients in the R variable MAT. Assuming the data for the four groups are stored in x, the command lincon(x,con=MAT) performs the three comparisons with $\alpha = 0.05$, and the results are

```
$test
       con.num       test       crit         se         df
[1,]         1  0.6099785   2.969545   43.97761   7.892383
[2,]         2  0.9975252   3.040172   54.38857   7.177902
[3,]         3  0.8926159   2.998945   47.65732   7.578376


$psihat
       con.num     psihat   ci.lower   ci.upper    p.value
[1,]         1   26.82540  -103.7681   157.4189  0.5590247
[2,]         2   54.25397  -111.0967   219.6046  0.3509425
[3,]         3   42.53968  -100.3820   185.4614  0.3995184


$test:
       con.num       test       crit         se         df
[1,]         1  0.6099785   2.969545   43.97761   7.892383
[2,]         2  0.9975252   3.040172   54.38857   7.177902
[3,]         3  0.8926159   2.998945   47.65732   7.578376


$psihat:
       con.num     psihat   ci.lower   ci.upper
[1,]         1   26.82540  -103.7681   157.4189
[2,]         2   54.25397  -111.0967   219.6046
[3,]         3   42.53968  -100.3820   185.4614
```

A difference between this output and the output of the previous example is that now, the contrasts are numbered under the column labeled con.num. The results for the first linear contrast, $\mu_{t1} - \mu_{t4}$, are stored in the first row of the matrices $test and $psihat. Thus, the test statistic for $H_0 : \mu_{t1} = \mu_{t4}$ is 0.61, the critical value is 2.97, the estimate of $\Psi_1 = \mu_{t1} - \mu_{t4}$ is 26.8, and the 0.95 confidence interval is $(-103.8, 157.4)$. Note that the critical values in this example are smaller than those in the previous example. This is because only three contrasts are being tested now, as opposed to six contrasts before.

7.4.3 Multiple Comparisons for Two-way and Three-Way Designs

Relevant multiple comparisons in a two-way design can be tested using appropriate linear contrasts. Consider, for example, a 3-by-3 design with the trimmed means depicted as follows:

		Factor B		
		1	2	3
Factor A	1	μ_{t1}	μ_{t2}	μ_{t3}
	2	μ_{t4}	μ_{t5}	μ_{t6}
	3	μ_{t7}	μ_{t8}	μ_{t9}

Let

$$\Psi_1 = \mu_{t1} + \mu_{t2} + \mu_{t3} - \mu_{t4} - \mu_{t5} - \mu_{t6},$$

$$\Psi_2 = \mu_{t1} + \mu_{t2} + \mu_{t3} - \mu_{t7} - \mu_{t8} - \mu_{t9},$$

$$\Psi_3 = \mu_{t4} + \mu_{t5} + \mu_{t6} - \mu_{t7} - \mu_{t8} - \mu_{t9}.$$

Then an approach to comparing the main effects for Factor A is to test H_0: $\Psi_\ell = 0$, for $\ell = 1$, 2, and 3. Roughly, the goal is to compare level 1 of Factor A to level 2 of Factor A, then compare levels 1 and 3, and finally compare levels 2 and 3. Main effects for Factor B, as well as interactions, can be examined in a similar manner. For interactions, this means that for any two levels of Factor A, say j and j' ($j < j'$), and any two levels of Factor B, k and k' ($k < k'$), linear contrast coefficients are generated with the goal of testing

$$H_0 : \mu_{tj} - \mu_{tj'} = \mu_{tk} - \mu_{tk'}.$$

For convenience, an R function (described in the next section and called con2way) is provided that generates the contrast coefficients typically used in a two-way design. Three-way designs are handled in a similar manner by generating linear contrast coefficients with the R function con3way, which is also described in the next section.

7.4.4 R Functions mcp2atm, mcp2med, mcp3atm, mcp3med, con2way, and con3way

The R function

$$\text{mcp2atm(J,K,tr=0.2,alpha=0.05,grp=NA,op=F)}$$

tests all of the usual pairwise comparisons associated with the levels of each factor, as well as all interactions associated with any two rows and columns, based on trimmed means. It does this by calling the R function

$$\text{con2way(J,K),}$$

which generates linear contrast coefficients, and then it calls the R function lincon. If op=F, the $(J^2 - J)/2$ hypotheses associated with Factor A (all pairwise comparisons of the J levels) are tested with the probability of one or more type I errors designated by the argument alpha, which defaults to 0.05. The same is done for Factor B and all relevant interactions. If op=T, the function is designed so that for all comparisons associated with Factor A, Factor B, and all interactions, the probability of one or more type I errors is alpha. So with op=T, power will be lower because the probability of one or more type I errors is being controlled for all hypotheses under consideration. For the special case where the goal is to compare medians, the function

$$\text{mcp2med(J,K,x,con=0,alpha=0.05,grp=NA,op=F)}$$

is supplied, which is based in part on the McKean–Schrader estimate of the standard error of the sample medians. As previously noted, this estimate of the standard error appears to perform reasonably well *with no tied values*, but otherwise a percentile bootstrap method is recommended for comparing medians.

■ Example

Consider a 2-by-3 design. A portion of the output from the R command con2way(2,3) is

```
$conAB
      [,1] [,2] [,3]
[1,]    1    1    0
[2,]   -1    0    1
[3,]    0   -1   -1
[4,]   -1   -1    0
[5,]    1    0   -1
[6,]    0    1    1
```

The three columns contain the linear contrast coefficients relevant to the three interactions associated with the six groups being compared. Assuming means are being compared and that they are arranged as indicated in Table 7.2, the first column indicates that the linear contrast of interest is

$$\Psi = \mu_1 - \mu_2 - \mu_4 + \mu_5.$$

The typical goal is to test H_0: $\Psi = 0$, which of course is the same as testing

$$H_0 : \mu_1 - \mu_2 = \mu_4 - \mu_5,$$

the hypothesis of no interaction for levels 1 and 2 of both factors. The second column deals with the interaction associated with levels 1 and 3 of factor B. And the third column deals with levels 2 and 3. If factor A had three levels, conAB would have nine columns. The first three would deal with levels 1 and 2 of factor A, the next three would deal with levels 1 and 3 of factor A. And the final three would deal with levels 2 and 3. Again the first three columns of conAB would deal with the three levels of factor B, namely, levels 1 and 2, 1 and 3, and finally levels 2 and 3.

The R function

$$\text{mcp3atm}(J,K,L,tr=0.2,alpha=0.05,grp=NA,op=F)$$

is like mcp2atm, only it is designed for a three-way design. The linear contrast coefficients are generated by the R function

$$\text{con3way}(J,K,L).$$

So all pairwise comparisons associated with the levels of each factor are performed, as well as all interactions associated with the levels of each Factor. For medians, use the R function

$$\text{mcp3med}(J,K,L,tr=0.2,alpha=0.05,grp=NA,op=F).$$

■ Example

To illustrate the use of the R function con3way when dealing with a three-way interaction, consider a 2-by-2-by-3 design and focus on the contrast coefficients returned by the command con3way(2,2,3), which are stored in the matrix conABC. The means are assumed to be arranged as described at the beginning of Section 7.3. The first set of contrast coefficients, stored in the first column of conABC, deal with the A-by-B interaction at levels 1 and 2 of factor C. The second set of contrast coefficients deal with the A-by-B interactions at levels 1 and 3 of factor C. The contrast coefficients stored in the third column of conABC deal with the A-by-B interactions at levels 2 and 3 of factor C. For a 2-by-3-by-3 design, there are nine linear contrasts associated with a three-way interaction. The first three deal with the interactions associated levels 1 and 2 of factors A and B, respectively. The first set of linear contrast coefficients (in column 1 of conABC) is relevant to levels 1 and 2 of factor C, the next is relevant to levels 1 and 3 of factor C, and the third is relevant to levels 2 and 3 of factor C. The next three sets of linear contrast coefficients repeat this pattern, only now the focus is on levels 1 and 3 factor B (and levels 1 and 2 of factor A). The final three sets of linear contrasts coefficients deal with levels 2 and 3 of factor B.

■

7.4.5 A Bootstrap-t Procedure

When comparing trimmed means, and the amount of trimming is relatively small, all indications are that an extension of the bootstrap-t method to multiple comparisons has practical value. So in particular, when comparing means and when the sample sizes are small, this approach appears to perform relatively well, with the understanding that all methods based on means can be unsatisfactory.

Table 7.8: Bootstrap-t Confidence Intervals for C Linear Contrasts.

The goal is to compute confidence intervals for each of C linear contrasts, $\Psi_1, \ldots, \Psi_C$, such that the simultaneous probability coverage is $1 - \alpha$. The kth linear contrast has contrast coefficients $c_{1k}, \ldots, c_{Jk}$.

Step 1. For each of the J groups, generate a bootstrap sample, X_{ij}^*, $i = 1, \ldots, n_j$; $j = 1, \ldots, J$. For each of the J bootstrap samples, compute the trimmed mean, $\bar{X}_j^*$, and d_j^*, Yuen's estimate of the squared standard error of $\bar{X}_j^*$, $j = 1, \ldots, J$

Step 2. For the kth linear contrast, compute

$$T_k^* = \frac{|\hat{\Psi}_k^* - \hat{\Psi}_k|}{\sqrt{A_k^*}},$$

where $\hat{\Psi}_k^* = \sum c_{jk} \bar{X}_k^*$ and $A_k^* = \sum c_{jk}^2 d_j^*$.

Step 3. Let

$$T_m^* = \max\{T_1^*, \ldots, T_C^*\}.$$

In words, T_m^* is the maximum of the C values, $T_1^*, \ldots, T_C^*$.

Step 4. Repeat steps 1–3 B times yielding T_{mb}^*, $b = 1, \ldots, B$.
Let $T_{m(1)}^* \leq \cdots \leq T_{m(B)}^*$ be the T_{mb}^* values written in ascending order, and let $a = (1 - \alpha)B$, rounded to the nearest integer. Then the confidence interval for Ψ_k is

$$\hat{\Psi}_k \pm T_{m(a)}^* \sqrt{A_k},$$

and the simultaneous probability coverage is approximately $1 - \alpha$.

Table 7.8 describes a bootstrap-t method for computing confidence intervals for each of C linear contrasts, Ψ_k, $k = 1, \ldots, C$, such that the simultaneous probability coverage is approximately $1 - \alpha$. The method is essentially the same as the symmetric two-sided confidence interval using the bootstrap-t method described in Table 5.6, only modified so that for the C linear contrasts, the probability of at least one type I error is approximately α.

When using the Studentized maximum modulus distribution to compare all pairs of trimmed means, it is known that probability coverage can be more satisfactory when using 20% trimmed means versus no trimming at all. However, concerns persist when any of the sample sizes are small. If the goal is to avoid having the probability of a type I error excessively higher than α, the bootstrap-t method is a good choice based on extant simulation studies when the amount of trimming is small. A criticism of the bootstrap-t is that when all of the sample sizes are less than or equal to 15, the probability of at least one type I error can drop below .025 when testing at the .05 level.

Table 7.9 shows estimates of α (which is one minus the simultaneous probability coverage) when sampling from exponential or lognormal distributions with four groups, $B = 599$, and

Table 7.9: Simulation Estimates of α for Some Light-Tailed Distributions.

Distribution	n	σ	Bootstrap	No Bootstrap
exponential	(11,11,11,11)	(1,1,1,1)	0.039	0.060
	(11,11,11,11)	(1,1,1,5)	0.039	0.096
	(15,15,15,15)	(1,1,1,1)	0.050	0.042
	(15,15,15,15)	(1,1,1,5)	0.050	0.083
	(10,15,20,25)	(1,1,1,1)	0.041	0.046
	(10,15,20,25)	(1,1,1,5)	0.040	0.061
	(10,15,20,25)	(5,1,1,1)	0.052	0.098
lognormal	(11,11,11,11)	(1,1,1,1)	0.015	0.030
	(11,11,11,11)	(1,1,1,5)	0.046	0.091
	(15,15,15,15)	(1,1,1,1)	0.023	0.028
	(15,15,15,15)	(1,1,1,5)	0.052	0.085
	(10,15,20,25)	(1,1,1,1)	0.030	0.033
	(10,15,20,25)	(1,1,1,5)	0.039	0.056
	(10,15,20,25)	(5,1,1,1)	0.063	0.104

various configurations of sample sizes, **n**, and standard deviations, σ, when using 20% trimmed means. (Additional simulations are reported by Wilcox, 1996h.) If the sample sizes are large enough, probability coverage will be satisfactory without using the bootstrap-t method, but it is unknown just how large the sample sizes must be. For heavy-tailed distributions, the bootstrap-t method offers less of an advantage, but it is difficult to tell whether it can be safely abandoned simply by looking at the data.

7.4.6 R Functions linconb, bbtrim, and bbbtrim

The R function

$$\text{linconb(x,con=0,tr=0.2,alpha=0.05,nboot=599)}$$

is provided for applying the bootstrap-t method when testing d linear contrasts using trimmed means. This function is used exactly like the function lincon in Section 7.4.2, the only difference being the additional argument, nboot, which is used to specify B, the number of bootstrap samples to be used. Again, if con is not specified, all pairwise comparisons are performed. The default value for nboot is 599 which appears to suffice, in terms of controlling the probability of a type I error, when $\alpha = 0.05$. However, a larger choice for nboot might result in more power. The extent to which accurate probability coverage can be obtained, when $\alpha < 0.05$, is not known.

■ Example

Again consider the data in Table 7.7, suppose it is desired to compare the first three groups to the fourth, only now a bootstrap-t method is used. The results from linconb are

```
$psihat
     con.num   psihat  ci.lower ci.upper
[1,]       1 26.82540 -146.2109 199.8617
[2,]       2 54.25397 -159.7458 268.2537
[3,]       3 42.53968 -144.9750 230.0544

$test
     con.num      test       se  p.value
[1,]       1 0.6099785 43.97761 0.5525876
[2,]       2 0.9975252 54.38857 0.3439065
[3,]       3 0.8926159 47.65732 0.4140234

$crit
[1] 3.934646
```

The contrast matrix is also returned in linconb$con. The estimates of Ψ for the three linear contrasts of interest, plus the corresponding standard deviations, are the same as before. However, the confidence intervals are longer using the bootstrap method. The critical value for all three contrasts, used by the first method in this section, is approximately 3, but the bootstrap estimate of the critical value is 3.99. This was expected because the observations in Table 7.7 were generated by first generating values from an exponential distribution, shifting them so that the trimmed mean is equal to zero, and multiplying by 100. That is, observations were generated from a skewed distribution with a relatively light tail. If the observations were generated from a normal distribution instead, the expectation is that there would be little difference between the confidence intervals.

■

For convenience, the R function

$$bbtrim(J,K,x,tr=0.2,alpha=0.05,nboot=599).$$

is supplied for performing the usual multiple comparisons associated with a J-by-K design using a bootstrap-t method with trimmed means. The function generates the linear contrast coefficients via the R function con2way and then uses linconb to test the relevant hypotheses.

Multiple comparisons associated with a J-by-K-by-L three-way design are performed with the R function

$$\text{bbbtrim(J,K,L,x,tr}=0.2,\text{alpha}=0.05,\text{nboot}=599).$$

The contrast coefficients are generated via the R function

$$\text{con3way(J,K,L)}.$$

7.4.7 Percentile Bootstrap Methods for Comparing Medians and Other Trimmed Means

When the goal is to compare trimmed means, and the amount of trimming is not too small, say at least 20%, a percentile bootstrap method appears to be a relatively effective way of performing multiple comparisons. With 15% trimming, and even 10% trimming, a percentile bootstrap performs reasonably well. For the special case where the goal is to compare medians, and when tied values occur, it is the only known method that performs well in simulations, in terms of controlling the probability of a type I error. This section describes methods designed to control the probability of one or more type I errors when using a percentile bootstrap method.

Rom's Method

Imagine the goal is to test C hypotheses such that the probability of one or more type I errors is at most α, a simple way of proceeding is to use the *Bonferroni* method, meaning that each test is performed at the α/C level. However, several improvements on the Bonferroni method have been published that are designed to ensure that the probability of at least one type I error does not exceed some specified value, α. One that stands out is a *sequentially rejective* method derived by Rom (1990), which has been found to have good power relative to several competing techniques (e.g., Olejnik, Li, Supattathum, & Huberty, 1997; cf. Goeman & Solari, 2010). To apply it, compute a p-value for each of the C tests to be performed and label them $P_1, \ldots, P_C$. Next, put the p-values in descending order, which are labeled $P_{[1]} \geq P_{[2]} \geq \cdots \geq P_{[C]}$. Proceed as follows:

1. Set k=1.
2. If $P_{[k]} \leq d_k$, where d_k is read from Table 7.10, stop and reject all C hypotheses; otherwise, go to step 3.
3. Increment k by 1. If $P_{[k]} \leq d_k$, stop and reject all hypotheses having a significance level less than or equal d_k
4. If $P_{[k]} > d_k$, repeat step 3.
5. Continue until you reject or all C hypotheses have been tested.

Table 7.10: Critical Values, d_k, for Rom's Method.

k	$\alpha = 0.05$	$\alpha = 0.01$
1	0.05000	0.01000
2	0.02500	0.00500
3	0.01690	0.00334
4	0.01270	0.00251
5	0.01020	0.00201
6	0.00851	0.00167
7	0.00730	0.00143
8	0.00639	0.00126
9	0.00568	0.00112
10	0.00511	0.00101

Hochberg's Method

Hochberg's (1988) method for controlling the probability of one or more type I errors is applied as follows. Again let $p_1, \ldots, p_C$ be the p-values associated with the C tests, put these p-values in descending order, and label the results $p_{[1]} \geq p_{[2]} \geq \cdots \geq p_{[C]}$. Beginning with $k = 1$ (step 1), reject all hypotheses if

$$p_{[k]} \leq \alpha / k.$$

That is, reject all hypotheses if the largest p-value is less than or equal to α. If $p_{[1]} > \alpha$, proceed as follows:

1. Increment k by 1. If

$$p_{[k]} \leq \frac{\alpha}{k},$$

    stop and reject all hypotheses having a p-value less than or equal $p_{[k]}$
2. If $p_{[k]} > \alpha / k$, repeat step 1.
3. Repeat steps 1 and 2 until you reject or all C hypotheses have been tested.

Rom's method offers a slight advantage over Hochberg's method in terms of power. But Rom's method is limited to testing at the 0.05 and 0.01 levels, and tables for performing $C > 10$ hypotheses are not available. Hochberg's method avoids these limitations.

Benjamini–Hochberg Method

Benjamini and Hochberg (1995) proposed a variation of Hochberg's method where in step 1 of Hochberg's method, $p_{[k]} \leq \alpha / k$ is replaced by

$$p_{[k]} \leq \frac{(C - k + 1)\alpha}{C}$$

(cf. Williams, Jones, & Tukey, 1999; Peña, Habiger, & Wu, 2010). A criticism of the Benjamini–Hochberg method is that situations can be found where some hypotheses are true, some are false, and the probability of at least one type I error will exceed α among the hypotheses that are true (Hommel, 1988). In contrast, Hochberg's method does not suffer from this problem (assuming the actual level of each individual test is equal to the nominal level). However, when C hypotheses are tested, let Q be the proportion of hypotheses that are true and rejected. That is, Q is the proportion of type I errors among the null hypotheses that are correct. The *false discovery rate* is the expected value of Q. That is, if a study is repeated infinitely many times, the false discovery rate is the average proportion of type I errors among the hypotheses that are true. Benjamini and Hochberg (1995) show that their method ensures that the false discovery rate is less than or equal to α (cf. Peña et al., 2010).

Method TPB20

An early method aimed specifically at comparing 20% trimmed means, given the goal of performing all pairwise comparisons, was studied by Wilcox (2001a), which will be called method TPB20. Briefly, for each pair of groups, compute a bootstrap (generalized) p-value as described in Section 5.4. When comparing group j to group $k (j < k)$, denote this p-value by $2\hat{p}^*_{jk}$. Then reject H_0: $\mu_{tj} = \mu_{tk}$ at the .05 level if

$$2\hat{p}^*_{jk} \leq 2p_{\text{crit}},$$

where

$$p_{\text{crit}} = \frac{.0268660714}{C} - .0003321429,$$

and C is the number of hypotheses to be tested. (Adjustments, when testing at the .01 level, are available as well.) The method also yields confidence intervals for each linear contrast, which are designed so that the simultaneous probability coverage is either .05 or .01. All of the confidence intervals have the same probability coverage, a feature that is not available when using Rom's method or Hochberg's methods. A limitation of this method is that it can handle only situations where the probability of one or more type I error is set at .05 or .01. Also, situations can be constructed where Rom's method and Hochberg's method have more power.

7.4.8 R Functions tmcppb, bbmcppb, bbbmcppb, medpb, med2mcp, med3mcp, and mcppb20

The R function

$$\text{tmcppb}(x, \text{alpha} = 0.05, \text{nboot} = \text{NA}, \text{grp} = \text{NA}, \text{est} = \text{tmean}, \text{con} = 0, \text{bhop} = F, ...),$$

compares trimmed means using a percentile bootstrap method. By default, the function uses a 20% trimmed mean and all pairwise comparisons are performed. Rom's method is used to control the probability of one or more type I errors. For $C > 10$ hypotheses, or when the goal is to test at some level other than .05 and .01, Hochberg's method is used. Setting the argument bhop = T, the Benjamini–Hochberg method is used instead. Linear contrasts can be tested by storing linear contrast coefficients in the argument con, which is assumed to be a matrix with rows corresponding to groups. By default, all pairwise comparisons are performed.

For convenience, the R function

$$\text{bbmcppb(J, K, x, tr} = 0.2, \text{JK} = \text{J} * \text{K}, \text{alpha} = 0.05, \text{grp} = c(1{:}JK), \text{nboot} = 500, \text{bhop} = F,$$
$$\text{SEED} = T)$$

performs multiple comparisons for a two-way ANOVA design based on trimmed means. The function creates linear contrasts via the R function con2way and then uses the function

$$\text{bbmcppb.sub(J, K, x, tr} = 0.2, \text{JK} = \text{J} * \text{K}, \text{con} = 0, \text{alpha} = 0.05, \text{grp} = c(1{:}JK),$$
$$\text{nboot} = 500, \text{bhop} = F, \text{SEED} = T, ...)$$

to test hypotheses based on the resulting linear contrasts. The R function

$$\text{bbbmcppb(J, K, L, x, tr} = 0.2, \text{JKL} = \text{J} * \text{K} * \text{L}, \text{alpha} = 0.05, \text{grp} = c(1{:}JKL), \text{nboot} = 500,$$
$$\text{bhop} = F, \text{SEED} = T)$$

performs multiple comparisons for a three-way design.

For the special case where medians are to be compared, use the R function

$$\text{medpb(x, alpha} = 0.05, \text{nboot} = NA, \text{grp} = NA, \text{est} = \text{median}, \text{con} = 0, \text{bhop} = F,$$
$$\text{SEED} = T).$$

It performs well when there are tied values, and even with no tied values, it appears to be an excellent choice relative to competing techniques for comparing medians. The R functions

$$\text{med2mcp(J, K, x, grp} = c(1{:}p), \text{p} = \text{J} * \text{K}, \text{tr} = 0.2, \text{nboot} = NA, \text{alpha} = 0.05, \text{SEED} = T,$$
$$\text{bhop} = F)$$

and

$$\text{med3mcp(J, K, L, data, tr} = 0.2, \text{grp} = c(1{:}p), \text{alpha} = 0.05, \text{p} = \text{J} * \text{K} * \text{L}, \text{nboot} = NA,$$
$$\text{SEED} = T, \text{bhop} = F)$$

are designed to handle two-way and three-way designs, respectively.

The R function

$$\text{mcppb20(x, crit} = NA, \text{con} = 0, \text{tr} = 0.2, \text{alpha} = 0.05, \text{nboot} = 2000, \text{grp} = NA)$$

performs method TPB20.

7.4.9 Judging Sample Sizes

Suppose that all pairs of groups are compared with the R function lincon in Section 7.4.1 and that one or more of the hypotheses are not rejected. This might occur because there is little or no difference between the groups. But another possibility is that there is an important difference that was missed. One way of trying to distinguish between these two possibilities is to determine how many observations are needed so that the length of the confidence intervals are reasonably short. This can be done with an extension of a two-stage method for means that was developed by Hochberg (1975); see Wilcox (2004a).

Imagine that for all $j < k$, the goal is to compute a confidence interval for $\mu_{tj} - \mu_{tk}$ such that the simultaneous probability coverage is $1 - \alpha$ and the length of each confidence interval is $2m$, some value specified by the researcher. Let h be the $1 - \alpha$ quantile of the range of J independent Student t variates having degrees of freedom $h_1 - 1, \ldots, h_J - 1$, respectively, where $h_j = n_j - 2g_j - 1$ is the number of observations in the jth group left after trimming. Table 7.11 reports the $1 - \alpha$ quantiles of this distribution for selected degrees of freedom, $\alpha = 0.05$ and 0.01, when $h_1 - 1 = \cdots = h_J - 1 = \nu$, say. (For $\nu > 59$, the quantiles can be approximated with the quantiles of a Studentized range statistic with ν degrees of freedom.) For unequal sample sizes, a good choice for the degrees of freedom is

$$\nu = J \left(\sum \frac{1}{h_j - 1} \right)^{-1}.$$

Let

$$d = \left(\frac{m}{h} \right)^2.$$

The total number of observations needed from the jth group is

$$N_j = \max \left[n_j, \left(\frac{s_{jw}^2}{(1 - 2\gamma)^2 d} \right) + 1 \right]. \tag{7.8}$$

So if $N_j - n_j$ is large, and if we fail to reject any hypothesis involving μ_{tj}, it seems unreasonable to accept the null hypothesis.

If the additional $N_j - n_j$ observations can be obtained, confidence intervals are computed as follows. Let $\hat{\mu}_{jt}$ be the trimmed mean associated with the jth group based on all N_j values. For all pairwise comparisons, the confidence interval for $\mu_{jt} - \mu_{kt}$, the difference between the population trimmed means corresponding to groups j and k, is

$$(\hat{\mu}_{jt} - \hat{\mu}_{kt}) \pm hb,$$

Table 7.11: Percentage Points, h, of the Range of J Independent t Variates.

α	$v=5$	$v=6$	$v=7$	$v=8$	$v=9$	$v=14$	$v=19$	$v=24$	$v=29$	$v=39$	$v=59$
					$J=2$ Groups						
0.05	3.63	3.45	3.33	3.24	3.18	3.01	2.94	2.91	2.89	2.85	2.82
0.01	5.37	4.96	4.73	4.51	4.38	4.11	3.98	3.86	3.83	3.78	3.73
					$J=3$ Groups						
0.05	4.49	4.23	4.07	3.95	3.87	3.65	3.55	3.50	3.46	3.42	3.39
0.01	6.32	5.84	5.48	5.23	5.07	4.69	5.54	4.43	4.36	4.29	4.23
					$J=4$ Groups						
0.05	5.05	4.74	4.54	4.40	4.30	4.03	3.92	3.85	3.81	3.76	3.72
0.01	7.06	6.40	6.01	5.73	5.56	5.05	4.89	4.74	4.71	4.61	4.54
					$J=5$ Groups						
0.05	5.47	5.12	4.89	4.73	4.61	4.31	4.18	4.11	4.06	4.01	3.95
0.01	7.58	6.76	6.35	6.05	5.87	5.33	5.12	5.01	4.93	4.82	4.74
					$J=6$ Groups						
0.05	5.82	5.42	5.17	4.99	4.86	4.52	4.38	4.30	4.25	4.19	4.14
0.01	8.00	7.14	6.70	6.39	6.09	5.53	5.32	5.20	5.12	4.99	4.91
					$J=7$ Groups						
0.05	6.12	5.68	5.40	5.21	5.07	4.70	4.55	4.46	4.41	4.34	4.28
0.01	8.27	7.50	6.92	6.60	6.30	5.72	5.46	5.33	5.25	5.16	5.05
					$J=8$ Groups						
0.05	6.37	5.90	5.60	5.40	5.25	4.86	4.69	4.60	4.54	4.47	4.41
0.01	8.52	7.73	7.14	6.81	6.49	5.89	5.62	5.45	5.36	5.28	5.16
					$J=9$ Groups						
0.05	6.60	6.09	5.78	5.56	5.40	4.99	4.81	4.72	4.66	4.58	4.51
0.01	8.92	7.96	7.35	6.95	6.68	6.01	5.74	5.56	5.47	5.37	5.28
					$J=10$ Groups						
0.05	6.81	6.28	5.94	5.71	5.54	5.10	4.92	4.82	4.76	4.68	4.61
0.01	9.13	8.14	7.51	7.11	6.83	6.10	5.82	5.68	5.59	5.46	5.37

Reprinted, with permission, from R. Wilcox, "A table of percentage points of the range of independent t variables", *Technometrics*, 1983, 25, 201–204.

where

$$b = \max\left(\frac{s_{jw}}{(1-2\gamma)\sqrt{N_j}}, \frac{s_{kw}}{(1-2\gamma)\sqrt{N_k}}\right).$$

7.4.10 R Function hochberg

The two-stage method for trimmed means, just described, is performed by the R function

hochberg(x,x2=NA,cil=NA,crit=NA,con=0,tr=0.2,alpha=0.05,iter=10,000)

The first stage data are assumed to be stored in x, and if the second stage data are available, they are stored in x2, either in a matrix (having J columns) or in list mode. The argument cil is $2m$, the desired length of the confidence intervals. The argument crit is the $1-\alpha$ quantile of the range of independent Student t variates, some of which are reported in Table 7.11. If crit is not specified, the appropriate quantile is approximated via simulations with the number of replications controlled by the argument iter. As usual, the default amount of trimming, indicated by the argument tr, is 20%. (The function can also handle linear contrasts specified by the argument con, which is used as described, for example, in Section 7.4.1.)

7.4.11 Explanatory Measure of Effect Size

There are various ways one might characterize the differences among groups when dealing with a two-way ANOVA. Of course, measures of location can be used. If it is desired to use a measure of effect size that is based in part on some measure of variation among the groups, some extension of the explanatory measure of effect size might be used. First focus on Factor A. A simple approach is to ignore the levels for Factor B and use the explanatory measure of effect size previously discussed in Section 5.3.4. That is, for each level of Factor A, pool the data over the levels of Factor B, in which case for any two levels of Factor A, the explanatory measure of effect size can be computed. Of course, the same can be done for Factor B.

As for interactions, first focus on the simplest case: a 2-by 2 design. One possibility is to use the explanatory measure of effect size applied to the two distributions corresponding to the differences associated with each row. That is, for level 1 of Factor A, let F_1 be the distribution of the difference between randomly sampled observations from levels 1 and 2 of Factor B. Define F_2 for level 2 of Factor A in a similar manner. Then compute the explanatory measure of effect size based on estimates of F_1 and F_2. To elaborate, let X_{ijk} ($i = 1, \ldots, n_{jk}$; $j = 1, 2$, $k = 1, 2$) be the ith observation corresponding to the jth level of Factor A and the kth level of Factor B. Let $D_{ii'j} = X_{ij1} - X_{i'j2}$, where now $i = 1, \ldots n_{j1}$, $i' = 1, \ldots n_{j2}$ and $j = 1, 2$. Then the magnitude of the interaction can be characterized via the difference scores for Level 1 of Factor A and Level 2 of Factor A by computing the explanatory measure of effect size based on the two sets of data, $D_{ii'1}$ and $D_{ii'2}$. For the general case of a J-by-K design, this measure of effect size can be computed for levels j and j' of Factor A, and levels k and k' of Factor B, for all $j < j'$ and $k < k'$.

7.4.12 R Functions ESmainMCP and eslmcp

For all j and j' such that $1 \leq j < j' \leq J$, the R function

$$\text{ESmainMCP(J,K,x,tr=0.2,nboot=100,SEED=T)}$$

computes the explanatory measure of effect size for levels j and j' for Factor A as described in the previous section. That is, all pairwise comparisons are made among the J levels. The same is done for Factor B. The R function

$$\text{eslmcp(J,K,x,tr=0.2,nboot=100,SEED=T)}$$

computes the explanatory measure of effect size for interactions. Briefly, the function generates all of the relevant linear contrasts via the R function con2way and then, for each column of the resulting matrix of contrast coefficients, the corresponding explanatory measure of effect size is estimated.

7.5 A Random Effects Model for Trimmed Means

This section describes two random effects models based on Winsorization and trimmed means. When there is no trimming, and there is homogeneity of variance, both models reduce to the usual model covered in a basic course on the analysis of variance. The first of the two models is convenient when comparing trimmed means, while the other is covenient when estimating a Winsorized analog of the intraclass correlation coefficient.

It is well known that the standard random effects model provides poor control over the probability of a type I error when the usual assumptions of normality and homogeneous variances are violated. For example, when testing at the $\alpha = 0.05$ level with four groups and equal sample sizes of 20 in each group, the actual probability of a type I error can exceed .3 (Wilcox, 1994b). A striking feature of the model based on 20% trimming is the extent to which this problem is reduced. Among the various distributions considered by Wilcox (1994b), the highest estimated probability of a type I error was .074.

It is assumed that there is a pool of treatment groups that are of interest, but not all groups can be examined. Instead, J randomly sampled groups are used to make inferences about the pool of treatment groups. Once the J groups are randomly sampled, it is assumed that n_j observations are randomly sampled from the jth group. Let X_{ij} be the ith observation randomly sampled from the jth group, and let μ_{tj} be the population trimmed mean. Generalizing the standard random effects model in a natural way, let $\bar{\mu}_w = E_w(\mu_{tj})$. In words, $\bar{\mu}_w$ is the Winsorized mean for the population of trimmed means being sampled. A generalization of the usual random effects model is

$$X_{ij} = \bar{\mu}_w + b_j + \epsilon_{ij},$$

where $b_j = \mu_{tj} - \bar{\mu}_w$, $E_w(b_j) = 0$, and $E_w(\epsilon_{ij}) = 0$. Let the Winsorized variance of b_j be σ_{wb}^2, and for fixed j, let σ_{wj}^2 be the Winsorized variance of ϵ_{ij}. Also let $\sigma_w^2 = E_w(\sigma_{wj}^2)$, where the Winsorized expectation is taken with respect to a randomly sampled group. When there are no differences among the trimmed means associated with the pool of treatment groups under investigation, $\sigma_{wb}^2 = 0$.

Jeyaratnam and Othman (1985) derived a heteroscedastic method for comparing means in a random effects model. It can be extended to trimmed means using Winsorized expected

Table 7.12: Comparing Trimmed Means in a Random Effects Model.

For the jth group, Winsorize the observations by computing Y_{ij} as described in Section 7.5.1. To test the hypothesis of no differences among the trimmed means, $H_0 : \sigma_{wb}^2 = 0$, let h_j the effective sample size of the jth group (the number of observations left after trimming), and compute

$$\bar{Y}_j = \frac{1}{n_j} \sum_{i=1}^{n_j} Y_{ij},$$

$$s_{wj}^2 = \frac{1}{n_j - 1} \sum (Y_{ij} - \bar{Y}_j)^2,$$

$$\bar{X}_t = \frac{1}{J} \sum \bar{X}_{tj},$$

$$\text{BSST} = \frac{1}{J-1} \sum_{j=1}^{J} (\bar{X}_{tj} - \bar{X}_t)^2,$$

$$\text{WSSW} = \frac{1}{J} \sum_{j=1}^{J} \sum_{i=1}^{n_j} \frac{(Y_{ij} - \bar{Y}_j)^2}{h_j(h_j - 1)},$$

$$D = \frac{\text{BSST}}{\text{WSSW}}.$$

Let

$$q_j = \frac{(n_j - 1)s_{wj}^2}{J(h_j)(h_j - 1)}.$$

The degrees of freedom are estimated to be

$$\hat{v}_1 = \frac{[(J-1)\sum q_j]^2}{(\sum q_j)^2 + (J-2)J\sum q_j^2}$$

$$\hat{v}_2 = \frac{(\sum q_j)^2}{\sum q_j^2/(h_j - 1)}.$$

Reject if $D > f$, the $1 - \alpha$ quantile of an F distribution with $\hat{v}_1$ and $\hat{v}_2$ degrees of freedom.

values (Wilcox, 1994b). The computations are summarized in Table 7.12. Advantages of using trimmed means over means are better control over the probability of a type I error and the potential of substantially higher power, particularly when distributions have heavier than normal tails. However, there are situations where comparing means might mean more power as well. For example, the variation among the means might be larger than the variation among the trimmed means, so the Jeyaratnam and Othman method might yield more power. This might happen, for example, when distributions are skewed. As usual, the optimal amount of trimming will vary from one situation to the next, but to avoid poor power and undesirable power characteristics (a biased test), 20% trimming is a good choice for general use.

7.5.1 A Winsorized Intraclass Correlation

This subsection describes a Winsorized analog of the usual intraclass correlation. Several estimators of this parameter have been considered (Wilcox, 1994c), one of which is described here.

It is convenient to switch from the ANOVA model in the previous section to one that is slightly different. As before, the model is written as

$$X_{ij} = \bar{\mu}_w + b_j + \epsilon_{ij},$$

but now $b_j = \mu_{wj} - \bar{\mu}_w$, the difference between the Winsorized mean of the jth group and $\bar{\mu}_w = E_w(\mu_{wj})$, the Winsorized expected value of μ_{wj} with respect to a randomly sampled group. It can be shown that the Winsorized covariance between any two observations in the jth group, X_{ij} and $X_{i'j}$, $i \neq i'$, is σ_{wb}^2. Also, the Winsorized variance of X_{ij} is $\sigma_{wb}^2 + \bar{\sigma}^2$, where $\bar{\sigma}^2 = E_w(\sigma_{wj}^2)$, the Winsorized expected value of the Winsorized variance associated with a randomly sampled group. Then a heteroscedastic, Winsorized analog of the usual intraclass correlation is

$$\rho_{WI} = \frac{\sigma_{wb}^2}{\sigma_{wb}^2 + \bar{\sigma}^2}.$$

From a technical point of view, Winsorization is a convenient way to proceed because there is a relatively simple way of dealing with estimation problems. To briefly review a result mentioned in Chapter 2, if $g(X_1, \ldots, X_n)$ is any function of the random sample $X_1, \ldots, X_n$, and

$$E_w\{g(X_1, \ldots, X_n)\} = \xi,$$

then $g(Y_1, \ldots Y_n)$ estimates ξ, where

$$Y_i = \begin{cases} X_{(g+1)}, & X_i \leq X_{(g+1)} \\ X_i, & X_{(g+1)} < X_i < X_{(n-g)} \\ X_{(n-g)}, & X_i \geq X_{(n-g)} \end{cases}$$

For the ith observation in the jth group, let

$$Y_{ij} = \begin{cases} X_{(g+1)j}, & X_{ij} \leq X_{(g+1)j} \\ X_{ij}, & X_{(g+1)j} < X_{ij} < X_{(n-g)j} \\ X_{(n-g)j}, & X_{ij} \geq X_{(n-g)j}. \end{cases}$$

That is, Winsorize the observations in the jth group. Let s_{wj}^2 be the Winsorized sample variance for the jth group, let $m = [\gamma J]$, where as usual, γ is the amount of Winsorization, let $s_{w(1)}^2 \leq \cdots \leq s_{w(J)}^2$ be the Winsorized sample variances written in ascending order, let

$$SW = (m+1)s_{w(m+1)}^2 + s_{w(m+2)}^2 + \cdots + s_{w(J-m-1)}^2 + (m+1)s_{w(J-m)}^2,$$

and

$$\bar{s}_w^2 = \frac{\text{SW}}{J}.$$

Results in Rao, Kaplan, & Cochran (1981) suggest estimating ρ_{WI} with

$$\hat{\rho}_{\text{WI}} = \frac{\hat{\sigma}_{wb}^2}{\hat{\sigma}_{wb}^2 + \bar{s}_w^2},$$

where

$$\hat{\sigma}_{wb}^2 = \frac{1}{J} \sum \ell_j (\bar{Y}_j - \tilde{Y})^2,$$

$$\ell_j = \frac{n_j}{n_j + 1}$$

and

$$\tilde{Y} = \frac{\sum \ell_j \bar{Y}_j}{\sum \ell_j}.$$

Wilcox (1994c) compared the bias and mean squared error of $\hat{\rho}_{\text{WI}}$ to two alternative estimators and recommended $\hat{\rho}_{\text{WI}}$. When ρ_{WI} is close to zero, some type of bias reduction method might have practical value, but this issue needs further study before a recommendation can be made.

7.5.2 R Function *rananova*

The R function

$$\text{rananova(x,tr=0.2,grp=NA)}$$

performs the computations for the random effects ANOVA, where x is any R variable that is a data frame, or matrix, or has list mode, tr is the amount of trimming, which defaults to 0.2, and grp can be used to specify some subset of the groups if desired. If grp is not specified, all groups stored in x are used. The function returns the value of the test statistic, D, which is stored in rananova$teststat, the significance level is stored in rananova$siglevel, and an estimate of the Winsorized intraclass correlation, which is computed as described in the previous subsection of this chapter, is returned in the R variable rananova$rho.

■ Example

Assuming the data in Table 7.7 are stored in the R variable data, the command rananova(data) returns.

```
$teststat:
[1] 0.33194
```

```
$df:
[1]   2.520394 18.693590

$siglevel:
[1] 0.7687909

$rho:
[1] 0.0576178
```

■

7.6 Global Tests Based on M-Measures of Location

This section describes two bootstrap methods for comparing J independent groups that appear to perform relatively well when the goal is to test global hypotheses about robust measures of location. Here the emphasis is on M-estimators, but the methods described here perform well when using a one-step M-estimator and trimmed means, provided the amount of trimming is not too small. An exception is situations where the goal is to compare medians and there are tied values. (Non-bootstrap methods, based in part on some estimate of the standard error of an M-estimator, can perform poorly when dealing with skewed distributions. But some bootstrap methods that use estimates of standard errors seem to have practical value.) In more formal terms, the goal is to test

$$H_0 : \theta_1 = \cdots = \theta_J, \tag{7.9}$$

there θ_j represents a (population) M-measure of location associated with the jth group.

The first method is based on a test statistic mentioned by Schrader and Hettmansperger (1980), and studied by He, Simpson, and Portnoy (1990). The test statistic is

$$H = \frac{1}{N} \sum n_j (\hat{\theta}_j - \bar{\theta})^2,$$

where $N = \sum n_j$, and

$$\bar{\theta} = \frac{1}{J} \sum \hat{\theta}_j.$$

To determine the critical value, shift the empirical distributions of each group so that the estimated measure of location is zero, generate bootstrap samples from each group in the usual way from each of the shifted distributions, and compute the test statistic based on the bootstrap samples yielding H^*, say. Repeat this B times resulting in $H_1^*, \ldots, H_B^*$, and put these B values in order yielding $H_{(1)}^* \leq \cdots \leq H_{(B)}^*$. Then an estimate of an appropriate critical value is $H_{(u)}^*$, where $u = (1 - \alpha)B$, rounded to the nearest integer, and H_0 is rejected if $H > H_{(u)}^*$. (For simulation results on how this method performs, see Wilcox, 1993d.) This method appears to give reasonably good results when using an M-estimator (with Huber's Ψ), as well as the Harrell–Davis estimate of the median, but it is not recommended when comparing means or even trimmed means.

The second bootstrap method described here is based on a slight variation of a general approach described by Liu and Singh (1997). Currently, it appears to be the better of the two methods described in this section when working with the modified one step M-estimator (MOM), described in Section 3.10 (Keselman, Wilcox, Othman, and Fradette, 2002), and it appears to perform reasonably well when using M-estimators (with Huber's Ψ) and even trimmed means if the amount of trimming is sufficiently high. Let

$$\delta_{jk} = \theta_j - \theta_k,$$

where for convenience it is assumed that $j < k$. That is, the δ_{jk} values represent all pairwise differences among the J groups. When working with means, for example, δ_{12} is the difference between the means of groups 1 and 2, and δ_{35} is the difference for groups 3 and 5. If all J groups have a common measure of location (i.e., $\theta_1 = \cdots = \theta_J$), then in particular

$$H_0 : \delta_{12} = \delta_{13} = \cdots = \delta_{J-1,J} = 0 \tag{7.10}$$

is true. The total number of δ's in Eq. (7.10) is $L = (J^2 - J)/2$.

For each group, generate bootstrap samples from the *original* values. That is, the observations are *not* centered as was done in the previous method. Instead bootstrap samples are generated from the X_{ij} values. For each group, compute the measure of location of interest based on a bootstrap sample and repeat this B times. The resulting estimates of location are represented by $\hat{\theta}^*_{jb}$ ($j = 1, \ldots, J; b = 1, \ldots, B$) and the corresponding estimates of δ are denoted by $\hat{\delta}^*_{jkb}$. (That is, $\hat{\delta}^*_{jkb} = \hat{\theta}^*_{jb} - \hat{\theta}^*_{kb}$.) The general strategy is to determine how deeply $\mathbf{0} = (0, \ldots, 0)$ is nested within the bootstrap values $\hat{\delta}^*_{jkb}$ (where $\mathbf{0}$ is a vector having length L). For the special case where only two groups are being compared, this is tantamount to determining the proportion of times $\hat{\theta}^*_{1b} > \hat{\theta}^*_{2b}$, among all B bootstrap samples, which is how we proceeded in Chapter 5. But here we need special techniques for comparing more than two groups.

There remains the problem of measuring how deeply $\mathbf{0}$ is nested within the bootstrap values. Several strategies were described in Chapter 6, but in terms of type I error probabilities and power, all indications are that, for the situation at hand, the choice of method is irrelevant. However, from a computational point of view, the choice of method can matter, for reasons indicated at the end of this section. For the moment, the focus is on using Mahalanobis distance.

Let $\hat{\delta}_{jk} = \hat{\theta}_j - \hat{\theta}_k$ be the estimate of δ_{jk} based on the original data and let $\hat{\delta}^*_{jkb} = \hat{\theta}^*_{jb} - \hat{\theta}^*_{kb}$ based on the bth bootstrap sample ($b = 1, \ldots, B$). (It is assumed that $j < k$.) For notational convenience, we rewrite the $L = (J^2 - J)/2$ differences $\hat{\delta}_{jk}$ as $\hat{\Delta}_1, \ldots, \hat{\Delta}_L$ and the corresponding bootstrap values are denoted by $\hat{\Delta}_{\ell b}^*$ $(\ell = 1, \ldots, L)$. Let

$$\bar{\Delta}_\ell^* = \frac{1}{B} \sum_{b=1}^{B} \hat{\Delta}_{\ell b}^*,$$

$$Y_{\ell b} = \hat{\Delta}_{\ell b}^* - \bar{\Delta}_\ell^* + \hat{\Delta}_\ell,$$

(so the $Y_{\ell b}$ values are the bootstrap values shifted to have mean $\hat{\Delta}_\ell$) and let

$$S_{\ell m} = \frac{1}{B-1} \sum_{b=1}^{B} (Y_{\ell b} - \bar{Y}_\ell)(Y_{mb} - \bar{Y}_m),$$

where

$$\bar{Y}_\ell = \frac{1}{B} \sum_{b=1}^{B} Y_{\ell b}.$$

(Note that in the bootstrap world, the bootstrap population mean of $\bar{\Delta}_\ell^*$ is known and is equal to $\hat{\Delta}_\ell$.) Next, compute

$$D_b = (\hat{\mathbf{\Delta}}_b^* - \hat{\mathbf{\Delta}})\mathbf{S}^{-1}(\hat{\mathbf{\Delta}}_b^* - \hat{\mathbf{\Delta}})',$$

where $\hat{\mathbf{\Delta}}_b^* = (\hat{\Delta}_{1b}^*, \ldots, \hat{\Delta}_{Lb}^*)$ and $\hat{\mathbf{\Delta}} = (\hat{\Delta}_1, \ldots, \hat{\Delta}_L)$. D_b measures how closely $\hat{\mathbf{\Delta}}_b^*$ is located to $\hat{\mathbf{\Delta}}$. If $\mathbf{0}$ (the null vector) is relatively far from $\hat{\mathbf{\Delta}}$, reject. In particular, put the D_b values in ascending order yielding $D_{(1)} \leq \cdots \leq D_{(B)}$ and let $u = (1 - \alpha)B$, rounded to the nearest integer. Then reject H_0 if

$$T \geq D_{(u)},$$

where

$$T = (\mathbf{0} - \hat{\mathbf{\Delta}})\mathbf{S}^{-1}(\mathbf{0} - \hat{\mathbf{\Delta}})'.$$

A *p*-value can be computed as well and is given by

$$\frac{1}{B} \sum_{b=1}^{B} I_b,$$

where the indicator function $I_b = 1$ if $T < D_b$; otherwise $I_b = 0$.

Notice that with three groups ($J = 3$), $\theta_1 = \theta_2 = \theta_3$ can be true if and only if $\theta_1 = \theta_2$ and $\theta_2 = \theta_3$. So in terms of type I errors, it suffices to test

$$H_0 : \theta_1 - \theta_2 = \theta_2 - \theta_3 = 0$$

as opposed to testing

$$H_0 : \theta_1 - \theta_2 = \theta_2 - \theta_3 = \theta_1 - \theta_3 = 0,$$

the hypothesis that all pairwise differences are zero. However, if groups differ, then rearranging the groups could alter the conclusions reached if the first of these hypotheses is tested. For example, if the groups have means 6, 4, and 2, then the difference between groups one and two, as well as two and three, is 2. But the difference between groups one and three is 4, so comparing groups one and three could mean more power. That is, we might not reject when comparing group one to two and two to three, but we might reject if instead we compare one to three and two to three. To help avoid different conclusions depending on how the groups are arranged, all pairwise differences among the groups were used. However, a consequence of using all pairwise differences is that situations are encountered where the covariance matrix, used when computing Mahalanobis distance, is singular. This problem can be avoided by replacing Mahalanobis distance with the projection distance described in Section 6.2.5.

7.6.1 R Functions b1way and pbadepth

The R function b1way performs the first of the percentile bootstrap methods described in the previous subsection. It has the general form

b1way(x,est=onestep,alpha=0.05,nboot=599)

where x is any R variable that is a matrix (with J columns) or has list mode, alpha defaults to 0.05, and nboot, the value of B, defaults to 599. The argument est is any R function that computes a measure of location. It defaults to onestep, the one-step M-estimator with Huber's Ψ.

The function

pbadepth(x,est=onestep,con=0,alpha=0.05,nboot=2000,grp=NA,op=1,allp=T,
         MM=F,MC=F,cop=3,SEED=T,na.rm=F,...)

performs the other percentile bootstrap method and uses the one-step M-estimator by default. As usual, the argument ... can be used to reset default settings associated with the estimator being used. The argument allp indicates how the null hypothesis is defined. Setting allp=T, all pairwise differences are used. Setting allp=F, the function tests

$$H_0 : \theta_1 - \theta_2 = \theta_2 - \theta_3 = \cdots = \theta_{J-1} - \theta_J = 0.$$

The argument op determines how the depth of a point is measured within a bootstrap cloud. The choices are:

- op=1, Mahalanobis depth.
- op=2, Mahalanobis depth but with the usual covariance matrix replaced by the MCD estimate.
- op=3, projection depth computed via the function pdis (That is, use the measure of depth described in Section 6.2.5.)

The default is op=1 to reduce execution time. Using op=3 avoids a computational error that can occur when the argument allp $= $ T: in some situations, $\mathbf{S}^{-1}$ (as described in the previous section) cannot be computed. This event appears to be rare with $J \leq 4$, but it can occur. This problem might be avoided by setting allp=F, but in terms of power, this has practical concerns already described. With op=3, a measure of depth is used that does not require inverting a matrix. Given the speed of modern computers, perhaps using op=3 routinely is reasonable. If access to a multicore processor is available, setting the argument MC=T will reduce execution time. (Another option is to use the multiple comparison procedure in Section 7.6.2, which again does not require inverting a matrix.)

■ Example

The command b1way(x,est=hd) would test the hypothesis of equal medians using the Harrell–Davis estimator. For the data in Table 7.7, it reports a test statistic of $H = 329$ and a critical value of 3627. The command b1way(x) would compare M-measures of location instead.

■

7.6.2 M-estimators and Multiple Comparisons

This section describes two bootstrap methods for performing multiple comparisons that perform relatively well when comparing M-estimators. The first is based in part on a variation of bootstrap-t method that uses bootstrap estimates of the standard errors. (For relevant simulation results, see Wilcox, 1993d.) The other does not use estimated standard errors but rather relies on a variation of the percentile bootstrap method.

Variation of a Bootstrap-t Method

For the jth group, generate a bootstrap sample in the usual way and compute $\hat{\mu}^*_{mj}$, the M-estimate of location. Repeat this B times yielding $\hat{\mu}^*_{mjb}$, $b = 1, \ldots B$. Let

$$\hat{\tau}^2_j = \frac{1}{B-1} \sum_{b=1}^{B} (\hat{\mu}^*_{mjb} - \bar{\mu}^*)^2,$$

where $\bar{\mu}^* = \sum_b \hat{\mu}^*_{mjb} / B$. Let

$$H^*_{jkb} = \frac{|\hat{\mu}^*_{mjb} - \hat{\mu}^*_{mkb} - (\hat{\mu}_{mj} - \hat{\mu}_{mk})|}{\sqrt{\hat{\tau}^2_j + \hat{\tau}^2_k}},$$

and

$$H^*_b = \max H^*_{jkb},$$

where the maximum is taken over all $j < k$. Put the H^*_b values in order yielding $H^*_{(1)} \leq \cdots \leq H^*_{(B)}$. Let $u = (1 - \alpha)B$, rounded to the nearest integer. Then a confidence interval for $\mu_{mj} - \mu_{mk}$ is

$$(\hat{\mu}_{mj} - \hat{\mu}_{mk}) \pm H^*_{(u)} \sqrt{\hat{\tau}^2_j + \hat{\tau}^2_k},$$

and the simultaneous probability coverage is approximately $1 - \alpha$. With $\alpha = 0.05$, it seems that $B = 399$ gives fairly good probability coverage when all of the sample sizes are greater than or equal to 21.

The same method appears to perform well when using the Harrell–Davis estimate of the median. The extent to which it performs well when estimating other quantiles has not been determined.

Linear contrasts can be examined using a simple extension of the method for performing all pairwise comparisons. Let

$$\Psi_k = \sum_{j=1}^{J} c_{jk} \mu_{mj},$$

$k = 1, \ldots, C$, be C linear combinations of the M-measures of location. As in Section 7.4, the constants c_{jk} are chosen to reflect linear contrasts that are of interest, and for fixed k, $\sum c_{jk} = 0$. Included as a special case is the situation where all pairwise comparisons are to be performed. As before, generate bootstrap samples yielding $\hat{\mu}^*_{mjb}$, $b = 1, \ldots, B$, and $\hat{\tau}^2_j$, $j = 1, \ldots, J$. Let

$$\hat{\Psi}_k = \sum_{j=1}^{J} c_{jk} \hat{\mu}_{mj},$$

$$\hat{\Psi}^*_{kb} = \sum_{j=1}^{J} c_{jk} \hat{\mu}^*_{mjb},$$

and

$$H^*_{kb} = \frac{|\hat{\Psi}^*_{kb} - \hat{\Psi}_k|}{\sum c_{jk}^2 \hat{\tau}^2_j}.$$

Let

$$H^*_b = \max H^*_{kb},$$

where the maximum is taken over all k, $k = 1, \ldots, C$. Then a confidence interval for Ψ_k is

$$\hat{\Psi}_k \pm H^*_{(u)} \sqrt{\sum c_{jk}^2 \hat{\tau}^2_k},$$

where again $u = (1 - \alpha)B$, rounded to the nearest integer.

A Percentile Bootstrap Method: Method SR

When comparing modified one-step M-estimators, or M-estimators, if the sample sizes are small, an alternative bootstrap method appears to compete well with the method just described. Imagine that hypotheses for each of C linear contrasts are to be tested. For the cth hypothesis, let $2\hat{p}^*_c$ be the usual percentile bootstrap estimate of the p-value. Put the $\hat{p}^*_c$ values in *descending* order yielding $\hat{p}^*_{[1]} \geq \hat{p}^*_{[2]} \geq \cdots \geq \hat{p}^*_{[C]}$. Decisions about the individual hypotheses are made as follows. If $\hat{p}^*_{[1]} \leq \alpha_1$, where α_1 is read from Table 7.13, reject all C of the hypotheses. Put another way, if the largest estimated p-value, $2\hat{p}^*_{[1]}$, is less than or equal to α, reject all C hypotheses. If $\hat{p}^*_{[1]} > \alpha_1$, but $\hat{p}^*_{[2]} \leq \alpha_2$, fail to reject the hypothesis associated with $\hat{p}^*_{[1]}$, but the remaining hypotheses are rejected. If $\hat{p}^*_{[1]} > \alpha_1$ and $\hat{p}^*_{[2]} > \alpha_2$, but $\hat{p}^*_{[3]} \leq \alpha_3$, fail to reject the hypotheses associated with $\hat{p}^*_{[1]}$ and $\hat{p}^*_{[2]}$, but reject the remaining hypotheses. In general, if $\hat{p}^*_{[c]} \leq \alpha_c$, reject the corresponding hypothesis and all other hypotheses having smaller $\hat{p}^*_m$ values. For other values of α (assuming $c > 1$) or for $c > 10$, use

$$\alpha_c = \frac{\alpha}{c}$$

(which corresponds to a slight modification of a sequentially rejective method derived by Hochberg, 1988.) This will be called *Method SR*.

Table 7.13: Values of α_c for $\alpha = 0.05$ and 0.01.

c	$\alpha = 0.05$	$\alpha = 0.01$
1	0.02500	0.00500
2	0.02500	0.00500
3	0.01690	0.00334
4	0.01270	0.00251
5	0.01020	0.00201
6	0.00851	0.00167
7	0.00730	0.00143
8	0.00639	0.00126
9	0.00568	0.00112
10	0.00511	0.00101

Method SR, just described, has the advantage of providing type I error probabilities close to the nominal level for a fairly wide range of distributions. It is *not* recommended, however, when the sample sizes are reasonably large, say greater than about 80. Method SR does not conform to any known multiple comparison procedure; it represents a slight modification of a method derived by Rom (1990) that was designed for small sample sizes. But as the sample sizes get large, the actual familywise error rate (FWE), appears to converge to a value greater than the nominal level. Consequently, if any sample size is greater than 80, use the first of the two methods outlined here, or a percentile bootstrap with Hochberg's method, or perhaps the Benjamini–Hochberg method, which are described in Section 7.4.7.

7.6.3 R Functions linconm and pbmcp

The R function

$$\text{linconm}(x,\text{con}=0,\text{est}=\text{mest},\text{alpha}=0.05,\text{nboot}=399,...)$$

computes confidence intervals for measures of location using the first of the methods described in the previous subsection. As usual, x is any R variable that is a matrix or has list mode, and con is a J-by-C matrix containing the contrast coefficients of interest. If con is not specified, all pairwise comparisons are performed. The argument est is any estimator which defaults to the one-step M-estimator if unspecified. Again alpha is α and defaults to 0.05, and nboot is B which defaults to 399. The final argument, ... , can be any additional arguments required by the argument est. For example, linconm(w) will use the data stored in the R variable w to compute confidence intervals for all pairwise differences between M-measures of location. The command linconm(w,est=hd) will compute confidence intervals for medians based on the Harrell–Davis estimator, while linconm(w,est=hd,q=0.7) computes confidence intervals for the difference between the 0.7 quantiles, again using the Harrell–Davis estimator.

The command linconm(w,bend=1.1) would use an M-estimator, but the default value for the bending constant in Huber's Ψ would be replaced by 1.1.

The function

$$\text{pbmcp}(x, \text{alpha} = 0.05, \text{nboot} = NA, \text{grp} = NA, \text{est} = mom, \text{con} = 0, \text{bhop} = F, ...)$$

performs multiple comparisons using method SR described in the previous section. (Method SR should not be used when comparing trimmed means.) By default, all pairwise comparisons are performed, but a collection of linear contrasts can be specified via the argument con which is used as illustrated in Section 7.4.1. With bhop=F, method SR is used, and setting bhop=T, the Benjamini–Hochberg method is applied instead, which is described in Section 7.4.7.

7.6.4 M-Estimators and the Random Effects Model

Little has been done to generalize the usual random effects model to M-estimators. The approach based on the Winsorized expected value does not readily extend to M-estimators unless restrictive assumptions are made. Bansal and Bhandry (1994) consider M-estimation of the intraclass correlation coefficient, but they assume sampling is from an elliptical and permutationally symmetric probability density function.

7.6.5 Other Methods for One-Way Designs

The methods described in this section are far from exhaustive. For completeness, it is noted that Keselman, Wilcox, Othman, and Fradette (2002) compared 56 methods based on means, trimmed means with various amount of trimming, and even asymmetric trimming, and two methods based on MOM. Some of the more successful methods, as measured by the ability to control the probability of a type I error, were based on trimmed means used in conjunction with transformations studied by Hall (1992) and Johnson (1978). More results supporting the use of these transformation can be found in Guo and Luh (2000) and Luh and Guo (1999).

7.7 M-Measures of Location and a Two-Way Design

As was the case when dealing with one-way designs, comparing M-measures of location in a two-way design requires, at the moment, some type of bootstrap method to control the probability of a type I error, at least when the sample sizes are small. (There are no results on how large the sample sizes must be to avoid the bootstrap.) The method described here was initially used with M-measures of location, but it can be applied when comparing any measure of location, including trimmed means and the modified one-step M-estimator (MOM).

Let θ be any measure of location and let

$$\Upsilon_1 = \frac{1}{K}(\theta_{11} + \theta_{12} + \cdots + \theta_{1K}),$$

$$\Upsilon_2 = \frac{1}{K}(\theta_{21} + \theta_{22} + \cdots + \theta_{2K}),$$

$$\vdots$$

$$\Upsilon_J = \frac{1}{K}(\theta_{J1} + \theta_{J2} + \cdots + \theta_{JK}).$$

So Υ_j is the average of the K measures of location associated with the jth level of Factor A. The hypothesis of no main effects for Factor A is

$$H_0 : \Upsilon_1 = \Upsilon_2 = \cdots = \Upsilon_J.$$

and one variation of the percentile bootstrap method is to test this hypothesis using a slight modification of the method in Section 7.6. For example, one possibility is to test

$$H_0 : \Delta_1 = \cdots = \Delta_{J-1} = 0, \tag{7.11}$$

where

$$\Delta_j = \Upsilon_j - \Upsilon_{j+1},$$

$j = 1, \ldots, J - 1$. Briefly, generate bootstrap samples in the usual manner yielding $\hat{\Delta}_j^*$, a bootstrap estimate of Δ_j. Then proceed as described in Section 7.6. That is, determine how deeply $\mathbf{0} = (0, \ldots, 0)$ is nested within the bootstrap samples. If $\mathbf{0}$ is relatively far from the center of the bootstrap samples, reject.

For reasons previously indicated, the method just described is satisfactory when dealing with the probability of a type I error, but when the groups differ, this approach might be unsatisfactory in terms of power depending on the pattern of differences among the Υ_j values. One way of dealing with this issue is to compare all pairs of the Υ_j instead. That is, for every $j < j'$, let

$$\Delta_{jj'} = \Upsilon_j - \Upsilon_{j'},$$

and then test

$$H_0 : \Delta_{12} = \Delta_{13} = \cdots = \Delta_{J-1,J} = 0. \tag{7.12}$$

Of course, a similar method can be used when dealing with Factor B.

Now a test of the hypothesis of no interaction is described. For convenience, label the JK measures of location as follows:

	Factor B		
	θ_1	θ_2	$\cdots$	θ_K
Factor A	θ_{K+1}	θ_{K+2}	$\cdots$	θ_{2K}
	$\vdots$	$\vdots$	$\cdots$	$\vdots$
	$\theta_{(J-1)K+1}$	$\theta_{(J-1)K+2}$	$\cdots$	θ_{JK}

Let $\mathbf{C}_J$ be a $(J-1)$-by-J matrix having the form

$$\begin{pmatrix} 1 & -1 & 0 & 0 & \cdots & 0 \\ 0 & 1 & -1 & 0 & \cdots & 0 \\ & & & \vdots & & \\ 0 & 0 & \cdots & 0 & 1 & -1 \end{pmatrix}.$$

That is, $c_{ii} = 1$ and $c_{i,i+1} = -1$; $i = 1, \ldots, J-1$ and $\mathbf{C}_K$ is defined in a similar fashion. One approach to testing the hypothesis of no interactions is to test

$$H_0 : \Psi_1 = \cdots = \Psi_{(J-1)(K-1)} = 0,$$

where

$$\Psi_L = \sum c_{L\ell}\theta_\ell,$$

$L = 1, \ldots, (J-1)(K-1)$, $\ell = 1, \ldots, JK$, and $c_{L\ell}$ is the entry in the Lth row and ℓth column of $\mathbf{C}_J \otimes \mathbf{C}_K$. So in effect we have a situation similar to that in Section 7.6. That is, generate bootstrap samples yielding $\hat{\Psi}_L^*$ values, do this B times, and then determine how deeply $\mathbf{0} = (0, \ldots, 0)$ is nested within these bootstrap samples.

A criticism of this approach is that when groups differ, not all relevant differences are being tested which might affect power. A strategy for dealing with this problem is for every $j < j'$ and $k < k'$, set

$$\Psi_{jj'kk'} = \theta_{jk} - \theta_{jk'} + \theta_{j'k} - \theta_{j'k'}$$

and then test

$$H_0 : \Psi_{1212} = \cdots = \Psi_{J-1,J,K-1,K} = 0. \tag{7.13}$$

7.7.1 R Functions pbad2way and mcp2a

The R function

$$\text{pbad2way}(J,K,x,\text{est}=\text{mom},\text{conall}=T,\text{alpha}=0.05,\text{nboot}=2000,\text{grp}=NA,...)$$

performs the percentile bootstrap method just described, where J and K indicate the number of levels associated with Factors A and B. The argument conall=T indicates that all possible pairs are to be tested, as described, for example, by Eq. (7.13), and conall=F means that the hypotheses having the form given by Eq. (7.11) will be used instead. The remaining arguments are the same as those used in the function pbadepth described in Section 7.6.1.

■ Example

The data in Exercise 12, at the end of this chapter, are used to illustrate the R function pbad2way. The study involves a 2-by-2 design with weight gain among rats the outcome of interest. The factors are source of protein (beef vs. cereal) and amount of protein (high vs. low). Storing the data in the R variable weight, the command

$$\text{pbad2way}(2,2,\text{weight},\text{est}=\text{median})$$

tests all relevant hypotheses using medians. It is left as an exercise to verify that when using R, the p-values for Factors A and B are .39 and .056, respectively. The test for no interaction has a p-value of .16.

■

For convenience, when working with a two-way design, the function

$$\text{mcp2a}(J,K,x,\text{est}=\text{mom},\text{con}=0,\text{alpha}=0.05,\text{nboot}=NA,\text{grp}=NA,...)$$

is supplied for performing all pairwise comparisons for both factors and all interactions. The arguments are generally the same as those used by pbad2way. One difference is that the number of bootstrap samples is determined by the function unless a value for nboot is specified. Another is that if con=0, all pairwise differences, and all tetrad differences when dealing with interactions, are tested. If a particular set of C linear contrasts is of interest, they can be specified by con, a JK-by-C matrix.

7.8 Ranked-Based Methods for a One-Way Design

This section describes some rank-based methods for a one-way design. The classic method is the Kruskall–Wallis test, which is satisfactory, in terms of controlling the probability of a type I error, when comparing groups having identical distributions. But when the distributions differ, under general conditions an incorrect estimate of the standard errors is being used, which might adversely affect power. A method aimed at improving the Kruskall–Wallis test was derived by Rust and Fligner (1984), assuming that tied values occur with probability zero. The explicit goal stated by Rust and Fligner is to test the hypothesis that J groups have a common median, but under general conditions it fails to do this in a satisfactory manner. Letting $p_{jk} = P(X_{ij} < X_{ik})$, their technique is appropriate for testing the hypothesis that for all J groups, $p_{jk} = .5$. However, their method is based on the assumption that the distributions of the J groups differ in location only. If this assumption is violated, in essence the Rust–Fligner method is testing the hypothesis that the groups have identical distributions. A possible appeal of their method is that it is asymptotically distribution free under weaker conditions than the Kruskall–Wallis test.

A rank-based method that can handle tied values was derived by Brunner, Dette, and Munk (1997). Extensive comparisons with the Rust–Fligner method, when ties occur with probability zero, have not been made. With small sample sizes, the choice of method might make a practical difference, but a detailed study of when this is the case has yet to be performed. Here it is merely remarked that situations can be constructed where, with a common sample size of 50, the choice of method makes a practical difference. For example, the Rust–Fligner method can reject at the .05 level, even though the Brunner et al. method has a p-value equal to .188. Even for normal distributions with unequal variances, the p-values resulting from these two methods can differ substantially.

7.8.1 The Rust–Fligner Method

The basic idea is that if for any x,

$$H_0 : F_1(x) = \cdots = F_J(x),$$

is true, meaning that all J groups have identical distributions, and if ranks are assigned based on the pooled data, then the average ranks among the groups should not differ by too much. Table 7.14 describes how to calculate the test statistic, Q. This method can have relatively good power when sampling from heavy-tailed distributions. Because hypothesis testing methods based on robust measures of location are sensitive to different situations, compared with a method based on the average ranks, the Rust–Fligner method can have more power than methods based on robust measures of location, but there are situations where the reverse is true. That is, in terms of maximizing power, the choice of method depends on how the groups differ, which is not known.

Table 7.14: How to Compute the Rust–Fligner Test Statistic.

Let $V(x) = 1$ if $x \geq 0$, otherwise $V(x) = 0$. Let

$$R_{ij} = \sum_{\ell=1}^{J} \sum_{m=1}^{n_\ell} V(X_{ij} - X_{m\ell})$$

be the rank of X_{ij} among the pooled observations. Let

$$\bar{R}_{.j} = \sum_i R_{ij}/n_j, \quad U_j = \frac{n_j}{N(N+1)}(\bar{R}_{.j} - \bar{R}),$$

where $\bar{R}$ is the average of all the ranks. Let

$$P_{ij\ell} = \sum_{m=1}^{n_\ell} V(X_{ij} - X_{m\ell}), \quad T_{ij} = \sum_{\ell, \ell \neq j}^{J} P_{ij\ell},$$

where the notation $\sum_{\ell,\ell \neq j}$ means summation over all values of ℓ not equal to j. Let $N = \sum n_j$ be the total number of observations. Compute the matrix $\mathbf{A} = (a_{jk})$, where

$$N^3 a_{jj} = \sum_{m=1}^{n_j} (T_{mj} - T_{.j})^2 + \sum_{\ell, \ell \neq j} \sum_{m=1}^{n_\ell} (P_{m\ell j} - \bar{P}_{.\ell j})^2,$$

and for $j \neq k$

$$N^3 a_{jk} = \sum_{\ell, j \neq \ell \neq k} \sum_m (P_{m\ell j} - \bar{P}_{.\ell j})(P_{m\ell k} - \bar{P}_{.\ell k})$$

$$- \sum_m (P_{mjk} - \bar{P}_{.jk})(T_{mj} - \bar{T}_{.j}) - \sum_m (P_{mkj} - \bar{P}_{.kj})(T_{mk} - \bar{T}_{.k}),$$

where $\bar{P}_{.j\ell} = \sum_i P_{ij\ell}/n_j$, and $\bar{T}_{.j} = \sum_i T_{ij}/n_j$. Letting $\mathbf{U} = (U_1, \ldots, U_J)$, the test statistic is

$$Q = N \left(\prod_{j=1}^{J} \frac{n_j - 1}{n_j} \right) \mathbf{U A^- U'},$$

where $\mathbf{A}^-$ is any generalized inverse of $\mathbf{A}$. (See e.g., Graybill, 1983, for information about the generalized inverse of a matrix.) When the null hypothesis is true, Q has, approximately, a chi-square distribution with $J-1$ degrees of freedom. That is, reject H_0 if Q exceeds the $1-\alpha$ quantile of a chi-square distribution having $J-1$ degrees of freedom.

7.8.2 R Function rfanova

The R function

$$\text{rfanova(x,grp)}$$

performs the calculations in Table 7.14, where x is any R variable that is a matrix (with J columns), or a data frame, or x has list mode. The argument grp indicates which groups are to be used. If grp is unspecified, all J groups are used. (If tied values are detected, the function prints a warning message.) The function returns the value of the test statistic, Q, and the p-value.

■ Example

Table 7.6 contains data for eight groups of participants. If the data for the first group are stored in the R variable film[[1]], the data for the second group in film[[2]], and so on, the function rfanova reports that the test statistic is $Q = 10.04$ and the p-value is .19. The command rfanova(film,grp=c(1,3,4)) would compare groups 1, 3, and 4 only.

■

7.8.3 A Heteroscedastic Rank-Based Method that Allows Tied Values

Brunner, Dette, and Munk (1997) derived a heteroscedastic analog of the Kruskal–Wallis test that allows tied values. Like the Rust–Fligner method, the basic idea is that if

$$H_0 : F_1(x) = \cdots = F_J(x),$$

is true, then the average ranks among the groups should not differ by too much. Again, pool the data and assign ranks. In the event there are tied values, midranks are used. Let R_{ij} be the resulting rank of X_{ij}. The remaining calculations are relegated to Table 7.15.

7.8.4 R Function bdm

The R function

$$bdm(x)$$

performs the BDM rank-based ANOVA described in Table 7.15. Here, x can have list mode or it can be a matrix with columns corresponding to groups. The function returns the value of the test statistic, the degrees of freedom, the vector of relative effects, which is labeled q.hat, and the p-value.

Table 7.15: How to Perform the Brunner–Dette–Munk Test.

Let

$$\bar{R}_j = \frac{1}{n_j} \sum_{i=1}^{n_j} R_{ij}.$$

$$\mathbf{Q} = \frac{1}{N} \left(\bar{R}_1 - \frac{1}{2}, \ldots, \bar{R}_J - \frac{1}{2} \right).$$

The vector $\mathbf{Q}$ contains what are called the *relative effects*. For the jth group, compute

$$s_j^2 = \frac{1}{N^2(n_j - 1)} \sum_{i=1}^{n_j} (R_{ij} - \bar{R}_j)^2,$$

and let

$$\mathbf{V} = N\text{diag}\left\{ \frac{s_1^2}{n_1}, \ldots, \frac{s_J^2}{n_J} \right\}.$$

Let $\mathbf{I}$ be a J-by-J identity matrix, let $\mathbf{J}$ be a J by J matrix of 1s, and set $\mathbf{M} = \mathbf{I} - \frac{1}{J}\mathbf{J}$ (The diagonal entries in $\mathbf{M}$ have a common value, a property required to satisfy certain theoretical restrictions.) The test statistic is

$$F = \frac{N}{\text{tr}(\mathbf{M}_{11}^2 \mathbf{V})} \mathbf{Q}\mathbf{M}\mathbf{Q}', \tag{7.14}$$

where tr indicates trace and $\mathbf{Q}'$ is the transpose of the matrix $\mathbf{Q}$. The null hypothesis is rejected if $F \geq f$, where f is the $1 - \alpha$ quantile of an F distribution with

$$\nu_1 = \frac{M_{11}[\text{tr}(\mathbf{V})]^2}{\text{tr}(\mathbf{M}\mathbf{V}\mathbf{M}\mathbf{V})},$$

and

$$\nu_2 = \frac{[\text{tr}(\mathbf{V})]^2}{\text{tr}(\mathbf{V}^2\Lambda)},$$

degrees of freedom and $\Lambda = \text{diag}\{(n_1 - 1)^{-1}, \ldots, (n_J - 1)^{-1}\}$.

■ Example

In schizophrenia research, an issue that has received some attention is whether groups of individuals differ in terms of skin resistance (measured in Ohms). In one such study, the groups of interest were no schizophrenic spectrum disorder, schizotypal or paranoid personality disorder, schizophrenia, predominantly negative symptoms, and schizophrenia, predominantly positive symptoms. For a portion of this study, the following results were obtained (after measures were transformed):

(No Schiz.)	(Schizotypal)	(Schiz. Neg.)	(Schiz. Pos.)
0.49959	0.24792	0.25089	0.37667
0.23457	0.00000	0.00000	0.43561
0.26505	0.00000	0.00000	0.72968
0.27910	0.39062	0.00000	0.26285
0.00000	0.34841	0.11459	0.22526
0.00000	0.00000	0.79480	0.34903
0.00000	0.20690	0.17655	0.24482
0.14109	0.44428	0.00000	0.41096
0.00000	0.00000	0.15860	0.08679
1.34099	0.31802	0.00000	0.87532

The function bdm returns a *p*-value of .040. The relative effect sizes (the **Q** values) are reported as

```
$output$q.hat:
        [,1]
[1,] 0.4725
[2,] 0.4725
[3,] 0.3550
[4,] 0.7000
```

So the average of the ranks in group 3 is smallest, and the average is highest for group 4. ∎

7.8.5 Inferences about a Probabilistic Measure of Effect Size

Method CHMCP

Consider J independent groups. For groups j and k $(1 \leq J < k \leq J)$, let

$$p_{jk} = P(X_{ij} < X_{ik}) + .5P(X_{ij} = X_{ik}).$$

As noted in Section 5.7.2, Cliff's method can be used to

$$H_0 : p_{jk} = .5 \tag{7.15}$$

and to compute a confidence interval for p_{jk}. But imagine that it is desired to test Eq. (7.15) for all $j < k$, with the goal that the probability of one or more type I errors is α. A simple method, that seems to be relatively effective, is to compute *p*-values for each test and use Hochberg's method, described in Section 7.4.7, to control the probability of one or more type

I errors. This will be called *method CHMCP*. It has been found to be generally preferable to using a Studentized maximum modulus distribution as suggested in Wilcox (2003); see Wilcox (2010d).

Method WMWAOV

It is briefly noted how one might test

$$H_0 : p_{12} = p_{13} = \cdots = p_{J-1,J} = .5. \tag{7.16}$$

Wilcox (2010d) examined several methods for accomplishing this goal, but only the method that performed well in simulations is described here. A limitation of the method is that it assumes tied values never occur; it can perform poorly, in terms of controlling the probability of a type I error, when this is not the case.

For the jth and kth groups, let D_{jk} be the distribution of $X_j - X_k$. Recall from Section 5.7.2 that $p_{jk} = .5$ corresponds to $\theta_{jk} = 0$, where θ_{jk} is the (population) median of D_{jk}. (As pointed out in Section 5.7.2, under general conditions, $\theta_{jk} \neq \theta_j - \theta_k$.) So testing Eq. (7.16) is tantamount to testing

$$H_0 : \theta_{12} = \theta_{13} = \cdots = \theta_{J-1,J} = 0. \tag{7.17}$$

Let X_{ij} ($i = 1, \ldots, n_j$; $j = 1, \ldots, J$) be a random sample of size n_j from the jth group. Generate a bootstrap sample from jth group by randomly sampling with replacement n_j observations from $X_{1j}, \ldots, X_{n_j j}$, which will be labeled $X_{1j}^*, \ldots, X_{n_j j}^*$. Let M_{jk}^*, $j < k$, be the the usual sample median based on the $n_j n_k$ differences $X_{ij}^* - X_{\ell k}^*$ ($i = 1, \ldots, n_j$; $\ell = 1, \ldots, n_k$). Repeat this process B times yielding M_{jkb}^*, $b = 1, \ldots, B$. So M_{jkb}^* represents B vectors, each having length $(J^2 - J)/2$. From Liu and Singh (1997), a p-value for testing Eq. (7.17) can be obtained by measuring how deeply $\mathbf{0} = (0, \ldots, 0)$ is nested within the bootstrap cloud of points. More precisely, let $\mathbf{G_0}$ be the depth of the null vector $(0, \ldots, 0)$ based on the notion of projection distance as described in Section 6.2.5. The projection distance of $\mathbf{M}_b^* = (M_{12b}^*, \ldots M_{J-1,Kb}^*)$ from the center of the bootstrap data cloud is denoted by G_b. Then a p-value is

$$\frac{1}{B} \sum I_b,$$

where the indicator function $I_b = 1$ if $G_0 < G_b$; otherwise $I_b = 0$. This will be called method WMWAOV. Though seemingly rare, it is possible for method WMWAOV to correctly reject even though method CHMCP finds no differences.

Method DBH

When performing all pairwise comparisons, there is a variation of method WMWAOV that should be mentioned. For each pair of groups, apply method WMWAOV and control the probability of one or more type I errors using Hochberg's method. This will be called method DBH. Note that with only two groups, a p-value can be computed as described, for example, in Section 4.4.1. (If $\hat{p}^*$ is the proportion of M_{12b}^* values less than 0, the p-value is $2\min(\hat{p}^*, 1 - \hat{p}^*)$.) Simulation results indicate that the actual probability of one or more type I errors will be closer to the nominal level compared with method CHMCP (Wilcox, 2010d). Moreover, method DBH might provide a bit more power. But DBH is not recommended when tied values can occur.

7.8.6 R Functions cidmulv2, wmwaov and cidM

The R function

$$\text{cidmulv2(x,alpha=0.05,g=NULL,dp=NULL,CI.FWE=F)}$$

tests Eq. (7.15). The output includes a column headed by p.crit, which indicates how small the p-value must be to reject using Hochberg's method. If the argument CI.FWE=F, the function returns confidence intervals for each p_{jk} having probability coverage $1 - \alpha$. If CI.FWE=T, the probability coverage corresponds to the "critical" p-value used to make decisions about rejecting Eq. (7.15) based on Hochberg's method. For example, if the goal is to have the probability of one or more type I errors equal to .05, and if the second largest p-value is less than or equal to .025, Hochberg's method rejects. The confidence interval returned by cidmulv2, for the two groups corresponding to the situation having the second largest p-value, will have probability coverage $1 - .025 = .975$. For the next largest p-value, the probability coverage will be $1 - .05/3$, and so on. If the argument g is specified, it is assumed that x is a matrix with the dependent variable stored in column dp and the levels of the factors stored in column g.

The R function

$$\text{wmwaov(x,nboot=500,MC=F,SEED=T,pro.dis=T,MM=F)}$$

performs the WMWAOV method. Setting the argument MC=T, the function takes advantage of a multicore processor, assuming one is available.

Finally, the R function

$$\text{cidM(x,nboot=1000,alpha=0.05,MC=F,SEED=T,g=NULL,dp=NULL)}$$

performs method DBH. (Both wmwaov and cidM check for tied values and print a warning message if any are found.)

7.9 A Rank-Based Method for a Two-Way Design

This section describes a rank-based method for a two-way design derived by Akritas, Arnold, and Brunner (1997). The basic idea stems from Akritas and Arnold (1994) and is based on the following point of view. For any value x, let

$$\bar{F}_{j.}(x) = \frac{1}{K} \sum_{k=1}^{K} F_{jk}(x)$$

be the average of the distributions among the K levels of Factor B corresponding to the jth level of Factor A. The hypothesis of no main effects for Factor A is

$$H_0 : \bar{F}_{1.}(x) = \bar{F}_{2.}(x) = \cdots = \bar{F}_{J.}(x).$$

for any x. Letting

$$\bar{F}_{.k}(x) = \frac{1}{J} \sum_{j=1}^{J} F_{jk}(x)$$

be the average of the distributions for the kth level of Factor B, the hypothesis of no main effects for Factor B is

$$H_0 : \bar{F}_{.1}(x) = \bar{F}_{.2}(x) = \cdots = \bar{F}_{.K}(x).$$

As for interactions, first consider a 2-by-2 design. Then no interaction is taken to mean that for any x,

$$F_{11}(x) - F_{12}(x) = F_{21}(x) - F_{22}(x),$$

which has a certain similarity to how no interaction based on means is defined. Here, no interaction in a J-by-K design means that for any two rows and any two columns, there is no interaction as just described. From a technical point of view, a convenient way of stating the hypothesis of no interactions among all JK groups is with

$$H_0 : F_{jk}(x) - \bar{F}_{j.}(x) - \bar{F}_{.k}(x) + \bar{F}_{..}(x) = 0,$$

for any x, all j $(j = 1, \ldots, J)$ and all k $(k = 1, \ldots, K)$, where

$$\bar{F}_{..}(x) = \frac{1}{JK} \sum_{j=1}^{J} \sum_{k=1}^{K} F_{jk}(x).$$

The computations begin by pooling all of the data and assigning ranks. For convenience, let $L = JK$ and let $R_{i\ell}$ be the ranks of the ℓth group $(\ell = 1, \ldots, L)$, where the first K groups correspond to the first level of the first factor, the next K correspond to the second level of the first factor, and so on. Let

$$\bar{R}_{\ell} = \frac{1}{n_{\ell}} \sum R_{i\ell}$$

and

$$s_\ell^2 = \frac{1}{N^2(n_\ell - 1)} \sum (R_{i\ell} - \bar{R}_\ell)^2,$$

where $N = \sum n_\ell$ is the total sample size. Set

$$\mathbf{V} = N \mathrm{diag}\left\{ \frac{s_1^2}{n_1}, \dots, \frac{s_L^2}{n_L} \right\}.$$

Let $\mathbf{I}_J$ be a J-by-J identity matrix, let $\mathbf{H}_J$ be a J-by-J matrix of ones, and let

$$\mathbf{P}_J = \mathbf{I}_J - \frac{1}{J}\mathbf{H}_J, \quad \mathbf{M}_A = \mathbf{P}_J \otimes \frac{1}{K}\mathbf{H}_K,$$

$$\mathbf{M}_B = \frac{1}{J}\mathbf{H}_J \otimes \mathbf{P}_K, \quad \mathbf{M}_{AB} = \mathbf{P}_J \otimes \mathbf{P}_K.$$

(The notation $\otimes$ refers to the right Kronecker product.)

The remaining calculations are summarized in Table 7.16.

Table 7.16: Two-Way, Heteroscedastic, Rank-Based ANOVA.

Let

$$\mathbf{Q} = \frac{1}{N}\left(\bar{R}_1 - \frac{1}{2}, \dots, \bar{R}_L - \frac{1}{2} \right)$$

be the *relative effects*. The test statistics are as follows:

$$F_A = \frac{N}{\mathrm{tr}(\mathbf{M}_{A11}\mathbf{V})}\mathbf{Q}\mathbf{M}_A\mathbf{Q}', \quad F_B = \frac{N}{\mathrm{tr}(\mathbf{M}_{B11}\mathbf{V})}\mathbf{Q}\mathbf{M}_B\mathbf{Q}',$$

$$F_{AB} = \frac{N}{\mathrm{tr}(\mathbf{M}_{AB11}\mathbf{V})}\mathbf{Q}\mathbf{M}_{AB}\mathbf{Q}'.$$

For Factor A, reject if $F_A \geq f$, where f is the $1 - \alpha$ quantile of an F distribution with degrees of freedom

$$\nu_1 = \frac{M_{A11}^2[\mathrm{tr}(\mathbf{V})]^2}{\mathrm{tr}(\mathbf{M}_A\mathbf{V}\mathbf{M}_A\mathbf{V})}, \quad \nu_2 = \frac{[\mathrm{tr}(\mathbf{V})]^2}{\mathrm{tr}(\mathbf{V}^2\Lambda)},$$

where $\Lambda = \mathrm{diag}\{(n_1 - 1)^{-1}, \dots, (n_L - 1)^{-1}\}$. Here M_{A11} is the first diagonal element of the matrix $\mathbf{M}_A$. (By design, all of the diagonal elements of $\mathbf{M}_A$ have a common value.) For Factor B, reject if $F_B \geq f$, where

$$\nu_1 = \frac{M_{B11}^2[\mathrm{tr}(\mathbf{V})]^2}{\mathrm{tr}(\mathbf{M}_B\mathbf{V}\mathbf{M}_B\mathbf{V})}.$$

(The value for ν_2 remains the same.) As for the hypothesis of no interactions, reject if $F_{AB} \geq f$, where now

$$\nu_1 = \frac{M_{AB11}^2[\mathrm{tr}(\mathbf{V})]^2}{\mathrm{tr}(\mathbf{M}_{AB}\mathbf{V}\mathbf{M}_{AB}\mathbf{V})}.$$

and ν_2 is the same value used to test for main effects.

7.9.1 R Function bdm2way

The R function

$$\text{bdm2way(J, K, x)}$$

performs the two-way ANOVA method described in Table 7.16.

7.9.2 The Patel–Hoel Approach to Interactions

Patel and Hoel (1973) proposed an alternative approach to interactions in a 2-by-2 design that can be extended to a multiple comparisons method for a J-by-K design, even when there are tied values. First consider a 2-by-2 design where X_{ijk} is the ith observation randomly sampled from the jth level of Factor A and the the kth level of Factor B. Temporarily assume ties occur with probability zero and let

$$p_{11,12} = P(X_{i11} < X_{i12}).$$

Note that ignoring level two of Factor A, levels one and two of Factor B can be compared by testing H_0: $p_{11,12} = 0$ as described in Sections 5.7. The Patel–Hoel definition of no interaction is that $p_{11,12} = p_{21,22}$. That is, the probability of an observation being smaller under level one of Factor B, versus level two, is the same for both levels of Factor A. In the event ties can occur, let

$$p_{11,12} = P(X_{i11} \leq X_{i12}) + \frac{1}{2}P(X_{i11} = X_{i12}),$$

$$p_{21,22} = P(X_{i21} \leq X_{i22}) + \frac{1}{2}P(X_{i21} = X_{i22}),$$

in which case the hypothesis of no interaction is

$$H_0 : p_{11,12} = p_{21,22}.$$

Again, temporarily ignore level two of Factor A and note that the two independent groups corresponding to the two levels of Factor B can be compared in terms of δ as described in Section 5.7.2. Let δ_1 represent δ when focusing on level one of Factor A with level two ignored and let $\hat{\delta}_1$ be the estimate of δ as given by Eq. (5.19). An estimate of the squared standard error of $\hat{\delta}_1$, $\hat{\sigma}_1^2$, is described in Section 5.7.2 as well. Similarly, let δ_2 be the estimate of δ_2 when focusing on level two of Factor A ,with level one ignored, and denote its estimate with $\hat{\delta}_2$. The estimated squared standard error of $\hat{\delta}_2$ is denoted by $\hat{\sigma}_2^2$. It can be seen that the null hypothesis of no interaction just defined corresponds to

$$H_0 : \Delta = \frac{\delta_2 - \delta_1}{2} = 0.$$

An estimate of $p_{11,12} - p_{21,22}$ is

$$\hat{\Delta} = \frac{\hat{\delta}_2 - \hat{\delta}_1}{2},$$

and the estimated squared standard error of $\hat{\Delta}$ is

$$S^2 = \frac{1}{4} \left(\hat{\sigma}_1^2 + \hat{\sigma}_2^2 \right),$$

and a $1 - \alpha$ confidence interval for Δ is

$$\hat{\Delta} \pm z_{1-\alpha/2} S,$$

where $z_{1-\alpha/2}$ is the $1 - \alpha/2$ quantile of a standard normal distribution. The hypothesis of no interaction is rejected if this confidence interval does not contain zero.

For the more general case of a J-by-K design, an analog of Dunnett's T3 method is used to control FWE. When working with levels j and j' of Factor A and levels k and k' of Factor B, we represent the parameter Δ by $\Delta_{jj'kk'}$, its estimate is labeled $\hat{\Delta}_{jj'kk'}$, and the estimated squared standard error is denoted by $S^2_{jj'kk'}$. For every $j < j'$ and $k < k'$, the goal is to test

$$H_0 : \Delta_{jj'kk'} = 0.$$

The total number of hypotheses to be tested is

$$C = \frac{J^2 - J}{2} \times \frac{K^2 - K}{2}.$$

The critical value, c, is the $1 - \alpha$ quantile of the C-variate Studentized maximum modulus distribution with degrees of freedom $\nu = \infty$. The confidence interval for $\Delta_{jj'kk'}$ is

$$\hat{\Delta}_{jj'kk'} \pm c S_{jj'kk'},$$

and the hypothesis of no interaction, corresponding to levels j and j' of Factor A and levels k and k' of Factor B, is rejected if this confidence interval does not contain zero.

7.9.3 R Function rimul

The R function

$$\text{rimul(J,K,x,p=J*K,grp=c(1:p),plotit=T,op=4)}$$

performs the test for interactions just described. (The argument, p=J*K, is not important in applied work; it is used to deal with certain conventions in R.) The groups are assumed to be arranged as in Section 7.2.1, and the argument grp is explained in Section 7.2.1 as well. If $J = K = 2$ and plotit=T, the function plots an estimate of the distribution of $D_{i1} = X_{i11} - X_{i12}$ and $D_{i2} = X_{i21} - X_{i12}$ via the function g2plot in Section 5.1.7. The argument op is relevant to g2plot and controls the type of plot that is created.

7.10 MANOVA Based on Trimmed Means

Multivariate analysis of variance, also known as *MANOVA*, deals with a generalization of ANOVA to situations where two or more measures are taken on each participant. More formally, consider J independent groups where, for each participant, p measures are taken. For the jth group, denote the p trimmed means by $\mu_j = (\mu_{tj1}, \ldots, \mu_{tjp})$. The goal is to test

$$H_0 : \mu_1 = \cdots = \mu_J. \tag{7.18}$$

Johansen (1980) derived a method for means that allows the covariances associated with the J groups to differ, in contrast to classic methods that assume the J groups have a common covariance matrix. The method represents a heteroscedastic approach to what is called the *general linear model*. Johansen assumed normality, but the method can be extended to trimmed means as described here. For the two-sample case, comparisons with a method derived by Kim (1992b) as well as several other methods, are reported by Wilcox (1995f). Lix, Keselman, & Hinds (2005) compared several methods based on both means and a 20% trimmed mean, again for the two-sample case. No single method dominated and it is unclear the extent the generalization of Johansen's method used here competes well with the methods compared by Lix et al.

The version of Johansen's method used here, extended to trimmed means, is applied as follows. For the jth group, there are n_j randomly sampled vectors of observations denoted by $(X_{ij1}, \ldots X_{ijp})$, $i = 1, \ldots, n_j$. Let $\bar{\mathbf{X}}_j = (\bar{X}_{tj1}, \ldots, \bar{X}_{tjp})$ denote the vector of trimmed means and let $\mathbf{V}_j$ be the Winsorized covariance matrix. Compute

$$\tilde{R}_j = \frac{n_j - 1}{(n_j - 2g_j)(n_j - 2g_j - 1)} \mathbf{V}_j,$$

where $g_j = \gamma n_j$, rounded down to the nearest integer, and γ is the amount of trimming,

$$\mathbf{W}_j = \tilde{R}_j^{-1},$$

$$\mathbf{W} = \sum \mathbf{W}_j$$

and

$$A = \frac{1}{2} \sum_{j=1}^{J} [\{tr(\mathbf{I} - \mathbf{W}^{-1}\mathbf{W}_j)\}^2 + tr\{(\mathbf{I} - \mathbf{W}^{-1}\mathbf{W}_j)^2\}]/f_j,$$

where $f_j = n_j - 2g_j - 1$. The estimate of the population trimmed means, assuming H_0 is true, is

$$\hat{\mu}_t = \mathbf{W}^{-1} \sum \mathbf{W}_j \bar{\mathbf{X}}_j.$$

The test statistic is

$$F = \sum_{j=1}^{J} \sum_{k=1}^{p} \sum_{m=1}^{p} w_{mkj}(\bar{X}_{mj} - \hat{\mu}_m)(\bar{X}_{kj} - \hat{\mu}_k), \tag{7.19}$$

where w_{mkj} is the mkth element of $\mathbf{W}_j$, $\bar{X}_{mj}$ is the mth element of $\bar{\mathbf{X}}_j$, and $\hat{\mu}_m$ is the mth element of $\hat{\mu}_t$. Reject the null hypothesis if

$$F \geq c + \frac{c}{2p(J-1)} \left\{ A + \frac{3cA}{p(J-1)+2} \right\},$$

where c is the $1 - \alpha$ quantile of a chi-squared distribution with $p(J-1)$ degrees of freedom.

Note that the MANOVA method based on trimmed means uses a measure of location that does not take into account the overall structure of the data. Todorov and Filzmoser (2010) derived a MANOVA method based on the MCD estimator, which does take into account the overall structure, but their method assumes that groups differ in location only.

For the special case where the goal is to compare two groups only, Yanagihara and Yuan (2005) derived a method for comparing means that compares well to several other heteroscedastic methods, in terms of controlling the probability of a type I error, when sampling from multivariate normal distributions. Currently, there are no published papers comparing the small-sample properties of the extended Yanagihara and Yuan method to the extension of Johansen's method. (A few simulations were run by the author using a 20% trimmed mean. Situations were found where the extended Yanagihara and Yuan method provides more satisfactory control over the probability of a type I error, no situation has been found where the reverse is true, but a more comprehensive study is needed.)

Let

$$T = (\bar{\mathbf{X}}_1 - \bar{\mathbf{X}}_2)'(\tilde{R}_1 + \tilde{R}_2)^{-1}(\bar{\mathbf{X}}_1 - \bar{\mathbf{X}}_2),$$

$$\bar{\mathbf{V}} = \frac{n_2}{n}\mathbf{V}_1 + \frac{n_1}{n}\mathbf{V}_2,$$

where $n = n_1 + n_2$,

$$P_1 = \frac{n_2^2(n-2)}{n^2(n_1-1)}\{\mathrm{tr}(\mathbf{V}_1\bar{\mathbf{V}}^{-1})\}^2 + \frac{n_1^2(n-2)}{n^2(n_2-1)}\{\mathrm{tr}(\mathbf{V}_2\bar{\mathbf{V}}^{-1})\}^2,$$

$$P_2 = \frac{n_2^2(n-2)}{n^2(n_1-1)}\mathrm{tr}(\mathbf{V}_1\bar{\mathbf{V}}^{-1}\mathbf{V}_1\bar{\mathbf{V}}^{-1}) + \frac{n_1^2(n-2)}{n^2(n_2-1)}\mathrm{tr}(\mathbf{V}_2\bar{\mathbf{V}}^{-1}\mathbf{V}_2\bar{\mathbf{V}}^{-1}),$$

and

$$\hat{v} = \frac{(h - 2 - P_1)^2}{(h - 2)P_2 - P_1},$$

where $h = h_1 + h_2$ and $h_j = n_j - 2g_j$ ($j = 1, 2$). The test statistic, based on an extension of the Yanagihara–Yuan method to trimmed means, is

$$T_f = \frac{n - 2 - P_1}{(n - 2)p} T,$$

which has, approximately, an F distribution with p and $\hat{v}$ degrees of freedom when the null hypothesis is true. When $p = 1$, this method reduces to Yuen's method described in Section 5.3.

7.10.1 R Functions MULtr.anova, MULAOVp, bw2list, and YYmanova

The R function

$$\text{MULtr.anova}(x, J = \text{NULL}, p - \text{NULL}, tr = 0.2, alpha = 0.05)$$

performs the robust MANOVA method based on the extension of Johansen's method to trimmed means. The argument J defaults to NULL, meaning that x is assumed to have list mode with length J, where x[[j]] contains a matrix with n_j rows and p columns, $j = 1, \ldots, J$. If the arguments J and p are specified, the data can be stored in list mode or a matrix. If stored in list mode, it is assumed that x[[1]] - x[[p]] contain p measures associated with the first group, x[[p+1]] - x[[2p]] contain the p measures for the next group, and so on. If the data are stored in a matrix or data frame, it is assumed the first p columns of x contain the p measures associated with the first group, the next p columns contain the p measures associated with the second groups, and so forth.

The R function

$$\text{MULAOVp}(x, J = \text{NULL}, p = \text{NULL}, tr = 0.2)$$

performs the same robust MANOVA method as the R function MULtr.anova, only it returns a p-value.

The R function

$$\text{bw2list}(x, grp.col, lev.col)$$

is provided to help with data management issues. If the data are stored in a matrix or data frame, the argument grp.col indicates which column contains the group identification information. The argument lev.col indicates the columns that contain the p measures that are to be compared.

■ Example

Imagine that two independent groups are to be compared based on measures taken at three different times. One way of comparing the groups is with a robust MANOVA method. If the data for the first group are stored in the R variable m1, a matrix having 3 columns, and if the data for the second group are stored in m2, also having 3 columns, the analysis can be performed as follows:

$$x=\text{list}()$$
$$x[[1]]=m1$$
$$x[[2]]=m2$$
$$\text{MULtr.anova}(x).$$

The function returns the test statistic and a critical value.

■

■ Example

The web page http://stat.cmu.edu/DASL/allsubjects.html, maintained by Carnegie Mellon University, includes a file called scents. The first five columns of the data, for the first row of the data, look like this:

```
ID Sex Smoker Opinion Age Order
1   M      N       pos  23     1
```

The next six columns for the first row look like this:

```
U.Trial.1 U.Trial.2 U.Trial.3 S.Trial.1  S.Trial.2 S.Trial.3
     38.4      27.7      25.7      53.1       30.6      30.2.
```

These last six columns contain the time participants required to complete a pencil and paper maze test when they were smelling a floral scent and when they were not. The columns headed by U.Trial.1 U.Trial.2 U.Trial.3 are the times for no scent, which were taken on three different occasions. Here we compare smokers (Y) and nonsmokers (N) based on all three of the no scent measures. So in the notation used here, $J = 2$ and $p = 3$. The first task is storing the data in a manner that can used by the R functions MULtr.anova and MULAOVp. Assuming the data have been stored in the R variable called scent, in a data frame or a matrix, this can be accomplished with the R command

$$z=\text{bw2list}(\text{scent},3,c(7{:}9)).$$

Then the R command

$$\text{MULAOVp(z,2,3)}$$

would compare the 20% trimmed means of smokers to nonsmokers. The p-value is .143.

■

The R function

$$\text{YYmanova(x1, x2, tr=0.2)}$$

performs the extension of the Yanagihara–Yuan MANOVA method to trimmed means, which is limited to $J = 2$ groups. The data for the first group are stored in the argument x1, which is assumed to be a matrix with p columns, as is the argument x2, which is assumed to be the data for group 2.

7.10.2 Linear Contrasts

Consider again J independent groups where for each participant, p measures are taken. This section deals with the goal of testing a set of linear contrasts in the context of multivariate data. There are two variations. The first uses some marginal measure of location, such as a trimmed or M-estimator, and the other uses some multivariate measure of location that takes into account the overall structure of the data such as those summarized in Section 6.3.

For convenience only, attention is focused on the marginal trimmed means with the understanding that any measure of location can be used. So now we let

$$\Psi = \sum c_j \mu_t,$$

where μ_t is a vector of p trimmed means and the goal is to test

$$H_0 : \Psi = \mathbf{0}.$$

With 20% trimming, currently a relatively good approach aimed at achieving this goal is to use a percentile bootstrap method. Generate a bootstrap sample by sampling with replacement n_j rows from p-variate data associated with the jth group. Compute the marginal trimmed means and label the result $\mathbf{X}_t^*$, followed by

$$\hat{\Psi}^* = \sum c_j \bar{\mathbf{X}}_t^*.$$

Repeat B times yielding $\Psi_1^*, \ldots, \Psi_B^*$. Next, compute the Mahalanobis distance of each Ψ_b^* ($b = 0, \ldots, B$), say d_b^*, where d_0^* is the distance of the null vector. The center of the bootstrap data cloud is taken to be $\hat{\Psi}$, the estimate of Ψ based on the observed data. And the covariance matrix when computing the Mahalanobis distances is just the sample covariance matrix based on the Ψ_b^* ($b = 0, \ldots, B$) values. Let P^* be the proportion of d_b values ($b = 0, \ldots, B$) such that $d_0 \geq d_b$. Then a p-value is $1 - P^*$.

This method is not recommended when using means. It appears to perform well when using a reasonably robust measure of location, but if the breakdown point is close to zero, it can be highly inaccurate.

Limited comparisons suggest that when comparing two groups, the bootstrap method described here performs about as well as the extension of Johansen's method when working with 20% trimmed means in terms of controlling the probability of a type error and when sampling from normal distributions. However, the p-values can differ substantially. In simulations, for example, when sampling from normal distributions, Johansen's method can have a substantially larger p-value, but situations where the reverse is true are encountered even though the ability of the two methods to control the type I error probability is similar. In practical terms, even under normality, the choice of method is not academic in terms of deciding whether to reject the null hypothesis.

Now consider a situation where $p = 3$, $J = 2$, $n_1 = 20$, $n_2 = 40$, the first group has a multivariate normal distribution with common correlation $\rho = 0$, but the other groups is generated from a g-and-h distribution with $g = 0.5$ and $h = 0$ (a skewed distribution with relatively light tails) and $\rho = 0.6$. Further imagine that for the second group, the marginal distributions are shifted so that they have a trimmed mean of zero. Then the probability of a type I error when testing at the 0.05 level, and using the bootstrap method described here, is .014 (based on 1000 replications). In contrast, the actual levels using the Yanagihara–Yuan and Johansen methods are 0.053 and 0.045, respectively. A similar result is obtained when the second group now has $g = 0$ and $h = 0.5$ (a symmetric distribution with relatively heavy tails). Of course, this is not convincing evidence that the Yanagihara–Yuan and Johansen methods are generally preferable to the bootstrap method. The only point is that there are situations where indeed they have a practical advantage.

■ Example

For the EEG data in Table 6.1, MULAOVp returns a p-value equal to .083. But using the percentile bootstrap method described here, the p-value is .789. Situations are encountered, however, where the percentile bootstrap method has a substantially smaller p-value. ■

Note that for $J > 2$ groups, the Yanagihara–Yuan method can be used to perform all pairwise comparisons with the probability of at least one type I error controlled by Rom's method. Limited simulations suggest that this approach performs relatively well in terms of type I errors and power, even with small sample sizes. However, the percentile bootstrap method can be used with any robust estimator. Moreover, situations are encountered where both Johansen's method and the Yanagihara–Yuan method cannot be applied because the Winsorized covariance matrix is singular. Because the percentile bootstrap method does not use any covariance matrix, this problem is avoided.

7.10.3 R Functions linconMpb, linconSpb, YYmcp, fac2Mlist, and fac2BBMlist

The R function

$$linconMpb(x, alpha = 0.05, nboot = 1000, grp = NA, est = tmean, con = 0, bhop = F,$$
$$SEED = T, PDIS = F, J = NULL, p = NULL,...)$$

tests hypotheses, based on linear contrasts, using the percentile bootstrap method described in the previous section, assuming that for each group, some marginal measure of location is used. The argument x is assumed to have list mode, where x[[1]] is a matrix with p columns associated with group 1, x[[2]] is a matrix with p columns associated with group 2, and so on. If x does not have list mode, but rather is a matrix or data frame with the first p columns corresponding group 1, the next p columns corresponding to group 2, and so forth, then specify how many groups there are via the argument J, or how specify how many variables there via the argument p. By default, all pairwise comparisons are performed based on the marginal 20% trimmed means, but M-estimators, for example, could be used by setting the argument est=onestep. The probability of at least one type I error is set via the argument alpha and is controlled using Rom's method. As usual, contrast coefficients can be specified via the argument con. Setting the argument PDIS=T, projection distances will be used the depth of the null vector in the bootstrap cloud of points.

For each group, it might be desired to use a multivariate measure of location that takes into account the overall structure of the data. That is, use one of the measures of location in Section 6.3. This can be done with a percentile bootstrap method via the R function

$$linconSpb(x, alpha = 0.05, nboot = 1000, grp = NA, est = smean, con = 0, bhop = F, SEED$$
$$= T, PDIS = F, J = NULL, p = NULL,...)$$

By default, the OP estimator of location is used, but this might result in relatively high execution time. (For an alternative approach based on an S-estimator or MM-estimator, see Van Aelst and Willems, 2011.)

The R function

$$YYmcp(x, alpha = 0.05, grp = NA, tr = 0.2, bhop = F, J = NULL, p = NULL, ...)$$

performs all pairwise comparisons via the extension of the Yanagihara–Yuan technique used in conjunction with Rom's method for controlling FWE. Setting the argument bhop=T, the

Benjamini–Hochberg method is used instead. The arguments J and p are used in the same manner as described in conjunction with the R function linconMpb.

Data Management

The following two R functions might help with data management. The R function

$$fac2Mlist(x,grp.col,lev.col,pr=T)$$

sorts *p*-variate data stored in the matrix (or data frame) x into groups based on the values stored in the column of x indicated by the argument grp.col. The results are stored in list mode in a manner that can be used by linconMpb and linconSpb. For example, the command

$$z=fac2Mlist(plasma,2,c(7:8))$$

will create groups based on the data in column 2. The result is that z[[1]] will contain the data for the first group stored as a matrix. The first column of this matrix corresponds to data stored in column 7 of the R variable plasma and the second column corresponds to data stored in column 8. Similarly, z[[2]] will contain the data for group 2, and so on. So the command

$$linconSpb(z)$$

would perform all pairwise comparisons.

The R function

$$fac2BBMlist(x,grp.col,lev.col,pr=T)$$

is like the function fac2Mlist, only it is designed to handle a between-by-between design. Now the argument grp.col is assumed to contain two values indicating the columns of x that contain the levels of the two factors. The multivariate data are stored in the columns indicated by the argument lev.col. For a J-by-K design, the result is an R variable having list mode with length JK.

■ Example

The command

$$z=fac2BBMlist(plasma,c(2,3),c(7,8))$$

will create groups based on the values in columns 2 and 3 of the R variable plasma. In this particular case, there are two levels for the first factor (meaning that column 2 of plasma has two unique values only) and three for the second. The result will be that z[[1]], ..., z[[6]] will each contain a matrix having two columns stemming from the bivariate data in columns 7 and 8 of plasma. Then the commands

$$con=con2way(2,3)$$

$$linconMpb(z,con=con\$conAB)$$

would test all hypotheses based on the linear contrast coefficients typically used when dealing with interactions.

∎

7.11 Nested Designs

Briefly, a two-way nested design refers to a situation where there is a hierarchy among the levels of two factors under study. This is in contrast to a completely crossed design as considered in Section 7.2. For example, a goal might be to compare the efficacy of two medical procedures. The first method is used in K randomly sampled hospitals, with n participants used within each hospital, and the same is done for another K randomly sampled hospitals for the second method. For various reasons, the efficacy of a method might depend on the hospital where it is used. Here, the factor hospital is nested within the two levels corresponding to the medical procedures. (There is no interaction term.) Similarly, the effectiveness of methods for teaching mathematics might depend on the school where they are used. If the goal is to compare J teaching strategies, this might be done based on K randomly sampled schools with n students within each school being taught based on a particular method. So the factor school is nested within the levels of J methods.

A simple way of dealing with nested designs, in a robust manner that allows heteroscedasticity, is to use the trimmed means from each level of factor B, which is nested within the levels of factor A. For the teaching strategies example, methods would be compared based on the trimmed means resulting from each school, where each trimmed mean is based on n participants within each school. For the special case where means are used, formal statements of this approach are given in Khuri (1992) where heteroscedastic methods are studied. (For the situation where the K levels of the nested factor are fixed, see Guo, Billard, & Luh, 2011.)

The goal when dealing with a nested design can be stated in a slightly more formal manner as follows. For the jth level of factor A, it is assumed that there are K randomly sampled levels of the nested factor. For j fixed, let μ_{tjk} be the population trimmed mean corresponding to level k of the nested factor B ($j = 1, \ldots, J; k = 1, \ldots, K$). Moreover, with j still fixed, μ_{tjk} is assumed to have some unknown distribution having a trimmed mean denoted by μ_{tj} and variance σ_j^2. There are two goals. The first is to test

$$H_0 : \mu_{t1} = \cdots = \mu_{tJ}. \tag{7.20}$$

The second is to perform all pairwise comparisons in a manner that controls the probability of at least one type I error. That is, the goal is to test

$$H_0 : \mu_{tj} = \mu_{tj'} \tag{7.21}$$

for each $j < j'$, such that the probability of at least one type I error is approximately equal to α.

Let X_{ijk} be the ith randomly sampled observation from kth randomly sampled level of factor B. For fixed j and k, let

$$X_{(1)jk} \leq \ldots \leq X_{(n)jk}$$

be the n observations written in ascending order. For some γ $(0 \leq \gamma < .5)$, let $g = [\gamma n]$, where $[\gamma n]$ is the value of γn rounded down to the nearest integer. Then the γ sample trimmed mean is

$$\bar{X}_{jk} = \sum_{g+1}^{n-g} X_{(i)jk}.$$

In essence, the unit of analysis becomes the $\bar{X}_{tjk}$ values when applying methods for trimmed means already covered. To elaborate, let $\bar{X}_j$ be the trimmed mean of the values $\bar{X}_{j1}, \ldots, \bar{X}_{jK}$ and let

$$W_{jk} = \begin{cases} \bar{X}_{(g+1)jk}, & \text{if } \bar{X}_{jk} \leq \bar{X}_{(g+1)jk} \\ \bar{X}_{jk}, & \text{if } \bar{X}_{(g+1)jk} < \bar{X}_{jk} < \bar{X}_{(n-g)jk} \\ \bar{X}_{(n-g)jk}, & \text{if } \bar{X}_{jk} \geq \bar{X}_{(n-g)jk}. \end{cases}$$

The Winsorized sample mean corresponding to $\bar{X}_{j1}, \ldots, \bar{X}_{jK}$ (j fixed) is

$$\bar{X}_{wj} = \frac{1}{K} \sum_{k=1}^{K} W_{jk}$$

and the Winsorized variance is

$$s_{wj}^2 = \frac{1}{K-1} \sum (W_{jk} - \bar{X}_{wj})^2.$$

Let

$$d_j = \frac{(K-1)s_{wj}^2}{h_j \times (h_j - 1)},$$

where $h_j = K - 2G$, $G = [\gamma K]$,

$$w_j = \frac{1}{d_j}$$

$$U = \sum w_j$$

$$\tilde{X} = \frac{1}{U} \sum w_j \bar{X}_j$$

$$A = \frac{1}{J-1} \sum w_j (\bar{X}_j - \tilde{X})^2$$

$$B = \frac{2(J-2)}{J^2-1} \sum \frac{(1 - \frac{w_j}{U})^2}{h_j - 1}.$$

For $J = 2$, a test of Eq. (7.20) can be performed using an analog of Yuen's (1974) method. The test statistic is

$$T_y = \frac{\bar{X}_1 - \bar{X}_2}{\sqrt{d_1 + d_2}}. \qquad (7.22)$$

When the null hypothesis is true, T_y has, approximately, a Student's t-distribution with degrees of freedom

$$\hat{v}_y = \frac{(d_1 + d_2)^2}{\frac{d_1^2}{h_1 - 1} + \frac{d_2^2}{h_2 - 1}}.$$

For $J \geq 2$, the test statistic

$$F_t = \frac{A}{1 + B}$$

can be used to test (7.20). When the null hypothesis is true, F_t has, approximately, an F distribution with degrees of freedom

$$v_1 = J - 1$$

$$v_2 = \left[\frac{3}{J^2 - 1} \sum \frac{(1 - w_j/U)^2}{h_j - 1} \right]^{-1}.$$

Finally, there is goal of testing Eq. (7.21) such that the probability of at least one type I error is approximately equal to α. When comparing groups j and j', reject if $|T_y| \geq c$, where c is the $1 - \alpha$ quantile of a Studentized maximum modulus distribution having degrees of freedom

$$\hat{v}_y = \frac{(d_j + d_{j'})^2}{\frac{d_j^2}{h_j - 1} + \frac{d_{j'}^2}{h_{j'} - 1}}.$$

Khuri (1992) derived another approach based on means that compares the means of factor A based in part on Hotelling's T^2. It is unknown whether an extension of the method to trimmed means, along the lines in Section 6.7.2, has any practical value.

Compared with the method for testing Eq. (7.20), the method aimed at testing Eq. (7.21) has been found to perform well in simulations for a broader range of situations in terms of controlling the probability of a type I error (Wilcox, 2011). When $K = 5$ or 6, the method for testing Eq. (7.20) can have an actual type I error probability exceeding .08 when testing at the 0.05 level. Extant results indicate that with $K > 6$, the actual type I error probability will not exceed .075. In contrast, when the method for testing Eq. (7.21) was used, the probability of at least one type I error never exceeded .064. If the distribution of the μ_{tjk} (j fixed) is heavy-tailed, both methods can have type I error probabilities less than .025.

7.11.1 R Functions anova.nestA, mcp.nestA, and anova.nestAP

The R function

$$\text{anova.nestA(x,tr=0.2)}$$

tests the hypothesis given by Eq. (7.20) and the R function

$$\text{mcp.nestA(x,tr=0.2)}$$

tests the hypothesis given by Eq. (7.21). Both of these functions assume the argument x has list mode with length J. Moreover, x[[j]] ($j = 1, \ldots J$) is assumed to contain a matrix with n rows and K columns.

The R function

$$\text{anova.nestAP(x,tr=0.2)}$$

compares the J levels of factor A after pooling the observations over the levels of the nested factor. The hypothesis that the J levels of factor A have a common trimmed mean is tested using the method in Section 7.1.1. Multiple comparisons, based on the pooled data, are performed by the function The R function

$$\text{mcp.nestAP(x,tr=0.2)}$$

7.12 Exercises

1. Describe how M-measures of location might be compared in a two-way design with a percentile bootstrap method. What practical problem might arise when using the bootstrap and sample sizes are small?

2. If data are generated from exponential distributions, what problems would you expect in terms of probability coverage when computing confidence intervals? What problems with power might arise?

3. From well-known results on the random effects model (e.g., Graybill, 1976; Jeyaratnam and Othman, 1985), it follows that

$$\text{BSSW} = \sum \frac{(\bar{Y}_j - \bar{Y})^2}{J-1}$$

estimates

$$\sigma_{wb}^2 + \sum \frac{\sigma_{wj}^2}{Jn_j},$$

and

$$\text{WSSW} = \sum \sum \frac{(Y_{ij} - \bar{Y}_j)^2}{Jn_j(n_j - 1)}$$

estimates

$$\sum \frac{\sigma_{wj}^2}{Jn_j}.$$

Use these result to derive an alternative estimate of ρ_{WI}.

4. Some psychologists have suggested that teachers' expectancies influence intellectual functioning. The file VIQ.dat contains pretest verbal IQ scores for students in grades 1 and 2 who were assigned to one of three ability tracks. (The data are from Elashoff and Snow, 1970, and originally collected by R. Rosenthal. See Section 1.8 on how to obtain this data.) The experimental group consisted of children for whom positive expectancies had been suggested to teachers. Compare the trimmed means of the control group with the experimental group taking into account grade and tracking ability. When examining tracking ability, combine ability levels 2 and 3 into one category, so a 2-by-2-by-2 design is being used.

5. Using the data in the previous exercise, use the function lincon to compare the experimental group with the control group taking into account grade and the two tracking abilities. (Again, tracking abilities 2 and 3 are combined.) Comment on whether the results support the conclusion that the experimental and control group have similar trimmed means.

6. Using the data from the previous two exercises, compare the 20% trimmed means of the experimental group to the control taking into account grade. Also test for no interactions using lincon and linconb. Is there reason to suspect that the confidence interval returned by linconb will be longer than the confidence interval returned by lincon?

7. Suppose three different drugs are being considered for treating some disorder, and it is desired to check for side effects related to liver damage. Further suppose that the following data are collected on 28 participants.

ID	Damage	ID	Damage	ID	Damage
1	92	2	88	3	110
1	91	2	83	3	112
1	84	2	82	3	101
1	78	2	68	3	119
1	82	2	83	3	89
1	90	2	86	3	99
1	84	2	92	3	108
11	91	2	101	3	107
1	78	2	89		
1	95	3	99		

The values under the columns headed by ID indicate which of the three drugs a subject received. Store this data in an R variable having matrix mode with 28 rows and 2 columns with the first column containing the subjects' ID number, and the second column containing the resulting measure of liver damage. For example, the first subject received the first drug and liver damage was rated as 92. Use the function selby to put the data in the second column into an R variable having list mode, then compare the groups using t1way.

8. For the data in the previous exercise, compare the groups using both the Rust–Fligner and Brunner–Dette–Munk methods.

9. For the data in the previous two exercises, perform all pairwise comparisons using the Harrell–Davis estimate of the median.

10. Snedecor and Cochran (1967) report weight gains for rats randomly assigned to one of four diets that varied in the amount and source of protein. The results were as follows:

Beef Low	Beef High	Cereal Low	Cereal High
90	73	107	98
76	102	95	75
90	118	97	56
64	104	80	111
86	81	98	95
51	107	74	88
72	100	74	82
90	87	67	77
95	117	89	86
78	111	58	92

Verify the results based on the R function pba2way mentioned in the example of Section 7.7.1.

11. Generate data for a 2-by-3 design and use the function pbad2way. Note the contrast coefficients for interactions. If you again use pbad2way, but with conall=F, what will happen to these contrast coefficients? Describe the relative merits of using conall=T.

12. For the schizophrenia data in Section 7.8.4, compare the groups with t1way and pbadepth.

CHAPTER 8

Comparing Multiple Dependent Groups

This chapter covers basic methods for comparing dependent groups, including both a between-by-within and a within-by-within design. Three-way designs are covered as well where one or more factors involve dependent groups.

As noted in Chapter 5, when comparing dependent groups based on some measure of location, there are three general approaches that might be used. The first is to compare measures of location associated with the marginal distributions. The second is to make inferences based on a measure of location associated with the difference scores. And the third focuses on measures of location associated with the distribution of the difference between two dependent random variables. When comparing means, it makes no difference which view is adopted, but when using robust measures of location, this is no longer the case. Methods relevant to all three approaches are described and comments on their relative merits are provided.

Note that when comparing measures of location associated with the marginal distributions, there are two types of estimators that might be used. The first estimates a measure of location for each marginal distribution, ignoring the other variables under study. That is, for p-variate data X_{ij} $(i = 1, \ldots, n; j = 1, \ldots, p)$, compute the trimmed mean or some other measure of location using the n values associated with each j. This is in contrast to using a location estimator that takes into account the overall structure of the data when dealing with outliers, such as the OP-estimator in Section 6.5. The bulk of the methods in this chapter are based on the former type of estimator. A multiple comparison procedure that deals with the latter type of estimator is described at the end of Section 8.2.7.

8.1 Comparing Trimmed Means

This section focuses on nonbootstrap methods for testing hypotheses about trimmed means. Methods that use other estimators based on the marginal distributions, such as robust M-estimators, are described in Section 8.2.

8.1.1 Omnibus Test Based on the Trimmed Means of the Marginal Distributions

For J dependent groups, let μ_{tj} be the population trimmed mean associated with the jth group. That is, μ_{tj} is the trimmed mean associated with the jth marginal distribution. The goal in this section is to test

$$H_0 : \mu_{t1} = \cdots = \mu_{tJ},$$

the hypothesis that the trimmed means of J dependent groups are equal. The method used here is based on a generalization of the Huynh–Feldt method for means which is designed to handle violations of the sphericity assumption associated with the standard F-test. (See Kirk, 1995, for details about sphericity. For simulation results on how the test for trimmed means performs, see Wilcox, 1993c.) The method begins by Winsorizing the values in essentially the same manner described in Section 5.9.3. That is, fix j, let $X_{(1)j} \leq X_{(2)j} \leq \cdots \leq X_{(n)j}$ be the n values in the jth group written in ascending order, and let

$$Y_{ij} = \begin{cases} X_{(g+1)j} & \text{if } X_{ij} \leq X_{(g+1)j} \\ X_{ij} & \text{if } X_{(g+1)j} < X_{ij} < X_{(n-g)j} \\ X_{(n-g)j} & \text{if } X_{ij} \geq X_{(n-g)j}, \end{cases}$$

where g is the number of observations trimmed or Winsorized from each end of the distribution corresponding to the jth group. The test statistic, F, is computed as described in Table 8.1, and Table 8.2 describes how to compute the degrees of freedom.

8.1.2 R Function rmanova

The R function

$$\text{rmanova(x,tr=.2,grp=c(1:length(x)))}$$

tests the hypothesis of equal population trimmed means among J dependent groups using the calculations in Tables 8.1 and 8.2. The data are stored in any variable x, which can be either an n-by-J matrix, the jth column containing the data for the jth group, or an R variable having list mode. In the latter case, x[[1]] contains the data for group 1, x[[2]] contains the data for group 2, and so on. As usual, tr indicates the amount of trimming which defaults to 0.2, and grp can be used to compare a subset of the groups. If the argument grp is not specified, the trimmed means of all J groups are compared. If, for example, there are five groups, but the goal is to test $H_0 : \mu_{t2} = \mu_{t4} = \mu_{t5}$, the command rmanova(x,grp=c(2,4,5)) accomplishes this goal using 20% trimming.

■ Example

Section 8.6.2 reports measures of hangover symptoms for participants belonging to one of two groups, with each participant consuming alcohol on three different occasions.

Table 8.1: Test Statistic for Comparing the Trimmed Means of Dependent Groups.

Winsorize the observations in the jth group, as described in this section, yielding Y_{ij}. Let $h = n - 2g$ be the effective sample size, where $g = [\gamma n]$, and γ is the amount of trimming. Compute

$$\bar{X}_t = \frac{1}{J} \sum \bar{X}_{tj}$$

$$Q_c = (n - 2g) \sum_{j=1}^{J} (\bar{X}_{tj} - \bar{X}_t)^2$$

$$Q_e = \sum_{j=1}^{J} \sum_{i=1}^{n} (Y_{ij} - \bar{Y}_{.j} - \bar{Y}_{i.} + \bar{Y}_{..})^2,$$

where

$$\bar{Y}_{.j} = \frac{1}{n} \sum_{i=1}^{n} Y_{ij}$$

$$\bar{Y}_{i.} = \frac{1}{J} \sum_{j=1}^{J} Y_{ij}$$

$$\bar{Y}_{..} = \frac{1}{nJ} \sum_{j=1}^{J} \sum_{i=1}^{n} Y_{ij}.$$

The test statistic is

$$F = \frac{R_c}{R_e},$$

where

$$R_c = \frac{Q_c}{J - 1}$$

$$R_e = \frac{Q_e}{(h - 1)(J - 1)}.$$

For present purposes, focus on group 1 (the control group) with the goal of comparing the responses on the three different occasions. The function rmanova reports a p-value of .09. ∎

8.1.3 Pairwise Comparisons and Linear Contrasts Based on Trimmed Means

Suppose that for J dependent groups, it is desired to compute a $1 - \alpha$ confidence interval for

$$\mu_{tj} - \mu_{tk},$$

Table 8.2: How to Compute Degrees of Freedom when Comparing Trimmed Means?

Let

$$v_{jk} = \frac{1}{n-1} \sum_{i=1}^{n} (Y_{ij} - \bar{Y}_{.j})(Y_{ik} - \bar{Y}_{.k})$$

for $j = 1, \ldots, J$ and $k = 1, \ldots, J$, where Y_{ij} is the Winsorized observation corresponding X_{ij}. When $j = k$, $v_{jk} = s_{wj}^2$, the Winsorized sample variance for the jth group, and when $j \neq k$, v_{jk} is a Winsorized analog of the sample covariance.

Let

$$\bar{v}_{..} = \frac{1}{J^2} \sum_{j=1}^{J} \sum_{k=1}^{J} v_{jk}$$

$$\bar{v}_d = \frac{1}{J} \sum_{j=1}^{J} v_{jj}$$

$$\bar{v}_{j.} = \frac{1}{J} \sum_{k=1}^{J} v_{jk}$$

$$A = \frac{J^2 (\bar{v}_d - \bar{v}_{..})^2}{J - 1}$$

$$B = \sum_{j=1}^{J} \sum_{k=1}^{J} v_{jk}^2 - 2J \sum_{j=1}^{J} \bar{v}_{j.}^2 + J^2 \bar{v}_{..}^2$$

$$\hat{\epsilon} = \frac{A}{B}$$

$$\tilde{\epsilon} = \frac{n(J-1)\hat{\epsilon} - 2}{(J-1)[n - 1 - (J-1)\hat{\epsilon}]}.$$

The degrees of freedom are

$$\nu_1 = (J-1)\tilde{\epsilon}$$

$$\nu_2 = (J-1)(h-1)\tilde{\epsilon},$$

where h is the effective sample size for each group.

for all $j < k$. That is, the goal is to compare all pairs of trimmed means. One possibility is to compare the jth trimmed mean to the kth trimmed mean using the R function yuend in Chapter 5, and control the familywise error (FWE) rate, (the probability of at least one type I error) with the Bonferroni inequality. That is, if C tests are to be performed, perform each test at the α/C level. A practical concern with this approach is that the actual probability of at least one type I error can be considerably less than the nominal level. For example, if $J = 4$, $\alpha = 0.05$, and sampling is from independent normal distributions, the actual probability of at least one type I error is approximately .019 when comparing 20% trimmed means with

$n = 15$. If each pair of random variables has correlation 0.1, the probability of at least one type I error drops to .014, and it drops even more as the correlations are increased. Part of the problem is that the individual tests for equal trimmed means tends to have type I error probabilities less than the nominal level, so performing each test at the α/C level makes matters worse. In fact, even when sampling from heavy-tailed distributions, power can be low compared to using means, even though the sample mean has a much larger standard error (Wilcox, 1997a). One way of improving on this approach is to use results in Rom (1990) to control FWE.

Momentarily consider a single linear contrast

$$\Psi = \sum_{j=1}^{J} c_j \mu_j,$$

where $\sum c_j = 0$ and the goal is to test

$$H_0 : \Psi = 0.$$

Let Y_{ij} ($i = 1, \ldots, n$; $j = 1, \ldots, J$) be the Winsorized values which are computed as described in Section 8.1.1. Let

$$A = \sum_{j-1}^{J} \sum_{k=1}^{J} c_j c_k d_{jk},$$

where

$$d_{jk} = \frac{1}{h(h-1)} \sum_{i=1}^{n} (Y_{ij} - \bar{Y}_j)(Y_{ik} - \bar{Y}_k),$$

and $h = n - 2g$ is the number of observations left in each group after trimming. Let

$$\hat{\Psi} = \sum_{j=1}^{J} c_j \bar{X}_{tj}.$$

The test statistic is

$$T = \frac{\hat{\Psi}}{\sqrt{A}}$$

and the null hypothesis is rejected if $|T| \geq t$, where t is the $1 - \alpha/2$ quantile of a Student's t-distribution with $\nu = h - 1$ degrees of freedom.

When testing C hypotheses, the following method, motivated by results in Rom (1990), appears to be relatively effective at controlling FWE. Let p_k be the p-value associated with the kth hypothesis and put these C p-values in descending order yielding $p_{[1]} \geq \cdots \geq p_{[C]}$. Then

Table 8.3: Critical Values, d_k, for Rom's Method.

k	$\alpha = 0.05$	$\alpha = 0.01$
1	0.05000	0.01000
2	0.02500	0.00500
3	0.01690	0.00334
4	0.01270	0.00251
5	0.01020	0.00201
6	0.00851	0.00167
7	0.00730	0.00143
8	0.00639	0.00126
9	0.00568	0.00112
10	0.00511	0.00101

1. Set $k=1$.
2. If $p_{[k]} \le d_k$, where d_k is read from Table 8.3, stop and reject all C hypotheses; otherwise, go to step 3. (When $k > 10$, then $d_k = \alpha/k$.)
3. Increment k by 1. If $p_{[k]} \le d_k$, stop and reject all hypotheses having p-values less than or equal to d_k
4. If $P_{[k]} > d_k$, repeat step 3.
5. Continue until you reject or all C hypotheses have been tested.

Note that Table 8.3 is limited to $k \le 10$. If $k > 10$, here FWE is controlled with Hochberg's (1988) method. That is, proceed as just indicated, but rather than use d_k read from Table 8.3, use $d_k = \alpha/k$.

8.1.4 Linear Contrasts Based on the Marginal Random Variables

The method just described is readily extended to a situation that contains comparisons based on difference scores as a special case. Let

$$D_{ik} = \sum_{j=1}^{J} c_{jk} X_{ij},$$

where for any k $(k = 1, \ldots, C)$, $\sum c_{jk} = 0$, and let μ_{tk} be the population trimmed mean of the distribution from which the random sample $D_{1k}, \ldots, D_{nk}$ was obtained. For example, if $c_{11} = 1$, $c_{21} = -1$, and $c_{31} = \cdots = c_{J1} = 0$, then

$$D_{i1} = X_{i1} - X_{i2},$$

the difference scores for groups 1 and 2, and μ_{t1} is the (population) trimmed mean associated with this difference. Similarly, if $c_{22} = 1$, $c_{32} = -1$, and $c_{12} = c_{41} = \cdots = c_{J1} = 0$, then

$$D_{i2} = X_{i2} - X_{i3}$$

and μ_{t2} is the corresponding (population) trimmed mean. The goal is to test

$$H_0 : \mu_{tk} = 0$$

for each $k = 1, \ldots, C$ such that FWE is approximately α. Each hypothesis can be tested using results in Chapter 4, but there is the added goal of controlling FWE. Here, Rom's method, described in Section 8.1.3, is used to accomplish this goal.

It should be noted that the multiple comparison procedures in this chapter are designed to control the probability of one or more type I errors. As was the case in Chapter 7, the expectation is that the actual probability of one more type I error will be reduced if the multiple comparison procedures in this chapter are used contingent on a global test rejecting at the α level. That is, power might be adversely affected (cf. Bernhardson, 1975).

Section 5.3.4 described ξ, a robust, heteroscedastic measure of effect size based on the notion of explanatory power. One way of characterizing the difference between two dependent groups is to again use this measure of effect size, which can be done for all pairs of groups via the R function esmcp in Section 7.1.2.

8.1.5 R Function rmmcp and rmmismcp

The R function

$$\text{rmmcp}(x, \text{con} = 0, \text{tr} = 0.2, \text{alpha} = 0.05, \text{dif}=T)$$

performs multiple comparisons among dependent groups using trimmed means and Rom's method for controlling FWE. By default, difference scores are used. Setting dif=F results in comparing the marginal trimmed means. When α differs from both 0.05 and 0.01, FWE is controlled with Hochberg's (1988) method. That is, proceed as indicated in Section 8.1.3 but rather than use d_k from Table 8.3, use $d_k = \alpha/k$.

When there are values missing at random, method M2 in Section 5.9.13 can be used to perform multiple comparisons via the R function

$$\text{rmmismcp}(x, y = NA, \text{alpha} = 0.05, \text{con} = 0, \text{est} = \text{tmean}, \text{plotit} = T, \text{grp} = NA, \text{nboot} =$$
$$500, \text{SEED} = T, \text{xlab} = \text{"Group 1"}, \text{ylab} = \text{"Group 2"}, \text{pr} = F, \ldots),$$

which was introduced in Section 5.9.14 and controls the probability of one or more type I errors using Hochberg's method. By default, 20% trimmed means are used, but other robust estimators can be used via the argument est.

8.1.6 Judging the Sample Size

Let $D_{ijk} = X_{ij} - X_{ik}$ and let μ_{tjk} be a trimmed mean corresponding to D_{ijk}. If when testing $H_0 : \mu_{tjk} = 0$ for any $j < k$, a non-significant result is obtained, this might be because the null hypothesis is true, or of course, a type II error might have been committed due to a sample size that is too small. To help determine whether the latter explanation is reasonable, an extension of Stein's (1945) two-stage method for means might be used. Suppose it is desired to have all-pairs power greater than or equal to $1 - \beta$ when for any $j < k$, $\mu_{tjk} = \delta$. That is, the probability of rejecting H_0 for all $j < k$ for which $\mu_{tjk} = \delta$ is to be at least $1 - \beta$. The goal here is to determine whether the sample size used, namely n, is large enough to accomplish this goal, and if not, the goal is to determine how many more observations are needed. The following method performs well in simulations (Wilcox, 2004b).

Let $C = (J^2 - J)/2$ and

$$d = \left(\frac{\delta}{t_\beta - t_{1-\alpha/(2C)}} \right)^2 ,$$

where t_β is the β quantile of Student's t distribution with $\nu = n - 2g - 1$ degrees of freedom, and g is the number of observations trimmed from each tail. (So $n - 2g$ is the number of observations not trimmed.) Let

$$N_{jk} = \max \left(n, \left[\frac{s_{wjk}^2}{(1 - 2\gamma)^2 d} \right] + 1 \right)$$

where s_{wjk}^2 is the Winsorized variance of the D_{ijk} values. Then the required sample size in the second stage is

$$N = \max N_{jk},$$

the maximum being taken over all $j < k$. So if $N = n$, the sample size used is judged to be adequate for the specified power requirement.

In the event the additional $N - n$ vectors of observations can be obtained, familiarity with Stein's (1945) original method suggests how H_0 should be tested, but in simulations, a slight modification performs a bit better in terms of power. Let S_{wjk} be the Winsorized variance based on all N of the observations, where the amount of Winsorizing is equal to the amount of trimming. Let $\hat{\mu}_{tjk}$ be the trimmed mean based on all N D_{ijk} differences and let

$$T_{jk} = \frac{\sqrt{N}(1 - 2\gamma)\hat{\mu}_{tjk}}{S_{wjk}} .$$

Then reject $H_0 : \mu_{tjk} = 0$ if $|T_{jk}| \geq t_{1-\alpha/(2C)}$. So as would be expected based on Stein's method, the degrees of freedom depend on the initial sample size, n, not the ultimate sample size, N. But contrary to what is expected based on Stein's method, the Winsorized variance when computing T_{jk} is based on all N observations. (All indications are that no adjustment for β is needed when computing d when multiple tests are performed and the goal is to have all-pairs power greater than or equal to $1 - \beta$. Also, a variation of the method aimed at comparing the marginal trimmed means has not been investigated.)

8.1.7 R Functions stein1.tr and stein2.tr

Using the method just described, the R function

$$\text{stein1.tr(x,del,alpha=0.05,pow=0.8,tr=0.2)}$$

determines the required sample size needed to achieve all-pairs power equal to the value indicated by the argument pow for a difference specified by the argument del which corresponds to δ. In the event additional data are needed to achieve the desired amount of power, and if these additional observations can be acquired,

$$\text{stein2.tr(x,y,alpha=0.05,tr=0.2)}$$

tests all pairwise differences. Here the first-stage data are stored in x (which is a vector or a matrix with J columns) and y contains the second-stage data.

8.2 Bootstrap Methods Based on Marginal Distributions

This section focuses on bootstrap methods aimed at making inferences about measures of location associated with the marginal distributions. (Section 8.3 takes up measures of location associated with difference scores.) As in previous chapters, two general types of bootstrap methods appear to deserve serious consideration in applied work. (As usual, this is not intended to suggest that all other variations of the bootstrap have no practical value for the problems considered here, only that based on extant studies, the methods covered here seem to perform relatively well.) The first type uses estimated standard errors and reflects extensions of the bootstrap-t methods in Chapter 5; they are useful when comparing trimmed means. The other is an extension of the percentile bootstrap method where estimated standard errors do not play a direct role. When comparing robust M-measures of location, this latter approach is the only known way of controlling the probability of a type I error for a fairly wide range of distributions.

8.2.1 Comparing Trimmed Means

Let μ_{tj} be the population trimmed mean associated with the jth marginal distribution and consider the goal of testing

$$H_0 : \mu_{t1} = \cdots = \mu_{tJ},$$

An extension of the bootstrap-t method to this problem is straightforward. Set

$$C_{ij} = X_{ij} - \bar{X}_{tj}$$

with the goal of estimating an appropriate critical value, based on the test statistic F in Table 8.1, when the null hypothesis is true. The remaining steps are as follows:

1. Generate a bootstrap sample by randomly sampling, with replacement, n rows of data from the matrix

$$\begin{pmatrix} C_{11}, \dots, C_{1J} \\ \vdots \\ C_{n1}, \dots, C_{nJ} \end{pmatrix}$$

yielding

$$\begin{pmatrix} C_{11}^*, \dots, C_{1J}^* \\ \vdots \\ C_{n1}^*, \dots, C_{nJ}^* \end{pmatrix}.$$

2. Compute the test statistic F in Table 8.1 based on the C_{ij}^* values generated in step 1, and label the result F^*.
3. Repeat steps 1 and 2 B times and label the results $F_1^*, \dots, F_B^*$.
4. Put these B values in ascending order and label the results $F_{(1)}^* \leq \cdots \leq F_{(B)}^*$.

The critical value is estimated to be $F_{(u)}^*$, where $u = (1 - \alpha)B$ rounded to the nearest integer. That is, reject the hypothesis of equal trimmed means if

$$F \geq F_{(u)}^*,$$

where F is the statistic given in Table 8.1 based on the X_{ij} values.

8.2.2 R Function rmanovab

The R function

$$\text{rmanovab(x, tr} = 0.2, \text{alpha} = 0.05, \text{grp} = 0, \text{nboot} = 599)$$

performs the bootstrap-t method just described.

8.2.3 Multiple Comparisons Based on Trimmed Means

This section describes bootstrap methods for performing multiple comparisons based on trimmed means. First consider the goal of performing all pairwise comparisons. That is, the goal is to test

$$H_0 : \mu_{tj} = \mu_{tk}$$

for all $j < k$. A bootstrap-t method is applied as follows. Generate bootstrap samples as was done in Section 8.2.1 yielding

$$\begin{pmatrix} C_{11}^*, \ldots, C_{1J}^* \\ \vdots \\ C_{n1}^*, \ldots, C_{nJ}^* \end{pmatrix}.$$

For every $j < k$, compute the test statistic T_y, given by Eq. (5.24), using the values in the jth and kth columns of the matrix just computed. That is, perform the test for trimmed means corresponding to two dependent groups using the data $C_{1j}^*, \ldots, C_{nj}^*$ and $C_{1k}^*, \ldots, C_{nk}^*$. Label the resulting test statistic T_{yjk}^*. Repeat this process B times yielding $T_{yjk1}^*, \ldots, T_{yjkB}^*$. Because these test statistics are based on data generated from a distribution for which the trimmed means are equal, they can be used to estimate an appropriate critical value. In particular, for each b, set

$$T_b^* = \max|T_{yjkb}^*|,$$

the maximum being taken over all $j < k$. Let $T_{(1)}^* \leq \cdots \leq T_{(B)}^*$ be the T_b^* values written in ascending order and let $u = (1 - \alpha)B$, rounded to the nearest integer. Then $H_0 : \mu_{tj} = \mu_{tk}$ is rejected if $T_{yjk} > T_{(u)}^*$. That is, for the jth and kth groups, test the hypothesis of equal trimmed means using the method in Section 5.9.5, only the critical value is $T_{(u)}^*$, which was determined so that the probability of a least one type I error is approximately equal to α. Alternatively, the confidence interval for $\mu_{tj} - \mu_{tk}$ is

$$(\bar{X}_{tj} - \bar{X}_{tk}) \pm T_{(u)}^* \sqrt{d_j + d_k - 2d_{jk}},$$

where $\sqrt{d_j + d_k - 2d_{jk}}$ is the estimate of the standard error of $\bar{X}_{tj} - \bar{X}_{tk}$, which is computed as described in Section 5.9.5. The simultaneous probability coverage is approximately $1 - \alpha$. Probability coverage appears to be reasonably good with n as small as 15 when using 20% trimming with $J = 4$, $\alpha = 0.05$, $B = 599$ (Wilcox, 1997a). When there is no trimming, probability coverage can be poor, and no method can be recommended. Also, the power of the bootstrap method, with 20% trimmed means, compares well to an approach based on means and the Bonferroni inequality.

The method is easily extended to situations where the goal is to test C linear contrasts, $\Psi_1, \ldots, \Psi_C$, where

$$\Psi_k = \sum c_{jk} \mu_{tj},$$

and c_{jk} ($j = 1, \ldots, J$ and $k = 1, \ldots, C$) are constants chosen to reflect some hypothesis of interest. As before, Ψ_k is estimated with $\hat{\Psi}_k = \sum c_{jk} \bar{X}_{tj}$, but now the squared standard error is estimated with

$$A_k = \sum_{j=1}^{J} \sum_{\ell=1}^{J} c_{jk} c_{\ell k} d_{j\ell},$$

where

$$d_{jk} = \frac{1}{h(h-1)} \sum (Y_{ij} - \bar{Y}_j)(Y_{ik} - \bar{Y}_k),$$

and Y_{ij} are the Winsorized observations for the jth group. (When $j = k$, $d_{jk} = d_j^2$.)

To compute a $1 - \alpha$ confidence interval for Ψ_k, generate a bootstrap sample yielding C_{ij}^* and let

$$T_{yk}^* = \frac{\hat{\Psi}_k^*}{\sqrt{A_k^*}},$$

where $\hat{\Psi}_k^*$ and A_k^* are computed with the bootstrap observations. Repeat this bootstrap process B times yielding T_{ykb}^*, $b = 1, \ldots B$. For each b, let $T_b^* = \max|T_{ykb}^*|$, the maximum being taken over $k = 1, \ldots, C$. Put the T_b^* values in order yielding $T_{(1)}^* \leq \cdots \leq T_{(B)}^*$, in which case an appropriate critical value is estimated to be $T_{(u)}^*$, where $u = (1 - \alpha)B$, rounded to the nearest integer. Then an approximate $1 - \alpha$ confidence interval for Ψ_k is

$$\hat{\Psi}_k \pm T_{(u)}^* \sqrt{A_k}.$$

8.2.4 R Functions pairdepb and bptd

The R function

$$\text{pairdepb(x,tr=0.2,alpha=0.05,grp=0,nboot=599)}$$

performs all pairwise comparisons among J dependent groups using the bootstrap method just described. The argument x can be an n-by-J matrix of data, or it can be an R variable having list mode. In the latter case, x[[1]] contains the data for group 1, x[[2]] contains the data for group 2, and so on. The argument tr indicates the amount of trimming, which, if unspecified, defaults to 0.2. The value for α defaults to alpha=0.05, and B defaults to nboot=599. The argument grp can be used to test the hypothesis of equal trimmed means using a subset of the groups. If missing values are detected, they are eliminated via the function elimna described in Section 1.9.1.

■ Example

For the alcohol data reported in Section 8.6.2, suppose it is desired to perform all pairwise comparisons using the time 1, time 2, and time 3 data for the control group. The R function pairdepb returns

```
$test:
        Group  Group        test         se
[1,]      1      2    -2.115985   1.693459
[2,]      1      3    -2.021208   1.484261
[3,]      2      3     0.327121   1.783234


$psihat:
        Group  Group       psihat    ci.lower     ci.upper
[1,]      1      2    -3.5833333   -7.194598   0.02793158
[2,]      1      3    -3.0000000   -6.165155   0.16515457
[3,]      2      3     0.5833333   -3.219376   4.38604218


$crit:
[1] 2.132479
```

Thus, none of the pairwise differences is significantly different at the 0.05 level.

Assuming the data are stored in the R variable dat, the command pairdepb(dat,grp=c(1,3)) would compare groups 1 and 3, ignoring group 2. It is left as an exercise to show that if the data are stored in list mode, the command ydbt(dat[[1]],dat[[3]]) returns the same confidence interval.

The function

$$bptd(x,tr=0,alpha=0.05,con=0,nboot=599)$$

computes confidence intervals for each of C linear contrasts, Ψ_k, $k = 1, \ldots, C$, such that the simultaneous probability coverage is approximately $1 - \alpha$. The only difference between bptd and pairedpb is that bptd can handle a set of specified linear contrasts via the argument con. The argument con is a J-by-C matrix containing the contrast coefficients. The kth column of con contains the contrast coefficients corresponding to Ψ_k. If con is not specified, all pairwise comparisons are performed. So for this special case, pairedpb and bptd always produce the same results.

■ Example

If there are three dependent groups, and con is a 3-by-1 matrix with the values 1, −1, and 0, and if the data are stored in the R variable xv, the command bptd(xv,con=con) will compute a confidence interval for $\Psi = \mu_{t1} - \mu_{t2}$, the difference between the 20% trimmed means corresponding to the first two groups. If xv has list mode, the command ydbt(xv[[1]],xv[[2]]) returns the same confidence interval. (The function ydbt was described in Section 5.9.8.)

8.2.5 Percentile Bootstrap Methods

This section describes two types of percentile bootstrap methods that can be used to compare J dependent groups based on any measures of location, θ, associated with the marginal distributions. Included as special cases are M-measures of location and trimmed means. The goal is to test

$$H_0 : \theta_1 = \cdots = \theta_J. \tag{8.1}$$

Method RMPB3

The first method uses the test statistic

$$Q = \sum (\hat{\theta}_j - \bar{\theta})^2,$$

where $\bar{\theta} = \sum \hat{\theta}_j / J$. An appropriate critical value is estimated using an approach similar to the bootstrap-t technique. First, set $C_{ij} = X_{ij} - \hat{\theta}_j$. That is, shift the empirical distributions so that the null hypothesis is true. Next a bootstrap sample is obtained by resampling, with replacement, as described in step 1 of Section 8.2.1. As usual, label the results

$$\begin{pmatrix} C_{11}^*, \ldots, C_{1J}^* \\ \vdots \\ C_{n1}^*, \ldots, C_{nJ}^* \end{pmatrix}.$$

For the jth column of the bootstrap data just generated, compute the measure of location that is of interest and label it $\hat{\theta}_j^*$. Compute

$$Q^* = \sum (\hat{\theta}_j^* - \bar{\theta}^*)^2,$$

where $\bar{\theta}^* = \sum \hat{\theta}_j^* / J$, and repeat this process B times yielding $Q_1^*, \ldots, Q_B^*$. Put these B values in ascending order yielding $Q_{(1)}^* \leq \cdots \leq Q_{(B)}^*$. Then reject the hypothesis of equal measures of location if $Q > Q_{(u)}^*$, where again $u = (1 - \alpha)B$ rounded to the nearest integer.

Method RMPB4

If the null hypothesis is true, then all J groups have a common measure of location, θ. The next method estimates this common measure of location and then checks to see how deeply it is nested within the bootstrap values obtained when resampling from the original values. That is, in contrast to method RMPB3, the data are not centered, and bootstrap samples are obtained by resampling rows of data from

$$\begin{pmatrix} X_{11}, \ldots, X_{1J} \\ \vdots \\ X_{n1}, \ldots, X_{nJ} \end{pmatrix}$$

yielding

$$\begin{pmatrix} X_{11}^*, \ldots, X_{1J}^* \\ \vdots \\ X_{n1}^*, \ldots, X_{nJ}^* \end{pmatrix}.$$

For the jth group (or column of bootstrap values) compute $\hat{\theta}_j^*$. Repeating this process B times yields $\hat{\theta}_{jb}^*$, $(j = 1, \ldots, J; b = 1, \ldots, B)$. The remaining calculations are performed as outlined in Table 8.4.

Table 8.4: Repeated Measures ANOVA Based on the Depth of the Grand Mean.

Goal: Test the hypothesis

$$H_0 : \theta_1 = \cdots = \theta_J.$$

1. Compute

$$S_{jk} = \frac{1}{B-1} \sum_{b=1}^{B} (\hat{\theta}_{jb}^* - \bar{\theta}_j^*)(\hat{\theta}_{kb}^* - \bar{\theta}_k^*),$$

where

$$\bar{\theta}_j^* = \frac{1}{B} \sum_{b=1}^{B} \hat{\theta}_{jb}^*.$$

(The quantity S_{jk} is the sample covariance of the bootstrap values corresponding to the jth and kth groups.)

2. Let

$$\hat{\theta}_b^* = (\hat{\theta}_{1b}^*, \ldots, \hat{\theta}_{Jb}^*)$$

and compute

$$d_b = (\hat{\theta}_b^* - \hat{\theta}) \mathbf{S}^{-1} (\hat{\theta}_b^* - \hat{\theta})',$$

where $\mathbf{S}$ is the matrix corresponding to S_{jk}, $\hat{\theta} = (\hat{\theta}_1, \ldots, \hat{\theta}_J)$, $\hat{\theta}_j$ is the estimate of θ based on the original data for the jth group (the X_{ij} values, $i = 1, \ldots, n$), and $\hat{\theta}_b = (\hat{\theta}_{1b}, \ldots, \hat{\theta}_{Jb})$. The value of d_b measures how far away the bth bootstrap vector of location estimators is from $\hat{\theta}$, which is roughly the center of all B bootstrap values.

3. Put the d_b values in ascending order: $d_{(1)} \leq \cdots \leq d_{(B)}$.
4. Let $\hat{\theta}_G = (\bar{\theta}, \ldots, \bar{\theta})$, where $\bar{\theta} = \sum \hat{\theta}_j / J$, and compute

$$D = (\hat{\theta}_G - \hat{\theta}) \mathbf{S}^{-1} (\hat{\theta}_G - \hat{\theta})'.$$

D measures how far away the estimated common value is from the observed measures of location (based on the original data).

5. Reject if $D \geq d_{(u)}$, where $u = (1 - \alpha)B$, rounded to the nearest integer.

For completeness, yet another approach to comparing dependent groups is to use a *mixed linear model* in conjunction with the regression MM-estimator introduced in Chapter 10. Heritier, Cantoni, Copt, and Victoria-Feser (2009, Section 4.5) summarize the relevant details and computations. The mixed linear model has the form

$$Y = \mathbf{X}\alpha + \sum Z_j \beta_j + \epsilon,$$

where Y is a vector of N measurements, $\mathbf{X}$ is an $n \times q$ design matrix for the fixed effects, $\mathbf{Z}_j$ is an $N \times q_j$ design matrix for the random effects β_j, and ϵ is an N-vector of independent residual errors. Evidently it is unknown what advantages this approach might have, in terms of type I errors and power, over the other methods covered in this chapter. (Some concerns about the regression MM-estimator are described in Chapter 10.) Copt and Heritier (2007) derived a (nonbootstrap) method for testing hypotheses that is based in part on an appropriate estimate of the standard errors. However, a general pattern regarding M-estimators seems to be that nonbootstrap methods that use a test statistic based on an estimate of the standard error can perform poorly in terms of type I errors and probability coverage when dealing with skewed distributions. Perhaps the MM-estimator, in the context of the mixed linear model, is an exception, but this has not been investigated.

8.2.6 R Functions bd1way and ddep

The R functions

$$bd1way(x, est = onestep, nboot = 599, alpha = 0.05)$$

and

$$ddep(x, alpha = 0.05, est = onestep, grp = NA, nboot = 500)$$

perform the percentile bootstrap methods just described. The first function performs method RMPB3; it uses by default the one-step M-estimator of location (based on Huber's Ψ), but any other estimator can be used via the argument est. As usual, x is any R variable that is a matrix or has list mode, nboot is B, the number of bootstrap samples to be used, and grp can be used to analyze a subset of the groups, with the other groups ignored. (That is, grp is used as illustrated in Section 8.1.2.) The function ddep performs method RMPB4 described in Table 8.4.

When there are values missing at random, method M2 in Section 5.9.13 can be used to perform multiple comparisons via the R function

$$rmmismcp(x,y = NA, alpha = 0.05, con = 0, est = tmean, plotit = T, grp = NA, nboot = 500, SEED = T, xlab = ``Group 1", ylab = ``Group 2", pr = F, \ldots).$$

By default, 20% trimmed means are used, but other robust estimators can be used via the argument est.

■ Example

Table 6.5 shows the weight of cork borings taken from north, east, south, and west sides of the 28 trees. Assuming the data are stored in the R matrix cork, the command bd1way(cork) returns:

```
$test:
17.08

$crit:
34.09
```

So comparing one-step M-estimators, we fail to reject the hypothesis that the typical weight of a cork boring is the same for all four sides of a tree. If we compare groups using MOM in conjunction with method RMPB4, the *p*-value is .385. (Compare this result to the Example in Section 8.2.8.)

■ Example

Again consider the hangover data used to illustrate rmanova in Section 8.1.2. (The data are listed in Section 8.6.2.) Comparing M-measures of location results in an error because there are too many tied values resulting in MAD=0 within the bootstrap. Assuming the data are stored in x, the command bd1way(x,est=hd) compares medians based on the Harrell–Davis estimator. The function reports that $Q = 9.96$ with a 0.05 critical value of 6.3, so the null hypothesis is rejected at the 0.05 level.

8.2.7 *Multiple Comparisons Using M-estimators or Skipped Estimators*

Next consider C linear contrasts involving M-measures of location where the kth linear contrast is

$$\Psi_k = \sum_{j=1}^{J} c_{jk}\mu_{mj},$$

and, as usual, the c_{jk} values are constants that reflect linear combinations of the M-measures of location that are of interest and for fixed k, $\sum c_{jk} = 0$. The goal is to compute a confidence interval for Ψ_k, $k = 1, \ldots, C$, such that the simultaneous probability coverage is approximately $1 - \alpha$. Alternatively, test $H_0 : \Psi_k = 0$ with the goal that the probability of at least one type I error is α.

First, set $C_{ij} = X_{ij} - \hat{\mu}_{mj}$. Next, obtain a bootstrap sample by sampling, with replacement, n rows of data from the matrix C_{ij}. Label the bootstrap values C^*_{ij}. Use the n values in the jth column of C^*_{ij} to compute $\hat{\mu}^*_{mj}$, $j = 1, \ldots, J$. Repeat this process B times yielding $\hat{\mu}^*_{mjb}$, $b = 1, \ldots, B$. Next, compute the J-by-J covariance matrix associated with the $\hat{\mu}^*_{mjb}$ values. That is, compute

$$\hat{\tau}_{jk} = \frac{1}{B-1} \sum_{b=1}^{B} (\hat{\mu}^*_{mjb} - \bar{\mu}^*_j)(\hat{\mu}^*_{mkb} - \bar{\mu}^*_k),$$

where $\bar{\mu}^*_j = \sum \hat{\mu}^*_{mjb} / B$. Let

$$\hat{\Psi}_k = \sum_{j=1}^{J} c_{jk} \hat{\mu}_{mj},$$

$$\hat{\Psi}^*_{kb} = \sum_{j=1}^{J} c_{jk} \hat{\mu}^*_{mjb},$$

$$S^2_k = \sum_j \sum_\ell c_{jk} c_{\ell k} \hat{\tau}_{j\ell},$$

$$T^*_{kb} = \frac{\hat{\Psi}^*_{kb}}{S_k},$$

and

$$T^*_b = \max |T^*_{kb}|,$$

the maximum being taken over $k = 1, \ldots, C$. Then a confidence interval for Ψ_k is

$$\hat{\Psi}_k \pm T^*_{(u)} S_k,$$

where as usual $u = (1 - \alpha)B$, rounded to the nearest integer, and $T^*_{(1)} \leq \cdots \leq T^*_{(B)}$ are the T^*_b values written in ascending order. The simultaneous probability coverage is approximately $1 - \alpha$, but for $n \leq 21$ and $B = 399$, the actual probability coverage might be unsatisfactory. Under normality, for example, there are situations where the probability of at least one type I error exceeds .08 with $J = 4$, $\alpha = 0.05$, and $n = 21$, and where all pairwise comparisons are performed. Increasing $B = 599$ does not correct this problem. It seems that $n > 30$ is required if the probability of at least one type I error is not to exceed .075 when testing at the 0.05 level (Wilcox, 1997a).

An alternative approach, which appears to have some practical advantages over the method just described, is to use a simple extension of the percentile bootstrap method in

Section 5.9.11. Let $\hat{p}_k^*$ be the proportion of times $\hat{\Psi}_{kb}^* > 0$ among the B bootstrap samples. Then a (generalized) p-value for H_0: $\Psi_k = 0$ is $2\min(\hat{p}_k^*, 1 - \hat{p}_k^*)$. When using M-estimators or MOM, however, a bias adjusted estimate of the p-value appears to be beneficial; see Section 5.9.11. (With trimmed means, this bias adjustment appears to be unnecessary; see Wilcox & Keselman, 2002). FWE can be controlled with method SR outlined in Section 7.6.2. Again, for large sample sizes, say greater than 80, Hochberg's method (mentioned in Section 8.1.3) appears to be preferable.

Note that all of the methods described so far are based on measures of location that do not take into account the overall structure of the data when dealing with outliers. The skipped estimators in Section 6.5 do take the overall structure of the data into account and situations might be encountered where this makes a practical difference. A basic percentile bootstrap method can be used to test H_0: $\Psi_k = 0$ and appears to control the probability of a type I error reasonably well when using the OP estimator in Section 6.5.

8.2.8 R Functions lindm and mcpOV

Using the first method described in the Section 8.2.7, the R function lindm computes confidence intervals for C linear contrasts involving M-measures of location corresponding to J dependent groups. (The second method is performed by the R function in Section 8.3.3.) The function has the form

$$\text{lindm(x,con=0,est=onestep,grp=0,alpha=0.05,nboot=399, \ldots).}$$

The argument x contains the data and can be any n-by-J matrix, or it can have list mode. In the latter case, x[[1]] contains the data for group 1, x[[2]] the data for group 2, and so on. The optional argument con is a J-by-C matrix containing the contrast coefficients. If not specified, all pairwise comparisons are performed. The argument est is any statistic of interest. If unspecified, a one-step M-estimator is used. The argument grp can be used to select a subset of the groups for analysis. As usual, alpha is α and defaults to 0.05, and nboot is B which defaults to 399. The argument $\ldots$ is any additional arguments that are relevant to the function est.

■ Example

Again consider the hangover data (reported in Section 8.6.2) where two groups of participants are measured at three different times. Suppose the first row of data is stored in DAT[[1]], the second row in DAT[[2]], the third in DAT[[3]], and so forth, but it is desired to perform all pairwise comparisons using only the group 1 data at times 1, 2, and 3. Then the command lindm(DAT,grp=c(1:3)) attempts to perform the comparisons, but eventually the function terminates with the error message "missing values in x not allowed." This error arises because there are so many tied values, bootstrap samples yield $\mathrm{MAD} = 0$ which in turn makes it impossible to compute $\hat{\mu}^*$. This error can also arise when there are no tied values but one or more of the sample sizes are smaller than 20.

∎

The command lindm(DAT,est=hd,gpr=c(1:3)) compares medians instead, using the Harrell–Davis estimator, and returns

```
        con.num      psihat    ci.lower    ci.upper          se
[1,]          1  -3.90507026   -7.880827  0.07068639    2.001571
[2,]          2  -3.82383677   -9.730700  2.08302610    2.973775
[3,]          3   0.08123349   -6.239784  6.40225079    3.182279

$crit:
[1] 1.986318

$con:
       [,1]  [,2]  [,3]
[1,]      1     1     0
[2,]     -1     0     1
[3,]      0    -1    -1
```

Because the argument con was not specified, the function creates its own set of linear contrasts assuming all pairwise comparisons are to be performed. The resulting contrast coefficients are returned in the R variable $con. Thus, the first column, containing the values $1, -1$, and 0, indicates that the first contrast corresponds to the difference between the medians for times 1 and 2. The results in the first row of $con.num indicate that the estimated difference between these medians is -3.91, and the confidence interval is $(-7.9, 0.07)$. In a similar fashion, the estimated difference between the medians at times 1 and 3 is -3.82, and for time 2 versus time 3 the estimate is 0.08. The command lindm(DAT,est=hd,gpr=c(1:3),q=.4) would compare 0.4 quantiles.

The R function

mcpOV(x,alpha=0.05,nboot=NA,grp=NA,est=smean,con=0,bhop=F,SEED=T, . . .).

is like the R function the function lindm, only it is designed to handle skipped estimators that take into account the overall structure of the data when checking for outliers. By default it uses the OP-estimator, which is based on the projection method for detecting outliers.

8.3 Bootstrap Methods Based on Difference Scores

The following method, based on difference scores, has been found to have practical value, particularly in terms of controlling type I error probabilities when sample sizes are very small.

First consider the goal of testing the hypothesis that a measure of location associated with the difference scores $D_{ij} = X_{ij} - X_{i,j+1}$ has the value zero. That is, use the difference between the ith observation in group j and the ith observation in group $j+1$, $j = 1, \ldots, J-1$. Let θ_j be any measure of location associated with the D_{ij} values. So, for example, θ_1 might be an M-measure of location corresponding to the difference scores between groups 1 and 2, and θ_2 might be the M-measure of location associated with difference scores between groups 2 and 3. A simple alternative to Eq. (8.1) is to test

$$H_0 : \theta_1 = \cdots = \theta_{J-1} = 0, \tag{8.2}$$

the hypothesis that the typical difference scores do not differ and are all equal to zero. However, a criticism of this approach is that the outcome can depend on how we order the groups. That is, rather than take differences between groups 1 and 2, we could just as easily take differences between groups 1 and 3, which might alter our conclusions about whether to reject. We can avoid this problem by instead taking differences among all pairs of groups. There are a total of

$$L = \frac{J^2 - J}{2}$$

such differences which are labeled $D_{i\ell}$, $i = 1, \ldots, n$; $\ell = 1, \ldots, L$.

■ Example

For four groups ($J = 4$), there are $L = 6$ differences given by

$$D_{i1} = X_{i1} - X_{i2},$$

$$D_{i2} = X_{i1} - X_{i3},$$

$$D_{i3} = X_{i1} - X_{i4},$$

$$D_{i4} = X_{i2} - X_{i3},$$

$$D_{i5} = X_{i2} - X_{i4},$$

$$D_{i6} = X_{i3} - X_{i4}.$$

■

The goal is to test

$$H_0 : \theta_1 = \cdots = \theta_L = 0, \tag{8.3}$$

where θ_ℓ is the population measure of location associated with the ℓth set of difference scores, $D_{i\ell}$ ($i = 1, \ldots, n$). To test H_0 given by Eq. (8.3), resample vectors of D values, but unlike the bootstrap-t, observations are not centered. That is, a bootstrap sample now consists of resampling with replacement n rows from the matrix

$$
\begin{pmatrix}
D_{11}, \ldots, D_{1L} \\
\vdots \\
D_{n1}, \ldots, D_{nL}
\end{pmatrix}.
$$

yielding

$$
\begin{pmatrix}
D_{11}^*, \ldots, D_{1L}^* \\
\vdots \\
D_{n1}^*, \ldots, D_{nL}^*
\end{pmatrix}.
$$

For each of the L columns of the D^* matrix, compute whatever measure of location is of interest, and for the ℓth column label the result $\hat{\theta}_\ell^*$ ($\ell = 1, \ldots, L$). Next, repeat this B times yielding $\hat{\theta}_{\ell b}^*$, $b = 1, \ldots, B$ and then determine how deeply the vector $\mathbf{0} = (0, \ldots, 0)$, having length L, is nested within the bootstrap values $\hat{\theta}_{\ell b}^*$. For two groups, this is tantamount to determining how many bootstrap values are greater than zero, which leads to the (generalized) p-value described in Section 5.4. The computational details when dealing with more than two groups are relegated to Table 8.5.

8.3.1 R Function rmdzero

The R function

$$
\text{rmdzero(x,est} = \text{mom, grp} = \text{NA, nboot} = \text{NA}, \ldots)
$$

performs the test on difference scores outlined in Table 8.5.

■ Example

For the cork data in Table 6.5, rmdzero returns a p-value of .044, so in particular reject with $\alpha = 0.05$. That is, conclude that the typical difference score is not equal to zero for all pairs of groups. This result is in sharp contrast to comparing marginal measures of location based on a robust M-estimator or MOM and the method in Table 8.4; see the Example in Section 8.2.6. ■

8.3.2 Multiple Comparisons

Multiple comparisons based on a percentile bootstrap method and difference scores can be addressed as follows. First generate a bootstrap sample as described at the beginning of this

Table 8.5: Repeated Measures ANOVA Based on Difference Scores

Goal: Test the hypothesis given by Eq. (8.3).

1. Let $\hat{\theta}_\ell$ be the estimate of θ_ℓ. Compute bootstrap estimates as described in Section 8.3 and label them $\hat{\theta}_{\ell b}^*$, $\ell = 1, \ldots, L$; $b = 1, \ldots, B$.

2. Compute the L-by-L matrix

$$S_{\ell\ell'} = \frac{1}{B-1} \sum_{b=1}^{B} (\hat{\theta}_{\ell b}^* - \hat{\theta}_\ell)(\hat{\theta}_{\ell' b}^* - \hat{\theta}_{\ell'}).$$

Readers familiar with multivariate statistical methods might notice that $S_{\ell\ell'}$ uses $\hat{\theta}_\ell$ (the estimate of θ_ℓ based on the original difference values) rather than the seemingly more natural $\bar{\theta}_\ell^*$, where

$$\bar{\theta}_\ell^* = \frac{1}{B} \sum_{b=1}^{B} \hat{\theta}_{\ell b}^*.$$

If $\bar{\theta}_\ell^*$ is used, unsatisfactory control over the probability of a type I error can result.

3. Let $\hat{\theta} = (\hat{\theta}_1, \ldots, \hat{\theta}_L)$, $\hat{\theta}_b^* = (\hat{\theta}_{1b}^*, \ldots, \hat{\theta}_{Lb}^*)$ and compute

$$d_b = (\hat{\theta}_b^* - \hat{\theta})\mathbf{S}^{-1}(\hat{\theta}_b^* - \hat{\theta})',$$

where $\mathbf{S}$ is the matrix corresponding to $S_{\ell\ell'}$.

4. Put the d_b values in ascending order. $d_{(1)} \leq \cdots \leq d_{(B)}$.

5. Let

$$\mathbf{0} = (0, \ldots, 0)$$

having length L.

6. Compute

$$D = (\mathbf{0} - \hat{\theta})\mathbf{S}^{-1}(\mathbf{0} - \hat{\theta})'.$$

D measures how far away the null hypothesis is from the observed measures of location (based on the original data). In effect, D measures how deeply $\mathbf{0}$ is nested within the cloud of bootstrap values.

7. Reject if $D \geq d_{(u)}$, where $u = (1 - \alpha)B$, rounded to the nearest integer.

section yielding $D_{i\ell}^*$, $\ell = 1, \ldots, L$. When all pairwise differences are to be tested, $L = (J^2 - J)/2$, $\ell = 1$ corresponds to comparing group 1 to group 2, $\ell = 2$ is comparing group 1 to group 3, and so on. Let $\hat{p}_\ell^*$ be the proportion of times among B bootstrap resamples that $D_{i\ell}^* > 0$. As usual, let

$$\hat{p}_{m\ell}^* = \min(\hat{p}_\ell^*, \ 1 - \hat{p}_\ell^*),$$

in which case $2\hat{p}_{m\ell}^*$ is the estimated (generalized) p-value for the ℓth comparison.

One approach to controlling FWE is to put the p-values in descending order and to make decisions about which hypotheses are to be rejected using method SR outlined in

Section 7.6.2. That is, once the $\hat{p}^*_{mc}$ are computed, reject the hypothesis corresponding to $\hat{p}^*_{mc}$ if $\hat{p}^*_{mc} \leq \alpha_c$, where α_c is read from Table 7.13.

As for linear contrasts, consider any specific linear contrast with contrast coefficients $c_1, \ldots, c_J$, set

$$D_i = \sum c_j X_{ij},$$

and let θ_d be some (population) measure of location associated with this sum. Then $H_0 : \theta_d = 0$ can be tested by generating a bootstrap sample from the D_i values, repeating this B times, computing $\hat{p}^*$, the proportion of bootstrap estimates that are greater than zero, in which case $2\min(\hat{p}^*, 1 - \hat{p}^*)$ is the estimated significance level. Then FWE can by controlled in the manner just outlined.

When comparing groups using MOM or M-estimators, at the moment it seems that the method based on difference scores often provides the best power versus testing hypotheses based on measures of location associated with the marginal distributions. Both approaches do an excellent job of avoiding type I error probabilities greater than the nominal α level. But when testing hypotheses about measures of location associated with the marginal distributions, the actual type I error probability can drop well below the nominal level in situations where the method based on difference scores avoids this problem. This suggests that the method based on difference scores will have more power, and indeed, there are situations where this is the case even when the two methods have comparable type I error probabilities. It is stressed, however, that a comparison of these methods, in terms of power, needs further study. Also the bias adjusted critical value mentioned in Section 5.9.7 appears to help increase power.

While comparing marginal measures of location based on MOM or an M-estimator seems to result in relatively low power, there is weak evidence that comparing marginal measures of location based on the OP-estimator, via a percentile bootstrap method as mentioned at the end of Section 8.2.7, performs relatively well. But the extent this is true needs additional study.

When dealing with trimmed means, again the percentile bootstrap method just described can be used and appears to be relatively good choice provided the amount of trimming is not too small. With a small amount of trimming, use a bootstrap-t method instead. Controlling the probability of at least one type I error can be done with Hochberg's method.

8.3.3 R Functions rmmcppb, wmcppb, dmedpb, and lindepbt

The R function

rmmcppb(x,y = NA, alpha = 0.05, con = 0, est = mom, plotit = T, dif = T, grp = NA, nboot = NA, BA=F,hoch=F, . . .)

performs multiple comparisons among dependent groups using the percentile bootstrap methods just described. The argument dif defaults to T (for true) indicating that difference scores will be used, in which case Hochberg's method is used to control FWE. If dif=F, measures of location associated with the marginal distributions are used instead. If dif=F and BA=T, the bias adjusted estimate of the generalized p-value (described in Section 5.9.7) is applied; using BA=T (when dif=F) is recommended when comparing groups with M-estimators and MOM, but it is not necessary when comparing 20% trimmed means (Wilcox & Keselman, 2002). If hoch=F, then FWE is controlled using method SR in Section 7.6.2 if the sample size is less than 80, otherwise Hochberg's method is used as described in Section 8.1.3. If hoch=T, Hochberg's method is used regardless of the sample size. If no value for con is specified, then all pairwise differences will be tested. As usual, if the goal is to test hypotheses other than all pairwise comparisons, con can be used to specify the linear contrast coefficients.

When comparing trimmed means, it appears that Hochberg's method is preferable to method SR in terms of controlling the probability of at least one type I error. For convenience, the R function

$$\text{wmcppb(x, alpha} = 0.05, \text{con} = 0, \text{est} = \text{tmean, plotit} = T, \text{dif} = T, \text{grp} = NA, \text{nboot} = NA,$$
$$\text{BA=F,hoch=T, ...)}$$

is supplied. It is the same as the R function rmmcppb, only it defaults to comparing 20% trimmed means, and by default it uses Hochberg's method rather than method SR. (The R function dtrimpb is the same as the function wmmcppb.)

The R function

$$\text{dmedpb(x,y=NA,alpha=0.05,con=0,est=median,plotit=T,dif=F,grp=NA,}$$
$$\text{hoch=T,nboot=NA,xlab="Group 1",ylab="Group 2",pr=T,SEED=T,BA=F, ...)}$$

is similar to the R function rmmcppb, only it defaults to comparing medians and it is designed to handle tied values. Hochberg's method is used to control FWE. With a small sample size, say less than 30, setting the argument BA=T seems advisable, meaning that the p-value is adjusted as described in 5.9.11 (Wilcox, 2006b).

The R function

$$\text{lindepbt(x, con} = \text{NULL, tr} = 0.2, \text{alpha} = 0.05, \text{nboot=599,dif=T,SEED=T)}$$

performs multiple comparisons based on trimmed means using a bootstrap-t method. When the amount of trimming is small, a bootstrap-t method is preferable to a percentile bootstrap method, but it is unclear at what point this will be case. The function reports critical p-values based on Rom's method for controlling the probability of one or more type I errors. The function returns confidence intervals, but they are not adjusted so that the simultaneously probability coverage is $1 - \alpha$. Rather, each confidence interval is designed to have probability coverage $1 - \alpha$.

■ **Example**

For the cork boring data in Table 6.5, the R function wmcppb (with the argument dif=F as well as dif=T) finds no significant results when the probability of at least one type I error is taken to be .05. But the R function mcpOV (in Section 8.2.8), which compares marginal measures of location via the OP-estimator, finds three significant results, the only point being that the choice of method can make a practical difference. Again it is stressed that little is known about the extent the OP-estimator might have higher power compared to the many other methods that might be used to compare dependent groups.

■

8.4 Comments on which Method to Use

No single method in this chapter dominates based on various criteria used to compare hypothesis testing techniques. However, a few comments can be made about their relative merits that might be useful. First, the expectation is that in many situations where groups differ, all methods based on means perform poorly, in terms of power, relative to approaches based on some robust measure of location such as MOM or a 20% trimmed mean. Currently, with a sample size as small as 21, the bootstrap-t method in Section 8.2.2, which is performed by the R function rmanovab, appears to provide excellent control over the probability of a type I error when used in conjunction with 20% trimmed means. Its power compares reasonably well to most other methods that could be used, but as noted in previous chapters, different methods are sensitive to different features of the data and arguments for some other measure of location, such as an M-estimator, have been made.

The percentile bootstrap methods in Section 8.2.5 also do an excellent job of avoiding type I errors greater than the nominal level, but there are indications that when using method RMPB3, and the sample size is small, the actual probability of a type I error can be substantially less than α suggesting that some other method might provide better power. Nevertheless, if there is specific interest in comparing M-estimators associated with the marginal distributions, it is suggested that method RMPB3 be used when the sample size is greater than 20. Also, it can be used to compare groups based on MOM, but with very small sample sizes power might be inadequate relative to other techniques that could be used. Given the goal of testing some omnibus hypothesis, currently, among the techniques covered in this chapter, it seems that the two best methods for controlling type I error probabilities and simultaneously providing relatively high power are the bootstrap-t method based on 20% trimmed means and the percentile bootstrap method in Table 8.5, which is based in part on difference scores; the computations are performed by the R function rmdzero. (But also consider the multiple comparison procedure in Section 8.1.4.) With near certainty, situations arise where some other technique is more optimal, but typically the improvement is small. But again, comparing groups with MOM is not the same as comparing means, trimmed means or M-estimators and certainly there will be situations where some other estimator has higher power than any method based on MOM or a 20% trimmed mean. If the goal is to maximize power, several methods are contenders for routine use. With sufficiently large sample sizes, trimmed means can be compared without resorting to the bootstrap-t method, but it remains unclear just how large the sample size must be. Roughly, as the amount of trimming increases from 0% to 20%, the smaller the sample size must be to control the probability of a type I error without resorting to a bootstrap technique. But if too much trimming is done, power might be relatively low. When comparing medians, a percentile bootstrap method is recommended when dealing with tied values.

As for the issue of whether to use difference scores rather than robust measures of location based on the marginal distributions, each approach provides a different perspective on how groups differ and they can give different results regarding whether groups are significantly different. There is some evidence that difference scores typically provide more power and better control over the probability of a type I error, but situations are encountered where the reverse is true. A more detailed study is needed to resolve this issue.

As previously mentioned, method RMPB4 performed by the R function ddep in Section 8.2.6 is very conservative in terms of type I errors, meaning that when testing at the 0.05 level, say, often the actual probability of a type I error will be less than or equal to α and typically smaller than any other method described in this chapter. So a concern is that power might be low relative to the many other methods that might be used.

Regarding methods designed for performing multiple comparisons in a manner that controls the probability of at least one type I error, currently it seems that using the R function wmcppb, which compares groups based on trimmed means, is a good choice, particularly in terms of maximizing power. But again, there are exceptions. Perhaps the method in Section 8.2.7, which is based on the OP-estimator and performed by the R function mcpOV, generally competes well with the R function wmcppb, but little is known about the extent to which this is true. In general, no single method is always best. As in previous chapters, in terms of controlling the probability of at least one type I error, the method used by wmcppb and mcpOV does not assume or require that one first test and reject the global hypothesis that all groups have identical population trimmed means. It is not necessary, for example, that the function rmanovab (described in Section 8.2.2) returns a significant result before using the function wmcppb.

8.5 Some Rank-Based Methods

This section describes two rank-based methods for testing

$$H_0 : F_1(x) = \cdots = F_J(x),$$

the hypothesis that J dependent groups have identical marginal distributions. Friedman's test is the best-known test of this hypothesis, but no details are given here. The first of the two methods was derived by Agresti and Pendergast (1986) and has higher power than Friedman's test when sampling from normal distributions, and their test can be expected to have good power when sampling from heavy-tailed distributions, so it is included here. The other method stems from Brunner, Domhof, and Langer (2002, Section 7.2.2), which appears to have an advantage over the Agresti–Pendergast method in terms of power (Tian & Wilcox, 2007).

Method AP

The calculations for the Agresti–Pendergast method are shown in Table 8.6.

Method BPRM

Let R_{ij} be defined as in Table 8.6 and let $\mathbf{R}_i = (R_{i1}, \ldots, R_{iJ})'$ be the vector of ranks for the ith participant, where $(R_{i1}, \ldots, R_{iJ})'$ is the transpose of $(R_{i1}, \ldots, R_{iJ})$. Let

$$\bar{\mathbf{R}} = \frac{1}{n} \sum_{i=1}^{n} \mathbf{R}_i$$

be the vector of ranked means, let

$$\bar{R}_{.j} = \frac{1}{n} \sum_{i=1}^{n} R_{ij}$$

denote the mean of the ranks for group j and let

$$\mathbf{V} = \frac{1}{N^2(n-1)} \sum_{i=1}^{n} (\mathbf{R}_i - \bar{\mathbf{R}})(\mathbf{R}_i - \bar{\mathbf{R}})'.$$

The test statistic is

$$F = \frac{n}{N^2 \text{tr}(\mathbf{PV})} \sum_{j=1}^{J} \left(\bar{R}_{.j} - \frac{N+1}{2} \right)^2, \qquad (8.4)$$

where

$$\mathbf{P} = \mathbf{I} - \frac{1}{J}\mathbf{J},$$

$\mathbf{J}$ is a $J \times J$ matrix of all ones, and $\mathbf{I}$ is the identity matrix.

Table 8.6: Computing the Agresti–Pendergast Test Statistic.

Pool all the observations and assign ranks. Let R_{ij} be the resulting rank of the ith observation in the jth group. Compute

$$\bar{R}_j = \frac{1}{n} \sum_{i=1}^{n} R_{ij}$$

$$s_{jk} = \frac{1}{n - J + 1} \sum_{i=1}^{n} (R_{ij} - \bar{R}_j)(R_{ik} - \bar{R}_k).$$

Let the vector $\mathbf{R}'$ be defined by

$$\mathbf{R}' = (\bar{R}_1, \ldots, \bar{R}_J),$$

and let $\mathbf{C}$ be the $(J - 1)$-by-J matrix given by

$$\begin{pmatrix} 1 & -1 & 0 & \cdots & 0 & 0 \\ 0 & 1 & -1 & \cdots & 0 & 0 \\ \cdot & \cdot & \cdot & \cdot & \cdot & \cdot \\ 0 & 0 & 0 & \cdots & 1 & -1 \end{pmatrix}.$$

The test statistic is

$$F = \frac{n}{J - 1} (\mathbf{CR})'(\mathbf{CSC}')^{-1} \mathbf{CR},$$

where

$$\mathbf{S} = (s_{jk}).$$

The degrees of freedom are $\nu_1 = J - 1$ and $\nu_2 = (J - 1)(n - 1)$, and you reject if $F > f_{1-\alpha}$, the $1 - \alpha$ quantile of an F distribution with ν_1 and ν_2 degrees of freedom.

Decision Rule

Reject the hypothesis of identical distributions if

$$F \geq f,$$

where f is the $1 - \alpha$ quantile of an F distribution with degrees of freedom

$$\nu_1 = \frac{[\text{tr}(\mathbf{PV})]^2}{\text{tr}(\mathbf{PVPV})}$$

and $\nu_2 = \infty$. Note that based on the test statistic, a crude description of method BPRM is that it is designed to be sensitive to differences among the average ranks.

8.5.1 R Functions apanova and bprm

The R function

$$\text{apanova(x,grp=0)}$$

performs the Agresti–Pendergast test of equal marginal distributions using the calculations in Table 8.6. As usual, x can have list mode, or x can be an n-by-J matrix, and the argument grp can be used to specify some subset of the groups. If grp is unspecified, all J groups are used. The function returns the value of the test statistic, the degrees of freedom, and the p-value. For example, the command apanova(dat,grp=c(1,2,4)) would compare groups 1, 2, and 4 using the data in the R variable dat.

The R function

$$\text{bprm(x)}$$

performs method BPRM; it returns a p-value.

8.6 Between-by-Within and Within-by-Within Designs

This section describes some methods for testing hypotheses in a between-by-within (or split-plot) design. That is, a J-by-K ANOVA design is being considered where the J levels of the first factor correspond to independent groups (between subjects), and the K levels of the second factor are dependent (within subjects). Within-by-within designs are covered as well.

8.6.1 Analyzing a Between-by-Within Design Based on Trimmed Means

We begin with a between-by-within design. For the jth level of factor A, let Σ_j be the K-by-K population Winsorized covariance matrix for the K dependent random variables associated with the second factor. The better-known methods for analyzing a split-plot design are based on the assumption that $\Sigma_1 = \cdots = \Sigma_J$, but violating this assumption can result in problems controlling the probability of a type I error. Keselman, Keselman, and Lix (1995) found that a method derived by Johansen (1980), that does not assume there is a common covariance matrix, gives better results, so a generalization of Johansen's method, to trimmed means, is described here. (For related results, see Keselman, Algina, Kowalchuk, & Wolfinger, 1999; Keselman, Carriere, & Lix, 1993; Livavcic-Rojas, Vallejo, & Fernández, 2010.)

As in the case of a two-way design for independent groups, it is easier to describe the method in terms of matrices. Main effects and interactions are examined by testing

$$H_0 : \mathbf{C}\mu_t = \mathbf{0},$$

where $\mathbf{C}$ is a k-by-JK contrast matrix having rank k that reflects the null hypothesis of interest. Let $\mathbf{C}_m$ and $\mathbf{j}'$ be defined as in Section 7.3. Then for factor A, $\mathbf{C} = \mathbf{C}_J \otimes \mathbf{j}'_K$, and $k = J - 1$. For factor B, $\mathbf{C} = \mathbf{j}'_J \otimes \mathbf{C}_K$ and $k = K - 1$, and the test for no interactions uses $\mathbf{C} = \mathbf{C}_J \otimes \mathbf{C}_K$.

For every level of factor A, there are K dependent random variables, and each pair of these dependent random variables has a Winsorized covariance that must be estimated. In symbols, let X_{ijk} be the ith observation randomly sampled from the jth level of factor A and the kth level of factor B. For fixed j, the Winsorized covariance between the mth and ℓth levels of factor B is estimated with

$$s_{jm\ell} = \frac{1}{n_j - 1} \sum_{i=1}^{n_j} (Y_{ijm} - \bar{Y}_{.jm})(Y_{ij\ell} - \bar{Y}_{.j\ell}),$$

where

$$Y_{ijk} = \begin{cases} X_{(g+1),jk} & \text{if } X_{ijk} \leq X_{(g+1),jk} \\ X_{ijk} & \text{if } X_{(g+1),jk} < X_{ij} < X_{(n-g),jk} \\ X_{(n-g),jk} & \text{if } X_{ijk} \geq X_{(n-g),jk}, \end{cases}$$

and

$$\bar{Y}_{.jm} = \frac{1}{n} \sum_{i=1}^{n} Y_{ijm}.$$

For fixed j, let $\mathbf{S}_j = (s_{jm\ell})$. That is, $\mathbf{S}_j$ estimates Σ_j, the K-by-K Winsorized covariance matrix for the jth level of factor A. Let

$$\mathbf{V}_j = \frac{(n_j - 1)\mathbf{S}_j}{h_j(h_j - 1)}, \quad j = 1, \ldots, J,$$

and let $\mathbf{V} = \text{diag}(\mathbf{V}_1, \ldots, \mathbf{V}_J)$ be a block diagonal matrix. The test statistic is

$$Q = \bar{\mathbf{X}}'\mathbf{C}'(\mathbf{C}\mathbf{V}\mathbf{C}')^{-1}\mathbf{C}\bar{\mathbf{X}}, \tag{8.5}$$

where $\bar{\mathbf{X}}' = (\bar{X}_{t11}, \ldots, \bar{X}_{tJK})$. Let $\mathbf{I}_{K \times K}$ be a K-by-K identity matrix, let $\mathbf{Q}_j$ be a JK by JK block diagonal matrix (consisting of J blocks, each block being a K-by-K matrix), where the tth block ($t = 1, \ldots, J$) along the diagonal of $\mathbf{Q}_j$ is $\mathbf{I}_{K \times K}$ if $t = j$, and all other elements are zero. (For example, if $J = 3$ and $K = 4$, then $\mathbf{Q}_1$ is a 12-by-12 matrix block diagonal matrix where the first block is a 4-by-4 identity matrix, and all other elements are zero. As for $\mathbf{Q}_2$, the second block is an identity matrix, and all other elements are zero.) Compute

$$A = \frac{1}{2} \sum_{j}^{J} \{ \text{tr}[(\mathbf{V}\mathbf{C}'(\mathbf{C}\mathbf{V}\mathbf{C}')^{-1}\mathbf{C}\mathbf{Q}_j)^2] + [\text{tr}(\mathbf{V}\mathbf{C}'(\mathbf{C}\mathbf{V}\mathbf{C}')^{-1}\mathbf{C}\mathbf{Q}_j)]^2 \}/(h_j - 1).$$

where tr indicates trace, and let

$$c = k + 2A - \frac{6A}{k+2}.$$

When the null hypothesis is true, Q/c has, approximately, an F distribution with $\nu_1 = k$ and $\nu_2 = k(k+2)/(3A)$ degrees of freedom, so reject if $Q/c > f_{1-\alpha}$, the $1-\alpha$ quantile. Recent simulation results reported by Livavcic-Rojas et al. (2010) indicate that this method performs relatively well, in terms of controlling the probability of a type I error, when comparing means. However, their results do not consider the effect of having different amounts of skewness.

■ Example

Consider a 2-by-3 design where for the first level of factor A, observations are generated from a multivariate normal distribution with all correlations equal to zero. For the second level of factor A, the marginal distributions are lognormal that have been shifted to have mean zero. Further suppose that the covariance matrix for the second level is three times larger that the covariance matrix for the first level. If the sample sizes are $n_1 = n_2 = 30$ and the hypothesis of no main effects for factor A, based on means, is tested at the 0.05 level, the actual level is approximately 0.088. If the sample sizes are $n_1 = 40$ and $n_2 = 70$ and for the first level of factor A the marginal distributions are g-and-h distributions with $g = h = 0.5$, the probability of a type I error, again testing at the 0.05 level, is approximately .188. Comparing 20% trimmed means instead, the actual type I error probability is approximately .035.

■

8.6.2 R Functions bwtrim and tsplit

The R function

$$bwtrim(J,K,x,tr=.2,grp=c(1:p),p=J*K)$$

tests the hypotheses of no main effects and no interactions in a between-by-within (split-plot) design, where J is the number of independent groups, K is the number of dependent groups, and the argument x contains the data stored in list mode, or a matrix, or a data frame. The optional argument, tr, indicates the amount of trimming, which defaults to 0.2 if unspecified.

The groups are assumed to be ordered as described in Section 7.2.1. If the data are not stored in the proper order, grp can be used to indicate how they are stored. For example, if a 2-by-2 design is being used, the R command

$$bwtrim(2,2,x,grp=c(3,1,2,4))$$

indicates that the data for the first level of both factors are stored in x[[3]], the data for level 1 of Factor A and level 2 of Factor B are in x[[1]], and so forth. If x is a matrix or a data frame, the first K columns correspond to the first level of Factor A and the K levels of Factor B, the next K columns correspond to the second level of Factor A, and so on.

The R function tsplit, which is described in earlier editions of this book, performs the same analysis. The function bwtrim was added to match naming conventions used by other functions to be described.

■ Example

In a study on the effect of consuming alcohol, hangover symptoms were measured for two independent groups, with each subject consuming alcohol and being measured on three different occasions. One group (group 2) consisted of sons of alcoholics and the other was a control group. Here, $J = 2$ and $K = 3$. The results were as follows.

Group 1, Time 1	0 32 9 0 2 0 41 0 0 0 6 18 3 3 0 11 11 2 0 11
Group 1, Time 2	4 15 26 4 2 0 17 0 12 4 20 1 3 7 1 11 43 13 4 11
Group 1, Time 3	0 25 10 11 2 0 17 0 3 6 16 9 1 4 0 14 7 5 11 14
Group 2, Time 1	0 0 0 0 0 0 0 0 1 8 0 3 0 0 32 12 2 0 0 0
Group 2, Time 2	2 0 7 0 4 2 9 0 1 14 0 0 0 0 15 14 0 0 7 2
Group 2, Time 3	1 0 3 0 3 0 15 0 6 10 1 1 0 2 24 42 0 0 0 2

Suppose the first row of data is stored in DAT[[1]], the second row in DAT[[2]], the third in DAT[[3]], the fourth in DAT[[4]], the fifth in DAT[[5]], and the sixth in DAT[[6]]. That is, the data are stored as assumed by the function bwtrim. Then the command bwtrim(2,3,DAT,tr=0) will compare the means and returns

```
$Qa:
[1] 3.277001

$Qa.siglevel:
[1] 0.149074

$Qb:
[1] 0.7692129

$Qb.siglevel:
[1] 0.5228961

$Qab:
[1] 0.917579

$Qab.siglevel:
[1] 0.4714423
```

The test statistic for factor A is $Qa = 3.28$, and the corresponding p-value is .159. For factor B the p-value is .52, and for the interaction it is .47.

∎

This section described one way of comparing independent groups in a between-by-within subjects design. Another approach is simply to compare the J independent groups for each level of Factor B. That is, do not sum or average the data over the levels of Factor B as was done here. So the goal is to test

$$H_0 : \mu_{t1k} = \cdots \mu_{tJk}$$

for each $k = 1, \ldots, K$. The next example illustrates that in applied work, the choice between these two methods can make a practical difference. (Yet another strategy for comparing the levels of Factor A is to apply the robust MANOVA method in Section 7.10.)

∎ Example

Section 7.8.4 reported data from a study comparing schizophrenics to a control group based on a measure taken at two different times. Analyzing the data with the function tsplit, no main effects for Factor A are found, the p-value being .245. So no difference between the schizophrenics and the control group was detected at the 0.05 level. But if the groups are compared using the first measurement only, using the function yuen (described in Chapter 5), the p-value is .012. For the second measurement, ignoring the first, the p-value is .89. (Recall that in Chapter 6, using some multivariate methods, again a difference between the schizophrenics and the control group was found.)

∎

8.6.3 Data Management: R Function bw2list

Imagine a situation where data are stored in a matrix or a data frame, say x, with one column indicating the levels of the between factor, but the K levels of the within group factor are stored in K columns of the matrix x. In order to use the R function bwtrim, it is necessary to store the data in the format that is allowed. The R function

bw2list(x, grp.col, lev.col)

is provided to help accomplish this goal. The argument grp.col indicates the column containing information about the levels of the independent groups. The values in this column can be numeric or character data. And the argument lev.col indicates the K columns where the within group data are stored. The function returns the data stored in list mode, which can then be used by bwtrim as well as other functions aimed at dealing with a between-by-within design. The function will store the data sorted in ascending (or alphabetical) order based on the values found in the column of x indicated by the argument grp.col. The next example illustrates this feature.

■ Example

Imagine that three medications are being investigated regarding their effectiveness to lower cholesterol and that column 3 of the matrix m indicates which medication a participant received. Moreover, columns 5 and 8 contain the participants' cholesterol level at times 1 and 2, respectively. The R command

$$z=bw2list(m,3,c(5,8))$$

will store the data in z in list mode. If column 3 contains the character values "P", "CH", and "BN", then z[[1]] and z[[2]] will contain the data for times 1 and 2, respectively, corresponding to level "BN" of Factor A, z[[3]] and z[[4]] will contain the data for level "CH", and z[[5]] and z[[6]] will contain the data for level "P". The R command

$$bwtrim(3,2,z)$$

will compare the groups based on 20% trimmed means.

■

8.6.4 Bootstrap-t Method for a Between-by-Within Design

To apply a bootstrap-t method, when working with trimmed means and dealing with a between-by-within design, first center the data in the usual way. In the present context, this means you compute

$$C_{ijk} = X_{ijk} - \bar{X}_{tjk},$$

$i = 1, \ldots, n_j$; $j = 1, \ldots, J$; and $k = 1, \ldots, K$. That is, for the group corresponding to the jth level of factor A and the kth level of factor B, subtract the corresponding trimmed mean from each of the observations. Next, for each j, generate a bootstrap sample based on the C_{ijk} values by resampling with replacement n_j vectors of observations from the n_j rows of data corresponding to level j of factor A. That is, for each level of factor A, you have an n_j-by-K matrix of data, and you generate a bootstrap sample from this matrix of data as described in Section 8.2.5 where for fixed j, resampling is based on the C_{ijk} values. Label the resulting bootstrap samples C_{ijk}^*. Compute the test statistic Q, based on the C_{ijk}^* values as described in Section 8.6.1 and label the result Q^*. Repeat this B times yielding $Q_1^*, \ldots, Q_B^*$ and then put these B values in ascending order yielding $Q_{(1)}^* \leq \cdots \leq Q_{(B)}^*$. Next, compute Q using the original data (the X_{ijk} values) and reject if $Q \geq Q_{(c)}^*$, where $c = (1 - \alpha)$ rounded to the nearest integer.

A crude rule that seems to apply to a wide variety of situations is: the more the distributions (associated with groups) differ, the more beneficial it is to use some type of bootstrap method, at least when sample sizes are small. Keselman et al. (2000) compared the bootstrap method just described to the nonbootstrap method for a split-plot design covered in Section 8.6.1. For the situations they examined, this rule did not apply; it was found that the bootstrap offered little or no advantage. Their study included situations where the correlations (or covariances) among the dependent groups differ across the independent groups being compared. However, the more complicated the design, the more difficult it becomes to consider all the factors that might influence the operating characteristics of a particular method. One limitation of their study was that the differences among the covariances were taken to be relatively small. Another issue that has not been addressed is how the bootstrap method performs when distributions differ in skewness. Having differences in skewness is known to be important when dealing with the simple problem of comparing two groups only. There is no reason to assume that this problem diminishes as the number of groups increases, and indeed there are reasons to suspect that it becomes a more serious problem. So currently, it seems that if groups do not differ in any manner, or the distributions differ slightly, it makes little difference whether you use a bootstrap versus a nonbootstrap method for comparing trimmed means. However, if distributions differ in shape, there is indirect evidence that a bootstrap method might offer an advantage when using a split-plot design, but the extent to which this is true is not well understood.

8.6.5 R Functions bwtrimbt and tsplitbt

The R function

$$\text{tsplitbt(J,K,x,tr=0.2,alpha=0.05,JK=J*K,grp=c(1:JK),nboot=599)}$$

performs a bootstrap-t method for a split-plot design as just described. The data are assumed to be arranged as indicated in conjunction with the R function tsplit (as described in Section 8.6.2), and the arguments J, K, tr, and alpha have the same meaning as before. The argument JK can be ignored, and grp can be used to rearrange the data if they are not stored as expected by the function. (For an R function that might help when dealing with organizing the data in a manner that is accepted by tsplitbt, see Section 8.6.3.)

The R function

$$\text{bwtrimbt(J,K,x,tr=0.2,JK=J*K,grp=c(1:JK),nboot=599)}$$

is the same as tsplitbt, only bwtrimbt reports p-values rather than α level critical values.

8.6.6 Percentile Bootstrap Methods for a Between-by-Within Design

Comparing groups based on MOMs, medians, and M-estimators in a between-by-within design is possible using extensions of percentile bootstrap methods already described. And they provide yet another way of comparing trimmed means.

Again consider a two-way design where factor A consists of J independent groups and Factor B corresponds to K dependent groups. First consider the dependent groups. One approach to comparing these K groups, ignoring Factor A, is to simply form difference scores and then apply the method in Section 8.3. More precisely, imagine you observe X_{ijk} ($i = 1, \ldots, n_j$; $j = 1, \ldots, J$; $k = 1, \ldots, K$). That is, X_{ijk} is the ith observation in level j of Factor A and level k of Factor B. Note that if we ignore the levels of Factor A, we can write the data as Y_{ik}, $i = 1, \ldots, N$; $k = 1, \ldots, K$, where $N = \sum n_j$. Now consider levels k and k' of Factor B ($k < k'$) and set

$$D_{ikk'} = Y_{ik} - Y_{ik'},$$

and let $\theta_{kk'}$ be some measure of location associated with $D_{ikk'}$. Then the levels of Factor B can be compared, ignoring Factor A, by testing

$$\theta_{12} = \cdots = \theta_{k-1,k} = 0 \tag{8.6}$$

using the method in Section 8.3. In words, the null hypothesis is that the typical difference score between any two levels of Factor B, ignoring Factor A, is zero.

As for Factor A, ignoring Factor B, one approach is as follows. Momentarily focus on the first level of Factor B and note that the levels of Factor A can be described as in Chapter 7. That is, the null hypothesis of no differences among the levels of Factor A is

$$H_0 : \theta_{11} = \theta_{21} = \cdots = \theta_{J1},$$

where of course these J groups are independent, and a percentile bootstrap method can be used. More generally, for any level of Factor B, say the kth, the hypothesis of no main effects is

$$H_0 : \theta_{1k} = \theta_{2k} = \cdots = \theta_{Jk},$$

($k = 1, \ldots, K$), and the goal is to determine whether these K hypotheses are simultaneously true. Here we take this to mean that we want to test

$$H_0 : \theta_{11} - \theta_{21} = \cdots \theta_{J-1,1} - \theta_{J1} = \cdots = \theta_{J-1,K} - \theta_{JK} = 0. \tag{8.7}$$

In this last equation, there are $C = K(J^2 - J)/2$ differences, all of which are hypothesized to be equal to zero. Proceeding along the lines in Chapter 7, for each level of Factor A, generate bootstrap samples as is appropriate for K dependent groups and then test Eq. (8.7). Label the C differences based on the observed data as $\delta_1, \ldots, \delta_C$ and then denote bootstrap estimates by $\hat{\delta}_c^*$ ($c = 1, \ldots, C$). For example, $\hat{\delta}_1^* = \theta_{11}^* - \theta_{21}^*$. Then we test Eq. (8.5) by determining how deeply the vector $(0, \ldots, 0)$, having length C, is nested within the B bootstrap values, which is done as described in Table 8.5. However, a criticism of this method is that control over the probability of a type I error can be unsatisfactory it can (exceed .075 when testing at the 0.05 level) when the sample size is small.

For Factor A, an alternative approach, which seems more satisfactory in terms of type I errors, is to base the analysis on the average measures of location across the K levels of Factor B. In symbols, let

$$\bar{\theta}_{j.} = \frac{1}{K} \sum_{k=1}^{K} \theta_{jk},$$

in which case the goal is to test

$$H_0 : \bar{\theta}_{1.} = \cdots = \bar{\theta}_{J.}.$$

Again for each level of Factor A, generate B samples for the K dependent groups as described in Section 8.2.5 in conjunction with method RMPB4. Let $\bar{\theta}_{j.}^{*}$ be the bootstrap estimate for the jth level of Factor A. For levels j and j' of Factor A, $j < j'$, set $\delta_{jj'}^{*} = \bar{\theta}_{j.}^{*} - \bar{\theta}_{j'.}^{*}$, Then you determine how deeply $\mathbf{0}$, having length $(J^2 - J)/2$, is nested within the B bootstrap values for $\delta_{jj'}^{*}$ using the method described in Table 8.5. When dealing with Factor A, this approach seems to be more satisfactory than the strategy described in the previous paragraph.

As for interactions, again there are several approaches one might adopt. Here an approach based on difference scores among the dependent groups is used. To explain, first consider a 2-by-2 design, and for the first level of Factor A let $D_{i1} = X_{i11} - X_{i12}$, $i = 1, \ldots, n_1$. Similarly, for level 2 of Factor A let $D_{i2} = X_{i21} - X_{i22}$, $i = 1, \ldots, n_2$, and let θ_{d1} and θ_{d2} be the population measure of location corresponding to the D_{i1} and D_{i1} values, respectively. Then the hypothesis of no interaction is taken to be

$$H_0 : \theta_{d1} = \theta_{d2},$$

which of course is the same as

$$H_0 : \theta_{d1} - \theta_{d2} = 0. \tag{8.8}$$

Again the basic strategy for testing hypotheses is generating bootstrap estimates and determining how deeply 0 is embedded in the B values that result. For the more general case of a J-by-K design, there are a total of

$$C = \frac{J^2 - J}{2} \times \frac{K^2 - K}{2}$$

equalities, one for each pairwise difference among the levels of Factor B and any two levels of Factor A.

8.6.7 R Functions sppba, sppbb, and sppbi

The R function

$$\text{sppba(J,K,x,est=onestep,grp = c(1:JK),avg=T,nboot=500,MC=F,MDIS=T, ...)}$$

argument avg to T (for true) indicates that the averages of the measures of location (the $\bar{\theta}_{j.}$ values) will be used. That is, $H_0 : \bar{\theta}_{1.} = \cdots = \bar{\theta}_{J.}$ is tested. Otherwise, the hypothesis given by Eq. (8.6) is tested. By default, the argument MDIS=T, meaning that the depths of the points in the bootstrap cloud are based on Mahalanobis distance. Otherwise a projection distance is used, which was described in Section 6.2.5. If MDIS=F and MC=T, a multi-core processor will be used if one is available. The remaining arguments have their usual meaning.

The R function

$$\text{sppbb(J,K,x,est=onestep,grp = c(1:JK),nboot=500, ...)}$$

tests the hypothesis of no main effects for Factor B (as described in the previous section) and

$$\text{sppbi(J,K,x,est=onestep,grp = c(1:JK),nboot=500, ...)}$$

tests the hypothesis of no interactions.

■ Example

We examine once more the EEG measures for murderers versus a control group reported in Table 6.1, only now we use the data for all four sites in the brain where measures were taken. If we label the typical measures for the control group as $\theta_{11}, \ldots, \theta_{14}$, and the typical measures for the murderers as $\theta_{21}, \ldots, \theta_{24}$, we have a 2-by-4, between-by-within design and a possible approach to comparing the groups is testing

$$H_0 : \theta_{11} - \theta_{21} = \theta_{12} - \theta_{22} = \theta_{13} - \theta_{23} = \theta_{14} - \theta_{24} = 0.$$

This can be done with the R function sppba with the argument avg set to F. If the data are stored in a matrix called eeg having eight columns, with the first four corresponding to the control group, then the command sppba(2,4,eeg,est=mom) performs the calculations based on the MOM measure of location and returns a significance level of 0.098. An alternative approach is to average the value of MOM over the four brain sites for each group, and then compare these averages. That is, test $H_0 : \bar{\theta}_{1.} = \bar{\theta}_{2.}$, where $\bar{\theta}_{j.} = \sum \theta_{jk}/4$. This can be done with the command sppba(2,4,eeg,avg=T). Now the p-value is .5 illustrating that the p-value can vary tremendously depending on how groups are compared.

8.6.8 Multiple Comparisons

When dealing with multiple comparisons associated with a between-by-within design, there are several approaches that might be taken that answer different questions. This section outlines some of the possibilities.

Method BWMCP

Focusing on trimmed means, multiple comparisons, when dealing with a between-by-within design, can be tested using linear contrasts, which are created in the same manner as outlined in Section 7.4.3. Consider any linear contrast Ψ, with the understanding that multiple linear contrasts are generally of interest. As usual, the goal is to test

$$H_0 : \Psi = 0,$$

and among the C linear contrasts of interest, often it is desired to have the probability of one or more type I errors equal to some specified value, α. Two methods for accomplishing this goal seem to be relatively effective: a bootstrap-t method and a percentile bootstrap method.

For convenience, we write the $L = JK$ trimmed means as $\bar{X}_{t1}, \dots, \bar{X}_{tL}$. The estimate of

$$\Psi = \sum c_\ell \mu_{t\ell}$$

is

$$\hat{\Psi} = \sum c_\ell \bar{X}_{t\ell},$$

where $c_1, \dots c_L$ are the linear contrast coefficients.

To test hypotheses using a bootstrap-t method, first note that the variances and covariances among the sample trimmed means can be estimated using results in Section 5.9.5. (Of course, when two sample trimmed means are independent, their covariance is taken to be zero.) Let $\mathbf{S}$ denote this L-by-L covariance matrix. (So the diagonal elements are the estimated squared standard errors.) Let $\mathbf{C}$ be a column matrix having length L that contains the contrast coefficients. Then the squared standard error of $\hat{\Psi}$ is estimated with

$$s_{\hat{\Psi}}^2 = \mathbf{C}'\mathbf{S}\mathbf{C}$$

and an appropriate test statistic is

$$W = \frac{|\hat{\Psi}|}{s_{\hat{\Psi}}}.$$

A bootstrap-t method is used to estimate the null distribution of W. First, compute

$$Y_{ijk} = X_{ijk} - \bar{X}_{tjk},$$

where $\bar{X}_{tjk}$ is the trimmed mean corresponding to level j of Factor A and level k of Factor B. Next, take bootstrap samples based on the Y_{ijk} values. So for level j of Factor A, n_j rows of data are sampled with replacement. Based on this bootstrap sample, compute the test statistic W, which is labeled W^*. Repeat this process B times yielding $W_1^*, \ldots, W_B^*$. Let $c = (1-\alpha)B$, rounded to the nearest integer. Then a $1-\alpha$ confidence interval for Ψ is

$$\hat{\Psi} \pm W_{(c)}^* \frac{s_{\hat{\Psi}}}{\sqrt{n}}.$$

Alternatively, reject the null hypothesis if $W \geq W_{(c)}^*$.

Section 4.4.3 made a distinction between a symmetric bootstrap-t confidence interval and an equal-tailed confidence interval. Here, a symmetric confidence interval is used. For the situation at hand, there are no results on whether an equal-tailed confidence interval ever offers a practical advantage.

A percentile bootstrap method can be applied as well. As usual, no standard errors are used. For each hypothesis to be tested, corresponding to some linear contrast, a p-value can be computed as indicated at the end of Section 8.2.7.

Method BWAMCP: Comparing Levels of Factor A for Each Level of Factor B

To provide more detail about how groups differ, another strategy is to focus on a particular level of Factor B and perform all pairwise comparisons among the levels of Factor A. Of course, this can be done for each level of Factor B.

■ Example

Consider again a 3-by-2 design where the means are arranged as follows:

		Factor B 1	2
Factor A	1	μ_1	μ_2
	2	μ_3	μ_4
	3	μ_5	μ_6

For level 1 of Factor B, method BWAMCP would test H_0: $\mu_1 = \mu_3$, H_0: $\mu_1 = \mu_5$ and H_0: $\mu_3 = \mu_5$. For level 2 of Factor B, the goal is to test H_0: $\mu_2 = \mu_4$, H_0: $\mu_2 = \mu_6$ and H_0: $\mu_4 = \mu_6$. These hypotheses can be tested by creating the appropriate linear contrasts and using the R function lincon, which can be done with the R function bwamcp described in the next section.

Method BWBMCP: Dealing with Factor B

When dealing with Factor B, there are four variations of method BWMCP that might be used, which are described here under the appellation method BWBMCP. The first two variations ignore the levels of Factor A and test hypotheses based on the trimmed means. The first variation uses difference scores and the second uses the marginal trimmed means. Both of these variations begin by pooling the data over the levels of Factor A. In essence, Factor A is ignored. The other two variations do not pool the data over the levels of Factor A, but rather perform an analysis based on difference scores or the marginal trimmed means for each level of Factor A. In more formal terms, consider the jth level of Factor A. Then there are $(K^2 - K)/2$ pairs of groups that can be compared. If for each of the J levels of Factor A, all pairwise comparisons are performed, the total number of comparisons is $J(K^2 - K)/2$.

■ Example

Consider a 2-by-2 design where the first level of Factor A has 10 pairs of observations and the second has 15. So we have a total of 25 pairs of observations with the first 10 corresponding to level 1 of Factor A. When analyzing Factor B, pooling the data means the goal is to compare either the difference scores corresponding to all 25 pairs of observations, or to compare the marginal trimmed means, again based on all 25 observations. Not pooling means that for level 1 of Factor A either test hypotheses based on difference scores or compare the marginal trimmed means. And the same could be done for level 2 of Factor A.

■

Method BWIMCP: Interactions

As for interactions, we focus on a 2-by-2 design with the understanding that the same analysis can be done for any two levels of Factor A and any two levels of Factor B. Rather than define interactions as done when using Method BWMCP, difference scores might be used instead. To elaborate, consider the first level of Factor A. There are n_1 pairs of observations corresponding to the two levels of Factor B. Form the difference scores, which for level j of Factor A are denoted by

$$D_{ij},$$

$(i = 1, \ldots, n_j)$ and let μ_{tj} be the population trimmed means associated with these difference scores. Then one way of stating the hypothesis of no interaction is

$$H_0 : \mu_{t1} = \mu_{t2}.$$

In words, the hypothesis of no interaction corresponds to the trimmed means of the difference scores associated with level 1 of Factor A being equal to the trimmed means of the difference scores associated with level 2 of Factor A. When either factor has more than two levels, a possible goal is to test all similar hypotheses (associated with any two levels of Factor A and Factor B) in a manner that controls FWE, which might be done using Rom's method or Hochberg's method.

Methods SPMCPA, SPMCPB, and SPMCPI

If it is desired to compare groups based on using a percentile bootstrap method, which appears to be the best method when comparing groups based on an M-estimator or MOM, analogs of methods BWAMCP, BWBMCP, and BWIMCP can be used, which are called methods SPMCPA, SPMCPB, and SPMCPI, respectively.

8.6.9 R Functions bwmcp, bwamcp, bwbmcp, bwimcp, spmcpa, spmcpb, and spmcpi

The R function

$$bwmcp(J, K, x, tr = 0.2, alpha = 0.05, con=0, nboot=599)$$

performs method BWMCP described in the previous section. By default, it creates all relevant linear contrasts for main effects and interactions by calling the R function con2way.

The R function

$$bwamcp(J, K, x, tr = 0.2, alpha = 0.05)$$

performs multiple comparisons associated with Factor A using the (bootstrap-t) method BWAMCP, described in the previous section. The function creates the appropriate set of linear contrasts and calls the R function lincon. The function returns three sets of results corresponding to Factor A, Factor B, and all interactions. The critical value reported for each of the three set of tests in designed to control the probability of at least one type I error.

The R function

$$spmcpa(J,K,x,est=tmean,JK=J*K,grp=c(1:JK),con=0,avg=F,alpha=.05,$$
$$nboot=NA,pr=T, \ldots)$$

is like the R function bwamcp, only a percentile bootstrap method is used.

The R function

$$bwbmcp(J, K, x, tr = 0.2, con = 0, alpha = 0.05, dif = T, pool=F)$$

uses method BWBMCP to compare the levels of Factor B. If the argument pool=T, the function pools the data for you and then calls the function rmmcp. If the argument dif=F, the marginal trimmed means are compared instead. By default, pool=F meaning that

$$H_0 : \mu_{tjk} = \mu_{tjk'}$$

is tested for all $k < k'$ and $j = 1, \ldots, J$. For each level of Factor A, the function simply selects data associated with the levels of Factor B and tests hypotheses via the R function rmmcp. "Critical *p*-values" are reported in the column headed by p.crit. That is, p.crit indicates how small the *p*-value must be in order to reject, given the goal that FWE be equal to some specified α value. The R function

spmcpb(J,K,x,est=tmean,JK=J*K,grp=c(1:JK),dif=T,alpha=0.05, nboot=NA,pr=T, ...)

is like bwbmcp, only a percentile bootstrap method is used.

As for interactions, the R function

$$\text{bwimcp(J, K, x, tr} = 0.2, \text{alpha} = 0.05)$$

compares trimmed means using a nonbootstrap method. The R function

spmcpi(J,K,x,est=tmean,JK=J*K,grp=c(1:JK),alpha=0.05,nboot=NA, SEED=T,pr=T, ...)

uses a percentile bootstrap technique instead.

The R function

$$\text{bwmcppb(J, K, x, tr} = 0.2, \text{alpha} = 0.05, \text{nboot} = 500, \text{bhop} = F)$$

simultaneously performs all multiple comparisons related to all main effects and interactions using a percentile bootstrap method. Unlike spmcpa, spmcpb, and spmcpi, the function bwmcppb is designed for trimmed means only and has an option for using the Benjamini–Hochberg method via the argument bhop.

The R function

$$\text{bwmcppb(J, K, x, tr} = 0.2, \text{alpha} = 0.05, \text{nboot} = 500, \text{bhop} = F)$$

tests the same hypotheses as done by the R function bwmcp, only a percentile bootstrap method used. Unlike spmcpa, spmcpb, and spmcpi, the function bwmcppb is designed for trimmed means only and has an option for using the Benjamini–Hochberg method via the argument bhop.

8.6.10 Within-by-Within Designs

The methods for dealing with a between-by-within design are readily extended to a within-by-within design. That is, all JK groups being compared are dependent. For example, the method in Section 8.6.1 can be modified to handle this situation by taking $\mathbf{V}$ in Eq. (8.5) to be the Winsorized variance–covariance of all JK variables under study. (That is, for a between-by-within design, $\mathbf{V}$ was a block diagonal matrix, but for the situation at hand, generally this is no longer the case.) A similar extension can be used when dealing with linear contrasts. Note that when dealing with linear contrasts, again there are two basic goals that might be of interest. The first is to test hypotheses about linear contrasts stated in terms of the measures of location associated with the marginal distributions. Section 8.1.3 provides explicit details when dealing with trimmed means that can be used to analyze a within-by-within design. The second strategy is to use an extension of methods based on difference scores. That is, now the hypotheses of interest take the form described in Section 8.1.4. R functions specifically designed for within-by-within design are described in the next section.

8.6.11 R Functions wwtrim, wwtrimbt, wwmcppb, and wwmcpbt

The R function

$$\text{wwtrim(J, K, x, grp} = c(1{:}p), p = J * K, tr = 0.2)$$

tests for main effects and interactions in a within-by-within design using a modification of the method for trimmed means described in Section 8.6.1. (The modification simply takes into account the possibility that all JK variables might be dependent.) The R function

$$\text{wwtrimbt(J, K, x, tr} = 0.2, JKL = J * K, grp = c(1{:}JK), nboot = 599, SEED = T, \ldots)$$

is the same as the R function wwtrim, only a bootstrap-t method is used. The R function

$$\text{wwmcp(J,K,x,tr=0.2,alpha=0.05,dif=T)}$$

performs multiple comparisons relevant to both main effects and interactions. (The function creates the appropriate linear contrasts and then uses the R function rmmcp.) By default, linear contrasts are created along the lines described in Section 8.1.4. To use linear contrasts based on the marginal trimmed means, set the argument dif=F. The R function

$$\text{wwmcppb(J,K,x, alpha} = 0.05, con = 0, est=tmean, plotit = F, dif = T, grp = NA, nboot = NA, BA = T, hoch = T, xlab = ``Group\ 1", ylab = ``Group\ 2", pr = T, SEED = T, \ldots)$$

is like the R function wwmcp, only a percentile bootstrap method is used. It defaults to using a 20% trimmed mean, but other measures of location can be used via the argument est. (When using an M-estimator, setting the argument hoch=F is suggested.) This function (using default settings) appears to be a relatively good choice, particularly when dealing with a small sample size. When the amount of trimming is small, use the R function

$$\text{wwmcpbt(J,K,x, tr=0.2, alpha} = 0.05, nboot = 599),$$

which uses a bootstrap-t method.

8.6.12 A Rank-Based Approach

This section describes a rank-based approach to a split-plot (or between-by-within subjects) design taken from Brunner et al. (2002, Chapter 8). There are other rank-based approaches (e.g., Beasley, 2000; Beasley & Zumbo, 2003), but it seems that the practical merits of these competing methods, versus the method described here, have not been explored.

Main effects for Factor A are expressed in terms of

$$\bar{F}_{j.}(x) = \frac{1}{K}\sum_{k=1}^{K} F_{jk}(x),$$

the average of the distributions among the K levels of Factor B corresponding to the jth level of Factor A. The hypothesis of no main effects for Factor A is

$$H_0 : \bar{F}_{1.}(x) = \bar{F}_{2.}(x) = \cdots = \bar{F}_{J.}(x).$$

for any x. Letting

$$\bar{F}_{.k}(x) = \frac{1}{J}\sum_{j=1}^{J} F_{jk}(x)$$

be the average of the distributions for the kth level of Factor B, the hypothesis of no main effects for Factor B is

$$H_0 : \bar{F}_{.1}(x) = \bar{F}_{.2}(x) = \cdots = \bar{F}_{.K}(x).$$

As for interactions, first consider a 2-by-2 design. Then no interaction is taken to mean that for any x,

$$F_{11}(x) - F_{12}(x) = F_{21}(x) - F_{22}(x).$$

More generally, the hypothesis of no interactions among all JK groups is

$$H_0 : F_{jk}(x) - \bar{F}_{j.}(x) - \bar{F}_{.k}(x) + \bar{F}_{..}(x) = 0,$$

for any x, all j ($j = 1, \ldots, J$) and all k ($k = 1, \ldots, K$), where

$$\bar{F}_{..}(x) = \frac{1}{JK}\sum_{j=1}^{J}\sum_{k=1}^{K} F_{jk}(x).$$

As usual, let X_{ijk} represent the ith observation for level j of Factor A and level k of Factor B. Here, $i = 1, \ldots, n_j$. That is, the jth level of Factor A has n_j vectors of observations, each vector containing K values. So for the jth level of Factor A there are a total of $n_j K$ observations, and among all the groups, the total number of observations is denoted by N. So the total number of vectors among the J groups is $n = \sum n_j$, and the total number of observations is $N = K \sum n_j = Kn$.

Pool all N observations and assign ranks. As usual, midranks are used if there are tied values. Let R_{ijk} represent the rank associated with X_{ijk}. Let

$$\bar{R}_{.jk} = \frac{1}{n_j} \sum_{i=1}^{n_j} R_{ijk},$$

$$\bar{R}_{.j.} = \frac{1}{K} \sum_{k=1}^{K} \bar{R}_{.jk},$$

$$\bar{R}_{ij.} = \frac{1}{K} \sum_{k=1}^{K} R_{ijk},$$

$$\hat{\sigma}_j^2 = \frac{1}{n_j - 1} \sum_{i=1}^{n_j} (\bar{R}_{ij.} - \bar{R}_{.j.})^2,$$

$$S = \sum_{j=1}^{J} \frac{\hat{\sigma}_j^2}{n_j},$$

$$U = \sum_{j=1}^{J} \left(\frac{\hat{\sigma}_j^2}{n_j} \right)^2,$$

$$D = \sum_{j=1}^{J} \frac{1}{n_j - 1} \left(\frac{\hat{\sigma}_j^2}{n_j} \right)^2.$$

Factor A: The test statistic is

$$F_A = \frac{J}{(J-1)S} \sum_{j=1}^{J} (\bar{R}_{.j.} - \bar{R}_{...})^2,$$

where $\bar{R}_{...} = \sum \bar{R}_{.j.}/J$. The degrees of freedom are

$$\nu_1 = \frac{(J-1)^2}{1 + J(J-2)U/S^2},$$

and

$$\nu_2 = \frac{S^2}{D}.$$

Reject if $F_A \geq f$, where f is the $1 - \alpha$ quantile of an F distribution with ν_1 and ν_2 degrees of freedom.

Factor B: Let

$$\mathbf{R}_{ij} = (R_{ij1}, \ldots, R_{ijK})',$$

$$\bar{\mathbf{R}}_{.j} = \frac{1}{n_j} \sum_{i=1}^{n_j} \mathbf{R}_{ij}, \ \bar{\mathbf{R}}_{..} = \frac{1}{J} \sum_{j=1}^{J} \bar{\mathbf{R}}_{.j},$$

$n = \sum n_j$ (so $N = nK$),

$$\mathbf{V}_j = \frac{n}{N^2 n_j(n_j - 1)} \sum_{i=1}^{n_j} (\mathbf{R}_{ij} - \bar{\mathbf{R}}_{.j})(\mathbf{R}_{ij} - \bar{\mathbf{R}}_{.j})'.$$

So $\mathbf{V}_j$ is a K-by-K matrix of covariances based on the ranks. Let

$$\mathbf{S} = \frac{1}{J^2} \sum_{j=1}^{J} \mathbf{V}_j$$

and let $\mathbf{P}_K$ be defined as in Section 7.9. The test statistic is

$$F_B = \frac{n}{N^2 \mathrm{tr}(\mathbf{P}_K \mathbf{S})} \sum_{k=1}^{K} (\bar{R}_{..k} - \bar{R}_{...})^2.$$

The degrees of freedom are

$$\nu_1 = \frac{(\mathrm{tr}(\mathbf{P}_K \mathbf{S}))^2}{\mathrm{tr}(\mathbf{P}_K \mathbf{S} \mathbf{P}_K \mathbf{S})}, \ \nu_2 = \infty,$$

and H_0 is rejected if $F_B \geq f$, where f is the $1 - \alpha$ quantile of an F distribution with ν_1 and ν_2 degrees of freedom.

Interactions: Let $\mathbf{V}$ be the block diagonal matrix based on the matrices $\mathbf{V}_j$, $j = 1, \ldots, J$. Letting $\mathbf{M}_{AB}$ be defined as in Section 7.9, the test statistic is

$$F_{AB} = \frac{n}{N^2 \mathrm{tr}(\mathbf{M}_{AB} \mathbf{V})} \sum_{j=1}^{J} \sum_{k=1}^{K} (\bar{R}_{.jk} - \bar{R}_{.j.} - \bar{R}_{..k} + \bar{R}_{...})^2.$$

The degrees of freedom are

$$\nu_1 = \frac{(\mathrm{tr}(\mathbf{M}_{AB} \mathbf{V}))^2}{\mathrm{tr}(\mathbf{M}_{AB} \mathbf{V} \mathbf{M}_{AB} \mathbf{V})}, \ \nu_2 = \infty.$$

Reject if $F_A \geq f$ (or if $F_{AB} \geq f$), where f is the $1 - \alpha$ quantile of an F distribution with ν_1 and ν_2 degrees of freedom.

8.6.13 R Function bwrank

The R function

$$bwrank(J,K,x)$$

performs a between-by-within ANOVA based on ranks using the method just described. In addition to testing hypotheses as just indicated, the function returns the average ranks ($\bar{R}_{.jk}$) associated with all JK groups as well as the relative effects, $(\bar{R}_{.jk} - 0.5)/N$.

■ Example

Lumley (1996) reports data on shoulder pain after surgery; the data are from a study by Jorgensen, Gilles, Hunt, Caplehorn, and Lumley (1995). Table 8.7 shows a portion of

Table 8.7: Shoulder Pain Data (1=low, 5=high).

Active Treatment			No Active Treatment		
Time 1	Time 2	Time 3	Time 1	Time 2	Time 3
1	1	1	5	2	3
3	2	1	1	5	3
3	2	2	4	4	4
1	1	1	4	4	4
1	1	1	2	3	4
1	2	1	3	4	3
3	2	1	3	3	4
2	2	1	1	1	1
1	1	1	1	1	1
3	1	1	1	5	5
1	1	1	1	3	2
2	1	1	2	2	3
1	2	2	2	2	1
3	1	1	1	1	1
2	1	1	1	1	1
1	1	1	5	5	5
1	1	1	3	3	3
2	1	1	5	4	4
4	4	2	1	3	3
4	4	4			
1	1	1			
1	1	1			

the results where two treatment methods are used and measures of pain are taken at three different times. The output from bwrank is

```
$test.A:
[1] 12.87017

$sig.A:
[1] 0.001043705

$test.B:
[1] 0.4604075

$sig.B:
[1] 0.5759393

$test.AB:
[1] 8.621151

$sig.AB:
[1] 0.0007548441

$avg.ranks:
          [,1]      [,2]      [,3]
[1,] 58.29545 48.40909 39.45455
[2,] 66.70455 82.36364 83.04545

$rel.effects:
           [,1]       [,2]       [,3]
[1,] 0.4698817 0.3895048 0.3167036
[2,] 0.5382483 0.6655580 0.6711013
```

So at approximately the 0.001 level, treatment methods are significantly different and there is a significant interaction, but no significant difference is found over time. Note that the average ranks and relative effects suggest that a disordinal interaction might exist. In particular, for group 1 (the active treatment group), time 1 has higher average ranks versus time 2, and the reverse is true for the second group. However, the Wilcoxon signed rank test fails to reject at the 0.05 level when comparing time 1 to time 2 for both groups. When comparing time 1 versus time 3 for the first group, again using the Wilcoxon signed rank test, we reject at the 0.05 level, but a nonsignificant result is obtained for group 2. So again a disordinal interaction appears to be a possibility, but the empirical evidence is not compelling.

■

■ Example

Section 6.11 illustrated a method for comparing multivariate data corresponding to two independent groups based on the extent that points from one group are nested within the other. For the data in Table 6.6, it was found that schizophrenics differed from the control group; also see Figure 6.10. If the two groups are compared based on the OP estimator (using the function smean2), again the two groups are found to differ. Comparing the groups with the method for means and trimmed means described in this section, no difference between the schizophrenics and control group is found at the 0.05 level. Using the rank-based method in this section, again no difference is found. (The *p*-value is .11.) The only point is that how we compare groups can make a practical difference about the conclusions reached.

∎

8.6.14 Rank-Based Multiple Comparisons

Multiple comparisons based on the rank-based methods covered here can be performed using simple combinations of methods already considered. When dealing with Factor A, for example, one can simply compare level j to level j', ignoring the other levels. When comparing all pairs of groups, FWE can be controlled with Rom's method or the Benjamini–Hochberg technique. Factor B and the collection of all interactions (corresponding to any two rows and any two columns) can be handled in a similar manner.

8.6.15 R Function bwrmcp

The R function

$$bwrmcp(J,K,x,grp=NA,alpha=0.05,bhop=F)$$

performs all pairwise multiple comparisons using the method of Section 8.6.14 with the FWE (familywise error) rate controlled using Rom's method of the Benjamini–Hochberg method. For example, when dealing with Factor A, the function simply compares level j to level j' ignoring the other levels. All pairwise comparisons among the J levels of Factor A are performed and the same is done for Factor B and all relevant interactions.

8.6.16 Multiple Comparisons when Using a Patel–Hoel Approach to Interactions

Rather than compare distributions when dealing with a between-by-within design, one could use a simple analog of the Patel–Hoel approach instead. First consider a 2-by-2 design and focus on level one of Factor A. Then the two levels of Factor B are dependent and can be compared with the sign test. In essence, inferences are being made about p_1, the probability that for a randomly sampled pair of observations, the observation from level one of Factor B is less than the corresponding observation from level two. Of course, for level two of Factor A, we can again compare levels one and two of Factor B with the sign test. Now we let p_2 be the probability that for a randomly sampled pair of observations, the observation from level one of Factor B is less than the corresponding observation from level two. Then no interaction can be defined as $p_1 = p_2$.

The hypothesis of no interaction,

$$H_0 : p_1 = p_2,$$

is just the hypothesis that two independent binomials have equal probabilities of success, which can be tested using one of the methods described in Section 5.8. Here, Beal's method is used rather than the Storer–Kim method because it currently seems that Beal's method provides more accurate control over FWE for the problem at hand, execution time can be much lower when sample sizes are large, and unlike the Storer–Kim procedure, Beal's method provides confidence intervals. Method KMS in Section 5.8.3 might be used as well, but there are no published results on how it performs for the situation at hand.

There are various ways FWE might be controlled. Among a collection of techniques considered by Wilcox (2001c), the following method was found to be relatively effective. Let q be the $1 - \alpha$ quantile of a C-variate Studentized maximum modulus distribution with degrees of freedom $\nu = \infty$, where C is the total number of hypotheses to be tested. Assuming that all pairs of rows and columns are to be considered when testing the hypothesis of no interactions,

$$C = \frac{J^2 - J}{2} \times \frac{K^2 - K}{2}.$$

(For a formal definition of a Studentized maximum modulus distribution, see Miller, 1966, p. 71. Some quantiles are reported in Wilcox, 2003a.) Let Z be a standard normal random variable. Then if FWE is to be α, test each of the C hypotheses at the α_a level where

- If $(J, K) = (5, 2)$, then $\alpha_a = 2[1 - P(Z \leq q)]$.
- If $(J, K) = (3, 2)$, $(4, 2)$ or $(2, 3)$, then $\alpha_a = 3[1 - P(Z \leq q)]$.
- For all other J and K values, $\alpha_a = 4[1 - P(Z \leq q)]$.

These adjusted α values appear to work well when the goal is to achieve FWE less than or equal to 0.05. Whether this remains the case with FWE equal to 0.01 is unknown. For $C > 28$ and FWE equal to 0.05, use

$$q = 2.383904C^{1/10} - 0.202.$$

(Of course, for $C = 1$, no adjustment is necessary; simply use Beal's method.)

Tied values are handled in the same manner as with the signed rank test: pairs of observations with identical values are simply discarded. So among the remaining observations, for every pair of observations, the observation from level one of Factor B, for example, is either less than or greater than the corresponding value from level two.

A criticism of this method is that power can be relatively low. However, it directly addresses an issue that might be deemed interesting and useful that is not directly addressed by other methods in this chapter.

A variation of the approach in this section is where, for level one of Factor B, p_1 is the probability that an observation from level one of Factor A is less than an observation from level 2. Similarly, p_2 is now defined in terms of the two levels of Factor A when working with level two of Factor B. However, the details of how to implement this approach have not been studied.

8.6.17 R Function sisplit

The method just described for interactions can be applied with the R function

$$sisplit(J,K,x)$$

This function assumes $\alpha = 0.05$; other values are not allowed. As usual, x is any R variable containing the data that is an n-by-JK matrix or has list mode.

8.7 Some Rank-Based Multivariate Methods

This section describes two rank-based methods for comparing J independent groups with K measures associated with each group.

8.7.1 The Munzel–Brunner Method

The first method was derived by Munzel and Brunner (2000). (For recent results regarding how the Munzel–Brunner method compares to several techniques not covered here, see Bathke, Solomon, & Madden, 2008. A variation of the Munzel–Brunner method can be used in place of the Agresti–Pendergast method, but the relative merits of these two techniques have not been explored.) Let n_j represent the number of randomly sampled vectors from the jth group, each vector containing K measures. Let $F_{jk}(x)$ be the distribution associated with the jth group and kth measure. So for example, $F_{32}(6)$ is the probability that for the third group, the second variable will be less than or equal to 6 for a randomly sampled individual. For the kth measure, the goal is to test the hypothesis that all J groups have identical distributions. And the more general goal is to test the hypothesis that simultaneously, all groups have identical distributions for each of the K measures under consideration. That is, the goal is to test

$$H_0 : F_{1k}(x) = \cdots = F_{Jk}(x) \text{ for all } k = 1, \ldots, K. \tag{8.9}$$

To apply the method, begin with the first of the K measures, pool all the observations among the J groups and assign ranks. Ties are handled in the manner described in Section 5.7.2. Repeat this process for all K measures and label the results R_{ijk}. That is, R_{ijk} is the rank of the ith observation in the jth group and for the kth measure. Let

$$\bar{R}_{jk} = \frac{1}{n_j} \sum_{i=1}^{n_j} R_{ijk},$$

be the average rank for the jth group corresponding to the kth measure. Set

$$\hat{Q}_{jk} = \frac{\bar{R}_{jk} - 0.5}{n},$$

where $n = \sum n_j$ is the total number of randomly sampled vectors among the J groups. The remaining calculations are summarized in Table 8.8. The $\hat{Q}$ values are called the *relative effects* and reflect the ordering of the average ranks. If, for example, $\hat{Q}_{11} < \hat{Q}_{21}$, the typical rank for variable one in group one is less than the typical rank for variable one in group two. More generally, if $\hat{Q}_{jk} < \hat{Q}_{j'k}$, then based on the kth measure, the typical rank (or observed value) for group j is less than the typical rank for group j'.

Table 8.8: The Munzel–Brunner One-Way Multivariate Method.

Let

$$\hat{\mathbf{Q}} = (\hat{Q}_{11}, \hat{Q}_{12}, \ldots, \hat{Q}_{1K}, \hat{Q}_{21}, \ldots, \hat{Q}_{JK})',$$

$$\mathbf{R}_{ij} = (R_{ij1}, \ldots, R_{ijK})', \; \bar{\mathbf{R}}_j = (\bar{R}_{j1}, \ldots, \bar{R}_{jK})',$$

$$\mathbf{V}_j = \frac{1}{nn_j(n_j - 1)} = \sum_{i=1}^{n_j} (\mathbf{R}_{ij} - \bar{\mathbf{R}}_j)(\mathbf{R}_{ij} - \bar{\mathbf{R}}_j)',$$

$n = \sum n_j$ and let

$$\mathbf{V} = \mathrm{diag}\{\mathbf{V}_1, \ldots, \mathbf{V}_J\}.$$

Compute the matrix $\mathbf{M}_A$ as described in Section 7.9. The test statistic is

$$F = \frac{n}{\mathrm{tr}(\mathbf{M}_A \mathbf{V})} \hat{\mathbf{Q}}' M_A \hat{\mathbf{Q}}.$$

Decision Rule: Reject if $F \geq f$, where f is the $1 - \alpha$ quantile of an F distribution with

$$\nu_1 = \frac{(\mathrm{tr}(\mathbf{M}_A \mathbf{V}))^2}{\mathrm{tr}(\mathbf{M}_A \mathbf{V} \mathbf{M}_A \mathbf{V})},$$

and $\nu_2 = \infty$ degrees of freedom.

8.7.2 R Function mulrank

The R function

$$\text{mulrank}(J, K, x)$$

performs the one-way multivariate method in Table 8.8. The data are stored in x which can be a matrix or have list mode. If x is a matrix, the first K columns correspond to the K measures for group 1, the second K correspond to group 2, and so forth. If stored in list mode, x[[1]], ..., x[[K]] contain the data for group 1, x[[K+1]], ..., x[[2K]] contain the data for group 2, and so on.

■ Example

Table 8.9 summarizes data (reported by Munzel & Brunner, 2000) from a psychiatric clinical trial where three methods are compared for treating individuals with panic disorder. The three methods are exercise, clomipramine and a placebo. The two measures of effectiveness were a clinical global impression (CGI) and the patient's global impression (PGI). The test statistic is $F = 12.7$ with $v_1 = 2.83$ and a significance level less than 0.001. The relative effects are:

```
$q.hat:
          [,1]       [,2]
[1,] 0.5074074 0.5096296
[2,] 0.2859259 0.2837037
[3,] 0.7066667 0.7066667
```

Table 8.9: CGI and PGI Scores After Four Weeks of Treatment.

Exercise		Clomipramine		Placebo	
CGI	PGI	CGI	PGI	CGI	PGI
4	3	1	2	5	4
1	1	1	1	5	5
2	2	2	0	5	6
2	3	2	1	5	4
2	3	2	3	2	6
1	2	2	3	4	6
3	3	3	4	1	1
2	3	1	4	4	5
5	5	1	1	2	1
2	2	2	0	4	4
5	5	2	3	5	5
2	4	1	0	4	4
2	1	1	1	5	4
2	4	1	1	5	4
6	5	2	1	3	4

So among the three groups, the second group, clomipramine, has the lowest relative effects. That is, the typical ranks were lowest for this group, and the placebo group had the highest ranks on average.

∎

8.7.3 The Choi–Marden Multivariate Rank Test

This section describes a multivariate analog of the Kruskal–Wallis test derived by Choi and Marden (1997). There are actually many variations of the approach they considered, but here attention is restricted to the version they focused on. As with the method in Section 8.7.1, we have K measures for each individual and there are J independent groups. For the jth group and any vector of constants $\mathbf{x} = (x_1, \ldots, x_K)$, let

$$F_j(\mathbf{x}) = P(X_{j1} \leq x_1, \ldots, X_{jK} \leq x_K).$$

So for example, $F_1(\mathbf{x})$ is the probability that for the first group, the first of the K measures is less than or equal to x_1, the second of the K measures is less than or equal to x_2, and so forth. The null hypothesis is that for any $\mathbf{x}$,

$$H_0 : F_1(\mathbf{x}) = \cdots = F_J(\mathbf{x}), \tag{8.10}$$

which is sometimes called the *multivariate hypothesis* to distinguish it from Eq. (8.8), which is called the *marginal hypothesis*. The multivariate hypothesis is a stronger hypothesis in the sense that if it is true, then by implication the marginal hypothesis is true as well. For example, if the marginal distributions for both groups are standard normal distributions, the marginal hypothesis is true, but if the groups have different correlations, the multivariate hypothesis is false.

The Choi–Marden method represents an extension of a technique derived by Möttönen & Oja (1995) and is based on a generalization of the notion of a rank to multivariate data which was also used by Chaudhuri (1996, Section 4). First consider a random sample of n observations with K measures for each individual or thing and denote the ith vector of observations by

$$\mathbf{X}_i = (X_{i1}, \ldots, X_{iK}).$$

Let

$$A_{ii'} = \sqrt{\sum_{k=1}^{K}(X_{ik} - X_{i',k})^2},$$

Table 8.10: The Choi–Marden Method.

Pool the data from all J groups and compute rank vectors as just described in the text. The resulting rank vectors are denoted by $\mathbf{R}_1, \ldots, \mathbf{R}_n$, where $n = \sum n_j$ is the total number of vectors among the J groups. For each of the J groups, average the rank vectors and denote the average of these vectors for the jth group by $\bar{\mathbf{R}}_j$.

Next, assign ranks to the vectors in the jth group, ignoring all other groups. We let $\mathbf{V}_{ij}$ (a column vector of length K) represent the rank vector corresponding to the ith vector of the jth group ($i = 1, \ldots, n_j$; $j = 1, \ldots, J$) to make a clear distinction with the ranks based on the pooled data. Compute

$$\mathbf{S} = \frac{1}{n-J} \sum_{j=1}^{J} \sum_{i=1}^{n_j} \mathbf{V}_{ij} \mathbf{V}'_{ij},$$

where $\mathbf{V}'_{ij}$ is the transpose of $\mathbf{V}_{ij}$ (so $\mathbf{S}$ is a K-by-K matrix). The test statistic is

$$H = \sum_{j=1}^{J} n_j \bar{\mathbf{R}}'_j \mathbf{S}^{-1} \bar{\mathbf{R}}_j. \tag{8.11}$$

(For $K = 1$, H does not quite reduce to the Kruskal–Wallis test statistic. In fact, H avoids a certain technical problem that is not addressed by the Kruskal–Wallis method.)

Decisions Rule: Reject if $H \geq c$, where c is the $1 - \alpha$ quantile of a chi-squared distribution with degrees of freedom $K(J - 1)$.

Here, the "rank" of the ith vector is itself a vector (having length K) given by

$$\mathbf{R}_i = \frac{1}{n} \sum_{i'=1}^{n} \frac{\mathbf{X}_i - \mathbf{X}_{i'}}{A_{ii'}},$$

where

$$\mathbf{X}_i - \mathbf{X}_{i'} = (X_{i1} - X_{i'1}, \ldots, X_{iK} - X_{i'K}).$$

The remaining calculations are summarized in Table 8.10. All indications are that this method provides good control over the probability of a type I error when ties never occur. There are no known problems when there are tied values, but this issue is in need of more research.

8.7.4 R Function cmanova

The R function

cmanova(J,K,x)

performs the Choi–Marden method just described. The data are assumed to be stored in x as described in Section 8.6.2.

8.8 Three-Way Designs

Generally, two-way designs can be extended to a three-way design where one or more factors involve dependent groups. This section outlines some of the methods that might be used. It is stressed, however, that simulation studies reporting the relative merits of the methods considered are extremely limited.

8.8.1 Global Tests Based on Trimmed Means

The method in Section 8.6.1 is readily extended to a three-way design. Note that the matrix V used in Eq. (8.4) reflects the variances and covariances among the trimmed means where the covariances are taken to be zero if the groups are independent. Here, V is computed in a similar manner. That is, for a J-by-K-by-L design, V is a JKL square matrix that contains the squared standard errors and covariances among the sample trimmed means, with independent trimmed means having a covariance of zero. Once V is available, compute the test statistic Q given by Eq. (8.4), where now the matrix C is computed as described in Table 7.5.

When dealing with situations where one or more factors involve dependent groups, comments should be made regarding the approximation of the null distribution using an F distribution. Unlike the method in Section 8.6.1 where the second degree of freedom, v_2, is estimated based on the data, the strategy here is to simply set $v_2 = 999$. The reason is that even with $v_2 = 999$, the actual level of the method can drop well below the nominal level when the sample size is small and 20% trimmed means are used. For example, when dealing with a between-by-between-by-within design, under normality with all correlations equal to zero and $J = 2$, $K = L = 3$, and $n = 25$, the actual type I error probability is approximately .003 when dealing with the A-by-C interaction and testing at the 0.05 level. Increasing n to 50, now the actual level is approximately 0.013, for $n = 100$ it is 0.037, and for $n = 900$ it is 0.04. For the main effect associated with factor A, the estimated level is 0.050 with $n = 25$ and 0.052 with $n = 900$. A bootstrap-t method appears to suffer from the same problem, but an extensive study of this issue has not been conducted. A similar problem occurs when dealing with a between-by-within-by-within design. Again $v_2 = 999$ is used, but even with $n = 100$, the actual level can drop as low as 0.025 under normality. Using instead a bootstrap-t method (via the R function bbwtrimbt), with $n = 25$, the actual level was estimated to be 0.067.

Evidently there are no published studies comparing methods, in terms of type I errors, when dealing with three-way designs with one or more within group factors. Very limited results suggest that perhaps a better approach, compared to the methods described here, is to use the R functions in Section 8.8.6. They test hypotheses about all of the usual linear contrasts associated with a three-way design using a percentile bootstrap method in conjunction with a trimmed mean. In terms of controlling the probability of one or more type I errors, performing one of the global tests described here is not required when using the percentile bootstrap methods via the R functions in Section 8.8.6. Moreover, limited results suggest that control over the type I error probability is more satisfactory when the amount of trimming is 20%. For the situation considered here, where $n = 25$ and the actual type I error is approximately .003, the probability of one or more type I errors was estimated to be .039 when using a percentile bootstrap method via the R function bbwmcppb, based on a simulation with 1000 replications. (And execution time can be substantially less when using a percentile bootstrap method rather than the bootstrap-t method.) Using instead the R function bwwmcppb, the probability of one or more type I errors was estimated to be .049. But again, a more comprehensive study is needed.

8.8.2 R Functions bbwtrim, bwwtrim, wwwtrim, bbwtrimbt, bwwtrimbt, and wwwtrimbt

The R function

$$bbwtrim(J,K,L,x,grp=c(1:p),tr=0.2)$$

tests all omnibus main effects and interactions associated with a between-by-between-by-within design. The data are assumed to be stored as described in Section 7.3.1. For a between-by-within-by-within design use

$$bwwtrim(J,K,L,x,grp=c(1:p),tr=0.2).$$

And for a within-by-within-by-within design use

$$wwwtrim(J,K,L,x,grp=c(1:p),tr=0.2).$$

The R functions

$$bbwtrimbt(J,K,L,x,grp=c(1:p),tr=0.2, nboot = 599, SEED = T)$$

$$bwwtrim(J,K,L,x,grp=c(1:p),tr=0.2, nboot = 599, SEED = T).$$

and

$$wwwtrimbt(J,K,L,x,grp=c(1:p),tr=0.2, nboot = 599, SEED = T).$$

are the same as the functions bbwtrim, bwwtrim, and wwwtrim, respectively, only a bootstrap-t method is used.

8.8.3 Data Management: R Functions bw2list and bbw2list

For a between-by-within-by-within design, the R function

$$bw2list(x, grp.col, lev.col),$$

which was introduced in Section 8.6.3, can be used when dealing with data that are stored in a matrix or a data frame with one column indicating the levels of the independent groups and other columns containing data corresponding to within group levels. For example, setting the argument grp.col=c(5) would indicate that the levels for Factor A are stored in column 5 and lev.col=c(3,9,10,12) indicates that the within levels data are stored in columns 3, 9, 10, and 12. Note that it must be the case that KL is equal to the number of values stored in lev.col. So lev.col=c(3,9,10,12) would be appropriate if the within factors have two levels each, with the data for two levels of Factor C being stored in columns 10 and 12.

The R function

$$\text{bbw2list(x, grp.col, lev.col),}$$

deals with a between-by-between-by-within design and assumes that the argument grp.col contains two values that indicate the columns of x that indicate the levels of Factors A and B. Now the argument lev.col indicates the columns containing the within data.

■ Example

Imagine that for a between-by-between-by-within design, column 14 of the R variable dis contains values indicating the levels of Factor A, column 10 has values that contain the levels of Factor B, and columns 2, 4, and 9 contain the outcomes values at times 1, 2, and 3, respectively. Then

$$\text{z=bbw2list(dis, grp.col=c(14,10), lev.col=c(2,4,9))}$$

would store the data in list mode in z, after which the command

$$\text{bbwtrim(3,4,3,z)}$$

would test the usual hypotheses, assuming that Factors A and B have 3 and 4 levels, respectively. The values in column 14 and 10 would be sorted in ascending order, or in alphabetical order if the values in these columns are character data. ■

8.8.4 Multiple Comparisons

Multiple comparisons in a three-way design can be performed using a straightforward extension of methods described in previous sections. The R function con3way, described in Section 7.4.4, can be used to generate the linear contrast coefficients that are often used. Here, when computing A in Section 8.1.3, we set $d_{jk} = 0$ whenever j and k correspond to independent groups. Otherwise, this term is computed as described in Section 8.1.3. The next two sections summarize some R functions aimed at facilitating the analysis.

8.8.5 R Function rm3mcp

When dealing with a within-by-within-by-within design, a nonbootstrap method can be used to test the hypotheses associated with all of the linear contrasts generated by the R function con3way. This can be done with the R function

$$\text{rm3mcp}(J, K, L, x, tr = 0.2, alpha = 0.05, dif = T, grp = NA).$$

(That is, it uses the R function con3way to generate the linear contrast coefficients and then it tests the corresponding hypotheses.) When dealing with designs where there are both between and within factors, use a bootstrap method via one of the R functions described in the next section. Another approach is to use the R function rmmcp in Section 8.1.5 in conjunction with the R function con3way. (For an illustration of how to interpret three-way interactions based on the contrast coefficients returned by con3way, see the example at the end of Section 7.4.4.)

8.8.6 R Functions bbwmcp, bwwmcp, bbwmcppb, bwwmcppb, and wwwmcppb

Bootstrap-t Methods

The R function

$$\text{bbwmcp}(J, K, L, x, tr = 0.2, JKL = J * K * L, con = 0, alpha = 0.05, grp = c(1\!:\!JKL), nboot}$$
$$= 599, SEED = T, \ldots)$$

performs all multiple comparisons associated with main effects and interactions using a bootstrap-t method in conjunction with trimmed means when analyzing a between-by-between-by-within design. The function uses con3way to generate all of the relevant linear contrasts and then uses the function lindep to test the hypotheses. The critical value is designed to control the probability of at least one type I error among all the linear contrasts associated with factor A. The same is done for factor B and factor C.

The R function

$$\text{bwwmcp}(J, K, L, x, tr = 0.2, JKL = J * K * L, con = 0, alpha = 0.05, grp = c(1\!:\!JKL),$$
$$\text{nboot} = 599, SEED = T, \ldots)$$

handles a between-by-within-by-within design.

Percentile Bootstrap Methods

For a between-by-between-by-within design, the R function

$$\text{bbwmcppb}(J, K, L, x, tr = 0.2, JKL = J * K * L, con = 0, alpha = 0.05, grp = c(1\!:\!JKL),$$
$$\text{nboot} = 599, SEED = T, \ldots)$$

tests hypotheses using a percentile bootstrap method. As for a between-by-within-by-within and within-by-within-by-within design, use the functions

bwwmcppb(J, K, L, x, tr = 0.2, JKL = J * K * L, con = 0, alpha = 0.05, grp = c(1:JKL), nboot = 599, SEED = T, ...)

and

wwwmcppb(J, K, L, x, tr = 0.2, JKL = J * K * L, con = 0, alpha = 0.05, grp = c(1:JKL), nboot = 599, SEED = T, ...),

respectively.

8.9 Exercises

1. Section 8.6.2 reports data on hangover symptoms. For group 2, use the R function rmanova to compare the trimmed means corresponding to times 1, 2, and 3.
2. For the data used in Exercise 1, compute confidence intervals for all pairs of trimmed means using the R function pairdepb.
3. Analyze the data for the control group reported in Table 6.1 using the methods in Sections 8.1 and 8.2. Compare and contrast the results.
4. Repeat Exercise 3 using the rank-based method in Section 8.5. How do the results compare to using a measure of location?
5. Repeat Exercises 3 and 4 using the data for the murderers in Table 6.1.
6. Analyze the data in Table 6.1 using the methods in Sections 8.6.1 and 8.6.4.
7. Repeat Exercise 6, only now use the rank-based method in Section 8.6.12.

CHAPTER 9

Correlation and Tests of Independence

There are many approaches to finding robust measures of correlation and covariance (e.g., Ammann, 1993; Davies, 1987; Devlin et al., 1981; Goldberg & Iglewicz, 1992; Hampel et al., 1986, Chapter 5; Huber, 1981, Chapter 8; Li & Chen, 1985; Lupuhaä, 1989; Maronna, 1976; Mosteller & Tukey, 1977, p. 211; Wang & Raftery, 2002; Wilcox, 1993b), but no attempt is made to give a detailed description of all the strategies that have been proposed. Some of these measures are difficult to compute, others are not always equal to zero under independence, and from a technical point of view, some do not have all the properties one might want. One of the main goals in this chapter is to describe some tests of zero correlation that have practical value relative to the standard test based on the usual product moment correlation, r. Some alternative methods for testing the hypothesis of independence, that are not based on some type of correlation coefficient, are described as well. (For a collection of alternative methods for detecting dependence, see Kallenberg & Ledwina, 1999.)

9.1 Problems with the Product Moment Correlation

The most common measure of covariance between any two random variables, X and Y, is

$$\text{COV}(X, Y) = \sigma_{xy}$$

$$= E[(X - \mu_x)(Y - \mu_y)],$$

and the corresponding (Pearson) measure of correlation is

$$\rho = \frac{\sigma_{xy}}{\sigma_x \sigma_y}.$$

A practical concern with ρ is that it is not robust. If one of the marginal distributions is altered slightly, as measured by Kolmogorov distance, but the other marginal distribution is left unaltered, the magnitude of ρ can be changed substantially. More formally, the influence

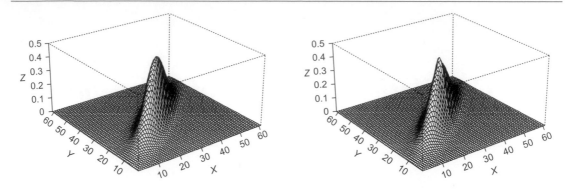

Figure 9.1: Pearson's correlation for the bivariate normal distribution shown on the left panel is 0.8. In the right panel, x has a contaminated normal distribution and Pearson's correlation is 0.2.

function of Pearson's correlation is

$$IF(x, y) = xy - \left(\frac{x^2 + y^2}{2}\right)\rho,$$

which is unbounded (Devlin et al., 1981). That is, Pearson's correlation does not have infinitesimal robustness.

The left panel of Figure 9.1 shows a bivariate normal distribution with $\rho = 0.8$. Suppose the marginal distribution of X is replaced by a contaminated normal given by Eq. (1.1) with $\epsilon = 0.9$ and $K = 10$. The right panel of Figure 9.1 shows the resulting joint distribution. As is evident, there is little visible difference between these two distributions, but in the right panel of Figure 9.1, $\rho = 0.2$. Put another way, even with an infinitely large sample size, the usual estimate of ρ, given by r in the next paragraph, can be misleading.

Let $(X_1, Y_1), \ldots, (X_n, Y_n)$ be a random sample from some bivariate distribution. The usual estimate of ρ is

$$r = \frac{\sum(X_i - \bar{X})(Y_i - \bar{Y})}{\sqrt{\sum(X_i - \bar{X})^2 \sum(Y_i - \bar{Y})^2}}.$$

A practical concern with r is that it is not resistant—a single unusual point can dominate its value.

■ Example

Figure 9.2 shows a scatterplot of data on the logarithm of the effective temperature at the surface of 47 stars versus the logarithm of its light intensity. (The data are reported

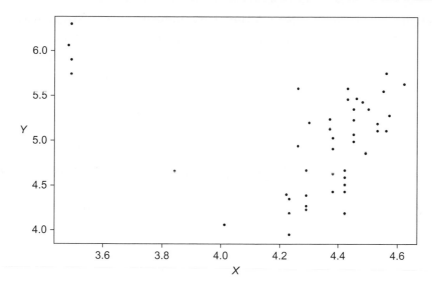

Figure 9.2: Scatterplot of the star data.

in Rousseeuw & Leroy, 1987, p. 27.) The scatterplot suggests that in general, there is a positive relationship between temperature and light, yet $r = -0.21$. The reason is that the four points in the upper left corner of Figure 9.2 (which are giant red stars) are outliers that dominate the value of r. (Two additional points are flagged as outliers by the R function out.)

From basic principles, if X and Y are independent, then $\rho = 0$. The best-known test of

$$H_0 : \rho = 0 \tag{9.1}$$

is based on the test statistic

$$T = r \sqrt{\frac{n-2}{1-r^2}}. \tag{9.2}$$

If H_0 is true, T has a Student's t-distribution with $\nu = n - 2$ degrees of freedom if at least one of the marginal distributions is normal (e.g., Muirhead, 1982, p. 146). In particular, reject $H_0 : \rho = 0$ if $|T| > t_{1-\alpha/2}$, the $1 - \alpha/2$ quantile of Student's t-distribution with $n - 2$ degrees of freedom. When X and Y are independent, there are general conditions under which $E(r) = 0$ and $E(r^2) = 1/(n-1)$ (Huber, 1981, p. 204). (All that is required is that the distribution of X or Y be invariant under permutations of the components.) This suggests that the test of independence, based on T, will be reasonably robust in terms of type I errors, and this seems to be the case for a variety of situations (Kowalski, 1972; Srivastava & Awan, 1984). However, problems arise in at least three situations: when $\rho = 0$ but X and Y are dependent (e.g., Edgell & Noon, 1984), when performing one-sided tests (Blair & Lawson, 1982), and when considering the more general goal of testing for independence among all pairs of p random variables. There is also the problem of computing a confidence interval for ρ. Many methods have been proposed, but simulations do not support their use, at least for small to moderate sample sizes, and it is unknown just how large of a sample size is needed before any particular method can be expected to give good probability coverage (Wilcox, 1991a). A modified percentile bootstrap method appears to perform reasonably well in terms of probability coverage provided ρ is reasonably close to zero. But as ρ gets close to one it begins to break down (Wilcox & Muska, 2001). Many books recommend *Fisher's r-to-Z transformation* when computing confidence intervals, but under general conditions, it is not even asymptotically correct when sampling from nonnormal distributions (Duncan & Layard, 1973).

9.1.1 Features of Data that Affect r and T

There are several features of data that affect the magnitude of Pearsons's correlation, as well as the magnitude of T, given by Eq. (9.2). These features are important when interpreting robust correlation coefficients, so they are described here.

Five features of data that affect r are as follows:

1. Outliers
2. The magnitude of the slope around which points are clustered (e.g., Barrett, 1974; Loh, 1987b). Put another way, rotating points can raise or lower r
3. Curvature
4. The magnitude of the residuals
5. Restriction of range

The effects of outliers have already been illustrated. The effect of curvature seems fairly evident, as does the magnitude of the residuals. It is well known that restricting the range of X or Y can lower r, and the star data in Figure 9.2 illustrate that a restriction in range can increase r as well. The effect of rotating points is illustrated by Figure 9.3. Thirty points were generated for X from a standard normal distribution, ϵ was taken to have a normal distribution with mean zero and standard deviation 0.25, and then $Y = X + \epsilon$ was computed. The least squares estimate of the slope is 1.00 and it was found that $r = 0.964$. Then the points were rotated clockwise by $35°$. The rotated points are indicated by the o's in Figure 9.3. Now the least squares estimate of the slope is 0.19 and $r = 0.81$. Rotating the points by $40°$, instead, $r = 0.61$. Continuing to rotate the points in the same direction, the correlation will decrease until both r and the least squares estimate of the slope are zero.

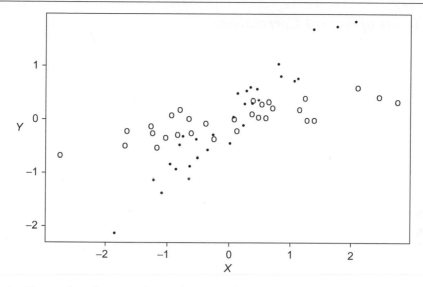

Figure 9.3: **An illustration that rotating points can alter Pearson's correlation. The dots have correlation** $r = 0.964$. **The points marked by an o are the dots rotated by 35 degrees; these rotated points have** $r = 0.81$.

9.1.2 Heteroscedasticity and the Classic Test that $\rho = 0$

Now consider the test of H_0: $\rho = 0$ based on the test statistic T given Eq. (9.2). As just pointed out, five features of data affect the magnitude of r, and hence T. There is, in fact, a sixth feature that affects T even when $\rho = 0$: heteroscedasticity. In regression, homoscedasticity refers to a situation where the conditional variance of Y, given X, does not depend on X. That is, $\text{VAR}(Y|X) = \sigma^2$. Heteroscedasticity refers to a situation where the conditional variance of Y varies with X. Independence implies homoscedasticity, but $\rho = 0$ does not necessarily mean there is homoscedasticity. The reason heteroscedasticity is relevant to the test of H_0 based on (9.2) is that the derivation of the test statistic, T, is based on the assumption that X and Y are independent. Even if $\rho = 0$, but there is heteroscedasticity, the wrong standard error is being used by T.

To illustrate what can happen, imagine that both X and Y have normal distributions with both means equal to zero. Further assume that both X and Y have variance 1 unless $|X| > 0.5$, in which case Y has standard deviation $|X|$. So there is dependence, but $\rho = 0$. With $n = 20$ and testing at the $\alpha = 0.05$ level with T, the actual probability of a type I error is .098. With $n = 40$, it is 0.125 and for $n = 200$ it is 0.159. Even though $\rho = 0$, the probability of rejecting is increasing as n gets large because the wrong standard error is being used. So when H_0 is rejected, it is reasonable to conclude dependence, but the nature of the dependence (the reason why H_0 was rejected) is unclear. (Section 9.3.13 describes two methods aimed at dealing with this problem.)

9.2 Two Types of Robust Correlations

Here, robust analogs of Pearson's correlation are classified into one of two types: those that protect against outliers among the marginal distributions without taking into account the overall structure of the data, and those that take into account the overall structure of the data when dealing with outliers. In terms of developing tests of the hypothesis of independence between two random variables, it is a bit easier working with the first type. Recently, however, some progress has been made when working with the second. For convenience, the first type will be called a *type M correlation* and the second will be called *type O*.

9.3 Some Type M-Measures of Correlation

This section describes four type M correlations and how they can be used to test the hypothesis of independence.

9.3.1 The Percentage Bend Correlation

The first type M correlation that has proven to be relatively successful, in terms of controlling type I error probabilities when testing the hypothesis of independence, is the so-called percentage bend correlation. It is estimated with r_{pb} using the computations described in Table 9.1. Table 9.2 describes how the population parameter corresponding to r_{pb}, ρ_{pb}, is defined. When X and Y are independent, $\rho_{pb} = 0$. Under normality, ρ and ρ_{pb} have very similar values, but ρ_{pb} is more robust, and their population values can differ substantially, even when there is little apparent difference between the bivariate distributions (Wilcox, 1994d).

Perhaps it should be emphasized that r_{pb} is not intended as an estimate of ρ. Rather, the goal is to estimate a measure of correlation, ρ_{pb}, that is not overly sensitive to slight changes in the distributions. There might be situations where ρ is of interest despite its lack of robustness, in which case r_{pb} has little or no value. The situation is similar to finding a robust measure of location. If there is direct interest in the population mean μ, the 20% sample trimmed mean does not estimate μ when distributions are skewed and would not be used. The problem is that μ is not robust, in which case some other measure of location might be of interest, such as a 20% trimmed mean. In a similar fashion, ρ_{pb} provides a robust measure of the linear association between two random variables that is designed so that its value is not overly sensitive to a relatively small proportion of the population under study.

Note that the definition of ρ_{pb} depends in part on a measure of scale, ω_x, which is a generalization of MAD. A technical point of some interest is that ω_x is a measure of dispersion. (See Section 2.3.) If in the definition of ρ_{pb}, $\Psi(x) = \max[-1, \min(1, x)]$ is

Table 9.1: Computing the Percentage Bend Correlation.

The goal is to estimate the percentage bend correlation, ρ_{pb}, based on the random sample $(X_1, Y_1), \ldots, (X_n, Y_n)$. For the observations $X_1, \ldots, X_n$, let M_x be the sample median. Select a value for $\beta, 0 \leq \beta \leq 0.5$. Compute

$$W_i = |X_i - M_x|,$$

$$m = [(1 - \beta)n].$$

Note that $[(1 - \beta)n]$ is $(1 - \beta)n$ rounded down to the nearest integer. Let $W_{(1)} \leq \cdots \leq W_{(n)}$ be the W_i values written in ascending order. Set

$$\hat{\omega}_x = W_{(m)}.$$

For example, if the observations are 4, 2, 7, 9, and 13, then the sample median is $M_x = 7$, so $W_1 = |4-7| = 3$, $W_2 = |2-7| = 5$, $W_3 = |7-7| = 0$, $W_4 = 2$, and $W_5 = 6$; so $W_{(1)} = 0$, $W_{(2)} = 2$, $W_{(3)} = 3$, $W_{(4)} = 5$, and $W_{(5)} = 6$. If $\beta = 0.1$, $m = [0.9(5)] = 4$, and $\hat{\omega} = W_{(4)} = 5$.

Let i_1 be the number of X_i values such that $(X_i - M_x)/\hat{\omega}_x < -1$. Let i_2 be the number of X_i values such that $(X_i - M_x)/\hat{\omega}_x > 1$. Compute

$$S_x = \sum_{i=i_1+1}^{n-i_2} X_{(i)}$$

$$\hat{\phi}_x = \frac{\hat{\omega}_x(i_2 - i_1) + S_x}{n - i_1 - i_2}.$$

Set $U_i = (X_i - \hat{\phi}_x)/\hat{\omega}_x$. Repeat these computations for the Y_i values yielding $V_i = (Y_i - \hat{\phi}_y)/\hat{\omega}_y$. Let

$$\Psi(x) = \max[-1, \min(1, x)].$$

Set $A_i = \Psi(U_i)$ and $B_i = \Psi(V_i)$. The percentage bend correlation between X and Y is estimated to be

$$r_{pb} = \frac{\sum A_i B_i}{\sqrt{\sum A_i^2 \sum B_i^2}}.$$

replaced by $\Psi(x) = \max[-K, \min(K, x)]$ for some $K > 1$, ω_x is no longer a measure of dispersion (Shoemaker & Hettmansperger, 1982).

9.3.2 A Test of Independence Based on ρ_{pb}

When X and Y are independent, $\rho_{pb} = 0$. To test the hypothesis H_0: $\rho_{pb} = 0$, assuming independence, compute

$$T_{pb} = r_{pb}\sqrt{\frac{n-2}{1-r_{pb}^2}}, \tag{9.3}$$

Table 9.2: Definition of the Population Percentage Bend Correlation.

Let

$$\Psi(x) = \max[-1, \min(1, x)],$$

which is a special case of Huber's Ψ. Let θ_x and θ_y be the population medians corresponding to the random variables X and Y, let ω_x be defined by the equation

$$P(|X - \theta_x| < \omega_x) = 1 - \beta.$$

Shoemaker and Hettmansperger (1982) use $\beta = 0.1$, but the resulting breakdown point might be too low in some cases. The percentage bend measure of location, corresponding to X, is the quantity ϕ_{pbx} such that

$$E[\Psi(U)] = 0,$$

where

$$U = \frac{X - \phi_{\text{pbx}}}{\omega_x}.$$

In terms of Chapter 2, ϕ_{pbx} is an M-measure of location for the particular form of Huber's Ψ being used here. Let

$$V = \frac{Y - \phi_{\text{pby}}}{\omega_y}.$$

Then the percentage bend correlation between X and Y is

$$\rho_{\text{pb}} = \frac{E[\Psi(U)\Psi(V)]}{\sqrt{E[\Psi^2(U)]E[\Psi^2(V)]}}.$$

Under independence, $\rho_{\text{pb}} = 0$, and $-1 \le \rho_{\text{pb}} \le 1$.

and reject H_0 if $|T_{\text{pb}}| > t_{1-\alpha}$, the $1 - \alpha$ quantile of Student's t-distribution with $\nu = n - 2$ degrees of freedom. All indications are that this test provides reasonably good control over the probability of a type I error for a broader range of situations than the test based on r (Wilcox, 1994d).

The breakdown point of the percentage bend correlation is at most β. However, if $\beta = 0.5$ is used, the power of the test for independence, based on T_{pb}, can be substantially less than the test based on r when sampling from a bivariate normal distribution (as will be illustrated by results in Table 9.5.) Here, the default value for β is 0.2. In exploratory studies, several values might be considered.

Like the conventional T test of H_0: $\rho = 0$, the method just described for testing H_0: $\rho_{\text{pb}} = 0$ is sensitive to heteroscedasticity. That is, even when H_0: $\rho_{\text{pb}} = 0$ is true, if there is heteroscedasticity, the wrong standard error is being used and the probability of rejecting can increase with the sample size. For a test of H_0: $\rho_{\text{pb}} = 0$ that is designed to be insensitive to heteroscedasticity, see Section 9.3.13.

9.3.3 R Function pbcor

The R function

$$\text{pbcor(x,y,beta}=0.2),$$

written for this book, estimates the percentage bend correlation for the data stored in any two vectors. If unspecified, the argument beta, the value for β when computing the measure of scale $W_{(m)}$, defaults to 0.2. The function returns the value of r_{pb} in pbcor$cor, the value of the test statistic, T_{pb}, in pbcor$test, and the p-value in pbcor$siglevel. It is noted that the function pbcor automatically removes any pair of observations for which one or both values are missing. (This is done via the function elimna mentioned in Section 1.9.)

■ **Example**

The example in Section 8.6.2 of Chapter 8 reports the results of drinking alcohol for two groups of subjects measured at three different times. Consider the measures at times 1 and 2 for the control group. Then $r = 0.37$, and $H_0: \rho = 0$ is not rejected with $\alpha = 0.05$. However, with $\beta = 0.1$, $r_{\text{pb}} = 0.5$, and $H_0: \rho_{\text{pb}} = 0$ is rejected, the p-value being .024.   ■

■ **Example**

Consider the star data in Figure 9.2. As previously noted, $r = -0.21$. In contrast, $r_{\text{pb}} = 0.06$ with $\beta = 0.1$. Increasing β to 0.2, $r_{\text{pb}} = 0.26$, and the p-value is .07. For $\beta = 0.3$, $r_{\text{pb}} = 0.3$ with a p-value of .04, and for $\beta = 0.5$, $r_{\text{pb}} = 0.328$.   ■

9.3.4 A Test of Zero Correlation among p Random Variables

Consider a random sample of n vectors from some p-variate distribution, $X_{i1}, \ldots, X_{ip}$, $i = 1, \ldots, n$. Let $\rho_{\text{pb}jk}$ be the percentage bend correlation between the jth and kth random variables, $1 \le j < k \le p$. This section considers the problem of testing

$$H_0 : \rho_{\text{pb}jk} = 0, \quad \text{for all } j < k.$$

Put another way, the hypothesis is that the matrix of percentage bend correlations among all p random variables is equal to the identity matrix.

Currently, the best method for testing this hypothesis, in terms of controlling the probability of a type I error under independence, begins by computing

$$c_{jk} = \sqrt{(n - 2.5) \times \ln\left(1 + \frac{T_{\mathrm{pb}jk}^2}{n - 2}\right)},$$

where $T_{\mathrm{pb}jk}$ is the statistic given by Eq. (9.3) for testing independence between the jth and kth random variables, and ln indicates the natural logarithm. Let

$$z_{jk} = c_{jk} + \frac{c_{jk}^3 + 3c_{jk}}{b} - \frac{4c_{jk}^7 + 33c_{jk}^5 + 240c_{jk}^3 + 855c_{jk}}{10b^2 + 8bc_{jk}^4 + 1000b}.$$

When H_0 is true,

$$H = \sum_{j < k} z_{jk}^2$$

has, approximately, a chi-squared distribution with $p(p - 1)/2$ degrees of freedom. Consequently, reject H_0 if $H > \chi_{1-\alpha}^2$, the $1 - \alpha$ quantile.

When the percentage bend correlation is replaced by r in the hypothesis testing procedure just described, the resulting test statistic will be labeled H_r. Gupta and Rathie (1983) suggest yet another test of the hypothesis that all pairs of random variables have zero correlations, again using r. Table 9.3 reports the estimated probability of a type I error for $p = 4$ and 10, and various g-and-h distributions, when using H, H_r, or the Gupta–Rathie (GR) method, and when $n = 10$ and 20. For $p = 10$ and $n \leq 20$, the GR method cannot always be computed, and the corresponding entry in Table 9.3 is left blank. As is evident, the test based on the

Table 9.3: Estimated Type I Error Probabilities, $\alpha = 0.05$.

			$p = 4$			$p = 10$		
g	h	n	H_r	GR	H	H_r	GR	H
0.0	0.0	10	0.050	0.070	0.053	0.054	—	0.056
		20	0.049	0.022	0.053	0.048	—	0.046
0.5	0.0	10	0.055	0.076	0.050	0.062	—	0.055
		20	0.058	0.025	0.053	0.059	—	0.051
1.0	0.0	10	0.092	0.111	0.054	0.126	—	0.062
		20	0.091	0.055	0.054	0.126	—	0.055
0.0	0.5	10	0.082	0.106	0.050	0.124	—	0.052
		20	0.099	0.062	0.050	0.152	—	0.054
0.5	0.5	10	0.097	0.118	0.051	0.157	—	0.053
		20	0.097	0.118	0.051	0.185	—	0.053
1.0	0.5	10	0.130	0.158	0.053	0.244	—	0.059
		20	0.135	0.105	0.053	0.269	—	0.055

percentage bend correlation is easily the most satisfactory, with the estimated probability of a type I error (based on simulations with 10,000 replications) ranging between .046 and .062. If the usual correlation, r, is used instead, the probability of type I error can exceed .2, and when using method GR, it exceeds .15.

9.3.5 R Function pball

The R function

$$pball(m, beta=0.2),$$

computes the percentage bend correlation for all pairs of random variables, and it tests the hypothesis that all of the correlations are equal to zero. Here, m is an n-by-p matrix of data. If the data are not stored in a matrix, the function prints an error message and terminates. (Use the R command matrix to store the data in the proper way. See Becker, Chambers, & Wilks, 1988, for details.) Again beta, which is β in Table 9.2, defaults to 0.2. The function returns a p-by-p matrix of correlations in pball\$pbcorm, another matrix indicating the p-values for the hypotheses that each correlation is zero, plus the test statistic H and its corresponding p-value.

■ Example

Again consider the alcohol data in Section 8.6.2 where measures of the effect of drinking alcohol are taken at three different times. If the data for the control group are stored in the R matrix amat, the command pball(amat) returns

```
$pbcorm:
          [,1]        [,2]        [,3]
[1,] 1.0000000  0.5028002  0.7152667
[2,] 0.5028002  1.0000000  0.5946712
[3,] 0.7152667  0.5946712  1.0000000


$siglevel:
             [,1]          [,2]          [,3]
[1,]           NA  0.023847557  0.0003925285
[2,] 0.0238475571           NA  0.0056840954
[3,] 0.0003925285  0.005684095            NA


$H:
[1] 5.478301e+192


$H.siglevel:
[1] 0
```

For example, the correlation between variables 1 and 2 is 0.5, between 1 and 3 it is 0.72, and between 2 and 3 it is 0.59. The corresponding p-values are .024, .0004, and .0057.

The test statistic, H, for testing the hypothesis that all three correlations are equal to zero, has a p-value approximately equal to 0. To use $\beta = 0.1$ instead, type the command pball(amat,0.1).

∎

9.3.6 The Winsorized Correlation

Another (type M) robust analog of ρ is the Winsorized correlation. The population Winsorized correlation between two random variables, X_1 and X_2, is given by

$$\rho_w = \frac{E_w[(X_1 - \mu_{w1})(X_2 - \mu_{w2})]}{\sigma_{w1}\sigma_{w2}},$$

where σ_{wj} is the population Winsorized standard deviation of X_j, and $E_w(X)$ is the Winsorized expected value of X, as defined in Section 2.5 of Chapter 2. The numerator of ρ_w is the population Winsorized covariance which is written simply as σ_{w12} when convenient. When X_1 and X_2 are independent, $\rho_w = 0$, and $-1 \le \rho_w \le 1$.

To estimate ρ_w, based on the random sample $(X_{11}, X_{12}), \ldots, (X_{n1}, X_{n2})$, first Winsorize the observations by computing the Y_{ij} values as described in Section 8.1.1. Then ρ_w is estimated by computing the Pearson's correlation with the Y_{ij} values. That is, estimate ρ_w with

$$r_w = \frac{\sum (Y_{i1} - \bar{Y}_1)(Y_{i2} - \bar{Y}_2)}{\sqrt{\sum (Y_{i1} - \bar{Y}_1)^2 \sum (Y_{i2} - \bar{Y}_2)^2}}.$$

Here, 20% Winsorization is assumed unless stated otherwise.

To test $H_0\colon \rho_w = 0$, compute

$$T_w = r_w \sqrt{\frac{n-2}{1-r_w^2}},$$

and reject if $|T_w| > t_{1-\alpha/2}$, the $1 - \alpha/2$ quantile of Student's t-distribution with $\nu = h - 2$ degrees of freedom, where h, the effective sample size, is the number of pairs of observations not Winsorized. (Equivalently, $h = n - 2g$, $g = [\gamma n]$, is the number of observations left after trimming.) Unless stated otherwise, $\gamma = 0.2$ is assumed. In terms of type I error probabilities when testing the hypothesis of zero correlation, the Winsorized correlation appears to compete well with the test based on r under independence, but the percentage bend correlation is better still, at least when $\beta = 0.1$. Like all of the hypothesis testing methods in this section, T_w is sensitive to heteroscedasticity. As for power, the best method depends in part on the values of ρ, ρ_w, and ρ_{pb}, which are unknown. Perhaps there are situations where using ρ_w will result in more power. This depends in part on how much ρ_w differs from ρ_{pb}.

Table 9.4: Estimated Type I Error Probabilities, $\alpha = 0.05$.

		$T_{\text{pb.1}}$		$T_{\text{pb.5}}$		T		T_w	
g	*h*	$n = 10$	$n = 20$	$n = 10$	$n = 20$	$n = 10$	$n = 20$	$n = 10$	$n = 20$
0.0	0.0	.050	.050	.053	.049	.049	.049	.040	.045
0.0	0.2	.050	.049	.054	.049	.054	.052	.043	.045
0.0	0.5	.047	.048	.053	.049	.062	.067	.038	.044
0.5	0.0	.050	.049	.053	.050	.039	.050	.037	.043
0.5	0.2	.048	.048	.053	.050	.055	.053	.044	.044
0.5	0.5	.047	.047	.053	.049	.064	.065	.037	.043
1.0	0.0	.045	.048	.053	.050	.054	.055	.037	.043
1.0	0.2	.046	.047	.053	.050	.062	.053	.041	.058
1.0	0.5	.045	.046	.053	.050	.070	.065	.035	.044
1.0	1.0	.044	.045	.052	.050	.081	.071	.034	.043

Table 9.4 compares type I error probabilities when testing for independence using T, T_w, and T_{pb}. The notation $T_{\text{pb.1}}$ means that $\beta = 0.1$ is used, and $T_{\text{pb.5}}$ means $\beta = 0.5$. (The first two columns in Table 9.4 indicate the *g*-and-h distribution associated with the marginal distributions.) As can be seen, the test based on *r* is the least stable in terms of type I errors.

9.3.7 R Functions wincor and winall

The R function

$$\text{wincor(x,y,tr=0.2)}$$

estimates the Winsorized correlation between two random variables. As usual, x and y can be any R variables containing data. The default amount of Winsorization, tr, is 0.2. The function returns the Winsorized correlation, r_w, the Winsorized covariance, plus the test statistic T_w, and the corresponding *p*-value.

The function

$$\text{winall(m,tr=0.2)}$$

estimates the correlation for all pairs of *p* random variables, assuming the data are stored in an *n*-by-*p* matrix. If m is not a matrix, the function prints an error message and terminates. The function returns the Winsorized correlations in winall$wcor, the covariances in winall$wcov, and the *p*-values associated with each correlation is returned in winall$siglevel.

Section 9.3.4 described a method for testing the hypothesis that all percentage bend correlations, among all pairs of random variables, are equal to zero. The method is easily extended to test the hypothesis that all Winsorized correlations are equal to zero, but there are no simulation results on how well this approach performs in terms of type I errors, so it is not recommended at this time.

■ Example

For the alcohol data in Section 8.6.2 used to illustrate the R function pball, winall returns

```
$wcor:
            [,1]        [,2]        [,3]
[1,] 1.0000000  0.5134198  0.6957740
[2,] 0.5134198  1.0000000  0.6267765
[3,] 0.6957740  0.6267765  1.0000000


$wcov:
            [,1]        [,2]        [,3]
[1,] 44.77895  24.12632  27.68421
[2,] 24.12632  49.31316  26.17105
[3,] 27.68421  26.17105  35.35526


$siglevel:
            [,1]          [,2]          [,3]
[1,]          NA  0.023645294  0.001061593
[2,] 0.023645294          NA  0.004205145
[3,] 0.001061593  0.004205145          NA
```

Thus, the estimated correlation between variables 1 and 2 is 0.51, the covariance is 24.1, and the *p*-value, when testing the hypothesis that the Winsorized correlation is zero, is 0.024. In this particular case, the results are very similar to those obtained with the percentage bend correlation.

■

9.3.8 The Biweight Midcovariance

It should be noted that the percentage bend covariance and correlation are a special case of a larger family of measures of association. Let Ψ be any odd function, such as those summarized in Section 2.2.4. Let μ_x be any measure of location for the random variable X, let τ_x be some measure of scale, let K be some constant, and let $U = (X - \mu_x)/(K\tau_x)$ and $V = (Y - \mu_y)/(K\tau_y)$. Then a measure of covariance between X and Y is

$$\gamma_{xy} = \frac{nK^2\tau_x\tau_y E[\Psi(U)\Psi(V)]}{E[\Psi'(U)\Psi'(V)]}$$

and a measure of correlation is $\gamma_{xy}/\sqrt{\gamma_{xx}\gamma_{yy}}$. If μ_x and μ_y are measures of location such that $E[\Psi(U)] = E[\Psi(V)] = 0$, then $\gamma_{xy} = 0$ when X and Y are independent.

Among the many choices for Ψ and K, the so-called biweight midcovariance has played a role in a regression method covered in Chapter 10, so for completeness an estimate of this parameter is described here. It is based on $K = 9$ and the biweight function described in Table 2.1 of Chapter 2. Let $(X_1, Y_1), \ldots, (X_n, Y_n)$ be a random sample from some bivariate distribution. Let

$$U_i = \frac{X_i - M_x}{9 \times \text{MAD}_x},$$

where M_x and MAD_x are the median and the value of MAD for the X values. Similarly, let

$$V_i = \frac{Y_i - M_y}{9 \times \text{MAD}_y}.$$

Set $a_i = 1$ if $-1 \le U_i \le 1$, otherwise $a_i = 0$. Similarly, set $b_i = 1$ if $-1 \le V_i \le 1$, otherwise $b_i = 0$. The sample biweight midcovariance between X and Y is

$$s_{bxy} = \frac{n \sum a_i (X_i - M_x)(1 - U_i^2)^2 b_i (Y_i - M_y)(1 - V_i^2)^2}{[\sum a_i (1 - U_i^2)(1 - 5U_i^2)][\sum b_i (1 - V_i^2)(1 - 5V_i^2)]}.$$

The statistic s_{bxx} is the biweight midcovariance mentioned in Chapter 3, and

$$r_b = \frac{s_{bxy}}{\sqrt{s_{bxx} s_{byy}}}$$

is an estimate of what is called the *biweight midcorrelation* between X and Y. The main reasons for considering this measure of covariance are that it is relatively easy to compute, and it appears to have a breakdown point of 0.5, but a formal proof has not been found.

9.3.9 R Functions bicov and bicovm

The R function

$$\text{bicov(x,y).}$$

computes the biweight midcovariance between two random variables. The function bicovm computes the biweight midcovariance and midcorrelation for all pairs of p random variables stored in some R variable, m, which can be either an n-by-p matrix or a variable having list mode. It has the form

$$\text{bicovm(m).}$$

■ Example

For the star data in Figure 9.2, bicovm reports that the biweight midcorrelation is 0.6. In contrast, the highest percentage bend correlation, among the choices 0.1, 0.2, 0.3,

0.4, and 0.5 for β, is 0.33, and a similar result is obtained when Winsorizing instead. As previously noted, these data have several outliers. The main point here is that r_{pb} and r_w offer more resistance than r, but they can differ substantially from other resistant estimators.

■

9.3.10 Kendall's tau

A well-known type M correlation is Kendall's tau. For completeness, it is briefly described here.

Consider two pairs of observations, (X_1, Y_1) and (X_2, Y_2). For convenience, assume tied values never occur and that $X_1 < X_2$. Then these two pairs of observations are said to be concordant if $Y_1 < Y_2$; otherwise they are discordant. For n pairs of points, let $K_{ij} = 1$ if the ith and jth points are concordant, and if they are discordant, $K_{ij} = -1$. Then Kendall's tau is given by

$$\hat{\tau} = \frac{2}{n(n-1)} \sum_{i<j} K_{ij} \tag{9.4}$$

Under independence, the population value of $\hat{\tau}$, τ, is zero. The usual test of $H_0\colon \tau = 0$ is to reject if

$$|Z| \geq z_{1-\frac{\alpha}{2}},$$

where

$$Z = \frac{\hat{\tau}}{\sigma_\tau},$$

and

$$\sigma_\tau^2 = \frac{2(2n+5)}{9n(n-1)}.$$

It is left as an exercise to show that heteroscedasticity affects the probability of rejecting when H_0 is true.

If X and Y have the bivariate distribution H, the influence function of Kendall's tau is

$$IF(x, y) = 2(2P_H[(X-x)(Y-y) > 0] - 1 - \tau)$$

(Croux & Dehon, 2010). So a positive feature of Kendall's tau is that it has infinitesimal robustness. (It's influence function is bounded.) However, although Kendall's tau provides protection against outliers among the X values ignoring Y, or among the Y values ignoring X, it can be seen that outliers can substantially alter its value. (Details are relegated to the exercises.)

9.3.11 Spearman's rho

Assign ranks to the X values, ignoring Y, and assign ranks to the Y values ignoring X. Then Spearman's rho, r_s, is just Pearson's correlation based on the resulting ranks. Like all of the correlations in this section it provides protection against outliers among the X values, ignoring Y, as well as outliers among the Y values, ignoring X, but outliers properly placed can alter its value substantially. Letting ρ_s be the population value of Spearman's rho, the influence function of ρ_s is

$$IF(x, y) = -3\rho_s - 9 + 12\{F(x)G(y) + E[F(X)I(Y \geq y)] + E[G(Y)I(X \geq x)]\},$$

where F and G are the marginal distributions of X and Y, respectively, and I is the indicator function (Croux & Dehon, 2010). In terms of asymptotic efficiency and other robustness considerations, results in Croux and Dehon (2010) indicate that Kendall's tau is preferable to Spearman's rho.

When X and Y are independent, $\rho_s = 0$. The usual test of H_0: $\rho_s = 0$ is to reject if $|T| \geq t$, where t is the $1 - \alpha/2$ quantile of Student's t-distribution with $\nu = n - 2$ degrees of freedom, and

$$T = \frac{r_s \sqrt{n - 2}}{\sqrt{1 - r_s^2}}.$$

Like all of the hypothesis testing methods in this section, heteroscedasticity affects the probability of rejecting, even when H_0 is true.

Table 9.5 shows estimated power when testing the hypothesis of a zero correlation and $\rho = 0.5$. Included is the power of the test based on Kendall's tau, under the column headed Kend., and Spearman's rho under the column Spear. The main point is that different methods can have more or less power than other methods, one reason being that the parameters being estimated can differ, so it is difficult to select a single method for general use based on the criterion of high power.

9.3.12 R Functions tau, spear, cor, and taureg

The function

$$\text{tau(x,y,alpha=0.05)}$$

computes Kendall's tau, and

$$\text{spear(x,y)}$$

Table 9.5: Estimated Power, $n = 20$, $\rho = .5$.

g	h	T_w	T	$T_{pb.1}$	$T_{pb.5}$	Kend.	Spear.
0.0	0.0	0.562	0.637	.620	.473	0.551	0.568
0.0	0.2	0.589	0.633	.638	.512	0.594	0.597
0.0	0.5	0.614	0.603	.658	.552	0.644	0.626
0.0	1.0	0.617	0.573	.658	.588	0.692	0.659
0.5	0.0	0.588	0.620	.629	.499	0.602	0.602
0.5	0.2	0.602	0.608	.643	.525	0.624	0.615
0.5	0.5	0.614	0.591	.653	.558	0.656	0.638
0.5	1.0	0.611	0.565	.608	.591	0.698	0.664
1.0	0.0	0.621	0.597	.650	.537	0.668	0.641
1.0	0.2	0.620	0.585	.652	.550	0.667	0.644
1.0	0.5	0.619	0.571	.652	.569	0.683	0.652
1.0	1.0	0.611	0.559	.653	.595	0.709	0.669

computes Spearman's rho. The built-in R function

cor(x, y = NULL, use = "everything", method = c("pearson", "kendall", "spearman"))

can be used to compute Kendall's by setting the argument method="kendall" and Spearman's rho by setting method="spearman". The R functions tau and spear automatically test the hypothesis of a zero correlation. The R function cor does not. For convenience, the function

taureg(m,y,corfun=tau)

computes the p correlations between every variable in the matrix m, having p columns, and the variable y. The argument corfun can be any function that computes a correlation between two variables only and returns the value in corfun$cor along with the p-value in corfun$siglevel. By default, Kendall's tau is used.

9.3.13 Heteroscedastic Tests of Zero Correlation

The tests of the independence based on type M correlations, including Pearson's correlation, are sensitive to heteroscedasticity. That is, even when these correlations are equal to zero, if there is heteroscedasticity, the probability of rejecting can increase as the sample size gets large. So when rejecting, it is reasonable to conclude that the variables under study are dependent, but the reason for rejecting might be due more to heteroscedasticity than to the correlation differing from zero. To test the hypothesis that a (type M) correlation is equal to zero in a manner that is insensitive to heteroscedasticity, a percentile bootstrap method can be used.

When using robust correlations, all indications are that a basic percentile bootstrap performs well in terms of controlling the probability of a type I error. For a random sample $(X_1, Y_1), \ldots (X_n, Y_n)$, generate a bootstrap sample by resampling with replacement n pairs of points yielding $(X_1^*, Y_1^*), \ldots (X_n^*, Y_n^*)$. Compute any of the robust estimators described in this section and label the result r^*. Repeat this B times yielding $r_1^*, \ldots, r_B^*$. Let $\ell = \alpha B / 2$, rounded to the nearest integer, and let $u = B - \ell$. Then reject the hypothesis of a zero correlation if $r_{(\ell+1)}^* > 0$ or $r_{(u)}^* < 0$, where $r_{(1)}^* \leq \cdots \leq r_{(B)}^*$ are the values $r_1^*, \ldots, r_B^*$ written in ascending order.

For the special case where r is Pearson's correlation, a modified percentile bootstrap method is required. When $B = 599$, an approximate a 0.95 confidence interval for ρ is

$$(r_{(a)}^*, r_{(c)}^*)$$

where again for $n < 40$, $a = 7$, and $c = 593$; for $40 \leq n < 80$, $a = 8$, and $c = 592$; for $80 \leq n < 180$, $a = 11$, and $c = 588$; for $180 \leq n < 250$, $a = 14$, and $c = 585$; while for $n \geq 250$, $a = 15$, and $c = 584$. As usual, if this interval does not contain zero, reject $H_0 : \rho = 0$.

Section 10.1.1 describes a heteroscedastic method for computing confidence intervals for the usual least squares regression slopes based on what is called the HC4 estimator of the standard errors. Here it is noted that the method is readily adapted to the problem of computing a confidence interval for Pearson's correlation, ρ, which can be applied with the R function pcorhc4 described in the next section. Limited studies suggest that it performs about as well as the modified percentile bootstrap method described here. Possible reasons for preferring the HC4 method is that it can be used when testing at any α value and a p-value is readily determined.

9.3.14 R Functions corb, pcorb, and pcorhc4

The R function

$$\text{corb(x,y,corfun=pbcor,nboot=599,...)}$$

tests the hypothesis of a zero correlation using the heteroscedastic bootstrap method just described. By default, it uses the percentage bend correlation, but any correlation can be specified by the argument corfun. For example, the command corb(x,y,corfun=wincor,tr=0.25) will use a 25% Winsorized correlation.

When working with Pearson's correlation, use the function

$$\text{pcorb(x,y),}$$

which applies the modified percentile bootstrap method described in the previous section. The R function

$$\text{pcorhc4(x,y,alpha=0.05)}$$

applies the HC4 method.

9.4 Some Type O Correlations

Type M correlations have the property that two properly placed outliers can substantially alter their value. (Illustrations are relegated to the exercises at the end of this chapter.) Type O correlations are an attempt to correct this problem. Section 6.2 described various measures that reflect how deeply a point is nested within a cloud of data, where the measures of depth take into account the overall structure of the data. Roughly, *type O correlations* are correlations that possibly downweight or eliminate one or more points that have low measures of depth. In essence, they are simple extensions of W-estimators described in Section 6.3.6. Included among this class of correlations coefficients are so-called *skipped correlations*, which remove any points flagged as outliers and then compute some correlation coefficient with the data that remain.

9.4.1 MVE and MCD Correlations

An example of a type O correlation has, in essence, already been described in connection with the MCD and MVE estimators of scatter described in Section 6.3. These measures search for the central half of the data and then use this half of the data to estimate location and scatter. As is evident, the covariance associated with this central half of the data readily yields a correlation coefficient. For example, simply compute Pearson's correlation based on the central half of the data.

9.4.2 Skipped Measures of Correlation

Skipped correlations are obtained by checking for any outliers using one of the methods described in Section 6.4, removing them, and applying some correlation coefficient to the remaining data. An example is to remove outliers using the MVE or MCD methods and compute Pearson's correlation after outliers are removed. It is noted that when using the R functions cov.mve and cov.mcd, already described, setting the optional argument cor to T (for true) causes this correlation to be reported. For example, when using cov.mve, the command

$$\text{cov.mve(m,cor=T)}$$

accomplishes this goal.

9.4.3 The OP Correlation

The so-called OP correlation coefficient begins by eliminating any outliers using the projection method in Section 6.4.9. Then it merely computes some correlation coefficient with the data that remain. Pearson's correlation is assumed unless stated otherwise.

Imagine that data are randomly sampled from some bivariate normal distribution. If the goal is to use a skipped correlation coefficient that gives a reasonably accurate estimate of Pearson's correlation, ρ, relative to r, then the OP estimator is the only skipped estimator known to be reasonably satisfactory.

Let r_p represent the skipped correlation coefficient and let m be the number of pairs of points left after outliers are removed. A seemingly simple method for testing the hypothesis of independence is to apply the usual T test for Pearson's correlation but with r replaced by r_p and n replaced by m. But this simple solution fails because it does not take into account the dependence among the points remaining after outliers are removed. If this problem is ignored, unsatisfactory control over the probability of a type I error results (Wilcox, 2010f). However, let

$$T_p = r_p \sqrt{\frac{n-2}{1-r_p^2}}$$

and suppose the hypothesis of independence is rejected at the $\alpha = 0.05$ level if $|T_p| \geq c$, where

$$c = \frac{6.947}{n} + 2.3197.$$

The critical value c was determined via simulations under normality by determining an appropriate critical value for n ranging between 10 and 200, and then a least squares regression line was fit to the data. For nonnormal distributions, all indications are that this hypothesis testing method has an actual type I error probability reasonably close to the nominal 0.05 level.

9.4.4 Inferences Based on Multiple Skipped Correlations

The hypothesis testing method just described has been extended to the problem of testing the hypothesis that $p \geq 2$ random variables are independent. However, rather than use Pearson's correlation after outliers are removed, Spearman's rho is used. When using Pearson's correlation, no method has been found that adequately controls the probability of a type I error. But switching to Spearman's rho corrects this problem among extant simulations (Wilcox, 2003e).

Let $\hat{\tau}_{cjk}$ be Spearman's correlation between variables j and k after points flagged as outliers by the projection method are removed. For convenience, it is assumed that

$$H_0 : \tau_{cjk} = 0 \tag{9.5}$$

is to be tested for all $j < k$. Let

$$T_{jk} = \hat{\tau}_{cjk} \sqrt{\frac{n-2}{1-\hat{\tau}_{cjk}^2}},$$

and let

$$T_{\max} = \max|T_{jk}|, \tag{9.6}$$

where the maximum is taken overall $j < k$. The strategy used here to control FWE is to approximate, via simulations, the distribution of $T_{\max}$ under normality when all correlations are zero and $p < 4$, determine the 0.95 quantile, say q, for $n = 10, 20, 30, 40, 60, 100$, and 200, And then reject $H_0 : \tau_{cjk} = 0$ if $|T_{jk}| \geq q$. For $p \geq 4$, normal distributions were replaced by a g-and-h distribution with $(g, h) = (0, 0.5)$. The reason is that for $p \geq 4$, the probability of at least one type I error was found to be largest for this special case among the situations considered in Wilcox (2003e). That is, the strategy for choosing an appropriate critical value was to determine q for a distribution that appears to maximize the probability of a type I error with the goal that FWE should not exceed .05 for any distribution that might be encountered in practice. Based on this strategy, the following approximations of q were determined:

$$p = 2, \ \hat{q} = 5.333n^{-1} + 2.374,$$
$$p = 3, \ \hat{q} = 8.800n^{-1} + 2.780,$$
$$p = 4, \ \hat{q} = 25.67n^{-1.2} + 3.030,$$
$$p = 5, \ \hat{q} = 32.83^{-1.2} + 3.208,$$
$$p = 6, \ \hat{q} = 51.53n^{-1.3} + 3.372,$$
$$p = 7, \ \hat{q} = 75.02n^{-1.4} + 3.502,$$
$$p = 8, \ \hat{q} = 111.34n^{-1.5} + 3.722,$$
$$p = 9, \ \hat{q} = 123.16n^{-1.5} + 3.825,$$
$$p = 10, \ \hat{q} = 126.72n^{-1.5} + 3.943.$$

Rather than test the hypothesis of a zero correlation among all pairs of random variables, the goal might be to test

$$H_0 : \tau_{c1k} = 0, \tag{9.7}$$

for each k, $k = 2, \ldots, p$. Now the approximate critical values are:

$$p = 2, \hat{q} \doteq 5.333n^{-1.0} + 2.374,$$

$$p = 3, \hat{q} = 8.811n^{-1.0} + 2.540,$$

$$p = 4, \hat{q} = 14.89n^{-1.2} + 2.666,$$

$$p = 5, \hat{q} = 20.59n^{-1.2} + 2.920,$$

$$p = 6, \hat{q} = 51.01n^{-1.5} + 2.999,$$

$$p = 7, \hat{q} = 52.15n^{-1.5} + 3.097,$$

$$p = 8, \hat{q} = 59.13n^{-1.5} + 3.258,$$

$$p = 9, \hat{q} = 64.93n^{-1.5} + 3.286,$$

$$p = 10, \hat{q} = 58.50n^{-1.5} + 3.414.$$

Again, these approximate critical values are designed so that FWE is approximately 0.05.

9.4.5 R Functions scor and mscor

The R function

$$\text{scor(x,y=NA,corfun=pcor,gval=NA,plotit=T,cop=1,op=T)}$$

computes the skipped correlation for the data in the variables x and y. If y is not specified, it is assumed that x is an n-by-p matrix in which case the skipped correlation is computed for each pair of variables. The argument corfun controls which correlation is computed after outliers are removed; by default, Pearson's correlation is used. The arguments gval, cop, and op are relevant to the projection outlier detection method; see Section 6.4.10.

The function

$$\text{mscor(m,corfun=spear,cop=1,gval=NA,ap=T,pw=T)}$$

also computes a skipped correlation coefficient, but it defaults to using Spearman's correlation, and when testing the hypotheses corresponding to Eq. (9.5) or Eq. (9.7), it controls FWE (the familywise error rate) using the approximations of an appropriate critical value outlined in the previous subsection, assuming FWE is to be 0.05. If ap=T, the hypothesis of a zero correlation for each pair of variables is tested. If ap=F, the hypotheses corresponding to Eq. (9.7) are tested instead.

9.5 A Test of Independence Sensitive to Curvature

Let $(\mathbf{X_1}, Y_1), \ldots, (\mathbf{X}_n, Y_n)$ be a random sample of n pairs of points where $\mathbf{X}$ is a vector having length p. The goal in this section is to test the hypothesis that $\mathbf{X}$ and Y are independent in a manner that is sensitive to curvature as well as any linear association that might exist.

Method INDT

The first method described here stems from general theoretical results derived by Stute, Gonzalez Manteiga, and Presedo Quindimil (1998). It does not assume or require homoscedasticity and is based in part on what is called a wild bootstrap method. Stute et al. establish that other types of bootstrap methods are not suitable when using the test statistic to be described. Essentially, the method in this section is designed to test the hypothesis that the regression surface for predicting Y, given $\mathbf{X}$, is a horizontal plane. Let $E(Y|\mathbf{X})$ represent the conditional mean of Y given $\mathbf{X}$. The goal is to test

$$H_0 : E(Y|\mathbf{X}) = \mu_y.$$

That is, the conditional mean of Y, given $\mathbf{X}$, does not depend on $\mathbf{X}$.

The test statistic is computed as follows. Let $\bar{Y}$ be the mean based on $Y_1, \ldots, Y_n$. (Using a trimmed mean or some other robust estimator can result in poor control over the probability of a Type I error when Y has a sufficiently skewed distribution.) Fix j and set $I_i = 1$ if $\mathbf{X}_i \leq \mathbf{X}_j$, otherwise $I_i = 0$. The notation $\mathbf{X}_i \leq \mathbf{X}_j$ means that for every k, $k = 1, \ldots, p$, $X_{ik} \leq X_{jk}$. Let

$$R_j = \frac{1}{\sqrt{n}} \sum I_i(Y_i - \bar{Y})$$
$$= \frac{1}{\sqrt{n}} \sum I_i r_i, \tag{9.8}$$

where

$$r_i = Y_i - \bar{Y}.$$

The test statistic is the maximum absolute value of all the R_j values. That is, the test statistic is

$$D = \max|R_j|. \tag{9.9}$$

An appropriate critical value is estimated with the *wild bootstrap method* as follows. Generate $U_1, \ldots, U_n$ from a uniform distribution and set

$$V_i = \sqrt{12}(U_i - 0.5),$$
$$r_i^* = r_i V_i,$$

and

$$Y_i^* = \bar{Y}_t + r_i^*.$$

Then based on the n pairs of points $(\mathbf{X}_1, Y_1^*)$, ..., $(\mathbf{X}_n, Y_n^*)$, compute the test statistic as described in the previous paragraph and label it D^*. Repeat this process B times and label the resulting (bootstrap) test statistics $D_1^*, \ldots, D_B^*$. Finally, put these B values in ascending order, which we label $D_{(1)}^* \leq \cdots \leq D_{(B)}^*$. Then the critical value is $D_{(u)}^*$, where $u = (1 - \alpha)B$ rounded to the nearest integer. That is, reject if

$$D \geq D_{(u)}^*.$$

An alternative test statistic has been studied where D is replaced by

$$W = \frac{1}{n}(R_1^2 + \cdots + R_n^2). \tag{9.10}$$

The critical value is determined in a similar manner as before. First, generate a wild bootstrap sample and compute W yielding W^*. Repeating this B times and reject if

$$W \geq W_{(u)}^*,$$

where again $u = (1 - \alpha)B$ rounded to the nearest integer, and $W_{(1)}^* \leq \cdots \leq W_{(B)}^*$ are the B W^* values written in ascending order. The test statistic D is called the *Kolmogorov-Smirnov* test statistic, and W is called the *Cramér-von Mises* test statistic. The choice between these two test statistics is not clear cut. For $p = 1$, currently it seems that there is little separating them in terms of controlling type I errors. The extent to which this remains true when $p > 1$ appears to have received little or no attention.

Method MEDIND

A seemingly natural way of generalizing the method to a robust measure of location is to replace $\bar{Y}$ with say the median or 20% trimmed mean. But when the distribution of Y is sufficiently skewed, control over the probability of a type I error can be highly unsatisfactory. There is, however, an alternative method that can be used with the median, which is based on a modification of a method derived by He and Zhu (2003); see Wilcox (2008e) for details.

Let $\mathbf{x}$ be the $n \times (p + 1)$ matrix with the first column containing all ones and the remaining p columns are the columns of $\mathbf{X}$. Following He and Zhu (2003), it is assumed that the design has been normalized so that $n^{-1} \sum \mathbf{x}_j \mathbf{x}_j' - I = o(1)$. Let $r_i = Y_i - \hat{Y}_\gamma$, where $\hat{Y}_\gamma$ is some estimate of the γth quantile of Y. Currently, simulation results on how well the method controls the probability of a type I error are limited to the quartiles. For the 0.5 quantile, $\hat{Y}_{0.5}$

is taken to be the usual sample median. Here the lower and upper quartiles are estimated via the ideal fourths. Let

$$\mathbf{W}_i = n^{-1/2} \sum_{k=1}^{n} \psi(r_k) \mathbf{x}_k I(\mathbf{x}_k \leq \mathbf{x}_i),$$

where $\psi(r) = \gamma I(r > 0) + (\gamma - 1) I(r < 0)$. For fixed j, let U_{ij} be the ranks of the n values in the jth column of $\mathbf{x}$, $j = 2, \ldots, q$. Let $F_i = \max U_{ij}$, the maximum being taken over $j = 2, \ldots, q$. If $\mathbf{x}_k \leq \mathbf{x}_i$, then $F_k \leq F_i$. The test statistic is D_n, the largest eigenvalue of

$$\mathbf{Z} = \frac{1}{n} \sum \mathbf{W}_i \mathbf{W}_i'.$$

The strategy for determining an appropriate critical value is to temporarily assume normality, use simulations to approximate the $1 - \alpha$ quantile of the null distribution, say c, and then reject the null hypothesis if $T_n \geq c$ even when sampling from a nonnormal distribution.

An advantage of the methods just described is that they are sensitive to a variety of ways two or more variables might be dependent. But a limitation is that when they reject, it is unclear why. That is, these tests do not provide any information about the nature of the association.

9.5.1 R Functions indt, indtall, and medind

The R function

$$\text{indt(x,y,nboot=500,alpha=0.05,flag=1)}$$

tests the hypothesis of independence using method INDT. As usual, x and y are R variables containing data, tr indicates the amount of trimming used when computing the Cramér–von Mises test statistic, and nboot is B. Here, x can be a single variable or a matrix having n rows and p columns. The argument flag indicates which test statistic will be used:

- flag=1 means the Kolmogorov–Smirnov test statistic, D, is used.
- flag=2 means the Cramér–von Mises test statistic, W, is used.
- flag=3 means both test statistics are computed.

■ Example

Sockett, Daneman, Carlson, and Ehrich (1987) report data from a study dealing with diabetes in children. One of the variables was the age of a child at diagnosis and another was a measure called base deficit. Using the conventional (Student's t) test of $H_0 : \rho = 0$ based on Pearson's correlation, we fail to reject at the 0.05 level. (The p-value is .135.) We again fail to reject with a skipped correlation, a 20% Winsorized correlation, and a percentage bend correlation. Using the bootstrap methods for Pearson's correlation or the percentage bend correlation, again no association is detected. But using the function indt, we reject at the 0.05 level. A possible explanation is that (based on methods covered in Chapter 11), the regression line between these two variables appears to have some curvature which might mask a true association when attention is restricted to one of the correlation coefficients covered in this chapter.

The function

$$\text{indtall}(x,y=\text{NULL},\text{nboot}=500,\text{alpha}=0.05)$$

performs all pairwise tests of independence for the variables in the matrix x if y=NA. If data are found in y, then the function performs p tests of independence between each of the p variables in x and y. The current version computes only the Kolmogorov–Smirnov test statistic. Each test is performed at the level indicated by the argument alpha.

The function

$$\text{medind}(x, y, \text{qval} = 0.5, \text{nboot} = 1000, \text{SEED} = T, \text{alpha} - 0.05, pr = T, \text{xout} = F, \\ \text{outfun} = \text{out}, ...)$$

tests the hypothesis of independence using method MEDIND. The function contains critical values for a range of situations. If a critical value is not available, one is determined via simulations, with the number of replications determined by the argument nboot.

9.6 Comparing Correlations: Independent Case

This section deals with comparing correlations associated with two independent groups.

9.6.1 Comparing Pearson Correlations

Numerous methods have been proposed for testing the hypothesis that two Pearson correlations are equal. More formally, the goal is to test

$$\rho_1 = \rho_2, \tag{9.11}$$

where r_1 and r_2, the estimates of ρ_1 and ρ_2, respectively, are independent. A comparison of various methods (Wilcox, 2009d) indicates that a modified percentile bootstrap method tends to be best in terms of controlling the probability of a type I error. Let $N = n_1 + n_2$ be the total number of pairs of observations. For the jth group, generate a bootstrap sample of n_j pairs of observations. Let r_1^* and r_2^* represent the resulting correlation coefficients and set

$$D^* = r_1^* - r_2^*.$$

Repeat this process 599 times yielding $D_1^*, \ldots, D_{599}^*$. Then a 0.95 confidence interval for the difference between the population correlation coefficients $(\rho_1 - \rho_2)$ is

$$(D_{(\ell)}^*, D_{(u)}^*),$$

where for $\ell = 7$ and $u = 593$ if $N < 40$; $\ell = 8$ and $u = 592$ if $40 \leq N < 80$; $\ell = 11$ and $u = 588$ if $80 \leq N < 180$; $\ell = 14$ and $u = 585$ if $180 \leq N < 250$; $\ell = 15$ and $u = 584$ if $N \geq 250$.

9.6.2 Comparing Robust Correlations

When comparing robust correlations, all indications are that a basic percentile bootstrap method performs reasonably well. That is, no modification, as described in the previous section, is necessary.

9.6.3 R Functions twopcor and twocor

The R function

$$\text{twopcor(x1, y1, x2, y2, SEED} = \text{T)}$$

computes a confidence interval for $\rho_1 - \rho_2$ using the modified bootstrap method just described. The R function

$$\text{twocor(x1, y1, x2, y2, corfun} = \text{pbcor, nboot} = 599, \text{alpha} = 0.05, \text{SEED} = \text{T, ...)}$$

tests the hypothesis that two robust correlation coefficients are equal. The choice of correlation is determined by the argument corfun, which defaults to the percentage bend correlation. The function returns a $1 - \alpha$ confidence interval and a p-value.

9.7 Exercises

1. Generate 20 observations from a standard normal distribution and store them in the R variable ep. Repeat this and store the values in x. Compute y=x+ep and compute Kendall's tau. Generally, what happens if two pairs of points are added at $(2.1, -2.4)$? Does this have a large impact on tau? What would you expect to happen to the p-value when testing $H_0\colon \tau = 0$?

2. Repeat Exercise 1 with Spearman's rho, the percentage bend correlation, and the Winsorized correlation.

3. Demonstrate that heteroscedasticity affects the probability of a type I error when testing the hypothesis of a zero correlation based on any type M correlation and nonbootstrap method covered in this chapter.

4. Use the function cov.mve(m,cor=T) to compute the MVE correlation for the star data in Figure 9.2. Compare the results with the Winsorized, percentage bend, skipped, and biweight correlations, as well the M-estimate of correlation returned by the R function relfun.

5. Using the Group 1 alcohol data in Section 8.6.2, compute the MVE estimate of correlation and compare the results with the biweight midcorrelation, the percentage bend correlation using $\beta = 0.1, 0.2, 0.3, 0.4$, and 0.5, Winsorized correlation using $\gamma = 0.1$ and 0.2, and the skipped correlation.

6. Repeat the previous problem using the data for Group 2.

7. The method of detecting outliers, described in Section 6.4.3, could be modified by replacing the MVE estimator with the Winsorized mean and covariance matrix. Discuss how this would be done and its relative merits.

8. Using the data in the file read.dat, test for independence using the data in columns 2, 3, and 10 and the R function pball. Try $\beta = 0.1, 0.3$, and 0.5. Comment on any discrepancies.

9. Examine the variables in the last exercise using the R functions mscor.

10. For the data used in the last two exercises, test the hypothesis of independence using the function indt. Why might indt find an association not detected by any of the correlations covered in this chapter?

11. For the data in the file read.dat, test for independence using the data in columns 4 and 5 and $\beta = 0.1$.

12. The definition of the percentage bend correlation coefficient, ρ_{pb}, involves a measure of scale, ω_x, that is estimated with $\hat{\omega} = W_{(m)}$, where $W_i = |X_i - M_x|$ and $m = [(1-\beta)n]$, and $0 \le \beta \le 0.5$. Note that this measure of scale is defined even when $0.5 < \beta < 1$ provided that $m > 0$. Argue that the finite sample breakdown point of this estimator is maximized when $\beta = 0.5$.

13. If in the definition of the biweight midcovariance, the median is replaced by the biweight measure of location, the biweight midcovariance is equal to zero under independence. Describe some negative consequences of replacing the median with the biweight measure of location.

14. Let X be a standard normal random variable, and suppose Y is contaminated normal with probability density function given by Eq. (1.1) of Chapter 1. Let $Q = \rho X + \sqrt{1-\rho^2}Y$, $-1 \le \rho \le 1$. Verify that the correlation between X and Q is

$$\frac{\rho}{\sqrt{\rho^2 + (1-\rho^2)(1-\epsilon+\epsilon K^2)}}.$$

Examine how the correlation changes as K gets large with $\epsilon = 0.1$. What does this illustrate about the robustness of ρ?

CHAPTER 10

Robust Regression

Suppose $(y_i, x_{i1}, \ldots, x_{ip})$, $i = 1, \ldots, n$, are n vectors of observations randomly sampled from some $p + 1$-variate distribution, where $(x_{i1}, \ldots, x_{ip})$ is a vector of predictor values.[1] In some situations the predictors are fixed, known constants, but this distinction is not particularly relevant for most of the results reported here. (Exceptions are noted when necessary.) A general goal is understanding how y is related to the p predictors, which includes finding a method of estimating a conditional measure of location associated with y given $(x_{i1}, \ldots, x_{ip})$, and there is now a vast arsenal of regression methods that might be used. Even when attention is restricted to robust methods, all relevant techniques would easily take up an entire book. In order to reduce the number of techniques to a reasonable size, attention is focused on estimation and hypothesis testing methods that perform reasonably well in simulation studies, in terms of efficiency and probability coverage, particularly when there is a heteroscedastic error term. For more information about robust regression, beyond the topics covered in the final two chapters of this book, see Belsley, Kuh, and Welsch (1980), Birkes and Dodge (1993), Carroll and Ruppert (1988), Cook and Weisberg (1992), Fox (1999), Hampel et al. (1986), Hettmansperger (1984), Hettmansperger and McKean (1998), Huber (1981), Li (1985), Maronna, Martin, and Yohai (2006), Montgomery and Peck (1992), Rousseeuw and Leroy (1987), and Staudte and Sheather (1990). Robust methods that have practical value when the error term is homoscedastic, but are unsatisfactory when the error term is heteroscedastic, are not described or only briefly mentioned. A reasonable suggestion for trying to salvage homoscedastic methods is to test the hypothesis that there is, indeed, homoscedasticity. But under what circumstances would such a test have enough power to detect situations where heteroscedasticity is a practical issue? There is no known way of adequately answering this question, so here attention is focused on heteroscedastic methods that appear to perform about as well as homoscedastic methods in the event there is homoscedasticity.

In regression, the most common assumption is that

$$y_i = \beta_0 + \beta_1 x_{i1} + \cdots + \beta_p x_{ip} + \epsilon_i, \qquad (10.1)$$

---

[1] When dealing with regression, the remaining two chapters write random variables as lowercase Roman letters.

where $\beta_0, \ldots, \beta_p$ are unknown parameters, $i = 1, \ldots, n$, the ϵ_i are independent random variables with $E(\epsilon_i) = 0$, $\text{VAR}(\epsilon_i) = \sigma^2$, and ϵ_i is independent of x_{ik} for each k, $k = 1, \ldots, p$. This model implies that the conditional mean of y_i, given $(x_{i1}, \ldots, x_{ip})$, is $\beta_0 + \sum \beta_k x_{ik}$, a linear combination of the predictors. Equation (10.1) is a *homoscedastic* model, meaning that the ϵ_i have a common variance. If the error term, ϵ_i, has variance σ_i^2, and $\sigma_i^2 \neq \sigma_j^2$, for some $i \neq j$, the model is said to be *heteroscedastic*. Even when ϵ has a normal distribution, heteroscedasticity can result in relatively low efficiency when using the conventional (ordinary least squares) estimator, meaning that the estimator of β can have a relatively large standard error. Also, probability coverage can be poor when computing confidence intervals, as will be illustrated. Dealing with these two problems is one of the major goals in this chapter. Other general goals are dealing with outliers and achieving high efficiency when sampling from heavy-tailed distributions.

Before continuing, some comments about notation might be useful. At times, standard vector and matrix notation will be used. In particular, let

$$\mathbf{x}_i = (x_{i1}, \ldots, x_{ip})$$

and

$$\beta = (\beta_1, \ldots, \beta_p)'$$

$$= \begin{pmatrix} \beta_1 \\ \vdots \\ \beta_p \end{pmatrix}.$$

Then

$$\mathbf{x}_i \beta = \beta_1 x_{i1} + \cdots + \beta_p x_{ip}$$

and Eq. (10.1) becomes

$$y_i = \beta_0 + \mathbf{x}_i \beta + \epsilon_i.$$

This chapter begins with a summary of practical problems associated with least squares regression. Then various robust estimators are described and some comments are made about their relative merits. Inferential techniques, based on the robust regression estimators introduced in this chapter, are described and illustrated in Chapter 11.

10.1 Problems with Ordinary Least Squares

Let b_j be any estimate of β_j, $j = 0, 1, \ldots, p$, and let

$$\hat{y}_i = b_0 + b_1 x_{i1} + \cdots + b_p x_{ip}.$$

From basic principles, the ordinary least squares (OLS) estimator arises without making any distributional assumptions. The estimates are the b_j values that minimize

$$\sum (y_i - \hat{y}_i)^2,$$

the sum of squared residuals. In order to test hypotheses or compute confidence intervals, typically the homoscedastic model given by Eq. (10.1) is assumed with the additional assumption that ϵ has a normal distribution with mean zero. Even when ϵ is normal, but heteroscedastic, problems with computing confidence intervals arise.

Consider, for example, simple regression where there is only one predictor ($p = 1$). Then the OLS estimate of the slope is

$$\hat{\beta}_1 = \frac{\sum (x_{i1} - \bar{x}_1)(y_i - \bar{y})}{\sum (x_{i1} - \bar{x}_1)^2},$$

where $\bar{x}_1 = \sum x_{i1}/n$ and $\bar{y} = \sum y_i/n$, and the OLS estimate of β_0 is

$$\hat{\beta}_0 = \bar{y} - \hat{\beta}_1 \bar{x}_1.$$

Suppose

$$y_i = \beta_0 + \beta_1 x_{i1} + \lambda(x_{i1})\epsilon_i, \tag{10.2}$$

where $VAR(\epsilon_i) = \sigma^2$ and λ is some unknown function of x_{i1} used to model heteroscedasticity. That is, the error term is now $\lambda(x_{i1})\epsilon_i$, and its variance varies with x_{i1}, so it is heteroscedastic. The usual homoscedastic model corresponds to $\lambda(x_{i1}) = 1$. To illustrate the effect of heterogeneity when computing confidence intervals, suppose both x_{i1} and ϵ_i are standard normal, $\beta_1 = 1$, $\beta_0 = 0$, and $\lambda(x_{i1}) = \sqrt{|x_{i1}|}$. Thus, the error term, $\lambda(x_{i1})\epsilon_i$, has a relatively small variance when x_{i1} is close to zero, and the variance increases as x_{i1} moves away from zero. The standard $1 - \alpha$ confidence interval for the slope, β_1, is

$$\hat{\beta}_1 \pm t_{1-\alpha/2} \sqrt{\frac{\hat{\sigma}^2}{\sum (x_{i1} - \bar{x}_1)^2}},$$

where $t_{1-\alpha/2}$ is the $1 - \alpha/2$ quantile of a Student's t-distribution with $n - 2$ degrees of freedom,

$$\hat{\sigma}^2 = \sum \frac{r_i^2}{n-2},$$

and $r_i = y_i - \hat{\beta}_1 x_{i1} - \hat{\beta}_0$ are the residuals. When the error term is homoscedastic and normal, the probability coverage is exactly $1 - \alpha$.

Let $\hat{\alpha}$ be a simulation estimate of one minus the probability coverage when computing a $1 - \alpha$ confidence interval for β_1. When using the conventional confidence interval for β_1, with $n = 20$, $\lambda(x_{i1}) = \sqrt{|x_{i1}|}$, and $\alpha = 0.05$, $\hat{\alpha} = 0.135$ based on a simulation with 1000 replications. If $\lambda(x_{i1}) = |x_{i1}|$, the estimate increases to 0.214. If instead x_{i1} has a g-and-h distribution (described in Section 4.2) with $g = 0$ and $h = 0.5$ (a symmetric, heavy-tailed distribution) $\hat{\alpha}$ increases to 0.52. Put another way, when testing H_0: $\beta_1 = 1$ with $\alpha = 0.05$, the actual probability of a type I error can be more than 10 times the nominal level. A similar problem arises when testing H_0: $\beta_0 = 0$. A natural strategy is to test the assumptions of normality and homogeneity, but as already noted, such tests might not have enough power to detect a situation where these assumptions yield poor probability coverage. Just how large the sample size should be, before standard assumptions can be adequately tested, is unknown. Currently, a better approach is simply to abandon it in favor of a method that performs relatively well under heteroscedasticity, and which competes well with the standard method when in fact the error term is homoscedastic.

Another problem with OLS is that it can be highly inefficient, and this can result in relatively low power. As will be illustrated, this problem arises even when ϵ_i is normal but $\lambda(x_{i1})$ is not equal to one. That is, the error term has a normal distribution but is heteroscedastic. Low efficiency also arises when the error term is homoscedastic but has a heavy-tailed distribution. As an illustration, again consider simple regression. When the error term is homoscedastic, $\hat{\beta}_1$, the OLS estimate of β_1, has variance

$$\frac{\sigma^2}{\sum(x_{i1} - \bar{x}_1)^2}.$$

But from results described and illustrated in Chapters 1–3, σ^2, the variance of the error term, becomes inflated if sampling is from a heavy-tailed distribution. That is, slight departures from normality, as measured by the Kolmogorov distance function, result in large increases in the standard error of $\hat{\beta}_1$. (This problem is well known and discussed by Hampel, 1973; He et al., 1990; Schrader & Hettmansperger, 1980, among others.) Note, however, that if the x_{i1} are sampled from a heavy-tailed distribution, this inflates the expected value of $\sum(x_{i1} - \bar{x}_1)^2$, relative to sampling from a normal distribution, in which case the standard error of $\hat{\beta}_1$ tends to be smaller versus the situation where x_{i1} is normal. In fact, a single outlier among the x_{i1} values inflates $\sum(x_{i1} - \bar{x}_1)^2$, causing the estimate of the squared standard error to decrease. Consequently, there is interest in searching for methods that have good efficiency when ϵ has a heavy-tailed distribution, and maintains relatively high efficiency when the x_i values are randomly sampled from a heavy-tailed distribution as well.

One strategy is to check for outliers, remove any that are found, and proceed with standard OLS methods using the data that remain. As was the case when dealing with measures of location, a problem with this approach is that the estimated standard error may not converge to the correct value as the sample size gets large (He & Portnoy, 1992). A bootstrap estimate of the standard error could be used instead, but little is known about the effectiveness of this approach. Also, simply removing outliers does not always deal effectively with low efficiency due to a heteroscedastic error term.

Yet another practical problem is that OLS has a breakdown point of only $1/n$. That is, a single point, properly placed, can cause the OLS estimator to have virtually any value. Not only do unusual y values cause problems, outlying x values, called *leverage points*, can have an inordinate influence on the estimated slopes and intercept.

It should be noted that two types of leverage points play a role in regression: good and bad. Roughly, leverage points are good or bad depending on whether they are reasonably consistent with the true regression line. A *regression outlier* is a point with a relatively large residual. A *bad leverage point* is a leverage point that is also a regression outlier. A *good leverage point* is a leverage point that is not a regression outlier. Leverage points can reduce the standard error of the OLS estimator, but a bad leverage point can result in a poor fit to the bulk of the data.

10.1.1 Computing Confidence Intervals under Heteroscedasticity

When using the OLS estimator, various methods have been proposed for computing confidence intervals for regression parameters when the error term is heteroscedastic. One strategy when dealing with heteroscedasticity is to transform the data (e.g., Carroll & Ruppert, 1988). The focus here is on methods that appear to perform well without relying on any transformation. Perhaps situations arise where transformations have practical value relative to the methods described here, but it seems that this issue has not been investigated. The methods described here compete well with homoscedastic methods when indeed the error term is homoscedastic. Attention is restricted to the seemingly better methods followed by comments regarding their relative merits.

Wilcox (1996c) found that for the special case $p = 1$ (one predictor only), only one method performed well among the situations he considered. For $p > 1$ a slight modification is recommended (Wilcox, 2003f) when the goal is to have simultaneous probability coverage equal to $1 - \alpha$ for all p slope parameters. The method begins by sampling, with replacement, n vectors of observations from $(y_i, \mathbf{x}_i)$, $i = 1, \ldots, n$. Put another way, a bootstrap sample is obtained by randomly sampling, with replacement, n rows of data from the n-by-$(p+1)$ matrix

$$\begin{pmatrix} y_1, x_{11}, \ldots, x_{1p} \\ y_2, x_{21}, \ldots, x_{2p} \\ \vdots \\ y_n, x_{n1}, \ldots, x_{np} \end{pmatrix}.$$

The resulting n vectors of observations are labeled

$$(y_1^*, x_{11}^*, \ldots, x_{1p}^*), \ldots, (y_n^*, x_{n1}^*, \ldots, x_{np}^*).$$

Let $\hat{\beta}_j^*$ be the OLS estimate of β_j, the jth slope parameter, $j = 1, \ldots, p$, based on the bootstrap sample just obtained. Repeat this bootstrap process B times yielding $\hat{\beta}_{j1}^*, \hat{\beta}_{j2}^*, \ldots, \hat{\beta}_{jB}^*$. Let $\hat{\beta}_{j(1)}^* \leq \hat{\beta}_{j(2)}^* \leq \cdots \leq \hat{\beta}_{j(B)}^*$ be the B bootstrap estimates written in ascending order.

For the special case $p = 1$, a slight modification of the standard percentile bootstrap method is used. When $B = 599$, the 0.95 confidence interval for β_1 is

$$(\hat{\beta}_{1(a+1)}^*, \hat{\beta}_{1(c)}^*),$$

where for $n < 40$, $a = 6$ and $c = 593$; for $40 \leq n < 80$, $a = 7$ and $c = 592$; for $80 \leq n < 180$, $a = 10$ and $c = 589$; for $180 \leq n < 250$, $a = 13$ and $c = 586$; whereas for $n \geq 250$, $a = 15$ and $c = 584$. Note that this method becomes the standard percentile bootstrap procedure when $n \geq 250$. If, for example, $n = 20$, the lower end of the 0.95 confidence interval is given by $\hat{\beta}_{1(7)}^*$. A confidence interval for the intercept is computed in the same manner, but currently it seems best to use $a = 15$ and $c = 584$ for any n. That is, use the usual percentile bootstrap confidence interval. From results described in Chapter 4, there are situations where the confidence interval for the intercept can be expected to have unsatisfactory probability coverage. In essence, the situation reduces to computing a percentile bootstrap confidence interval for the mean when the slope parameters are all equal to zero. A criticism of this method is that it is limited to $\alpha = 0.05$.

As for $p > 1$, if the goal is to achieve simultaneous probability coverage equal to $1 - \alpha$, it currently appears that the best approach is to use a standard percentile bootstrap method in conjunction with the Bonferroni inequality (Wilcox, 2003f). So, set

$$\ell = \frac{\alpha B}{2p},$$

round ℓ to the nearest integer, let $u = B - \ell$, in which case the confidence interval for the jth predictor ($j = 1, \ldots, p$) is

$$(\hat{\beta}_{j(\ell+1)}^*, \hat{\beta}_{j(u)}^*).$$

For $p = 1$, the bootstrap confidence interval just described is based on the strategy of finding a method that gives good results under normality and homoscedasticity, and then using this method when there is heteroscedasticity or sampling is from a nonnormal distribution. Simulations were then used to see whether the method continues to perform well when sampling from nonnormal distributions, or when there is heteroscedasticity. Relative to other methods that have been proposed, the modified bootstrap procedure has a clear advantage. This result is somewhat unexpected because in general, when working with nonrobust measures of location and scale, this strategy performs rather poorly. If, for example, the percentile bootstrap method is adjusted so that the resulting confidence interval for the mean has probability coverage close to the nominal level when sampling from a normal distribution, probability coverage can be poor when sampling from nonnormal distributions instead.

Another strategy is to obtain bootstrap samples by resampling residuals, as opposed to vectors of observations as is done here. When dealing with heteroscedasticity, theoretical results do not support this approach (Wu, 1986), and simulations indicate that unsatisfactory probability coverage can result. Of course, one could check for homoscedasticity in an attempt to justify resampling residuals, but there is no known way of being reasonably certain that the error term is sufficiently homoscedastic. Again, any test of the assumption of homoscedasticity might not have enough power to detect heteroscedasticity in situations where the assumption should be discarded.

Nanayakkara and Cressie (1991) derived another method for computing a confidence interval for the regression parameters when the error term is heteroscedastic. When the x_{i1} values are fixed and evenly spaced, their method appears to give good probability coverage, but otherwise probability coverage can be unsatisfactory.

Long and Ervin (2000) compared several nonbootstrap methods for dealing with heteroscedasticity and recommended one particular method for general use, which is based on what is called the HC3 estimate of the standard errors. The HC3 estimator is

$$\text{HC3} = (\mathbf{X}'\mathbf{X})^{-1}\mathbf{X}'\text{diag}\left[\frac{r_i^2}{(1 - h_{ii})^2}\right]\mathbf{X}(\mathbf{X}'\mathbf{X})^{-1},$$

where r_i, $i = 1, \ldots, n$ are the usual residuals,

$$h_{ii} = \mathbf{x}_i(\mathbf{X}'\mathbf{X})^{-1}\mathbf{x}_i',$$

and

$$\mathbf{X} = \begin{pmatrix} 1 & x_{11} \cdots & x_{1p} \\ 1 & x_{21} \cdots & x_{2p} \\ \vdots & \vdots & \vdots \\ 1 & x_{n1} \cdots & x_{np} \end{pmatrix}$$

and $\mathbf{x}_i$ is the ith row of $\mathbf{X}_i$ (e.g., MacKinnon & White, 1985). If $b_0, \ldots, b_p$ are the least squares estimates of the $p + 1$ parameters, the diagonal elements of the matrix HC3 represent the estimated squared standard errors. So if S_j^2 ($j = 0, \ldots, p$) is the jth diagonal element

HC3, the $1 - \alpha$ confidence interval for β_j is taken to be

$$b_j \pm t S_j,$$

where t is the $1 - \alpha/2$ quantile of a Student's t-distribution with $\nu = n - p - 1$ degrees of freedom. But it is unknown how large n must be to ensure reasonably accurate confidence intervals. For a single predictor, it is known that $n = 60$ might not suffice (Wilcox, 2001b).

More recently, an alternative to the HC3 estimator has been studied that appears to be better for general use (Godfrey, 2006). Let h_{ii} be defined as done when using the HC3 estimator. Let $\bar{h} = \sum h_{ii}/n$, $e_{ii} = h_{ii}/\bar{h}$, and $d_{ii} = \min(4, e_{ii})$. The HC4 estimator is

$$S = (\mathbf{X}'\mathbf{X})^{-1}\mathbf{X}'\text{diag}\left[\frac{\mathbf{r_i^2}}{(\mathbf{1 - h_{ii}})^{\mathbf{d_{ii}}}}\right]\mathbf{X}(\mathbf{X}'\mathbf{X})^{-1}.$$

The diagonal elements of the matrix $\mathbf{S}$, which we denote by $S_0^2, S_1^2, \ldots, S_p^2$, are the estimated squared standard errors of $b_0, b_1, \ldots b_p$, respectively. Following Ng and Wilcox (2009), the $1 - \alpha$ confidence interval for β_j is taken to be

$$b_j \pm t S_j,$$

where t is the $1 - \alpha/2$ quantile of a Student's t-distribution with $\nu = n - p - 1$ degrees of freedom. (Cribari-Neto, Souza, & Vasconcellos, 2007, suggest an alternative to the HC4 estimator, but for the situation at hand it seems to offer no practical advantage; see Ng, 2009b.)

Wald-Type Statistics Used in Conjunction with a Wild Bootstrap

Method HC4WB-D

Two wild bootstrap methods should be mentioned, both of which are based in part on what is called a Wald-type statistic. The first, which is labeled the HC4WB-D is performed as follows:

1. Again let b_j be the ordinary least squares estimate of β_j, compute S_j, the HC4 estimate of the standard error.
2. Compute the Wald test statistic

$$W = (b_j - 0)S_j^{-1}(b_j - 0).$$

3. Generate $D_1, \ldots, D_n$ from a two-point (lattice) distribution. That is,

$$D_i = \begin{cases} -1 & \text{with probability .5} \\ 1 & \text{with probability .5.} \end{cases}$$

A bootstrap sample $(y_i^*, \mathbf{x}_i)$, assuming the null hypothesis is true, is given by $y_i^* = \bar{y} + D_i r_i$, $i = 1, \ldots n$, where $\hat{\beta} = (\hat{\beta}_1, \ldots, \hat{\beta}_p)'$ are the ordinary least squares estimate under the assumption that the null hypothesis is true.

4. Compute the ordinary least squares estimate (b_j^*) based on this bootstrap sample as well as the HC4 estimate of the standard error (S_j^*). Compute the Wald test statistic

$$W^* = (b_j^* - 0)S^{*-1}(b_j^* - 0).$$

based on the bootstrap sample.

5. Repeat steps 2 – 4 B times yielding W_b^*, $b = 1, \ldots, B$.
6. A p-value for H_0: $\beta_j = 0$ is given by

$$p = \frac{\#\{W_b^* \geq W\}}{B}.$$

Reject H_0 if $p \leq \alpha$.

Method HC4WB-C

Method HC4WB-C is exactly like method HC4WB-D, only now

$$D_i = \sqrt{12}(U_i - .5),$$

where U has a uniform distribution over the unit interval.

In contrast to Godfrey (2006), the more extensive simulations by Ng and Wilcox (2009) indicate that the wild bootstrap methods do not have a striking advantage over the nonbootstrap HC4 method in terms of achieving a type I error probability reasonably close to the nominal level. However, in terms of minimizing the variability of the type I error probabilities among the situations that were considered, HC4WB-C was found to be best when testing at the 0.05 level. Although the methods based on the HC4 estimator perform relatively well, there are situations where all methods based on the HC4 estimator fail to control the type I error probability reasonably accurate manner, even with $n = 100$.

10.1.2 An Omnibus Test

Rather than test hypotheses about the individual parameters, a common goal is to test

$$H_0 : \beta_1 = \cdots = \beta_p = 0,$$

the hypothesis that all p slope parameters are zero. When using the OLS estimator and $p > 1$, it seems that no method has been found to be effective, in terms of controlling the probability of a type I error, when there is heteroscedasticity, nonnormality, or both. Mammen (1993) studied a method based in part on a so-called wild bootstrap technique. While preparing this chapter, the author ran some simulations as a partial check on this approach and found that it generally performed reasonable well with $n = 30$ and $p = 4$. However, a situation was found where, with a heteroscedastic error term, the actual probability of a type I error was estimated to be .29 when testing at the 0.05 level. In fairness, perhaps for nearly all practical situations, the method performs reasonably well, but resolving this issue is difficult at best.

With the understanding that, when dealing with least squares regression, no single method is always satisfactory in terms of controlling the probability of a type I error, two methods currently seem best for general use. Both are based on a more general form of the HC4 estimate of the standard error, which was derived by Cribari-Neto (2004).

Let $\mathbf{V}$ be the HC4 estimate of the variances and covariances of $\mathbf{b} = (b_1, \ldots, b_p)'$, the least squares estimate of the slope parameters. A test statistic for testing the hypothesis that all slopes are zero is

$$W = n\mathbf{b}'\mathbf{V}\mathbf{b},$$

which has, approximately, a chi-squared distribution with p degrees of freedom. However, for $p > 1$, this method is unsatisfactory in terms of controlling the probability of a type I error.

A better approach is to use a wild bootstrap method. That is, generate wild bootstrap values y_i^* as was done in conjunction with the HC4WB-C or the HC4WB-D methods. Based on this bootstrap sample, compute the test statistic W yielding W^*. Repeat this B times yielding $W_1^*, \ldots, W_B^*$. A p-value is given by

$$\frac{1}{B}\sum I_i,$$

where the indicator function $I_i = 1$ if $W_i \leq W$; otherwise $I_i = 0$.

10.1.3 R Functions lsfitNci, lsfitci, olshc4, hc4test, and hc4wtest

The R function

$$\text{lsfitci(x,y,nboot=599)}$$

is supplied for computing 0.95 confidence intervals for regression parameters, based on the OLS estimator, using the percentile bootstrap method described in Section 10.1.1. As usual, x is an n-by-p matrix of predictors. This function is designed for $\alpha = 0.05$ only.

For large sample sizes, the bootstrap can be avoided by using the estimate of the squared standard errors given by HC3. The computations are performed by the function

$$\text{lsfitNci(x,y,alpha=0.05).}$$

But again, it is unclear how large the sample size must be in order for this approach to achieve the same control over the type I error probability achieved by the percentile bootstrap method described here.

With the understanding that no single estimator is always best, it appears that using the HC4 estimator is preferable to the HC3 estimator. HC4 does not dominate HC3, but it is difficult to know when HC3 gives more accurate results. The R function

$$\text{olshc4(x,y,alpha=0.05,xout=F,outfun=out)}$$

computes $1 - \alpha$ confidence intervals using the HC4 estimator, and p-values are returned as well. By default, 0.95 confidence intervals are returned. Setting the argument alpha equal to 0.1, for example, will result in 0.9 confidence intervals. Leverage points are removed if the argument xout=T using the R function specified by the argument outfun, which defaults to the MVE method.

The function

$$\text{hc4test(x,y,xout=F,outfun=out)}$$

tests the hypothesis that all slope parameters are equal to zero. With a sufficiently large sample size, this method will perform well in terms of controlling the probability of a type I error. But it is unclear just how large the sample size needs to be. With a small to moderate sample size all indications are that it is safer to use the R function

$$\text{hc4wtest(x, y, nboot = 500, SEED = T, RAD = T),}$$

which uses a wild bootstrap method. When the argument RAD=T, method HC4WB-D is used. Otherwise METHOD HC4WB-C is used.

■ Example

Assuming both x and ϵ have standard normal distributions, 30 pairs of observations were generated according to the model $y = (|x| + 1)\epsilon$. The standard F-test for $H_0 : \beta_1 = 0$ was applied and this process was repeated 1000 times. Testing at the 0.05 level, the proportion of type I errors was .144. So the standard F-test correctly detects an association about 14% of the time, but simultaneously provides an inaccurate assessment of β_1. This again illustrates that under heteroscedasticity, the standard F test does not control the probability of a type I error. Using instead the R function olshc4, the proportion of rejections was 0.06, which is reasonably close to the nominal 0.05 level.

■

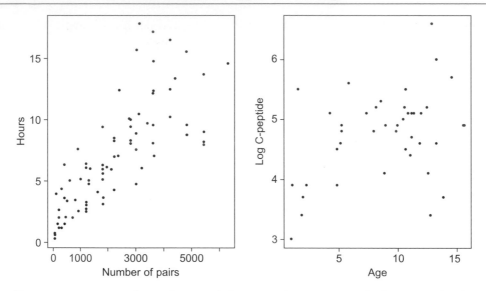

Figure 10.1: Scatterplots of two real data sets that appear to be heteroscedastic.

■ Example

Cohen, Dalal, and Tukey (1993) report data on the number of hours, y, needed to splice x pairs of wires for a particular type of telephone cable. The left panel of Figure 10.1 shows a scatterplot of the data. (The data are stored in the file splice.dat, which can be obtained as described in Section 1.8 of Chapter 1.) Note that the data appear to be heteroscedastic. The usual 0.95 confidence interval (multiplied by 1000 for convenience), based on the assumption of normality and homogeneity, is (1.68, 2.44). If the y values are stored in the R vector yvec, and the x values are stored in the R variable splice, the command lsfitci(splice,y) reports that the 0.95 bootstrap confidence interval is (1.64, 2.57). The ratio of the lengths is $(2.44 - 1.68)/(2.57 - 1.64) = 0.82$. It might be argued that the lengths are reasonably similar. However, the probability coverage of the usual method can be less than the nominal level, it is unclear whether this problem can be ignored for the data being examined, and all indications are that the bootstrap method provides better probability coverage under heteroscedasticity. Consequently, using the bootstrap confidence interval seems more satisfactory.

■

■ Example

Sockett et al. (1987) collected data with the goal of understanding how various factors are related to the patterns of residual insulin secretion in children. (The data can be found in the file diabetes.dat.) One of the response measurements is the logarithm of C-peptide concentration (pmol/ml) at diagnosis, and one of the predictors considered is age. The right panel of Figure 10.1 shows a scatterplot of the data. The scatterplot suggests that the error term is heteroscedastic, with the smallest variance near age 7. (Results in Chapter 11 lend support for this speculation.) The 0.95 confidence interval for the slope, using the standard OLS method, is $(0.0042, 0.0263)$, the estimate of the slope being 0.015. In contrast, lsfitci returns a 0.95 confidence interval of $(-0.00098, 0.029)$, and the ratio of the lengths is $(0.0263 - 0.0042)/(0.029 + 0.00098) = 0.74$. Again there is concern that the standard confidence interval is too short and that its actual probability coverage is less than the nominal level. Note that the standard confidence interval rejects $H_0 : \beta_1 = 0$, but lsfitci does not.

10.1.4 Comments on Comparing Means via Dummy Coding

A well-known approach to comparing the means of multiple groups is via least squares regression coupled with dummy coding (e.g., Montgomery & Peck, 1992) Yet another way of dealing with heteroscedasticity when comparing means, beyond the methods in Chapter 7, is to use dummy coding in conjunction with the HC4 estimator. But results in Ng (2009b) do not support this approach. Control over the probability of a type I error can be unsatisfactory.

10.1.5 Comments on Trying to Salvage the Homoscedasticity Assumption

One strategy for trying to salvage the homoscedasticity assumption, when using classic inferential methods associated with the least squares regression estimator, is to simply test the hypothesis that there is homoscedasticity. Most methods for testing this hypothesis have been found to be unsatisfactory in terms of controlling the probability of a type I error. That is, there is a high probability of rejecting the hypothesis that there is homoscedasticity when indeed there is homoscedasticity. Two methods that have been found to perform well in simulations are described in Section 11.4. However, imagine that the classic homoscedastic methods are used if these methods fail to reject the hypothesis that the error term is homoscedastic. A basic issue is whether these methods have enough power to detect situations where there is heteroscedasticity that invalidates the homoscedastic methods that are typically used. With $n \le 100$, Ng and Wilcox (2011) found that the answer is no. The usual homoscedastic methods can still have actual type I error probabilities well above the nominal level. To the extent it is desired to make inferences about the regression parameters, without being sensitive to heteroscedasticity, all indications are that it is best to abandon homoscedastic methods and always use one of the heteroscedastic methods in this section.

10.2 Theil–Sen Estimator

This section describes an estimator first proposed by Theil (1950) and later extended by Sen (1968) that is restricted to the case of a single predictor ($p = 1$). Then various extensions to $p > 1$ are discussed.

Temporarily focusing on $p = 1$, one view of the Theil–Sen estimator is that it attempts to find a value for the slope that makes Kendall's correlation tau, between $y_i - bx_i$ and x_i, (approximately) equal to zero. This can be seen to be tantamount to the following method. For any $i < i'$, for which $x_i \neq x_{i'}$, let

$$S_{ii'} = \frac{y_i - y_{i'}}{x_i - x_{i'}}.$$

The Theil–Sen estimate of the slope is b_{1ts}, the median of all the slopes represented by $S_{ii'}$. The intercept is estimated with

$$M_y - b_1 M_x,$$

where M_y and M_x are the usual sample medians of the y and x values, respectively. Sen (1968) derived the asymptotic standard error of the slope estimator, but it plays no role here and therefore is not reported. Dietz (1987) showed that the Theil–Sen estimator has an asymptotic breakdown point of 0.293. For results on its small-sample efficiency, see Dietz (1989), Talwar (1991) and Wilcox (1998a, 1998b). Sievers (1978) and Scholz (1978) proposed a generalization of the Theil–Sen estimator by attaching weights to the pairwise slopes, $S_{ii'}$, but in terms of efficiency it seems to offer little or no advantage, and in some cases its bias is considerably larger, so it is not described here. For results when the median of the slopes is replaced by a trimmed mean, see Luh and Guo (2000).

It is noted that the Theil–Sen estimator has close similarities to a regression estimator studied by Maronna and Yohai (1993, Section 3.3). This alternative estimator belongs to what are called projection estimators; see Section 10.13.11. Martin, Yohai, and Zamar (1989) proved that this alternative estimate of the slope is minimax in the class of all regression equivariant estimates if the error term has a symmetric and unimodal distribution.

There are at least three general ways the Theil–Sen estimator might be extended to two or more predictors (cf. Hussain & Sprent, 1983). The first, which will be called method *TS*, is to apply the back-fitting, Gauss–Seidel method as described in Hastie and Tibshirani (1990, pp. 106–108). (For a general description of the Gauss–Seidel method and its properties, see e.g., Dahlquist & Björck, 1974; Golub & van Loan, 1983.) For the problem at hand, the method is applied as follows:

1. Set $k = 0$ and choose an initial estimate for β_j, say $b_j^{(0)}$, $j = 0, \dots, p$. Here, the initial estimate is taken to be the Theil–Sen estimate of the slope based on the jth regressor

only. That is, simply ignore the other predictors to obtain an initial estimate of β_j. The initial estimate of β_0 is the median of

$$y_i - \sum_{j=1}^{p} b_j^{(0)} x_{ij}, \quad i = 1, \ldots, n.$$

2. Increment k and take the kth estimate of β_j $(j = 1, \ldots, p)$ to be $b_j^{(k)}$, the Theil-Sen estimate of the slope based on the regression estimate of x_{ij} with

$$r_i = y_i - b_0^{(k-1)} - \sum_{\ell=1, \ell \neq j}^{p} b_\ell^{(k-1)} x_{i\ell}.$$

The updated estimate of the intercept, $b_0^{(k)}$, is the median of

$$y_i - \sum_{j=1}^{p} b_j^{(k)} x_{ij}, \quad i = 1, \ldots, n.$$

3. Repeat step 2 until convergence.

The second general approach toward extending the Theil–Sen estimator to multiple predictors is based on so-called *elemental subsets*, an idea that appears to have been first suggested in unpublished work by Oja and Niimimaa; see Rousseeuw and Leroy (1987, p. 146, cf. Hawkins & Olive, 2002). In a regression data set, an elemental subset consists of the minimum number of cases required to estimate the unknown parameters of a regression model. With p regressors, the Oja and Niimimaa extension is to use all $N = n!/[(p+1)!(n-p-1)!]$ elemental subsets. That is, for each elemental subset, estimate the slope parameters using OLS. At this point, a simple strategy is to use the median of the N resulting estimates. That is, letting $\hat{\beta}_{ji}$ be the estimate of β_j based on the ith elemental subset, $i = 1, \ldots, N$, the final estimate of β_j would be the median of these N values. One practical problem is that the number of elemental subsets increases rapidly with n and p. For example, with $n = 100$ and $p = 4$, the number of elemental subsets is 106,657,320. Also note that intuitively, some elemental subsets will yield a highly inaccurate estimate of the slopes. An alternative strategy is to use $(n^2 - n)/2$ randomly sampled elemental subsets, the same number of elemental subsets used when $p = 1$. In terms of efficiency, results in Wilcox (1998b) support this approach over using all N elemental subsets instead. For convenience, this method is labeled *TSG*. Note that the back-fitting, Gauss–Seidel method eliminates the random component associated with the method just described, and Gauss-Seidel also offers faster execution time.

Let $\hat{\tau}_j$ be Kendall's tau between the jth predictor, x_j, and $y - b_1x_1 - \cdots - b_px_p$. A third approach, when generalizing the Theil–Sen estimator to $p > 1$ predictors, is to determine

$b_1, \ldots, b_p$ so that $\sum |\hat{\tau}_j|$ is approximately equal to zero. Note that this approach can be used to generalize the Theil–Sen estimator by replacing Kendall's tau with any reasonable correlation coefficient. The relative merits of this third way of extending the Theil–Sen estimator to multiple predictors have not been explored.

Results regarding the small-sample efficiency of the Gauss–Seidel method, versus using randomly sampled elemental subsets, are reported in Wilcox (2004d). The choice of method can make a practical difference, but currently there is no compelling reason to prefer one method over the other based solely on efficiency.

A criticism of method TSG is that as p increases, its finite sample breakdown point decreases (Rousseeuw & Leroy, 1987, p. 148). Another possible concern is that the marginal medians are location equivariant but not affine equivariant. (See Eq. (6.9) for a definition of affine equivariance when referring to a multivariate location estimator.) A regression estimator T is *affine equivariant* if for any nonsingular matrix $\mathbf{A}$,

$$T(\mathbf{x_i A}, \ y_i; i = 1, \ldots, n) = \mathbf{A}^{-1} \mathbf{T}(\mathbf{x_i}, \ y_i; i = 1, \ldots, n).$$

Because the marginal medians are not affine equivariant, TSG is not affine equivariant either. Yet one more criticism is that with only $(n^2 - n)/2$ randomly sampled elemental subsets, if n is small, rather unstable results can be obtained, meaning that if a different set of $(n^2 - n)/2$ elemental subsets is used, the estimates can change substantially. If, for example, $n = 20$ and $p = 2$, only 190 resamples are used from among the 1140 elemental subsets. (Of course, when n is small, it is a simple matter to increase the number of sampled elemental subsets, but just how many additional samples should be taken has not been investigated.)

A regression estimator T is *regression equivariant* if for any vector $\mathbf{v}$,

$$T(\mathbf{x_i}, \ y_i + \mathbf{x_i v}; i = 1, \ldots, n) = \mathbf{T}(\mathbf{x_i}, \ y_i; i = 1, \ldots, n) + \mathbf{v}.$$

And T is said to be *scale equivariant* if

$$T(\mathbf{x_i}, \ cy_i; i = 1, \ldots, n) = \mathbf{c T}(\mathbf{x_i}, \ y_i; i = 1, \ldots, n).$$

It is noted that method TS also fails to achieve affine equivariance, but it does achieve regression equivariance and scale equivariance.

10.2.1 R Functions tsreg, correg, and regplot

The R function

$$tsreg(x,y,iter=10)$$

computes the TS regression estimator just described. When $p > 1$, the default maximum number of iterations when using the Gauss–Seidel method, indicated by the argument iter, is 10.

The function

$$correg(x,y,corfun=tau)$$

computes the Theil–Sen estimate by determining $b_1, \ldots, b_p$ so that $\sum |\hat{\tau}_j|$ is approximately equal to zero, where $\hat{\tau}_j$ is Kendall's tau between the jth predictor, x_j, and $y - b_1 x_1 - \cdots - b_p x_p$. It generalizes the function tsreg by allowing Kendall's tau to be replaced by some other correlation. For example, correg(x,y,corfun=pbcor) would use the percentage bend correlation coefficient.

When dealing with a single predictor, only two R commands are needed to create a scatterplot that includes a regression line. (The two R functions are plot and abline.) To make this task even easier, the R function

$$regplot(x,y,regfun=tsreg,xlab="X",ylab="Y")$$

is supplied, which by default plots the Theil–Sen regression line. Other regression lines can be specified via the argument regfun, which assumes that the estimated slope and intercept, returned by the R function specified, are stored in $coef.

10.3 Least Median of Squares

The *least median of squares* (LMS) regression estimator appears to have been first proposed by Hampel (1975) and further developed by Rousseeuw's (1984). That is, the regression estimates are taken to be the values that minimize

$$\text{MED}(r_1^2, \ldots, r_n^2),$$

the median of the squared residuals. (Also see Davies, 1993; Hawkins & Simonoff, 1993; Rousseeuw & Leroy, 1987.) It was the first equivariant estimator to attain a breakdown point of approximately 0.5, but its efficiency relative to the ordinary least squares estimator is 0. (And its rate of convergence is $n^{-1/3}$ rather than the usual rate of $n^{-1/2}$.) Despite these negative properties, the LMS estimator is often suggested as a diagnostic tool or as preliminary fit to data.

10.3.1 R Function lmsreg

The R function

$$lmsreg(x,y)$$

computes the least median of squares regression estimator. To gain access to this function, first enter the command library(lqs).

10.4 Least Trimmed Squares Estimator

Rousseeuw's (1984) *least trimmed squares* (LTS) estimator is based on minimizing

$$\sum_{i=1}^{h} r_{(i)}^2,$$

where $r_{(1)}^2 \leq \cdots \leq r_{(h)}^2$ are the squared residuals written in ascending order. With $h = [n/2] + 1$, the same breakdown point as LMS is achieved. However, $h = [n/2] + [(p+1)/2]$ is often used to maintain regression equivariance. LTS has a relatively low asymptotic efficiency (Croux, Rousseeuw, & Hössjer, 1994), but it seems to have practical value. For example, it plays a role in the asymptotically efficient M-estimator described in Section 10.9.

10.4.1 R Functions ltsreg and ltsgreg

The R function

ltsreg(x,y)

computes the LTS regression estimate. Access to this function is obtained via the command library(lqs). The function

ltsgreg(x,y,tr=0.2,h=NA),

written for this book, also computes the LTS estimate but allows you to set the amount of trimming via the argument tr. (The parameters are estimated with the Nelder–Mead method.) If no value for h is specified, the value for h is taken to be $n-[\text{tr}(n)]$, where tr defaults to .2.

10.5 Least Trimmed Absolute Value Estimator

A close variation of the LTS estimator is the least trimmed absolute (LTA) value estimator. Now the strategy is to choose the intercept and slope so as to minimize

$$\sum_{i=1}^{h} |r|_{(i)}, \tag{10.3}$$

where $|r|_{(i)}$ is the ith smallest absolute residual and h is defined as in Section 10.4. (For recent results on the LTA estimator, see Hawkins & Olive, 1999. For asymptotic results, see

Tableman, 1994. For asymptotic results when $h = n$ and the error term is homoscedastic, see Knight, 1998.) Like LTS, the LTA estimator can have a much smaller standard error than the least squares estimator, but its improvement over the LTS estimator seems to be marginal at best, at least based on what is currently known (Wilcox, 2001b.)

10.5.1 R Function ltareg

The R function

$$\text{ltareg(x,y,tr=0.2,h=NA)}$$

computes the LTA estimate using the Nelder–Mead method for minimizing a function (which was mentioned in Chapter 6 in connection with the spatial median). If no value for h is specified, the value for h is taken to be $n-[\text{tr}(n)]$, where tr defaults to .2.

10.6 M-Estimators

Regression M-estimators represent a generalization of the location M-estimator described in Chapter 3. There are, in fact, many variations of M-estimators when dealing with regression, some of which appear to be particularly important in applied work. We begin, however, with some of the early and fairly simple methods. They represent an important improvement on ordinary least squares, but by today's standards they are relatively unsatisfactory. Nevertheless, some of these early methods have become fairly well known, so they are included here for completeness.

Generally, M-estimators of location can be extended to regression estimators by choosing a function ξ and then estimating β_j, $(j = 0, \ldots, p)$ with the b_j values that minimize

$$\sum \xi(r_i).$$

Typically, ξ is some symmetric function chosen so that it has desirable properties plus a unique minimum at zero. Table 2.1 lists some choices.

As explained in Chapter 2, a measure of scale needs to be used with M-estimators of location so that they are scale equivariant, and a measure of scale is needed for the more general situation considered here. In the present context, this means that if the y values are multiplied by some constant, c, the estimated slope parameters should be multiplied by the same value. For example, if the estimated slope is 0.304, and if $y_1, \ldots, y_n$ are multiplied by 10, then the estimated slope should become 3.04.

Letting τ be any measure of scale, M-estimators minimize

$$\sum \xi\left(\frac{r_i}{\tau}\right).$$

Following Hill and Holland (1977), τ is estimated with

$$\hat{\tau} = \frac{\text{median of the largest } n - p - 1 \text{ of the } |r_i|}{0.6745}. \qquad (10.4)$$

Note that $\hat{\tau}$ is resistant, which is needed so that the M-estimator is resistant against unusual y values.

Let Ψ be the derivative of ξ, some of which are given in Table 2.1. Here the focus of attention is on Huber's Ψ given by

$$\Psi(x) = \max[-K, \min(K, x)].$$

For the problem at hand, $K = 2\sqrt{(p+1)/n}$ is used following the suggestion of Belsley et al. (1980). M-estimators solve the system of $p + 1$ equations

$$\sum_{i=1}^{n} x_{ij} w_i r_i = 0, \quad j = 0, \ldots, p, \qquad (10.5)$$

where $x_{i0} = 1$ and

$$w_i = \begin{cases} \frac{\Psi(r_i/\hat{\tau})}{r_i/\hat{\tau}}, & \text{if } y_i \neq \hat{y}_i \\ 1, & \text{if } y_i = \hat{y}_i \end{cases}$$

As was the case when dealing with M-estimators of location, there is no explicit equation that gives the estimate of the regression parameters, an iterative estimation method must be used instead.

Equation (10.5) represents a problem in weighted least squares. That is, Eq. (10.5) is equivalent to estimating the parameters by minimizing

$$\sum w_i r_i^2.$$

As with W-estimators of location, described in Chapter 3, a technical problem here is that the weights, w_i, depend on unknown parameters. The iterative method used with the W-estimator suggests an iterative estimation procedure for the problem at hand, but the details are postponed until Section 10.8.

10.7 The Hat Matrix

The M-estimator described in Section 10.6 provides resistance against unusual or outlying y values, but a criticism is that it is not resistant against leverage points. In fact the breakdown point is only $1/n$. That is, a single unusual point can completely dominate the estimate of the parameters. Moreover, its influence function is unbounded. An early approach to these problems is based in part on the so-called hat matrix, which is described here. As will become evident, the hat matrix yields an M-estimator that has certain practical advantages – it competes very well with OLS in terms of efficiency and computing accurate confidence intervals under heteroscedasticity and nonnormality – but there are situations where it is not as resistant as one might want. In particular, its breakdown point is only $2/n$, meaning that it can handle a single outlier, but two outliers might destroy it. But methods based on the hat matrix have become increasingly well known, so a description of some of them seems in order.

It is a bit easier to convey the idea of the hat matrix in terms of simple regression, so this is done first, after which attention is turned to the more general case where the number of predictors is $p \geq 1$. One strategy for determining whether the point (y_i, x_i) is having an inordinate effect on $\hat{\beta}_1$ and $\hat{\beta}_0$, the OLS estimates of the slope and intercept, is to consider how much the estimates change when this point is eliminated. It turns out that this strategy provides a method for judging whether x_i is a leverage point.

Let

$$h_j = \frac{1}{n} + \frac{(x_j - \bar{x})^2}{\sum (x_i - \bar{x})^2}, \tag{10.6}$$

$j = 1, \ldots, n$. Let $\hat{\beta}_1(i)$, read beta hat sub 1 not i, be the OLS estimate of the slope when the ith pair of observations is removed from the data. Let

$$\hat{y}_i = \hat{\beta}_0 + \hat{\beta}_1 x_i,$$

$$A_j = \frac{\sum x_i^2}{n \sum (x_i - \bar{x})^2} - \frac{x_j \bar{x}}{\sum (x_i - \bar{x})^2},$$

and

$$B_j = \frac{x_j - \bar{x}}{\sum (x_i - \bar{x})^2}.$$

Then the change from $\hat{\beta}_1$ which is based on all of the data, versus the situation where the ith pair of values is removed, can be shown to be

$$\hat{\beta}_1 - \hat{\beta}_1(i) = B_i \frac{r_i}{1 - h_i},$$

the change in the intercept is

$$\hat{\beta}_0 - \hat{\beta}_0(i) = A_i \frac{r_i}{1 - h_i},$$

and the change in the predicted value of y_i, based on x_i, is

$$\hat{y}_i - \hat{y}_i(i) = \frac{h_i}{1 - h_i} r_i.$$

In particular, the bigger h_i happens to be, the more impact there is on the slope, the intercept, and the predicted value of y. Moreover, the change in $\hat{y}_i$ depends only on the residual, r_i, and h_i. But h_i reflects the amount x_i differs from the typical predictor value, $\bar{x}$, because the numerator of the second fraction in Eq. (10.6) is $(x_i - \bar{x})^2$. That is, h_i provides a measure of how much x_i influences the estimated regression equation, which is related to how far x_i is from $\bar{x}$, so h_i provides a method for judging whether x_i is unusually large or small relative to all the predictor values being used.

The result just described can be extended to the more general case where $p \geq 1$. Let

$$\mathbf{X} = \begin{pmatrix} 1 & x_{11} \cdots & x_{1p} \\ 1 & x_{21} \cdots & x_{2p} \\ \vdots & \vdots & \vdots \\ 1 & x_{n1} \cdots & x_{np} \end{pmatrix}.$$

Let $\hat{\beta}$ be the vector of OLS estimates, and let $\hat{\beta}(i)$ be the estimate when $(y_i, \mathbf{x}_i)$ is removed. Then the change in the estimates is

$$\hat{\beta} - \hat{\beta}(i) = (\mathbf{X}'\mathbf{X})^{-1}\mathbf{x}_i' \frac{r_i}{1 - h_{ii}},$$

where h_{ii} is the ith diagonal element of

$$\mathbf{H} = \mathbf{X}(\mathbf{X}'\mathbf{X})^{-1}\mathbf{X}'.$$

The matrix $\mathbf{H}$ is called the *hat matrix* because the vector of predicted y values is given by

$$\hat{\mathbf{y}} = \mathbf{H}\mathbf{y}.$$

In particular,

$$\hat{y}_i = \sum_{j=1}^{n} h_{ij} y_i.$$

In words, the predicted value of y, based on $\mathbf{x}_i$, is obtained by multiplying each element in the ith row of the hat matrix by y_i and adding the results. Furthermore, when $(y_i, \mathbf{x}_i)$ is removed, the vector of predicted values, $\hat{\mathbf{y}}$, is changed by

$$\hat{\mathbf{y}} - \hat{\mathbf{y}}(i) = \mathbf{X}(\mathbf{X}'\mathbf{X})^{-1}(1, x_{i1}, \ldots, x_{ip})' \frac{r_i}{1 - h_{ii}},$$

and the change in the ith predicted value, $\hat{y}_i$, is

$$\hat{y}_i - \hat{y}_i(i) = \frac{h_{ii}}{1 - h_{ii}} r_i.$$

The main point here is that the bigger h_{ii} happens to be, the more impact it has on the OLS estimator and the predicted values. In terms of identifying leverage points, Hoaglin and Welsch (1978) suggest regarding h_{ii} as being large if it exceeds $2(p+1)/n$. (For more information regarding the hat matrix, and the derivation of relevant results, see Belsley et al., 1980; Cook & Weisberg, 1992; Huber, 1981; Li, 1985; and Staudte and Sheather, 1990. The R function hat computes $h_{ii}, i = 1, \dots, n$.)

■ Example

Consider the observations

x	1	2	3	3	4	4	15	5	6	7
y	21	19	23	20	25	30	40	35	30	26.

The h_i values are 0.214, 0.164, 0.129, 0.129, 0.107, 0.107, 0.814, 0.100, 0.107, and 0.129, and $2(p+1)/n = 2(1+1)/10 = 0.4$. Because $h_7 = 0.814 > 0.4$, $x_7 = 15$ would be flagged as a leverage point. The OLS estimate of the slope is 1.43, but if the seventh point is eliminated, the estimate increases to 1.88.

■

There is an interesting connection between the hat matrix and the conditions under which the OLS estimator is asymptotically normal. In particular, a necessary condition for asymptotic normality is that $h_{ii} \to 0$ as $n \to \infty$ (Huber, 1981). A related result turns out to be relevant in the search for an appropriate M-estimator.

10.8 Generalized M-Estimators

This section takes up the problem of finding a regression estimator that guards against leverage points. A natural strategy is to attach some weight, w_i, to x_i with the idea that the more outlying or unusual x_i happens to be, relative to all the x_i values available, the less weight it is given. In general, it is natural to look for some function of the predictor values that reflects in some sense the extent to which the point x_i influences the OLS estimator. From the previous section, the leverage point, h_{ii}, is one way to measure how unusual x_i happens to be. Its value satisfies $0 < h_{ii} \le 1$. If $h_{ii} > h_{jj}$, this suggests that x_i is having a larger impact on the OLS estimator versus x_j.

However, there is a concern. Consider, for example, the simple regression model where $h_{ii} = 1/n + (x_i - \bar{x})^2 / \sum (x_i - \bar{x})^2$. The problem is that $(x_i - \bar{x})^2$ is not a robust measure of the extent to which $\mathbf{x}_i$ is an outlier. Practical problems do indeed arise, but there are some advantages to incorporating the leverage points in an estimation procedure, as will be seen.

One way of attaching a weight to $\mathbf{x}_i$ is with $w_i = \sqrt{1 - h_{ii}}$, which satisfies the requirement that if $h_{ii} > h_{jj}$, $w_i < w_j$. In other words, high leverage points get a relatively low weight. One specific possibility is to estimate the regression parameters as those values solving the $p + 1$ equations

$$\sum_{i=1}^{n} w_i \Psi(r_i/\hat{\tau}) x_{ij} = 0, \tag{10.7}$$

where again $x_{i0} = 1$ and $j = 0, \dots, p$. Equation (10.7) is generally referred to as an M-estimator using Mallows weights which derives its name from Mallows (1975).

Schweppe (see Hill, 1977) took this one step further with the goal of getting a more efficient estimator. The basic idea is to give more weight to the residual r_i if $\mathbf{x}_i$ has a relatively small weight, w_i. (Recall that outlying x values reduce the standard error of the OLS estimator.) Put another way, Mallows weights can result in a loss of efficiency if there are any outlying $\mathbf{x}$ values, and Schweppe's approach is an attempt at dealing with this problem by dividing r_i by w_i (Krasker & Welsch, 1982). The resulting M-estimator is now the solution to $p + 1$ equations

$$\sum_{i=1}^{n} w_i \Psi(r_i/(w_i \hat{\tau})) x_{ij} = 0. \tag{10.8}$$

$j = 0, \dots, p$. (For some technical details related to the choice of Ψ and w_i, see Hampel, 1968; Krasker, 1980; Krasker & Welsch, 1982.) Hill (1977) compared the efficiency of the Mallows and Schweppe estimators to several others and found that they dominate, with the Schweppe method having an advantage. Solving Eq. (10.8), which includes a choice for the measure of scale, τ, is based on a simple iterative procedure described in Table 10.1. For convenience, the estimator will be labeled $\hat{\beta}_m$.

An even more general framework is to consider $w_i = u(\mathbf{x}_i)$, where $u(\mathbf{x}_i)$ is some function of $\mathbf{x}_i$, chosen to supply some desirable property such as high efficiency. Two choices for $u(\mathbf{x}_i)$, discussed by Markatou and Hettmansperger (1990), are $\sqrt{1 - h_i}$ and $(1 - h_i)/\sqrt{h_i}$. The first choice, already mentioned, is due to Schweppe and was introduced in Handschin, Schweppe, Kohlas, and Fiechter (1975). The second choice is due to Welsch (1980).

In the context of testing hypotheses, and assuming ϵ has a symmetric distribution, Markatou and Hettmansperger (1990) recommend $w_i = (1 - h_i)/\sqrt{h_i}$. However, when ϵ has an asymmetric distribution, results in Carroll and Welsh (1988) indicate using $w_i = \sqrt{1 - h_i}$ or Mallows weights. The reason is that otherwise, under general conditions, the estimate of β is

Table 10.1: Iteratively Reweighted Least Squares for M Regression, $\hat{\beta}_m$.

To compute an M regression estimator with Schweppe weights, begin by setting $k = 0$ and computing the OLS estimate of the intercept and slope parameters, $\hat{\beta}_{0k}, \ldots, \hat{\beta}_{pk}$. Proceed as follows:

1. Compute the residuals, $r_{i,k} = y_i - \hat{\beta}_{0k} - \hat{\beta}_{1k}x_{i1} - \cdots - \hat{\beta}_{pk}x_{ip}$, let M_k be equal to the median of the largest $n - p$ of the $|r_{i,k}|$, $\hat{\tau}_k = 1.48M_k$, and let $e_{i,k} = r_{i,k}/\hat{\tau}_k$
2. Form weights,

$$w_{i,k} = \frac{\sqrt{1 - h_{ii}}}{e_{i,k}} \Psi\left(\frac{e_{i,k}}{\sqrt{1 - h_{ii}}}\right),$$

where

$$\Psi(x) = \max[-K, \min(K, x)]$$

is Huber's Ψ with $K = 2\sqrt{(p + 1)/n}$.

3. Compute the residuals, $r_{i,k} = y_i - \hat{\beta}_{0k} - \hat{\beta}_{1k}x_{i1} - \ldots - \hat{\beta}_{pk}x_{ip}$, let M_k be equal to the median of the largest $n - p$ of the $|r_{i,k}|$, $\hat{\tau}_k = 1.48M_k$, and let $e_{i,k} = r_{i,k}/\hat{\tau}_k$
4. Form weights,

$$w_{i,k} = \frac{\sqrt{1 - h_{ii}}}{e_{i,k}} \Psi\left(\frac{e_{i,k}}{\sqrt{1 - h_{ii}}}\right),$$

where

$$\Psi(x) = \max[-K, \min(K, x)]$$

is Huber's Ψ with $K = 2\sqrt{(p + 1)/n}$.

5. Use these weights to obtain a weighted least squares estimates, $\hat{\beta}_{0,k+1}, \ldots, \hat{\beta}_{p,k+1}$. Increase k by 1.
6. Repeat steps 1–3 until convergence. That is, iterate until the change in the estimated parameters is small.

not consistent, meaning that it does not converge to the correct value as the sample size gets large. In particular, if Eq. (10.8) is written in the more general form

$$\sum w_i \Psi(r_i/(u(\mathbf{x}_i)\hat{\tau}))x_{ij} = 0,$$

Carroll and Welsh (1988) show that if $u(\mathbf{x}_i) = 1$, which corresponds to using Mallows weights, the estimate is consistent, but it is not if $u(\mathbf{x}_i) \neq 1$. Thus, this suggests using $w_i = \sqrt{1 - h_i}$ versus $w_i = (1 - h_i)/\sqrt{h_i}$ because for almost all random sequences, $h_i \to 0$ as $n \to \infty$, for any i. In the one-predictor case, it is easy to see that $h_i \to 0$ if the x_i are bounded. In fact, as previously indicated, $h_i \to 0$ is a necessary condition for the least squares estimator to be asymptotically normal. (An example where h_i does not converge to zero is $x_i = 2^i$, as noted by Staudte & Sheather, 1990.) For completeness, it is pointed out that there are also *Mallows type estimators* where weights are given by the leverage points. In particular, perform weighted least squares with weights

$$w_i = \min\left[1, \left(\frac{b}{h_i}\right)^{j/2}\right],$$

where $b = h_{(mn)}$, $h_{(r)}$ is the rth ordered leverage value, and j and m are specified; see Hamilton (1992) as well as McKean, Sheather, and Hettmansperger (1993). The choices $j = 1, 2$, and 4 are sometimes labeled GMM1, GMM2, and GMM4 estimators. Little or nothing is known about how these estimators perform under heteroscedasticity, so they are not discussed further.

To provide at least some indication of the efficiency of $\hat{\beta}_m$, the M-estimator with Schweppe weights, suppose $n = 20$ observations are randomly sampled from the model $y_i = x_i + \lambda(x_i)\epsilon_i$ where both x_i and ϵ_i have standard normal distributions. First consider $\lambda(x_i) = 1$, which corresponds to the usual homoscedastic model, and suppose efficiency is measured with R, the estimated standard error of the OLS estimator divided by the estimated standard error of the M-estimator. Then $R < 1$ indicates that OLS is more efficient, and $R > 1$ indicates that the reverse is true. For the situation at hand, a simulation estimate of R, based on 1000 replications, is 0.89, so OLS gives better results. If instead $\lambda(x) = |x|$, meaning that the variance of y increases as x moves away from its mean, zero, $R = 1.09$. For $\lambda(x) = x^2$, $R = 1.9$, and for $\lambda(x) = 1 + 2/(|x| + 1)$, $R = 910$, meaning that the OLS estimator is highly unsatisfactory. In the latter case, the variance of y, given x, is relatively large when x is close to its mean.

Suppose instead ϵ has a symmetric heavy-tailed distribution (a g-and-h distribution with $g = 0$ and $h = 0.5$). Then the estimated efficiency, R, for the four λ functions considered here, are 3.02, 3.87, 6.0, and 226. Thus, for these four situations, OLS performs poorly, particularly for the last situation considered.

It should be noted that in some instances, R can be relatively large. Wilcox (1996d) reports additional values of R, again based on 1000 replications, but using a different set of random numbers (a different seed in the random number generator) yielding $R = 721$ versus 226 for $\lambda(x) = 1 + 2/(|x| + 1)$, and the distributions considered in the previous paragraph. Even if the expected value of R is 10, surely OLS is unsatisfactory. In particular, a large value for R reflects that OLS can yield a wildly inaccurate estimate of the slope when there is a heteroscedastic error term. For example, among the 1000 replications reported in Wilcox (1996d), where $\beta_1 = 1$ and $R = 721$, there is one case where the OLS estimate of the slope is $-3,140$.

If x has a heavy-tailed distribution, this will favor OLS if ϵ is normal and homoscedastic, but when ϵ is heteroscedastic, OLS can be less efficient than $\hat{\beta}_m$. Suppose for example that x has a g-and-h distribution with $g = 0$ and $h = 0.5$. Then estimates of R, for the four choices for λ

considered here, are 0.89, 1.45, 5.48, and 36. Evidently, there are situations where OLS is slightly more efficient than $\hat{\beta}_m$, but there are situations where $\hat{\beta}_m$ is substantially more efficient than OLS, so $\hat{\beta}_m$ appears to have considerable practical value.

10.8.1 R Function bmreg

The R function

$$\text{bmreg(x,y,iter=20,bend=2*sqrt((ncol(x)+1)/nrow(x))).}$$

computes the bounded influence M regression with Huber's Ψ and Schweppe weights using the iterative estimation procedure described in Table 10.1. The argument x is any n-by-p matrix of predictors. If there is only one predictor, x can be stored in an R vector as opposed to an R variable having matrix mode. The argument iter controls the maximum number of iterations allowed, which defaults to 20 if unspecified. The argument bend is K in Huber's Ψ which defaults to $2\sqrt{(p+1)/n}$. The estimate of the regression parameters is returned in bmreg$coef. The function also returns the residuals in bmreg$residuals, and the final weights (the w_i values) are returned in bmreg$w.

■ **Example**

The file read.dat, which accompanies the R functions written for this book, contains data from a reading study conducted by L. Doi. (See Section 1.8 for instructions on how to obtain this data.) One of the goals was to predict WWISST2, a word identification score (stored in column 8 of the file read.dat). One of the predictors is TAAST1, a measure of phonological awareness (stored in column 2). The OLS estimates of the slope and intercept are 1.72 and 73.56, respectively. The corresponding estimates returned by bmreg are 1.38 and 80.28.

■ **Example**

Although $\hat{\beta}_m$ offers more resistance than OLS, it is important to keep in mind that $\hat{\beta}_m$ might not be resistant enough. The star data in Figure 6.3 illustrate that problems might arise. The OLS estimates are $\hat{\beta}_1 = -0.41$ and $\hat{\beta}_0 = 6.79$. In contrast, if the data are stored in the R variables x and y, bmreg(x,y) returns $\hat{\beta}_1 = -0.1$ and $\hat{\beta}_0 = 5.53$. As is evident from Figure 6.3, $\hat{\beta}_m$ does a poor job of capturing the relationship among the majority of points, although it is less influenced by the outliers than is OLS. As noted in Chapter 6, there are several outliers, and there is the practical problem that more than one outlier can cause $\hat{\beta}_m$ to be misleading.

10.9 The Coakley–Hettmansperger and Yohai Estimators

The M-estimator $\hat{\beta}_m$, described in the previous section, has a bounded influence function, but a criticism is that its finite sample breakdown point is only $2/n$. That is, it can handle one outlier, but two outliers might destroy it. Coakley and Hettmansperger (1993) derived an M-estimator that has a breakdown point nearly equal to 0.5, a bounded influence function, and high asymptotic efficiency for the normal model. Their strategy is to start with the LTS estimator and then adjust it, the adjustment being a function of empirically determined weights. More formally, letting $\hat{\beta}_0$ (a vector having length $p+1$) be the LTS estimator, their estimator is

$$\hat{\beta}_{ch} = \hat{\beta}_0 + (\mathbf{X}'\mathbf{BX})^{-1}\mathbf{X}'\mathbf{W}\Psi(r_i/(w_i\hat{\tau}))\hat{\tau},$$

where $\mathbf{W} = \operatorname{diag}(w_i)$,

$$\mathbf{B} = \operatorname{diag}(\Psi'(r_i/\hat{\tau}w_i)),$$

$\mathbf{X}$ is the n-by-$(p+1)$ design matrix (described in Section 10.7), and $\Psi'(x)$ is the derivative of Huber's Ψ. They suggest using $K = 1.345$ in Huber's Ψ if uncertain about which value to use, so this value is assumed here unless stated otherwise. As an estimate of scale, they use $\hat{\tau} = 1.4826(1 + 5/(n-p)) \times \operatorname{med}\{|r_i|\}$. The weight given to $\mathbf{x_i}$, the ith row of predictor values, is

$$w_i = \min\{1, [b/(\mathbf{x_i} - \mathbf{m_x})'\mathbf{C}^{-1}(\mathbf{x_i} - \mathbf{m_x})]^{a/2}\},$$

where the quantities $\mathbf{m_x}$ and $\mathbf{C}$ are the minimum volume ellipsoid (MVE) estimators of location and covariance associated with the predictors. (See Section 6.3.1.) These estimators have a breakdown point approximately equal to 0.5. When computing w_i, Coakley and Hettmansperger suggest setting b equal to the 0.95 quantile of a chi-squared distribution with p degrees of freedom and using $a = 2$.

The Coakley–Hettmansperger regression method has high asymptotic efficiency when the error term, ϵ, has a normal distribution. That is, as the sample size gets large, the standard error of the OLS estimator will not be substantially smaller than the standard error of $\hat{\beta}_{ch}$. To provide some indication of how the standard error of $\hat{\beta}_{ch}$ compares to the standard error of OLS when n is small, consider $p = 1$, $a = 2$, $K = 1.345$, and suppose both x and ϵ have standard normal distributions. When there is homoscedasticity, R, the standard error of OLS divided by the standard error of $\hat{\beta}_{ch}$, is 0.63 based on a simulation with 1000 replications. If x has a symmetric, heavy-tailed distribution instead (a g-and-h distribution with $g = 0$ and $h = 0.5$), then $R = 0.31$. In contrast, if $\hat{\beta}_m$ is used, which is computed as described in Table 10.1, the ratio is 0.89 for both of the situations considered here. Of course, the lower efficiency of $\hat{\beta}_{ch}$ must be weighed against the lower breakdown point of $\hat{\beta}_m$. Moreover, if x has a light-tailed distribution, and the distribution of ϵ is heavy-tailed, $\hat{\beta}_{ch}$ offers an advantage over OLS. For example, if x is standard normal, but ϵ has a g-and-h distribution with $g = 0$ and $h = 0.5$, $R = 1.71$. However, replacing $\hat{\beta}_{ch}$ with $\hat{\beta}_m$, $R = 3$.

It might be thought that the choice for the bending constant in Huber's Ψ, $K = 1.345$, is partly responsible for the low efficiency of $\hat{\beta}_{ch}$. If $\hat{\beta}_{ch}$ is modified by increasing the bending constant to $K = 2$, $R = 0.62$ when both x and ϵ are standard normal. If x has the symmetric, heavy-tailed distribution considered in the previous paragraph, again $R = 0.31$. It appears that $\hat{\beta}_{ch}$ is inefficient because of its reliance on LTS regression, and because it does not take sufficient advantage of good leverage points.

10.9.1 MM-Estimator

Yet another robust regression estimator that should be mentioned is the MM-estimator derived by Yohai (1987), which has certain similarities to the generalized M-estimators in Section 10.8. It has the highest possible breakdown point, 0.5, and high efficiency under normality. The parameters are estimated by solving an equation similar to Eq. (10.8), with Ψ taken to be some redescending function. A popular choice is Tukey's biweight, given in Table 2.1, which will be used here unless stated otherwise. That is, the regression parameters are estimated by determining the solution to $p + 1$ equations

$$\sum_{i=1}^{n} \Psi(r_i/\hat{\tau})x_{ij} = 0. \tag{10.9}$$

$j = 0, \ldots, p$, where again $\hat{\tau}$ is a robust measure of variation based on the residuals and

$$\Psi(r_i; c) = \frac{r_i}{\hat{\tau}}\left[\left(\frac{r_i}{c\hat{\tau}}\right)^2 - 1\right]^2, \quad \text{if } |r_i/\hat{\tau}| \leq c;$$

otherwise $\Psi(r_i; c) = 0$. The choice $c = 4.685$ leads to an MM-estimator with 95% efficiency compared to the least squares estimator and is the default value used here. The ratio p/n is relevant to the efficiency of this estimator; see Maronna and Yohai (2010) for details. In addition to having excellent theoretical properties, the small-sample efficiency of the MM-estimator appears to compare well with other robust estimators, but like several other robust estimators it can be sensitive to what is called contamination bias, as described in Section 10.14.1. Also, situations are encountered where the iterative estimation scheme used to compute the MM-estimator does not converge. For an extension of this estimator, in the context of ridge regression, see Maronna (2011).

The R function lmrob, which can be accessed via the R package robustbase, can be used to compute confidence intervals and test hypotheses when using the MM-estimator. The function computes estimates of the standard errors and assumes the null distribution of the test statistics has, approximately, a Student's t-distribution with $n - p - 1$ degrees of freedom.

However, even under normality and homoscedasticity, control over the probability of a type I error is poor when n is small. When testing at the 0.05 level, the actual level exceeds 0.05 by a substantial amount. With $n = 100$ it performs reasonably well, still assuming normality and homoscedasticity. Evidently there are no results regarding how well the method performs under nonnormality and heteroscedasticity. (For a recent extension of this estimator, see Koller & Stahel, 2011.) If there is interest in testing hypotheses based on the MM-estimator, currently the best approach appears to be to use the percentile bootstrap methods in Sections 11.1.1 and 11.1.3.

For completeness, it is noted that there are additional M regression methods not covered in this chapter (e.g., Jureckova & Portnoy, 1987).

10.9.2 R Functions chreg and MMreg

The R function

$$\text{chreg(x,y,bend=1.345)}.$$

computes the Coakley–Hettmansperger regression estimator. As will all regression functions, the argument x can be a vector or an n-by-p matrix of predictor values. The argument bend is the value of K used in Huber's Ψ.

The R function

$$\text{MMreg(x,y)}$$

computes Yohai's MM-estimator, assuming that the R package robustbase has been installed.

■ **Example**

If the star data in Figure 6.3 are stored in the R variables starx and stary, the command

$$\text{chreg(starx,stary)}$$

returns an estimate of the slope equal to 4.0. The R function MMreg estimates the slope to be 2.25. In contrast, the OLS estimate is -0.41, and the M-regression estimate, based on the method in Table 10.1, yields $\hat{\beta}_m = -0.1$.

  ■

10.10 Skipped Estimators

Skipped regression estimators generally refer to the strategy of checking for outliers using one of the multivariate outlier detection methods in Chapter 6, discarding any that are found, and applying some estimator to the data that remain. A natural issue is whether the ordinary least squares estimator might be used after outliers are removed, but checks on this approach by the author found that the small-sample efficiency of this method is rather poor, compared to other estimators, when the error term is normal and homoscedastic.

Consider the goal of achieving good small-sample efficiency, relative to OLS, when the error term is both normal and homoscedastic, and $\mathbf{x}'_i$ is multivariate normal, while simultaneously providing protection against outliers. Then a relatively effective method is to first check for outliers among the $(p+1)$-variate data $(\mathbf{x}'_i, y_i)$ $(i = 1, \ldots, n)$ using the MGV or projection method (described in Sections 6.4.7 and 6.4.9), remove any outliers that are found, and then apply the Theil–Sen estimator to the data that remain. Replacing the Theil–Sen estimator with one of the M-estimators previously described has been found to be rather unsatisfactory. (Checks on using the MM-estimator have not been made.) Generally, when using the MGV outlier detection in conjunction with any regression estimator, this will be called an *MGV estimator*. Here, it is assumed that the Theil–Sen estimator is used unless stated otherwise. When using the projection method for detecting outliers, this will be called an *OP estimator*, and again the Theil–Sen estimator is assumed.

10.10.1 R Functions mgvreg and opreg

The R function

$$\text{mgvreg(x,y,regfun=tsreg,outfun=outbox)}$$

computes a skipped regression estimator where outliers are identified and removed using the MGV outlier detection method in Section 6.4.7. The argument outfun controls the decision rule used when checking for outliers. The default is a boxplot rule based on Eq. (6.18). Setting outfun=out results in using the MAD-median rule corresponding to Eq. (6.20). Once outliers are eliminated, the regression estimator indicated by the argument regfun is applied and defaults to the Theil–Sen estimator.

The R function

$$\text{opreg(x,y,regfun=tsreg,cop=3,MC=F)}$$

is like mgvreg, only the OP outlier detection method in Section 6.4.9 is used. The argument cop determines the measure of location used when checking for outliers using the projection method in Section 6.4.9. (See Section 6.4.10 for details about the argument cop.) Unlike mgvreg, opreg can take advantage of a multi-core processor by setting the argument MC=T.

10.11   Deepest Regression Line

Rousseeuw and Hubert (1999) derived a method of fitting a line to data that searches for the deepest line embedded within a scatterplot. First consider simple regression. Let b_1 and b_0 be any choice for the slope and intercept, respectively, and let r_i $(i = 1, \ldots, n)$ be the corresponding residuals. The candidate fit, (b_0, b_1), is called a nonfit if a partition of the x values can be found such that all of the residuals for the lower x values are negative (positive), but for all of the higher x values the residuals are positive (negative). So, for example, if all of the points lie above a particular straight line, in which case all of the residuals are positive, this line is called a nonfit. More formally, a candidate fit is called a nonfit if and only if a value for v can be found such that

$$r_i < 0 \quad \text{for all } x_i < v$$

and

$$r_i > 0 \quad \text{for all } x_i > v$$

or

$$r_i > 0 \quad \text{for all } x_i < v$$

and

$$r_i < 0 \quad \text{for all } x_i > v.$$

The regression depth of a fit (b_1, b_0), relative to $(x_1, y_1), \ldots, (x_n, y_n)$, is the smallest number of observations that need to be removed to make (b_1, b_0) a nonfit. The deepest regression estimator corresponds to the values of b_1 and b_0 that maximize regression depth. Bai and He (1999) derived the limiting distribution of this estimator. When the $\mathbf{x}_i$ are distinct, the breakdown point is approximately 0.33.

The idea can be extended to multiple predictors. Let $r_i(b_0, \ldots, b_p)$ be the ith residual based on the candidate fit $b_0, \ldots, b_p$. This candidate fit is called a nonfit if there exists a hyperplane V in $\mathbf{x}$ space, such that no $\mathbf{x}_i$ belongs to V and such that $r_i(b_0, \ldots, b_p) > 0$ for all $\mathbf{x}_i$ in one of the open halfspaces corresponding to V and $r_i(b_0, \ldots, b_p) < 0$ in the other open halfspace. Regression depth is defined as in the $p = 1$ case.

10.11.1   R Function mdepreg

The R function

$$\text{mdepreg(x,y)}$$

computes the deepest regression line. For $p > 1$ predictors it uses an approximation of the depth of a hyperplane.

10.12 A Criticism of Methods with a High Breakdown Point

It seems that no single method is free from criticism, and regression methods that have a high breakdown point are no exception. A potential problem with these methods is that standard diagnostic tools for detecting curvature, by examining the residuals, might fail (Cook, Hawkins, & Weisberg, 1992; McKean et al., 1993). Thus, it is recommended that if a regression method with a high breakdown point is used, possible problems with curvature be examined using some alternative technique. Some possible ways of checking for curvature, beyond the standard methods covered in an introductory regression course, are described in Chapter 11.

10.13 Some Additional Estimators

Some additional regression estimators should be outlined. Although some of these estimators have theoretical properties that do not compete well with some of the estimators already described, they might have practical value. For instance, some of the estimators listed here have been suggested as an initial estimate that is refined in some manner or is used as a preliminary screening device for detecting regression outliers. Graphical checks suggest that these estimators sometimes provide a more reasonable summary of the data versus other estimators covered in the previous section. Also, it is not being suggested that by listing an estimator in this section, it necessarily should be excluded from consideration in applied work. In some cases, certain comparisons with estimators already covered, such as efficiency under heteroscedasticity, have not been explored. (For an extension to the so-called general linear model, see Cantoni & Ronchetti, 2001.)

10.13.1 S-Estimators and τ-Estimators

S-estimators of regression parameters, proposed by Rousseeuw and Yohai (1984), search for the slope and intercept values that minimize some measure of scale associated with the residuals. Least squares, for example, minimizes the variance of the residuals and is a special case of S-estimators. The hope is that by replacing the variance with some measure of scale that is relatively insensitive to outliers, we will obtain estimates of the slope and intercept that are relatively insensitive to outliers as well. As noted in Chapter 3, there are many measures of scale. The main point is that if, for example, we use the percentage bend midvariance (described in Section 3.12.3), situations arise where the resulting estimate of the slope and intercept has advantages over other regression estimators we might use. This is not to say that other measures of scale never provide a more satisfactory estimate of the regression parameters. But for general use, it currently seems that the percentage bend midvariance is a good choice. For relevant asymptotic results, see Davies (1990). Hössjer (1992) showed that S-estimators cannot achieve simultaneously both a high breakdown point and a high efficiency under the normal model. Also, Davies (1993) reports results on the inherit instability of S-estimators. Despite this, it may have practical value as preliminary fit to data; see Section 10.13.3.

Here a simple approximation of the S-estimator is used. (There are other ways of computing S-estimators, e.g., Croux et al., 1994; Ferretti, Kelmansky, Yohai, & Zamar, 1999, perhaps they have practical advantages, but it seems that this possibility has not been explored.) Let

$$R_i = y_i - b_1 x_{1i} - \cdots b_p x_{ip},$$

and use the Nelder–Mead method (mentioned in Chapter 6) to find the values $b_1, \ldots, b_p$ that minimize S, some measure of scale based on the values $R_1, \ldots, R_n$. The intercept is taken to be

$$b_0 = M_y - b_1 M_1 \cdots - b_p M_p,$$

where M_j and M_y are the medians of the x_{ij} ($i = 1, \ldots, n$) and y values, respectively. This will be called method *SNM*. (Again, for details motivating the Nelder–Mead method, see Olsson & Nelson, 1975.)

A related approach in the one-predictor case, called the *STS* estimator, is to compute the slope between points j and j', $S_{jj'}$ and take the estimate of the slope to be the value of $S_{jj'}$ that minimizes some measure of scale applied to the values $Y_1 - S_{jj'} X_1, \ldots, Y_n - S_{jj'} X_n$ values. Here, the back-fitting method is used to handle multiple predictors.

For completeness, τ-estimators proposed by Yohai and Zamar (1988) generalize S-estimators by using a broader class of scale estimates. Gervini and Yohai (2002, p. 584) note that tuning these estimators for high efficiency will result in an increase in bias. An extension of τ-estimators is the class of generalized τ-estimators proposed by Ferretti et al. (1999). Briefly, residuals are weighted, with high leverage values resulting small weights, and a measure of scale based on these weighted residuals is used to judge a fit to data. For recent results on computing τ-estimators, see Flores (2010).

10.13.2 R Functions snmreg and stsreg

The R function

snmreg(x,y)

computes the SNM estimate as just described. The measure of scale, S, is taken to be the percentage bend midvariance. (When using the Nelder–Mead method, the initial estimate of the parameters is based on the Coakley–Hettmansperger estimator.) The R function

stsreg(x,y,sc=pbvar)

computes the STS estimator, where sc is the measure of scale to be used. By default, the percentage bend midvariance is used.

10.13.3 E-Type Skipped Estimators

Skipped estimators remove any outliers among the cloud of data $(x_{i1}, \ldots, x_{ip}, y_i)$, $i = 1, \ldots n$, and then fit a regression line to the data that remain. E-type skipped estimators (where E stands for error term) look for outliers among the residuals based on some preliminary fit, remove (or downweight) the corresponding points, and then compute a new fit to the data. Rousseeuw and Leroy (1987) suggested using least median of squares (LMS) to obtain an initial fit, remove any points for which the corresponding standardized residuals are large, and then apply least squares to the data that remain. But He and Portnoy (1992) showed that the asymptotic efficiency is 0.

Another E-type skipped estimator is to apply one of the outlier detection methods in Chapter 3 to the residuals. For example, first fit a line to the data using the the STS estimator described in Section 10.13.1. Let r_i $(i = 1, \ldots, n)$ be the usual residuals. Let M_r be the median of the residuals and let MAD_r be the median of the values $|r_1 - M_r|, \ldots, |r_n - M_r|$. Then the ith point (x_i, y_i) is declared a regression outlier if

$$|r_i - M_r| > \frac{2(\text{MAD}_r)}{0.6745}. \tag{10.10}$$

The final estimate of the slope and intercept is obtained by applying the Theil–Sen estimator to those points not declared regression outliers. When there are p predictors, again compute the residuals based on STS and use Eq. (10.10) to eliminate any points with large residuals. This will be called method *TSTS*.

Another variation, called an *adjusted M-estimator*, is to proceed as in Table 10.1, but in step 2, set $w_{i,k} = 0$ if $(y_i, \mathbf{x}_i)$ is a regression outlier based, for example, on the regression outlier detection method in Rousseeuw and van Zomeren (1990); see Section 10.15. This estimator will be labeled $\hat{\beta}_{\text{ad}}$.

Gervini and Yohai (2002) used an alternative approach for determining whether any residuals are outliers and they derived some general theoretical results for this class of estimators. In particular, they describe conditions under which the asymptotic breakdown point is not less than the initial estimator, and they find that good efficiency is obtained under normality and homoscedasticity.

The method begins by obtaining an initial fit to the data; Gervini and Yohai focus on LMS and S-estimators to obtain this initial fit. Then the absolute value of the residuals are checked for outliers, any such points are eliminated, and the least squares estimator is applied to the remaining data. But unlike other estimators in this section, an adaptive method for detecting outliers, which is based on the empirical distribution of the residuals, is used to detect outliers. An interesting result is that under general conditions, if the errors are normally distributed, the estimator has full asymptotic efficiency.

To outline the details, let R_i be the residuals based on an initial estimator, and let

$$r_i = \frac{R_i}{S}$$

be the standardized residuals, where S is some measure of scale applied to the R_i values. Following Gervini and Yohai, S is taken to be MADN (MAD divided by 0.6745). Let $|r|_{(1)} \le \cdots \le |r|_{(n)}$ be the absolute values of the standardized residuals written in ascending order and let $i_0 = \max\{|r|_{(i)} < \eta\}$ for some constant η; Gervini and Yohai use $\eta = 2.5$. Let

$$d_n = \max\left\{\Phi(|r|_{(i)}) - \frac{i-1}{n}\right\},$$

where the maximum is taken over all $i > i_0$, and where Φ is the cumulative standard normal distribution. In the event $d_n < 0$, set $d_n = 0$, and let $i_n = n - [d_n]$, where $[d_n]$ is the greatest integer less than or equal to d_n. The point corresponding to $|r|_{(i)}$ is eliminated (is given zero weight) if $i > i_n$, and the least squares estimator is applied to the data that remain.

For the situations in Table 10.2, the Gervini–Yohai estimator does not compete well in terms of efficiency when using LMS regression as the initial estimate when checking for regression outliers. Switching to the LTS estimator as the initial estimator does not improve efficiency for the situations considered. (Also see Section 10.14.)

10.13.4 R Functions mbmreg, tstsreg, and gyreg

The function

$$tstsreg(x,y,sc=pbvar,...)$$

computes the E-type estimator described in Section 10.13.3. The argument sc indicates which measure of scale will be used when method STS is employed to detect regression outliers, and the default measure of scale is the percentage bend midvariance. The function

$$mbmreg(x,y,iter = 20, bend = (2 * sqrt(ncol(x) + 1))/nrow(x)))$$

computes the so-called adjusted M-estimator. Finally, the function

$$gyreg(x,y,rinit = lmsreg, K = 2.5)$$

computes the Gervini–Yohai estimator where the argument rinit indicates which initial estimator will be used to detect regression outliers. By default, the LMS estimator is used. The argument K corresponds to η.

10.13.5 Methods Based on Robust Covariances

A general approach to regression, briefly discussed by Huber (1981), is based on estimating a robust measure of covariance which in turn can be used to estimate the parameters of the model. For the one-predictor case, the slope of the OLS regression line is

$$\beta_1 = \frac{\sigma_{xy}}{\sigma_x^2},$$

where σ_{xy} is the usual covariance between x and y. This suggests estimating the slope with

$$\hat{\beta}_1 = \frac{\hat{\tau}_{xy}}{\hat{\tau}_x^2},$$

where $\hat{\tau}_{xy}$ estimates τ_{xy}, some measure of covariance chosen to have good robustness properties, and $\hat{\tau}_x^2$ is an estimate of some measure of scale. The intercept can be estimated with

$$\hat{\beta}_0 = \hat{\theta}_y - \hat{\beta}_1 \hat{\theta}_x$$

for some appropriate estimate of location, θ. As noted in Chapter 9, there are many measures of covariance. Here, the biweight midcovariance is employed, which is described in Section 9.3.8 of Chapter 9 and is motivated in part by results in Lax (1985) who found that the biweight midvariance is relatively efficient.

For the more general case where $p \geq 1$, let

$$\mathbf{A} = (s_{byx_1}, \ldots, s_{byx_p})'$$

be the vector of sample biweight midcovariances between y and the p predictors. The quantity s_{byx_j} estimates the biweight midcovariance between y and the jth predictor, x_j. Let

$$\mathbf{C} = (s_{bx_j x_k})$$

be the p-by-p matrix of estimated biweight midcovariances among the p predictors. By analogy with OLS, the regression parameters $(\beta_1, \ldots, \beta_p)'$ are estimated with

$$(\hat{\beta}_1, \ldots, \hat{\beta}_p)' = \mathbf{C}^{-1} \mathbf{A}, \tag{10.11}$$

and an estimate of the intercept is

$$\hat{\beta}_0 = \hat{\mu}_{my} - \hat{\beta}_1 \hat{\mu}_{m1} - \cdots - \hat{\beta}_p \hat{\mu}_{mp}, \tag{10.12}$$

where $\hat{\mu}_{my}$ is taken to be the one-step M-estimator based on the y values, and $\hat{\mu}_{mj}$ is the one-step M-estimator based on the n values corresponding to the jth predictor.

The estimation procedure just described performs reasonably well when there is a homoscedastic error term, but it can give poor results when the error term is heteroscedastic.

For example, Wilcox (1996e) reports a situation where the error term is heteroscedastic, $\beta_1 = 1$, yet with $n = 100,000$, the estimated slope is approximately equal to 0.5. Apparently the simple estimation procedure described in the previous paragraph is not even consistent when the error term is heteroscedastic. However, a simple iterative procedure corrects this problem.

Set $k = 0$, let $\hat{\beta}_k$ be the $p + 1$ vector of estimated slopes and intercept using Eqs (10.10) and (10.11), and let r_{ki} be the resulting residuals. Let $\hat{\delta}_k$ be the $p + 1$ vector of estimated slopes and intercept when using $\mathbf{x}$ to predict the residuals. That is, replace y_i with r_{ki} and compute the regression slopes and intercepts using Eqs (10.10) and (10.11). Then an updated estimate of β is

$$\hat{\beta}_{k+1} = \hat{\beta}_k + \hat{\delta}_k, \tag{10.13}$$

and this process can be repeated until convergence. That is, compute a new set of residuals using the estimates just computed, increase k by 1, and use Eqs (10.10) and (10.11) to get a new adjustment, $\hat{\delta}_k$. The iterations stop when all of the $p + 1$ values in $\hat{\delta}_k$ are close to zero, say within .0001. The final estimate of the regression parameters is denoted by $\hat{\beta}_{\mathrm{mid}}$.

It is noted that for certain measures of scatter, this iteration scheme does not appear to be necessary. For example, Zu and Yuan (2010) used the multivariate measure of scatter derived by Maronna (1976), and checks on this approach indicate that when there is heteroscedasticity, using Eqs (10.10) and (10.11) suffices.

A concern about some robust covariances is that they do not take into account the overall structure of the data and so might be influenced by properly placed outliers. One could use the skipped correlations described in Chapter 6 for the situation at hand, but there are no results regarding the small-sample properties of this approach.

Another approach to estimating regression parameters is to replace the biweight midcovariance with the Winsorized covariance in the biweight midregression method. This will be called *Winsorized regression*. An argument for this approach is that the goal is to estimate the Winsorized mean of y, given x, and the Winsorized mean satisfies the Bickel–Lehmann condition described in Chapter 2. In terms of probability coverage, it seems that there is little or no reason to prefer Winsorized regression over the biweight midregression procedure.

It should be noted that Srivastava, Pan, Sarkar, and Mudholkar (2010) report results on what they call Winsorized regression where all variables are Winsorized as described in Section 9.3.6, after which they apply the usual least squares estimator using the Winsorized values. They demonstrate via simulations that the resulting estimator can be more efficient than least squares. Evidently there are no results on how the efficiency of this estimator compares to other robust estimators in this chapter. And no results were reported regarding the effects of heteroscedasticity.

Table 10.2: Estimates of R Using Covariance Methods, $n = 20$.

VP	x and ϵ normal			x normal, ϵ heavy-tailed			x heavy-tailed, ϵ normal		
	b_{mid}	$\gamma = 0.1$	$\gamma = 0.2$	b_{mid}	$\gamma = 0.1$	$\gamma = 0.2$	b_{mid}	$\gamma = 0.1$	$\gamma = 0.2$
1	0.94	0.92	0.81	1.10	2.55	2.64	0.61	0.78	0.57
2	1.80	1.69	2.17	2.28	4.49	6.46	24.41	9.08	19.58
3	18.25	13.82	10.26	13.79	9.62	9.64	13.20	2.83	3.57

To provide some indication of how the efficiency of the biweight midregression and Winsorized regression methods compare to OLS regression, Table 10.2 shows estimates of R, the standard error of OLS regression divided by the standard error of the competing method. The columns headed by $\gamma = 0.1$ are the values when 10% Winsorization is used, and $\gamma = 0.2$ is 20%. These estimates correspond to three types of error terms: $\lambda(x) = 1$, $\lambda(x) = x^2$, and $\lambda(x) = 1/|x|$. For convenience, these three choices are labeled VP1, VP2, and VP3. In general, the biweight and Winsorized regression methods compare well to OLS, and in some cases they offer a substantial advantage. Note, however, that when x has a heavy-tailed distribution, and ϵ is normal, OLS offers better efficiency when the error term is homoscedastic. In some cases, $\hat{\beta}_{mid}$ has better efficiency versus $\hat{\beta}_{ad}$, but in other situations the reverse is true.

10.13.6 R Functions bireg, winreg, and COVreg

The R function

$$\text{bireg(x,y,iter=20,bend=1.28)}$$

is supplied for performing the biweight midregression method just described. As usual, x can be a vector when dealing with simple regression ($p = 1$), or it is an n-by-p matrix for the more general case where there are $p \geq 1$ predictors. The argument iter indicates the maximum number of iterations allowed. It defaults to 20 which is more than sufficient for most practical situations. If convergence is not achieved, the function prints a warning message. The argument bend is the bending constant, K, used in Huber's Ψ when computing the one-step M-estimator. If unspecified, $K = 1.28$ is used. The function returns estimates of the coefficients in bireg$coef, and the residuals are returned in bireg$resid.

The R function

$$\text{winreg(x,y,iter=20,tr=0.2)}$$

performs Winsorized regression where tr indicates the amount of Winsorizing, which defaults to 20%. Again, iter is the maximum number of iterations allowed, which defaults to 20.

The R function

$$\text{COVreg(x,y,cov.fun=MARest,loc.fun=MARest,xout=F,outfun=out,...)}$$

estimates the slopes and intercept via Eqs (10.11) and (10.12) without iterations and defaults to using Marrona's M-estimator in Section 6.3.13.

■ **Example**

For the star data shown in Figure 6.3 of Chapter 6, bireg estimates the slope and intercept to be 2.66 and −6.7, respectively. The OLS estimates are −0.41 and 6.79. The function winreg estimates the slope to be 0.31 using 10% Winsorization (tr=0.1), and this is considerably smaller than the estimate of 2.66 returned by bireg. Also, winreg reports a warning message that convergence was not obtained in 20 iterations. This problem seems to be very rare. Increasing iter to 50, convergence is obtained, and again the slope is estimated to be 0.31, but the estimate of the intercept drops from 3.62 to 3.61. Using the default 20% Winsorization (tr=0.2), the slope is now estimated to be 2.1 and convergence problems are eliminated.

■

10.13.7 L-Estimators

A reasonable approach to regression is to compute some initial estimate of the parameters, compute the residuals, and then re-estimate the parameters based in part on the trimmed residuals. This strategy was employed by Welsh (1987a,b) and expanded upon by De Jongh, De Wet, and Welsh (1988). The small-sample efficiency of this approach does not compare well with other estimators such as $\hat{\beta}_m$ or M regression with Schweppe weights (Wilcox, 1996d). In terms of achieving high efficiency when there is heteroscedasticity, comparisons with the better estimators in this chapter have not been made. So, even though the details of the method are not described here, it is not being suggested that Welsh's estimator be abandoned.

10.13.8 L_1 and Quantile Regression

Yet another approach is to estimate the regression parameters by minimizing $\sum |r_i|$, the so-called L_1 norm, which is just the sum of the absolute values of the residuals. This approach predates OLS by 50 years. This is, of course, a special case of the LTA estimator in Section 10.5. The potential advantage of L_1 (or least absolute value) regression over OLS, in terms of efficiency, was known by Laplace (1818). The L_1 approach reduces the influence of outliers, but the breakdown point is still $1/n$. More precisely, L_1 regression protects against unusual y values, but not leverage points, which can have a large influence on the fit to data.

Another concern is that a relatively large weight is being given to observations with the smallest residuals (Mosteller & Tukey, 1977, p. 366). For these reasons, further details are omitted. (For a review of L_1 regression, see Narula, 1987, as well as Dielman & Pfaffenberger, 1982.) Hypothesis testing procedures are described by Birkes and Dodge (1993), but no results are given on how the method performs when the error term is heteroscedastic.

An interesting generalization of the L_1 estimator was proposed by Koenker and Bassett (1978), which is aimed at estimating the qth quantile of y given x. Let

$$\rho_q(u) = u(q - I_{u<0}),$$

where I is the indicator function. Then the regression line is determined by minimizing

$$\sum \rho_q(r_i).$$

So $q = 0.5$ corresponds to the least absolute value (or L_1) estimator and yields an estimate of the median of y, given x.

A (rank inversion) method for testing hypotheses about the individual parameters was studied by Koenker (1994) and can be applied via the R package quantreg. (Also see Koenker & Xiao, 2002.) The method is limited to testing at the $\alpha = 0.05$ level. The R function rqfit, described in the next section, applies the method. When the goal is to test at some other level, or if an omnibus test is to be performed, see Section 11.1.6.

10.13.9 R Functions qreg and rqfit

The R function

$$\text{qreg(x,y,qval=0.5)}$$

computes the Koenker–Bassett quantile regression estimator. The argument qval determines the quantile to be used.

The R function

$$\text{rqfit(x,y,qval=0.5,alpha=0.05,xout=F,outfun=out,res=T,...)}$$

tests hypotheses about the individual parameters. As usual, setting the argument xout=T eliminates leverage points. If the argument alpha is not equal to 0.05, an error message is printed indicating that the R function qregci (described in Section 11.1.6) should be used.

10.13.10 Methods Based on Estimates of the Optimal Weights

It is well known that when using weighted least squares, the optimal (most efficient) method of estimating regression parameters, when there is heteroscedasticity, is to use weights $w_i = 1/\sigma_i^2$, where σ_i^2 is the variance of ϵ_i. Several estimation procedures have been designed to handle heteroscedastic error terms based on this result. One such procedure was proposed by Cohen et al. (1993). The general strategy is to start with some robust estimator and then use the residuals to estimate appropriate weights based on an estimate of how the variance of ϵ_i varies with the predictor. Only a single predictor has been considered so far. The method can have high efficiency compared to OLS, and efficiency is very close to OLS when both x and ϵ have standard normal distributions. It remains unclear, in terms of efficiency, whether the method ever offers a practical advantage over various alternative estimators covered here. The method might be particularly effective when the predictor values are fixed and evenly spaced. Also, unlike many robust estimators, the percentile bootstrap method does not provide reasonably accurate confidence intervals for the parameters when n is small (Wilcox, 1996d). Because the practical value of the method needs more research, the lengthy computational details are not given here.

Wilcox (1996d) suggests a method of estimating σ_i^2 using a running interval smoother. (Smoothers are described in Chapter 11.) The efficiency of the resulting estimator compares well to OLS, and in various situations it offers a substantial advantage. At the moment, there seems to be little practical advantage to using this approach over other robust estimators that have good efficiency under heteroscedasticity.

Robinson (1987) suggests yet another estimator that uses a smoother to estimate σ_i^2. All indications are that it offers little advantage over OLS (Wilcox, 1996d), so no details are given. For a method based on the assumption that σ_i is given by some *known* function depending of x, β_1 and perhaps some additional unknown parameters, see Carroll and Ruppert (1982). For results on how the method performs when the function is incorrectly specified, see Mak (1992).

10.13.11 Projection Estimators

Maronna and Yohai (1993) derived yet another regression estimator called *projection regression*. Let $T(\mathbf{x}, y)$ be any estimating functional through the origin that is scale and affine equivariant. Let $s(\lambda'\mathbf{x})$ be a measure of scale based on the projection $\lambda'\mathbf{x}$. For any vectors β and λ having length p, let

$$A(\beta, \lambda) = |T(\lambda'\mathbf{x}, y - \beta\lambda')| s(\lambda'\mathbf{x}).$$
$$C(\beta) = \sup A(\beta, \lambda),$$

where the supremum is taken over all λ satisfying $\|\lambda\| = 1$. The projection estimate is the vector β that minimizes $C(\beta)$. Several variations of this estimator are considered by Maronna and Yohai. The variation given by their Eq. (3.11) was considered here but found to have relatively unsatisfactory efficiency under heteroscedasticity. Other variations have not been considered.

10.13.12 Methods Based on Ranks

Naranjo and Hettmansperger (1994) suggest estimating regression coefficients by minimizing

$$\sum_{i<j} c_{ij}|r_i - r_j|,$$

where the c_{ij} are (Mallows) weights given by $c_{ij} = h(\mathbf{x}_i)h(\mathbf{x}_j)$,

$$h(\mathbf{x}_i) = \min\{1, [c/(\mathbf{x_i} - \mathbf{m_x})'\mathbf{C}^{-1}(\mathbf{x_i} - \mathbf{m_x})]^{a/2}\}.$$

Letting $d_i = (\mathbf{x_i} - \mathbf{m_x})'\mathbf{C}^{-1}(\mathbf{x_i} - \mathbf{m_x})$, they suggest using $c = \text{med}\{d_i\} + 3\text{MAD}\{d_i\}$, where $\text{MAD}\{d_i\}$ means that MAD is computed using the d_i values, and med is the median. They report that $a = 2$ is effective in uncovering outliers. The quantities $\mathbf{m_x}$ and $\mathbf{C}$ are the minimum volume ellipsoid estimators of location and scale described in Chapter 6. When $c_{ij} \equiv 1$, their method reduces to Jaeckel's (1972) method, which minimizes a sum that is a function of the ranks of the residuals. For the one-predictor case they replace the minimum volume ellipsoid estimators $\mathbf{m_x}$ and $\mathbf{C}$ with $M_x = \text{med}\{x_i\}$ and $C = (1.483\text{MAD}\{x_i\})^2$, where $\text{MAD}\{x_i\}$ is the median of $|x_1 - M_x|, \ldots |x_n - M_x|$, and M_x is the median of the x_i values. The resulting breakdown point appears to be at least 0.15. Evidently, this method can be used to get good control over the probability of a type I error when testing hypotheses about the regression parameters, even when the error term is heteroscedastic, but its power can be poor (Wilcox, 1995e). For rank-based diagnostic tools, see McKean, Sheather, and Hettmansperger (1990). For other results and methods based on ranks, see Cliff (1994), Dixon and McKean (1996), Hettmansperger (1984), Hettmansperger and McKean (1977), and Tableman (1990). (For results on a multivariate linear model, see Davis & McKean, 1993.)

The method derived by Hettmansperger and McKean (1977) does not protect against leverage points, but their method can be of interest when trying to detect curvature (McKean et al., 1990, 1993). Consequently, a brief discussion of their method seems warranted. Let $R(y_i - \mathbf{x}_i'\beta)$ be the rank associated with the ith residual. They determine β by minimizing Jaeckel's (1972) dispersion function given by

$$D(\beta) = \sum a(R(y_i - \mathbf{x}_i'\beta))R(y_i - \mathbf{x}_i'\beta),$$

where

$$a(i) = \phi(i/(n+1))$$

for a nondecreasing function ϕ defined on $(0,1)$ such that $\int \phi(u)du = 0$ and $\int \phi^2(u)du = 1$. Two common choices for ϕ are $\phi(u) = \sqrt{12}(u - 0.5)$ and $\phi(u) = \text{sign}(u - 0.5)$. The slope parameters can be estimated by minimizing $D(\beta)$, but the intercept cannot. One approach to estimating the intercept, which can be used when the error term has a skewed distribution, is to use the median of the residuals after the slope parameters have been estimated.

Another approach is to estimate β to be the vector of values that minimizes

$$\frac{1}{n}\sum a(R(y_i - \mathbf{x}_i'\beta))|r_i|$$

(Hössjer, 1994). Hössjer shows that this estimator can be chosen with a breakdown point of 0.5, and he establishes asymptotic normality. However, Hössjer notes that poor efficiency can result with a breakdown point of 0.5 and suggests designing the method so that its breakdown point is between 0.2 and 0.3, but under normality, the asymptotic relative efficiency is only 0.56. Despite this, the method might have a practical advantage when the error term is heteroscedastic, but this has not been determined. Yet another rank-based method was derived by Chang, McKean, Naranjo, and Sheather (1999). It can have a breakdown point of 0.5, but direct comparisons with some of the better estimators in this chapter have not been made. (For some results on R estimators, see McKean & Sheather, 1991).

10.14 Comments About Various Estimators

A few additional comments about the various regression estimators in this chapter might help. We have seen illustrations that certain estimators can have high efficiency versus OLS when the error term is heteroscedastic. Here, some additional results relevant to this issue are summarized. Let R be the standard error of the OLS estimator divided by the standard error of some competing estimator. So if R is less than one, least squares tends to be more accurate, while $R > 1$ indicates the opposite. Suppose observations are generated according to the model $y = x + \lambda(x)\epsilon$, where the function $\lambda(x)$ reflects heteroscedasticity. Setting $\lambda(x) = 1$ corresponds to the homoscedastic case. Table 10.3 shows estimates of R (based on simulations with 5000 replications) for the estimators TS, MGV (described in Section 10.10), the deepest regression line estimator (T^*), TSTS (in Section 10.13.3) and the MM-estimator, where VP1 corresponds to $\lambda(X) = 1$, VP2 is where $\lambda(x) = x^2$, and VP3 is $\lambda(x) = 1/|x|$. So for VP2, the error term has more variance corresponding to extreme x values and VP3 is a situation where the opposite is true. The results are limited to situations where the distribution for x is symmetric, but very similar results are obtained when x has an asymmetric distribution instead. In Table 10.3, the distributions for ϵ are taken to be normal (N), a g-and-h distribution with $(g, h) = (0, 0.5)$, which is symmetric and heavy-tailed (SH), a g-and-h distribution with $(g, h) = (0.5, 0)$, which is asymmetric and relatively light-tailed (AL), and a g-and-h distribution with $(g, h) = (0.5, 0.5)$, which is asymmetric and relatively heavy-tailed (AH). Note that all five estimators in Table 10.3 generally compete well with the OLS estimator, the main exception being a situation in which x has a symmetric, heavy-tailed distribution and ϵ has a normal distribution. The MMreg estimator is a popular choice among some statisticians and it performs relatively well for the situations considered in Table 10.3. But some caution is needed when using this estimator for reasons described in

Table 10.3: Estimated Ratios of Standard Errors, x Distribution Symmetric, $n = 20$.

x	ϵ	VP	TS	MGV	T^*	TSTS	MMreg
N	N	1	0.91	0.91	0.76	0.88	0.98
		2	2.64	2.62	3.11	2.36	1.95
		3	96.53	77.70	67.72	100.86	116.27
N	SH	1	4.28	4.27	4.42	3.51	1.13
		2	10.67	10.94	11.03	8.66	1.97
		3	70.96	65.20	57.84	82.019	83.61
N	AL	1	1.13	1.13	0.92	1.05	0.95
		2	3.21	3.21	3.69	2.84	1.96
		3	96.52	77.69	67.72	100.86	116.26
N	AH	1	8.89	8.85	16.41	7.05	1.18
		2	26.66	27.07	25.81	20.89	1.94
		3	70.96	65.20	57.84	82.01	83.61
SH	N	1	0.81	0.80	0.61	0.76	0.95
		2	40.57	42.30	55.47	27.91	7.28
		3	106.50	49.79	58.82	106.28	151.61
SH	SH	1	3.09	2.78	2.88	2.41	1.12
		2	78.43	83.56	90.84	47.64	5.94
		3	106.50	49.79	58.82	106.28	151.61
SH	AL	1	0.99	0.87	0.73	0.90	1.20
		2	46.77	49.18	63.60	31.46	2.54
		3	106.50	49.79	58.82	106.28	151.61
SH	AH	1	6.34	5.64	6.75	4.62	1.19
		2	138.53	146.76	108.86	78.35	2.47
		3	81.95	42.27	53.54	92.65	107.26

N=normal; SH=symmetric, heavy-tailed
AL=asymmetric, light-tailed, AH=asymmetric, heavy-tailed.

Section 10.14.1. The OP estimator, described in Section 10.10, is not included in Table 10.3, but it is noted that it performs in a manner very similar to the MGV estimator.

It is not being suggested that if there is heteroscedasticity, it necessarily follows that the robust regression estimators considered here will have a substantially smaller standard error than the least squares estimator. In some situations these robust estimators offer little or no advantage in terms of efficiency. Also, there is a connection between the types of heteroscedasticity considered here and regression outliers. Variance pattern VP3, for example, has a tendency to generate regression outliers, which can have a relatively large effect on the standard error of the least squares estimator.

10.14.1 Contamination Bias

Clearly, one goal underlying robust regression is to avoid situations where a small number of points can completely dominate an estimator. In particular, a goal is to avoid getting a poor fit to the bulk of the points. An approach toward achieving this goal is to require that an estimator have a reasonably high finite sample breakdown point. But there are some regression estimators that, despite having a breakdown point reasonably close to 0.5, can be greatly influenced by a few outliers.

■ Example

Twenty points were generated where both x and ϵ have a standard normal distribution, and $y = x + \epsilon$ was computed, so the true slope is one. Then two aberrant points were added to the data at $(x, y) = (2.1, -2.4)$. Figure 10.2 shows a scatterplot of the points plus the LTS regression line which has an estimated slope of -0.94. So in this case, LTS is a complete disaster in terms of detecting how the majority of the points were generated. The least squares estimate is -0.63. The Coakley–Hettmansperger estimator relies on LTS as an initial estimate of the slope, and despite its high breakdown point, the estimate of the slope is -0.65. The MM-estimator, described in Section 10.9.1, estimates the slope to be -0.12. In contrast, the MGV estimate of the slope is 0.97 and the OP estimate is 0.89. The deepest regression line estimate is 0.66, and the STS estimator (described in Section 10.13.1) performs poorly in this particular case, the estimate being -0.98. The LMS estimate of the slope is 1.7, so it performs poorly as well in this instance. The TSTS estimator, which is an E-type estimator described in Section 10.13.3, yields an estimate of -0.05, and the Gervini–Yohai estimator, described

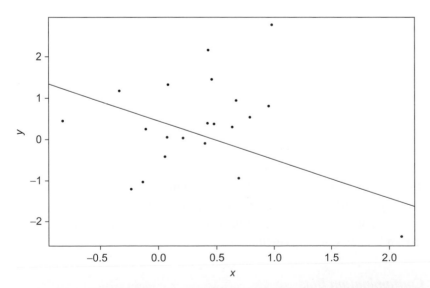

Figure 10.2: Scatterplot of 20 points, where $y = x + \epsilon$, with x and ϵ having independent, standard normal distributions, plus two outliers at $(2.1, -2.4)$. The straight line is the LTS regression line, which poorly estimates the slope used to generate the majority of the points.

in the same section, estimates the slope to be 1.49. The main point is that the choice of robust estimator is not an academic issue, but this one example is not intended as an argument that the estimators that perform poorly in this particular case should be excluded from consideration. Rather, the point is that despite the robust properties they enjoy, they can perform poorly in some situations where other methods do well. Also, although both the OP and MGV estimators do very well here, this is not to suggest that they be used to the exclusion of all other methods.

To add perspective, the process used to generate the data in Figure 10.2 was repeated 500 times, and estimates of the slope were computed using least squares, the M-estimator with Schweppe weights (using the R function bmreg), the Coakley–Hettmansperger estimator (chreg), the Theil–Sen estimator (tsreg), and the deepest regression line (depreg). Boxplots of the results are shown in Figure 10.3. Notice that the median of all these estimators differs from one, the value being estimated. This illustrates that these estimators can be sensitive to a type of *contamination bias*. That is, despite having a reasonably high finite sample breakdown point, it is possible for a few unusual points to result in a poor fit to the bulk of the observations. So these estimators, plus many other robust estimators, can provide substantial advantages versus least squares, but they do not eliminate all practical concerns.

Figure 10.4 shows the results when using the LTS, LTA, MGV, and OP estimators (with default settings). The LTA estimator gives results similar to LTS, and the OP estimator gives

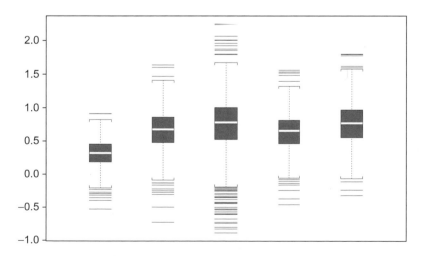

Figure 10.3: Boxplots of 500 estimates of the slope when data are generated as in Figure 10.2. From left to right, the boxplots are based on the ordinary least squares estimator, an M-estimator with Schweppe weights, the Coakley–Hettmansperger estimator, Theil–Sen, and the deepest regression line. All five estimators suffer from contamination bias.

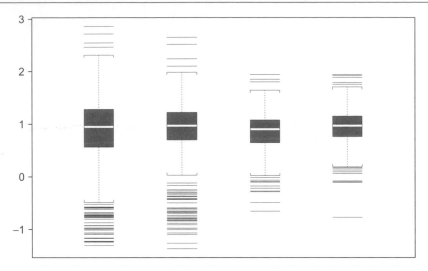

Figure 10.4: **Boxplots of 500 estimates of the slope. From left to right, the boxplots are based on the LTS estimator, the LTA estimator, the OP estimator, and the MGV estimator. In contrast to the estimators used in Figure 10.3, all four estimators avoid the contamination bias problem; each is an approximately median unbiased estimator of the slope.**

results similar to MGV. In contrast to the estimators in Figure 10.3, the median of all the estimators is approximately 1. So in this particular situation, these estimators do a better job of avoiding contamination bias. Note that there is considerably more variation among the LTS estimates based on a breakdown point of 0.5.

There is some evidence that generally the STS estimator gives a better fit to the majority of points versus LTS and LMS. In particular, it seems common to encounter situations where STS is less affected by a few aberrant points. However, exceptions occur, as is illustrated next, so again it seems that multiple methods should be considered.

■ Example

Figure 10.5 shows 20 points that were generated in the same manner as in Figure 10.2. So the two aberrant points located at $(x, y) = (2.1, -2.4)$ are positioned relatively far from the true regression line which again has a slope of one and an intercept of zero. Also shown in Figure 10.5 are the STS, LMS, and MGV estimates of the regression line. In this particular case, STS performs poorly; the estimated slope is -0.23. The LMS estimate of the slope is 1.3 and the estimated slope and intercept based on the MGV estimator are 0.96 and -0.03, respectively, which are closer to the true values versus the other estimates considered here. The OP estimates are 1.26 and -0.34. The estimated

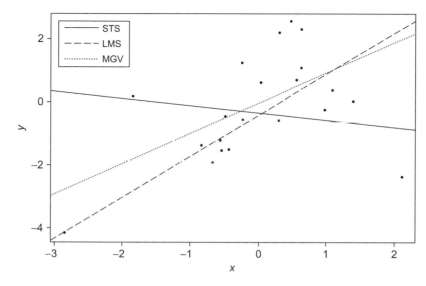

Figure 10.5: Scatterplot of 20 points where $y = x + \epsilon$, with x and ϵ having independent, standard normal distributions, plus two outliers at (2.1, −2.4). This illustrates that only two outliers can have a substantial impact on the STS estimator. The LMS estimate of the slope is 1.3 and MGV estimate is 0.96.

slope based on the TSTS estimator is 0.58 and least squares returns an estimate of 0.4. So once again we see that the choice of which robust estimator to use can make a substantial difference in how the association between x and y is summarized.

It is stressed, however, that while several estimators compete well with least squares, it seems to be easy to find fault with any estimator that has been proposed. For example, both the MGV and OP estimators have several practical advantages over many other estimators that might be used. But using software written exclusively in R, the execution time required for both of these estimators can be relatively high when using a bootstrap method to test hypotheses. But access to a multi-core processor can substantially reduce execution time.

It should be noted that in the theoretical literature, the term contamination bias is used in a different manner. To begin with a simple case, first focus on location estimators. Let T be any functional as described in Chapter 2 and let $T(F) = \theta$. The *contamination bias* associated with T is

$$\sup |T((1 - \epsilon)F + \epsilon G) - \theta|,$$

where $0 \leq \epsilon \leq 1$, and the supremum is taken overall all distributions G. Huber (1964) established results on the contamination bias of the median and various extension have appeared in the literature (e.g., He & Simpson, 1993).

The idea can be extended to regression estimators. Following, for example, Maronna and Yohai (1993), let $\mathbf{V}$ be any affine equivariant scatter matrix associated with F, let $H = (1 - \epsilon)F + \epsilon G$, and let $T(F) = \beta$ be the vector of regression parameters. The bias at G is

$$[(T(H) - \beta)' \mathbf{V}^{-1} (T(H) - \beta)]^{1/2}.$$

Maronna and Yohai (1993) derive results related to the maximum bias of the projection estimator. Variations on this approach are described, for example, by He and Simpson (1993).

10.15 Outlier Detection Based on a Robust Fit

It should be mentioned that several outlier detection methods have been proposed that are based in part on first fitting a robust regression model assuming that Eq. (10.1) is true. Typically these methods assume a homoscedastic error term. Given the issue of contamination bias, it would seem that they should be used with caution. And the issue of how to deal with a heteroscedastic error term seems to warrant consideration. Billor and Kiral (2008) compare several techniques. No single method dominates and the most effective method depends to some extent on where the outliers occur.

10.15.1 Detecting Regression Outliers

Rousseeuw and van Zomeren (1990) suggest using the LMS estimator to detect what are called *regression outliers*. Roughly, these are points that deviate substantially from the linear pattern for the bulk of the points under study. Their method begins by computing the residuals associated with LMS regression, $r_1, \ldots, r_n$. Next, let M_r be the median of $r_1^2, \ldots, r_n^2$, the squared residuals, and let

$$\hat{\tau} = 1.4826 \left(1 + \frac{5}{n - p - 1} \right) \sqrt{M_r}.$$

The point $(y_i, x_{i1}, \ldots, x_{ip})$ is labeled a regression outlier if the corresponding standardized residual is large. In particular, Rousseeuw and van Zomeren label the ith vector of observations a *regression outlier* if $|r_i|/\hat{\tau} > 2.5$.

10.15.2 R Function reglev

The R function

$$\text{reglev(x,y,plotit=T)}.$$

is provided for detecting regression outliers and leverage points using the method described in the previous section. If the ith vector of observations is a regression outlier, the function stores the value of i in the R variable reglev\$regout. If $\mathbf{x}_i$ is an outlier based on the method in Section 6.4.3, it is declared a leverage point and the function stores the value of i in reglev\$levpoints. The plot created by this function can be suppressed by setting plotit=F.

■ Example

If the reading data in the first example of Section 10.8.1 are stored in the R variables x and y, the command reglev(x,y) returns

```
$levpoints:
[1] 8

$regout:
[1]  12 44 46 48 59 80
```

This says that x_8 is flagged as a leverage point (it is an outlier among the x values), and the points (y_{12}, x_{12}), (y_{44}, x_{44}), (y_{46}, x_{46}), (y_{48}, x_{48}), (y_{59}, x_{59}), and (y_{80}, x_{80}) are regression outliers. Note that even though x_8 is an outlier, the point (y_8, x_8) is not a regression outlier. For this reason, x_8 is called a *good leverage point*. (Recall that extreme x values can lower the standard error of an estimator.) If (y_8, x_8) had been a regression outlier, x_8 would be called a *bad leverage point*. Regression outliers, for which x is not a leverage point, are called *vertical outliers*. In the illustration all of the regression outliers are vertical outliers as well.

■

■ Example

The reading data used in the last example is considered again, only the predictor is now taken to be the data in column 3 of the file read.dat, which is another measure of phonological awareness called sound blending. The plot created by reglev is shown in Figure 10.6. Points below the horizontal line that intersects the y-axis at -2.24 are declared regression outliers, as are points above the horizontal line that intersects the y-axis at 2.24. There are three points that lie to the right of the vertical line that intersects the x-axis at $\sqrt{\chi^2_{.975,p}} = 2.24$; theses points are flagged as leverage points.

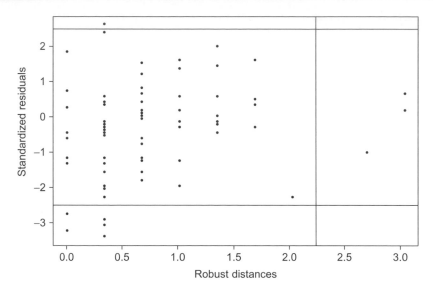

Figure 10.6: **The plot created by the function reglev based on the reading data. Points to the right of the vertical line located at 2.24 on the x-axis are declared leverage points. Points outside the two horizontal lines are declared regression outliers.**

These three points are not flagged as regression outliers, so they are deemed to be good leverage points.

■

10.16 *Logistic Regression and the General Linear Model*

A common situation is where the outcome variable y is binary. In the context of regression, a general approach is to assume that

$$P(y = 1 | \mathbf{X} = \mathbf{x}) = F(\mathbf{x}'\beta), \qquad (10.14)$$

where F is some strictly increasing cumulative distribution function and β is a vector of unknown parameters. The best-known choice for F is

$$F(t) = \frac{\exp(t)}{1 + \exp(t)},$$

which yields what is generally known as the logistic regression model. That is, assume that

$$P(y = 1 | \mathbf{X} = \mathbf{x}) = \frac{\exp(\beta_0 + \beta_1 x_1 + \cdots + \beta_p x_p)}{1 + \exp(\beta_0 + \beta_1 x_1 + \cdots + \beta_p x_p)}. \qquad (10.15)$$

The maximum likelihood estimator of $\beta = (\beta_0, \ldots, \beta_p)$ is the vector $\mathbf{b} = (\mathbf{b_0}, \ldots, \mathbf{b_p})$ that minimizes

$$\sum D(y_i, \mathbf{x}'_i \mathbf{b}),$$

where

$$D(y_i, \mathbf{x}'_i \mathbf{b}) = -y_i \ln(F(\mathbf{x}'_i \mathbf{b})) - (1 - y_i) \ln(1 - F(\mathbf{x}'_i \mathbf{b})).$$

This maximum likelihood estimator is routinely used, but it is not robust. Roughly, leverage points can have an inordinate influence on the estimates. Croux, Flandre, and Haesbroeck (2002) discuss its breakdown point. Here the focus is on a variation of a robust estimator derived by Bianco and Yohai (1996), which is motivated by results in Croux and Haesbroeck (2000). Also see Bianco and Martínez (2009). The robust estimate is the value of $\mathbf{b}$ that minimizes

$$\sum w_i \phi(y_i, \mathbf{x}_i \mathbf{b}), \tag{10.16}$$

where

$$\phi(y, t) = y\rho(-\ln(F(t))) + (1 - y)\rho(-\ln(1 - F(t))) + G(F(t)) + G(1 - F(t)) - G(1),$$

$$G(t) = \int_0^t \psi(-\ln u)du,$$

$\psi(t) = \rho'(t)$, and

$$\rho(t) = \begin{cases} te^{-\sqrt{c}}, & \text{if } t \leq c \\ -2e^{-\sqrt{c}}(1 + \sqrt{t}) + e^{-\sqrt{c}}(2(1 + \sqrt{c}) + c), & \text{if } t > c, \end{cases}$$

and c is a constant. Following the suggestion by Croux and Haesbroeck, $c = 0.5$ is used here. (Croux and Haesbroeck also provide an analytic form for $G(t)$.) For additional results related to robust estimators for the logistic regression model, some of which are derived in the more general framework of the general linear model outlined in Section 10.16.2, see Bondell (2005, 2008), Carroll and Pedersen (1993), Christmann (1994), Künsch, Stefanski, and Carroll (1989), Morgenthaler (1992), Pregibon (1987), Rousseeuw and Christmann (2003), and Stefanski, Carroll, and Ruppert (1986). When computing confidence intervals for the parameters in this model, the percentile bootstrap method in Section 11.1.3 is recommended. The R function wlogregci in Section 11.1.4 performs the calculations.

10.16.1 R Functions glm, logreg, wlogreg, and logreg.plot

The built-in R function glm can be used to compute the maximum likelihood estimate of the parameters in the logistic regression model. And the R function summary tests hypotheses. If, for example, the data are stored in the R variables x and y, use the commands

$$\text{fit=glm(formula=y\~x,family=binomial)}$$
$$\text{summary(fit)}$$

For convenience, the R function

$$\text{logreg(x, y, xout = F, outfun = outpro, plotit = F)}$$

is provided, which performs both of the R commands glm and summary. The function also removes any leverage points if the argument xout=T; it will use the outlier detection method indicated by the argument outfun. By default, the projection method in Section 6.4.9 is used. For a single predictor, if the argument plotit = T, the regression line will be plotted.

The Bianco–Yohai estimator is applied with the R function

$$\text{wlogreg(x, y)}$$

The function wlogreg returns estimates of the standard errors, but using them to compute confidence intervals and test hypotheses is not recommended. (Use the R function wlogregci, which is described in Chapter 11.) Finally, the R function

$$\text{logreg.plot(x, y, MLE = F, ROB = T, xlab = "X", ylab = "P(X)")}$$

plots the robust estimate of the regression line, assuming $p = 1$ predictor. To plot the usual (maximum likelihood) estimate simultaneously, set the argument MLE=T. If ROB=F, the robust regression line is not plotted.

Note that there are two ways of dealing with leverage points. Use the R function logreg with xout=T, or use the Bianco–Yohai estimator via the R function wlogreg. In terms of achieving a relatively small standard error, all indications are that the Bianco–Yohai estimator is preferable to using the R function logreg with xout=T. However, each method reacts differently to outliers and it is not completely clear which is preferable for general use in terms of achieving relatively high power and short confidence intervals.

10.16.2 The General Linear Model

Briefly, the *general linear model* model consists of three components. The first is the assumption that an outcome variable y has a distribution that belongs to the exponential family. This family of distributions includes the normal, binomial, Poisson, and gamma distributions as special cases. (In practice, one specifies which of these distributions will be assumed.) It is further assumed that the independent random variables $y_1, \ldots, y_n$ have the same distribution. In the context of regression, typically homoscedasticity is assumed. (But some generalized linear models are designed to allow heteroscedasticity.) The second component is a set of p predictors $\mathbf{x}$ and associated parameters β. And the third component is a monotone link function g that satisfies

$$g(\mu) = \mathbf{x}\beta.$$

If the link function is taken to be the identity function, we get the usual linear model given by Eq. (10.1). A class of M-estimators for the generalized linear model has been derived, a summary of which can be found in Heritier et al. (2009, Section 5.3). For results on testing hypotheses, see Cantoni and Ronchetti (2001). Here it is noted that the generalized linear model provides yet another approach to logistic regression, and it has the advantage of being able to handle discrete data assuming that y has a Poisson distribution.

10.16.3 R Function glmrob

Robust estimation and hypothesis testing can be performed via the generalized linear model just described using the R function

glmrob(formula, family, data),

which belongs to the R package robustbase. (Hypothesis testing is accomplished with the R function summary.) Mallows or Huber type robust estimators, as described in Cantoni and Ronchetti (2001), are used. In principle, this class of M-estimators can handle continuous outcomes, but currently the R function glmrob only allows discrete outcomes where y has a binomial or Poisson distribution. That is, the argument family can be equal to "binomial" or "poisson". (When dealing with continuous outcomes, methods for testing hypotheses that perform well when there is heteroscedasticity are described in Chapter 11.)

10.17 Multivariate Regression

Consider a regression problem where there are p predictors $\mathbf{x}' = (x_1, \ldots, x_p)$ and q responses $\mathbf{y} = (y_1, \ldots, y_q)$. The usual multivariate regression model is

$$\mathbf{y} = \mathbf{B}'\mathbf{x} + \mathbf{a} + \epsilon, \tag{10.17}$$

where $\mathbf{B}$ is a $(p \times q)$ slope matrix, $\mathbf{a}$ is a q-dimensional intercept vector, and the errors $\epsilon = (\epsilon_1, \ldots, \epsilon_q)$ are independent and identically distributed with mean $\mathbf{0}$ and covariance matrix Σ_ϵ, a positive definite matrix of size q. Let μ be some measure of location associated with the joint distribution of $(\mathbf{x}, \mathbf{y})$ and let Σ be some measure of scatter. Partitioning $(\mathbf{x}, \mathbf{y})$ and Σ in an obvious way yields

$$\mu = \begin{pmatrix} \mu_x \\ \mu_y \end{pmatrix} \quad \text{and} \quad \Sigma = \begin{pmatrix} \Sigma_{xx} & \Sigma_{xy} \\ \Sigma_{yx} & \Sigma_{yy} \end{pmatrix}.$$

In practice, Eq. (10.17) is typically assumed and the most common choice for μ is the population mean, which is estimated with the usual sample mean, say $\hat{\mu}$, and the estimate of Σ is typically taken to be the usual covariance matrix, say $\hat{\Sigma}$. The resulting estimates of **B** and **a** are

$$\hat{\mathbf{B}} = \hat{\Sigma}_{xx}^{-1}\hat{\Sigma}_{xy} \tag{10.18}$$

and

$$\hat{\mathbf{a}} = \hat{\mu}_y - \hat{\mathbf{B}}'\hat{\mu}_x, \tag{10.19}$$

respectively. The estimate of the covariance matrix associated with the error term, ϵ, is

$$\hat{\Sigma}_\epsilon = \hat{\Sigma}_{yy} - \hat{\mathbf{B}}'\hat{\Sigma}_{xx}\hat{\mathbf{B}} \tag{10.20}$$

It is well known, however, that this classic estimator is extremely sensitive to outliers. Another concern is that when $q = 1$, it is known that the efficiency of the least squares estimator can be poor relative to other estimators that might be used.

Another point worth mentioning is that the estimator just described does not take into account the overall structure of the y values. Indeed, it is tantamount to simply computing the least squares regression line for each of the q response variables $y_1, \ldots, y_q$ (e.g., Jhun & Choi, 2009).

10.17.1 The RADA Estimator

Let $\mathbf{z} = (\mathbf{x}, \mathbf{y})$ represent the joint $(\mathbf{x}, \mathbf{y})$ variables and let $\mathbf{z}_i$ $(i = 1, \ldots, n)$ be a random sample of size n. Rousseeuw, Van Aelst, Van Driessen, & Agulló (2004) proposed three robust alternatives to Eqs (10.18) and (10.19), and they recommended one for general use based on simulation estimates of its efficiency. They begin by computing the MCD estimate based on $\mathbf{z}_i$ $(i = 1, \ldots, n)$. Recall that the MCD estimator searches for a subset $\{\mathbf{z}_{i_1}, \ldots, \mathbf{z}_{i_h}\}$ of size h whose covariance matrix has the smallest determinant, where $\lceil n/2 \rceil \le h \le n$. Let $\gamma = (n - h)/n$, so $0 \le \gamma \le 0.5$. The estimated center is

$$\hat{\theta} = \sum_{j=1}^{h} \mathbf{z}_{i_j}/h,$$

and the estimated scatter is

$$\hat{\Xi} = c_n c_\gamma \frac{1}{h} \sum_{j=1}^{h} (\mathbf{z}_{i_j} - \hat{\theta})(\mathbf{z}_{i_j} - \hat{\theta})',$$

where c_n is a small-sample correction factor and c_γ is a consistency factor (Pison et al., 2002). For the problem at hand, Rousseeuw et al. found that $h \approx 3n/4$ provides relatively good efficiency and this choice is used here unless stated otherwise. Once the MCD estimates of location and scatter, based on $\mathbf{z}$, are available, their values are used in Eqs (10.18) and (10.19) to get estimates of the slopes and intercept. But efficiency can be relatively low.

Rousseeuw et al. consider two strategies for improving efficiency. Briefly, their first strategy uses weighted measures of location and scatter, with the weights computed as follows. Let $d(\mathbf{z}) = [(\mathbf{z}_i - \hat{\theta})' \hat{\Xi}^{-1} (\mathbf{z}_i - \hat{\theta})]^{1/2}$ and $w_i = I(d^2(\mathbf{z}) \leq q)$, where q is the 0.975 quantile of a chi-squared distribution with $p + q$ degrees of freedom. Then the weighted measure of location and scatter (omitting a consistency factor) are

$$\hat{\theta}_1 = \frac{\sum w_i \mathbf{z}_i}{\sum w_i}, \tag{10.21}$$

and

$$\hat{\Xi}_1 = \frac{\sum w_i (\mathbf{z}_i - \hat{\theta}_1)(\mathbf{z}_i - \hat{\theta}_1)'}{\sum w_i}, \tag{10.22}$$

respectively. Their second and recommended method uses updated weights based on the residuals associated with Eqs (10.21) and (10.22). Let $\mathbf{r}_i$ be the residuals. Then the weights are taken to be $w_i = (\mathbf{r}_i' \hat{\Sigma}_\epsilon \mathbf{r}_i)^{1/2}$. One appealing feature of this reweighting scheme is that good leverage points (outliers among the $\mathbf{x}$ values for which the corresponding residual is not an outlier) are not downweighted. This will be called the RADA estimator henceforth.

Wilcox (2009b) compared the RADA estimator to several other estimators, which included situations where the error term is heteroscedastic. The RADA estimator did not dominate, but it performed reasonably well, particularly when there is dependence among the $\mathbf{y}$ values.

10.17.2 The Least Distance Estimator

Bai, Chen, Miao, and Rao (1990) proposed another estimator that takes into account the dependence among the outcome variables, $\mathbf{y}$. Called the *least distance estimator*, the regression parameters are estimated with the matrix $\mathbf{B}$ that minimizes

$$\sum_{i=1}^{n} \| \mathbf{y}_i - \mathbf{B}' \mathbf{x}_i \|, \tag{10.23}$$

where now the design matrix $\mathbf{x}$ is assumed to have a column of 1's when the model includes an intercept term. The least distance estimator generalizes the spatial median estimator of multivariate location. Jhun and Choi (2009) confirm that the efficiency of the least distance estimator compares well to the least absolute estimator, meaning that the univariate least absolute regression estimator is applied for each of the q outcome variables. In particular, the efficiency of the least distance estimator, relative to the least absolute regression estimator, improves under normality as the correlation among the outcome variables, **y**, increases.

There is some indication that the least distance estimator competes well with the RADA estimator, in terms of mean squared error, when the error term is homoscedastic. When the error term is heteroscedastic, the reverse might be true. A negative feature of the least distance estimator is that it can be a bit biased with $n = 40$, while bias is negligible when using RADA. It is stressed, however, that a systematic comparison of these two estimators has not been made.

10.17.3 R Functions mlrreg and Mreglde

The R function

$$\text{mlrreg(x,y,cov.fun=cov.mcd)}$$

computes the RADA multivariate regression estimator. By default, it uses the MCD estimator, but some other covariance matrix can be used via the argument cov.fun. The function assumes the argument y is a matrix with two or more columns. The R function

$$\text{Mreglde(x,y,xout=F,eout=F,outfun=outpro)}$$

computes the least distance estimator. If the argument eout=T, the function combines the columns of data in the arguments x and y into a single matrix and then removes all outliers detected by the method indicated by the argument outfun. If xout=T, the function removes any row of data from both x and y for which the row in x is declared an outlier. By default the projection-type outlier detection method is used.

■ Example

A practical issue is whether situations are encountered where the choice of a robust multivariate regression estimator can result in estimates that appear to differ substantially. This can indeed occur as illustrated here using the reading data mentioned in Section 10.8.1. Consider the first two predictors (stored in columns 2 and 3) and the first two outcome variables of interest stored in columns 8 and 9. The estimates returned by the R function mlrreg (the RADA estimator) are

```
                   Y1          Y2
   Intercept   66.000739   68.2289879
   V2           1.027754    0.6587633
   V3           2.587086    2.1111645
```

The estimates returned by Mreglde (the least distance estimator) are

```
                 Y            Y
INTER   95.444444   89.987654
SLOPE    7.344828    3.444856
SLOPE    7.045528    6.303742
```

■

10.17.4 Multivariate Least Trimmed Squares Estimator

Agulló, Croux, and Van Aelst (2008) suggest another approach to multivariate regression based on what they call the least trimmed squares estimators. For the usual least squares estimate of $\mathbf{B}$, say $\hat{\mathbf{B}}_{\mathrm{LS}}$, let

$$\hat{\Sigma}_{\mathrm{LS}} = \frac{1}{n-p}(\mathbf{Y} - \mathbf{X}\hat{\mathbf{B}}_{\mathrm{LS}})'(\mathbf{Y} - \mathbf{X}\hat{\mathbf{B}}_{\mathrm{LS}}).$$

Consider any subset of the $\mathbf{z}_i$ vectors (defined as in Section 10.17.1) having cardinality h. For this subset of the data, and some choice for $\mathbf{B}$, let $\mathbf{r}_i = \mathbf{y}_i - \mathbf{B}'\mathbf{x}_i$ be the matrix of residuals and

$$\mathrm{cov}(\mathbf{B}) = \frac{1}{h}\sum(\mathbf{r}_i - \bar{\mathbf{r}})(\mathbf{r}_i - \bar{\mathbf{r}})',$$

where $\bar{\mathbf{r}} = \sum \mathbf{r}_i / h$. Their strategy is to first search for the subset of the data that minimizes $|\hat{\Sigma}_{\mathrm{LS}}|$. Their multivariate least trimmed squares (MLTS) estimator, $\hat{\mathbf{B}}_{\mathrm{MLTS}}$ is the least squares estimate based on this subset of the data. They establish that this is tantamount to choosing $\mathbf{B}$ so as to minimize the determinant of the MCD scatter matrix estimate based on the residuals from $\mathbf{B}$.

A criticism is that the efficiency of this estimator can be relatively low. Agulló et al. deal with this issue by using a one-step reweighted estimator. Let

$$\hat{\Sigma}_{\mathrm{MLTS}} = \frac{1}{n-p}(\mathbf{Y} - \mathbf{X}\hat{\mathbf{B}}_{\mathrm{MLTS}})'(\mathbf{Y} - \mathbf{X}\hat{\mathbf{B}}_{\mathrm{MLTS}}).$$

Let $J = [j : d_j^2(\hat{\mathbf{B}}_{\mathrm{MLTS}}, \hat{\Sigma}_{\mathrm{MLTS}}) \le q_\delta]$, where

$$d_j^2(\mathbf{B}, \Sigma) = \mathbf{r}_i' \Sigma^{-1} \mathbf{r}_i.$$

Agulló et al. take $\delta = 0.01$ and c_δ equal to the $1 - \delta$ quantile of a chi-squared distribution with q degrees of freedom. The reweighted estimate is taken to be $\hat{\mathbf{B}}_{RMLTS}$, the least squares estimate based on the vectors of observations corresponding to the set J.

10.17.5 R Function MULTtsreg

The R function

$$\text{MULtsreg}(x, y, \text{tr} = 0.2, \text{RMLTS} = T)$$

computes the multivariate least trimmed squares estimator. The argument RMLTS=T means the reweighted estimator is returned; otherwise the MLTS estimator is returned.

10.17.6 Other Robust Estimators

Not all multivariate regression estimators, which have been proposed, are listed here. But in case it helps, two others are mentioned. The first uses Eqs (10.18) and (10.19) to estimate the slopes and intercept but with the usual mean and covariance matrices replaced by some robust analog. Zhou (2008) has studied this approach when using the projection estimate of location and scatter in Section 6.3.7. (The form of the Stahel–Donoho W-estimator suggested by Zuo, Hengjian, & He, 2004; Zuo, Hengjian, & Young, 2004, was used.) Currently, execution time can be an issue and little is known about how this approach compares to the estimators in Sections 10.17.1 and 10.17.2. Yet another approach was derived by Ben, Martínez, and Yohai (2006). The regression coefficients and the covariance matrix of the errors are estimated simultaneously by minimizing the determinant of the covariance matrix, subject to a constraint on a robust scale of the Mahalanobis norms of the residuals. They use a τ-estimate of scale. Ben et al. report simulation results indicating that their estimator compares favorably to S-estimates.

10.18 Exercises

1. The average LSAT scores (x) for the 1973 entering classes of 15 American law schools, and the corresponding grade point averages (y), are as follows.

    x: 576 635 558 578 666 580 555 661 651 605 653 575 545 572 594
    y: 3.39 3.30 2.81 3.03 3.44 3.07 3.00 3.43 3.36 3.13 3.12 2.74 2.76 2.88 2.96

    Using the R function lsfitci, verify that the 0.95 confidence interval for the slope, based on the least squares regression line, is $(0.0022, 0.0062)$.

2. Discuss the relative merits of $\hat{\beta}_{ch}$.

3. Using the data in Exercise 1, show that the estimate of the slope given by $\hat{\beta}_{ch}$ is 0.0057. In contrast, the OLS estimate is 0.0045, and $\hat{\beta}_m = 0.0042$. Comment on the difference among the three estimates.

4. Let T be any regression estimator that is affine equivariant. Let $\mathbf{A}$ be any nonsingular square matrix. Argue that the predicted y values, $\hat{y}_i$, remain unchanged when $\mathbf{x}_i$ is replaced by $\mathbf{x}_i \mathbf{A}$.

5. For the data in Exercise 1, use the R function reglev to comment on the advisability of using M regression with Schweppe weights.

6. Compute the hat matrix for the data in Exercise 1. Which x values are identified as leverage points? Relate the result to the previous exercise.

7. The example in Section 6.6.1 reports the results of drinking alcohol for two groups of subjects measured at three different times. Using the group 1 data, compute an OLS estimate of the regression parameters for predicting the time 1 data using the data based on times 2 and 3. Compare the results to the estimates given by $\hat{\beta}_m$ and $\hat{\beta}_{ch}$.

8. For the data used in the previous exercise, compute 0.95 confidence intervals for the parameters using OLS as well as M regression with Schweppe weights.

9. Referring to Exercise 6, how do the results compare to the results obtained with the R function reglev?

10. For the data in Exercise 6, verify that the 0.95 confidence interval for the regression parameters, using the R function regci with M regression and Schweppe weights, are $(-0.2357, 0.3761)$ and $(-0.0231, 1.2454)$. Also verify that if regci is used with OLS, the confidence intervals are $(-0.4041, 0.6378)$ and $(0.2966, 1.7367)$. How do the results compare to the confidence intervals returned by lsfitci? What might be wrong with confidence intervals based on regci when the OLS estimator is used?

11. The file read.dat contains reading data collected by L. Doi. Of interest is predicting WWISST2, a word identification score (stored in column 8), using TAAST1, a measure of phonological awareness stored in column 2, and SBT1 (stored in column 3), another measure of phonological awareness. Compare the OLS estimates to the estimates given by $\hat{\beta}_m$, $\hat{\beta}_{ad}$, and $\hat{\beta}_{mid}$.

12. For the data used in Exercise 11, compute the hat matrix and identify any leverage points. Also check for leverage points with the R function reglev. How do the results compare?

13. For the data used in Exercise 11, RAN1T1 and RAN2T1 (stored in columns 4 and 5) are measures of digit naming speed and letter naming speed. Use M regression with Schweppe weights to estimate the regression parameters when predicting WWISST2. Use the function elimna, described in Chapter 1, to remove missing values. Compare the results with the OLS estimates and $\hat{\beta}_{ch}$.

14. For the data in Exercise 13, identify any leverage points using the hat matrix. Next, identify leverage points with the function reglev. How do the results compare?

15. Graphically illustrate the difference between a regression outlier and a good leverage point. That is, plot some points for which $y = \beta_1 x + \beta_0$, and then add some points that represent regression outliers and good leverage points.

16. Describe the relative merits of the OP and MGV estimators in Section 10.10.

17. For the star data in Figure 6.3, which are stored in the file star.dat, eliminate the four outliers in the upper left corner of the plot by restricting the range of the x values. Then using the remaining data, estimate the standard error of the least squares estimator, the M-estimator with Schweppe weights, as well as the OP and MGV estimators. Comment on the results.

CHAPTER 11

More Regression Methods

This final chapter describes some additional robust regression methods that have been found to have practical value, including some inferential techniques that perform well in simulation studies even when the error term is heteroscedastic. Also covered are methods for testing the hypothesis that two or more of the regression parameters are equal to zero, a method for comparing the slope parameters of independent groups, measures of association based on a given fit to the data, methods for dealing with curvilinear relationships, and methods for performing an analysis of covariance.

11.1 Inferences About Robust Regression Parameters

This section deals with testing hypotheses about the parameters in the regression model

$$y_i = \beta_0 + \beta_1 x_{i1} + \cdots + \beta_p x_{ip} + \epsilon_i,$$

where the error term might be heteroscedastic and some robust regression estimator is used. (Section 10.1 described methods designed specifically for the situation where the least squares estimator is used.) A special case that is commonly of interest is testing

$$H_0 : \beta_1 = \cdots = \beta_p = 0, \tag{11.1}$$

the hypothesis that all of the slope parameters are equal to zero, but the methods described here can also be used to test

$$H_0 : \beta_1 = \cdots = \beta_q = 0,$$

the hypothesis that $q < p$ of the parameters are equal to zero. A more general goal is to test the hypothesis that q parameters are equal to some specified value, and the method described here accomplishes this goal as well. And there is the goal of computing confidence intervals for the individual parameters.

11.1.1 Omnibus Tests for Regression Parameters

Currently, the best methods for testing hypotheses based on robust regression estimators are based on some type of bootstrap method. This section begins with testing Eq. (11.1) and then the goal of computing confidence intervals for the individual parameters is addressed.

When working with robust regression, three strategies for testing hypotheses have received attention in the statistical literature and should be mentioned. The first is based on the so-called *Wald scores*, the second is a *likelihood ratio test*, and the third is based on a measure of *drop in dispersion*. Details about these methods can be found in Markatou, Stahel, and Ronchetti (1991), as well as Heritier and Ronchetti (1994). Coakley and Hettmansperger (1993) suggest using a Wald scores test in conjunction with their estimation procedure, assuming that the error term is homoscedastic. The method estimates the standard error of their estimator which can be used to get an appropriate test statistic for which the null distribution is chi-square. When both x and ϵ are normal, and the error term is homoscedastic, the method provides reasonably good control over the probability of a type I error when $n = 50$. However, if ϵ is nonnormal, the actual probability of a type I error can exceed .1 when testing at the $\alpha = 0.05$ level, even with $n = 100$ (Wilcox, 1994e). Consequently, details about the method are not described. Birkes and Dodge (1993) describe drop-in-dispersion methods when working with M regression methods that do not protect against leverage points. Little is known about how this approach performs under heteroscedasticity, so it is not discussed either. Instead, attention is focused on a method that has been found to perform well when there is a heteroscedastic error term. It is not being suggested that the method described here outperforms all other methods that might be used, only that it gives good results over a relatively wide range of situations, and based on extant simulation studies, it is the best method available.

The basic strategy is to generate B bootstrap estimates of the parameters and then determine whether the vector of values specified by the null hypothesis is far enough away from the bootstrap samples to warrant rejecting H_0. This strategy is illustrated with the tree data in the Minitab handbook (Ryan, Joiner, & Ryan, 1985), p. 329). The data consist of tree volume (V), tree diameter (d), and tree height (h). If the trees are cylindrical or cone shaped, then a reasonable model for the data is $y = \beta_1 x_1 + \beta_2 x_2 + \beta_0$, where $y = \ln(V)$, $x_1 = \ln(d)$, $x_2 = \ln(h)$ with $\beta_1 = 2$ and $\beta_2 = 1$ (Fairley, 1986). The OLS estimate of the intercept is $\hat{\beta}_0 = -6.632$, $\hat{\beta}_1 = 1.98$, and $\hat{\beta}_2 = 1.12$. Using M regression with Schweppe weights (the R function bmreg), the estimates are -6.59, 1.97, and 1.11, respectively.

Suppose bootstrap samples are generated as described in Section 10.1.1 (cf. Salibian-Barrera & Zamar, 2002). That is, rows of data are sampled with replacement. Figure 11.1 shows a scatterplot of 300 bootstrap estimates, using M regression with Schweppe weights, of the intercept, β_0, and β_2, the slope associated with log height. The square marks the hypothesized

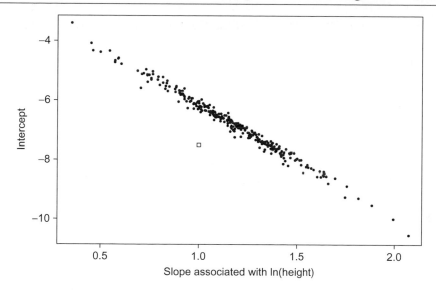

Figure 11.1: Scatterplot of bootstrap estimates using the tree data. The square marks the null values.

values, $(\beta_0, \beta_2) = (-7.5, 1)$. These bootstrap values provide an estimate of a confidence region for (β_0, β_2) that is centered at the estimated values $\hat{\beta}_0 = -6.59$ and $\hat{\beta}_2 = 1.11$. Figure 11.1 suggests that the hypothesized values might not be reasonable. That is, the point $(-7.5, 1)$ might be far enough away from the bootstrap values to suggest that it is unlikely that β_0 and β_2 simultaneously have the values -7.5 and 1, respectively. The problem is measuring the distance between the hypothesized values and the estimated values, and then finding a decision rule that rejects the null hypothesis with probability α when H_0 is true.

For convenience, temporarily assume the goal is to test the hypothesis given by Eq. (11.1). A simple modification of the method in Section 8.2.5 can be used where bootstrap samples are obtained by resampling with replacement rows from

$$\begin{pmatrix} y_1, x_{11}, \ldots, x_{1J} \\ \vdots \\ y_n, x_{n1}, \ldots, x_{nJ} \end{pmatrix}$$

yielding

$$\begin{pmatrix} y_1^*, x_{11}^*, \ldots, x_{1J}^* \\ \vdots \\ y_n^*, x_{n1}^*, \ldots, x_{nJ}^* \end{pmatrix}.$$

Let $\hat{\beta}_{jb}^*$, $j = 1, \ldots, p$; $b = 1, \ldots, B$ be an estimate of the jth parameter based on the bth bootstrap sample and any robust estimator described in Chapter 10. Then an estimate of the covariance between $\hat{\beta}_j$ and $\hat{\beta}_k$ is

$$v_{jk} = \frac{1}{B-1} \sum_{b=1}^{B} (\hat{\beta}_{jb}^* - \bar{\beta}_j^*)(\hat{\beta}_{kb}^* - \bar{\beta}_k^*),$$

where $\bar{\beta}_j^* = \sum \hat{\beta}_{jb}^* / B$. Here $\hat{\beta}_j$ can be any estimator of interest. Now the distance between the bth bootstrap estimate of the parameters, and the estimate based on the original observations, can be measured with

$$d_b^2 = (\hat{\beta}_{1b}^* - \hat{\beta}_1, \ldots, \hat{\beta}_{pb}^* - \hat{\beta}_p)\mathbf{V}^{-1}(\hat{\beta}_{1b}^* - \hat{\beta}_1, \ldots, \hat{\beta}_{pb}^* - \hat{\beta}_p)',$$

where $\mathbf{V}$ is the p-by-p covariance matrix with the element in the jth row and kth column equal to v_{jk}. That is, $\mathbf{V} = (v_{jk})$ is the sample covariance matrix based on the bootstrap estimates of the parameters. The square root of d_b^2, d_b, represents a simple generalization of the Mahalonobis distance. If the point corresponding to the vector of hypothesized values is sufficiently far from the estimated values, relative to the distances d_b, reject H_0. This strategy is implemented by sorting the d_b values yielding $d_{(1)} \leq \cdots \leq d_{(B)}$, setting $M = [(1-\alpha)B]$, and letting m be the value of M rounded to the nearest integer. The null hypothesis is rejected if

$$D > d_{(m)}, \tag{11.2}$$

where

$$D = \sqrt{(\hat{\beta}_1, \ldots, \hat{\beta}_p)\mathbf{V}^{-1}(\hat{\beta}_1, \ldots, \hat{\beta}_p)'}.$$

The method just described is easily generalized to testing

$$H_0 : \beta_1 = \beta_{10}, \beta_2 = \beta_{20}, \ldots, \beta_q = \beta_{q0},$$

the hypothesis that q of the $p+1$ parameters are equal to specified constants, $\beta_{10}, \ldots, \beta_{q0}$. Proceed as before, only now

$$d_b = \sqrt{(\hat{\beta}_{1b}^* - \hat{\beta}_1, \ldots, \hat{\beta}_{qb}^* - \hat{\beta}_q)\mathbf{V}^{-1}(\hat{\beta}_{1b}^* - \hat{\beta}_1, \cdots, \hat{\beta}_{qb}^* - \hat{\beta}_q)'},$$

and $\mathbf{V}$ is a q-by-q matrix of estimated covariances based on the B bootstrap estimates of the q parameters being tested. The critical value, $d_{(m)}$, is computed as before, and the test statistic is

$$D = \sqrt{(\hat{\beta}_1 - \beta_{10}, \ldots, \hat{\beta}_q - \beta_{q0})\mathbf{V}^{-1}(\hat{\beta}_1 - \beta_{10}, \ldots, \hat{\beta}_q - \beta_{q0})'}.$$

The (generalized) p-value is

$$\hat{p}^* = \frac{1}{B} \sum I(D \le d_b),$$

where $I(D \le d_b) = 1$ if $D \le d_b$, and $I(D \le d_b) = 0$ if $D > d_b$.

The hypothesis testing method just described can be used with any regression estimator. When using the OLS estimator, it has advantages over the conventional F-test, but problems remain. This is illustrated in Table 11.1, which shows the estimated probability of a type I error for various situations when testing $H_0 : \beta_1 = \beta_2 = 0$, $\alpha = 0.05$, and where

$$y = \beta_1 x_1 + \beta_2 x_2 + \lambda(x_1, x_2)\epsilon. \tag{11.3}$$

(For recent results on testing hypotheses when the function λ is known, see Zhao & Wang, 2009.) In Table 11.1, VP1 corresponds to $\lambda(x_1, x_2) = 1$ (a homoscedastic error term), VP2 is $\lambda(x_1, x_2) = |x_1|$, and VP3 is $\lambda(x_1, x_2) = 1/(|x_1| + 1)$. Both x_1 and x_2 have identical g-and-h distributions, with the g and h values specified by the first two columns. In some cases the conventional F-test performs well, but it performs poorly for VP2. The bootstrap method improves matters considerably, but the probability of a type I error exceeds .075 in various situations. In practical terms, when testing hypotheses using OLS, use the methods in Section 10.1.1 rather than the bootstrap method described here.

Table 11.2 shows $\hat{\alpha}$, the estimated probability of a type I error when using $\hat{\beta}_{\text{mid}}$, the biweight midregression estimator with $n = 20$, and the goal is to test $H_0 : \beta_1 = \beta_2 = 0$ with $\alpha = 0.05$.

Table 11.1: Estimated Type I Error Probabilities Using OLS, $\alpha = 0.05$, $n = 20$.

\(x \)		\(\epsilon \)		VP1		VP2		VP3	
g	h	g	h	Boot	F	Boot	F	Boot	F
0.0	0.0	0.0	0.0	.072	.050	.097	.181	.009	.015
0.0	0.0	0.0	0.5	.028	.047	.046	.135	.004	.018
0.0	0.0	0.5	0.0	.052	.049	.084	.174	.009	.018
0.0	0.0	0.5	0.5	.028	.043	.042	.129	.005	.019
0.0	0.5	0.0	0.0	.022	.055	.078	.464	.003	.033
0.0	0.5	0.0	0.5	.014	.074	.042	.371	.002	.038
0.0	0.5	0.5	0.0	.017	.048	.072	.456	.005	.032
0.0	0.5	0.5	0.5	.011	.070	.039	.372	.005	.040
0.5	0.0	0.0	0.0	.054	.044	.100	.300	.013	.032
0.5	0.0	0.0	0.5	.024	.057	.049	.236	.010	.038
0.5	0.0	0.5	0.0	.039	.048	.080	.286	.010	.033
0.5	0.0	0.5	0.5	.017	.058	.046	.217	.010	.040
0.5	0.5	0.0	0.0	.013	.054	.083	.513	.006	.040
0.5	0.5	0.0	0.5	.009	.073	.043	.416	.002	.048
0.5	0.5	0.5	0.0	.013	.053	.079	.505	.005	.043
0.5	0.5	0.5	0.5	.006	.067	.036	.414	.005	.050

Table 11.2: Values of $\hat{\alpha}$ Using Biweight Midregression, $\alpha = 0.05$, $n = 20$

X		ϵ				
g	h	g	h	VP1	VP2	VP3
0.0	0.0	0.0	0.0	.047	.039	.015
0.0	0.0	0.0	0.5	.018	.024	.008
0.0	0.0	0.5	0.0	.038	.037	.011
0.0	0.0	0.5	0.5	.021	.025	.003
0.0	0.5	0.0	0.0	.016	.018	.002
0.0	0.5	0.0	0.5	.009	.018	.002
0.0	0.5	0.5	0.0	.015	.016	.002
0.0	0.5	0.5	0.5	.009	.012	.003
0.5	0.0	0.0	0.0	.033	.037	.012
0.5	0.0	0.0	0.5	.020	.020	.006
0.5	0.0	0.5	0.0	.024	.031	.009
0.5	0.0	0.5	0.5	.015	.021	.005
0.5	0.5	0.0	0.0	.015	.021	.002
0.5	0.5	0.0	0.5	.008	.011	.002
0.5	0.5	0.5	0.0	.014	.017	.002
0.5	0.5	0.5	0.5	.006	.007	.002

Now the probability of a type I error is less than or equal to the nominal level, but in some cases it is too low, particularly for VP3 where it drops as low as .002. Perhaps better results are obtained when using some other robust regression method, but this has not been determined. (For more details about the simulations used to create Tables 11.1 and 11.2, see Wilcox, 1996f.)

11.1.2 R Function regtest

The R function

$$regtest(x,y,regfun=tsreg,nboot=600,alpha=0.05,plotit=T,grp=c(1:ncol(x)), nullvec = c(rep(0, length(grp))))$$

tests hypotheses with the bootstrap method described in the previous section. As usual, x is an n-by-p matrix containing the predictors. The argument regfun is any R function that estimates regression coefficients and returns the estimates in the vector regfun$coef. If unspecified, regfun defaults to tsreg which computes the Theil–Sen estimate. The assumption is that the first element of regfun$coef contains the estimated intercept, the second contains the estimate of β_1, and so on. The arguments nboot and alpha are B, the number of bootstrap samples to be used, and α, respectively. The default values are $B = 599$ and $\alpha = 0.05$. The argument grp indicates which parameters are to be tested. By default, the null hypothesis is that all p slope parameters are equal to zero (the hypothesis given by Eq. 9.1). If, for example, the goal is to test $H_0 : \beta_2 = \beta_4 = 0$, type the R command grp=c(2,4) in which case the command regtest(x,y,regfun=bmreg,grp=grp) would test H_0 for the data in x and y using M regression with Schweppe weights. That is, grp is a vector containing the subscripts of the parameters to be tested. Alternatively, use the command regtest(x,y,grp=c(2,4)). To test $H_0 : \beta_0 = \beta_4 = 0$, the hypothesis that the intercept and fourth slope parameter are equal to zero, use the command regtest(x,y,grp=c(0,4)). The optional argument nullvec contains the null values. If unspecified, nullvec defaults to a vector of zeros. The vectors nullvec and grp must have the same length. If they do not, the function returns an error message and terminates.

■ Example

For the tree data used to create Figure 11.1, suppose there is reason to believe that $\beta_0 = -7.5$ and $\beta_2 = 1$. If the logarithm of the predictor values are stored in mtree, and the logarithm of the volume (the y values) are stored in ytree, then the command

regtest(mtree,ytree,regfun=bmreg,grp=c(0,2),nullvec=c(-7.5,1))

will test the hypothesis that $H_0 : \beta_0 = -7.5$ and $\beta_2 = 1$ using the R function bmreg to estimate the parameters. The function regtest reports a test statistic of 97.24, with a 0.05 critical value of 7.98, so H_0 is rejected. Using the Theil–Sen estimator, the test statistic is 93.8, the 0.05 critical value is 7.06, and the (generalized) p-value is 0.

■

11.1.3 Inferences About Individual Parameters

When the goal is to compute a confidence interval for the individual parameters in a regression model, a simple percentile bootstrap method appears to perform well, in terms of probability coverage, with most robust regression estimators. Essentially, proceed as in Section 10.1.1, but with the least squares estimator replaced by any robust estimator, and when applying the percentile bootstrap method, no adjustment is made when the sample size is small. That is, bootstrap samples are obtained by randomly sampling n vectors of observations, with replacement, which is in contrast to bootstrap methods that resample residuals. Let $\hat{\beta}_j^*$ be any robust estimate of β_j based on the bootstrap sample just obtained. Repeat this process B times yielding $\hat{\beta}_{j1}^*, \ldots \hat{\beta}_{jB}^*$. Then for fixed j, the $1 - \alpha$ confidence interval for β_j is

$$(\hat{\beta}_{j(\ell+1)}^*, \hat{\beta}_{j(u)}^*), \tag{11.4}$$

where $\ell = \alpha B/2$, rounded to the nearest integer, $u = B - \ell$, and $\hat{\beta}_{j(1)}^* \leq \cdots \leq \hat{\beta}_{j(B)}^*$ are the B bootstrap estimates of β_j written in ascending order. In other words, use the standard percentile bootstrap method, as opposed to the modified method used when working with OLS. A (generalized) p-value can be computed in the usual way. Let $\hat{p}^*$ be the proportion of bootstrap estimates greater than zero. Then the p-value is

$$\hat{p}_m^* = 2\min(\hat{p}^*, \, 1 - \hat{p}^*).$$

(Under certain circumstances, when there is one predictor only, an alternative approach to computing a confidence interval for the slope that might have practical value is described by Adrover & Salibian-Barrera, 2010.)

To provide some indication of how well the method performs when using the M-estimator $\hat{\beta}_m$, Table 11.3 shows values of $\hat{\alpha}$, simulation estimates of one minus the actual probability coverage, when $n = 20$ and $\alpha = 0.05$. The notation VP1 indicates a homoscedastic error term ($\lambda(x) = 1$ in Eq. (10.2)), VP2 is a heteroscedastic error term where the variance of the error term increases as x moves away from its median value ($\lambda(x) = x^2$), and VP3 is where the variance decreases as x moves away from its median ($\lambda(x) = 1 + 2/(|x| + 1)$). The $\hat{\alpha}$ values never exceed .075, but for VP2 they can exceed .070. (For results when using the Theil–Sen estimator, see Wilcox, 1998a,b. For results related to the OP-estimator in Section 10.10, see Wilcox, 2004d.)

To provide a bit more perspective, Table 11.4 shows simulation estimates of the probability of a type I error when using the Theil–Sen estimator with $p = 2$. Under VP1, the first entry is the estimated probability of a type I error using Eq. (11.2), and the second entry is the probability of at least one type I error when using Eq. (11.4). The same is true for the columns headed by VP2 and VP3. For brevity, only results where x has a symmetric distribution are shown. So

Table 11.3: Values of $\hat{\alpha}$ Using $\hat{\beta}_m$, $\alpha = 0.05$, $n = 20$.

\(x\)		\(\epsilon\)				
g	h	g	h	VP1	VP2	VP3
0.0	0.0	0.0	0.0	.054	.065	.050
0.0	0.0	0.0	0.5	.051	.064	.051
0.0	0.0	0.5	0.0	.057	.066	.066
0.0	0.0	0.5	0.5	.055	.065	.049
0.0	0.5	0.0	0.0	.058	.070	.034
0.0	0.5	0.0	0.5	.057	.067	.035
0.0	0.5	0.5	0.0	.058	.069	.036
0.0	0.5	0.5	0.5	.059	.069	.037
0.5	0.0	0.0	0.0	.049	.071	.051
0.5	0.0	0.0	0.5	.049	.064	.047
0.5	0.0	0.5	0.0	.051	.068	.053
0.5	0.0	0.5	0.5	.050	.065	.050
0.5	0.5	0.0	0.0	.054	.072	.043
0.5	0.5	0.0	0.5	.054	.071	.047
0.5	0.5	0.5	0.0	.056	.071	.044
0.5	0.5	0.5	0.5	.056	.071	.044

Table 11.4: Values of $\hat{\alpha}$ Using Eqs. (11.2), (11.4) and the Theil–Sen Estimator, $\alpha = 0.05$, $n = 20$.

x	ϵ		VP1		VP2		VP3	
h	g	h	(11.2)	(11.4)	(11.2)	(11.4)	(11.2)	(11.4)
0.0	0.0	0.0	.036	.033	.037	.043	.017	.030
0.0	0.0	0.0	.010	.031	.013	.038	.007	.029
0.0	0.0	0.5	.020	.033	.034	.041	.013	.030
0.0	0.0	0.5	.008	.032	.010	.037	.001	.031
0.5	0.0	0.0	.015	.033	.036	.039	.007	.026
0.5	0.0	0.0	.008	.032	.029	.036	.004	.032
0.5	0.0	0.5	.008	.035	.029	.036	.004	.032
0.5	0.0	0.5	.004	.031	.013	.039	.002	.032

for n small, using Eq. (11.2) can result in type I error probabilities considerably smaller than the nominal level.

11.1.4 R Functions regci and wlogregci

The R function

$$\text{regci(x,y,regfun=tsreg,nboot=599,alpha=0.05)}$$

is supplied for computing confidence intervals for regression parameters with the percentile bootstrap method just described. Here, x can be a vector or a matrix having n rows and p columns. The optional argument regfun can be any R function that estimates regression parameters and returns the results in regfun$coef. The first element of regfun$coef is assumed to be the estimated intercept, the second element is the estimate of β_1, and so on. Regression methods that come with R follow this convention, as do all of the R regression functions written for this book. For example, bmreg returns the estimated values in bmreg$coef. If unspecified, regfun is tsreg which is the Theil–Sen estimator. The default value for nboot, which is the number of bootstrap samples to be used, B, is 599.

■ Example

As a simple illustration, consider the data

x	-80 79 -90 11 137 141 116 -54 92 -58 -9 -96 -27 -135
	76 -56 19 -93 -19 -158
y	7 56 -84 -69 88 103 -102 -82 25 84 -69 -78 -127 50
	210 -51 120 -212 174 -72

The 0.95 confidence interval for β_1, returned by the command regci(x,y), is $(-0.0225,$ 1.21). If the function lsfitci is used instead, the 0.95 confidence interval is $(-0.0147,$ 1.08). Note that this interval is shorter than the interval based on regci. This was expected because both x and ϵ were generated from normal distributions with $\beta_1 = 1$. ∎

The command regci(x,y,regfun=lsfit) would return a confidence interval based on the OLS estimator, using the standard percentile bootstrap method, but this is not recommended for reasons already explained. However, regci appears to give good results when working with nearly all of the robust regression methods described in this chapter. (Exceptions are noted in Sections 11.1.5 and 11.1.7)

∎ Example

For the star data in Figure 6.3, the 0.95 confidence interval for the slope returned by the R function regci, using the default estimator, is $(-0.78, 5.03)$. If the bounded influence M regression method is used instead, $\hat{\beta}_m$, the 0.95 confidence interval for the slope is $(-1.076, 2.436)$. So even among robust estimators, the estimator used can make a practical difference when computing confidence intervals. ∎

∎ Example

For the tree data used in the last example of Section 11.1.2, the hypothesis H_0: $(\beta_0, \beta_2) = (-7.5, 1)$ was rejected using M regression with Schweppe weights. The 0.95 confidence intervals for these two parameters returned by regci, again using M regression with Schweppe weights (i.e., setting regfun=bmreg when using regci) are $(-9.1, -4.9)$ and $(0.65, 1.76)$, respectively, suggesting that the hypothesized values for β_0 and β_2 are reasonable. This illustrates the well-known result that confidence intervals can fail to reject when an omnibus test rejects. (The reason is that the confidence region used by the omnibus test is an ellipse, versus a rectangular confidence region when computing confidence intervals for the individual parameters. See Fairley, 1986, for more details.) ∎

For the special case where y is binary and the robust estimator given by Eq. (10.16) is used, the R function

wlogregci(x,y,nboot=400,alpha=0.05,SEED=T,MC=F, xout=F,outfun=out, ...)

can be used.

11.1.5 Methods Based on the Quantile Regression Estimator

Section 10.13.8 described a nonbootstrap R function for making inferences about the parameters associated with a quantile regression estimator. Two limitations of the method are that it can be used only when testing at the $\alpha = 0.05$ level, and it does not provide a way of testing the omnibus hypothesis that two or more parameters are equal to zero. Switching to a percentile bootstrap method, simulations indicate that type I error probabilities greater than the nominal level are avoided. But the actual level can drop well below the nominal level when the sample size is small. A slightly better approach appears to be one based in part on a bootstrap estimate of the standard errors, but again the actual level can be lower than intended. Here, an adjustment is made for dealing with this problem that was suggested by Wilcox and Costa (2009).

Generate B bootstrap estimates of the slope yielding $b_1^*, \ldots, b_B^*$. Then an estimate of the squared standard error of b_1 is

$$S^2 = \frac{1}{B-1} \sum_{b=1}^{B} (b_b^* - \bar{b})^2,$$

where $\bar{b} = \sum b_b^*/B$. So an approximate $1 - \alpha$ confidence interval for β_1 is

$$b_1 \pm z_{1-\alpha/2} S,$$

where $z_{1-\alpha/2}$ is the $1 - \alpha/2$ quantile of a standard normal distribution.

To avoid type I error probabilities well below the nominal level when the sample sizes are small, Wilcox and Costa found that the following adjusted critical values perform reasonably well in simulations:

1. if $\alpha = 0.1$, $z_a = 1.645 - 1.19/\sqrt{(n)}$
2. if $\alpha = 0.05$, $z_a < -1.96 - 1.37/\sqrt{(n)}$
3. if $\alpha = 0.025$, $z_a = 2.24 - 1.18/\sqrt{(n)}$
4. if $\alpha = 0.01$, $z_a = 2.58 - 1.69/\sqrt{(n)}$.

That is, an approximate $1 - \alpha$ confidence interval for β_1 is taken to be

$$b_1 \pm z_a S.$$

This approximation appears to work well when estimating the γth quantile regression line when $0.2 \leq \gamma \leq 0.8$.

As for testing the global hypothesis given by Eq. (11.1), that all slope parameters are equal to zero, take a bootstrap sample in the usual manner and label the resulting estimate of the slopes

b_k^*, $k = 1, \ldots, p$. Repeat this process B times yielding $b_{1k}^*, \ldots b_{Bk}^*$. An estimate of the variances and covariances associated with $b_1, \ldots, b_p$ is

$$S = \frac{1}{B-1} \sum_{c=1}^{B} (\mathbf{b}_c^* - \bar{\mathbf{b}})^2,$$

where $\mathbf{b}_\mathbf{c}^* = (b_{c1}^*, \ldots, b_{cp}^*)$, $\bar{\mathbf{b}} = (\bar{b}_1^*, \ldots, \bar{b}_p^*)$ and $\bar{b}_k^* = \sum b_{bk}^*/B$. A reasonable test statistic is

$$T^2 = n\bar{\mathbf{b}}'\mathbf{S}^{-1}\bar{\mathbf{b}}. \tag{11.5}$$

And from basic principles, a natural strategy is to reject if

$$T^2 \geq \frac{n-1}{n-p} f_{p,n-p},$$

where $f_{p,n-p}$ is the $1 - \alpha$ quantile of an F distribution with p and $n - p$ degrees of freedom. All indications are that the actual probability of type I error is less than the nominal level when the sample size is small, particularly as the number of predictors increases. For example, when $\gamma = 0.5$, $p = 2$, $n = 20$, $\alpha = 0.05$, and x_1 and x_2 have a bivariate normal distribution with correlation $\rho = 0$, the actual type I error probability is approximately .026. Increasing p to 6, the estimate is now .001. But with $n = 60$, the actual probability of a type I error has been found to be reasonably close to .05 (Wilcox, 2007). Adjusted critical values, when $n < 60$ and $\alpha = 0.1, 0.05, 0.025$ and 0.01 are reported by the R function rqtest described in the next section.

It is briefly noted that He and Zhu (2003) derived a method for testing the hypothesis that a specified family of quantile regression models fits the data. In particular, one can test the hypothesis that for some choice for $\beta_0, \ldots, \beta_p$, the γ quantile of y, given $x_1, \ldots, x_p$, is given by

$$y = \beta_0 + \beta_1 x_1 + \cdots + \beta_p x_p.$$

A simple variation of their method has been found to reduce execution time considerably (Wilcox, 2008b). The details are omitted, but an R function (called qrchk) is supplied for performing the analysis.

11.1.6 R Functions rqtest, qregci, and qrchk

The R function

```
rqtest(x,y,qval=0.5,nboot=200,alpha=0.05,SEED=T,xout=F,outfun=out, ...)
```

tests the hypothesis that p slope parameters are equal to zero, assuming the parameters are estimated via the quantile regression method. Reject if the reported p-value is less than or equal to the value stored in adjusted.alpha. For situations where an adjusted critical value cannot be computed, or when $n > 60$, the function sets adjusted.alpha equal to alpha.

The R function

$$\text{qregci(x,y,qval=0.5,nboot=200,alpha=0.05,SEED=T,xout=F,outfun=out, ...)}$$

computes confidence intervals for the individual parameters.

The R function

$$\text{qrchk(x, y, qval} = 0.5, \text{nboot} = 1000, \text{com.pval} = F, \text{SEED} = T, \text{alpha} = 0.05, \text{pr} = T,$$
$$\text{xout} = F, \text{outfun} = \text{out, ...)}$$

tests the hypothesis that for some $\beta_0, \ldots, \beta_p$, the γ quantile of y, given $x_1, \ldots, x_p$, is given by

$$y = \beta_0 + \beta_1 x_1 + \cdots + \beta_p x_p.$$

The function contains appropriate critical values when testing at the 0.1, 0.05, 0.025, and 0.01 levels. A p-value can be computed by setting the argument com.pval=T, which will increase execution time considerably. Reject the null hypothesis if the test statistic exceeds the critical value.

11.1.7 Inferences Based on the OP-Estimator

When using the skipped estimators in Section 10.10 and when $p > 1$, the bootstrap methods in Sections 11.1.1 and 11.1.3 tend to be too conservative in terms of type I errors when the sample size is small. That is, when testing at the 0.05 level, the actual probability of a type I error tends to be considerably less than 0.05 when the sample size is less than 60 (Wilcox, 2004d). Accordingly, the following modifications are suggested when using the OP-estimator. When testing (11.1), the hypothesis that all slope parameters are zero, let $\hat{p}^*$ be the bootstrap estimate of the p-value given in Section 11.1.1. Let $n_a = n$ if $20 \leq n \leq 60$. If $n < 20$, $n_a = 20$, and if $n > 60$, $n_a = 60$. Then the adjusted p-value used here is

$$\hat{p}_a^* = \frac{\hat{p}^*}{2} + \left(\frac{n_a - 20}{40}\right) \frac{\hat{p}^*}{2},$$

and the null hypothesis is rejected if $\hat{p}_a^* \leq \alpha$.

As for testing hypotheses about the individual slope parameters, let

$$C = 1 - \frac{60 - n_a}{80},$$

and let $\hat{p}_m^*$ be computed as in Section 11.1.3. Then the adjusted p-value is

$$\hat{p}_a^* = C \hat{p}_m^*.$$

To control FWE (the probability of at least one type I error), Hochberg's (1988) method is used. For convenience, let Q_j be the adjusted p-value associated with the bootstrap test of H_0: $\beta_j = 0$. Put the Q_j values in descending order yielding $Q_{[1]} \geq Q_{[2]} \geq \cdots \geq Q_{[p]}$. Beginning with $k = 1$, reject all hypotheses if

$$Q_{[k]} \leq \alpha/k.$$

That is, reject all hypotheses if the largest p-value is less than or equal to α. If $Q_{[1]} > \alpha$, proceed as follows:

1.   Increment k by 1. If

$$Q_{[k]} \leq \frac{\alpha}{k},$$

stop and reject all hypotheses having a p-value less than or equal $Q_{[k]}$.
2.   If $Q_{[k]} > \alpha/k$, repeat step 1.
3.   Repeat steps 1 and 2 until a significant result is obtained or all p hypotheses have been tested.

Table 11.5 shows simulation estimates of the actual probability of a type I error using the adjusted p-values just described, where BD indicates the omnibus test using the bootstrap depth method, H indicates Hochberg's method as just described, and $n = 20$.

■ Example

Suppose x_1, x_2, and ϵ are independent and have standard normal distributions, and that the goal is to test H_0: $\beta_2 = 0$ at the 0.05 level with $n = 20$ assuming that $y = x_1 + x_2 + \epsilon$. Further imagine that unknown to us, $y = x_1 + x_1 x_2 + \epsilon$. Using the conventional Student's t-test of H_0: $\beta_2 = 0$, the actual probability of rejecting is approximately .16 (based on a simulation with 1000 replications using the built-in R functions lm and summary). Increasing n to 100, the actual probability of rejecting is again approximately .16. Using instead the R function opregpb (described in the next section), with $n = 20$, the probability of rejecting is approximately .049. ■

11.1.8 R Functions opregpb and opregpbMC

The R function

opregpb(x,y,nboot=1000,alpha=0.05,om=T,ADJ=T,nullvec=rep(0, ncol(x) + 1), plotit=T,
    gval = sqrt(qchisq(0.95,ncol(x) + 1)))

Table 11.5: Values of $\hat{\alpha}$ Using the Method in Section 11.1.5.

x		ε		VP	$\hat{\alpha}$		x		ε		VP	$\hat{\alpha}$	
g	*h*	*g*	*h*		BD	H	*g*	*h*	*g*	*h*		BD	H
0.0	0.0	0.0	0.0	1	.039	.027	0.5	0.0	0.0	0.0	1	.037	.023
0.0	0.0	0.0	0.0	2	.035	.024	0.5	0.0	0.0	0.0	2	.021	.021
0.0	0.0	0.0	0.0	3	.043	.027	0.5	0.0	0.0	0.0	3	.033	.022
0.0	0.0	0.0	0.5	1	.026	.024	0.5	0.0	0.0	0.5	1	.023	.018
0.0	0.0	0.0	0.5	2	.024	.016	0.5	0.0	0.0	0.5	2	.022	.018
0.0	0.0	0.0	0.5	3	.025	.025	0.5	0.0	0.0	0.5	3	.021	.020
0.0	0.0	0.5	0.0	1	.026	.024	0.5	0.0	0.5	0.0	1	.026	.018
0.0	0.0	0.5	0.0	2	.031	.019	0.5	0.0	0.5	0.0	2	.022	.022
0.0	0.0	0.5	0.0	3	.029	.014	0.5	0.0	0.5	0.0	3	.023	.017
0.0	0.0	0.5	0.5	1	.014	.017	0.5	0.0	0.5	0.5	1	.018	.017
0.0	0.0	0.5	0.5	2	.020	.020	0.5	0.0	0.5	0.5	2	.021	.018
0.0	0.0	0.5	0.5	3	.019	.020	0.5	0.0	0.5	0.5	3	.019	.016
0.0	0.5	0.0	0.0	1	.024	.023	0.5	0.5	0.0	0.0	1	.019	.019
0.0	0.5	0.0	0.0	2	.017	.014	0.5	0.5	0.0	0.0	2	.024	.014
0.0	0.5	0.0	0.0	3	.021	.020	0.5	0.5	0.0	0.0	3	.018	.020
0.0	0.5	0.0	0.5	1	.015	.017	0.5	0.5	0.0	0.5	1	.012	.019
0.0	0.5	0.0	0.5	2	.017	.013	0.5	0.5	0.0	0.5	2	.015	.013
0.0	0.5	0.0	0.5	3	.015	.018	0.5	0.5	0.0	0.5	3	.013	.017
0.0	0.5	0.5	0.0	1	.019	.021	0.5	0.5	0.5	0.0	1	.018	.017
0.0	0.5	0.5	0.0	2	.020	.017	0.5	0.5	0.5	0.0	2	.021	.018
0.0	0.5	0.5	0.0	3	.010	.016	0.5	0.5	0.5	0.0	3	.019	.016
0.0	0.5	0.5	0.5	1	.010	.012	0.5	0.5	0.5	0.5	1	.006	.013
0.0	0.5	0.5	0.5	2	.022	.012	0.5	0.5	0.5	0.5	2	.018	.015
0.0	0.5	0.5	0.5	3	.010	.013	0.5	0.5	0.5	0.5	3	.006	.016

tests hypotheses based on the OP-estimator and modified bootstrap method just described. Both an omnibus test and confidence intervals for the individual parameters are reported. To avoid the omnibus test, set om=F. The argument gval is the critical value used by the projection-type outlier detection method. Setting ADJ=F, the adjustments of the *p*-values, described in Section 11.1.5, are not made. The function

$$\text{opregpbMC(x,y,nboot=1000,alpha=0.05,om=T,ADJ=T,nullvec=rep(0, ncol(x) + 1),}$$
$$\text{plotit=T, gval = sqrt(qchisq(0.95,ncol(x) + 1)))}$$

is the same as opregpb, only it uses a multi-core processor, assuming one is available and that the R package multicore has been installed.

11.1.9 Hypothesis Testing when Using the Multivariate Regression Estimator RADA

Consider again the multivariate regression model in Section 10.17.1 where

$$\mathbf{y} = \mathbf{B}'\mathbf{x} + \mathbf{a} + \epsilon, \tag{11.6}$$

B is a $(p \times q)$ slope matrix, **a** is a q-dimensional intercept vector, and the errors $\epsilon = (\epsilon_1, \ldots, \epsilon_q)$ are independent and identically distributed with mean **0** and covariance matrix Σ_ϵ, a positive definite matrix of size q. (That is, for any nonzero vector **x**, $\mathbf{x}'\Sigma_\epsilon\mathbf{x} > 0$.) When using the multivariate regression estimator RADA, in Section 10.17.1, consider the issue of testing

$$H_0 : \mathbf{B} = \mathbf{0}.$$

A natural guess is to proceed along the lines in Section 11.1.1 and use a simple modification of the R function regtest. But checks on this method have found it to be unsatisfactory in simulations. A percentile bootstrap method appears to avoid type I error probabilities above the nominal level. However, the actual level can be substantially smaller than the nominal level suggesting that power might be relatively poor. Imagine, for example, that with $n = 40$, $p = 2$, and $q = 3$, that a p-value is computed in the usual way. Under normality, an actual type I error probability of .05 is achieved if the null hypothesis is rejected when the estimated p-value is less than or equal to .16.

Currently, only one method has been found that performs reasonably well in terms of controlling the type I error probability, including situations where there is heteroscedasticity. Briefly, let **C** be a row vector of length pq containing the pq slope estimates. Let $\hat{\Sigma}$, a pq-by-pq matrix, be a bootstrap estimate of the variances and covariances associated with **C**. The test statistic is

$$\frac{1}{pq}\mathbf{C}\hat{\Sigma}^{-1}\mathbf{C}',$$

with the null distribution taken to be an F distribution with $\nu_1 = pq - 1$ and $\nu_2 = n - pq$ degrees of freedom.

11.1.10 R Function mlrGtest

The R function

$$\text{mlrGtest(x,y,regfun=mlrreg,nboot=300,SEED=T)}$$

tests the hypothesis

$$H_0 : \mathbf{B} = \mathbf{0},$$

using the method just described. By default, the RADA estimator is used. Any multivariate regression estimator could be used via the argument regfun. Currently, however, simulation results regarding the ability of the method to control the probability of a type I error are limited to the RADA estimator.

11.1.11 Robust ANOVA via Dummy Coding

As noted in Section 10.1.4, a well-known approach to comparing the means of multiple groups is via least squares regression coupled with dummy coding (e.g., Montgomery & Peck, 1992). An issue of interest is whether this approach might be generalized by replacing the least squares regression estimator with one of the robust estimators described in Chapter 10. When using the Theil–Sen estimator, simulations do not support this strategy: control over the type I error probability can be poor (Ng, 2009b). Whether a similar problem occurs when using some other robust regression estimator has not been investigated.

11.2 Comparing the Parameters of Two Independent Groups

For two independent groups, let $(y_{ij}, \mathbf{x}_{ij})$ be the ith vector of observations in the jth group, $i = 1, \ldots, n_j$. Suppose

$$y_{ij} = \beta_{0j} + \mathbf{x}'_{1j}\beta_j + \lambda_j(\mathbf{x}_j)\epsilon_{ij},$$

where $\beta_j = (\beta_{1j}, \ldots, \beta_{pj})'$ is a vector of slope parameters for the jth group, $\lambda_j(\mathbf{x}_j)$ is some unknown function of $\mathbf{x}_j$, and ϵ_{ij} has variance σ_j^2. This section considers the problem of computing a $1 - \alpha$ confidence interval for $\beta_{k1} - \beta_{k2}$, $k = 1, \ldots, p$, the difference between the slope parameters associated with the kth predictor. It is not assumed that ϵ_{i1} and ϵ_{i2} have a common variance. Moreover, the goal is to get an accurate confidence interval without specifying what λ might be. That is, for each group, the error term can be heteroscedastic, and nothing is assumed about how the variance of the error terms, corresponding to each group, are related. The conventional method for comparing slope parameters (e.g., Huitema, 1980) performs poorly when standard assumptions are violated. Conerly and Mansfield (1988) provide references to other solutions. Included is Chow's (1960) likelihood ratio test, which is also known to fail. (For results related to the method described here, see Wilcox, 1997c.)

When using any robust estimator with a reasonably high breakdown point, the percentile bootstrap technique appears to give reasonably accurate confidence intervals for a fairly broad range of nonnormal distributions and heteroscedastic error terms. This suggests a method for addressing the goal considered here, and simulations support its use. Briefly, the procedure begins by generating a bootstrap sample from the jth group as described, for example, in Section 11.1.1 That is, for the jth group, randomly sample n_j vectors of observations, with replacement, from $(y_{1j}, \mathbf{x}_{1j}), \ldots, (y_{n_j j}, \mathbf{x}_{n_j j})$. Let $d_k^* = \hat{\beta}_{k1}^* - \hat{\beta}_{k2}^*$ be the difference between the resulting estimates of the kth predictor, $k = 1, \ldots, p$. Repeat this process B times yielding $d_{k1}^*, \ldots, d_{kB}^*$. Put these B values in ascending order yielding $d_{k(1)}^* \leq \cdots \leq d_{k(B)}^*$. Let $\ell = \alpha B/2$, $u = (1 - \alpha/2)B$, rounded to the nearest integer, in which case an approximate $1 - \alpha$ confidence interval for $\beta_{k1} - \beta_{k2}$ is $(d_{k(\ell+1)}^*, d_{k(u)}^*)$.

To provide some indication of how well the method performs when computing a 0.95 confidence interval, Tables 11.6 and 11.7 show $\hat{\alpha}$, an estimate of one minus the probability coverage, when $n = 20$, $p = 1$, and M regression with Schweppe weights is used. In these tables, VP refers to three types of error terms: $\lambda(x) = 1$, $\lambda(x) = x^2$, and $\lambda(x) = 1 + 2/(|x| + 1)$. For convenience, these three variance patterns are labeled VP1, VP2, and VP3, respectively. The situation VP2 corresponds to large error variances when the value of x is in the tails of its distribution, and VP3 is the reverse. Three additional conditions are considered as well. The first, called C1, is where x_{i1} and x_{i2}, as well as ϵ_{i1} and ϵ_{i2}, have identical distributions. The second condition, C2, is the same as the first condition only $\epsilon_{i2} = 4\epsilon_{i1}$. The third condition, C3, is where for the first group, both x_{i1} and ϵ_{i1} have standard normal distributions, but for the second group, both x_{i2} and ϵ_{i2} have a g-and-h distribution.

Notice that $\hat{\alpha}$ never exceeds .06, and in general it is less than .05. There is room for improvement, however, because in some situations $\hat{\alpha}$ drops below .020. This happens when ϵ has a heavy-tailed distribution, as would be expected based on results in Chapters 4 and 5.

Table 11.6: Values of $\hat{\alpha}$, Using the Method in Section 11.2 when x has a Symmetric Distribution, $n = 20$.

x		ϵ			Condition		
g	h	g	h	VP	C1	C2	C3
0.0	0.0	0.0	0.0	1	.029	.040	.040
				2	.042	.045	.045
				3	.028	.039	.039
0.0	0.0	0.0	0.5	1	.029	.036	.036
				2	.045	.039	.039
				3	.025	.037	.037
0.0	0.0	0.5	0.0	1	.026	.040	.041
				2	.043	.043	.043
				3	.029	.040	.040
0.0	0.0	0.5	0.5	1	.028	.036	.036
				2	.042	.040	.040
				3	.023	.037	.037
0.0	0.5	0.0	0.0	1	.024	.035	.040
				2	.051	.058	.046
				3	.014	.023	.039
0.0	0.5	0.0	0.5	1	.023	.035	.036
				2	.049	.054	.039
				3	.013	.020	.037
0.0	0.5	0.5	0.0	1	.022	.039	.040
				2	.050	.039	.043
				3	.014	.022	.040
0.0	0.5	0.5	0.5	1	.024	.037	.036
				2	.052	.058	.040
				3	.013	.020	.037

Table 11.7: Values of $\hat{\alpha}$, x has an Asymmetric Distribution, $n = 20$.

X		ϵ			Condition		
g	h	g	h	VP	C1	C2	C3
0.5	0.0	0.0	0.0	1	.026	.040	.040
				2	.044	.048	.046
				3	.032	.037	.039
0.5	0.0	0.0	0.5	1	.028	.039	.036
				2	.041	.047	.039
				3	.030	.031	.037
0.5	0.0	0.5	0.0	1	.025	.040	.040
				2	.046	.052	.040
				3	.032	.038	.043
0.5	0.0	0.5	0.5	1	.024	.038	.036
				2	.041	.045	.040
				3	.031	.034	.037
0.5	0.5	0.0	0.0	1	.018	.032	.040
				2	.049	.050	.046
				3	.019	.020	.039
0.5	0.5	0.0	0.5	1	.019	.031	.036
				2	.045	.049	.039
				3	.015	.018	.037
0.5	0.5	0.5	0.0	1	.019	.031	.040
				2	.050	.050	.043
				3	.014	.020	.040
0.5	0.5	0.5	0.5	1	.022	.027	.036
				2	.046	.051	.040
				3	.016	.019	.037

Also, VP3, which corresponds to large error variances when x is near the center of its distribution, plays a role. Despite this, all indications are that, in terms of probability coverage, the bootstrap method in conjunction with $\hat{\beta}_m$ (M regression with Schweppe weights) performs reasonably well over a broader range of situations than any other method that has been proposed, and using M regression offers the additional advantage of a relatively efficient estimator for the situations considered. It appears that when using other robust estimators such as Theil–Sen, good probability coverage is again obtained, but extensive simulations have not yet been performed.

11.2.1 R Function reg2ci

The R function reg2ci

```
reg2ci(x1,y1,x2,y2,regfun=tsreg,nboot=599,alpha=0.05,plotit=T)
```

computes a 0.95 confidence interval for the difference between regression slope parameters corresponding to two independent groups. The first two arguments contain the data for the first group, and the data for group 2 are contained in x2 and y2. The optional argument regfun indicates the regression method to be used. If not specified, tsreg, the Theil–Sen estimator, is used. Setting regfun=bmreg results in using $\hat{\beta}_m$, M regression with Schweppe weights. The default number of bootstrap samples, nboot, is $B = 599$, and alpha, which is α, defaults to 0.05 if unspecified. When the argument plotit equals T (for true), the function also creates a scatterplot that includes the regression lines for both groups.

■ Example

A controversial issue is whether teachers' expectancies influence intellectual functioning. A generic title for studies that address this issue is Pygmalion in the classroom. Rosenthal and Jacobson (1968) argue that teachers' expectancies influence intellectual functioning, and others argue that it does not. A brief summary of some of the counterarguments can be found in Snow (1995). Snow illustrates his concerns with data collected by Rosenthal where children in grades 1 and 2 were used. Here, other issues are examined using robust regression methods.

■

One of the analyses performed by Rosenthal involved comparing an experimental group of children, for whom positive expectancies had been suggested to teachers, to a control group for whom no expectancies had been suggested. One measure was a reasoning IQ pretest score, and a second was a reasoning IQ posttest score. The data are given in Elashoff and Snow (1970) and they are stored in the files pyge.dat and pygc.dat. (The file pyge.dat contains the results for the experimental group, and pgyc.dat contains data for the control.) If the posttest scores are compared using Yuen's method for trimmed means, the 0.95 confidence interval for the difference between the 20% trimmed means is $(-30.76, -5.04)$ with a p-value of .009. Comparing means instead, the p-value is .013.

Suppose that the control data are stored in pygcx and pygcy, and the data for the experimental group are stored in pygex and pygey. The command

$$\text{reg2ci(pygcx,pygcy,pygex,pygey,regfun=bmreg)}$$

indicates that a 0.95 confidence interval for the difference between the slopes, using M regression with Schweppe weights, is $(-1.5, 0.40)$. That is, there might be little or no difference between the slopes. The 0.95 confidence interval for the difference between the intercepts is $(-47.4, 97.8)$ suggesting that there might be little or no difference between the groups when the pretest scores are taken into account. Switching to the (default) Theil–Sen estimator, the 0.95 confidence interval for the difference between the slopes is $(-1.34, 0.35)$.

There are many concerns about this data that go beyond the scope of this book, some of which deal with psychometric issues, so it is not being suggested that the analysis just presented resolves the controversy associated with this topic. The only goal is to illustrate the method for comparing slopes and intercepts, and to demonstrate that a different perspective can result compared to ignoring the pretest scores.

11.3 Detecting Heteroscedasticity

Yet another way of establishing dependence between some outcome variable y and some predictor x is to test the hypothesis that the (conditional) variation of y, given x, does not vary with x. Among the many methods that have been proposed, most do not perform well in simulations. Two that do perform well are described in this section.

The better-known regression models, which are routinely used, assume homoscedasticity. It is *not* being suggested that the methods in this section be used to justify homoscedastic techniques. That is, if a test of the assumption of homoscedasticity fails to reject, it is not recommended that a homoscedastic regression model would then be used. The reason is that it is unclear when the power of the methods in this section will be high enough to detect a departure from the usual homoscedastic regression model that is important. Rather, the methods in this section are intended as a method for establishing that a certain type of dependence is present. Situations are encountered where the methods in this section reject, yet methods aimed at testing the hypothesis that the slope parameters are equal to zero fail to reject. And methods for testing the hypothesis of a zero correlation can fail to reject as well.

11.3.1 A Quantile Regression Approach

Let

$$y_\gamma = \alpha_\gamma + \beta_\gamma x$$

be the regression line for predicting the γth quantile of y, given x. The goal is to test

$$H_0 : \beta_{.2} = \beta_{.8}$$

which represents a type of homoscedasticity. Of course, other quantiles might be used. The strategy is to choose quantiles different enough to help achieve relatively high power. But if the quantiles are too close to 0 and 1, controlling the type I error probability can be difficult. Wilcox and Keselman (2006b) considered several methods and found the following to be best among those that were considered.

Compute a bootstrap estimate of the standard error of $d = b_{0.2} - b_{0.8}$ and label the result s_d^*. Here, $B = 100$ is used. Then an appropriate test statistic is

$$T = \frac{d}{s_d^*}.$$

Assuming that this test statistic has approximately a standard normal distribution, the actual type I error probability was found to be less than the nominal level among the situations considered by Wilcox and Keselman (2006b). But even for $n = 100$, the actual level can drop well below the nominal level. (The same problem occurs when using a percentile bootstrap method.) When testing at the 0.05 level, Wilcox and Keselman suggest using an approximate critical value given by

$$q = \Phi^{-1}\left(\frac{-0.104}{\sqrt{n}} + 0.975\right),$$

where Φ^{-1} is the inverse of the cumulative standard normal distribution. That is, reject if $|T| \geq q$.

11.3.2 Koenker's Method

In terms of controlling the probability of a type I error when testing the hypothesis of homoscedasticity, Koenker's (1981) method has been found to perform well in simulations by Lyon and Tsai (1996) as well as Wilcox and Keselman (2006b). The method begins by fitting an ordinary least squares regression line. Let r_i be the usual residuals ($i = 1, \ldots, n$). If the null hypothesis is true, then

$$\hat{\sigma}^2 = \frac{1}{n}\sum r_i^2$$

provides an estimate of the common variance. Let $A = \sum(r_i^2 - \hat{\sigma}^2)^2/n$ and $\tilde{Y} = \sum \hat{Y}_i/n$. The test statistic is

$$V = \frac{\{\sum r_i^2(\hat{Y}_i - \tilde{Y})\}^2}{A\sum(\hat{Y}_i - \tilde{Y})^2},$$

which has, approximately, a chi-squared distribution with 1 degree of freedom when the null hypothesis is true.

11.3.3 R Functions qhomt and khomreg

The R function

qhomt(x, y, nboot = 100, alpha = 0.05, qval = c(0.2, 0.8), plotit = T, SEED = T, xlab = "X",
ylab = "Y")

tests the hypothesis

$$H_0 : \beta_{0.2} - \beta_{0.8}.$$

The quantiles that are used can be altered via the argument qval. For example, qval=c(0.25, 0.75) would test H_0: $\beta_{0.25} = \beta_{0.75}$. The R function

khomreg(x, y)

performs Koenker's test.

11.4 Curvature and Half-Slope Ratios

This section describes an approach to dealing with curvature where the strategy is to attempt to straighten a regression line by replacing the x values with x^a for some a to be determined. Here, the so-called half-slope ratio is used to help suggest an appropriate choice for a. Another general approach when dealing with curvature is to use some type of nonparametric regression method described in Section 11.5.

Temporarily consider the situation where there is only one predictor ($p = 1$), and let $m = n/2$, rounded down to the nearest integer. Suppose the x values are divided into two groups: x_L, the m smallest x values, and x_R, the $n - m$ largest. Let y_L and y_R be the corresponding y values. For example, if the (x, y) points are $(1, 6)$, $(8, 4)$, $(12, 9)$, $(2, 23)$, $(11, 33)$, and $(10, 24)$, then the (x_L, y_L) values are $(1, 6)$, $(2, 23)$, and $(8, 4)$; and the (x_R, y_R) values are $(10, 24)$, $(11, 33)$, and $(12, 9)$. That is, the x values are sorted into two groups containing the lowest half and the highest half of the values, and the y values are carried along.

Suppose some regression method is used to estimate the slope using the (x_L, y_L) values yielding say $\hat{\beta}_L$. Similarly, let $\hat{\beta}_R$ be the slope corresponding to the other half of the data. The *half-slope ratio* is $H = \hat{\beta}_R/\hat{\beta}_L$. If the regression line is straight, then H should have a value reasonably close to 1. In principle, the method can be extended to $p \geq 1$ predictors. But the practical utility of the method is unclear. One simply chooses a particular predictor, say the kth, and then divides the vectors of observations into two groups, the first containing the lowest m values of the kth predictor, and the second containing the $n - m$ highest. Simultaneously the y values and all of the remaining predictor values are carried along. That is, rows of data are sorted according to the values of the predictor being considered.

Next, estimate the p slope parameters for the first group, yielding $\hat{\beta}_{Lk1}, \ldots, \hat{\beta}_{Lkp}$, where the second subscript, k, indicates that the data are divided into two groups based on the kth predictor. Do the same for the second group of observations yielding $\hat{\beta}_{Rk1}, \ldots, \hat{\beta}_{Rkp}$, and the half-slope ratios are

$$H_{k\ell} = \hat{\beta}_{Rk\ell}/\hat{\beta}_{Lk\ell},$$

$k = 1, \ldots, p$ and $\ell = 1, \ldots, p$. That is, $H_{k\ell}$ is the ratio of the estimated regression slopes for the ℓth predictor when the data are split using the kth predictor.

It is stressed that the half-slope ratio is an exploratory tool that should be used in conjunction with other techniques such as the smoothers described in Section 11.5. A practical advantage is that it might suggest a method of straightening the regression line. In simple regression ($p = 1$), it can suggest a choice for a such that the regression line $y = \beta_1 x^a + \beta_0$ gives a better fit to data, as will be illustrated. However, even when the half-slope ratio appears to be substantially different from 1, replacing x with x^a might have little practical advantage, as will be seen. Also, the half-slope ratio can be highly misleading when, for example, the usual linear model is correct but the slope parameters are close to zero.

11.4.1 R Function hratio

The R function

$$\text{hratio(x,y,regfun=bmreg)}$$

computes the half-slope ratios as just described. As usual, x is an n-by-p matrix containing the predictors. The optional argument regfun can be any R regression function that returns the estimated coefficients in regfun$coef. If regfun is not specified, bmreg (the bounded influence M regression estimator with Schweppe weights) is used.

The function returns a p-by-p matrix. The first row reports the half-slope ratios when the data are divided into two groups using the first predictor. The first column is the half-slope ratio for the first predictor, the second column is the half-slope ratio for the second predictor, and so forth. The second row contains the half-slope ratios when the data are divided into two groups using the second predictor, and so on.

■ Example

Table 11.8 shows some data on the rate of breast cancer per 100,000 women and the amount of solar radiation received (in calories per square centimeter) within the indicated city. (These data are also stored in the file cancer.dat. See Section 1.8.) The half-slope ratio is estimated to be 0.64 using the default regression function, bmreg. The estimates of the slope and intercept, based on all of the data, are -0.030 and 35.3.

Table 11.8: Breast Cancer Rate Versus Solar Radiation

City	Rate	Daily Calories	City	Rate	Daily Calories
New York	32.75	300	Chicago	30.75	275
Pittsburgh	28.00	280	Seattle	27.25	270
Boston	30.75	305	Cleveland	31.00	335
Columbus	29.00	340	Indianapolis	26.50	342
New Orleans	27.00	348	Nashville	23.50	354
Washington, DC	31.20	357	Salt Lake City	22.70	394
Omaha	27.00	380	San Diego	25.80	383
Atlanta	27.00	397	Los Angeles	27.80	450
Miami	23.50	453	Fort Worth	21.50	446
Tampa	21.00	456	Albuquerque	22.50	513
Las Vegas	21.50	510	Honolulu	20.60	520
El Paso	22.80	535	Phoenix	21.00	520

When the half-slope ratio is between 0 and 1, it might be possible to straighten the regression line by replacing x with x^a, where $0 < a < 1$, but there is no explicit equation for determining what a should be. However, it is a simple matter to try a few values and see what effect they have on the half-slope ratio. When the half-slope ratio is greater than one, try $a > 1$. If $a < 0$, often it is impossible to find an a that gives a better fit to the data (e.g., Velleman and Hoaglin, 1981). In the illustration, the half-slope ratio is less than one, so try $a = 0.5$. Replacing x with x^a, the half-slope ratio increases to 0.8, and $a = 0.2$ increases it to 0.9. In this latter case, the slope and intercept are now estimated to be -1.142 and 46.3. However, a check of the residuals indicates that using $a = 0.2$, rather than $a = 1$, offers little advantage for the data at hand. This is not surprising, based on a cursory examination of a scatterplot of the data, and the smoothers in the Section 11.5 also suggest that $a = 1$ will provide a reasonable fit to the data. However, in some situations, this process proves to be valuable.

■ Example

L. Doi conducted a study on variables that predict reading ability. Two predictors of interest were TAAST1, a measure of phonological awareness (auditory analysis), and SBT1, another measure of phonological awareness (sound blending). (The data are stored in the file read.dat in columns 2 and 3.) One goal is to predict an individual's score on a word identification test, WWISST2 (column 8 in the file read.dat), and more generally to understand the association among the three variables. Published papers typically assume that $y = \beta_1 x_1 + \beta_2 x_2 + \epsilon$. The function hratio returns

```
            [,1]         [,2]
[1,] -0.03739762   0.8340422
[2,]  0.34679647  -0.9352855
```

That is, dividing the data into two groups using the first predictor, the half-slope ratio for the first predictor is estimated to be -0.037, and the second predictor has an estimated half-slope ratio of 0.83. Dividing the data into two groups using the second predictor, the estimates are 0.35 and -0.93. This suggests that a regression plane might be an unsatisfactory representation of how the variables are related. (A method for testing the hypothesis that the regression surface is a plane is described in Section 11.6.1.) ∎

11.5 Curvature and Nonparametric Regression

Roughly, nonparametric regression deals with the problem of estimating a conditional measure of location associated with y, given the p predictors $x_1, \ldots, x_p$, assuming only that this conditional measure of location is given by some unknown function $m(x_1, \ldots, x_p)$. The problem, then, is estimating the function m. This is in contrast to specifying m in terms of unknown parameters, where the best-known approach assumes

$$y_i = \beta_0 + \beta_1 x_{i1} + \cdots + \beta_p x_{ip} + \epsilon_i.$$

There is now a vast literature on estimating m nonparametrically with most methods assuming that the measure of location of interest is the mean (e.g.. Efromovich, 1999; Eubank, 1999; Fan & Gijbels, 1996; Fox, 2001; Green & Silverman, 1993; Györfi, Kohler, Krzyzk, & Walk, 2002; Härdle, 1990; Hastie & Tibshirani, 1990). For $p = 1$ and 2, these techniques provide useful graphical methods for studying curvature. Complete details about all methods cannot be covered here. Instead the focus is on a few methods that appear to have considerable practical value when the focus is on robust measures of location.

11.5.1 Smoothers

Methods for estimating the unknown function m are generally based on what are called smoothing techniques. The basic idea is that if $m(x_1, \ldots, x_p)$ is a smooth function, then among n observations, those points near $(x_1, \ldots, x_p)$ should contain information about the value of m at $(x_1, \ldots, x_p)$. So a crude description of smoothing techniques is that they identify which points, among n vectors of observations, are close to $(x_1, \ldots, x_p)$, and then some measure of location is computed based on the corresponding y values. The result is $\hat{m}(x_1, \ldots, x_p)$, an estimate of the measure of location associated with y at the point $(x_1, \ldots, x_p)$. The estimator $\hat{m}$ is called a *smoother*, and the outcome of a smoothing procedure is called a *smooth* (Tukey, 1977). A slightly more precise description of smoothers is that they are weighted averages of the y values with the weights a function of how close the vector of predictor values is to the point of interest.

11.5.2 Kernel Estimators and Cleveland's LOWESS

To elaborate, first consider the case of a single predictor ($p = 1$) and suppose it is desired to estimate some measure of location for y given x. Let w_i be some measure of how close x_i is to x. Then generally, the estimate of $m(x)$ is taken to be

$$\hat{m}(x) = \sum w_i y_i, \tag{11.7}$$

and the goal is to choose the w_i in some reasonable manner.

Kernel Smoothing

Although many methods aimed at estimating the conditional mean of y, given x, cannot be covered here, a brief description of a few of these methods might help. The first is called *kernel smoothing*. Let the kernel $K(u)$ be a continuous, bounded, and symmetric real function such that

$$\int K(u)du = 1$$

An example is the Epanechnikov kernel in Section 3.2.4. Then an estimate of $m(x)$ is given by Eq. (11.7) where

$$w_i = \frac{1}{W_s} K\left(\frac{x - x_i}{h}\right),$$

$$W_s = \sum K\left(\frac{x - x_i}{h}\right),$$

and h is the span described in Section 3.2.4. Even within the class of kernel smoothers, many variations are possible. In case it is useful, one of these variations is outlined here; it represents a very slight modification of the kernel regression estimator in Fan (1993). (In essence, the description given by Bjerve & Doksum, 1993, is used, but with the span taken to be min $\{s, \text{IQR}/1.34\}$.)

Again let $K(u)$ be the Epanechnikov kernel given in Section 3.2.4 and let

$$h = \min(s, \text{IQR}/1.34).$$

Then given x, $m(x)$ is estimated with $\hat{m}(x) = b_0 + b_1 x$, where b_0 and b_1 are estimated via weighted least squares with weights $w_i = K[(x_i - x)/h]$. A smooth can be created by taking x to be a grid of points and plotting the results. (The method can be extended to multiple predictors using the multivariate extension of the Epanechnikov kernel described at the end of Section 6.9, but often the smooth seems to be a bit lumpy. Altering the span might improve matters, but this has not been investigated.)

Cleveland's LOWESS

Another approach to smoothing was developed by Cleveland (1979) and is generally known as locally weighted scatter plot smoothing (LOWESS). Briefly, let

$$\delta_i = |x_i - x|.$$

Next, sort the δ_i values and retain the fn pairs of points that have the smallest δ_i values, where f is a number between 0 and 1 and plays the role of a span. Let δ_m be the largest δ_i value among the retained points. Let

$$Q_i = \frac{|x - x_i|}{\delta_m},$$

and if $0 \le Q_i < 1$, set

$$w_i = (1 - Q_i^3)^3,$$

otherwise set

$$w_i = 0.$$

Next, use weighted least squares to predict y using w_i as weights (cf. Fan, 1993). That is, determine the values b_1 and b_0 that minimize

$$\sum w_i (y_i - b_0 - b_1 x_i)^2$$

and estimate the mean of y corresponding to x to be $\hat{y} = b_0 + b_1 x$. Because the weights (the w_i values) change with x, generally a different regression estimate of y is used when x is altered. Finally, let $\hat{y}_i$ be the estimated mean of y given that $x = x_i$ based on the method just described. Then an estimate of the regression line is obtained by the line connecting the points $(x_i, \hat{y}_i)$ $(i = 1, \ldots, n)$. (For some interesting comments relevant to lowess versus kernel regression methods, see Hastie & Loader, 1993.)

Cleveland (1979) also discussed a robust version of this method. In effect, extreme y values get little or no weight, the result being that multiple outliers among the y values have little or no impact on the smooth. R provides access to a function, called lowess, that performs the computations. (An outline of the computations can be found in Härdle, 1990, p. 192.) For a smoothing method that is based in part on L_1 regression, see Wang and Scott (1994). (Related results are given by Fan & Hall, 1994.) For a method based in part on M-estimators, see Verboon (1993).

11.5.3 R Functions lplot and kerreg

It is a fairly simple matter to create a plot of the smooth returned by the built-in R function lowess, which applies Cleveland's method described in the previous section. But to facilitate the use of lowess, a function is supplied that creates the plot automatically. It has the form

lplot(x, y, span=0.75, pyhat=F, eout=F, xout=F, outfun=out, plotit=T, expand=0.5,
low.span=2/3, varfun=pbvar, cor.op=F, cor.fun=pbcor, pr=T, scale=F, xlab="X",
ylab="Y", zlab="", theta=50, phi=25, family="gaussian", duplicate="error",
pc="*",ticktype="simple"),

where the argument span is f. (More than one predictor can be handled using a method outlined in Section 11.5.13. With two predictors, setting the argument ticktype="detailed" will create ticks as done when using a two-dimensional plot.) If the argument pyhat=T and the number or predictors is less than or equal to 4, the function returns the $\hat{m}(x_i)$ values, $i = 1, \ldots, n$. If eout=T, the function first eliminates any outliers among the (x_i, y_i) values using the outlier detection method specified by the argument outfun. If xout=T instead, the function removes outliers (leverage points) among the x_i values only. To suppress the plot, set plotit=F. (The argument family is relevant only when $p = 2$; see Section 11.5.13.) The arguments theta and phi can be used to rotate a three-dimensional plot. The arguments xlab, ylab, and zlab indicate labels for the x-axis, y-axis, and z-axis, respectively. The argument varfun is explained in Section 11.9. When dealing with a single predictor, the argument pc controls how points will be represented in the scatterplot. By default, an * is used. But with large n , this can make it difficult seeing the regression line, in which case pc="." might be more satisfactory.

The function

kerreg(x,y,pyhat=F,pts=NA,plotit=T,theta=50,phi=25,expand=0.5,
scale=F,zscale=F,eout=F,xout=F,outfun=out,np=100,xlab="X",ylab="Y",zlab="Z",
varfun=pbvar,e.pow=T,pr=T,ticktype="simple")

creates a smooth using a slight modification of the method derived by Fan (1993). The arguments are the same as those used by the function lplot, except the argument np, which determines how many x values are used when creating the smooth. (Details about the argument np can be found in the R function locreg.)

11.5.4 The Running Interval Smoother

Now we consider how smoothers might be generalized to robust measures of location. To help fix ideas, we momentarily focus on the single predictor case ($p = 1$). One approach to estimating m and exploring curvilinearity is with the so-called running interval smoother. To be concrete, suppose the goal is to use the data in Table 11.8 to estimate the 20% trimmed mean of the breast cancer rate, given that solar radiation is 390. The strategy behind the running interval smoother is to compute the 20% trimmed mean using all of the y_i values for which the corresponding x_i values are close to the x value of interest, 390. The immediate problem is finding a rule for determining which y values satisfy this criterion.

Let f be some constant that is chosen in a manner to be described and illustrated. Then the point x is said to be close to x_i if

$$|x_i - x| \leq f \times \text{MADN},$$

where MADN is computed using $x_1, \ldots, x_n$. So for normal distributions, x is close to x_i if x is within f standard deviations of x_i. Let

$$N(x_i) = \{j : |x_j - x_i| \leq f \times \text{MADN}\}.$$

That is, $N(x_i)$ indexes the set of all x_j values that are close to x_i. Let $\hat{\theta}_i$ be an estimate of some parameter of interest, based on the y_j values such that $j \in N(x_i)$. That is, use all of the y_j values for which x_j is close to x_i. For example, if x_3, x_8, x_{12}, x_{19}, and x_{21} are the only values close to $x = 390$, then the 20% trimmed mean of y, given that $x = 390$, is estimated by computing the 20% sample trimmed mean using the corresponding y values y_3, y_8, y_{12}, y_{19}, and y_{21}. To get a graphical representation of the regression line, compute $\hat{\theta}_i$, the estimated value of y given that $x = x_i$, $i = 1, \ldots, n$, and then plot the points $(x_1, \hat{\theta}_1), \ldots, (x_n, \hat{\theta}_n)$ to gain some indication of how x and y are related. This process will be called a *running interval smoother*. (For an alternative approach for creating smooths using M-estimators and trimmed means based on generalizations of kernel smoothers for means, see Härdle, 1990, Chapter 6. Also see Hall & Jones, 1990.)

A practical problem is choosing f. If there are no ties among the x values, and if f is chosen small enough, the running interval smoother produces a scatterplot of the points. If f is too large, the horizontal line $\hat{y} = \hat{\theta}$ is obtained where $\hat{\theta}$ is the estimate of θ using all n of the y values. The problem, then, is to choose f large enough so that the resulting plot is reasonably smooth, but not too large so as to mask any nonlinear relationship between x and y. Often the choice $f = 1$ gives good results, but both larger and smaller values might be better, particularly when n is small. As with all smoothers, a good method is to try some values within an interactive-graphics environment, the general strategy being to find the smallest f so that the plot of points is reasonably smooth.

The smoother is first illustrated with some data generated from a known model, to demonstrate how well it performs, and then some additional illustrations are given using data from actual studies. For convenience, it is again assumed that the goal is to predict the 20% trimmed mean of y given x. First consider the situation where both x and ϵ are standard

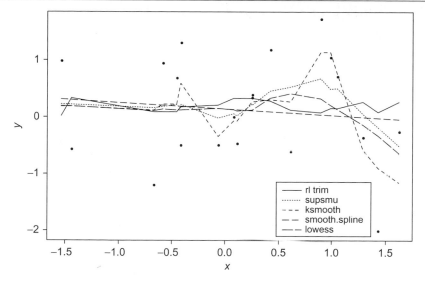

Figure 11.2: Various smoothers, $n = 20$, $f = 1$. The straight line is based on a spline method and gives the best results in this instance. But in other situations, alternative smoothers give superior results.

normal, and $\beta_1 = \beta_0 = 0$. Then the correct regression line is $y = 0$. Figure 11.2 shows the running interval smoother, plus several other smoothers for $n = 20$ points, where both x and ϵ were generated from standard normal distributions. The solid line (labeled rl trim) is the running interval smoother. Note that the running interval smoother does a relatively good job of capturing the true regression line. The additional smoothers that come with R include a kernel smoother, a super smoother (labeled supsmu), Cleveland's method described in Section 11.5.2, and a smoothing spline. (Super smoothers and smoothing splines are discussed in manuals, but the details go beyond the scope of this book.) Of course, this one example is not convincing evidence that the running interval smoother has practical value.

A challenge for any smoother is correctly identifying a straight line when in fact the regression line is straight and n is small. Figure 11.3 illustrates some of the problems that can arise using both the running interval smoother and lowess, described in Section 11.5.2. The upper left panel of Figure 11.3 is based on the same data used in Figure 11.2, but only lowess and the running interval smoother are shown. The upper right panel of Figure 11.3 is the same as the upper left, but two of the y values were altered so that they are now outliers. Note that lowess suggests a curved regression line, although the curvature is small enough that it might be discounted. The lower left panel of Figure 11.3 shows the same data as in the upper right panel, only the furthest point to the right is moved to $(x, y) = (-2.5, -2.5)$. That is, both x and y are outliers. The curvature in lowess is more pronounced, but even the running interval smoother suggests that there is curvature. This is because the largest x value is so far removed from the other x values, the corresponding trimmed mean of y is based on only one value,

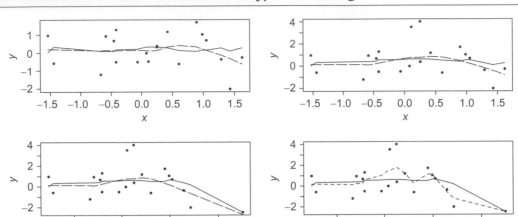

Figure 11.3: **A comparison of a running interval smooth versus some smoothers for means,** $n = 20$. **The upper left panel shows the running interval smooth versus lowess. The upper right panel is the same as the upper left, but with two of the** y **values increased so that they are outliers. The lower left panel is the same as the upper right, but now with three outliers and one leverage point. The lower right panel shows the same points as the lower left with lowess replaced by a kernel smooth.**

$y = -2.5$. One obvious way of dealing with this problem is to check for any outlying or isolated x values, remove them, and see what effect this has on the smoother. The lower right panel of Figure 11.3 shows what happens when lowess is replaced with a kernel smoother used by R.

The left panel of Figure 11.4 shows a running interval smooth, with $f = 0.75$, based on $n = 20$ points generated from the model $y = x^2 + \epsilon$, with both x and ϵ having standard normal distributions. (Using the default $f = 1$ is a bit less satisfactory.) The right panel is based on $n = 40$ and $f = 1$. The dashed line in both panels is the true regression line, $y = x^2$. The solid, ragged line is the estimate of the regression line using the running interval smoother. (For more about the running interval smoother, see Wilcox, 1995f.)

Figure 11.5 shows the results of applying the smoother to various data sets. The first scatterplot is based on data from a study of diabetes in children (Sockett et al., 1987). The upper left panel of Figure 11.5 shows the age in months versus the logarithm of serum C-peptide. Also shown is the smoother resulting from the lowess command in R. As is evident, they give similar results. One interesting feature of the data is that the half-slope ratio is approximately zero, so a regression model of the form $\hat{y} = \beta_1 x^a + \beta_0$, for some appropriately chosen a, is not very satisfactory.

The upper right panel of Figure 11.5 shows data from a reading study where one of the goals is to consider how well a measure of phonological awareness (sound blending as measured by

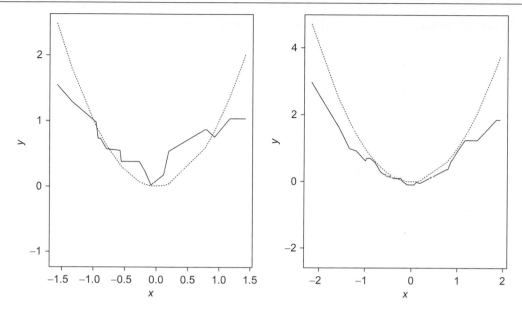

Figure 11.4: Two smooths where data were generated according to the model $y = x^2 + \epsilon$. The left panel is with a span of $f = 0.75$ and the right is with $f = 1$.

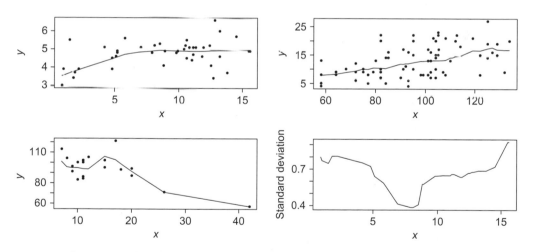

Figure 11.5: The upper left panel is a smooth for predicting the logarithm of C-peptide with age. The upper right panel is a smooth for predicting WWISST2 with SBT1. The lower left panel is a smooth of a Gesell score with age. And the lower right panel is a smooth of the standard deviation of log(C-peptide) versus age.

the variable SBT1 in the data file read.dat) predicts a word identification score (WWISST2). It appears that a straight line does a reasonably good job of capturing the relationship between the two random variables being investigated.

The lower-left panel of Figure 11.5 shows a scatterplot of data reported by Mickey, Dunn, and Clark (1967) for $n = 21$ children where the goal is to predict a child's Gesell adaptive score based on age in months when a child utters its first word. There is a suggestion that the regression line decreases sharply for older children, but there are too few observations to be sure. Clearly the two largest x values are outliers. If the outliers are eliminated, the running interval smoother returns a nearly flat line for children 12 months old and younger, but for older children a decreasing regression line appears again. (The details are left as an exercise.) Of course, with only 19 observations left, most of which correspond to children under the age of 12 months, more data are needed to resolve this issue.

An appeal of the running interval smoother is its versatility. Consider the diabetes data in the upper left panel of Figure 11.5. The lower right panel of Figure 11.5 shows a running interval smoother where the goal is to predict the standard deviation of the log C-peptide values based on the child's age. Note that the standard deviation drops dramatically, until about the age of 7, and then increases rapidly. Based on results in Chapter 10, this suggests that even if the error term has a normal distribution, OLS regression might be relatively inefficient compared to various robust estimators.

In some situations, particularly when the sample size is small, the running interval smooth can be somewhat ragged compared to other smoothers. An approach that might be used to help correct this problem is a bootstrap method called *bagging* (e.g., Breiman, 1996a,b; Bühlmann & Yu, 2002; Davison et al., 2003). In the present context, the method begins by applying the running interval smoother yielding, say, $m(x|d_n)$, where $d_n = (x_i, y_i)$, $i = 1, \ldots, n$. That is, $m(x|d_n)$ is some measure of location for y, given x, that is based on the n pairs of observations that are available. Generate a bootstrap sample by randomly sampling, with replacement, n pairs of points from d_n. Label the results d^*. Repeat this B times yielding $d_1^*, \ldots, d_B^*$. Then the bagged estimate of $m(x)$ is

$$\hat{m}(x|d) = \frac{1}{B} \sum_{b=1}^{B} m(x|d_b^*).$$

That is, use the average of the bootstrap estimates of $m(x)$.

11.5.5 R Functions runmean, rungen, runmbo, and runhat

Four R functions are supplied for applying the running interval smoother. The first has the form

    runmean(x, y, fr=1, tr=.2, pyhat=F, eout=F, outfun=out, xout=F, xlab="x", ylab="y")

and is designed to estimate the trimmed mean of y corresponding to x_i, $i = 1, \ldots, n$. The argument fr is f which, if unspecified, defaults to 1, and the argument tr is the amount of trimming which defaults to 0.2. (Also see the R function rplot in Section 11.5.12.) The function automatically creates a scatterplot of the data plus a smooth using the running interval method. If you want to use something other than the default value for f, which is used to determine which of the x_i values is close to a given point, set fr to the desired value. For example, the R command runmean(x,y,fr=0.75) would use $f = 0.75$ when creating the running interval smoother. If unsure, first try fr$=1$, and if the line seems smooth and straight, try fr$= 0.75$ to see what happens. Similarly, the command runmean(x,y,tr=0.1) would cause the running interval smoother to use the 10% trimmed mean. The command runmean (x,y,tr=0.1,plotit=F) would suppress the plot of the smooth. Most of the smooths in Figures 11.1–11.4 were created in this manner. If the argument pyhat is set to T (for true), the function returns the n values $\hat{m}(x_i)$, $i = 1, \ldots, n$. If eout=T is used, the function first eliminates outliers among the (x_i, y_i) values using the outlier detecting method specified by the argument outfun, and then a smooth is created based on the data that remain. If xout=T, the function checks for outliers among the x values only.

■ Example

Figure 11.6 shows the plot created by runmean using $f = 1$ with the experimental group of the Pygmalion study described in Section 11.2.1. Notice that the regression line is fairly straight for the bulk of the data, but the left end of the line curves up. It is evident that this is due to the two lowest x values. Because there are so few x values in this region, the smooth might be misleading for $x \leq 50$. It is left as an exercise to try $f = 0.75$.

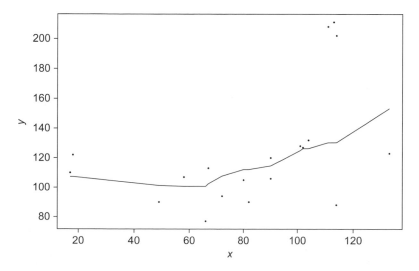

Figure 11.6: A smooth of the pygmalion data for the experimental group.

The function rungen is supplied in case it is desired to estimate some other measure of location associated with y. It has the form

rungen(x,y,est=onestep,fr=1,plotit=T,scat=T,pyhat=F,eout=F,xout=F, outfun=out, . . .).

The argument est can be any R function. If unspecified, it defaults to the modified one-step M-estimator. For example, to use the Harrell–Davis estimate of the median, use the command rungen(x,y,est=hd), while the command rungen(x,y) would use the one-step M-estimator instead. Again, the argument fr is the span f which defaults to 1. The last argument, . . . , can be any additional arguments required by the function est. For example, the command rungen(x,y,est=hd,q=0.25) would result in a running interval smoother that predicts the 0.25 quantile of y given x. The command rungen(x,y,est=mean,tr=0.2) would result in a running interval smoother based on the 20% trimmed mean. That is, rungen can create the smooth produced by runmean. The function runmean is supplied merely for convenience. The function creates a scatterplot of the data plus a smooth unless plotit=F is used. The argument scat defaults to T for true, meaning that a scatterplot of the data will be created. This adds perspective. To avoid the scatterplot, use scat=F. This might be done, for example, when the goal is to see how a measure of scale associated with y varies with x. Care must be taken because scat=F means that a plot of the smoothed values versus x is created, and this might affect one's perspective on the degree of curvature. (See Exercise 14 at the end of this chapter.)

A smoother does not provide an explicit equation for predicting y given x. The functions runmean and rungen can be used to compute $\hat{m}(x_i)$, but they do not report values of $\hat{m}(x)$ for some $x \neq x_i$, for all $i = 1, \ldots, n$. So the function

runhat(x,y,pts=x,est=onestep,fr=1,. . .)

is provided for computing $\hat{m}(x)$ for each of the values stored in the vector pts. The argument est defaults to the function onestep, which computes the one-step M-estimator. If, for example, it is desired to compute $\hat{m}(x)$ for $x = 1$ and 3, using a 20% trimmed mean, the command runhat(x,y,pts=c(1,3),mean,tr=0.2) accomplishes this goal.

The function

runmbo(x,y,fr=1,est=tmean,pyhat=F,eout=F,outfun=out,plotit=T,xout=F,
nboot=40,SEED=T, . . .)

can be used to produce a bagged version of the running interval smooth. For small sample sizes, this version of the running interval smooth seems preferable to using the function runmean.

11.5.6 Skipped Smoothers

Consider any smooth derived with the goal of estimating the conditional mean of y given x. Of course, another approach to robust smoothing is simply to eliminate any outliers and then apply this smooth to the data that remain. One variation is to search for outliers among the x values only, eliminate any point (x_i, y_i) where x_i is flagged an outlier, and then compute a smooth with the data that remain. Another approach is to eliminate any point that is an outlier based on one of the multivariate outlier detection methods in Section 6.4. That is, search for outliers among the points (x_i, y_i), $i = 1, \ldots, n$, in contrast to searching for outliers among the x_i values only. The functions lplot and kerreg, in Section 11.5.3, contain both approaches. Setting the argument xout to T (for true), these functions eliminate any points where x_i is flagged an outlier. Setting eout=T, now the function searches for points (x_i, y_i) that are outliers.

11.5.7 Smoothers for Estimating Quantiles via Splines

Another approach to nonparametric regression is based on what are called splines. They are a compromise between polynomial regression, which has been criticized due to the global nature of its fit, and other smoothers that have an explicit local nature. Regression splines compromise by employing a piecewise polynomial. The region that defines the pieces are separated by a sequence of knots or breakpoints. (For a summary of data-driven methods for choosing the knots, see for example Hastie & Tibshirani, 1990, Chapter 9.) A common goal is to force the piecewise polynomials to join smoothly at the knots. One popular choice consists of piecewise cubic polynomials constrained to be continuous and to have continuous first and second derivatives at the knots. Informal comparisons with other smoothers suggest that certain variations of methods based on splines are not quite as satisfactory as other smoothers that might be used (Härdle, 1990). However, some variations seem to have practical value. One such variation, called constrained B-spline smoothing (COBS), provides a way of dealing with quantiles (e.g., He & Ng, 1999; Koenker & Ng, 2005; Ng, 1996). A brief outline of the strategy behind COBS is provided here. Readers interested in the many computation details are referred to Koenker and Ng (2005); see in particular Section 4 of their paper. (For other approaches when estimating quantiles, see for example Doksum & Koo, 2000.) The Koenker–Ng method improves on a computational method studied by He and Ng (1999) and builds upon results in Koenker (1994).

Let $\rho_\tau(u) = u(\tau - I(u < 0))$, where the indicator function $I(u < 0) = 1$ if $u < 0$; otherwise $I(u < 0) = 0$. The goal is to estimate the τ quantile of y given x by finding a function $g(x)$ that minimizes

$$\sum \rho_\tau(y_i - g(x_i)) + \lambda \int |g''| dx$$

based on the random sample $(x_1, y_1), \ldots, (x_n, y_n)$, where λ is a scalar that controls smoothness. By default, COBS uses $\lambda = 0$ with quadratic B-splines and the number of knots chosen via a Schwartz-type information criterion. B-splines refer to a particular class of basis functions that offer numerical advantages. For general results on B-splines, see for example de Boor (1978).

11.5.8 R Function qsmcobs

The R package COBS performs the computations associated with the smoother just described. In case it helps, the R function

qsmcobs(x, y, qval = 0.5, xlab = "X", ylab = "Y", FIT = T, pc = ".", plotit = T, xout = F, outfun = out, ...)

is provided that plots the smooth (assuming that the R package cobs has been installed). The argument qval determines the quantile that will be used and defaults to the median. There are two options regarding how the plot is created. The default approach, when FIT=T, is to estimate the quantiles of y, given x, using the predict command associated with the R command cobs. The second, when the argument FIT=F, estimates the quantile of y for each observed x_i and plots the results.

It is noted that when the goal is to predict some quantile of y, rather than use qsmcobs, another possibility is to use the R functions rplot or rplotsm, which are described in Section 11.5.12. The relative merits of using these two functions, rather than qsmcobs, have not been studied. Yet another option when dealing with $p > 1$ predictors, when dealing with quantiles, is to use the R function runpd, which is described in Section 11.5.12 as well.

11.5.9 Special Methods for Binary Outcomes

When y is binary, now $m(x)$ is taken to be the (conditional) probability that $y = 1$ given x. Smoothers based on means can again be used, but some smoothers cannot be recommended. Examples are Cleveland's LOWESS estimator and the kernel estimator in Section 11.5.2. Both of these estimators can yield an estimate of $m(x)$ that is substantially smaller than 0 or larger than 1. However, there are estimators that deal explicitly with binary outcomes that guarantee that $0 \leq m(x) \leq 1$. One relevant study is by Copas (1983). Hosmer & Lemeshow (1989, p. 85) suggest using an estimator that is motivated in part by general results in Kay and Little (1987). Here, a slight modification of the Hosmer–Lemeshow estimator is used. The estimate of $m(x)$ is taken to be

$$\hat{m}(x) = \frac{\sum w_i y_i}{\sum w_i}, \qquad (11.8)$$

where

$$w_i = I_h e^{-(x_i - x)^2},$$

and $I_h = 1$ if $|x_i - x| < h$, otherwise $I_h = 0$. Also, unless stated otherwise, it is assumed that the x values have been standardized by subtracting the median and dividing by MADN. That is, if the observed predictors are $X_1, \ldots, X_n$, use $x_i = (X_i - M)/\text{MADN}$. If the predictors are not standardized, a change in scale can have a major impact on $\hat{m}$ yielding highly inaccurate and misleading results. The choice $h = 1.2$ appears to perform relatively well. Yet another approach is to use the running interval smoother in Section 11.5.4 with the amount of trimming set equal to zero.

Other variations have been studied by Signorini and Jones (2004) that are based in part on kernel density estimators. Let $f(x)$ be the probability density function of x, given that $y = 1$, and let $g(x)$ be the density given that $y = 0$. One of the estimators they studied has the form

$$\hat{m}(x) = \frac{n_1 f(x)}{n_1 \hat{f}(x) + n_0 \hat{g}(x)}, \tag{11.9}$$

where n_j is the number of times $y = j$, $j = 0, 1$. (So, for example, n_1 is the observed number of successes.) Here, $\hat{f}(x)$ and $\hat{g}(x)$ are taken to be adaptive kernel estimators described in Section 3.2.4.

A limitation of Eq. (11.8) is that it can handle only a single predictor. A slight variation of this estimator, which can handle more than one predictor, is to take

$$w_i = I_h e^{-d_i}, \tag{11.10}$$

where d_i is the squared Mahalanobis distance between $\mathbf{x}_i$ and $\mathbf{x}$, but with the usual covariance matrix replaced by the MVE estimator. That is,

$$d_i = (\mathbf{x}_i - \mathbf{x})' \mathbf{S}^{-1} (\mathbf{x}_i - \mathbf{x}),$$

where $\mathbf{S}$ is the MVE measure of scatter. When using Eq. (11.10), now $h = 2$ appears to be good choice for general use. Of course, the MVE estimator could be replaced by some other robust measure of scatter, but the practical advantages of doing so are unknown.

None of the estimators listed in this section dominate in terms of mean squared error and bias, but the estimator given by Eq. (11.10) appears to perform relatively well with the running interval another good choice (Wilcox, 2010e).

11.5.10 R Functions logrsm bkreg, logSM, and rplot.bin

The R functions in this section are designed with the explicit goal of creating a smooth when the outcome variable y is binary. The function

$$logrsm(x, y, fr = 1.2, plotit = T, pyhat = F, xlab="X", ylab="Y", STAND=T,$$
$$xout=F, outfun=outpro, \ldots)$$

computes the smooth given by Eq. (11.8), where the argument fr is h. The argument STAND defaults to T for true, meaning that x will be standardized by subtracting the median and then dividing my MADN.

The R function

$$bkreg(x, y, kerfun = akerd, pyhat = F, plotit = T, xlab ="X", ylab = "Y", zlab = "Z",$$
$$xout = F, outfun = outpro, pr = T, theta = 50, phi = 25, duplicate = "error", expand = 0.5,$$
$$scale = F, \ldots)$$

uses a variation of the estimator given by Eq. (11.9). By default, the adaptive kernel density estimator is used, but other kernel density estimators can be used via the argument kerfun. Unlike logrsm, bkreg can be used with more than one predictor. Limited results suggest that bkreg offers little advantage over other estimators in terms of mean squared error and bias, and situations arise where the reverse is true. With more than one predictor, the function

$$logSM(x,y,pyhat=F,plotit=T,xlab="X",ylab="Y",$$
$$zlab="Z",xout=F,outfun=outpro,pr=T,theta=50,phi=25,duplicate="error",$$
$$expand=0.5,scale=F,fr=2,\ldots)$$

applies the method based on Eq. (11.10) and appears to be a relatively good choice when y is binary. The argument fr corresponds to h in Eq. (11.10).

Finally, the R function

$$rplot.bin(x, y, est = mean, scat = T, fr = 1.2, plotit = T, pyhat = F, efr = 0.5, theta = 50,$$
$$phi = 25, scale = F, expand = 0.5, SEED = T, nmin = 0, xout = F, outfun = out, eout = F,$$
$$xlab = "X", ylab = "Y", zlab =" ", pr = T, duplicate = "error", zscale = T, \ldots)$$

uses the running interval smoother. It is essentially the same as the R function rplot, but for convenience it is designed specifically for situations where y is binary.

11.5.11 Smoothing with More than One Predictor

The running interval smoother can be generalized to more than one predictor by replacing MADN with the minimum volume ellipsoid estimate of scatter, $\mathbf{M}$, introduced in Chapter 6, and by measuring the distance between $\mathbf{x}_i$ and $\mathbf{x}_j$ with

$$D_{ij} = \sqrt{(\mathbf{x}_i - \mathbf{x}_j)'\mathbf{M}^{-1}(\mathbf{x}_i - \mathbf{x}_j)}.$$

When trying to predict y, given $\mathbf{x}_i$, simply compute the trimmed mean of all y_j values such that $\mathbf{x}_j$ is close to $\mathbf{x}_i$. More formally, compute the trimmed mean of all the y_j values for which the subscript j satisfies $D_{ij} \leq f$. The choice $f = 1$ or 0.8 often gives good results. When there are only two predictors, adjustments can be made as in the previous subsection. That is, start with $f = 1$, generate a graph of the three-dimensional smooth, and try other choices for f to see how the graph is affected. (For $p = 2$ and when estimating quantiles, also see He, Ng, & Portnoy, 1998.)

To provide some indication of how well the method performs, first suppose $y = x_1 + x_2 + \epsilon$. The left panel of Figure 11.7 shows a smooth based on $f = 1$ and $n = 20$ observations, where x_1, x_2, and ϵ all have a standard normal distribution. As can be seen, the shape of the regression plane is captured reasonably well. The right panel of Figure 11.7 shows a smooth when $y = x_1^2 + x_2 + \epsilon$, otherwise the situation is the same as before. Again the shape of the regression surface is captured. Of course, it is not being suggested that the correct surface is always reflected with only 20 points. Even with only one predictor, a smooth might suggest there is some curvature when data are generated from a straight line. Also, any smooth might be unreliable for extreme x_1 and x_2 values simply because there might be few points available for estimating the trimmed mean of y.

A possible concern with using D_{ij}, a robust analog of Mahalanobis distance, is that an ellipsoid is being used to identify the points close to $\mathbf{x}_i$. This might suffice, but a more flexible

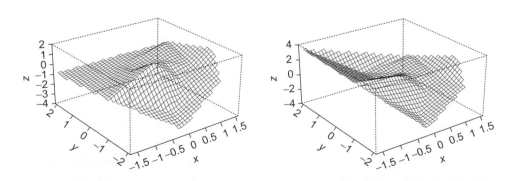

Figure 11.7: Illustrations of how runm3d performs under normality with a small sample size.

approach is to use projection distances instead. That is, use Approximation A1 in Section 6.2.3.

A criticism of the running interval smoother is that with a small sample size, the regression surface can be relatively ragged when it should be smooth. One way of improving the method is to apply the bootstrap bagging method as described at the end of Section 11.5.4.

11.5.12 R Functions runm3d, run3hat, rung3d, run3bo, rung3hat, rplot, rplotsm, and runpd

R functions are supplied for applying the running interval smoother when there is more than one predictor. The R function designed for trimmed means has the form

$$runm3d(x,y,theta=50,phi=25,fr=0.8,tr=0.2,plotit=T,pyhat=F,nmin=0,$$
$$scale=F,xout=F,outfun=out,ticktype=``simple").$$

Again, fr is the value of the span, which defaults to 1, and tr is the amount of trimming which defaults to 0.2. For a three-dimensional plot, setting the argument ticktype="detailed" will create ticks as done when creating a two-dimensional plot. The function returns the estimated trimmed mean of y for each of the n vectors of predictors stored in the n-by-p matrix, x. If the data are not stored in an R variable having matrix mode, the function prints an error message and terminates. When x is an n-by-2 matrix, the function automatically plots the estimated regression surface. To avoid the plot, set the argument plotit=F.

The argument nmin can be used to modify how the regression surface is estimated. By default, nmin is 0 meaning that the regression surface is estimated using all n rows of x. If, for example, nmin=2, the regression surface is estimated using only those points $\mathbf{x}_i$ for which the number of points close to $\mathbf{x}_i$ is greater than 2. Put another way, the regression surface is estimated using only those points for which the sample trimmed mean of y is based on more than nmin values. Setting the argument xout=T eliminates outliers among the $\mathbf{x}$ values before creating the plot, and eout=T causes outliers among the $(\mathbf{x}, y)$ to be removed.

When there is no association, and the regression surface is a flat, horizontal plane, using the default scale=F typically gives the best visual representation. But when there is an association, often scale=T provides a better perspective. The arguments theta and phi control the orientation of the plot. The argument theta controls the azimuthal direction and phi the colatitude. The left graph in Figure 11.8 shows a plot of $y = x_1 + x_2$ using the default values for theta and phi. The right panel is the same plot but with theta=20. (Changing the argument phi tilts the plot forward or backwards.)

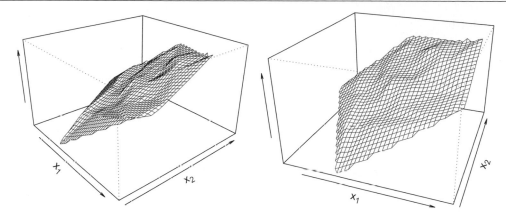

Figure 11.8: **An illustration of what happens when the argument theta is altered in the R function runm3d.**

Like runmean, runm3d can return $m(\mathbf{x}_i)$, $i = 1, \ldots, n$, but it is not set up to return $m(\mathbf{x})$ for some $\mathbf{x} \neq \mathbf{x}_i$, for all $i = 1, \ldots, n$. To evaluate $m(\mathbf{x})$ for any $\mathbf{x}$, the function

$$\text{run3hat(x,y,pts,fr=0.8,tr=0.2)}$$

is provided. The arguments are the same as those used by runm3d except for the argument pts which indicates the $\mathbf{x}$ values for which $m(\mathbf{x})$ is to be computed; pts is a matrix having p columns. So if pts contains ℓ rows, ℓ predicted values are returned in the R variable \$rmd. The number of values used to estimate y is stored in \$nval. That is, when predicting y given $\mathbf{x}$, \$nval is the number of y_j values for which the corresponding $\mathbf{x}_j$ value satisfies $D_j = \sqrt{(\mathbf{x} - \mathbf{x}_j)'\mathbf{M}^{-1}(\mathbf{x} - \mathbf{x}_j)} \leq f$.

The R function

rung3d(x, y, est = onestep, fr = 1, plotit = T, theta = 50, phi = 25, pyhat = F, expand = 0.5, scale = F, zscale = T, nmin = 0, xout = F, outfun = out, SEED = T, xlab = "X", ylab = "Y", zlab = " ", pr = T, duplicate = "error", ticktype = "simple", ...).

applies the running interval smoother using any location estimator specified by the argument est. If unspecified, it defaults to the one-step M-estimator using Huber's Ψ. The arguments x, y, fr, plotit, and nmin are the same as those in runm3d. The final argument, ..., can be any additional arguments required by est. For example, the command rung3d(x,y,est=hd,q=.4) would use the Harrell–Davis estimate of the 0.4 quantile.

The function rung3hat can be used to estimate some measure of location associated with y, given $\mathbf{x}$, when there is interest in some measure of location other than the trimmed mean.

It has the form

$$\text{rung3hat(x,y,est=onestep,pts,fr=1, ...)}.$$

The arguments are the same as those used by run3hat except for the argument est, which indicates the measure of location to be used, and the argument ..., which can be any additional arguments required by est. Like run3hat, est defaults to the modified one-step M-estimator.

■ Example

If tp is a 2-by-3 matrix with the first row equal to zero and the second equal to 1, the command run3hat(x,y,est=onstep,pts=tp) returns two values in $rmd: the predicted one-step M-estimate of y given that x is equal to (0,0,0), and the predicted value when x is equal to (1,1,1). The function also returns, in the R variable $nval, the number of y values used to compute the measure of location. The first value in $nval is the number of predictors that are close to (0,0,0), and the second value is the number of predictors close to (1,1,1). For example, if the first value in nval is 8, there were eight points close to (0,0,0), which in turn means that the predicted value of y is based on eight values as well.

■

The function

run3bo(x, y, fr = 1, est = tmean, theta = 50, phi = 25, nmin = 0, pyhat = F, eout = F, outfun = out, plotit = T, xout = F, nboot = 40, SEED = T, expand = 0.5, scale = F, xlab = "X", ylab = "Y", zlab = "", ticktype = "simple", ...)

can be used to create a bagged version of the running interval smoother; see the end of Section 11.5.4. This function can give substantially better results, compared to runm3d, when the sample size is relatively small.

The functions just described assume $p > 1$. For convenience, the functions

rplot(x, y, est = tmean, scat = T, fr = NA, plotit = T, pyhat = F, efr = 0.5, theta = 50, phi = 25, scale = F, expand = 0.5, SEED = T, varfun = pbvar, nmin = 0, xout = F, outfun = out, eout = F, xlab = "X", ylab = "Y", zlab = " " pr = T, duplicate = "error", ticktype="simple", ...)

and

rplotsm(x, y, est = tmean, fr = 1, plotit = T, pyhat = F, nboot = 40, atr = 0, nmin = 0, outfun = out, eout = F, xlab = "X", ylab = "Y", scat = T, SEED = T, expand = 0.5, scale = F, varfun = pbvar, pr = T, ticktype="simple", ...)

are supplied to handle the general case $p \geq 1$. The function rplot calls rungen if $p = 1$ or rung3d otherwise. The arguments for rplot are the same as rungen or rung3d. The function rplotsm also handles $p \geq 1$, but it computes a bagged version of the smooth by calling runmbo when $p = 1$ and run3bo otherwise. The only difference from the separate functions is that both rplot and rplotsm use a 20% trimmed mean by default. (That is, the argument est=tmean is used.) To use a one-step M-estimator, for example, set est=onestep.

■ Example

This example illustrates that a smooth, based on bagging, can make a practical difference. Using R, 50 values were generated from a standard normal distribution for both x and ϵ and $y = x + \epsilon$ was computed. Then a smooth of the conditional variance of y, given x, was created with the command rplot(x,y,est=var,scat=F). The result is shown in the left panel of Figure 11.9. Then a bagged version of the smooth was created with the command rplotsm(x,y,est=var,scat=F) and the result is shown in the right panel of Figure 11.9. As is evident, the bagged version gives a much more accurate indication of the conditional variance of y given x.

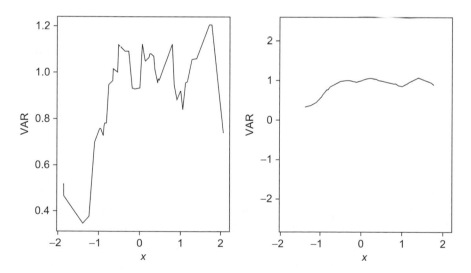

Figure 11.9: The left panel shows a smooth of y, given x, where the conditional variance of y, given any x, is one. The right panel shows a bagged version of the smooth created by the function rplotsm.

When dealing with $p > 1$ predictors, the R functions previously described in this section determine which points are close to some specified **x** using a robust analog of Mahalanobis distance based on the MVE covariance matrix. So the closest points to **x** are based on ellipsoids. As noted in the previous section, a more flexible approach to identifying the closest points is to use projection distances instead. This is done by the R function

runpd(x, y, pts = x, est = tmean, fr = 0.8, plotit = T, pyhat = F, nmin = 0, scale = F, expand = 0.5, xout = F, outfun = out, pr = T, xlab = "X1", ylab = "X2", zlab = " ", theta = 50, phi = 25, duplicate = "error", MC = F, ...).

The function runpd uses the R function

pdclose(x, pts = x, fr = 1, MM = F, MC = F)

to determine which points stored in x are close to the points stored in the argument pts.

It was noted in Section 11.5.8 that an alternative to COBS, when the goal is to predict some quantile of y, given x, can be accomplished with the R functions rplot and rplotsm. For instance, setting the argument est=hd would use the Harrel–Davis estimator, or est=qest would use a single order statistic to estimate the quantile of interest. For example, the command

rplot(x,y,est=hd,q=.25)

would plot the smooth for estimating the 0.25 quantile of y given x. Note that these two functions are capable of handling situations where there are $p > 1$ predictors, in contrast to COBS, which is limited to $p = 1$.

11.5.13 LOESS

There is an extension of the smoother lowess (described in Section 11.5.2) to multiple predictors that was derived by Cleveland and Devlin (1988); it can be applied with the function loess which comes with R. The R function lplot, described in Section 11.5.3, uses loess to create a plot when $p = 2$. Like lowess, the goal is to estimate the conditional mean of y, but unlike lowess (which handles $p = 1$ only), when using the default settings of the function, a single outlier can grossly distort the estimate of the regression surface and nonnormality can greatly influence the plot. One way of addressing this problem is to set the argument family="symmetric" when using the function lplot. Another possibility is to eliminate all outliers by setting the argument eout=T and use the default value for the argument family.

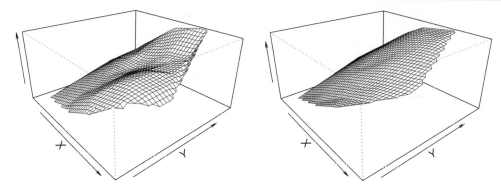

Figure 11.10: An illustration of how nonnormality might affect smooths created by lowess. The left panel shows a smooth using the default settings of the function lplot. The right panel is a plot of the same data, but with outliers removed by setting the argument eout=T.

■ Example

As an illustration, $n = 100$ points were generated from the model $y = x_1 + x_2 + \epsilon$, where x_1 and x_2 are independent standard normal random variables and ϵ has a g-and-h distribution with $g = h = 0.5$. The left panel of Figure 11.10 shows the plot created by lplot using the default settings, and the right panel is the plot with eout=T, which eliminates all outliers before creating the plot.

■

■ Example

Using the reading data, with the two independent variables taken to be TAAST1 and SBT1 (which are measures of phonological awareness and stored in columns 2 and 3 of the file read.dat), and the dependent variable taken to be OCT2 (a measure of orthographic ability and stored in column 10), Figure 11.11 shows an estimate of the regression surface using four different smoothers. The upper left graph was created by lplot using the default values for the arguments. The upper right graph was created by runm3d, again using the default values. The lower left graph was created by lplot but with eout=T so that outliers are eliminated before creating the smooth. The lower right graph was created by runm3d but with fr=1.1, to get a smoother estimate, and xout=T to eliminate any outliers among the independent variables.

■

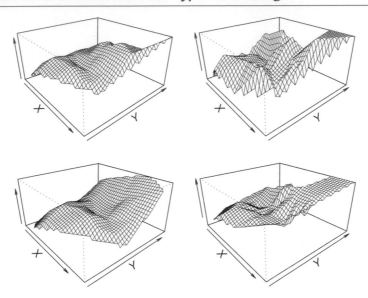

Figure 11.11: Four different smooths based on the reading data.

■ Example

To illustrate run3hat, again using the reading data, suppose it is desired to estimate WWISST2 (a word identification score stored in column 8) when TASST1 is 15 and SBT1 is 8. Then there is $\ell = 1$ point of interest, so store the values 15 and 8 in any 1-by-2 matrix. For example, the R command val=matrix(c(15,8),1,2) could be used. Assuming the values of the predictors are stored in the R variable x, and the WWISST2 values are stored in y, the command run3hat(x,y,val) returns the value 106.2 in the R variable \$rmd. That is, the estimated 20% trimmed mean of WWISST2, given that TASST1 is 15 and SBT1 is 8, is equal to 106.2. If instead it is desired to compute $\hat{y}$ for the points (15, 8) and (15, 9), enter the command val=matrix(c(15,8,15,9),2,2, byrow=T). Then the first row of the matrix val contains (15, 8), the second row contains (15, 9), and the command runm3hat(x,y,val) returns the values 106.2 and 114.0, which are stored in the R variable \$rmd.

■

■ Example

Kyphosis is a postoperative spinal deformity. R has built-in data, stored in the R variable kyphosis, reporting the presence or absence of kyphosis versus the age of the patient, in months, the number of vertebrae involved in the spinal operation, and a variable called

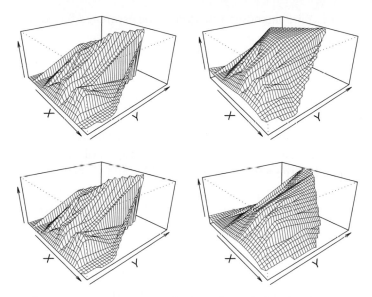

Figure 11.12: Four smooths based on the kyphosis data.

start, which is the beginning of the range of vertebrae involved. Suppose it is desired to estimate the probability of kyphosis based on age and the number of vertebrae involved. The function runm3d accomplishes this goal by setting the argument tr equal to zero, or the function rplot.bin could be used. The top two graphs in Figure 11.12 show the resulting estimate of the regression surface using runm3d (with tr=0) and lplot (shown on the right). The bottom two graphs were again created by runm3d and lplot, but both functions used xout=T to eliminate any outliers among the independent variables. (Three outliers were found using the MVE method.) Also, runm3d used fr=1.1 to smooth the plot. (Standard logistic regression is typically used when y is binary. See Section 10.16 for some robust alternatives.)

■

11.5.14 Other Approaches

Yet another approach when dealing with two or more predictors is to use what is called a *generalized additive model*. That is, assume that

$$y = \beta_0 + \sum_{j=1}^{p} g_j(x_j) + \epsilon \tag{11.11}$$

where $g_1(x_1), \ldots, g_p(x_p)$ are unknown functions to be estimated based on the available data. This is in contrast to assuming

$$y = g(x_1, \ldots, x_p) + \epsilon. \tag{11.12}$$

A concern about the more general model given by Eq. (11.12) is that, when using various kernel smoothers to estimate the (conditional) mean of y, the so-called *curse of dimensionality* comes into play: neighborhoods with a fixed number of points become less local as the dimensions increase (Bellman, 1961). Regardless of the extent Eq. (11.12) improves upon Eq. (11.11), the additive model provides an interesting generalization of the usual linear model $y_i = \beta_0 + \beta_1 x_{i1} + \ldots + \beta_p x_{ip} + \epsilon_i$, when testing hypotheses and trying to gain insight into any associations that might exist. (Illustrations are given in Section 11.6.)

When dealing with robust measures of location, the generalized additive model given by Eq. (11.11) can be fit to data using the running interval smoother in conjunction with the so-called *backfitting algorithm* (e.g., Friedman & Stuetzle, 1981). More generally, virtually any smoother can be used, including the many smoothers designed specifically for means. The backfitting algorithm is applied as follows. Set $k = 0$ and let g_j^0 be some initial estimate of g_j. Here, $g_j^0 = S_j(y|x_j)$, where $S_j(y|x_j)$ is the running interval smooth based on the jth predictor, ignoring the other $p - 1$ predictors that are available. Next, iterate as follows.

1. Increment k by 1.
2. For each j, $j = 1, \ldots, p$, let

$$g_j^k = S_j(y - \sum_{\ell \neq j} g_\ell^{k-1} | x_j).$$

3. Repeat steps 1 and 2 until convergence.

(For general theoretical results on the backfitting algorithm, see Buja, Hastie, & Tibshirani, 1989.)

Finally, estimate β_0 with

$$b_0 = m(y - \sum g_j^k),$$

where m indicates the measure of location used when computing the smooths. R contains functions that estimate the generalized additive model given by Eq. (11.11) when the goal is to estimate the mean of y. Again, when the goal is to get a more robust version of these methods, a simple approach is to remove any outliers before using these R functions.

Methods have been derived that are blend of both a parametric model and a nonparametric smoother (e.g., Ruppert, Wand, & Carroll, 2003). For recent results on how this approach might be implemented in a robust manner, see Boente and Rodriguez (2010).

11.5.15 R Function adrun, adrunl, gamplot, and gamplotINT

The R function

$$\text{adrun}(x, y, est = tmean, iter = 10, pyhat = F, plotit = T, fr = 1, xlab = \text{``X'}, ylab = \text{``Y},$$
$$zlab = \text{`` ''}, theta = 50, phi = 25, expand = 0.5, scale = F, zscale = T, xout = F,$$
$$eout = xout, outfun = out, ticktype = \text{``simple''}, \ldots),$$

fits the additive model given by Eq. (11.11) in conjunction with the running interval smoother. As usual, the arguments theta and phi control the orientation of the plot; see Section 11.5.12. (At each iteration, the individual smooths are obtained by calling the function rungen in Section 11.5.5.) The measure of location is specified by the argument est and defaults to a 20% trimmed mean. The command adrun(x,y,est=mean,tr=0.1), for example, would result in a smooth based on a 10% trimmed mean instead. Setting the argument pyhat=T causes the function to return the estimates of y for each design point, and fr specifies the span. For bivariate data, the function plots the smooth if plotit=T (for true). To avoid the plot, set plotit=F. As p, the number of predictors, gets large, caution must be exercised. Situations can arise where the fit to data is wildly inaccurate due to the span being too small. So at a minimum it is suggested to check the output with pyhat=T to make sure the function is returning reasonable results. The function

$$\text{adrunl}(x, y, est = tmean, iter = 10, pyhat = F, plotit = T, fr = 0.8, xlab = \text{``x1''}, ylab = \text{``x2''},$$
$$zlab = \text{`` ''}, theta = 50, phi = 25, expand = 0.5, scale = F, zscale = T, xout = F, outfun = out,$$
$$ticktype = \text{``simple''}, \ldots)$$

is like the function adrun, only the the running interval smoother is replaced by lowess.

The R function

$$\text{gamplot}(x,y,pyhat=F,sop=T,plotit=T,theta=50,phi=25,scale=F,eout=F,$$
$$outfun=out,ticktype=\text{``simple''},)$$

creates a plot based on an additive fit for means that is computed via a call to the built-in R function gam. (Splines are used to create the smooth. With the argument sop=F, the usual linear model is used.) The R function gam has many more options for modeling the regression surface than are used by the function gamplot. For two predictors, the function gamplot is intended as way of graphing the regression surface assuming that Eq. (11.11) holds.[1] The current version is limited to $p = 4$. The other arguments are the same as those described in

---

[1] The R function gam differs in fundamental ways from the S-PLUS function gam.

Section 11.5.12 in conjunction with runm3d. (When y is binary, the function logadr fits a generalized additive model in conjunction with Copas's method previously described.) The R function

$$\text{gamplotINT}(x, y, \text{pyhat} = F, \text{plotit} = T, \text{theta} = 50, \text{phi} = 25, \text{expand} = 0.5, \text{scale} = F,$$
$$\text{zscale} = T, \text{eout} = F, \text{outfun} = \text{out}, \text{ticktype} = \text{"simple"},)$$

is like gamplot, only it is limited to $p = 2$ predictors and is based on the model $y = g_1(x_1) + g_2(x_2) + g_3(x_1, x_2) + \epsilon$ rather than $y = g_1(x_1) + g_2(x_2) + \epsilon$. This is useful when checking for interactions as described in Section 11.7.

11.6 Checking the Specification of a Regression Model

Typically, when testing hypotheses, a particular parametric form for a regression model is specified and inferences are made about the parameters assuming that the model is correct. A practical concern is that the assumed parametric form might be wrong, which in turn can lead to erroneous conclusions. As a simple example, values for x were generated by the author from a bivariate normal distribution with $\rho = 0$, the marginal distributions as well as ϵ had a standard normal distribution, $n = 20$, and the error term was homoscedastic. Now imagine we assume that $y = \beta_0 + \beta_1 x_1 + \beta_2 x_2 + \epsilon$ and the goal is to test H_0: $\beta_2 = 0$. Furthermore, based on how the data were generated, power is approximately 0.26 when testing at the 0.05 level. Is it reasonable to conclude that the model is a good approximation of how the data were generated and that indeed, $\beta_2 \neq 0$? Here, such a conclusion would be erroneous; the data were generated using the model $y = \beta_1 x_1 + \beta_2 x_2^2 + \epsilon$. So an issue is whether it is reasonable to assume that for some β_0, β_1 and β_2, $y = \beta_0 + \beta_1 x_1 + \beta_2 x_2 + \epsilon$. Of course, exploratory graphical methods, already covered, help address this issue. Here the goal is to describe some additional tools for dealing this problem.

There are, in fact, many methods for testing the hypothesis that a regression equation has a particular parametric form. Typically these methods are based on estimates of the conditional mean of y given $\mathbf{x}$. Included are methods that begin with a kernel-type smooth and then compare the fitted y values to those obtained by an assumed parametric model. Miles and Mora (2003) summarize and compare a variety of these methods assuming normality. More generally, there is the problem of testing the hypothesis that a regression surface belongs to some particular family of models. For example, can we rule out the possibility that a generalized additive model generated the data? Samarov (1993) provides an interesting overview of various models and how they might be investigated. It seems that few results are available on how extensions of these methods to robust estimators perform.

11.6.1 Testing the Hypothesis of a Linear Association

Given p predictors, $x_1, \ldots, x_p$, let $\mathcal{M}$ be the family of all regression equations having the form $y = \beta_0 + \beta_1 x_1 + \cdots + \beta_p x_p + \epsilon$, where the error term may be heteroscedastic. This section describes a test of the hypothesis

$$H_0 : m(\mathbf{x}) \in \mathcal{M} \tag{11.13}$$

where as usual, $m(\mathbf{x})$ represents some conditional measure of location given $\mathbf{x}$. That is, the null hypothesis is that the data are generated from the model $y = \beta_0 + \beta_1 x_1 + \cdots + \beta_p x_p + \epsilon$. If, for example, $y = \beta_0 + \beta_1 x_1^2 + \epsilon$, the null hypothesis is false. The method described here stems from Stute, Gonzalez Manteiga, and Presedo-Quindimil (1998).

Let $\hat{y}$ be some regression estimate of y. Least squares could be used, but it has been shown that this can lead to problems in terms of controlling the probability of a type I error (Wilcox, 1999), so it is suggested that some robust estimator be used instead. For fixed j $(1 \le j \le n)$, set $I_i = 1$ if $\mathbf{x}_i < \mathbf{x}_j$, otherwise $I_i = 0$, and let

$$
\begin{aligned}
R_j &= \frac{1}{\sqrt{n}} \sum I_i (y_i - \hat{y}_i) \\
&= \frac{1}{\sqrt{n}} \sum I_i r_i,
\end{aligned} \tag{11.14}
$$

where $r_i = y_i - \hat{y}_i$ are the usual residuals. The (Kolmogorov) test statistic is the maximum absolute value of all the R_j values. That is, the test statistic is

$$D = \max |R_j|, \tag{11.15}$$

where max means that D is equal to the largest of the $|R_j|$ values. As in Section 9.5, a Cramér–von Mises test statistic can be used instead, where now

$$D = \frac{1}{n} \sum R_j^2. \tag{11.16}$$

A critical value is determined using the wild bootstrap method. Generate n observations from a uniform distribution and label the results $U_1, \ldots, U_n$. Next, for $i = 1, \ldots, n$, set

$$V_i = \sqrt{12}(U_i - 0.5),$$

$$r_i^* = r_i V_i,$$

and

$$y_i^* = \hat{y}_i + r_i^*.$$

Then based on the n pairs of points $(\mathbf{x}_1, y_1^*), \ldots, (\mathbf{x}_n, y_n^*)$, compute the test statistic and label it D^*. Repeat this process B times and label the resulting test statistics $D_1^*, \ldots, D_B^*$. Finally, put these B values in ascending order yielding $D_{(1)}^* \leq \cdots \leq D_{(B)}^*$. The critical value is $D_{(u)}^*$, where $u = (1 - \alpha)B$ rounded to the nearest integer. That is, reject if

$$D \geq D_{(u)}^*.$$

(Wang & Qu, 2007, propose another approach, but it is unknown how it compares to the method covered here. For a method aimed specifically at L_1 regression, see Horowitz & Spokoiny, 2002. For yet another method dealing with quantile regression, see He & Zhu, 2003.)

11.6.2 R Function lintest

The R function

$$\text{lintest(x,y,regfun=tsreg,nboot=500,alpha=0.05)}$$

tests the hypothesis that a regression surface is a plane (more generally that the regression surface corresponds to a linear model) using the method just described. (Execution time is fairly fast with one predictor, but on some computers it might be slow when there are multiple predictors. This problem can be greatly reduced by using regfun=chreg, which uses the Coakley–Hettmansperger M-estimator.) When reading the output, the Kolmogorov test statistic is labeled dstat and its critical value is labeled critd. The Cramér–von Mises test statistic is labeled wstat. The default regression method (indicated by the argument regfun) is Theil–Sen.

■ Example

For the diabetes data shown in Figure 11.5, suppose the goal is to test the hypothesis that there is a linear association between the logarithm of the C-peptide values and age. That is, the hypothesis is that for some β_0 and β_1, $y = \beta_0 + \beta_1 x + \epsilon$, where x is age. The Kolmogorov test statistic returned by the R version of lintest is $D = 0.179$, it reports a 0.05 critical value of 0.269, so fail to reject. If both predictors (age and base deficit) are used, again we fail to reject at the 0.05 level.   ■

11.6.3 Testing the Hypothesis of a Generalized Additive Model

This section describes a variation and extension of the test of linearity given in Section 11.6.1. Here, rather than test the hypothesis of a linear association, the goal is to test the hypothesis that the data were generated from a generalized additive model. More formally, given p predictors, $x_1, \ldots, x_p$, now let $\mathcal{M}$ be the family of all regression equations having the form given by Eq. (11.11). The goal is to test the hypothesis

$$H_0 : m(\mathbf{x}) \in \mathcal{M} \tag{11.17}$$

where as usual, $m(\mathbf{x})$ represents some conditional measure of location given $\mathbf{x}$. There are various ways this problem might be addressed. For example, some obvious extension of the method in Dette (1999) might be used, but so far no such variation has been found that performs well in simulations. Another approach is suggested by results in Samarov (1993), but again there are no simulation results supporting this strategy. Another possibility is to fit the additive model and test the hypothesis that the regression surface for the residuals, versus $\mathbf{x}$, is a horizontal plane, which can be done along the lines in Section 9.5, or one might compare the fit of the additive model to the fit obtained by the method in Section 11.5.11. Wild bootstrap methods based on these last two strategies have, so far, proven to be rather unsatisfactory in simulations.

Currently, the only method that performs well in simulations, when the sample size is small, is applied exactly as in Section 11.6.1, only rather than compute $\hat{y}$ based on some robust regression estimator, use $\hat{y} - \hat{m}(\mathbf{x})$ based on the additive fit described in Section 11.5.14 (Wilcox, 2003e). Here it is assumed that the additive fit is obtained using the 20% trimmed mean. The method does not perform well when using means and nothing is known about how it performs when using an M-estimator.

There is, however, a practical concern about the choice of the span when applying the running interval smoother to get the additive fit. If $p = 2$ and the span is too large, the actual type I error probability can drop well below the nominal level. For this special case, and when testing at the 0.05 level, approximations of a good choice for the span corresponding to the sample sizes 20, 30, 50, 80, and 150 are .4, .36, .18, .15, and .09, respectively. It is suggested that when $20 \leq n \leq 150$, interpolation based on these values be used, and for $n > 150$ simply use a span equal to .09. So for n sufficiently large, perhaps the actual type I error probability might be well below the nominal level, but exactly how the span should be modified when $n > 150$ is an issue that is in need of further investigation. For $p = 3$, the choice of the span seems less sensitive to the sample size, with a span of $f = 0.8$ being a reasonable choice for $n < 100$. What happens when $p > 3$ has not been investigated.

11.6.4 R Function adtest

The R function

```
adtest(x,y,est=tmean,nboot=100,alpha=0.05,fr=NA,xout=F,outfun=out,SEED=T,...)
```

tests the hypothesis given by Eq. (11.17). If xout=T, outliers among the **x** values are first identified and $(y_i, \mathbf{x}_i)$ is eliminated if $\mathbf{x}_i$ is flagged an outlier.

11.6.5 Inferences About the Components of a Generalized Additive Model

Inferences about the components of a generalized additive model, based on the running interval smoother, can be made as follows. For convenience, assume the goal is to test

$$H_0 : g_1(x_1) = 0.$$

Fit the generalized additive model yielding

$$\hat{y}_i = b_0 + \hat{g}_2(x_{i2}) + \cdots + \hat{g}_p(x_{ip}).$$

Let $r_i = y_i - \hat{y}_i$, $i = 1, \ldots, n$. The strategy is to test the hypothesis that the association between the residuals and x_1 is a straight horizontal line, and this can be done with the wild bootstrap method in Section 9.5 (cf. Härdle & Korostelev, 1996).

When using the running interval smoother, the choice of the span can be crucial in terms of controlling the probability of a type I error (Wilcox, 2006a). Letting f be the span used in Section 11.5.4. The choice for f when using means or a 20% trimmed mean are as follows:

n	20% trimming	Mean
20	1.20	0.80
40	1.0	0.70
60	0.85	0.55
80	0.75	0.50
120	0.65	0.50
160	0.65	0.50

So, for example, if $n = 60$ and a generalized additive model based on the running interval smoother and a 20% trimmed mean is to be used to test H_0, choose the span to be $f = 0.85$.

In principle, the method is readily extended to situations where something other than the running interval smoother is used to fit the generalized additive model, but currently there are no results on the resulting probability of a type I error.

11.6.6 R Function adcom

The R function

adcom(x, y, est = mean, tr = 0, nboot = 600, alpha = 0.05, fr = NA, jv = NA, ...)

tests hypotheses about the components of a generalized additive model using the method just described. With the argument fr=NA, the function chooses the appropriate span, as a function of the sample size, using linear interpolation where necessary. By default, all components are tested. The argument jv can be used to limit which components are tested. For example, jv=2 would test only H_0: $g_2(x_2) = 0$.

11.7 Regression Interactions and Moderator Analysis

As an application of the method in Section 11.6.3, note that it provides a flexible approach to the so-called regression interaction problem. Consider the two predictor case and let c_1 and c_2 be two distinct values for the second predictor, x_2. Roughly, no interaction refers to a situation where the regression line between y and x_1, given that $x_2 = c_1$, is parallel to regression line between y and x_1, given that $x_2 = c_2$. An early approach to modeling interactions assumes that

$$y = \beta_0 + \beta_1 x_1 + \beta_2 x_2 + \beta_3 x_1 x_2 + \epsilon, \tag{11.18}$$

where an interaction is said to exist if $\beta_3 \neq 0$ (e.g., Saunders, 1956). This model often plays a role in what is called a *moderator analysis*, roughly meaning that the goal is to determine the extent to which knowing the value of one variable, x_2 here, alters the association between y and x_1. Note that Eq. (11.18) can be written as

$$y = (\beta_0 + \beta_2 x_2) + (\beta_1 + \beta_3 x_2) x_1 + \epsilon,$$

so the slope for x_1 changes as a linear function of x_2. (An R function, called ols.plot.inter, described in Section 11.7.1, plots the regression surface when using the least squares estimate of the parameters.) Currently, a commonly used method for testing the hypothesis of no interaction is to test H_0: $\beta_3 = 0$, meaning that the slope for x_1 does not depend on x_2.

A more general approach to testing the hypothesis of no interaction is to use a variation of the method in Section 11.5.1 to test the hypothesis that for some functions g_1 and g_2, $y = g_1(x_1) + g_2(x_2) + \epsilon$. This can be done with the function adtest in Section 11.5.4. Another way of stating the problem is described, for example, by Barry (1993) who uses an ANOVA-type decomposition. Essentially, write

$$m(x_1, x_2) = \beta_0 + g_1(x_1) + g_2(x_2) + g_3(x_1, x_2) + \epsilon,$$

in which case the hypothesis of no interaction is

$$H_0 : g_3(x_1, x_2) \equiv 0.$$

Barry (1993) derived a Bayesian-type test of this hypothesis assuming the mean of y is to be estimated and that prior distributions for g_1, g_2, and g_3 can be specified. Another approach is outlined by Samarov (1993), but when dealing with robust measures of location, the details have not been investigated. Note that this last hypothesis can be tested with the function adcom in Section 11.6.6. How this approach compares to using the function adtest is unknown.

Now we describe graphical methods that might be useful when studying interactions. The first simply plots a smooth of y versus x_1 given a particular value for x_2. So if there is no interaction, and this plot is created at say $x_2 = c_1$ and $x_2 = c_2$, $c_1 \neq c_2$, the regression lines should be parallel. Here, creating this plot is tackled using a simple extension of the kernel estimator (the modification of Fan's method) described in Section 11.5.2. (Many alternative versions are possible and might have practical value.)

Momentarily consider a single predictor x. In Section 11.5.2, an estimate of the conditional mean of y at x is obtained using weighted least squares with weights $K[(x - x_i)/h]$. One possibility for extending this method to estimating $m(x_1)$, given that $x_2 = c$, which is used here, begins with a bivariate Epanechnikov kernel, where, if $1 - x_1^2 - x_2^2 < 1$,

$$K(x_1, x_2) = \frac{2}{\pi}(1 - x_1^2 - x_2^2),$$

otherwise $K(x_1, x_2) = 0$. An estimate of the bivariate density $f(\mathbf{x})$, based on (x_{i1}, x_{i2}), $i = 1, \ldots, n$, is

$$\hat{f}(\mathbf{x}) = \frac{1}{nh^2} \sum_{i=1}^{n} K\left[\frac{1}{h}(\mathbf{x} - \mathbf{x}_i)\right],$$

where as usual, h is the span. For the jth predictor, let $u_j = \min(s_j, \text{IQR}_j/1.34)$, where s_j and IQR_j are the sample standard deviation and interquartile range (estimated with the ideal fourths) based on $x_{1j}, \ldots, x_{nj}$. Here the span is taken to be

$$h = 1.77n^{-1/6}\sqrt{u_1^2 + u_2^2}.$$

(See Silverman, 1986, pp. 86–87.) Then an estimate of $m(x_{i1})$, given that $x_{i2} = c$, is obtained via weighted least squares applied to (y_i, x_{i1}), $i = 1, \ldots, n$, with weights

$$w_i = \frac{K(x_{i1}, x_{i2} = c)}{K_2(x_{i2} = c)},$$

where K_2 is the Epanechnikov kernel used to estimate the probability density function of x_2.

Let $\hat{y}_i$ be the estimate of y based on (x_{i1}, x_{i2}) and the generalized additive model given by Eq. (11.11). Another approach to gaining insight regarding any interaction is to plot (x_{i1}, x_{i2}) versus the residuals $y_i - \hat{y}_i$, $i = 1, \ldots, n$.

Yet one more possibility is to split the data into two groups according to whether x_{i2} is less than some constant. For example, one might let M_2 be the median of the x_{i2} values, then take the first group to be the (x_{i1}, y_i) values for which $x_{i2} < M_2$, and the second group would be the (x_{i1}, y_i) values for which $x_{i2} \geq M_2$, and then a smooth for both groups could be created. If there is no interaction, the two smooths should be reasonably parallel.

11.7.1 R Functions kercon, riplot, runsm2g, ols.plot.inter, and reg.plot.inter

The R functions in Section 11.5.14 can be used to get some graphical information about how regression surfaces compare when no interaction is assumed versus situations where an interaction term is included. This section summarizes some additional R functions that might be useful.

The R function

ols.plot.inter(x, y, pyhat = F, eout = F, xout = F, outfun = out, plotit = T, expand = 0.5, scale = F, xlab = "X", ylab = "Y", zlab = " ", theta = 50, phi = 25, family = "gaussian", duplicate = "error", ticktype = "simple",)

plots the regression surface assuming that Eq. (11.18) is true and that the least squares estimates of the parameters are used. Because this model is often used, an issue of interest is how the estimated regression surface compares to other plots that are based on a more flexible nonparametric estimator.

The R function

reg.plot.inter(x, y, regfun=tsreg, pyhat = F, eout = F, xout = F, outfun = out, plotit = T, expand = 0.5, scale = F, xlab = "X", ylab = "Y", zlab = " ", theta = 50, phi = 25, family = "gaussian", duplicate = "error", ticktype = "simple",)

is exactly like the function ols.plot.inter, only it can be used with any regression estimator that returns the residuals in $residuals. By default, the Theil–Sen estimator is used.

■ Example

A portion of a study conducted by Shelley Tom and David Schwartz dealt with the association between a Totagg score and two predictors: grade point average (GPA) and a measure of academic engagement. The Totagg score was a sum of peer nomination items that were based on an inventory that included descriptors focusing on adolescents' behaviors and social standing. (The peer nomination items were obtained by giving children a roster sheet and asking them to nominate a certain amount of peers who fit particular behavioral descriptors.) The sample size is $n = 336$. The left panel of Figure 11.13 shows the plot of the regression surface created with the R function ols.plot.inter. Compare this to the right panel, which is an estimate of the regression surface using LOESS and created by the R function lplot. This suggests that using the

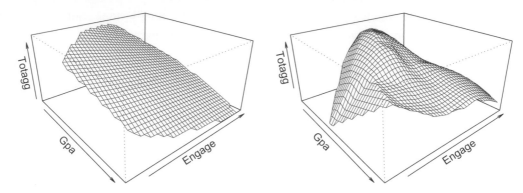

Figure 11.13: Plots of the estimated regression surface based on the peer nomination data. The left panel shows the plot created by ols.plot.inter, which assumes that an interaction can be modeled with $y = \beta_0 + \beta_1 x_1 + \beta_2 x_2 + \beta_3 x_1 x_2 + \epsilon$ and where the least squares estimate of the parameters is used. The right panel shows an approximation of the regression surface based on the R function lplot.

usual interaction model is unsatisfactory for the situation at hand. Testing H_0: $\beta_3 = 0$, assuming Eq. (11.18) is true and using ordinary least squares, the resulting p-value returned by the R function olshc4 is .64. The R function adtest returns a p-value less than .01 indicating that an interaction exists.

■

The R function

$$kercon(x,y,cval=NA,eout=F,xout=F, outfun=out,xlab=``X",ylab=``Y")$$

creates a plot using the first of the two methods described in the previous section. It assumes there are two predictors and terminates with an error message if this is not the case. For convenience, let x1 and x2 represent the data in columns one and two of the R variable x. The function estimates the quartiles of the data stored in x2 using the ideal fourths, and then creates three smooths between y and x1. By default, the smooths correspond to the regression lines between y and $x1$ given that x2 is equal to the estimated lower quartile, the median, and the upper quartile. If it is desired to use other values for $x2$, this can be done via the argument cval. The arguments are used in the same manner as described, for example, in Section 11.5.8.

The R function

$$riplot(x,y,adfun=adrun,plotfun=lplot,eout=T,xout=T)$$

fits a model to data using the function specified by the argument adfun, which defaults to the generalized additive model given by Eq. (11.11). It then computes the residuals and plots them versus the data in x. Again, x must be a matrix with two columns of data.

The R function

$$runsm2g(x1,y1,x2,val=median(x2),est=tmean,sm=F,\ldots)$$

splits the x1 and y1 values into two groups according to whether x2 is less than the value stored in the argument val. By default, val is the median of the values stored in x2. It then creates a smooth for both groups (via the function rungen in Section 11.5.5). Setting the argument sm=T results in a bagged version of the smooths. With small sample sizes, setting sm=T can be beneficial.

■ Example

Two hundred values were generated for x_1, x_2, and ϵ, where x_1, x_2, and ϵ are independent and have standard normal distributions. The left panel of Figure 11.14 shows the output from kercon when $y = x_1 + x_2 + \epsilon$. The solid line is the smooth for y and x_1 given that x_2 is equal to the estimate of its lower quartile. The middle line is the smooth given that x_2 is equal to its estimated median, and the upper line is the smooth for the upper quartile. The right panel shows the output where now $y = x_1 + x_2 + x_1 x_2 + \epsilon$.

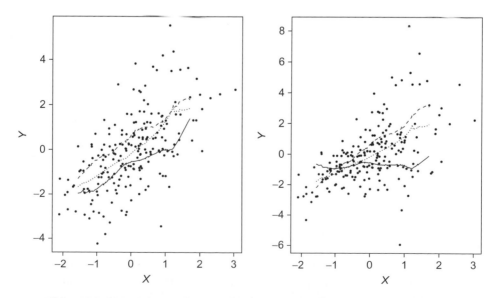

Figure 11.14: An illustration of the plot created by the function kercon.

11.7.2 Mediation Analysis

This section provides some very brief comments about what is generally known as *mediation analysis*. (For a book dedicated to this topic, see MacKinnon, 2008.) Mediation analysis is similar to a moderator analysis in the sense that the goal is to understand how the association between two variables is related to a third (mediating) variable. (Blends of the two methods, yielding what are called moderated-mediation analyses, have been proposed as well. See, e.g., Preacher et al., 2007.) In the parlance of researchers working on this problem, an *indirect effect*, also known as a *mediation effect*, refers to a situation where two variables of interest are associated via a third variable. For example, stress and obesity are believed to be associated through cortisol secretion (Rosmond, Dallman, & Björntorp, 1998). The strategy behind a mediation analysis is to assume that the three variables of interest satisfy three linear models. The first is that two primary variables of interest x and y (e.g., stress and obesity) are related via the usual linear model

$$y = \beta_{01} + \beta_{11}x + \epsilon_1. \tag{11.19}$$

The second assumption is that the mediating variable (cortisol in the example), which here is labeled x_m, is related to x via

$$x_m = \beta_{02} + \beta_{12}x + \epsilon_2. \tag{11.20}$$

And finally, it is assumed that

$$y = \beta_{03} + \beta_{13}x + \beta_{23}x_m + \epsilon_3. \tag{11.21}$$

Roughly, if $\beta_{13} = 0$, this is said to constitute full mediation (Judd & Kenny, 1981a, 1981b). If the strength of the association between x and y is reduced when the mediator is included, in the sense that $\beta_{13} < \beta_{11}$, there is said to be partial mediation.

Various strategies have been proposed for assessing whether x_m mediates the association between y and x. (For recent discussion of the issues and strategies when dealing with mediation analyses, see Zhao, 2010.) One is to focus on testing $H_0: \beta_{11} = \beta_{13}$. Another is to focus on the product $\beta_{12}\beta_{23}$, which has been called the *mediated effect* or *indirect effect*. This latter approach arises by noting that if Eq. (11.20) is substituted into Eq. (11.21), the total effect represented by the slope in Eq. (11.19) satisfies $\beta_{11} = \beta_{12}\beta_{23} + \beta_{13}$. (See MacKinnon et al., 1995, for more details.) Consequently, a common goal is testing

$$H_0 : \beta_{12}\beta_{23} = 0. \tag{11.22}$$

Under normality and homoscedasticity, a bootstrap method for testing this hypothesis, using the least squares estimator, has been found to perform reasonably well in simulations. But under nonnormality, or when there is heteroscedasticity, this is no longer the case (Ng, 2009a). Replacing the least squares estimator with the Theil–Sen estimator, Ng (2009a) found that a percentile bootstrap method performs well in simulations when $\beta_{12} = \beta_{23} = 0$. But otherwise, control over the probability of a type I error can be unsatisfactory in some situations. Biesanz, Falk, and Savalei (2010) compared several alternative methods. But the results relevant to nonnormality were limited to a single nonnormal distribution that is skewed with a relatively light tail. No results on the effects of heteroscedasticity were reported.

Another approach when performing a mediation analysis is to compute a confidence interval for $\beta_{11} - \beta_{13}$ using some robust regression estimator and a percentile bootstrap method. Briefly, take a bootstrap sample in the usual way assuming Eq. (11.21) is true, which yields a bootstrap estimate of β_{13}, say b_{13}^*. Using this same bootstrap sample, compute a bootstrap estimate of β_{11} assuming that Eq. (11.19) is true, yielding b_{11}^*. Let $d^* = b_{11}^* - b_{13}^*$. Repeat this process B times yields a confidence interval for $\beta_{11} - \beta_{13}$, and a p-value when testing H_0: $\beta_{11} = \beta_{13}$, by proceeding along the lines in Section 11.2. Limited simulation studies suggest that when testing at the 0.05 level, the actual level can drop well below 0.05 when the sample size is less than or equal to 40. With $n = 80$, this does not seem to be an issue.

Zu and Yuan (2010) derived an approach to testing Eq. (11.22) based on a Huber-type M-estimator that is used in conjunction with a percentile bootstrap method. Briefly, their method begins by computing the multivariate measure of location and scatter derived by Maronna (1976) based on (x_i, x_{mi}, y_i), $i = 1, \ldots, n$, yielding say $\hat{\mu}$ and $\hat{\Sigma}$. They then estimate the regression parameters via the method in Section 10.13.5. Finally, Eq. (11.22) is tested via a percentile bootstrap method. (Zu and Yuan also consider hypothesis testing techniques based on an estimate of the standard errors.) The percentile bootstrap method appears to perform relatively well in terms of controlling the probability of a type I error, but situations are encountered where it can be unsatisfactory. For example, under normality with $n = 40$, $\beta_{23} = 0.5$ and $\beta_{12} = 0$, if there is heteroscedasticity in the form where the error term is $\epsilon_3/(|x| + 1)$, the actual level of the test is approximately .09 when testing at the 0.05 level. Increasing n to 60, the actual level drops to about .056. But with $n = 60$ and a homoscedastic error term, if two additional points are added at $(x, x_m, y) = (3, -2, -3)$, the actual level is again approximately .09. Using instead the Theil-sen estimator in conjunction with a percentile bootstrap method for testing H_0: $\beta_{11} = \beta_{13}$, the actual level is approximately .025. But a criticism of this latter approach is that in various situations, the actual level can drop well below the nominal level. Currently, the best method for dealing with these problems is to modify slightly the Zu and Yuan method. In particular, use their method after excluding any (x_i, x_{mi}, y_i) for which x_i is an outlier among the values $x_1, \ldots, x_n$. Another seemingly natural strategy is to instead eliminate any (x_i, x_{mi}, y_i) for which (x_i, x_{mi}) is an outlier. But this can result in poor control over the probability of a type I error.

It should be noted that Green, Ha, and Bullock (2010) argue that mediation analyses have been based on regression models that rest on naive assumptions. The stated goal in the abstract of their paper is "to puncture the widely held view that it is a relatively simple matter to establish the mechanism by which causality is transmitted. This means puncturing the faith that has been placed in commonly used statistical methods of establishing mediation."

11.7.3 R functions ZYmediate, regmed2, and regmediate

The R function

$$\text{ZYmediate}(x, y, \text{nboot} = 2000, \text{alpha} = 0.05, \text{kappa} = 0.05, \text{SEED} = T, \text{xout} = F,$$
$$\text{outfun} = \text{out})$$

tests the hypothesis given by Eq. (11.22) using the method derived by Zu and Yuan (2010), which was outlined in the previous section. By default, the functions eliminate any point for which x_i is an outlier. This improves control over the probability of a type I error when there is heteroscedasticity. Currently, it seems to be one of the better methods when the sample size is small.

In case it helps, the R function

$$\text{regmed2}(x, y, \text{regfun} = \text{tsreg}, \text{nboot} = 400, \text{alpha} = 0.05, \text{xout} = F, \text{outfun} = \text{out}, \text{MC} = F,$$
$$\text{SEED} = T, \text{pr} = T, \ldots)$$

tests the two hypotheses $H_0: \beta_{12} = 0$ and $H_0: \beta_{22} = 0$, which are relevant to a mediation analysis as explained in the previous section. By default the Theil–Sen estimator is used, but other regression estimators can be used via the argument regfun. As usual, setting the argument xout=T results in leverage points being removed.

The R function

$$\text{regmediate}(x, y, \text{regfun=tsreg}, \text{nboot=400}, \text{alpha=0.05}, \text{xout=F}, \text{outfun=out}, \text{MC=F},$$
$$\text{SEED=T}, \ldots)$$

computes a confidence interval for $\beta_{11} - \beta_{13}$, and a p-value when testing $H_0: \beta_{11} = \beta_{13}$ is returned as well. Again by default, the Theil–sen estimator is used.

11.8 Comparing Parametric, Additive, and Nonparametric Fits

One way of comparing two different fits to data is to simply compute $m(\mathbf{x}_i)$, $i = 1, \ldots, n$ using both methods and then plot the results. That is, if $\hat{y}_{i1}$ is $m(\mathbf{x}_i)$ based on the fit using the first method, and $\hat{y}_{i2}$ is $m(\mathbf{x}_i)$ based on the second fit, plot $\hat{y}_{i1}$ versus $\hat{y}_{i2}$. So, for example, if data are generated according to a generalized additive model, then a plot of $\hat{y}_{i1}$ obtained by a method that assumes a generalized additive model generated the data, versus $\hat{y}_{i2}$ obtained by the running interval smooth in Section 11.5.11, should consist of points that are reasonably close to a line having slope one and intercept zero.

11.8.1 R Functions adpchk and pmodchk

The R function

$$\text{adpchk(x,y,adfun=adrun,gfun=runm3d,xout=T,outfun=out,} \ldots)$$

computes the $\hat{y}_{i1}$ values using the method specified by the argument adfun, which defaults to the generalized additive model given by Eq. (11.11) in Section 11.5.14. It then computes $\hat{y}_{i2}$ using the method specified by the argument gfun, which defaults to the running interval smoother. It then plots $\hat{y}_{i1}$ versus $\hat{y}_{i2}$. So if the two methods agree, the plotted points should be centered around a line having slope one and intercept zero. Here, $p > 2$ is allowed.

The R function

$$\text{pmodchk(x,y,regun=tsreg,gfun=runm3d,op=1,xout=F,eout=F)}$$

is like adpchk, only the third argument is now regfun, which is assumed to be some parametric fit. The default method is the Theil–Sen estimator.

■ Example

Values for x_1, x_2, and ϵ were generated from a g-and-h distribution with x_1, x_2, and ϵ independent and $g = h = 0.5$. The upper two panels of Figure 11.15 show the plots created by adpchk and pmodchk when $y = x_1 + x_2 + \epsilon$. (The graph created by adpchk is in the left panel.) In this case the model assumed by the third argument (adfun and tsreg) is correct and all points are tightly clustered around the line having slope 1. The lower panels show the plots where now $y = x_1 + x_1 x_2^2 + \epsilon$. So now the models, assumed

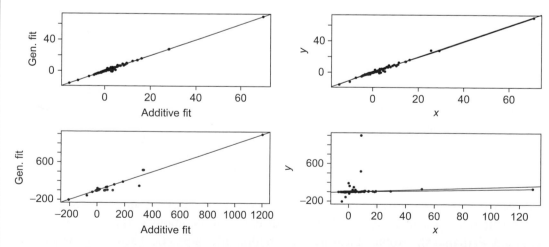

Figure 11.15: Illustration of the plots created by the functions adpchk and pmodchk.

by the third argument, are wrong, and the graphs indicate that in some instances, there are relatively large discrepancies between the assumed models and the more general model indicated by the fourth argument, gfun.

∎

11.9 Measuring the Strength of an Association Given a Fit to the Data

The measures of association, covered in Chapter 9, are not based on any particular regression model or fit to the data. Pearson's correlation has a well-known connection to the least squares regression line, but for the bulk of the robust correlations, there is no explicit connection to any of the robust regression methods covered in Chapter 10. This section is aimed at filling this gap. There are, in fact, various ways one might proceed. The immediate goal is to describe how this might be done based on simple generalizations of the notion of explanatory power, which was studied in a general context by Doksum and Samarov (1995).

Let $\hat{y}$ be some predicted value of y, given the values of p predictors $x_1, \ldots, x_p$. Explanatory power is

$$\frac{\sigma^2(\hat{y})}{\sigma^2(y)},$$

the usual variance of the predicted values divided by the variance of the observed y values. If $\hat{y}$ is based on the usual least squares regression line, and when there is $p = 1$ predictor, explanatory power reduces to ρ^2, the coefficient of determination. To see this, note that from basic principles, the least squares regression line can be written as

$$\hat{y} = \beta_0 + \rho \frac{\sigma_y}{\sigma_x} x.$$

So $\sigma^2(\hat{y}) = \rho^2(\sigma_y^2/\sigma_x^2)\sigma_x^2 = \rho^2\sigma_y^2$. Dividing this last quantity by σ_y^2 yields ρ^2.

A robust generalization of explanatory power consists of simply replacing the usual variance with some robust analog and taking $\hat{y}$ to be the predicted value of y based on any regression estimator or smoother. In symbols, let $\tau^2(y)$ be any measure of variation. Then a robust analog of explanatory power is

$$\eta^2 = \frac{\tau^2(\hat{y})}{\tau^2(y)}. \qquad (11.23)$$

The explanatory strength of the association is the (positive) square root of explanatory power, η. From Chapter 3, there are several reasonable choices for τ^2. Here, unless stated otherwise, τ^2 is taken to be the percentage bend midvariance, which is computed as described in Table 3.9. Perhaps other robust measures of variation offer a practical advantage when measuring the strength of association, but this has not been explored. R functions previously described that report the explanatory strength of association include lplot (lowess) and tsreg (the Theil–sen estimator).

In principle, explanatory power can be estimated when using any regression method or smoother. First, compute the percentage bend midvariance based on predicted y values, say $\hat{\tau}^2(\hat{y})$, compute the percentage bend midvariance based on the observed y values, $\hat{\tau}^2(y)$, in which case the estimate of η^2 is

$$\hat{\eta}^2 = \frac{\hat{\tau}^2(\hat{y})}{\hat{\tau}^2(y)}. \tag{11.24}$$

But a fundamental issue is whether the choice of method for obtaining the predicted y values make a practical difference when estimating η^2. For small to moderate sample sizes, it has been found that it does (e.g., Wilcox, 2010b). Two regression estimators that seem to perform relatively well, given the goal of estimating η^2, are the Theil–Sen estimator when the regression surface is a plane, and Cleveland's smoother (LOWESS), described in Section 11.5.2, when there is curvature.

Section 11.5.3 described an R function, lplot, for plotting Cleveland's nonparametric regression line (LOESS). One of the arguments is varfun, which can now be explained. It indicates the measure of variation used when estimating explanatory power and defaults to the percentage bend midvariance. The R function tsreg, which computes the Theil–Sen estimator, also contains the argument varfun, which again indicates how explanatory power is computed.

Renaud and Victoria-Feser (2010) compared several other robust analogs of R^2, the coefficient of determination, which are based in part on a fit to the data obtained via the MM-estimator in Section 10.9.1. Their approach represents a generalization of a measure of association suggested by Maronna, Martin, and Yohai (2006, p. 171). For yet another approach to getting a robust version of R^2, see Croux and Dehon (2003).

Let $\Psi(r_i; c)$ be defined as in Section 10.9.1. In principle, some other choice for Ψ, associated with some M-estimator, could be used, but the focus here is on the choice used by the MM-estimator. The measure of association proposed by Maronna et al. (2006) is

$$R^2_{MM} = 1 - \frac{\sum \Psi\left(\frac{r_i}{\hat{\tau}}\right)}{\sum \Psi\left(\frac{y_i - \hat{\mu}}{\hat{\tau}}\right)},$$

where $\hat{\mu}$ is some robust measures of location, taken here to be the M-measure of location associated with Ψ.

For convenience, write $w_i = \Psi(r_i; c)$. The generalization of R^2_{MM}, suggested by Renaud and Victoria-Feser is

$$R^2_w = \frac{\sum w_i(\hat{y} - \tilde{y})^2}{\sum w_i(\hat{y}_i - \tilde{y})^2 + a\sum w_i(y_i - \hat{y})^2},$$

where a is a correction factor for achieving consistency, $\tilde{y} = (1/\sum w_i) \sum w_i \hat{y}_i$ and $\hat{y}_i$ are the predicted y values produced by the MM-estimator. The motivation for this generalization is that it reduces the bias associated with R_{MM}^2. Following Renaud and Victoria-Feser, $a = 1.2067$ is used.

11.9.1 R Function RobRsq

The R function

$$RobRsq(x,y)$$

computes R_w^2, the measure of association derived by Renaud and Victoria-Feser (2010).

11.9.2 Comparing Two Independent Groups via Explanatory Power

For two independent groups, let η_j^2 be the explanatory power associated with the jth group ($j = 1, 2$). This section describes a modified percentile bootstrap method for testing

$$H_0 : \eta_1^2 = \eta_2^2. \tag{11.25}$$

A simple strategy is to use a percentile bootstrap method. That is, generate bootstrap samples from the jth group, estimate η_j^2 yielding say $\tilde{\eta}_j^2$, repeat this B times yielding $\tilde{\eta}_{jb}^2$ ($b = 1, \ldots, B$), in which case a p-value is $p = 2\min(P, 1 - P)$, where P is the proportion of times $\tilde{\eta}_1^2 > \tilde{\eta}_2^2$. Imagine that the goal is to test at the $\alpha = 0.05$ level, in which case H_0 is rejected if $p \leq .05$. Then the actual level of the percentile bootstrap method just described is very close to 0.05 with sample sizes $n_1 = n_2 = 200$ (Wilcox, 2009c). But for smaller sample sizes the actual level is substantially smaller than 0.05, particularly when both sample sizes are less than 100. However, Wilcox (2009c) found that the actual level of the test was fairly stable among the nonnormal distributions that were considered, which suggests a simple modification: determine an adjusted level α_a with the goal of achieving a .05 type I error probability if the null hypothesis is rejected when $p \leq \alpha_a$. First consider $n_1 = n_2 = n$. For standard normal distributions it was found that for $n = 30, 50$, and 100, $\alpha_a = .3, .21$, and $.08$, respectively. For other sample sizes, simple linear interpolation is suggested. More precisely, if $30 < n < 50$, use linear interpolation based on n and the α_a values .3 and .21. For $50 < n < 100$ interpolate using the α_a values .21 and .08, and for $100 < n < 200$ use the values .08 and .05. As for $n_1 \neq n_2$, let α_1 and α_2 be the values of α_a corresponding to n_1 and n_2, respectively. Then the adjusted level is taken to be $(n_2\alpha_1 + n_1\alpha_2)/(n_1 + n_2)$. For example, with $n_1 = 30$ and $n_2 = 100$, this yields, $\alpha_a = .249$, which is nearly equal to the estimate of α_a based on simulations, namely .24. And the simulation estimate of the actual type I error probability remains .05. For $n_1 = 30$ and $n_2 = 50$, this yields .266, the simulation estimate is .26, and the level of the test using .266 is again 0.05.

A related goal is testing

$$H_0 : \eta_1 = \eta_2, \tag{11.26}$$

which generalizes methods for testing the hypothesis that two independent groups have equal Pearson correlations. When there is curvature, an obvious way of attaching a sign to the square root of $\hat{\eta}^2$ is to use the positive square root if the association is monotonic increasing, otherwise use the negative square root. If the regression line is not monotonic, a possibility is to attach a sign indicating whether in general the regression line is increasing. For convenience, assume $x_1 \leq \cdots \leq x_n$, let

$$S = \sum_{i=2}^{n} \text{sign}(\hat{y}_i - \hat{y}_{i-1}),$$

and let I=1 if $S \geq 0$, otherwise I= -1. Then use $I\hat{\eta}$ as the measure of association. (Choosing the sign in this manner has similarities to Kendall's tau.)

As for testing Eq. (11.26), a slight modification of the method for testing Eq. (11.25) is needed to avoid type I error probabilities well above the nominal level when the sample size is small. For $n \geq 50$, determine α_u exactly as done when testing (11.25). But for $n < 50$, extrapolate using the α_a values .21 and .08, which correspond to the sample sizes 50 and 100, respectively.

11.9.3 R Functions smcorcom and smstrcom

The R function

   smcorcom(x1, y1, x2, y2, nboot = 200, pts = NA, plotit = T, SEED = T, varfun = pbvar)

tests Eq. (11.25). If the argument plotit=T, the two regression lines are plotted (by calling the R function lplot2g in Section 11.11.2). The R function

   smstrcom(x1, y1, x2, y2, nboot = 200, plotit = T, SEED = T, varfun = pbvar, xout=F,
                           outfun=out, . . .)

tests the hypothesis given by Eq. (11.26).

11.10 Comparing Predictors

When dealing with two or more predictors, an issue that has received considerable attention is determining which predictors, or which collection of predictors, is best. Numerous methods have been proposed, many of which are known to be unsatisfactory. Relatively well-known methods that have proven to be unsatisfactory include stepwise regression (e.g., Montgomery & Peck, 1992, Section 7.2.3; Derksen & Keselman, 1992), a related (forward selection) method (see Kuo & Mallick, 1998; Huberty, 1989; Chatterjee & Hadi, 1988; cf. Miller, 1990), methods based on R^2 (the squared multiple correlation), and the classic F statistic that tests the hypothesis that all slopes are zero. A homoscedastic approach based on

$$C_p = \frac{1}{\hat{\sigma}^2} \sum (Y_i - \hat{Y}_i)^2 - n + 2p,$$

called Mallows (1973) C_p criterion, cannot be recommended either (Miller, 1990). Another approach is based on what is called ridge regression, but it suffers from problems listed by Breiman (1995). Briefly, ridge regression is not scale invariant. If the scale of the predictors is changed, the ridge coefficients do not change inversely proportional to the changes in the variable scale. An approach to this criticism is to standardize each predictor so that they each have mean 0 and variance 1. Breiman notes, for example, that if the interquartile range were used instead of the usual variance to normalize the predictors, this would give different regression predictors (cf. Smith & Cambell, 1980). Three alternative approaches, versions of which are described later in this section, are cross-validation, bootstrap methods, namely the .632 estimator used here, and the so-called nonnegative garrote technique derived by Breiman (1995). Efron and Tibshirani (1993) provide additional details regarding the .632 estimator.

11.10.1 Comparing Pearson Correlations

This section describes a method for comparing predictors via Pearson's correlation. As previously stressed, Pearson's correlation is not robust, but perhaps there are situations where comparing Pearson correlations has practical value. Many such methods have been derived, comparisons of which are reported in Wilcox (2009d). The method described here combines a method derived by Zou (2007) with the HC4 method for computing a confidence interval for ρ, which was described in Section 9.3.14.

For notational convenience, let ρ_{jk} be the correlation between x_j and x_k, $j = 1, 2, 3$; $k = 1, 2, 3$. The goal is to compute a confidence interval for $\rho_{12} - \rho_{13}$. In a regression context, x_1 corresponds to the the outcome variable y. Let (l_1, u_1) and (l_2, u_2) be $1 - \alpha$ confidence intervals for ρ_{12} and ρ_{13}, respectively, which are based on the HC4 method. Then a $1 - \alpha$ confidence interval for $\rho_{12} - \rho_{13}$ is

$$(L, U),$$

where

$$L = r_{12} - r_{13} - \sqrt{(r_{12} - l_1)^2 + (u_2 - r_{13})^2 - 2\widehat{corr}(r_{12}, r_{13})(r_{12} - l_1)(u_2 - r_{13})},$$

$$U = r_{12} - r_{23} + \sqrt{(u_1 - r_{12})^2 + (r_{23} - l_2)^2 - 2\widehat{corr}(r_{12}, r_{13})(u_1 - r_{12})(r_{23} - l_2)},$$

and

$$\widehat{corr}(r_{12}, r_{13}) = \frac{(r_{23} - .5r_{12}r_{23})(1 - r_{12}^2 - r_{13}^2 - r_{23}^2) + r_{23}^2}{(1 - r_{12}^2)(1 - r_{13})^2}.$$

11.10.2 Methods Based on Estimating Prediction Error

This section describes two methods for comparing predictors based on the notion of prediction error. The first approach is called the .632 bootstrap method, which allows heteroscedasticity. The other uses a leave-one-out cross-validation method.

Imagine that the n pairs of values $(x_1, y_1), \ldots, (x_n, y_n)$ are used to determine the regression line $\hat{y} = b_0 + b_1 x$. Now imagine that a new x value is observed, which is labeled x_0, in which case the predicted value of y, based on the original n pairs of points, is $\hat{y}_0 = b_0 + b_1 x_0$. *Prediction error* refers to the discrepancy between the predicted value of y, $\hat{y}_0$, and the actual value of y, y_0, if it could be observed. One way of measuring the typical amount of prediction error is with

$$E[(y_0 - \hat{y}_0)^2],$$

the expected squared difference between the observed and predicted value of Y. And another possibility is

$$E[|y_0 - \hat{y}_0|],$$

the expected absolute error. As is evident, the notion of prediction error is easily generalized to multiple predictors. The basic idea is that via some method we get a predicted value for y, which we label $\hat{y}$, and the goal is to measure the discrepancy between $\hat{y}_0$ (the predicted value of y based on a future collection of x values) and the actual value of y, y_0, if it could be observed.

A simple estimate of prediction error is the so-called apparent error rate, which is just the average error when predicting the observed y values with $\hat{y}$. More formally, let $Q(y, \hat{y})$ be some measure of the discrepancy between an observation, y, and its predicted value, $\hat{y}$. So squared error corresponds to

$$Q(y, \hat{y}) = (y - \hat{y})^2.$$

The goal is to estimate the typical amount of error for future observations. In symbols, the goal is to estimate

$$\eta = E[Q(y_0, \hat{y}_0)],$$

the expected error between a predicted value for y, based on a future value of x, and the actual value of y, y_0, if it could be observed. A simple estimate of η is the *apparent error*:

$$\hat{\eta}_{\text{ap}} = \frac{1}{n} \sum Q(y_i, \hat{y}_i).$$

So for squared error, the apparent error is

$$\hat{\eta}_{\text{ap}} = \frac{1}{n} \sum (y_i - \hat{y}_i)^2,$$

the average of the squared residuals. (For results on estimating prediction error when using the MM-estimator, see Khan, van Aelst, & Zamar, 2010.)

A practical concern is that the apparent error is biased downward because the data used to come up with a prediction rule ($\hat{y}$) are also being used to estimate error (Efron & Tibshirani, 1993). That is, it tends to underestimate the true error rate, η. The so-called .632 bootstrap estimator is designed to address this problem and currently seems to be a relatively good choice for identifying the best predictors.

The .632 Estimator

The .632 estimator is applied as follows. Generate a bootstrap sample, only rather than sample n vectors of observations with replacement, as is typically done, sample $m < n$ vectors of observations instead. (Setting $m = n$, Shao, 1996, shows that the probability of selecting the correct model may not converge to one as n gets large.) Here, $m = 5\log(n)$ is used, which was derived from results reported by Shao (1996). Let $\hat{y}_i^*$ be the estimate of y_i based on the bootstrap sample, $i = 1, \ldots, n$. Repeat this process B times yielding $\hat{y}_{ib}^*$, $b = 1, \ldots, B$. Then an estimate of η is

$$\hat{\eta}_{\text{Boot}} = \frac{1}{nB} \sum_{b=1}^{B} \sum_{i=1}^{n} Q(y_i, \hat{y}_{ib}^*).$$

A refinement of $\hat{\eta}_{\text{Boot}}$ is to take into account whether a y_i value is contained in the bootstrap sample used to compute $\hat{y}_{ib}^*$. Let

$$\hat{\epsilon}_0 = \frac{1}{n} \sum_{i=1}^{n} \frac{1}{B_i} \sum_{b \in C_i} Q(y_i, \hat{y}_{ib}^*),$$

where C_i is the set of indices of the bth bootstrap sample not containing y_i and B_i is the number of such bootstrap samples. Then the .632 estimate of the prediction error is

$$\hat{\eta}_{.632} = .368\hat{\eta}_{\text{ap}} + .632\hat{\epsilon}_0. \tag{11.27}$$

This estimator arises in part from a theoretical argument showing that .632 is approximately the probability that a given observation appears in a bootstrap sample of size n.

The Leave-One-Out Cross-Validation Method.

Prediction error using the leave-one-out cross-validation method is applied as follows. Momentarily omit the ith point $(\mathbf{x}_i, y_i)$ and fit a regression model to the data. Based on this fit, let $\hat{y}_{-i}$ be the estimate of y using $\mathbf{x}_i$. Let $e_i = y_i - \hat{y}_{-i}$. Then prediction error is measured via some measure of variation applied to the e_i values ($i = 1, \ldots, n$). Here the percentage bend midvariance is used unless stated otherwise, but this is not to suggest that alternative measures of variation should be ruled out.

A practical issue is how this leave-one-out cross-validation method compares to the .632 estimator. This issue has not been explored as yet.

11.10.3 R Functions TWOpov, regpre, and regpreCV

The R function

$$\text{TWOpov}(x, y, \text{alpha} = 0.05)$$

computes a confidence interval for the difference between Pearson correlations, corresponding to two predictors, using the method in Section 11.10.1. The argument x is assumed to be a matrix with two columns corresponding to two predictors.

The R function

regpre(x, y, regfun=lsfit, error=absfun, nboot=100, adz=T, mval=round(5*log(length(y))), model=NULL, locfun=mean, pr=T, xout F, outfun=out, plotit=T, xlab="Model Number", ylab="Prediction Error", SEED=T, ...)

estimates prediction error using the .632 bootstrap method. By default, least squares regression is used, but results in Wilcox (2008d) indicate that the Theil–Sen estimator is better for general use. This can be done by setting the argument regfun=tsreg. With adz=T, the function includes an estimate of prediction error based on using only the measure of location indicated by the argument locfun. That is, no predictors are used. The argument mval is m, the number of observations sampled when generating bootstrap samples. The argument error=absfun means that absolute error is used by default. Setting error=sqfun would use squared error. Other robust measures of variation might be used as well. For example, error=winvar would use the 20% Winsorized variance.

The R function

regpreCV(x, y, regfun=tsreg, varfun=pbvar, adz=T, model=NULL, locfun=mean, xout=F, outfun=out, plotit=T, xlab="Model Number", ylab = "Prediction Error", ...)

performs the leave-one-out cross-validation estimate of prediction error.

■ Example

The R function regpre is illustrated with the reading data described in the first example of Section 10.8.1. Here we consider how well the first three predictors, stored in columns 2–4, compare when predicting a word identification score (stored in column 8 of the file read.dat). Assuming the data are stored in the R variable read, the command regpre(read[,2:4],read[,8],regfun=tsreg,locfun=median) returns

```
$estimates
      apparent.error   boot.est   err.632   var.used   rank
[1,]       12.48052    13.36920   13.25599         1      4
[2,]       11.99610    12.74614   12.70693         2      2
[3,]       14.87278    16.12612   16.08528         3      8
[4,]       11.47575    12.51257   12.55914        12      1
[5,]       12.76048    14.33530   14.27367        13      6
[6,]       11.90024    13.24456   13.28122        23      5
[7,]       11.30447    13.30425   13.21919       123      3
[8,]             NA          NA   14.30223         0      7
```

The column headed by var.used indicates the predictors used in the model. The entry 12 means that both predictors 1 and 2 were used, ignoring predictor 3. The entry 123 is the case where all three predictors are used. The last column provides an easy way of identifying which combination of predictors produced the lowest prediction error. Here, using both predictors 1 and 2 performed best. The worst model was using predictor 3, ignoring the other predictors. The last row is for the case where all predictors are ignored. So here, using predictor 3 is worse than using no predictors at all, meaning that one simply uses the median of the *y* values to predict future observations. Using instead the R function regpreCV, the results are:

```
      apparent.error   boot.est   err.632   var.used   rank
[1,]       12.54924    13.05253   13.06939         1      5
[2,]       11.56878    12.23330   12.23345         2      1
[3,]       14.13608    15.15757   15.15858         3      8
[4,]       11.51583    12.24795   12.32174        12      2
[5,]       12.56508    13.63793   13.65641        13      6
[6,]       11.56411    12.75894   12.77784        23      4
[7,]       11.45809    12.63197   12.72554       123      3
[8,]             NA          NA   14.27227         0      7
```

The R functions automatically generate all possible combinations of predictors, assuming the number of predictors is at most 5. The argument model, which is assumed to have list mode, can be used to analyze only models that are of specific interest. For example, setting model[[1]]=1 and model[[2]]=c(1,2,3), and then setting the argument model=model, prediction error would be estimated when using predictor 1, as well as using predictors 1, 2, and 3 simultaneously, but no other models would be considered.

11.10.4 R Function larsR

Another method for identifying the best predictors that is based on what is called the lasso (Tibshirani, 1996). And a related approach is least angle regression; see Efron, Hastie, Johnstone, and Tibshirani (2004). Also see Wang and Leng (2007), Owen (2006), as well as Radchenko and James (2011).

Both the lasso and least angle regression can be applied with the R function

$$larsR(x,y,type=``lasso",xout=F,outfun=outpro).$$

By default, the lasso method is used. To use least angle regression, set the argument type="lar". To eliminate leverage points via the function indicated by the argument outfun, set the argument xout=T. The function returns estimates of which estimates are best, in descending order. Unlike the R functions regpre and regpreCV, larsR does not provide information about which subsets of variables are best. That is, it does not indicate, for example, whether predictors 1 and 2, taken together, are better in some sense than using predictor 1 only.

11.10.5 Comparing Predictors via Explanatory Power and a Robust Fit

Method BTS

Yet another approach to comparing predictors is to estimate the strength of the association based on the Theil–Sen estimator and then use a percentile bootstrap method to test

$$H_0 : \eta_1^2 = \eta_2^2,$$

where now η_j^2 is explanatory power when using predictor x_j. But as the correlation between x_1 and x_2 increases, the actual level of this method can drop well below the nominal level, even with a sample size of $n = 100$. Taking independent bootstrap samples, the first from (x_{i1}, y_i) and the second from (x_{i2}, y_i), has been found to improve matters (Wilcox, in press b). Imagine that this is done yielding $D^* = \tilde{\eta}_1^2 - \tilde{\eta}_2^2$, the difference between the two estimates of η^2. Repeating this process B times yields $D_1^*, \ldots, D_B^*$, which can be used to estimate $P = P(D < 0)$ in the manner already described, which in turn yields the generalized p-value. But again, control over the probability of a type I error has been found to be not quite satisfactory: when testing at the 0.05 level, the actual level can be substantially smaller than 0.05. Some improvement is obtained if rather than estimate $P = P(D < 0)$ with the bootstrap samples in the usual way, a kernel density estimate is used instead, a strategy motivated by results in Racine and MacKinnon (2007b). Here the adaptive kernel estimator in Section 3.2.4

is used, in which case the distribution of D is estimated with

$$\hat{f}(d) = \frac{1}{nh} \sum_{i=1}^{B} K\left(\frac{d - D_i}{h}\right),$$

where K is taken to be the Epanechnikov kernel and h is the span. An estimate of $P(D < 0)$ is

$$\hat{P}(D < 0) = \frac{1}{nh} \sum_{i=1}^{n} \int_{\ell}^{0} K\left(\frac{t - D_i}{h}\right) dt.$$

The method just described performs well in simulations when $\rho_{12} = 0$, but again the actual level can drop well below the nominal level when $\rho_{12} \neq 0$. Let ρ_{k12} be Kendall's tau for x_1 and x_2. Compute a 0.95 confidence interval ρ_{k12} using the method in Section 9.3.14. If this interval contains 0, let $\tilde{p} = 0$; otherwise

$$\tilde{p} = .352|r_{k12}| + .049. \tag{11.28}$$

For $n \leq 100$, reject at the 0.05 level if the p-value is less than or equal to $\tilde{p}$. For $n > 100$, use the p-value in the usual manner.

Method SM

Method BTS can be extended to the situation where explanatory power is estimated via LOWESS, described in Section 11.5.2, when testing at the 0.05 level. Let

$$\check{p} = .25|r_{k12}| + .05 + (100 - n)/10000, \tag{11.29}$$

$\check{p} = \max(.05, \check{p})$, and reject if $p \leq \check{p}$. For $n > 200$, $\check{p}$ is taken to be .05.

11.10.6 R Functions ts2str and sm2strv7

The R function

$$\text{ts2str(x, y, nboot = 400, SEED = T)}$$

performs method BTS and the function

$$\text{sm2strv7(x, y, nboot = 100, SEED = T, xout = F, outfun = outpro, ...)}$$

performs method SM.

11.11 ANCOVA

A common problem is comparing two independent groups of participants in terms of some measure of location while taking into account a covariate. More generally, a common goal is to compare the regression curves corresponding to two or more groups. The Pygmalion data in Section 11.2.1 provides an illustration. If the experimental and control groups are compared using post IQ reasoning scores, the 20% trimmed means are found to be significantly different at the $\alpha = 0.05$ level. A general issue is whether there continues to be a difference when the pretest IQ scores are taken into account. Section 11.2 describes one approach to this problem where the slopes of the regression lines are compared.

Before continuing, it is noted that there is a vast literature on the analysis of covariance (ANCOVA) where the goal is to compare groups in terms of some measure of location, usually means, while taking into account some covariate. For an entire book devoted to the subject, see Huitema (1980). For a review of some recent developments, see Rutherford (1992) and Harwell (2003). (For some recent results when using means, see Ceyhan and Goad, 2009.) Obviously all relevant methods cannot be described here. Rather, attention is restricted to situations where comparisons are to be made using some robust measure of location.

For the jth group, let $m_j(x)$ be some population measure of location associated with y given x. Given x, the problem is determining how the typical value of y in the first group compares to the typical value in the second. In the Pygmalion study, for example, the goal might be to determine how the 20% trimmed mean of the experimental group compares to the trimmed mean of the control group, given that a student's IQ reasoning pretest score is $x = 90$. Of course, a more general goal is to determine how the trimmed means compare as x varies. The most common strategy is to assume that a straight regression line can be used to predict y given x. That is, assume that for the jth group, $m_j(x) = \beta_{1j}x_{1j} + \beta_{0j}$, $j = 1, 2$. Next, estimate the slope and intercept for each group, and use the estimates to compute a confidence interval for $m_1(x) - m_2(x)$. Based on results covered in this chapter, as well as Chapter 5, a reasonable speculation is that a percentile bootstrap procedure will provide fairly accurate probability coverage when working with some robust regression method. However, simulations do not support this approach. In fact, probability coverage can be poor, at least when $n \leq 50$.

Another strategy is to assume that the regression slopes are parallel, as is done using a standard ANCOVA method for means. If this assumption is true, then the groups can be compared simply by comparing the intercepts. However, this approach is not very satisfying. The assumption of parallel regression lines can be tested using the method in Section 11.2, but how much power should this test have in order to be reasonably certain that the slopes are, for all practical purposes, sufficiently parallel? Also, if the slopes are not parallel what should be done instead? There is a solution based on means and the assumption that the error term within each group is homoscedastic (Wilcox, 1987b), but one of the goals here is to allow the error term to be heteroscedastic. Yet another problem is determining what to do if the regression line is curvilinear. In some cases it might help to replace x with x^a, for some constant a, but as already noted, this method of straightening a regression is not always effective, and even if a good choice for a could be found, there remains the problem of finding an effective method for computing a confidence interval for $m_1(x) - m_2(x)$.

In recent years, a number of nonparametric methods have been proposed for testing

$$H_0 : m_1(x) = m_2(x),$$

for any x (e.g., Bowman & Young, 1996; Delgado, 1993; Dette & Neumeyer, 2001; Ferreira & Stute, 2004; Härdle & Marron, 1990; Hall & Hart, 1990; Hall, Huber, & Speckman, 1997; King, Hart, & Wherly, 1991; Kulasekera, 1995; Kulasekera & Wang, 1997; Munk & Dette, 1998; Neumeyer & Dette, 2003; Srihera & Stute, 2010; Young & Bowman, 1995; Zou et al., 2010). Typically they are based on kernel-type regression estimators where $m(x)$ is the conditional mean of y given x. Many of these methods make rather restrictive assumptions, such as homoscedasticity or equal design points, but recent efforts have yielded methods that remove these restrictions (e.g., Dette & Neumeyer, 2001; Neumeyer & Dette, 2003). The method derived by Srihera and Stute (2010) allows heteroscedasticity, but it is unknown how well it performs under nonnormality. There are various ways these methods might be extended to robust measures of location, so far simulations do not support their use, but many variations have yet to be investigated.

11.11.1 *Methods Based on Specific Design Points*

This section describes a method for computing a $1 - \alpha$ confidence interval for $m_1(x) - m_2(x)$ that makes no parametric assumption about how y is related to x. In particular, a straight regression line is not assumed. Furthermore, complete heteroscedasticity is allowed, meaning that the error term for each group can be heteroscedastic, and nothing is assumed about how the variance of the error term in the first group is related to the second. There are many variations of the method that might prove to be useful, as will become evident. The technique used here is one of the few methods that has been found to perform well in simulations under nonnormality and heteroscedasticity. The goal is to impart the general flavor of the method with the understanding that some researchers might want to make modifications depending on the situation. In principle, $m_j(x)$ could be any measure of location or scale, but for now attention is focused on the 20% trimmed mean.

The general strategy is to approximate the regression lines with a running interval smoother and then use the components of the smoother to make comparisons at appropriate design points. To explain, first assume that an x has been chosen with the goal of computing a confidence interval for $m_1(x) - m_2(x)$. For the jth group, let $x_{ij}, i = 1, \ldots, n_j$ be values of the predictors that are available. As previously indicated, $m_j(x)$ is estimated with the trimmed mean of the y_{ij} values such that i is an element of the set

$$N_j(x) = \{i : |x_{ij} - x| \le f_j \times \text{MADN}_j\}.$$

That is, for fixed j, estimate $m_j(x)$ using the y_{ij} values corresponding to the x_{ij} values that are close to x. As already noted, the choice $f_j = 0.8$ or $f_j = 1$ generally gives good results, but some other value might be desirable. Let $M_j(x)$ be the cardinality of the set $N_j(x)$. That is, $M_j(x)$ is the number of points in the jth group that are close to x, which in turn is the number of y_{ij} values used to estimate $m_j(x)$. When $m_j(x)$ is the 20% trimmed mean of y, given x, the two regression lines are defined to be *comparable* at x if $M_1(x) \ge 12$ and $M_2(x) \ge 12$. The idea is that if the sample sizes used to estimate $m_1(x)$ and $m_2(x)$ are sufficiently large, then a reasonably accurate confidence interval for $m_1(x) - m_2(x)$ can be computed using the methods in Chapter 5. Yuen's method often gives satisfactory results, and the the bootstrap-t is even better, albeit more costly in terms of computer time.

When comparing the regression lines at more than one design point, confidence intervals for $m_1(x) - m_2(x)$, having simultaneous probability coverage approximately equal to $1 - \alpha$, can be computed as described in Chapter 7. When this is done for the problem at hand, the value for f that is used is related to how close the actual simultaneous probability coverage is to the nominal level.

To illustrate what can happen, suppose it is desired to compare the regression lines at five x values: z_1, z_2, z_3, z_4, and z_5. Of course, in practice, an investigator might have some substantive reason for picking certain design points, but this process is difficult to study via simulations. For illustrative purposes, suppose the design points are chosen using the following process. First, for notational convenience, assume that for fixed j, the x_{ij} values are in ascending order. That is, $x_{1j} \le \cdots \le x_{n_j j}$. Suppose z_1 is taken to be the smallest x_{i1} value for which the regression lines are comparable. That is, search the first group for the smallest x_{i1} such that $M_1(x_{i1}) \ge 12$. If $M_2(x_{i1}) \ge 12$, in which case the two regression lines are comparable at x_{i1}, set $z_1 = x_{i1}$. If $M_2(x_{i1}) < 12$, consider the next largest x_{i1} value and continue until it is simultaneously true that $M_1(x_{i1}) \ge 12$ and $M_2(x_{i1}) \ge 12$. Let i_1 be the value of i. That is, i_1 is the smallest integer such that $M_1(x_{i_1 1}) \ge 12$ and $M_2(x_{i_1 1}) \ge 12$. Similarly, let z_5 be the largest x value in the first group for which the regression lines are comparable. That is, z_5 is the largest x_{i1} value such that $M_1(x_{i1}) \ge 12$ and $M_2(x_{i1}) \ge 12$. Let i_5 be the corresponding value of i. Let $i_3 = (i_1 + i_5)/2$, $i_2 = (i_1 + i_3)/2$, and $i_4 = (i_3 + i_5)/2$. Round i_2, i_3, and i_4 down to the nearest integer and set $z_2 = x_{i_2 1}$, $z_3 = x_{i_3 1}$, and $z_4 = x_{i_4 1}$. Finally, consider computing confidence intervals for $m_1(z_q) - m_2(z_q)$, $q = 1, \ldots, 5$ by applying the methods for trimmed means described in Chapter 5. In particular, perform Yuen's test using the y values for which the corresponding x values are close to the design point z_q, and control the probability of at least one type I error among the five tests by using the critical value given by the five-variate Studentized maximum modulus distribution.

Table 11.9: Estimated Type I Error Probabilities, $\alpha = 0.05$.

X		ϵ		$Y = X$ $n = 30$ $f = 1$	$Y = X^2$ $n = 40$ $f = 1$	$Y = X^2$ $n = 30$ $f = 0.75$	$Y = X^2$ $n = 40$ $f = 0.75$	$Y = X^2$ $n = 40$ $f = 0.5$
g	h	g	h					
0	0	0	0	.046	.049	.045	.045	.039
0	0	0	0.5	.024	.034	.030	.027	.023
0	0	0.5	0	.034	.055	.045	.043	.037
0	0	0.5	0.5	.022	.038	.032	.031	.024
0	0.5	0	0	.042	.065	.045	.059	.041
0	0.5	0	0.5	.027	.049	.030	.037	.031
0	0.5	0.5	0	.041	.073	.043	.052	.031
0	0.5	0.5	0.5	.027	.041	.035	.026	.021
0.5	0	0	0	.042	.071	.029	.059	.039
0.5	0	0	0.5	.027	.046	.021	.037	.031
0.5	0	0.5	0	.041	.066	.026	.052	.035
0.5	0	0.5	0.5	.026	.043	.018	.026	.021
0.5	0.5	0	0	.044	.082	.039	.060	.044
0.5	0.5	0	0.5	.033	.056	.024	.036	.033
0.5	0.5	0.5	0	.045	.078	.041	.057	.035
0.5	0.5	0.5	0.5	.032	.053	.028	.031	.023

Table 11.9 shows some simulation results when x and ϵ are generated from various g-and-h distributions. Column five shows a simulation estimate of the actual probability of at least one type I error, $\hat{\alpha}$, when $\alpha = 0.05$, $y = x + \epsilon$, $n = 30$, and $f = 1$. The control over the probability of a type I error is reasonably good, the main problem being that the actual probability of a type I error can drop slightly below .025 when ϵ has a heavy-tailed distribution. Simulations are not reported for $n = 20$ because situations arise where five design points cannot always be found for which the the regression lines are comparable.

Column six shows the results when $y = x^2 + \epsilon$, $n = 40$, and $f = 1$. When x is highly skewed and heavy tailed ($g = h = 0.5$), $\hat{\alpha}$ can exceed .075, the highest estimate being equal to .082. With $n = 30$, not shown in Table 11.9, the estimate goes as high as .089. There is the additional problem that $f = 1$ might not be sufficiently small, as previously illustrated. If $f = 0.75$ is used, $\hat{\alpha}$ never exceeds .05, but in a few cases it drops below .025, the lowest estimate being equal to .018. Increasing n to 40, the lowest estimate is .026. Results in Chapter 5 suggest that even better probability coverage can be obtained using a bootstrap method, but the extent to which the probability coverage is improved for the problem at hand has not been determined. (For more details about the simulations, see Wilcox, 1997b.)

Another positive feature of the method described here is that its power compares well with the conventional approach to ANCOVA when the standard assumptions of normality, homogeneity of variance, and parallel regression lines are true. For example, if both groups have sample sizes of 40, $y_{i1} = x_{i1} + \epsilon_{i1}$, but $y_{i2} = x_{i2} + \epsilon_{i2} + 1$, the conventional approach to

ANCOVA has power approximately equal to .867 when testing at the 0.05 level. The method described here has power .828, where power is the probability that the null hypothesis is rejected for at least one of the empirically chosen design points, $z_1, \ldots, z_5$. If the error terms have heavy-tailed distributions, the traditional approach has relatively low power, as is evident from results described in previous chapters. Even under normality, standard ANCOVA can have relatively low power. For example, if $y_{i1} = x_{i1} + \epsilon_{i1}$, but $y_{i2} = .5x_{i2} + \epsilon_{i2}$, standard ANCOVA has power .039 versus .225 for the method based on trimmed means, again with sample sizes of 40. Even when ANCOVA has relatively good power, an advantage of using trimmed means with a running interval smoother is that the goal is to determine where the regression lines differ and by how much, and this is done without assuming that the regression lines are straight.

Bootstrap Bagging

A possible way of increasing power is to combine the running interval smoother with bootstrap bagging. In terms of hypothesis testing, a percentile bootstrap method is currently the only known method that performs well in simulations (Wilcox, 2009a). So in effect, a nested bootstrap method is used where for each bootstrap sample, bootstrap bagging is applied. Let $D = \hat{m}_1^*(x) - \hat{m}_2^*(x)$, where $\hat{m}_1^*(x)$ and $\hat{m}_2^*(x)$ are estimates of $m_1(x)$ and $m_2(x)$, respectively, based on bootstrap bagging. Repeat this process B times yielding $D_1, \ldots, D_B$. Then a (generalized) p-value when testing H_0: $m_1(x) = m_2(x)$ is

$$P = \frac{1}{B} \sum_{b=1}^{B} (I_{D_b < 0} + .5 I_{D_b = 0}),$$ (11.30)

where the indicator function $I_{D_b < 0} = 1$ if $D_b < 0$; otherwise $I_{D_b < 0} = 0$.

11.11.2 R Functions ancova, ancpb, runmean2g, lplot2g, ancboot, ancbbpb, and cobs2g

Several R functions are supplied with the hope that one of them matches the needs of the reader when dealing with ANCOVA. The first is

```
ancova(x1,y1,x2,y2,fr1=1,fr2=1,tr=0.2,alpha=0.05,plotit=T,
       pts=NA,sm=F,xout=F,outfun=out, ...)
```

which compares trimmed means. The data for group 1 are stored in x1 and y1, and for group two they are stored in x2 and y2. The arguments fr1 and fr2 are the values of f (the span) for the first and second group, used by the running interval smoother, which default to 1 if unspecified, and tr is the amount of trimming which defaults to 0.2. The default value for alpha (α), the probability of at least one type I error, is .05. If the argument pts=NA, five design points are chosen as described in the previous section. The results are returned in the matrix ancova$output, as illustrated in the next example. If values are stored in pts, the function compares groups at the values specified. So if pts contains the values, 5, 8, and 12, the function will test H_0: $m_1(x) = m_2(x)$ at $x = 5, 8$, and 12, and it controls the probability of a type I error by determining a critical value based on the Studentized maximum modulus distribution (as described in Chapter 7). When plotit=T, the function creates a scatterplot and smooth for both groups by calling the function

$$\text{runmean2g}(x1,y1,x2,y2,fr=0.8,est=tmean,xlab=``x'',ylab=``y'',}$$
$$\text{sm=F,nboot=40,SEED=T,eout=F,xout=F,outfun=out, \ldots),}$$

which creates a separate smooth for each group. Setting the argument sm=T results in using a bagged version of the smooth, which can be useful when the sample size is small. If the argument xout=T, leverage points are removed when plotting the regression lines.

■ Example

The Pygmalion data in Section 11.2.1 is used to illustrate the function ancova. The goal is to compare posttest scores taking into account pretest scores. Suppose the data for the experimental group are stored in the R variables conx1 and cony1, and the control data are in x2 and y2. The command ancova(conx1,cony1,x2,y2) returns

```
$output
         X   n1   n2      DIF       TEST         se      ci.low      ci.hi     p.value
[1,]    72   12   63  13.39103   1.848819   7.243016   -9.015851   35.79790   0.09387996
[2,]    82   16   68  14.79524   1.926801   7.678655   -8.211174   37.80165   0.07732813
[3,]   101   14   59  22.43243   1.431114  15.674806  -26.244186   71.10905   0.18315241
[4,]   111   12   47  23.78879   1.321946  17.995286  -35.644021   83.22161   0.22452259
[5,]   114   12   43  21.59722   1.198906  18.014112  -37.832791   81.02724   0.26640590
      crit.val
[1,] 3.093584
[2,] 2.996151
[3,] 3.105405
[4,] 3.302688
[5,] 3.299081
```

The first column of the matrix $output indicates the x values at which the two groups are compared, and the next two columns indicate the sample sizes being used [the values of $M_1(x)$ and $M_2(x)$]. For example, the first row has 72 under the column headed X, 12 under the column headed by n1, and 63 under the column headed n2. This means that a confidence interval for $m_1(72) - m_2(72)$ is being computed based on sample sizes of 12 and 63. That is, subjects in the experimental group, with an IQ reasoning pretest score of 72, are being compared to subjects in the control group who also have an IQ reasoning pretest score of 72. The estimated difference between the 20% trimmed

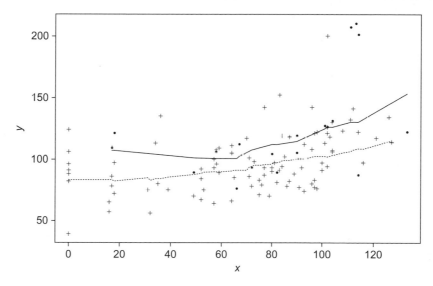

Figure 11.16: The plot created by ancova based on the Pygmalion data.

means is in the column headed by DIF, the test statistic for testing $H_0 : m_1(x) = m_2(x)$ is in the column headed by TEST, and the estimated standard error of DIF is in the column se. The final two columns give the lower and upper ends of the 0.95 confidence interval. The critical value used by all five tests is stored in the R variable ancova$crit and is approximately equal to 3.3. For $H_0 : m_1(72) = m_2(72)$, the test statistic is 1.84, this is less than the critical value, so do not reject. Figure 11.16 shows the plot created by the function ancova.

■

If both fr1 and fr2 are decreased to 0.75, all five comparisons are done at $x = 90$ because no other design points can be found for which the regression lines are comparable. If fewer than five design points are found for which the regression lines are comparable, the critical value is adjusted appropriately.

It should be noted that the critical values are designed so that the probability of one or more type I errors is approximately α. The confidence intervals are not adjusted accordingly. Each confidence is designed to have probability coverage $1 - \alpha$.

■ Example

Again consider the Pygmalion data, only suppose it is desired to compare the groups at $x = 70, 80$, and 90. Then the command

ancova(conx1,cony1,x2,y2,pts=c(70,80,90))

returns

```
          X n1 n2      DIF     TEST       se   ci.low    ci.hi    p.value crit.val
[1,] 70 12 58 16.41667 2.316398 7.087152 -3.926717 36.76005 0.04486589 2.870460
[2,] 80 16 68 15.74762 2.052538 7.672266 -5.270017 36.76525 0.06200997 2.739430
[3,] 90 15 62 14.32456 1.866934 7.672774 -6.846733 35.49586 0.08721893 2.759275
```

That is, the output is the same as before, but the critical value is computed based on the number of tests being performed, which in this case is three.

■

The function

$$\text{ancpb}(x1, y1, x2, y2, \text{est=mom}, \text{pts=NA}, \text{nboot=599}, \text{plotit=T}, \dots)$$

is like the function ancova, only any measure of location can be used, which is specified by the argument est, and a percentile bootstrap method is used to compute confidence intervals. The arguments are the same as before, only the argument nboot is added to indicate B, how many bootstrap samples to use.

The function

$$\text{ancboot}(x1, y1, x2, y2, \text{fr1=1}, \text{fr2=1}, \text{tr=0.2}, \text{nboot=599}, \text{pts=NA}, \text{plotit=T}).$$

is exactly like the function ancova, only a bootstrap-t method is used to compute confidence intervals and test hypotheses based on trimmed means.

■ Example

For the Pygmalion data, ancboot (using the default values for fr1, fr2, and tr) returns

```
$output:
          X  n1  n2      DIF     TEST    ci.low     ci.hi
[1,]  72  12  63  13.39103  1.848819  -11.58043  38.36248
[2,]  82  16  68  14.79524  1.926801  -11.67815  41.26863
[3,] 101  14  59  22.43243  1.431114  -31.60897  76.47384
[4,] 111  12  47  23.78879  1.321946  -38.25284  85.83043
[5,] 114  12  43  21.59722  1.198906  -40.50931  83.70376

$crit:
[1] 3.44766
```

■

■ Example

Consider the data used to illustrate anctgen, only bootstrap confidence intervals are computed instead with ancbootg. The function returns

```
$output:
          X  n1  n2        DIF       TEST         se      ci.low       ci.hi
[1,]  70  12  58   16.41667   2.316398   7.087152   -4.323841   37.15717
[2,]  80  16  68   15.74762   2.052538   7.672266   -6.705222   38.20046
[3,]  90  15  62   14.32456   1.866934   7.672774   -8.129766   36.77889

$crit:
[1] 2.926494
```

The critical value is slightly larger using the bootstrap method, compared to the critical value used by anctgen, and this results in slightly longer confidence intervals.

■

The R function

$$ancbbpb(x1, y1, x2, y2, fr1 = 1, fr2 = 1, nboot = 200, pts = NA, plotit = T, SEED = T,$$
$$alpha = 0.05, RNA = T)$$

is like the R function ancova, only bootstrap bagging is used. So it has the potential of more power at the cost of higher execution time.

■ Example

The first example in this section, based on the default settings of the R function ancova and applied to the the Pygmalion data, found no differences between the groups. Using instead the R function ancbbpb, the result are

```
$output
           X  n1  n2        DIF       ci.low       ci.hi      p.value
[1,]   72  12  63   12.03672   -2.7430218   22.37841   0.16528926
[2,]   82  16  68   16.24183    0.8489948   24.69427   0.03007519
[3,]  101  14  59   28.32713    3.5099184   48.23010   0.01398601
[4,]  111  12  47   31.94660    8.9667976   70.64249   0.01273885
[5,]  114  12  43   34.23546    7.4661331   71.24803   0.01149425
```

So now differences are found at four of the five design points when testing at the 0.05 level.

■

The R function

lplot2g(x1,y1,x2,y2,fr=0.8,est=tmean,xlab="X",ylab="Y",xout=F,eout=F,outfun=out, . . .)

is like runmean2g, only it plots the regression lines corresponding to two groups using Cleveland's smoother instead. The R function

cobs2g(x1,y1,x2,y2,xlab="X",ylab="Y",qval=0.5,xout=F,outfun=out, . . .)

plots two regression lines using the quantile the regression estimator COBS in Section 11.5.7.

11.11.3 Multiple Covariates

There are various ways the method described in Section 11.11.1 might be extended to the case of multiple covariates. Consider any point among the covariates. Momentarily focus on the ith value of the covariate in the first group, $\mathbf{x}_{i1}$. Then proceeding along the lines in Section 11.5.11, it is a simple matter to determine the set $N_1(\mathbf{x}_{i1}) = \{j : D_{1ij} \le f\}$, where

$$D_{1ij} = \sqrt{(\mathbf{x}_{i1} - \mathbf{x}_{j1})'\mathbf{M}^{-1}(\mathbf{x}_{i1} - \mathbf{x}_{j1})},$$

where as before, $\mathbf{M}$ is some measure of covariance. The set N_1 identifies all $\mathbf{x}_{j1}$ values such that $\mathbf{x}_{j1}$ is close to $\mathbf{x}_{i1}$. The same can be done for the second group. That is, $N_2(\mathbf{x}_{i1}) = \{j : D_{2ij} \le f\}$, where

$$D_{2ij} = \sqrt{(\mathbf{x}_{i1} - \mathbf{x}_{j2})'\mathbf{M}^{-1}(\mathbf{x}_{i1} - \mathbf{x}_{j2})},$$

Then the y_{j1} values, such that $j \in N_1$, can be compared to the y_{j2} values, $j \in N_2$, using some measure of location. If there is interest in some particular $\mathbf{x}_{i1}$, this approach is readily implemented, but otherwise, how should $\mathbf{x}_{i1}$ be chosen? One possibility is to take $\mathbf{x}_{i1}$ to be the point having the largest halfspace depth. Another possibility is to pool the $\mathbf{x}_{i1}$ and $\mathbf{x}_{i2}$ and replace $\mathbf{x}_{i1}$ in the method just described with the point having the largest halfspace depth among the pooled data. Of course, multiple choices for $\mathbf{x}_{i1}$ might be used. For example, in addition to the deepest point, one might consider those points whose depth is equal to the median depth among all points in the first group, or among the pooled values. That is, for each point that lies on the 0.5 depth contour, compare group 1 and group 2 as just described.

Notice that the method for handling multiple covariates is readily extended to comparing more than two groups. Again pick design points for each of the J groups to be compared, determine observed covariate values that are close to the chosen design points, then use methods in Chapter 7 to test hypotheses about the corresponding y values.

11.11.4 R Functions ancdes, ancovamp, ancmppb, and ancmg

The R function

$$\text{ancovamp}(x1,y1,x2,y2,fr1=1,fr2=1,tr=0.2,alpha=0.05,pts=NA)$$

compares two groups based on trimmed means and takes into account multiple covariates using the method outlined in Section 11.11.3. The arguments are the same as ancova in Section 11.11.2. By default, pts=NA meaning that the points among the covariates at which the groups will be compared are determined by the function; it finds a point among the x1 values that has the deepest halfspace depth, plus the points on the 0.5 depth contour, and the groups are compared at these points provided that the corresponding sample sizes are at least 10. Should one want to pool the data and then find the deepest point, plus the points on the 0.5 depth contour, this can be done as indicated by an Example later in this section. The function controls the familywise error rate by determining a critical value via the Studentized maximum modulus distribution (as described in Chapter 7).

The function

$$\text{ancmppb}(x1,y1,x2,y2,fr1=1,fr2=1,tr=0.2,alpha=0.05,pts=NA,est=tmean,}$$
$$\text{nboot=NA,bhop=F,}\ldots)$$

is like ancovamp, only a percentile bootstrap method is used and any measure of location can be employed. By default, a 20% trimmed mean is used. In essence, the function determines groups as before, then it compares the corresponding *y* values by calling the function pbmcp in Section 7.6.3, where the argument bhop is explained. (It determines the approach used to control the probability of at least one type I error among the tests performed.)

It is noted that both of the functions just described determine the points among the covariates that will be used by calling the function

$$\text{ancdes}(x,depfun=fdepth).$$

This function determines the halfspace depth of the points in x, where the depth of points is determined using the function indicated by the argument depfun (see Chapter 6), and then ancdes determines a point having the maximum depth plus those points on the 0.5 depth contour. So if one wanted to use ancmppb, but pool the data in x1 and x2 when determining the points among the covariates when deciding where the groups are to be compared, the command

$$\text{ancmppb}(x1,y1,x2,y2,pts=ancdes(rbind(x1,x2)))$$

would accomplish this goal.

The R function

$$\text{ancmg}(x, y, \text{pool} = T, \text{jcen} = 1, \text{fr} = 1, \text{depfun} = \text{fdepth}, \text{nmin} = 8, \text{op} = 3, \text{tr} = 0.2,$$
$$\text{pts} = NA, \text{SEED} = T, \text{pr} = T, \text{cop} = 3, \text{con} = 0, \text{nboot} = NA, \text{alpha} = 0.05, \text{bhop} = F)$$

can be used to compare multiple groups when there are multiple covariates. The argument op determines how the groups are compared. The default op=3 means that multiple comparisons are performed based on trimmed means via the R function pbmcp. This is done for each value of the covariates that is of interest. To perform a global test instead, set op=1, which means the R function t1way will be used. The choices op=2 and 4 result in comparing medians, where op=4 performs multiple comparisons by calling by calling medpb. If there are tied values, op=2 is not recommended; use op=4 to compare medians.

One way of storing the data, when using ancmg, is in list mode. Imagine that three groups are to be compared based on two covariates. Then x[[1]] would contain a matrix of data with n_1 rows and $p = 2$ columns, x[[2]] would contain a matrix of data with n_2 rows and $p = 2$ columns, x[[3]] would contain a matrix of data with n_3 rows and $p = 2$ columns, and y[[1]], y[[2]] and y[[3]] would contain the outcome measures to be compared. Another option is to store the covariate data in a matrix with Jp columns, assuming all J groups have the same sample size. And y could be a matrix with J columns.

11.11.5 Some Global Tests

The methods in the previous section are aimed at comparing groups at specific design points. This section describes some methods where the goal is to test the hypothesis that the groups do not differ for any design point.

Method TG

Method TG is a global test based on trimmed means and is limited to a single covariate. It is assumed that x has been standardized based on some robust measure of location and scale. Unless stated otherwise, the median and median absolute deviation are used. That is, if we begin with the covariate z, we work with

$$x = \frac{z - M}{\text{MADN}}.$$

The goal is to test

$$H_0 : m_1(x) = m_2(x) \ \forall x. \tag{11.31}$$

The method is based in part on a simple generalization of the notion of regression depth. Recall from Section 10.11 that when estimating the slope and intercept of the usual linear model, a candidate fit, (b_0, b_1), is called a nonfit if a partition of the x values can be found such that all of the residuals for the lower x values are negative (positive), but for all of the higher x values the residuals are positive (negative). For the random sample $(x_1, y_1), \ldots, (x_n, y_n)$, and letting $r_i = y_i - b_0 - b_1 x_i$, a candidate fit is called a nonfit if and only if a value for v can be found such that

$$r_i < 0 \quad \text{for all } x_i < v$$

and

$$r_i > 0 \quad \text{for all } x_i > v$$

or

$$r_i > 0 \quad \text{for all } x_i < v$$

and

$$r_i < 0 \quad \text{for all } x_i > v.$$

Rousseeuw and Hubert define the regression depth of a fit (b_1, b_0), relative to $(x_1, y_1), \ldots, (x_n, y_n)$, as the smallest number of observations that need to be removed to make (b_1, b_0) a nonfit. Their deepest regression line estimator corresponds to the values of b_1 and b_0 that maximize regression depth.

Now consider any fit $\hat{y}_i = m(x_i)$, which might be obtained via any of the nonparametric regression methods previously described. Given a fit, the depth of the fit can be measured using a simple extension of the Rousseeuw and Hubert approach because their notion of depth is based entirely on the residuals and the values of the covariate. In particular, it does not require that the regression line be straight. That is, now $r_i = y_i - \hat{y}_i$, and given $x_1, \ldots, x_n$, depth is defined as before.

Using a simple modification of the computational algorithm in Rousseeuw and Hubert (1999), the depth of a nonparametric regression line can be computed as follows. First, reorder the covariate values so that $x_1 \leq \cdots \leq x_n$. Then the regression depth of $m(x)$ is

$$D = \min_{1 \leq i \leq n} (\min\{L^+(x_i) + R^-(x_i), R^+(x_i) + L^-(x_i)\}),$$

where

$$L^+(v) = \#\{j; x_j \leq v \text{ and } r_j \geq 0\},$$

$$R^-(v) = \#\{j; x_j > v \text{ and } r_j \leq 0\},$$

and L^- and R^+ are defined accordingly. Note that regression depth is scale invariant. From Theorem 1 in Rousseeuw and Hubert, it follows that the maximum possible value for D is greater than or equal to $n/3$ and less than or equal to n, where the ceiling z is the smallest integer $\geq z$.

Let (y_{ij}, x_{ij}) be a random sample from the jth group ($i = 1, \ldots, n_j$), and imagine that a nonparametric regression line is fitted to the data in the first group. Then in general, this fit can be used to estimate y given any value for x simply by computing the trimmed mean of the y_{i1} values for which the corresponding x_{i1} values are close to x. In particular, an estimate can be computed for the covariate values corresponding to the second group: $x_{i2}, i = 1, \ldots, n_2$. This assumes, of course, that given x_{i2}, there are one or more x_{i1} values that are close to x_{i2}; otherwise an estimate is not defined and is ignored. Let

$$\hat{y}_{ijk} = m_j(x_{ik}; x_{1j}, \ldots, x_{n_j j})$$

be the predicted value of y corresponding to the ith observation in the kth group using the fit obtained from the jth group. That is, $\hat{y}_{ijk}$ is the 20% trimmed mean of the y_{ij} values for which x_{ij} is close to x_{ik}. Let $r_{ijk} = y_{ijk} - \hat{y}_{ijk}$, and let D_{jk} be the resulting regression depth. So D_{11}, for example, is the depth of the first smooth relative to the points in the first group, and D_{12} is the depth of the first smooth relative to the second group. If H_0 is true, it should be the case that $D_{11} - D_{21}$, as well as $D_{12} - D_{22}$, are relatively small, suggesting the test statistic

$$T = D_{11} - D_{21} + D_{12} - D_{22}. \tag{11.32}$$

Because regression depth is scale invariant, T is scale invariant as well. If H_0 is rejected when $T \geq t$, the problem is determining t so as to control the probability of a type I error. Note that T has a discrete distribution, so in general choosing t so that the type I error probability is exactly α cannot be accomplished in most cases.

Momentarily assume the running interval smoother is used. The only known method that has been found to perform well in simulations begins by pooling the data from both groups and using bootstrap samples to estimate the null distribution of T (Wilcox, 2010a). Note that if the null hypothesis is true, the individual smooths estimate the same regression line estimated by the pooled estimate. Let $N = n_1 + n_2$ and generate a bootstrap sample by sampling with replacement N pairs of points from the pooled data. Based on this bootstrap sample, use the first n_1 pairs of points to compute the depths D_{11} and D_{21}, and label the results D_{11}^* and D_{21}^*. In a similar manner, the remaining n_2 points are used to compute D_{12} and D_{22} and the results are labeled D_{12}^* and D_{22}^*. Let

$$T^* = D_{11}^* - D_{21}^* + D_{12}^* - D_{22}^*.$$

Repeat this process B times yielding $T_1^*, \ldots, T_B^*$ and let

$$P = \frac{1}{B} \sum I_{T > T^*},$$

where $I_{T > T^*} = 1$ if $T > T^*$; otherwise $I_{T > T^*} = 0$. Then a (generalized) p-value is

$$p = 1 - P.$$

If H_0 is rejected when $p \leq .05$, simulations indicate that, generally, the actual level is reasonably close to 0.05 when the span is $f = 1$ and both sample sizes are between 40 and 150. However, for smaller or larger sample sizes, an alternative choice for f is required. If the smallest sample size is greater than 150, $f = 0.2$ was found to give good results with sample sizes as large as 800. If $\max(n_1, n_2) < 35$, use $f = 0.5$. For sample sizes between 150 and 180, both $f = 0.2$ and $f = 1$ perform well. But for $\min(n_1, n_2) > 200$, $f = 0.2$ should be used.

The choice of smoother is important. It is unknown, for example, how to control type I errors reasonably well, in simulations, when the running interval smoother is replaced by LOWESS. However, simulations do indicate that when using the quantile smoother COBS, the probability of a type I error will not exceed the nominal level when using sample sizes of at least 30. But when testing at the 0.05 level, the actual level can be as low as 0.01 in some situations. The choice between COBS and the running interval smoother can make a substantial difference in terms of power.

Also, it might seem that regression depth could be used to determine a nonparametric regression line, but this strategy is unsatisfactory. Note that if $x_i \neq x_j$ for all $i \neq j$, then the maximum possible depth, $D = n$, is achieved by taking $m(x_i) = y_i$. The point here is that given a nonparametric fit, based on some appropriate choice for the span, its regression depth can be computed, which in turn can be used to test the hypothesis given by Eq. (11.31).

There is a feature of the global tests in this section that should be stressed, which is relevant to the classic ANCOVA method as well. Imagine that for the first group, the range of the covariate values, x, is 0–10, and for the second group the range is 30–40. Classic ANCOVA would assume that the regression lines are parallel and compare the intercepts. If the usual linear model holds, the methods in this section are aimed at testing the hypothesis that the slopes, as well as the intercepts, are equal. More generally, the methods in this section ignore the fact that the two groups do not have any covariate values in common. But based on the ANCOVA methods in Section 11.11.1, comparisons would not be made. For instance, it might be of interest to determine whether the groups differ when the covariate $x = 5$. Because data are not available for the second group when $x = 5$, comparisons cannot be made based on the methods in Section 11.11.1. Perhaps comparisons should not be made by imposing assumptions such as those made by the classic ANCOVA model.

11.11.6 R Functions ancsm and Qancsm

The R function

ancsm(x1, y1, x2, y2, nboot = 200, SEED = T, est = tmean, fr = NULL, plotit = T, sm = F, tr = 0.2, xout=F, outfun=out, ...)

applies method TG using the running interval smoother. By default, a 20% trimmed mean is used. Setting sm=T, the plots of the regression lines will be based on bootstrap bagging. (But the test of the null hypothesis of identical regression lines is based on the running interval smoother without bootstrap bagging.) The function

Qancsm(x1,y1,x2,y2,crit.mat=NULL,nboot=200,SEED=T,REP.CRIT=F, qval=.5,xlab="X",ylab="Y",plotit=T,pr=T,xout=F,outfun=out, ...)

is like ancsm only COBS is used to estimate the quantile regression lines. The argument qval determines the quantile that is used and defaults to 0.5, the median.

11.12 Marginal Longitudinal Data Analysis: Comments on Comparing Groups

There is a vast literature on what is generally known as *longitudinal data* analysis (e.g., Diggle, Heagerty, Liang, & Zeger, 2002; Molenberghs & Verbeke, 2005). Roughly, the goal is to deal with situations where measures are taken over time. As a concrete example, consider again the data in Section 1.9 dealing with an orthodontic growth study. The measure of interest is the distance between the pituitary and pterygomaxillary fissure, which was measured at 8, 10, 12, and 14 years of age. As noted in Section 1.9, the first 10 rows of the data are:

```
    Distance  Age  Subject   Sex
1     26.0     8     M01    Male
2     25.0    10     M01    Male
3     29.0    12     M01    Male
4     31.0    14     M01    Male
5     21.5     8     M02    Male
6     22.5    10     M02    Male
7     23.0    12     M02    Male
8     26.5    14     M02    Male
9     23.0     8     M03    Male
10    22.5    10     M03    Male
```

There are 16 males and 11 females.

Many goals arise when dealing with longitudinal data. And often one of three models is used. The first is a marginal regression model where the goal is to understand the typical outcome (distance in the example) given the value of some explanatory variable, which here is age. The other two models are a random effects model and a transitional model where the covariate effects and the within-subject association are modeled through a single equation. Transitional models go beyond the scope of this book and are not discussed. Indeed, only a few comments are made about a narrow range of problems that are relevant to longitudinal data. From a robustness point of view, progress has been made, but more research in this area is needed for reasons outlined at the end of this section.

A common approach to longitudinal data is to fit some type of linear regression model that takes into account in some manner the time at which measures were taken. A simple example is to assume

$$y_{ij} = \beta_0 + \beta_1 t_j + \epsilon_{ij},$$

where t_j is some measure taken at the jth time point. In the orthodontic example, $t_1, \ldots, t_4$ correspond to ages 8, 10, 12 and 14, respectively. So a single slope and intercept are used to characterize the association between y and t for the population of individuals under study. A semiparametric regression model for longitudinal data was studied by Chen and Zhong (2010) where an empirical likelihood method is used to test hypotheses. Their simulations indicate that the method performs well under normality, but it seems that a robust version of this approach has not been derived. Recall from Section 4.7 that in the one-sample case, the empirical likelihood method can be relatively unsatisfactory when dealing with heavy-tailed distributions.

A variation of this approach fits a regression line for each individual. Diggle et al. (2002) describe a number of situations where this approach appears to be reasonable. For example, for each participant in the orthodontic growth data, the strategy would be to fit a regression model that relates distance to the age of the child. In more formal terms, assume that for the ith participant

$$y_{ij} = \beta_{0i} + \beta_{1i} t_{ij} + \epsilon_{ij}.$$

So each individual is characterized by a slope and intercept, with the slopes and intercepts possibly varying among the population of participants.

Imagine that the goal is to compare males and females. One approach is to use the between-by-within ANOVA method described in Section 8.6. Another approach is to compare the groups using a multivariate measure of location as described, for example, in Sections 6.8, 6.9, or 6.11. In terms of the orthodontic data, we have four measures for each individual, and so the groups could be compared, for example, based on the multivariate OP measure of location. For the kth group, let $(\theta_{0k}, \theta_{1k})$ represent some measure of location associated with $(\beta_{01k}, \beta_{11k}), \ldots, (\beta_{0n_k k}, \beta_{1n_k k})$ where β_{0ik} and β_{1ik} are the intercept and slope, respectively, associated with the kth group ($k = 1, 2$). For example, θ_{1k} might be the median of the slopes associated with group k. Yet another approach is to test

$$H_0 : (\theta_{01}, \theta_{11}) = (\theta_{02}, \theta_{12}). \tag{11.33}$$

So the p-variate data has been reduced to two variables, and these two variables could be compared, for example, using the methods in Sections 6.8 or 6.9.

A broader, more involved approach toward longitudinal data, based on a marginal model and the MM-estimator in Section 10.9.1, is summarized by Heritier et al. (2009, Section 6.2). Included is an inferential technique that is based in part on appropriate estimates of the standard errors; the test statistic is assumed to be approximately standard normal. Evidently, there are no results on the ability of this approach to control type I errors when dealing with skewed distributions or heteroscedasticity. In simpler situations, skewness is a serious concern when using M-estimators and a hypothesis testing method is based on a (nonbootstrap) technique that is a function of estimated standard errors. So caution seems warranted for the situation at hand. For robust methods based on a random effects model, see Mills, Field, and Dupuis (2002), Sinha (2004), and Noh and Lee (2007). The basic strategy is to estimate parameters assuming observations are randomly sampled from a class of distributions that includes normal distributions as a special case. For example, Mills et al. assume that sampling is from a mixture of normal and t distributions, which results in a bounded influence function. Certainly these methods are an improvement on methods that assume normality. Again, what is unclear is the extent skewness and heteroscedasticity affect efficiency and type I error probabilities. How well do these methods handle contamination bias as described in Section 10.14.1? If practical problems are found, perhaps some bootstrap methods can provide more satisfactory results, but this remains to be determined.

11.12.1 R Functions long2g, longreg, longreg.plot, and xyplot

The R function

$$\text{long2g(x, x.col, y.col, s.id, grp.id, regfun} = \text{tsreg, MAR} = \text{T, tr} = 0.2)$$

compares two groups based on estimates of the slope and intercept for each participant. The data are assumed to be stored in a matrix or data frame as illustrated by the orthodontic data in the previous section. The arguments x.col and y.col indicate the columns of x where the covariate and outcome variables are stored, respectively. The argument s.id is the column containing the subject's identification and grp.id is the column indicating group membership, which is assumed to have two possible values only. The regression line for each participant is fitted with the regression estimator indicated by the argument regfun, which defaults to tsreg, the Theil–Sen estimator. If MAR=T, the slopes and intercepts are compared using Yuen's method for trimmed means, which is described in Section 5.3. If MAR=F, the OP-estimator is used via the method in Section 6.8.

The R function

$$longreg(x, x.col, y.col, s.id, regfun = tsreg, est = tmean)$$

computes the slope and intercept for each participant, using the regression estimator indicated by the argument regfun, and returns the results in a matrix labeled S.est. The typical slope and intercept are returned as well, which is based on the estimator indicated by the argument est. The R function

$$longreg.plot((x,x.col,y.col,s.id,regfun=tsreg,scat=T,xlab="X", ylab="Y")$$

plots the regression lines based on the R function longreg. A scatterplot of the points can be created with the R function xyplot, which is in the R package lattice. If the orthodontic data are stored in the R variable x, the R command library(lattice) followed by

$$xyplot(x[,1]\~x[,2],group=x[,3])$$

accomplishes this goal. To add line segments connecting the responses for each participant, include the argument type="b". The command

$$long2g(x,2,1,3,4)$$

would compare the slopes and intercepts using Yuen's method.

11.13 Exercises

1. For the data in Exercise 1 of Chapter 10, the 0.95 confidence interval for the slope, based on the least squares regression line, is $(0.0022, 0.0062)$. Using R, the 0.95 confidence interval for the slope returned by lsfitci is $(0.003, 0.006)$. The 0.95 confidence interval returned by the R function regci (using the Theil–Sen estimator) is $(0.003, 0.006)$. Verify this result.

2. Section 8.6.2 reports data on the effects of consuming alcohol on three different occasions. Using the data for group 1, suppose it is desired to predict the response at time 1 using the responses at times 2 and 3. Test $H_0 : \beta_1 = \beta_2 = 0$ using the R function regtest and $\hat{\beta}_m$.

3. For the data in Exercise 1, test $H_0 : \beta_1 = 0$ with the functions regci and regtest. Comment on the results.

4. Use the function winreg to estimate the slope and intercept of the star data using 20% Winsorization. (The data are stored in the file star.dat. See Section 1.8 on how to obtain the data.)

5. For the Pygmalion data in Section 11.2.1, use the function reglev to determine which points, if any, are regression outliers. (The data for the control group are stored in pygc.dat, and the data for the experimental group are stored in pyge.dat.)

6. Use rplot to plot a smooth of the Pygmalion data using $f = 0.75$ and 20% trimmed means. Create a plot for both the control and experimental groups when the goal is to predict post IQ scores with pretest scores. Comment on how the results compare to using $f = 1$.

7. Based on the results of Exercise 6, speculate about what a nonrobust smoother might look like. Check your answer with the smoother lowess using the R lplot.

8. Use the function ancova and the Pygmalion data to compare the control group to the experimental group using means. What might be affecting power?

9. For the reading data in file read.dat, let x be the data in column 2 (TAAST1), and suppose it is desired to predict y, the data in column 8 (WWISST2). Speculate on whether there are situations where it would be beneficial to use x^2 to predict y taking into account the value stored in column 3 (SBT1). Use the functions in this chapter to address this issue.

10. For the reading data in the file read.dat, use the R function rplot to investigate the shape of the regression surface when predicting the 20% trimmed mean of WWISST2 (the data in column 8) with RAN1T1 and RAN2T1 (the data in columns 4 and 5).

11. The data in the lower left panel of Figure 11.5 are stored in the file agegesell.dat. Remove the two pairs of points having the largest x value and create a running interval smoother using the data that remain.

12. Using the Pygmalion data, compare the slope of the regression line of the experimental group to the control group using the biweight midregression estimator.

13. For the reading data in the upper right panel of Figure 11.5, recreate the smooth. If you wanted to find a parametric regression equation, what might be tried? Examine how well your suggestions perform.

14. For the experimental group of the Pygmalion data in Section 11.2.1, create a plot of the smooth using $f = 1$ and the function rplot. Recreate the plot, but this time omit the scatterplot of the points by setting the argument scat to F for false. What does this illustrate?

15. Generate 25 observations from a standard normal distribution and store the results in the R variable x. Generate 25 more observations and store them in y. Use rungen to plot a smooth based on the Harrell–Davis estimator of the median. Also create a smooth with the argument scat=F. Comment on how the two smooths differ.

16. Generate 25 pairs of observations from a bivariate normal distribution having correlation zero and store them in x. (The R function rmul, written for this book, can be used.) Generate 25 more observations and store them in y. Create a smooth using rplot using scale=T and compare it to the smooth when scale=F.

17. Generate data from a bivariate normal distribution with the R command x=rmul(200). Then enter the R command y=x[,1]+x[,2]+x[,1]*x[,2]+rnorm(200) and examine the plot returned by the R command gamplot(x,y,scale=T). Compare this to the plot returned by R command gamplotINT(x,y,scale=T).

References

Acion, L., Peterson, J. J., Temple, S., & Arndt, S. (2006). Probabilistic index: An intuitive non-parametric approach to measuring the size of treatment effects. *Statistics in Medicine, 25,* 591–602.

Adrover, J., & Salibian-Barrera, M. (2010). Globally robust confidence intervals for simple linear regression. *Computational Statistics and Data Analysis, 54,* 2899–2913.

Adrover, J., & Yohai, V. (2002). Projection estimates of multivariate location. *Annals of Statistics, 30,* 1760–1781.

Agresti, A., & Caffo, B. (2000). Simple and effective confidence intervals for the proportions and differences of proportions result from adding two successes and two failures. *American Statistician, 54,* 280–288.

Agresti, A., & Coull, B. A. (1998). Approximate is better than "exact" for interval estimation of binomial proportions. *American Statistician, 52,* 119–126.

Agresti, A., & Pendergast, J. (1986). Comparing mean ranks for repeated measures data. *Communications in Statistics–Theory and Methods, 15,* 1417–1433.

Agulló, J., Croux, C., & Van Aelst, S. (2008). The multivariate least trimmed squares estimator. *Journal of Multivariate Analysis, 99,* 311–338.

Akritas, M. G. (1990). The rank transform method in some two-factor designs. *Journal of the American Statistical Association, 85,* 73–78.

Akritas, M. G., & Arnold, S. F. (1994). Fully nonparametric hypotheses for factorial designs I: Multivariate repeated measures designs. *Journal of the American Statistical Association, 89,* 336–343.

Akritas, M. G., Murphy, S. A., & LaValley, M. P. (1995). The Theil-Sen estimator with doubly censored data and applications to astronomy. *Journal of the American Statistical Association, 90,* 170–177.

Akritas, M. G., Arnold, S. F., & Brunner, E. (1997). Nonparametric hypotheses and rank statistics for unbalanced factorial designs. *Journal of the American Statistical Association, 92*, 258–265.

Alba-Fernández, V., & Jiménez-Gamero, M. D. (2009). Bootstrapping divergence statistics for testing homogeneity in multinomial populations. *Mathematics and Computers in Simulation, 79*, 3375–3384.

Algina, J., Keselman, H. J., & Penfield, R. D. (2005). An alternative to Cohen's standardized mean difference effect size: A robust parameter and confidence interval in the two independent groups case. *Psychological Methods, 10*, 317–328.

Algina, J., & Olejnik, S. F. (1984). Implementing the Welch-James procedure with factorial designs. *Educational and Psychological Measurement, 44*, 39–48.

Algina, J., Oshima, T. C., & Lin, W.-Y. (1994). Type I error rates for Welch's test and James's second-order test under nonnormality and inequality of variance when there are two groups. *Journal of Educational and Behavioral Statistics, 19*, 275–291.

Ammann, L. P. (1993). Robust singular value decompositions: A new approach to projection pursuit. *Journal of the American Statistical Association, 88*, 505–514.

Anaya-Izquierdo, K., Critchley, F., & Vines, K. (2011). Orthogonal simple component analysis: A new, exploratory approach. *Annals of Applied Statistics, 5*, 486–522.

Andrews, D. F., Bickel, P. J., Hampel, F. R., Huber, P. J., Rogers, W. H., & Tukey, J. W. (1972). *Robust estimates of location*. Princeton, NJ: Princeton University Press.

Arcones, M. A., Chen, Z., & Gine, E. (1994). Estimators related to U-processes with applications to multivariate medians: Asymptotic normality. *Annals of Statistics, 44*, 587–601.

Arnold, B. C., Balakrishnan, N., & Nagaraja, H. N. (1992). *A first course in order statistics*. New York, NY: Wiley.

Atkinson, A. C. (1994). Fast very robust methods for the detection of multiple outliers. *Journal of the American Statistical Association, 89*, 1329–1339.

Bai, Z.-D., & He, X. (1999). Asymptotic distributions of the maximal depth estimators for regression and multivariate location. *Annals of Statistics, 27*, 1616–1637.

Bai, Z.-D., Chen, X. R., Miao, B. Q., & Rao, C. R. (1990). Asymptotic theory of least distance estimate in multivariate linear model. *Statistics, 21*, 503–519.

Balakrishnan, N., & Kannan, N. (2003). Variance of a Winsorized mean when the sample contains multiple outliers. *Communications in Statistics—Theory and Methods, 32*, 139–150.

Banik, S., & Kibria, B. M. G. (2010). Comparison of some parametric and nonparametric type one sample confidence intervals for estimating the mean of a positively skewed distribution. *Communications in Statistics–Simulation and Computation, 39*, 361–380.

Bansal, N. K., & Bhandry, M. (1994). Robust M-estimation of the intraclass correlation coefficient. *Australian Journal of Statistics, 36*, 287–301.

Barrett, J. P. (1974). The coefficient of determination—Some limitations. *Annals of Statistics, 28*, 19–20.

Barry, D. (1993). Testing for additivity of a regression function. *Annals of Statistics, 21*, 235–254.

Basu, S., & DasGupta, A. (1995). Robustness of standard confidence intervals for location parameters under departures from normality. *Annals of Statistics, 23*, 1433–1442.

Bathke, A. C., Solomon, W. H., & Madden, L. V. (2008). How to compare small multivariate samples using nonparametric tests. *Computational Statistics and Data Analysis, 52*, 4951–4965.

Baumgartner, W., Weiss, P., & Schindler, H. (1998). A nonparametric test for the general two-sample problem. *Biometrics, 54*, 1129–1135.

Beal, S. L. (1987). Asymptotic confidence intervals for the difference between two binomial parameters for use with small samples. *Biometrics, 43*, 941–950.

Beasley, T. M. (2000). Nonparametric tests for analyzing interactions among intra-block ranks in multiple group repeated measures designs. *Journal of Educational and Behavioral Statistics, 25*, 20–59.

Beasley, T. M., & Zumbo, B. D. (2003). Comparison of aligned Friedman rank and parametric methods for testing interactions in split-plot designs. *Computational Statistics and Data Analysis, 42*, 569–593.

Beasley, W. H., DeShea, L., Toothaker, L. E., Mendoza, J. L., Bard, D. E., & Rodgers, J. L. (2007). Bootstrapping to test for nonzero population correlation coefficients using univariate sampling. *Psychological Methods, 12*, 414–433.

Bechhofer, R. E., & Dunnett, C. W. (1982). Multiple comparisons for orthogonal contrasts. *Technometrics, 24*, 213–222.

Becker, C., & Gather, U. (1999). The masking breakdown point of multivariate outlier detection rules. *Journal of the American Statistical Association, 94,* 947–955.

Becker, R. A., Chambers, J. M., & Wilks, A. R. (1988). *The new S language.* Pacific Grove, CA: Wadsworth & Brooks/Cole.

Bedall, P. J., & Zimmermann, H. (1979). AS 143: the median centre. *Applied Statistics, 28,* 325–328.

Bellman, R. E. (1961). *Adaptive control processes.* Princeton, NJ: Princeton University Press.

Belsley, D. A., Kuh, E., & Welsch, R. E. (1980). *Regression diagnostics*: *Identifying influential data and sources of collinearity.* New York, NY: Wiley.

Ben, M. G., Martínez, E., & Yohai, V. J. (2006). Robust estimation for the multivariate linear model based on a τ-scale. *Journal of Multivariate Analysis, 90,* 1600–1622.

Benjamini, Y. (1983). Is the t test really conservative when the parent distribution is long-tailed? *Journal of the American Statistical Association, 78,* 645–654.

Benjamini, Y., & Hochberg, Y. (1995). Controlling the false discovery rate: A practical and powerful approach to multiple testing. *Journal of the Royal Statistical Society B, 57,* 289–300.

Benjamini, Y., & Yekutieli, D. (2001). The control of the false discovery rate in multiple testing under dependency. *Annals of Statistics, 29,* 1165–1188.

Berger, R. L. (1996). More powerful tests from confidence interval p values. *American Statistician, 50,* 314–318.

Berk, K. N., & Booth, D. E. (1995). Seeing a curve in multiple regression. *Technometrics, 37,* 385–398.

Bernhardson, C. (1975). Type I error rates when multiple comparison procedures follow a significant F test of ANOVA. *Biometrics, 31,* 719–724.

Bernholt, T., & Fischer, P. (2004). The complexity of computing the MCD-estimator. *Theoretical Computer Science, 326,* 383–393.

Bianco, A. M. & Martínez, E. (2009). Robust testing in the logistic regression model. *Computational Statistics & Data Analysis, 53,* 4095–4105.

Bianco, A. M., & Yohai, V. J. (1996). Robust estimation in the logistic regression model. In H. Reider (Ed.), *Robust statistics, data analysis, and computer intensive methods, lecture notes in statistics,* (Vol. 109, pp. 17–34). New York, NY: Springer.

Bickel, P. J., & Lehmann, E. L. (1975). Descriptive statistics for nonparametric models II. Location. *Annals of Statistics, 3,* 1045–1069.

Bickel, P. J., & Lehmann, E. L. (1976). Descriptive statistics for nonparametric models III. Dispersion. *Annals of Statistics, 4*, 1139–1158.

Biesanz, J. C., Falk, C. F., & Savalei, V. (2010). Assessing mediational models: Testing and interval estimation for indirect effects. *Multivariate Behavioral Research, 45*, 661–701.

Billor, N., & Kiral, G. (2008). A comparison of multiple outlier detection methods for regression data. *Communications in Statistics–Simulation and Computation, 37*, 521–545.

Birkes, D., & Dodge, Y. (1993). *Alternative methods of regression*. New York, NY: Wiley.

Bjerve, S., & Doksum, K. (1993). Correlation curves: Measures of association as functions of covariate values. *Annals of Statistics, 21*, 890–902.

Blair, R. C., & Lawson, S. B. (1982). Another look at the robustness of the product-moment correlation coefficient to population non-normality. *Florida Journal of Educational Research, 24*, 11–15.

Blyth, C. R. (1986). Approximate binomial confidence limits. *Journal of the American Statistical Association, 81*, 843–855.

Boente, G., & Rodriguez, D. (2010). Robust inference in generalized partially linear models. *Computational Statistics and Data Analysis, 54*, 2942–2966.

Boik, R. J. (1987). The Fisher-Pitman permutation test: A non-robust alternative to the normal theory *F* test when variances are heterogeneous. *British Journal of Mathematical and Statistical Psychology, 40*, 26–42.

Bondell, H. D. (2005). Minimum distance estimation for the logistic regression model. *Biometrika, 92*, 724–731.

Bondell, H. D. (2008). A characteristic function approach to the biased sampling model, with application to robust logistic regression. *Journal of Statistical Planning and Inference, 138*, 742–755.

Bonett, D. G., & Price, R. M. (2002). Statistical inference for a linear function of medians: Confidence intervals, hypothesis testing, and sample size requirements. *Psychological Methods, 7*, 370–383.

Booth, J. G., & Sarkar, S. (1998). Monte carlo approximation of bootstrap variances. *American Statistician, 52*, 354–357.

Borkowf, C. B. (2002). Computing the nonnull asymptotic variance and the asymptotic relative efficiency of Spearman's rank correlation. *Computational Statistics & Data Analysis, 39*, 271–286.

Bowman, A., & Young, S. (1996). Graphical comparison of nonparametric curves. *Applied Statistics, 45*, 83–98.

Box, G. E. P. (1954). Some theorems on quadratic forms applied in the study of analysis of variance problems, I. Effect of inequality of variance in the one-way model. *Annals of Mathematical Statistics, 25*, 290–302.

Bradley, J. V. (1978). Robustness? *British Journal of Mathematical and Statistical Psychology, 31*, 144–152.

Breiman, L. (1995). Better subset regression using the nonnegative garrote. *Technometrics, 37*, 373–384.

Breiman, L. (1996a). Heuristics of instability and stabilization in model selection. *Annals of Statistics, 24*, 2350–2383.

Breiman, L. (1996b). Bagging predictors. *Machine Learning, 24*, 123–140.

Brown, B. M. (1983). Statistical uses of the spatial median. *Journal of the Royal Statistical Society, Series B, 45*, 25–30.

Brown, L., & Li, X. (2005). Confidence intervals for two sample binomial distribution. *Journal of Statistical Planning and Inference, 130*, 359–375.

Brown, L. D., Cai, T. T., & DasGupta, A. (2002). Confidence intervals for a binomial proportion and asymptotic expansions. *Annals of Statistics, 30*, 160–201.

Brown, M. B., & Forsythe, A. (1974). The small sample behavior of some statistics which test the equality of several means. *Technometrics, 16*, 129–132

Brunner, E., Dette, H., & Munk, A. (1997). Box-type approximations in non-parametric factorial designs. *Journal of the American Statistical Association, 92*, 1494–1502.

Brunner, E., Domhof, S., & Langer, F. (2002). *Nonparametric analysis of longitudinal data in factorial experiments*. New York, NY: Wiley.

Brunner, E., & Munzel, U. (2000). The nonparametric Behrens-Fisher problem: Asymptotic theory and small-sample approximation. *Biometrical Journal, 42*, 17–25.

Brunner, E., Munzel, U., & Puri, M. L. (1999). Rank-score tests in factorial designs with repeated measures. *Journal of Multivariate Analysis, 70*, 286–317.

Brunner, E., & Puri, M. L. (2002). A class of rank-score tests in factorial designs. *Journal of Statistical Planning and Inference, 103*, 331–360.

Brys, G., Hubert, M., & Struyf, A. (2004). A robust measure of skewness. *Journal of Computational and Graphical Statistics, 13*, 996–1017.

Bühlmann, P., & Yu, B. (2002). Analyzing bagging. *Annals of Statistics, 30*, 927–961.

Buja, A., Hastie, T., & Tibshirani, R. (1989). Linear smoothers and additive models (with discussion). *Annals of Statistics, 17*, 453–555.

Butler, R. W., Davies, P. L., & Jhun, M. (1993). Asymptotics for the minimum covariance determinant estimator. *Annals of Statistics, 21*, 1385–1400.

Büning, H. (2001). Kolmogorov-Smirnov and Cramer-von Mises type two-sample tests with various weights. *Communications in Statistics— Theory and Methods, 30*, 847–866.

Campbell, N. A. (1980). Robust Procedures in Multivariate Analysis I: Robust Covariance Estimation. *Applied Statistics, 29*, 231237.

Cantoni, E., & Ronchetti, E. (2001). Robust inference for generalized linear models. *Journal of the American Statistical Association, 96*, 1022–1030.

Carling, K. (2000). Resistant outlier rules and the non-Gaussian case. *Computational Statistics & Data Analysis, 33*, 249–258.

Carroll, R. J., & Pedersen, S. (1993). On robustness in the logistic regression model. *Journal of the Royal Statistical Society, B, 55*, 693–706.

Carroll, R. J., & Ruppert, D. (1982). Robust estimation in heteroscedastic linear models. *Annals of Statistics, 10*, 429–441.

Carroll, R. J., & Ruppert, D. (1988). *Transformation and weighting in regression*. New York, NY: Chapman and Hall.

Carroll, R. J., & Welsh, A. H. (1988). A note on asymmetry and robustness in linear regression. *American Statistician, 42*, 285–287.

Cerioli, A. (2010). Multivariate outlier detection with high-breakdown estimators. *Journal of the American Statistical Association, 105*, 147–156.

Ceyhan, E., & Goad, C. L. (2009). A comparison of analysis of covariate-adjusted residuals and analysis of covariance. *Communications in Statistics–Simulation and Computation, 38*, 2019–2038

Chakraborty, B. (1999). On multivariate median regression. *Bernoulli, 5*, 683–703.

Chakraborty, B. (2001). On affine equivariant multivariate quantiles. *Annals of the Institute of Statistical Mathematics, 53*, 380–403.

Chambers, J. M. *Programming with data. A guide to the S language*. New York, NY: Springer-Verlag.

Chambers, J. M., & Hastie, T. J. (1992). *Statistical models in S* New York, NY: Chapman & Hall.

Chang, W. H., McKean, J. W., Naranjo, J. D., & Sheather, S. J. (1999). High-breakdown rank regression. *Journal of the American Statistical Association, 94*, 205–219.

Chaouch, M., & Goga, C. (2010). Design-based estimation for geometric quantiles with applications to outlier detection. *Computational Statistics & Data Analysis, 54*, 2214–2229.

Chatterjee, A., & Lahiri, S. N. (2011). Bootstrapping lasso estimators. *Journal of the American Association, 106*, 608–625.

Chatterjee, S., & Hadi, A. S. (1988). *Sensitivity analysis in linear regression analysis.* New York, NY: Wiley.

Chaudhuri, P. (1996). On a geometric notion of quantiles for multivariate data. *Journal of the American Association, 91*, 862–872.

Chen, H. (1990). The accuracy of approximate intervals for a binomial parameter. *Journal of the American Statistical Association, 85*, 514–518.

Chen, L. (1995). Testing the mean of skewed distributions. *Journal of the American Statistical Association, 90*, 767–772.

Chen, S. X., & Van Keilegom, I. (2009). A review of empirical likelihood for regression. *Test, 18*, 415–447.

Chen, S. X., & Zhong, P.-S. (2010). ANOVA for longitudinal data with missing values. *Annals of Statistics, 38*, 3630–3659.

Chen, T., Martin, E., & Montague, G. (2009). Robust probabilistic PCA with missing data and contribution analysis for outlier detection. *Computational Statistics and Data Analysis, 53*, 3706–3716.

Chen, T.-C., & Victoria-Feser, M.-P. (2002). High-breakdown estimation of multivariate mean and covariance with missing observations. *British Journal of Mathematical and Statistical Psychology, 55*, 317–336.

Chen, Z., & Tyler, D. E. (2002). The influence function and maximum bias of Tukey's median. *Annals of Statistics, 30*, 1737–1760.

Chernick, M. R. (1999). *Bootstrap methods: A practitioner's guide.* New York, NY: Wiley.

Choi, K., & Marden, J. (1997). An approach to multivariate rank tests in multivariate analysis of variance. *Journal of the American Statistical Association, 92*, 1581–1590.

Chow, G. C. (1960). Tests of equality between sets of coefficients in two linear regressions. *Econometrika, 28*, 591–606.

Christmann, A. (1994). Least median of weighted squares in logistic regression with large strata. *Biometrika, 81*, 413–417.

Clements, A., Hurn, S., & Lindsay, K. (2003). Mobius-like mappings and their use in kernel density estimation. *Journal of the American Statistical Association, 98*, 993–1000.

Cleveland, W. S. (1979). Robust locally weighted regression and smoothing scatterplots. *Journal of the American Statistical Association, 74*, 829–836.

Cleveland, W. S. (1985). *The elements of graphing data.* New York, NY: Chapman & Hall.

Cleveland, W. S., & Devlin, S. J. (1988). Locally-weighted regression: An approach to regression analysis by local fitting. *Journal of the American Statistical Association, 83*, 596–610.

Cliff, N. (1993). Dominance statistics: Ordinal analyses to answer ordinal questions. *Psychological Bulletin, 114*, 494–509.

Cliff, N. (1994). Predicting ordinal relations. *British Journal of Mathematical and Statistical Psychology, 47*, 127–150.

Cliff, N. (1996). *Ordinal methods for behavioral data analysis.* Mahwah, NJ: Erlbaum.

Coakley, C. W., & Hettmansperger, T. P. (1993). A bounded influence, high breakdown, efficient regression estimator. *Journal of the American Statistical Association, 88*, 872–880.

Coe, P. R., & Tamhane, A. C. (1993). Small sample confidence intervals for the difference, ratio, and odds ratio of two success probabilities. *Communications in Statistics—Simulation and Computation, 22*, 925–938.

Cohen, J. (1988). *Statistical power analysis for the behavioral sciences* (2nd ed.). New York, NY: Academic Press.

Cohen, M., Dalal, S. R., & Tukey, J. W. (1993). Robust, smoothly heterogeneous variance regression. *Applied Statistics, 42*, 339–354.

Conerly, M. D., & Mansfield, E. R. (1988). An approximate test for comparing heteroscedastic regression models. *Journal of the American Statistical Association, 83*, 811–817.

Cook, R. D., & Hawkins, D. M. (1990). Discussion of unmasking multivariate outliers and leverage points by P. Rousseuw and B. van Zomeren. *Journal of the American Statistical Association, 85*, 640–644.

Cook, R. D., Hawkins, D. M., & Weisberg, S. (1992). Comparison of model misspecification diagnostics using residuals from least mean of squares and least median of squares fit. *Journal of the American Statistical, 87*, 419–424.

Cook, R. D., & Weisberg, S. (1983). Diagnostics for heteroscedasticity in regression. *Biometrika, 70*, 1–10.

Cook, R. D., & Weisberg, S. (1992). *Residuals and influence in regression*. New York, NY: Chapman and Hall.

Cook, R. D., & Weisberg, S. (1994). *An introduction the regression graphics*. New York, NY: Wiley.

Copas, J. B. (1983). Plotting p against x. *Applied Statistics, 32*, 25–31.

Copt, S., & Heritier, S. (2007). Robust alternatives to the F-Test in mixed linear models based on MM-estimates. *Biometrics, 63*, 1045–1052.

Cramer, H. (1946). *Mathematical methods of statistics*. Princeton, NJ: Princeton University Press.

Crawley, M. J. (2007). *The R book*. New York, NY: Wiley.

Cressie, N. A. C., & Whitford, H. J. (1986). How to use the two sample t-test. *Biometrical Journal, 28*, 131–148.

Cribari-Neto, F. (2004). Asymptotic inference under heteroscedasticity of unknown form. *Computational Statistics & Data Analysis, 45*, 215–233.

Cribari-Neto, F., Souza, T. C., & Vasconcellos, A. L. P. (2007). Inference under heteroskedasticity and leveraged data. *Communication in Statistics – Theory and Methods, 36*, 1877–1888.

Cribbie, R. A., Fiksenbaum, L., Keselman, H. J., & Wilcox, R. R. (in press). Effects of nonnormality on test statistics for one-way independent groups designs. *British Journal of Mathematical and Statistical Psychology*.

Croux, C. (1994). Efficient high-breakdown M-estimators of scale. *Statistics and Probability Letters, 19*, 371–379.

Croux, C., & Dehon, C. (2003). Estimators of the multiple correlation coefficient: local robustness and confidence intervals. *Statistical Papers, 44*, 315–334.

Croux, C., & Dehon, C. (2010). Influence functions of the Spearman and Kendall Correlation measures. *Statistical Methods and Applications, 19*, 497–515.

Croux, C., Filzmoser, P., & Oliveira, M. R. (2007). Algorithms for projection-pursuit robust principal component analysis. *Chemometrics and Intelligent Laboratory Systems, 87,* 218–225.

Croux, C., Flandre, C., & Haesbroeck, G. (2002). The breakdown behavior of the maximum likelihood estimator in the logistic regression model. *Statistics and Probability Letters, 60,* 377–386.

Croux, C., & Haesbroeck, G. (2000). Principal component analysis based on robust estimators of the covariance or correlation matrix: Influence functions and efficiencies. *Biometrika, 87,* 603–618.

Croux, C., Rousseeuw, P. J., & Hössjer, O. (1994). Generalized S-estimators. *Journal of the American Statistical Association, 89,* 1271–1281.

Croux, C., & Ruiz-Gazen, A. (2005). High breakdown estimators for principal components: the projection-pursuit approach revisited. *Journal of Multivariate Analysis, 95,* 206–226.

Crumpacker, D. W., Cederlof, R., Friberg, L., Kimberling, W. J., Sorensen, S., Vandenberg, S. G., ... Roulette, I. (1979). A twin methodology for the study of genetic and environmental control of variation in human smoking behavior. *Acta Genet Med Gemellol, 28,* 173–195.

Cuesta-Albertos, J. A., Gordaliza, A., & Matran, C. (1997). Trimmed k-Means: An attempt to robustify quantizers. *Annals of Statistics, 25,* 553–576.

Cuesta-Albertos, J. A., & Nieto-Reyes, A. (2008). The random Tukey depth. *Computational Statistics & Data Analysis, 52,* 4979–4988.

Cushny, A. R., & Peebles, A. R. (1904). The action of optical isomers II. Hyoscines. *Journal of Physiology, 32,* 501–510.

Dahlquist, G., & Björck, A. (1974). *Numerical methods.* Englewood Cliffs, NJ: Prentice Hall.

Dana, E. (1990). *Salience of the self and salience of standards: Attempts to match self to standard.* Unpublished PhD dissertation, Department of Psychology, University of Southern California.

Daniell, P. J. (1920). Observations wighted according to order. *American Journal of Mathematics, 42,* 222–236.

Davidson, R., & MacKinnon, J. G. (2000). Bootstrap tests: How many bootstraps? *Econometric Reviews, 19,* 55–68.

Davies, L., & Gather, U. (1993). The identification of multiple outliers (with discussion). *Journal of the American Statistical Association, 88,* 782–792.

Davies, P. L. (1987). Asymptotic behavior of S-estimates of multivariate location parameters and dispersion matrices. *Annals of Statistics, 15,* 1269–1292.

Davies, P. L. (1990). The asymptotics of S-estimators in the linear regression model. *Annals of Statistics, 18,* 1651–1675.

Davies, P. L. (1993). Aspects of robust linear regression. *Annals of Statistics, 21,* 1843–1899.

Davis, J. B., & McKean, J. W. (1993). Rank-based method for multivariate linear models. *Journal of the American Statistical Association, 88,* 245–251.

Davison, A. C., & Hinkley, D. V. (1997). *Bootstrap methods and their application.* Cambridge, UK: Cambridge University Press.

Davison, A. C., Hinkley, D. V., & Young, G. A. (2003). Recent developments in bootstrap methodology. *Statistical Science, 18,* 141–157.

Dawkins, B. P. (1995). Investigating the geometry of a p-dimensional data set. *Journal of the American Statistical Association, 90,* 350–359.

de Boor, C. (1978). *A practical guide to splines.* New York: Springer-Verlag.

Debruyne, M., Hubert, M., & van Horebeek, J. V. (2010). Detecting influential observations in kernel PCA. *Computational Statistics & Data Analysis, 54,* 3007–3019.

De Jongh, P. J., De Wet, T., & Welsh, A. H. (1988). Mallows-type bounded-influence-regression trimmed means. *Journal of the American Statistical Association, 83,* 805–810.

Delgado, M. A. (1993). Testing the equality of nonparametric regression curves. *Statistics and Probability Letters, 17,* 199–204.

Derksen, S., & Keselman, H. J. (1992). Backward, forward and stepwise automated subset selection algorithms: Frequency of obtaining authentic and noise variables. *British Journal of Mathematical and Statistical Psychology, 45,* 265–282.

Devlin, S. J., Gnanadesikan, R., & Kettenring, J. R. (1981). Robust estimation of dispersion matrices and principal components. *Journal of the American Statistical Association, 76,* 354–362.

Dette, H. (1999). A consistent test for the functional form of a regression based on a difference of variances estimator. *Annals of Statistics, 27,* 1012–1040.

Dette, H., & Neumeyer, N. (2001). Nonparametric analysis of covariance. *Annals of Statistics, 29,* 1361–1400.

Devroye, L., & Lugosi, G. (2001). *Combinatorial methods in density estimation*. New York, NY: Springer-Verlag.

DiCiccio, T., Hall, P., & Romano, J. (1991). Empirical likelihood is Bartlett-correctable. *Annals of Statistics, 19*, 1053–1061.

Dielman, T., & Pfaffenberger, R. (1982). LAV (least absolute value) estimation in linear regression: A review. In S. H. Zanakis & J. Rustagi (Eds.), *Optimization in statistics* (pp. 31–52). New York, NY: North Holland.

Dielman, T., & Pfaffenberger, R. (1988a). Bootstrapping in least absolute value regression: An application to hypothesis testing. *Communications in Statistics–Simulation and Computation, 17*, 843–856.

Dielman, T., & Pfaffenberger, R. (1988b). Least absolute value regression: Necessary sample sizes to use normal theory inference procedures. *Decision Sciences, 19*, 734–743.

Dielman, T., & Pfaffenberger, R. (1990). Tests of linear hypotheses and LAV estimation: A Monte Carlo comparison. *Communications in Statistics–Simulation and Computation, 19*, 1179–1199.

Dielman, T., Lowry, C., & Pfaffenberger, R. (1994). A comparison of quantile estimators. *Communications in Statistics–Simulation and Computation, 23*, 355–371.

Dietz, E. J. (1987). A comparison of robust estimators in simple linear regression. *Communications in Statistics–Simulation and Computation, 16*, 1209–1227.

Dietz, E. J. (1989). Teaching regression in a nonparametric statistics course. *American Statistician, 43*, 35–40.

Diggle, P. J., Heagerty, P. J., Liang, K.-Y., & Zeger, S. L. (2002). *Analysis of longitudinal data* (2nd ed.). Oxford, UK: Oxford University Press.

Dixon, S. L., & McKean, J. W. (1996). Rank-based analysis of the heteroscedastic linear model. *Journal of the American Statistical Association, 91*, 699–712.

Dixon, W. J., & Tukey, J. W. (1968). Approximate behavior of the distribution of Winsorized t (trimming/Winsorization 2). *Technometrics, 10*, 83–98.

Doksum, K. A. (1974). Empirical probability plots and statistical inference for nonlinear models in the two-sample case. *Annals of Statistics, 2*, 267–277.

Doksum, K. A. (1977). Some graphical methods in statistics. A review and some extensions. *Statistica Neerlandica, 31*, 53–68.

Doksum, K. A., & Koo, J.-Y. (2000). On spline estimators and prediction intervals in nonparametric regression. *Computational Statistics & Data Analysis, 35*, 67–82.

Doksum, K. A. & Samarov, A. (1995). Nonparametric estimation of global functionals and a measure of the explanatory power of covariates in regression. *Annals of Statistics, 23*, 1443–1473.

Doksum, K. A., & Sievers, G. L. (1976). Plotting with confidence: graphical comparisons of two populations. *Biometrika, 63*, 421–434.

Doksum, K. A., Blyth, S., Bradlow, E., Meng, X., & Zhao, H. (1994). Correlation curves as local measures of variance explained by regression. *Journal of the American Statistical Association, 89*, 571–582.

Doksum, K. A., & Wong, C.-W. (1983). Statistical tests based on transformed data. *Journal of the American Statistical Association, 78*, 411–417.

Donoho, D. L. (1982). Breakdown properties of multivariate location estimators. PhD qualifying paper, Department of Statistics, Harvard University.

Donoho, D. L., & Gasko, M. (1992). Breakdown properties of the location estimates based on halfspace depth and projected outlyingness. *Annals of Statistics, 20*, 1803–1827.

Dorfman, A. H. (1991). Sound confidence intervals in the heteroscedastic linear model through releveraging. *Journal of the Royal Statistical Society, B, 53*, 441–452.

Draper, N. R. (1994). Applied regression analysis bibliography update 1992–93. *Communications in Statistics–Theory and Methods, 23*, 2701–2731.

Ducharme, G. R. (1995). Uniqueness of the least-distance estimator in regression with multivariate response. *Canadian Journal of Statistics, 23*, 421–424.

Duncan, G. T., & Layard, M. W. (1973). A Monte-Carlo study of asymptotically robust tests for correlation. *Biometrika, 60*, 551–558.

Dunnett, C. W. (1980). Pairwise multiple comparisons in the unequal variance case. *Journal of the American Statistical Association, 75*, 796–800.

Edgell, S. E., & Noon, S. M. (1984). Effect of violation of normality on the t test of the correlation coefficient. *Psychological Bulletin, 95*, 576–583.

Efromovich, S. (1999). *Nonparametric curve estimation: Methods, theory and applications.* New York, NY: Springer-Verlag.

Efron, B. (1969). Student's t-test under symmetric conditions. *Journal of the American Statistical Association, 64*, 1278–1302.

Efron, B. (1987). Better bootstrap confidence intervals. *Journal of the American Statistical Association, 82*, 171–185.

Efron, B., Hastie, T., Johnstone, I., & Tibshirani, R. (2004). Least angle regression (with discussion and rejoinder). *Annals of Statistics, 32*, 407–499.

Efron, B., & Tibshirani, R. J. (1993). *An introduction to the bootstrap*. New York, NY: Chapman & Hall.

Efron, B., & Tibshirani, R. J. (1997). Improvements on cross-validation: The .632+ bootstrap method. *Journal of the American Statistical Association, 92*, 548–560.

Elashoff, J. D., & Snow, R. E. (1970). A case study in statistical inference: Reconsideration of the Rosenthal-Jacobson data on teacher expectancy. Tech. Rep. No. 15, School of Education, Stanford University.

Emerson, J. D., & Hoaglin, D. C. (1983a). Resistant lines for y versus x. In D. Hoaglin, F. Mosteller, & J. W. Tukey (Eds.), *Understanding robust and exploratory data analysis*. New York, NY: Wiley.

Emerson, J. D., & Hoaglin, D. C. (1983b). Resistant multiple regression, one variable at a time. In D. Hoaglin, F. Mosteller, & J. W. Tukey (Eds.), *Understanding robust and exploratory data analysis* (pp. 241–280). New York, NY: Wiley.

Emerson, J. D., & Stoto, M. A. (1983). Transforming data. In D. C. Hoaglin, F. Mosteller, & J. W. Tukey (Eds.), *Exploring data tables, trends and shapes* (pp. 97–128). New York, NY: Wiley.

Engelen, S., Hubert, M., & Vanden Branden, K. (2005). A comparison of three procedures for robust PCA in high dimensions. *Australian Journal of Statistics, 2*, 117–126.

Eubank, R. L. (1999). *Nonparametric regression and spline smoothing*. New York, NY: Marcel Dekker.

Everitt, B. S., Landau, S., Leese, M., & Stahl, D. (2011). *Cluster analysis* (5th ed.). New York, NY: Wiley.

Fairley, D. (1986). Cherry trees with cones? *American Statistician, 40*, 138–139.

Fan, J. (1993). Local linear smoothers and their minimax efficiencies. *The Annals of Statistics, 21*, 196–216.

Fan, J. (1996). Test of significance based on wavelet thresholding and Neyman's truncation. *Journal of the American Statistical Association, 91*, 674–688.

Fan, J., & Gijbels, I. (1996). *Local polynomial modeling and its applications*. Boca Raton, FL: CRC Press.

Fan, J., & Hall, P. (1994). On curve estimation by minimizing mean absolute deviation and its implications. *The Annals of Statistics, 22*, 867–885.

Fenstad, G. U. (1983). A comparison between U and V tests in the Behrens-Fisher problem. *Biometrika, 70*, 300–302.

Ferreira, E., & Stute, W. (2004). Testing for differences between conditional means in a time series context. *Journal of the American Statistical Association, 99*, 169–174.

Ferretti, N., Kelmansky, D., Yohai, V. J., & Zamar, R. H. (1999). A class of locally and globally robust regression estimates. *Journal of the American Statistical Association, 94*, 174–188.

Filzmoser, P., Maronna, R., & Werner, M. (2008). Outlier identification in high dimensions. *Computational Statistics & Data Analysis, 52*, 1694–1711.

Fisher, R. A. (1922). On the mathematical foundations of theoretical statistics. *Philosophical Transactions of the Royal Astronomical Society of London, Series A 222*, 309–368.

Filzmoser, P. P., Maronna, R., & Werner, M. (2008). Outlier identification in high dimensions. *Computational Statistics & Data Analysis, 52*, 1694–1711.

Fligner, M. A., & Policello II, G. E. (1981). Robust rank procedures for the Behrens-Fisher problem. *Journal of the American Statistical Association, 76*, 162–168.

Flores, S. (2010). On the efficient computation of robust regression estimators. *Computational Statistics & Data Analysis, 54*, 3044–3056.

Fox, J. (1999). *Applied regression analysis, linear models, and related methods*. Thousands Oaks, CA: Sage.

Fox, J. (2001). *Multiple and generalized nonparametric regression*. Thousands Oaks, CA: Sage.

Fox, J. (2002). *An R and S-PLUS companion to applied regression*. Thousand Oaks, CA: Sage.

Freedman, D., & Diaconis, P. (1981). On the histogram as density estimator: L_2 theory. *Zeitschrift fr Wahrscheinlichkeitstheorie und verwandte Gebiete, 57*, 453–476.

Freedman, D., & Diaconis, P. (1982). On inconsistent M-estimators. *Annals of Statistics, 10*, 454–461.

Freidlin, B. & Gastwirth, J. L. (2000). Should the median test be retired from general use? *American Statistician, 54*, 161–164.

Friedman, J. H., & Stuetzle, W. (1981). Projection pursuit regression. *Journal of the American Statistical Association, 76*, 817–823.

Frigge, M., Hoaglin, D. C., & Iglewicz, B. (1989). Some implementations of the boxplot. *American Statistician, 43*, 50–54.

Fung, K. Y. (1980). Small sample behaviour of some nonparametric multi-sample location tests in the presence of dispersion differences. *Statistica Neerlandica, 34*, 189–196.

Fung, W.-K. (1993). Unmasking outliers and leverage points: A confirmation. *Journal of the American Statistical Association, 88*, 515–519.

Gail, M. H., Santner, T. J., & Brown, C. C. (1980). An analysis of comparative carcinogenesis experiments with multiple times to tumor. *Biometrics, 36* 255–266.

Gather, U., & Hilker, T. (1997). A note on Tyler's modification of the MAD for the Stahel-Donoho estimator. *Annals of Statistics, 25*, 2024–2026.

Gatto, R., & Ronchetti, E. (1996). General saddlepoint approximations of marginal densities and tail probabilities. *Journal of the American Statistical Association, 91*, 666–673.

Genton, M. G., & Lucas, A. (2003). Comprehensive definitions of breakdown points for independent and dependent observations. *Journal of the Royal Statistical Society, B, 65*, 81–94.

Gervini, D. (2002). The influence function of the Stahel-Donoho estimator of multivariate location and scatter. *Statistics & Probability letters, 60*, 425–435.

Gervini, D., & Yohai, V. J. (2002). A class of robust and fully efficient regression estimators. *Annals of Statistics, 30*, 583 616.

Gleason, J. R. (1993). Understanding elongation: The scale contaminated normal family. *Journal of the American Statistical Association, 88*, 327–337.

Glenn, N. (2002). Robust empirical likelihood. Unpublished PhD dissertation, Department of Statistics, Rice University.

Glenn, N., & Zhao, Y. (2007). Weighted empirical likelihood estimates and their robustness properties. *Computational Statistics & Data Analysis, 51*, 5130–5141.

Gnanadesikan, R., & Kettenring, J. R. (1972). Robust estimates, residuals and outlier detection with multiresponse data. *Biometrics, 28*, 81–124.

Godfrey, L. G. (2006). Tests for regression models with heteroskedasticity of unknown form. *Computational Statistics & Data Analysis, 50*, 2715–2733.

Goeman, J. J., & Solari, A. (2010). The sequential rejection principle of familywise error control. *Annals of Statistics, 38*, 3782–3810.

Goldberg, K. M. & Iglewicz, B. (1992). Bivariate extensions of the boxplot. *Technometrics, 34*, 307–320.

Goldberg, K. M. & Iglewicz, B. (1992). Bivariate extensions of the boxplot. *Technometrics, 34*, 307–320.

Golub, G. H., & van Loan, C. F. (1983). *Matrix computations*. Baltimore, MD: Johns Hopkins University Press.

Good, P. (2000). *Permutation tests*. New York, NY: Springer-Verlag.

Graybill, F. A. (1976). *Theory and application of the linear model*. Belmont, CA: Wadsworth.

Graybill, F. A. (1983). *Matrices with applications in statistics*. Belmont, CA: Wadsworth.

Green, D. P., Ha, S. E., & Bullock, J. G. (2010). Enough already about 'Black Box' experiments: Studying mediation is more difficult than most scholars suppose. *Annals of the American Academy of Political and Social Science, 628*, 200–208.

Green, P. J., & Silverman, B. W. (1993). *Nonparametric regression and generalized linear models: A roughness penalty approach*. Boca Raton, FL: CRC Press.

Grissom, R. J. (2000). Heterogeneity of variance in clinical data. *Journal of Consulting and Clinical Psychology, 68*, 155–165.

Gupta, A. K., & Rathie, P. N. (1983). On the distribution of the determinant of sample correlation matrix from multivariate Gaussian population. *Metron, 61*, 43–56.

Guo, J. H., & Luh, W. M. (2000). An invertible transformation two-sample trimmed t-statistic under heterogeneity and nonnormality. *Statistics & Probability Letters, 49*, 1–7.

Guo, J. H., Billard, L., & Luh, W.-M. (2011). New heterogeneous test statistics for the unbalanced fixed-effect nested design. *British Journal of Mathematical and Statistical Psychology, 64*, 259–276.

Gutenbrunner, C., Jureckova, J., Koenker, R., & Portnoy, S. (1993). Tests of linear hypotheses based on regression rank scores. *Journal of Nonparametrics, 2*, 307–331.

Györfi, L., Kohler, M., Krzyzk, A., & Walk, H. (2002). *A Distribution-Free theory of nonparametric regression*. New York, NY: Springer Verlag.

Haldane, J. B. S. (1948). Note on the median multivariate distribution. *Biometrika, 35*, 414–415.

Hall, P. (1986). On the number of bootstrap simulations required to construct a confidence interval. *Annals of Statistics, 14*, 1431–1452.

Hall, P. (1988a). On symmetric bootstrap confidence intervals. *Journal of the Royal Statistical Society, Series B, 50*, 35–45.

Hall, P. (1988b). Theoretical comparison of bootstrap confidence intervals. *Annals of Statistics, 16*, 927–953.

Hall, P. (1992). On the removal of skewness by transformation. *Journal of the Royal Statistical Society, Series B, 54*, 221–228.

Hall, P. G., & Hall, D. (1995). *The bootstrap and edgeworth expansion*. New York, NY: Springer Verlag.

Hall, P., & Hart, J. D. (1990). Bootstrap test for difference between means in nonparametric regression. *Journal of the American Statistical Association, 85*, 1039–1049.

Hall, P., Huber, C., & Speckman, P. L. (1997). Covariate-matched one-sided tests for the difference between functional means. *Journal of the American Statistical Association, 92*, 1074–1083.

Hall, P., & Jones, M. C. (1990). Adaptive M-estimation in nonparametric regression. *Annals of Statistics, 18*, 1712–1728.

Hall, P., & Padmanabhan, A. R. (1992). On the bootstrap and the trimmed mean. *Journal of Multivariate Analysis, 41*, 132–153.

Hall, P., & Presnell, B. (1999). Biased bootstrap methods for reducing the effects of contamination. *Journal of the Royal Statistical Society, B, 61*, 661–680.

Hall, P., & Sheather, S. J. (1988). On the distribution of a studentized quantile. *Journal of the Royal Statistical Society, B, 50*, 380–391.

Hall, P., & Welsh, A. H. (1985). Limit theorems for the median deviation. *Annals of the Institute of Statistical Mathematics, 37, A*, 27–36.

Hamilton, L. C. (1992). *Regression with graphics: A second course in applied statistics*. Pacific Grove, CA: Brooks/Cole.

Hampel, F. R. (1968). *Contributions to the theory of robust estimation*. Unpublished PhD dissertation, Department of Statistics, University of California, Berkeley.

Hampel, F. R. (1973). Robust estimation: A condensed partial survey. *Z. Wahrscheinlichkeitstheorie and Verw. Gebiete, 27*, 87–104.

Hampel, F. R. (1974). The influence curve and its role in robust estimation. *Journal of the American Statistical Association, 62*, 1179–1186.

Hampel, F. R. (1975). Beyond location parameters: Robust concepts and methods (with discussion). *Bulletin of the ISI, 46*, 375–391.

Hampel, F. R., Ronchetti, E. M., Rousseeuw, P. J., & Stahel, W. A. (1986). *Robust statistics*. New York, NY: Wiley.

Hand, A. (1998). *A history of mathematical statistics from 1750 to 1930.* New York, NY: Wiley.

Handschin, E., Schweppe, F. C., Kohlas, J., & Fiechter, A. (1975). Bad data analysis for power system state estimation. *IEEE Transactions of Power Apparatus and Systems, PAS-94,* 329–337.

Härdle, W. (1990). *Applied nonparametric regression.* Econometric Society Monographs No. 19. Cambridge, UK: Cambridge University Press.

Härdle, W., & Korostelev, A. (1996). Search for significant variables in nonparametric additive regression. *Biometrika, 83,* 541–549.

Härdle, W., & Marron, J. S. (1990). Semiparametric comparison of regression curves. *Annals of Statistics, 18,* 63–89.

Harrell, F. E., & Davis, C. E. (1982). A new distribution-free quantile estimator. *Biometrika, 69,* 635–640.

Harwell, M. (2003). Summarizing monte carlo results in methodological research: The single-factor, fixed effects ANCOVA case. *Journal of Educational and Behavioral Statistics, 28,* 45–70.

Hastie, T. J., & Loader, C. (1993). Local regression: Automatic kernel carpentry. *Statistical Science, 8,* 120–143.

Hastie, T. J., & Tibshirani, R. J. (1990). *Generalized additive models.* New York, NY: Chapman and Hall.

Hawkins, D. G., & Simonoff, J. S. (1993). Algorithm AS 282: High breakdown regression and multivariate estimation. *Applied Statistics, 42,* 423–431.

Hawkins, D. M., & Olive, D. (1999). Applications and algorithms for least trimmed sum of absolute deviations regression. *Computational Statistics & Data Analysis, 32,* 119–134.

Hawkins, D. M., & Olive, D. (2002). Inconsistency of resampling algorithms for high-breakdown regression estimators and a new algorithm. *Journal of the American Statistical Association, 97,* 136–147.

Hayes, A. F., & Cai, L. (2007). Further evaluating the conditional decision rule for comparing two independent means. *British Journal of Mathematical and Statistical Psychology, 60,* 217–244.

Hayes, A. F., & Preacher, K. J. (2010). Quantifying and testing indirect effects in simple mediation models when the constituent paths are nonlinear. *Multivariate Behavioral Research, 45,* 627–660.

He, X., & Portnoy, S. (1992). Reweighted LS estimators converge at the same rate as the initial estimator. *Annals of Statistics, 20*, 2161–2167.

He, X., Ng, P., & Portnoy, S. (1998). Bivariate quantile smoothing splines. *Journal of the Royal Statistical Society, B, 60*, 537–550.

He, X., & Ng, P. (1999). Quantile splines with several covariates. *Journal of Statistical Planning and Inference, 75*, 343–352.

He, X., & Simpson, D. G. (1993). Lower bounds for contamination bias: Global minimax versus locally linear estimation. *Annals of Statistics, 21*, 314–337.

He, X., Simpson, D. G., & Portnoy, S. L. (1990). Breakdown robustness of tests. *Journal of the American Statistical Association, 85*, 446–452.

He, X., & Wang, G. (1997). Convergence of depth contours for multivariate data sets. *Annals of Statistics, 25*, 495–504.

He, X., & Zhu, L.-X. (2003). A lack-of-fit test for quantile regression. *Journal of the American Statistical Association, 98*, 1013–1022.

Herbert, R. D., Hayen, A., Macaskill, P., & Walter, S. D. (2011). Interval estimation for the difference of two independent variances. *Communications in Statistics–Simulation and Computation, 40*, 744–758.

Heritier, S., & Ronchetti, E. (1994). Robust bounded-influence tests in general linear models. *Journal of the American Statistical Association, 89*, 897–904.

Heritier, S., Cantoni, E., Copt, S., & Victoria-Feser, M.-P. (2009). *Robust methods in biostatistics*. New York, NY: Wiley.

Herwindiati, D. E., Djauhari, M. A., & Mashuri, M. (2007). Robust multivariate outlier labeling. *Communications in Statistics–Simulation and Computation, 36*, 1287–1294.

Hettmansperger, T. P. (1984). *Statistical inference based on ranks*. New York, NY: Wiley.

Hettmansperger, T. P., & McKean, J. W. (1977). A robust alternative based on ranks to least squares in analyzing linear models. *Technometrics, 19*, 275–284.

Hettmansperger, T. P., & McKean, J. W. (1998). *Robust nonparametric statistical methods*. London, UK: Arnold.

Hettmansperger, T. P., & Sheather, S. J. (1986). Confidence interval based on interpolated order statistics. *Statistical Probability Letters, 4*, 75–79.

Hettmansperger, T. P., & Sheather, S. J. (1992). A cautionary note on the method of least median of squares. *The American Statistician, 46*, 79–83.

Hill, M., & Dixon, W. J. (1982). Robustness in real life: A study of clincial laboratory data. *Biometrics, 38*, 377–396.

Hill, R. W. (1977). *Robust regression when there are outliers in the carriers*. Unpublished PhD dissertation, Department of Statistics, Harvard University.

Hill, R. W., & Holland, P. W. (1977). Two robust alternatives to robust regression. *Journal of the American Statistical Association, 72*, 828–833.

Hilton, J. F., Mehta, C. R., & Patel, N. R. (1994). An algorithm for conducting exact Smirnov tests. *Computational Statistics and Data Analysis, 19*, 351–361.

Hoaglin, D. C. (1985). Summarizing shape numerically: The g-and-h distribution. In D. Hoaglin, F. Mosteller, & J. Tukey (Eds.), *Exploring data tables trends and shapes* (pp. 461–515). New York, NY: Wiley.

Hoaglin, D. C., & Iglewicz, B. (1987). Fine-tuning some resistant rules for outlier labeling. *Journal of the American Statistical Association, 82*, 1147–1149.

Hoaglin, D. C., & Welsch, R. (1978). The hat matrix in regression and ANOVA. *American Statistician, 32*, 17–22.

Hoaglin, D. C., Mosteller, F., & Tukey, J. W. (1991). *Fundamentals of exploratory analysis of variance*. New York, NY: Wiley.

Hochberg, Y. (1975). Simultaneous inference under Behrens-Fisher conditions: A two sample approach. *Communications in Statistics, 4*, 1109–1119.

Hochberg, Y. (1988). A sharper Bonferroni procedure for multiple tests of significance. *Biometrika, 75*, 800–802.

Hochberg, Y., & Tamhane, A. C. (1987). *Multiple comparison procedures*. New York, NY: Wiley.

Hodges, J. L., & Lehmann, E. L. (1963). Estimates of location based on rank tests. *Annals of Mathematical Statistics, 34*, 598–611.

Hogg, R. V. (1974). Adaptive robust procedures: A partial review and some suggestions for future applications and theory. *Journal of the American Statistical Association, 69*, 909–922.

Hollander, M., & Wolfe, D. A. (1973). *Nonparametric statistical methods* New York, NY: Wiley.

Hommel, G. (1988). A stagewise rejective multiple test procedure based on a modified Bonferroni test. *Biometrika, 75*, 383–386.

Horowitz, J. L., & Spokoiny, V. G. (2002). An adaptive, rate-optimal test of linearity for median regression models. *Journal of the American Statistical Association, 97*, 822–835.

Hosmer, D. W., & Lemeshow, S. (1989). *Applied logistic regression*. New York, NY: Wiley.

Hössjer, O. (1992). On the optimality of S-estimators. *Statistics and Probability Letters, 14*, 413–419.

Hössjer, O. (1994). Rank-based estimates in the linear model with high breakdown point. *Journal of the American Statistical Association, 89*, 149–158.

Hössjer, O., & Croux, C. (1995). Generalizing univariate signed rank statistics for testing and estimating a multivariate location parameter. *Non-parametric Statistics, 4*, 293–308.

Huber, P. J. (1964). Robust estimation of location parameters. *Annals of Mathematical Statistics, 35*, 73–101.

Huber, P. J. (1981). *Robust statistics*. New York, NY: Wiley.

Huber, P. J., & Ronchetti, E. (2009). *Robust statistics* (2nd Ed.). New York, NY: Wiley.

Huber, P. (1993). Projection pursuit and robustness. In S. Morgenthaler, E. Ronchetti, & W. Stahel (Eds.), *New directions in statistical data analysis and robustness* (pp. 139–146). Boston, MA: Birkhäuser Verlag.

Hubert, M., & Vandervieren, E. (2008). An adjusted boxplot for skewed distributions. *Computational statistics & data analysis, 52*, 5186–5201.

Hubert, M., Rousseeuw, P. J. & Vanden Branden, K. (2005). ROBPCA: A new approach to robust principal component analysis. *Technometrics, 47*, 64–79.

Hubert, M., Rousseeuw, P. J., & Verboven, S. (2002). A fast method for robust principal components with applications to chemometrics. *Chemometrics and Intelligent Laboratory Systems, 60*, 101–111.

Huberty, C. J. (1989). Problems with stepwise methods—better alternatives. *Advances in Social Science Methodology, 1*, 43–70.

Huitema, B. E. (1980). *The analysis of covariance and alternatives*. New York, NY: Wiley.

Hussain, S. S., & Sprent, P. (1983). Non-parametric regression. *Journal of the Royal Statistical Society, 146*, 182–191.

Hwang, J., Jorn, H., & Kim, J. (2004). On the performance of bivariate robust location estimators under contamination. *Computational Statistics & Data Analysis, 44*, 587–601.

Hyndman, R. J., & Fan, Y. (1996). Sample quantiles in statistical packages. *The American Statistician, 50*, 361–365.

Iglewicz, B. (1983). Robust scale estimators and confidence intervals for location. In D. Hoaglin, F. Mosteller, & J. Tukey (Eds.), *Understanding robust and exploratory data analysis* (pp. 404–431). New York, NY: Wiley

Jaeckel, L. A. (1972). Estimating regression coefficients by minimizing the dispersion of residuals. *Annals of Mathematical Statistics, 43*, 1449–1458.

Jeyaratnam, S., & Othman, A. R. (1985). Test of hypothesis in one-way random effects model with unequal error variances. *Journal of Statistical Computation and Simulation, 21*, 51–57.

Jhun, M., & Choi, I. (2009). Bootstrapping least distance estimator in the multivariate regression model. *Computational Statistics & Data Analysis, 53*, 4221–4227.

Jöckel, K.-H. (1986). Finite sample properties and asymptotic efficiency of Monte Carlo tests. *Annals of Statistics, 14*, 336–347.

Johansen, S. (1980). The Welch-James approximation of the distribution of the residual sum of squares in weighted linear regression. *Biometrika, 67*, 85–92.

Johansen, S. (1982). Amendments and corrections: The Welch-James approximation to the distribution of the residual sum of squares in a weighted linear regression. *Biometrika, 69*, 491.

Johnson, N. J. (1978). Modified t tests and confidence intervals for asymmetrical populations. *Journal of the American Statistical Association, 73*, 536–544.

Johnson, N. L., & Kotz, S. (1970). *Distributions in statistics: Continuous univariate distributions-2*. New York, NY: Wiley.

Johnstone, I. M., & Velleman, P. F. (1985). The resistant line and related regression methods. *Journal of the American Statistical Association, 80*, 1041–1054.

Jorgensen, J. O., Gilles, R. B., Hunt, D. R., Caplehorn, J. R. M., & Lumley, T. (1995). A simple and effective way to reduce postoperative pain after laparoscopic cholecystectomy. *Australian and New Zealand Jouranl of Surgery, 65*, 466–469.

Judd, C. M., & Kenny, D. A. (1981a). *Estimating the effects of social interventions*. New York, NY: Cambridge University Press.

Judd, C. M., & Kenny, D. A. (1981b). Process analysis: Estimating mediation in treatment evaluations. *Evaluation Review, 5*, 602–619.

Jureckova, J., & Portnoy, S. (1987). Asymptotics for one-step M-estimators with application to combining efficiency and high breakdown point. *Communications in Statistics–Theory and Methods, 16*, 2187–2199.

Kallenberg, W. C., & Ledwina, T. (1999). Data-driven rank tests for independence. *Journal of the American Statistical Association, 94*, 285–310.

Kay, R., & Little, S. (1987). Transformation of the explanatory variables in the logistic regression model for binary data. *Biometrika, 74*, 495–501.

Kent, J. T., & Tyler, D. E. (1996). Constrained M-estimation for multivariate location and scatter. *Annals of Statistics, 24*, 1346–1370.

Keppel, G. (1991). *Design and analysis: A researcher's handbook*. Englewood Cliffs, NJ: Prentice Hall.

Keselman, H. J., Algina, J., Kowalchuk, R. K., & Wolfinger, R. D. (1999). A comparison of recent approaches to the analysis of repeated measurements. *British Journal of Mathematical and Statistical Psychology, 52*, 62–78.

Keselman, H. J., Algina, J., Wilcox, R. R., & Kowalchuk, R. K. (2000). Testing repeated measures hypotheses when covariance matrices are heterogeneous: Revisiting the robustness of the Welch-James test again. *Educational and Psychological Measurement, 60*, 925–938.

Keselman, H. J., Carriere, K. C., & Lix, L. M. (1993). Testing repeated measures hypotheses when covariance matrices are heterogeneous. *Journal of Educational Statistics, 18*, 305–319.

Keselman, H. C., Keselman, J. C., & Lix, L. M. (1995). The analysis of repeated measurements: Univariate tests, multivariate tests, or both? *British Journal of Mathematical and Statistical Psychology, 48*, 319–338.

Keselman, H. C., Huberty, C. J., Lix, L. M., Olejnik, S., Cribbie, R. A., Donahue, B., . . . Levin, J. R. (1998). Statistical practices of educational researchers: An analysis of their ANOVA, MANOVA and ANCOVA analyses. *Review of Educational Research, 68*, 350–386.

Keselman, H. J., Othman, A. R., Wilcox, R. R., & Fradette, K. (2004). The new and improved two-sample t test. *Psychological Science, 15*, 47–51.

Keselman, H. C., Wilcox, R. R., Lix, L. M., Algina, J., & Fradette, K. (2003). Adaptive robust estimation and testing. Unpublished technical report, Department of Psychology, University of Manitoba.

Keselman, H. C., Wilcox, R. R., Othman, A. R., & Fradette, K. (2002). Trimming, transforming statistics, and bootstrapping: Circumventing the biasing effects of heteroscedasticity and nonnormality. *Journal of Modern Applied Statistical Methods, 1*, 288–309.

Khan, J. A., van Aelst, S., & Zamar, R. (2010). Fast robust estimation of prediction error based on resampling. *Computational Statistics and Data Analysis, 54*, 3121–3130.

Khuri, A. I. (1992). Tests concerning a nested mixed model with heteroscedastic random effects. *Journal of Statistical Planning and Inference, 30*, 33–44.

Kim, J., & Hwang, J. (2001). Asymptotic properties of location estimators based on projection depth. *Statistics and Probability Letters, 49*, 293–299.

Kim, P. J., & Jennrich, R. I. (1973). Tables of the exact sampling distribution of the two-sample Kolmogorov-Smirnov criterion, D_{mn}, $m \leq n$. In H. L. Harter & D. B. Owen (Eds.), *Selected tables in mathematical statistics* (Vol. 1, pp. 79–170). Providence, Rhode Island: American Mathematical Society.

Kim, S.-J. (1992a). A practical solution to the multivariate Behrens-Fisher problem. *Biometrika, 79*, 171–176.

Kim, S.-J. (1992b). The metrically trimmed mean as a robust estimator of location. *Annals of Statistics, 20*, 1534–1547.

King, E. C., Hart, J. D., & Wherly, T. E. (1991). Testing the equality of two regression curves using linear smoothers. *Statistics and Probability Letters, 12*, 239–247.

Kirk, R. E. (1995). *Experimental design*. Pacific Grove, CA: Brooks/Cole.

Knight, K. (1998). Limiting distributions for L_1 regression estimators under general conditions. *Annals of Statistics, 26*, 755–770.

Koenker, R. (1981). A note on studentizing a test for heteroscedasticity *Journal of Econometrics, 17*, 107–112.

Koenker, R. (1994). Confidence intervals for regression quantiles. In P. Mandl & M. Huskova (Eds.), *Asymptotic statistics. Proceedings of the fifth prague symposium* (pp. 349–359). Heidelberg: Physica-Verlag.

Koenker, R., & Bassett, G. (1978). Regression quantiles. *Econometrika, 46*, 33–50.

Koenker, R., & Ng, P. (2005). Inequality constrained quantile regression. *Sankhya, The Indian Journal of Statistics, 67*, 418–440.

Koenker, R., Ng, P., & Portnoy, S. (1994). Quantile smoothing splines. *Biometrika, 81*, 673–680.

Koenker, R., & Xiao, Z. J. (2002). Inference on the quantile regression process. *Econometrica, 70*, 1583–1612.

Koller, M., & Stahel, W. A. (2011). Sharpening Wald-type inference in robust regression for small samples. *Computational Statistics and Data Analysis, 55*, 2504–2515.

Kosinski, A. (1999). A procedure for the detection of multivariate outliers. *Computational Statistics & Data Analysis, 29*, 145–161.

Kowalski, C. J. (1972). On the effects of non-normality on the distribution of the sample product-moment correlation coefficient. *Applied Statistics, 21*, 1–12.

Kraemer, H. C., & Kupfer, D. J. (2006). Size of treatment effects and their importance to clinical research and practice. *Biological Psychiatry, 59*, 990–996.

Krasker, W. S. (1980). Estimation in linear regression models with disparate data points. *Econometrika, 48*, 1333–1346.

Krasker, W. S., & Welsch, R. E. (1982). Efficient bounded influence regression estimation. *Journal of the American Statistical, 77*, 595–604.

Krause, A., & Olson, M. (2002). *The basics of S-PLUS*. New York, NY: Springer-Verlag.

Krishnamoorthy, K., Lu, F., & Mathew, T. (2007). A parametric bootstrap approach for ANOVA with unequal variances: Fixed and random models. *Computational Statistics and Data Analysis, 51*, 5731–5742.

Kulasekera, K. B. (1995) Comparison of regression curves using quasi-residuals. *Journal of the American Statistical Association, 90*, 1085–1093.

Kulasekera, K. B., & Wang, J. (1997). Smoothing parameter selection for power optimality in testing of regression curves. *Journal of the American Statistical, 92*, 500 511.

Kulinskaya, E., & Staudte, R. G. (2006). Interval estimates of weighted effect sizes in the one-way heteroscedastic ANOVA. *British Journal of Mathematical and Statistical Psychology, 59*, 97–111.

Kulinskaya, E., Morgenthaler, S., & Staudte, R. G. (2010). Variance stabilizing the difference of two binomial proportions. *American Statistician, 64*, 350–356.

Künsch, H., Stefanski, L., & Carroll, R. (1989). Conditionally unbiased bounded influence estimation in general regression models, with applications to generalized linear models. *Journal of the American Statistical, 84*, 460–466.

Kuonen, D. (2005). Studentized bootstrap confidence intervals based on M-estimates. *Journal of Applied Statistics, 32*, 443–460.

Kuo, L., & Mallick, B. (1998). Variable selection for regression models. *Sankhya, Series B, 60*, 65–81.

Kuonen, D. (2005). Studentized bootstrap confidence intervals based on M-estimates. *Journal of Applied Statistics, 32*, 443–460.

Lambert, D. (1985). Robust two-sample permutation test. *Annals of Statistics, 13*, 606–625.

Laplace, P. S., de (1818). *Deuxieme supplement a la theorie analytique des probabilites*. Paris, France: Courcier.

Lax, D. A. (1985). Robust estimators of scale: Finite-sample performance in long-tailed symmetric distributions. *Journal of the American Statistical Association, 80*, 736–741.

Leger, C., & Romano, J. P. (1990a). Bootstrap adaptive estimation: The trimmed mean example. *The Canadian Journal of Statistics, 18*, 297–314.

Leger, C., & Romano, J. P. (1990b). Bootstrap choice of tuning parameters. *Annals of the Institute of Mathematical Statistics, 42*, 709–735.

Leger, C., Politis, D. N., & Romano, J. P. (1992). Bootstrap technology and applications. *Technometrics, 34*, 378–398.

Lee, H., & Fung, K. F. (1985). Behavior of trimmed F and sine-wave F statistics in one-way ANOVA. *Sankhya: The Indian Journal of Statistics, 47*, 186–201.

Li, G. (1985). Robust regression. In D. Hoaglin, F. Mosteller, & J. Tukey (Eds.), *Exploring data tables, trends and shapes*. New York, NY: Wiley.

Li, G., & Chen, Z. (1985). Projection-pursuit approach to robust dispersion and principal components: Primary theory and monte carlo. *Journal of the American Statistical Association, 80*, 759–766.

Li, G., Tiwari, R. C., & Wells, M. T. (1996). Quantile comparison functions in two-sample problems, with application to comparisons of diagonal markers. *Journal of the American Statistical Association, 91*, 689–698.

Lin, P., & Stivers, L. (1974). On the difference of means with missing values. *Journal of the American Statistical Association, 61*, 634–636.

Little, R. J. A., & Rubin, D. B. (2002). *Statistical analysis with missing data*. New York, NY: Wiley.

Liu, R. C., & Brown, L. D. (1993). Nonexistence of informative unbiased estimators in singular problems. *Annals of Statistics, 21*, 1–14.

Liu, R. Y. (1988). Bootstrap procedure under some non-i.d.d. models. *Annals of Statistics, 16*, 1696–1708.

Liu, R. Y. (1990). On a notion of data depth based on random simplices. *Annals of Statistics, 18*, 405–414.

Liu, R. Y., & Singh, K. (1993). A quality index based on data depth and multivariate rank tests. *Journal of the American Statistical Association, 88*, 252–260.

Liu, R. G., & Singh, K. (1997). Notions of limiting P values based on data depth and bootstrap. *Journal of the American Statistical Association, 92*, 266–277.

Liu, R. Y., Parelius, J. M., & Singh, K. (1999). Multivariate analysis by data depth: Descriptive statistics, graphics and inference. *Annals of Statistics, 27*, 783–858.

Livavcic-Rojas, P., Vallejo, G., & Fernández, P. (2010). Analysis of type I error rates of univariate and multivariate procedures in repeated measures designs. *Communications in Statistics–Simulation and Computation, 39*, 624–640.

Lix, L. M., & Keselman, H. J. (1998). To trim or not to trim: Tests of mean equality under heteroscedasticity and nonnormality. *Educational and Psychological Measurement, 58*, 409–429.

Lix, L. M, Keselman, H. J., & Hinds, A. M. (2005). Robust tests for the multivariate Behrens-Fisher problem. *Computer Methods and Programs Biomedicine, 77*, 129–139.

Locantore, N., Marron, J. S., Simpson, D. G., Tripoli, N., & Zhang, J. T. (1999). Robust principal components for functional data. *Test, 8*, 1–28.

Loh, W.-Y. (1987a). Calibrating confidence coefficients. *Journal of the American Statistical Association, 82*, 155–162.

Loh, W. Y. (1987b). Does the correlation coefficient really measure the degree of clustering around a line? *Journal of Educational Statistics, 12*, 235–239.

Lombard, F. (2005). Nonparametric confidence bands for a quantile comparison function. *Technometrics, 47*, 364–369.

Long, J. S., & Ervin, L. H. (2000). Using heteroscedasticity consistent standard errors in the linear regression model. *American Statistician, 54*, 217–224.

Luh, W. M., & Guo, J. H. (1999). A powerful transformation trimmed mean method for one-way fixed effects ANOVA model under non-normality and inequality of variance. *British Journal of Mathematical and Statistical Psychology, 52*, 303–320.

Luh, W., & Guo, J. (2000). Approximate transformation trimmed mean methods to the test of simple linear regression slope equality. *Journal of Applied Statistics, 27*, 843–857.

Lumley, T. (1996). Generalized estimating equations for ordinal data: A note on working correlation structures. *Biometrics, 52*, 354–361.

Lunneborg, C. E. (2000). *Data analysis by resampling: Concepts and applications*. Pacific Grove, CA: Duxbury.

Lopuhaä, H. P. (1989). On the relation between S-estimators and M-estimators of multivariate location and covariance. *Annals of Statistics, 17*, 1662–1683.

Lupuhaä, H. P. (1991). τ-estimators for location and scatter. *Canadian Journal of Statistics, 19*, 307–321.

Lupuhaä, H. P., & Rousseeuw, P. J. (1991). Breakdown points of affine equivariant estimators of multivariate location and covariance matrices. *Annals of Statistics, 19*, 229–248.

Lupuhaä, H. P. (1999). Asymptotics of reweighted estimators of multivariate location and scatter. *Annals of Statistics, 27*, 1638–1665.

Lyon, J. D., & Tsai, C.-L. (1996). A comparison of tests for homogeneity. *Statistician, 45*, 337–350.

MacKinnon, D. P. (2008). *Introduction to statistical mediation analysis.* Clifton, New Jersey: Psychology Press.

MacKinnon, D. P., Warsi, G., & Dwyer, J. H. (1995). A simulation study of mediated effect measures. *Multivariate Behavioral Research, 30*, 41–62.

MacKinnon, J. G., & White, H. (1985). Some heteroskedasticity consistent covariance matrix estimators with improved finite sample properties. *Journal of Econometrics, 29*, 53–57.

Mak, T. K. (1992). Estimation of parameters in heteroscedastic linear models. *Journal of the Royal Statistical Society, B, 54*, 649–655.

Mallows, C. L. (1973). Some comments on C_p. *Technometrics, 15*, 661–675.

Mallows, C. L. (1975). On some topics in robustness. Technical memorandum, Bell Telephone Laboratories.

Mammen, E. (1993). Bootstrap and wild bootstrap for high dimensional linear models. *Annals of Statistics, 21*, 255–285.

Marazzi, A. (1993). *Algorithms, routines, and S functions for robust statistics.* New York, NY: Chapman and Hall.

Mardia, K. V., Kent, J. T., & Bibby, J. M. (1979). *Multivariate analysis.* San Diego, CA: Academic Press.

Maritz, J. S., & Jarrett, R. G. (1978). A note on estimating the variance of the sample median. *Journal of the American Statistical Association, 73*, 194–196.

Markatou, M., & Hettmansperger, T. P. (1990). Robust bounded-influence tests in linear models. *Journal of the American Statistical Association, 85*, 187–190.

Markatou, M., Stahel, W. A., & Ronchetti, E. (1991). Robust M-type testing procedures for linear models. In W. Stahel & S. Weisberg (Eds.), *Directions in robust statistics diagnostics* (Pt. I, pp. 201–220). New York, NY: Springer-Verlag.

Markatou, M., & He, X. (1994). Bounded influence and high breakdown point testing procedures in linear models. *Journal of the American Statistical Association, 89*, 543–549.

Markowski, C. A., & Markowski, E. P. (1990). Conditions for the effectiveness of a preliminary test of variance. *American Statistician, 44*, 322–326.

Marmolejo-Ramos, F., & Tian, T. S. (2010). The shifting boxplot. A boxplot based on essential summary statistics around the mean. *International Journal of Psychological Research, 3*, 37–45.

Maronna, R. A. (1976). Robust M-estimators of multivariate location and scatter. *Annals of Statistics, 4*, 51–67.

Maronna, R. A. (2005). Principal components and orthogonal regression based on robust scales. *Technometrics, 47*, 264–273.

Maronna, R. A. (2011). Robust ridge regression for high-dimensional data. *Technometrics, 53*, 44–53.

Maronna, R. A., Martin, D. R., & Yohai, V. J. (2006). *Robust statistics: Theory and methods*. New York, NY: Wiley.

Maronna, R. A., & Morgenthaler, S. (1986). Robust regression through robust covariances. *Communications in Statistics–Theory and Methods, 15*, 1347–1365.

Maronna, R. A., & Yohai, V. J. (1993). Bias-robust estimates of regression based on projections. *Annals of Statistics, 21*, 965–990.

Maronna, R. A., & Yohai, V. J. (1995). The behavior of the Stahel-Donoho robust estimator. *Journal of the American Statistical Association, 90*, 330–341.

Maronna, R. A. & Yohai, V. J. (2010). Correcting MM estimates for "fat" data sets. *Journal of Computational Statistics and Data Analysis, 54*, 3168–3173.

Maronna, R. A., & Zamar, R. H. (2002). Robust estimates of location and dispersion for high-dimensional data sets. *Technometrics, 44*, 307–317

Maronna, R. A., & Zamar, R. H. (2010). Correcting MM estimates for "fat" data sets. *Computational Statistics and Data Analysis, 54*, 3168–3173.

Martin, R. A., & Zamar, R. H. (1993). Efficiency-constrained bias-robust estimation of location. *Annals of Statistics, 21*, 338–354.

Martin, R. A., Yohai, V. J., & Zamar, R. H. (1989). Asymptotically min-max bias robust regression. *Annals of Statistics, 17*, 1608–1630.

Masse, J.-C., & Plante, J.-F. (2003). A Monte Carlo study of the accuracy and robustness of ten bivariate location estimators. *Computational Statistics & Data Analysis, 42*, 1–26.

Maxwell, S. E., & Delaney, H. D. (1990). *Designing experiments and analyzing data: A model comparison perspective*. Belmont, CA: Wadsworth.

McKean, J. W., & Schrader, R. M. (1984). A comparison of methods for studentizing the sample median. *Communications in Statistics—Simulation and Computation, 13*, 751–773.

McKean, J. W., Sheather, S. J., & Hettmansperger, T. P. (1990). Regression diagnostics for rank-based methods. *Journal of the American Statistical Association, 85*, 1018–1029.

McKean, J. W., & Sheather, S. J. (1991). Small sample properties of robust analyses of linear models based on R-estimates: A survey. In W. Stahel & S. Weisberg (Eds.), *Directions in robust statistics and diagnostics* (Pt. II, pp. 1–20). New York, NY: Springer-Verlag.

McKean, J. W., Sheather, S. J., & Hettmansperger, T. P. (1993). The use and interpretation of residuals based on robust estimation. *Journal of the American Statistical Association, 88*, 1254–1263.

Mee, R. W. (1990). Confidence intervals for probabilities and tolerance regions based on a generalization of the Mann-Whitney statistic. *Journal of the American Statistical Association, 85*, 793–800.

Messer, K., & Goldstein, L. (1993). A new class of kernels for nonparametric curve estimation. *Annals of Statistics, 21*, 179–195.

Micceri, T. (1989). The unicorn, the normal curve, and other improbable creatures. *Psychological Bulletin, 105*, 156–166.

Mickey, M. R., Dunn, O. J., & Clark, V. (1967). Note on the use of stepwise regression in detecting outliers. *Computational Biomedical Research, 1*, 105–111.

Miles, D., & Mora, J. (2003). On the performance of nonparametric specification tests in regression models. *Computational Statistics & Data Analysis, 42*, 477–490.

Miller, A. J. (1990). *Subset selection in regression*. London, UK: Chapman and Hall.

Miller, R. G. (1966). *Simultaneous statistical inference*. New York, NY: McGraw-Hill.

Mills, J. E., Field, C. A., & Dupuis, D. J. (2002). Marginally specified generalized mixed models: A robust approach. *Biometrics, 58*, 727–734.

Mizera, I. (2002). On depth and deep points: A calculus. *Annals of Statistics, 30*, 1681–1736.

Molenberghs, G., & Verbeke, G. (2005). *Models for discrete longitudinal data*. New York, NY: Springer.

Montgomery, D. C., & Peck, E. A. (1992). *Introduction to linear regression analysis*. New York, NY: Wiley.

Mooney, C. Z., & Duval, R. D. (1993). *Bootstrapping: A nonparametric approach to statistical inference*. Newbury Park, CA: Sage.

Morgenthaler, S. (1992). Least-absolute deviations fit for generalized linear models. *Biometrika, 79*, 747–754.

Morgenthaler, S., & Tukey, J. W. (1991). *Configural polysampling*. New York, NY: Wiley.

Moser, B. K., Stevens, G. R., & Watts, C. L. (1989). The two-sample t-test versus Satterthwaite's approximate F test. *Communications in Statistics-Theory and Methods, 18*, 3963–3975.

Mosteller, F., & Tukey, J. W. (1977). *Data analysis and regression*. Reading, MA: Addison-Wesley.

Möttönen, J., & Oja, H. (1995). Multivariate spatial sign and rank methods. *Nonparametric Statistics, 5*, 201–213.

Muirhead, R. J. (1982). *Aspects of multivariate statistical theory*. New York, NY: Wiley.

Munk, A., & Dette, H. (1998). Nonparametric comparison of several regression functions: Exact and asymptotic theory. *Annals of Statistics, 26*, 2339–2386.

Munzel, U., & Brunner, E. (2000). Nonparametric test in the unbalanced multivariate one-way design. *Biometrical Journal, 42*, 837–854.

Myers, J. L. (1979). *Fundamentals of experimental design*. Boston, MA: Allyn and Bacon.

Nanayakkara, N., & Cressie, N. (1991). Robustness to unequal scale and other departures from the classical linear model. In W. Stahel & S. Weisberg (Eds.), *Directions in robust statistics and diagnostics* (Part II, pp. 65–113). New York, NY: Springer-Verlag.

Naranjo, J. D., & Hettmansperger, T. P. (1994). Bounded influence rank regression. *Journal of the Royal Statistical Society, B, 56*, 209–220.

Narula, S. C. (1987). The minimum sum of absolute errors regression. *Journal of Quality Technology, 19*, 37–45.

Nelder, J. A., & Mead, R. (1965). A simplex method for function minimization. *Computer Journal, 7*, 308–313.

Neuhäuser, M. (2003). A note on the exact test based on the Baumgartner-Weiss-Schindler statistic in the presence of ties. *Computational Statistics & Data Analysis, 42*, 561–568.

Neuhäuser, M., Lösch, C., & Jöckel, K.-H. (2007). The Chen-Luo test in case of heteroscedasticity. *Computational Statistics & Data Analysis, 51*, 5055–5060.

Neumeyer, N., & Dette, H. (2003). Nonparametric comparison of regression curves: An empirical process approach. *Annals of Statistics, 31*, 880–920.

Newcomb, S. (1896). A generalized theory of the combination of observations so as to obtain the best result. *American Journal of Mathematics, 8*, 343–366.

Newcombe, R. G. (1998). Interval estimation for the difference between independent proportions: Comparisons of eleven methods. *Statistics in Medicine, 17*, 873–890.

Newcombe, R. G. (2006a). Confidence intervals for an effect size measure based on the Mann-Whitney statistic. Part 1: General issues and tail-area-based methods. *Statistics in Medicine, 25*, 543–557.

Newcombe, R. G. (2006b). Confidence intervals for an effect size measure based on the Mann-Whitney statistic. Part 2: Asymptotic methods and evaluation. *Statistics in Medicine, 25*, 559–573.

Ng, M. (2009a). A comparison of bootstrap techniques for evaluating indirect effect. Unpublished technical report, Department of Psychology, University of Southern Californian.

Ng, M. (2009b). *Significance testing in regression analyses*. Unpublished doctoral dissertation, Department of Psychology, University of Southern California.

Ng, M., & Wilcox, R. R. (2009). Level robust methods based on the least squares regression estimator. *Journal of Modern and Applied Statistical Methods, 8*, 384–395.

Ng, M., & Wilcox, R. R. (2010a). Comparing the slopes of regression lines. *British Journal of Mathematical and Statistical Psychology, 63*, 319–340.

Ng, M., & Wilcox, R. R. (2010b). The small-sample efficiency of some recently proposed multivariate measures of location. *Journal of Modern and Applied Statistical Methods, 9*, 28–42.

Ng, M., & Wilcox, R. R. (2011). A comparison of two-stage procedures for testing least-squares coefficients under heteroscedasticity. *British Journal of Mathematical and Statistical Psychology, 64*, 244–258.

Ng, M., & Wilcox, R. R. (in press). A bootstrap method for comparing independent regression slopes. *British Journal of Mathematical and Statistical Psychology*.

Ng, P. (1996). An algorithm for quantile smoothing splines. *Computational Statistics & Data Analysis, 22*, 99–118.

Noh, M., & Lee, Y. (2007). Robust modeling for inference from generalized linear model classes. *Journal of the American Statistical Association, 102*, 1059–1072.

Oberhelman, D., & Kadiyala, R. (2007). A test for the equality of parameters of separate regression models in the presence of heteroskedasticity. *Communications in Statistics–Simulation and Computation, 36*, 99–121.

Olejnik, S., Li, J., Supattathum, S. and Huberty, C. J. (1997). Multiple testing and statistical power with modified Bonferroni procedures. *Journal of Educational and Behavioral Statistics, 22*, 389–406.

Olive, D. J. (2004). A resistant estimator of multivariate location and dispersion. *Computational Statistics & Data Analysis, 46*, 93–102.

Olive, D. J. (2010). The number of samples for resampling algorithms. Preprint www.math.siu.edu/olive/preprints.htm

Olive, D. J., & Hawkins, D. M. (2010). Robust multivariate location and dispersion. Preprint www.math.siu.edu/olive/preprints.htm

Olsson, D. M. (1974). A sequential simplex program for solving minimization problems. *Journal of Quality Technology, 6*, 53–57.

Olsson, D. M., & Nelson, L. S. (1975). The Nelder-Mead simplex procedure for function minimization. *Technometrics, 17*, 45–51.

Othman, A. R., Keselman, H. J., Wilcox, R. R., Fradette, K., & Padmanabhan, A. R. (2002). A test of symmetry. *Journal of Modern Applied Statistical Methods, 1*, 310–315.

Owen, A. B. (2001). *Empirical likelihood*. New York, NY: Chapman and Hall.

Owen, A. B. (2006). A robust hybrid of lasso and ridge regression. http://www-stat.stanford.edu/owen/reports/

Özdemir, A. F., & Wilcox, R. R. (2010). New results on the small-sample properties of some robust estimators. Technical Report, Department of Statistics, Dokuz Eylul University, Izmir, Turkey.

Özdemir, A. F., Wilcox, R. R., & Yildiztepe, E. (2010). Comparing measures of location: Some small-sample results when distributions differ in skewness. Technical Report, Department of Statistics, Dokuz Eylul University, Izmir, Turkey.

Parrish, R. S. (1990). Comparison of quantile estimators in normal sampling. *Biometrics, 46*, 247–257.

Patel, K. M., & Hoel, D. G. (1973). A nonparametric test for interaction in factorial experiments. *Journal of the American Statistical Association, 68*, 615–620.

Patel, K. R., Mudholkar, G. S., & Fernando, J. L. I. (1988). Student's t approximations for three simple robust estimators. *Journal of the American Statistical Association, 83*, 1203–1210.

Pearson, E. S., & Please, N. W. (1975). Relation between the shape of the population distribution and the robustness of four simple statistics. *Biometrika, 62*, 223–241.

Pedersen, W. C., Miller, L. C., Putcha-Bhagavatula, A. D. and Yang, Y. (2002). Evolved sex differences in sexual strategies: The long and the short of it. *Psychological Science, 13*, 157–161.

Peña, D., & Prieto, F. J. (2001). Multivariate outlier detection and robust covariance matrix estimation. *Technometrics, 43*, 286–299.

Peña, D., & Yohai, V. (1999). A fast procedure for outlier diagnostics in large regression problems. *Journal of the American Statistical Association, 94*, 434–445.

Peña, E. A., Habiger, J. D., & Wu, W. (2010). Power-enhanced multiple decision functions controlling family-wise error and false discovery rates. *Annals of Statistics, 39*, 556–583.

Pesarin, F. (2001). *Multivariate permutation tests*. New York, NY: Wiley.

Pison, G., Van Aelst, S., & Willems, G. (2002). Small-Sample Corrections for LTS and MCD. *Metrika, 55*, 111–123.

Politis, D. N., & Romano, J. P. (1997). Multivariate density estimation with general flat-top kernels of infinite order. *Journal of Multivariate Analysis, 68*, 1–25.

Poon, W.-Y., Lew, S-F., & Poon, Y. S. (2000). A local influence approach to identifying multiple outliers. *British Journal of Mathematical and Statistical Psychology, 53*, 255–273.

Potthoff, R. F., & Roy, S. N. (1964). A generalized multivariate analysis of variance model useful especially for growth curve problem. *Biometrika, 51*, 313–326.

Pratt, J. W. (1964). Robustness of some procedures for the two-sample location problem. *Journal of the American Statistical Association, 59*, 665–680.

Pratt, J. W. (1968). A normal approximation for binomial, F, beta, and other common, related tail probabilities, I. *Journal of the American Statistical Association, 63*, 1457–1483.

Preacher, K. J., Rucker, D. D. & Hayes, A. F. (2007). Addressing moderated mediation hypotheses: Theory, methods, and prescriptions. *Multivariate Behavioral Research, 42*, 185–227.

Pregibon, D. (1987). Resistant fits for some commonly used logistic models with medical applications. *Biometrics, 38*, 485–498.

Price, R. M., & Bonett, D. G. (2001). Estimating the variance of the median. *Journal of Statistical Computation and Simulation, 68*, 295–305.

R Development Core Team. (2010). *R: A language and environment for statistical computing.* R Foundation for Statistical Computing, Vienna, Austria. ISBN 3-900051-07-0, Retrieved from http://www.R-project.org.

Racine, J., & MacKinnon, J. G. (2007a). Simulation-based tests than can use any number of simulations. *Communications in Statistics–Simulation and Computation, 36*, 357–365.

Racine, J., & MacKinnon, J. G. (2007b). Inference via kernel smoothing of bootstrap P values. *Computational Statistics & Data Analysis, 51*, 5949–5957.

Radchenko, P., & James, G. M. (2011). Improved variable selection with Forward-LASSO adaptive shrinkage. *Annals of Applied Statistics, 5*, 427–448.

Raine, A., Buchsbaum, M., & LaCasse, L. (1997). Brain abnormalities in murderers indicated by positron emission tomography. *Biological Psychiatry, 42*, 495–508.

Ramsey, P. H. (1980). Exact type I error rates for robustness of Student's t test with unequal variances. *Journal of Educational Statistics, 5*, 337–349.

Randal, J. A. (2008). A reinvestigation of robust scale estimation in finite samples. *Computational Statistics & Data Analysis, 52*, 5014–5021.

Randles, R. H., & Wolfe, D. A. (1979). *Introduction to the theory of nonparametric statistics.* New York, NY: Wiley.

Rao, C. R. (1948). Tests of significance in multivariate analysis. *Biometrika, 35*, 58–79.

Rao, P. S. R. S., Kaplan, J., & Cochran, W. G. (1981). Estimators for one-way random effects model with unequal error variances. *Journal of the American Statistical Association, 76*, 89–97.

Rasch, D., Teuscher, F., & Guiard, V. (2007). How robust are tests for two independent samples? *Journal of Statistical Planning and Inference, 137*, 2706–2720.

Rasmussen, J. L. (1989). Data transformation, type I error rate and power. *British Journal of Mathematical and Statistical Psychology, 42*, 203–211.

Reed, J. F. (1998). Contributions to adaptive estimation. *Journal of Applied Statistics, 25*, 651–669.

Reed, J. F. (2009). Improved confidence intervals for the difference between two proportions. *Journal of Modern Applied Statistical Methods, 8*, 208–214.

Reed, J. F., & Stark, D. B. (1996). Hinge estimators of location: Robust to asymmetry. *Computer Methods and Programs in Biomedicine, 49*, 11–17.

Reiczigel, J., Abonyi-Tóth, Z., & Singer, J. (2008). An exact confidence set for two binomial proportions and exact unconditional confidence intervals for the difference and ratio of proportions. *Computational Statistics & Data Analysis, 52*, 5046–5053.

Reider, H. (1994). *Robust asymptotic statistics*. New York, NY: Springer-Verlag.

Renaud, O., & Victoria-Feser, M.-P. (2010). A robust coefficient of determination for regression. *Journal of Statistical Planning and Inference, 140*, 1852–1862.

Rivest, L.-P. (1994). Statistical properties of Winsorized means for skewed distributions. *Biometrika, 81*, 373–383.

Rizzo, M. L., & Székely, G. (2010). DISCO analysis: A nonparametric extension of analysis of variance. *Annals of Applied Statistics, 4*, 1034–1055.

Robinson, J., Ronchetti, E., & Young, G. A. (2003). Saddlepoint approximations and tests based on multivariate M-estimates. *Annals of Statistics, 31*, 1154–1169.

Robinson, P. M. (1987). Asymptotically efficient estimation in the presence of heteroskedasticity of unknown form. *Econometrica, 55*, 875–891.

Rocke, D. M., & Woodruff, D. L. (1993). Computation of robust estimates of multivariate location and shape. *Statistica Neerlandica, 47*, 27–42.

Rocke, D. M. (1996). Robustness properties of S-estimators of multivariate location and shape in high dimension. *Annals of Statistics, 24*, 1327–1345.

Rocke, D. M., & Woodruff, D. L. (1996). Identification of outliers in multivariate data. *Journal of the American Statistical Association, 91*, 1047–1061.

Rom, D. M. (1990). A sequentially rejective test procedure based on a modified Bonferroni inequality. *Biometrika, 77*, 663–666.

Romanazzi, M. (1997). A schematic plot for bivariate data. *Student, 2*, 149–158.

Romano, J. P. (1990). On the behavior of randomization tests without a group invariance assumption. *Journal of the American Statistical Association, 85*, 686–692.

Rosenberger, J. L., & Gasko, M. (1983). Comparing location estimators: Trimmed means, medians, and trimean. In D. Hoaglin, F. Mosteller, & J. Tukey (Eds.), *Understanding robust and exploratory data analysis* (pp. 297–336). New York, NY: Wiley.

Rosenthal, R., & Jacobson, L. (1968). *Pygmalion in the classroom: Teacher expectations and Pupil's intellectual development*. New York, NY: Holt, Rinehart and Winston.

Rosmond, R., Dallman, M. F., & Björntorp, P. (1998). Stress-related cortisol secretion in men: Relationships with abdominal obesity and endocrine, metabolic and hemodynamic abnormalities. *Journal of Clinical Endocrinology & Metabolism, 83*, 1853–1859.

Rousseeuw, P. J. (1984). Least median of squares regression. *Journal of the American Statistical Association, 79*, 871–880.

Rousseeuw, P. J., & Christmann, A. (2003). Robustness against separation and outliers in logistic regression. *Computational Statistics & Data Analysis, 43*, 315–332.

Rousseeuw, P. J., & Croux, C. (1993). Alternative to the median absolute deviation. *Journal of the American Statistical Association, 88*, 1273–1283.

Rousseeuw, P. J., & Hubert, M. (1999). Regression depth. *Journal of the American Statistical Association, 94*, 388–402.

Rousseeuw, P. J., & Leroy, A. M. (1987). *Robust regression & outlier detection*. New York, NY: Wiley.

Rousseeuw, P. J., & Ruts, I. (1996). AS 307: Bivariate location depth. *Applied Statistics, 45*, 516–526.

Rousseeuw, P. J., & Struyf, A. (1998). Computing location depth and regression depth in higher dimensions. *Statistics and Computing, 8*, 193–203.

Rousseeuw, P. J., Van Aelst, S., Van Driessen, K., & Agulló, J. (2004). Robust multivariate regression. *Technometrics, 46*, 293–305.

Rousseeuw, P. J., & van Driessen, K. (1999). A fast algorithm for the minimum covariance determinant estimator. *Technometrics, 41*, 212–223.

Rousseeuw, P. J., & van Zomeren, B. C. (1990). Unmasking multivariate outliers and leverage points. *Journal of the American Statistical Association, 85*, 633–639.

Rousseeuw, P. J., & Verboven, S. (2002). Robust estimation in very small samples. *Computational Statistics & Data Analysis, 40*, 741–758.

Rousseeuw, P. J., & Yohai, V. (1984). Robust regression by means of S-estimators. *Nonlinear Time Series Analysis. Lecture Notes in Statistics, 26*, 256–272. New York, NY: Springer.

Rubin, A. S. (1983). The use of weighted contrasts in analysis of models with heterogeneity of variance. *Proceedings of the business and economics statistics section, American Statistical Association*, Alexandria, VA, 347–352.

Ruppert, D. (1992). Computing S estimators for regression and multivariate location/dispersion. *Journal of Computational and Graphical Statistics, 1*, 253–270.

Ruppert, D., & Carroll, R. J. (1980). Trimmed least squares estimation in the linear model. *Journal of the American Statistical Association, 75*, 828–838.

Ruppert, D., Wand, M. P., & Carroll, R. J. (2003). *Semiparametric regression*. Cambridge, MA: Cambridge University Press.

Rust, S. W., & Fligner, M. A. (1984). A modification of the Kruskal-Wallis statistic for the generalized Behrens-Fisher problem. *Communications in Statistics–Theory and Methods, 13*, 2013–2027.

Rutherford, A. (1992). Alternatives to traditional analysis of covariance. *British Journal of Mathematical and Statistical Psychology, 45*, 197–224.

Ryan, B. F., Joiner, B. L., & Ryan, T. A. (1985). *MINITAB handbook*. Boston, MA: PWS-KENT.

Ryu, E., & Agresti, A. (2008). Modeling and inference for an ordinal effect size measure. *Statistics in Medicine, 27*, 1703–1717.

Sakaori, F. (2002). Permutation test for equality of correlation coefficients in two populations. *Communications in Statistics–Simulation and Computation, 31*, 641–652.

Salibian-Barrera, M., & Zamar, R. H. (2002). Bootstrapping robust estimates of regression. *Annals of Statistics, 30*, 556–582.

Salibin-Barrera, M., Van Aelst, S., & Willems, G. (2006). PCA based on multivariate MM-estimators with fast and robust bootstrap. *Journal of the American Statistical Association, 101*, 1198–1211.

Salk, L. (1973). The role of the heartbeat in the relations between mother and infant. *Scientific American, 235*, 26–29.

Samarov, A. M. (1993). Exploring regression structure using nonparametric functional estimation. *Journal of the American Statistical Association, 88*, 836–847.

Santner, T. J., Pradhan, V., Senchaudhuri, P., Mehta, C. R., & Tamhane, A. (2007). Small-sample comparisons of confidence intervals for the difference of two independent binomial proportions. *Computational Statistics & Data Analysis, 51*, 5791–5799.

Sawilowsky, S. S. (2002). The probable difference between two means when $\sigma_1 \neq \sigma_2$: The Behrens-Fisher problem. *Journal of Modern Applied Statistical Methods, 1*, 461–472.

Sawilowsky, S. S., & Blair, R. C. (1992). A more realistic look at the robustness and type II error properties of the t test to departures from normality. *Psychological Bulletin, 111,* 352–360.

Saunders, D. R. (1956). Moderator variables in prediction. *Educational and Psychological Measurement, 16,* 209–222.

Scheffé, H. (1959). *The analysis of variance.* New York, NY: Wiley.

Schlölkopf, S., Smola, Λ., & Müller, K. R. (1998). Nonlinear component analysis as a kernel eigenvalue problem. *Neural Computation, 10,* 1299–1319.

Schnys, M., Haesbroeck, G., & Critchley, F. (2010). RelaxMCD: Smooth optimisation for the minimum covariance determinant estimator. *Computational Statistics & Data Analysis, 54,* 843–857.

Scholz, F. W. (1978). Weighted median regression estimates. *Annals of Statistics, 6,* 603–609.

Schrader, R. M., & Hettmansperger, T. P. (1980). Robust analysis of variance. *Biometrika, 67,* 93–101.

Schroër, G., & Trenkler, D. (1995). Exact and randomization distributions of Kolmogorov-Smirnov tests two or three samples. *Computational Statistics and Data Analysis, 20,* 185–202.

Scott, D. W. (1979). On optimal and data-based histograms. *Biometrika, 66,* 605–610.

Scott, D. W. (1992). *Multivariate density estimation. theory, practice, and visualization.* New York, NY: Wiley.

Sen, P. K. (1968). Estimate of the regression coefficient based on Kendall's tau. *Journal of the American Statistical Association, 63,* 1379–1389.

Serfling, R. J. (1980). *Approximation theorems of mathematical statistics.* New York, NY: Wiley.

Serneels, S., & Verdonck, T. (2008). Principal components analysis for data containing outliers and missing elements. *Computational Statistics and Data Analysis, 52,* 1712–1727.

Serroyen, J., Molenbergs, G., Verbeke, G., & Davidian, M. (2009). Nonlinear models for longitudinal data. *American Statistician, 63,* 378–388.

Sfakianakis, M. E., & Verginis, D. G. (2008). A new family of nonparametric quantile estimators. *Communications in Statistics–Simulation and Computation, 37,* 337–345.

Shao, J. (1996). Bootstrap model selection. *Journal of the American Statistical Association,* *91*, 655–665.

Shao, J., & Tu, D. (1995). *The jackknife and the bootstrap*. New York, NY: Springer-Verlag.

Sheather, S. J., & Marron, J. S. (1990). Kernel quantile estimators. *Journal of the American Statistical Association, 85*, 410–416.

Sheather, S. J., & McKean, J. W. (1987). A comparison of testing and confidence intervals for the median. *Statistical Probability Letters, 6*, 31–36.

Shoemaker, L. H. (2003). Fixing the F test for equal variances. *American Statistician, 57*, 105–114.

Shoemaker, L. H., & Hettmansperger, T. P. (1982). Robust estimates and tests for the one- and two-sample scale models. *Biometrika, 69*, 47–54.

Sievers, G. L. (1978). Weighted rank statitics for simple linear regression. *Journal of the American Statistical Association, 73*, 628–631.

Signorini, D. F., & Jones, M. C. (2004). Kernel estimators for univariate binary regression. *Journal of the American Statistical Association, 99*, 119–126.

Silvapulle, M. J. (1992). Robust Wald-type tests of one-sided hypotheses in the linear model. *Journal of the American Statistical Association, 87*, 156–161.

Silverman, B. W. (1986). *Density estimation for statistics and data analysis*. New York, NY: Chapman and Hall.

Simonoff, J. S. (1996). *Smoothing methods in statistics*. New York, NY: Springer.

Simpson, D. G., Ruppert, D., & Carroll, R. J. (1992). Bounded-influence regression on one-step GM estimates and stability of inferences in linear regression. *Journal of the American Statistical Association, 87*, 439–450.

Singh, K. (1998). Breakdown theory for bootstrap quantiles. *Annals of Statistics, 26*, 1719–1732.

Sinha, S. K. (2004). Robust analysis of generalized linear mixed models. *Journal of the American Statistical Association, 99*, 451–460.

Smith, G., & Cambell, F. (1980). A critique of some ridge regression methods (with discussion). *Journal of the American Statistical Association, 70*, 74–103.

Snedecor, G. W., & Cochran, W. (1967). *Statistical methods* (6th ed.). Ames, IA: University Press.

Snow, R. E. (1995). Pygmalion and intelligence? *Current Directions in Psychological Science, 4*, 169–172.

Sockett, E. B., Daneman, D., Carlson, C., & Ehrich, R. M. (1987). Factors affecting and patterns of residual insulin secretion during the first year of type I (insulin dependent) diabetes mellitus in children. *Diabetes, 30*, 453–459.

Srihera, R., & Stute, W. (2010). Nonparametric comparison of regression functions. *Journal of Multivariate Analysis, 101*, 2039–2059.

Srivastava, D. K., Pan, J., Sarkar, I., & Mudholkar, G. S. (2010). Robust winsorized regression using bootstrap approach. *Communications in Statistics–Simulation and Computation, 39*, 45–67.

Srivastava, M. S., & Awan, H. M. (1984). On the robustness of the correlation coefficient in sampling from a mixture of two bivariate normals. *Communications in Statistics–Theory and Methods, 13*, 371–382.

Stahel, W. A. (1981). Breakdown of covariance estimators. Research report 31, Fachgruppe für Statistik, E.T.H. Zürich.

Staudte, R. G., & Sheather, S. J. (1990). *Robust estimation and testing*. New York, NY: Wiley.

Steele, C. M., & Aronson, J. (1995). Stereotype threat and the intellectual test performance of African Americans. *Journal of Personality and Social Psychology, 69*, 797–811.

Stefanski, L., Carroll, D., & Ruppert, D. (1986). Optimally bounded score functions for generalized linear models with applications to logistic regression. *Biometrika, 73*, 413–424.

Stein, C. (1945). A two-sample test for a linear hypothesis whose power is independent of the variance. *Annals of Statistics, 16*, 243–258.

Sterne, T. E. (1954). Some remarks on confidence or fiducial limits. *Biometrika, 41*, 275–278.

Stigler, S. M. (1973). Simon Newcomb, Percy Daniel, and the history of robust estimation 1885–1920. *Journal of the American Statistical Association, 68*, 872–879.

Storer, B. E., & Kim, C. (1990). Exact properties of some exact test statistics for comparing two binomial proportions. *Journal of the American Statistical Association, 85*, 146–155.

Stromberg, A. J. (1993). Computation of high breakdown nonlinear regression parameters. *Journal of the American Statistical Association, 88*, 237–244.

Struyf, A., & Rousseeuw, P. J. (2000). High-dimensional computation of the deepest location. *Computational Statistics & Data Analysis, 34,* 415–426.

Stuart, V. M. (2009). *Exploring robust alternatives to Pearson's r through secondary analysis of published behavioral science data.* Unpublished PhD dissertation, Department of Psychology, University of Southern California.

Stute, W., Gonzalez Manteiga, W. G., & Presedo Quindimil, M. P. (1998). Bootstrap approximations in model checks for regression. *Journal of the American Statistical Association, 93,* 141–149.

Tableman, M. (1990). Bounded-influence rank regression: A one-step estimator based on Wilcoxon scores. *Journal of the American Statistical Association, 85,* 508–513.

Tableman, M. (1994). The asymptotics of the least trimmed absolute deviation (LTAD) estimators. *Statistics and Probability Letters, 19,* 387–398.

Talib, B. A., & Midi, H. (2009). Robust estimator to deal with regression models having both continuous and categorical regressors: A simulation study. *Malaysian Journal of Mathematical Sciences, 3,* 161–181.

Talwar, P. P. (1991). A simulation study of some non-parametric regression estimators. *Computational Statistics & Data Analysis, 15,* 309–327.

Tamura, R., & Boos, D. (1986). Minimum Hellinger distance estimation for multivariate location and covariance. *Journal of the American Statistical Association, 81,* 223–229.

Tan, W. Y. (1982). Sampling distributions and robustness of t, F, and variance-ratio of two samples and ANOVA models with respect to departure from normality. *Communications in Statistics–Theory and Methods, 11,* 2485–2511.

Theil, H. (1950). A rank-invariant method of linear and polynomial regression analysis. *Indagationes Mathematicae, 12,* 85–91.

Thompson, G. L., & Amman, L. P. (1990). Efficiencies of interblock rank statistics for repeated measures designs. *Journal of the American Association, 85,* 519–528.

Thomson, A., & Randall-Maciver, R. (1905). *Ancient races of the Thebaid.* Oxford, UK: Oxford University Press.

Tian, T., & Wilcox, R. R. (2007). A comparison of two ranked tests for repeated measures designs. *Journal of Modern and Applied Statistical Methods, 6,* 331–335.

Tibshirani, R. (1996). Regression shrinkage and selection via the lasso. *Journal of the Royal Statistical Society, B, 58,* 267–288.

Tingley, M., & Field, C. (1990). Small-sample confidence intervals. *Journal of the American Statistical Association, 85*, 427–434.

Todorov, V., & Filzmoser, P. (2010). Robust statistic for the one-way MANOVA *Computational Statistics and Data Analysis, 54*, 37–48.

Tomarken, A., & Serlin, R. (1986). Comparison of ANOVA alternatives under variance heterogeneity and specific noncentrality structures. *Psychological Bulletin, 99*, 90–99.

Troendle, J. F. (1990). A stepwise resample method of multiple hypothesis testing. *Journal of the American Statistical Association, 90*, 370–378.

Tukey, J. W. (1960). A survey of sampling from contaminated normal distributions. In I. Olkin, S. Ghurye, W. Hoeffding, W. Madow, & H. Mann (Eds.), *Contributions to probability and statistics* (pp. 448–503). Stanford, CA: Stanford University Press.

Tukey, J. W., & McLaughlin, D. H. (1963). Less vulnerable confidence and significance procedures for location based on a single sample: Trimming/Winsorization 1. *Sankhya A, 25*, 331–352.

Tukey, J. W. (1975). Mathematics and the picturing of data. *Proceedings of the international congress of mathematicians, 2*, 523–531.

Tukey, J. W. (1977). *Exploratory data analysis*. Reading, MA: Addison-Wesley.

Tyler, D. E. (1991). Some issues in the robust estimation of multivariate location and scatter. In W. Stahel & S Weisberg (Eds.), *Directions in Robust statistics and diagnostics* (Pt. II, pp. 327–336). New York, NY: Springer–Verlag.

Tyler, D. E. (1994). Finite sample breakdown points of projection based multivariate location and scatter statistics. *Annals of Statistics, 22*, 1024–1044.

Van Aelst, S., & Willems, G. (2011). Robust and efficient one-way MANOVA tests. *Journal of the American Statistical Association, 106*, 706–718.

Vargha A., & Delaney, H. D. (2000). A critique and improvement of the CL common language effect size statistics of McGraw and Wong. *Journal of Educational and Behavioral Statistics, 25*, 101–132.

Velleman, P. F., & Hoaglin, D. C. (1981). *Applications, basics, and computing of exploratory data analysis*. Boston, MA: Duxbury Press.

Venables, W. N., & Ripley, B. D. (2000). *S programming*. New York, NY: Springer.

Venables, W. N., & Ripley, B. D. (2002). *Modern applied statistics with S*. New York, NY: Springer.

Venables, W. N., & Smith, D. M. (2002). *An introduction to R*. Bristol, UK: Network Theory Ltd.

Verboon, P. (1993). Robust nonlinear regression analysis. *British Journal of Mathematical and Statistical Psychology, 46*, 77–94.

Verzani, J. (2004). *Using R for introductory statistics*. Boca Raton, FL: CRC Press.

Vexler, A., Liu, S., Kang, L., & Hutson, A. D. (2009). Modifications of the empirical likelihood interval estimation with improved coverage probabilities. *Communications in Statistics– Simulation and Computation, 38*, 2171–2183.

Victoroff, J., Quota, S., Adelman, J. R., Celinska, M. A., Stern, N., Wilcox, R., & Sapolsky, R. M. (2010). Support for Religio-Political aggression among teenaged boys in Gaza: Pt. I: Psychological findings. *Aggressive Behavior, 36*, 219–231.

Wand, M. P., & Jones, M. C. (1995). *Kernel smoothing*. London, UK: Chapman & Hall.

Wang, F. T., & Scott, D. W. (1994). The L_1 method for robust nonparametric regression. *Journal of the American Statistical Association, 89*, 65–76.

Wang, H., & Leng, C. (2007). Unified LASSO estimation by least squares approximation. *Journal of the American Statistics Association, 102*, 1039–1048.

Wang, L., & Qu, A. (2007). Robust tests in regression models with omnibus alternatives and bounded influence. *Journal of the American Statistical Association, 102*, 347–358.

Wang, N., & Raftery, A. E. (2002). Nearest-neighbor variance estimation (NNVE): Robust covariance estimation via nearest-neighbor cleaning. *Journal of the American Statistical Association, 97*, 994–1006.

Welch, B. L. (1938). The significance of the difference between two means when the population variances are unequal. *Biometrika, 29*, 350–362.

Welch, B. L. (1951). On the comparison of several mean values: An alternative approach. *Biometrika, 38*, 330–336.

Welsch, R. E. (1980). Regression sensitivity analysis and bounded-influence estimation. In J. Kmenta & J. B. Ramsey (Eds.), *Evaluation of econometric models* (pp. 153–167). New York, NY: Academic Press.

Welsh, A. H. (1987a). One-Step L-estimators for the linear model. *The Annals of Statistics, 15*, 626–641.

Welsh, A. H. (1987b). The trimmed mean in the linear model (with discussion). *The Annals of Statistics, 15*, 20–45.

Welsh, A. H., Carroll, R. J., & Ruppert, D. (1994). Fitting heteroscedastic regression models. *Journal of the American Statistical Association, 89*, 100–116.

Welsh, A. H., & Morrison, H. L. (1990). Robust L estimation of scale with an application in astronomy. *Journal of the American Statistical Association, 85*, 729–743.

Westfall, P. H., & Young, S. S. (1993). *Resampling based multiple testing*. New York, NY: Wiley.

Wilcox, R. R. (1983). A table of percentage points of the range of independent t variables. *Technometrics, 25*, 201–204.

Wilcox, R. R. (1986). Improved simultaneous confidence intervals for linear contrasts and regression parameters. *Communications in Statistics–Simulation and Computation, 15*, 917–932.

Wilcox, R. R. (1987a). New designs in analysis of variance. *Annual Review of Psychology, 38*, 29–60.

Wilcox, R. R. (1987b). Pairwise comparisons of J independent regression lines over a finite interval, simultaneous comparison of their parameters, and the Johnson-Neyman technique. *British Journal of Mathematical and Statistical Psychology, 40*, 80–93.

Wilcox, R. R. (1989). Percentage points of a weighted Kolmogorov-Smirnov statistics. *Communications in Statistics–Simulation and Computation, 18*, 237–244.

Wilcox, R. R. (1990a). Comparing the means of two independent groups. *Biometrical Journal, 32*, 771–780.

Wilcox, R. R. (1990b). Determining whether an experimental group is stochastically larger than a control. *British Journal of Mathematical and Statistical Psychology, 43*, 327–333.

Wilcox, R. R. (1990c). Comparing variances and means when distributions have non-identical shapes. *Communications in Statistics–Simulation and Computation, 19*, 155–173.

Wilcox, R. R. (1991a). Bootstrap inferences about the correlation and variance of paired data. *British Journal of Mathematical and Statistical Psychology, 44*, 379–382.

Wilcox, R. R. (1991b). Testing whether independent treatment groups have equal medians. *Psychometrika, 56*, 381–396.

Wilcox, R. R. (1992a). Comparing one-step M-estimators of location corresponding to two independent groups. *Psychometrika, 57*, 141–154.

Wilcox, R. R. (1992b). Comparing the medians of dependent groups. *British Journal of Mathematical and Statistical Psychology, 45*, 151–162.

Wilcox, R. R. (1993a). Comparing the biweight midvariances of two independent groups. *The Statistician, 42*, 29–35.

Wilcox, R. R. (1993b). Some results on a Winsorized correlation coefficient. *British Journal of Mathematical and Statistical Psychology, 46*, 339–349.

Wilcox, R. R. (1993c). Analyzing repeated measures or randomized block designs using trimmed means. *British Journal of Mathematical and Statistical Psychology, 46*, 63–76.

Wilcox, R. R. (1993d). Comparing one-step M-estimators of location when there are more than two groups. *Psychometrika, 58*, 71–78.

Wilcox, R. R. (1994a). Some results on the Tukey-McLaughlin and Yuen methods for trimmed means when distributions are skewed. *Biometrical Journal, 36*, 259–306.

Wilcox, R. R. (1994b). A one-way random effects model for trimmed means. *Psychometrika, 59*, 289–306.

Wilcox, R. R. (1994c). Estimating Winsorized correlations in a univariate or bivariate random effects model. *British Journal of Mathematical and Statistical Psychology, 47*, 167–183.

Wilcox, R. R. (1994d). The percentage bend correlation coefficient. *Psychometrika, 59*, 601–616.

Wilcox, R. R. (1994e). Computing confidence intervals for the slope of the biweight midregression and Winsorized regression lines. *British Journal of Mathematical and Statistical Psychology, 47*, 355–372.

Wilcox, R. R. (1995a). Comparing two independent groups via multiple quantiles. *The Statistician, 44*, 91–99.

Wilcox, R. R. (1995b). Comparing the deciles of two dependent groups. Unpublished technical report, Department of Psychology, University of Southern California.

Wilcox, R. R. (1995c). Three multiple comparison procedures for trimmed means. *Biometrical Journal, 37*, 643–656.

Wilcox, R. R. (1995d). Some small-sample results on a bounded influence rank regression method. *Communications in Statistics–Theory and Methods, 24*, 881–888.

Wilcox, R. R. (1995e). A regression smoother for resistant measures of location. *British Journal of Mathematical and Statistical Psychology, 48*, 189–204.

Wilcox, R. R. (1995f). Simulation results on solutions to the multivariate Behrens-Fisher problem via trimmed means. *Statistician, 44*, 213–225.

Wilcox, R. R. (1996a). *Statistics for the social sciences*. San Diego, CA: Academic Press.

Wilcox, R. R. (1996b). A note on testing hypotheses about trimmed means. *Biometrical Journal, 38*, 173–180.

Wilcox, R. R. (1996c). Confidence intervals for the slope of a regression line when the error term has non-constant variance. *Computational Statistics & Data Analysis, 22*, 89–98.

Wilcox, R. R. (1996d). Estimation in the simple linear regression model when there is heteroscedasticity of unknown form. *Communications in Statistics–Theory and Methods, 25*, 1305–1324.

Wilcox, R. R. (1996e). Confidence intervals for two robust regression lines with a heteroscedastic error term. *British Journal of Mathematical and Statistical Psychology, 49*, 163–170.

Wilcox, R. R. (1996f). Comparing the variances of dependent groups. Unpublished technical report, Department of Psychology, University of Southern California.

Wilcox, R. R. (1996g) Testing hypotheses about regression parameters when the error term is heteroscedastic. Unpublished technical report, Department of Psychology, University of Southern California.

Wilcox, R. R. (1996h). Simulation results on performing pairwise comparisons of trimmed means. Unpublished technical report, Department of Psychology, University of Southern California.

Wilcox, R. R. (1997a). Pairwise comparisons using trimmed means or M-estimators when working with dependent groups. *Biometrical Journal, 39*, 677–688.

Wilcox, R. R. (1997b). ANCOVA based on comparing a robust measure of location at empirically determined design points. *British Journal of Mathematical and Statistical Psychology, 50*, 93–103.

Wilcox, R. R. (1997c). Comparing the slopes of two independent regression lines when there is complete heteroscedasticity. *British Journal of Mathematical and Statistical Psychology, 50*, 309–317.

Wilcox, R. R. (1997d). Tests of independence and zero correlations among p random variables. *Biometrical Journal, 39*, 183–193.

Wilcox, R. R. (1998a). A note on the Theil-Sen regression estimator when the regressor is random and the error term is heteroscedastic. *Biometrical Journal, 40*, 261–268.

Wilcox, R. R. (1998b). Simulation results on extensions of the Theil-Sen regression estimator. *Communications in Statistics–Simulation and Computation, 27*, 1117–1126.

Wilcox, R. R. (1999). Comments on Stute, Manteiga, and Quindimil. *Journal of the American Statistical Association, 94*, 659–660.

Wilcox, R. R. (2001a). Pairwise comparisons of trimmed means for two or more groups. *Psychometrika, 66*, 343–356.

Wilcox, R. R. (2001b). Comments on long and Ervin. *American Statistician, 55*, 374–375.

Wilcox, R. R. (2001c). Rank-based tests for interactions in a two-way design when there are ties. *British Journal of Mathematical and Statistical, 53*, 145–153.

Wilcox, R. R. (2002). Comparing the variances of independent groups. *British Journal of Mathematical and Statistical Psychology, 55*, 169–176.

Wilcox, R. R. (2003a). Approximating Tukey's depth. *Communications in Statistics–Simulations and Computations, 32*, 977–985.

Wilcox, R. R. (2003b). Two-sample, bivariate hypothesis testing methods based on Tukey's depth. *Multivariate Behavioral Research, 38*, 225–246 .

Wilcox, R. R. (2003c). *Applying contemporary statistical techniques*. San Diego, CA: Academic Press.

Wilcox, R. R. (2003d). Inferences based on multiple skipped correlations. *Computational Statistics & Data Analysis, 44*, 223–236.

Wilcox, R. R. (2003e). Testing the hypothesis that a regression model is additive. Unpublished technical report, Department of Psychology, University of Southern California.

Wilcox, R. R. (2003f). Multiple hypothesis testing based on the ordinary least squares estimator when there is heteroscedasticity. *Educational and Psychological Measurement, 63*, 758–764.

Wilcox, R. R. (2004a). Extension of Hochberg's two-stage multiple comparison method. In N. Mukhopadhyay, S. Datta, & S. Chattopoadhyay (Eds.), *Applied sequential methodologies*: *Real world examples with data analysis*. New York, NY: Dekker.

Wilcox, R. R. (2004b). An extension of Stein's two-stage method to pairwise comparisons among dependent groups based on trimmed means. *Sequential Analysis, 23*, 63–74.

Wilcox, R. R. (2004c). Inferences based on a skipped correlation coefficient. *Journal of Applied Statistics, 31*, 131–144.

Wilcox, R. R. (2004d). Some results on extensions and modifications of the Theil-Sen regression estimator. *British Journal of Mathematical and Statistical Psychology, 57,* 265–280.

Wilcox, R. R. (2005a). Depth and a multivariate generalization of the Wilcoxon-Mann-Whitney test. *American Journal of Mathematical and Management Sciences, 25,* 343–364.

Wilcox, R. R. (2005b). An affine invariant rank-based method for comparing dependent groups. *British Journal of Mathematical and Statistical Psychology, 58,* 33–42.

Wilcox, R. R. (2006a). Inference about the components of a generalized additive model. *Journal of Modern Applied Statistical Methods, 5,* 309–316.

Wilcox, R. R. (2006b). Pairwise comparisons of dependent groups based on medians. *Computational Statistics & Data Analysis, 50,* 2933–2941.

Wilcox, R. R. (2006c). Comparing medians. *Computational Statistics & Data Analysis, 51,* 1934–1943.

Wilcox, R. R. (2006d). A note on inferences about the median of difference scores. *Educational and Psychological Measurement, 66,* 624–630.

Wilcox, R. R. (2006e). Comparing robust generalized variances and comments on efficiency. *Statistical Methodology, 3,* 211–223.

Wilcox, R. R. (2006f). Some results on comparing the quantiles of dependent groups. *Communications in Statistics–Simulation and Computation, 35,* 893–900.

Wilcox, R. R. (2007). An omnibus test when using a quantile regression estimator with multiple predictors. *Journal of Modern and Applied Statistical Methods, 6,* 361–366.

Wilcox, R. R. (2008a). Some small-sample properties of some recently proposed multivariate outlier detection techniques. *Journal of Statistical Computation and Simulation, 78,* 701–712.

Wilcox, R. R. (2008b). Quantile regression: A simplified approach to a lack-of-fit test. *Journal of Data Science, 6,* 547–556.

Wilcox, R. R. (2008c). Robust principal components: A generalized variance perspective. *Behavioral Research Methods, 40,* 102–108.

Wilcox, R. R. (2008d). Post-hoc analyses in multiple regression based on prediction error. *Journal of Applied Statistics, 35,* 9–17.

Wilcox, R. R. (2008e). A test of independence via quantiles that is sensitive to curvature. *Journal of Modern and Applied Statistics, 7*, 11–20.

Wilcox, R. R. (2009a). Robust ancova using a smoother with bootstrap bagging. *British Journal of Mathematical and Statistical Psychology, 62*, 427–437.

Wilcox, R. R. (2009b). Robust multivariate regression when there is heteroscedasticity. *Communications in Statistics–Simulation and Computation, 38*, 1–13.

Wilcox, R. R. (2009c). Comparing robust measures of association estimated via a smoother. *Communications in Statistics–Simulation and Computation, 38*, 1969–1979.

Wilcox, R. R. (2009d). Comparing Pearson correlations: Dealing with heteroscedasticity and non-normality. *Communications in Statistics–Simulation and Computation, 38*, 2220–2234.

Wilcox, R. R. (2010a). Comparing robust nonparametric regression lines via regression depth. *Journal of Statistical Computation and Simulation, 80*, 379–387.

Wilcox, R. R. (2010b). Measuring and detecting associations: Methods based on robust regression estimators or smoothers that allow curvature. *British Journal of Mathematical and Statistical Psychology, 63*, 379–393.

Wilcox, R. R. (2010c). A note on principal components via a robust generalized variance. Unpublished technical report, Department of Psychology, University of Southern California.

Wilcox, R. R. (2010d). Inferences about a probabilistic measure of effect size when dealing with more than two groups. Unpublished technical report, Department of Psychology, University of Southern California.

Wilcox, R. R. (2010e). Nonparametric regression when estimating the probability of success. Unpublished technical report, Department of Psychology, University of Southern California.

Wilcox, R. R. (2010f). Regression: Comparing predictors and groups of predictors based on robust measures of association. *Journal of Data Science, 8*, 429–441.

Wilcox, R. R. (2010g). Inferences about the population mean: Empirical likelihood versus bootstrap-t. *Journal of Modern and Applied Statistical Methods, 9*, 9–14.

Wilcox, R. R. (2011). Nested ANOVA design: Methods that are robust and allow heteroscedasticity. Unpublished technical report, Department of Psychology, University of Southern California.

Wilcox, R. R. (in press a). Comparing two dependent groups: Dealing with missing values. *Journal of Data Science*.

Wilcox, R. R. (in press b). Comparing the strength of association of two predictors via smoothers or robust regression estimators. *Journal of Modern and Applied Statistical Methods*.

Wilcox, R. R., Charlin, V. L., & Thompson, K. L. (1986). New Monte Carlo results on the robustness of the ANOVA F, W, and F^* statistics. *Communications in Statistics–Simulation and Computation, 15*, 933–944.

Wilcox, R. R., & Costa, K. (2009). Quantile regression: On inferences about the slopes corresponding to one, two or three quantiles. *Journal of Modern and Applied Statistical Methods, 8*, 368–375.

Wilcox, R. R., & Keselman, H. J. (2002). Within groups multiple comparisons based on robust measures of location. *Journal of Modern Applied Statistical Methods, 1*, 281–287.

Wilcox, R. R., & Keselman, H. J. (2006a). A skipped multivariate measure of location: One- and two-sample hypothesis testing. In S. Sawilowsky (Ed.), *Real data analysis* (pp. 125–138). Charlotte, NC: IAP.

Wilcox, R. R., & Keselman, H. J. (2006b). Detecting heteroscedasticity in a simple regression model via quantile regression slopes. *Journal of Statistical Computation and Simulation, 76*, 705–712.

Wilcox, R. R., & Muska, J. (2001). Inferences about correlations when there is heteroscedasticity. *British Journal of Mathematical and Statistical Psychology, 54*, 39–47.

Wilcox, R. R. & Tian, T. (2011). Measuring effect size: a robust heteroscedastic approach for two or more groups. *Journal of Applied Statistics, 38*, 1359–1368.

Wilcox, R. R., & Vigen, C. (2011). Comparing discrete distributions when the sample space is small. Unpublished technical report, Occupational Science and Occupational Therapy, University of Southern California.

Williams, V. S. L., Jones, L. V., & Tukey, J. W. (1999). Controlling error in multiple comparisons, with examples from state-to-state differences in educational achievement. *Journal of Educational and Behavioral Statistics, 24*, 42–69.

Woodruff, D. L., & Rocke, D. M. (1994). Computable robust estimation of multivariate location and shape in high dimension using compound estimators. *Journal of the American Statistical Association, 89*, 888–896.

Wu, C. F. J. (1986). Jackknife, bootstrap, and other resampling methods in regression analysis. *The Annals of Statistics, 14*, 1261–1295.

Wu, P.-C. (2002). *Central limit theorem and comparing means, trimmed means one-step M-estimators and modified one-step M–estimators under non-normality.* Unpublished doctoral dissertation, Department of Education, University of Southern California.

Yanagihara, H., & Yuan, K. H. (2005). Three approximate solutions to the multivariate Behrens-Fisher problem. *Communications in Statistics–Simulation and Computation, 34*, 975–988.

Yang, L., & Marron, J. S. (1999). Iterated transformation-kernel density estimation. *Journal of the American Statistical Association, 94*, 580–589.

Yohai, V. J. (1987). High breakdown point and high efficiency robust estimates for regression. *Annals of Statistics, 15*, 642–656.

Yohai, V. J., & Zamar, R. H. (1988). High breakdown point estimates of regression by means of the minimization of an efficient scale. *Journal of the American Statistical Association, 83*, 406–414.

Yohai, V. J., & Zamar, R. H. (2004). Robust non-parametric inference for the median. *Annals of Statistics, 32*, 1841–1857.

Yoshizawa, C. N., Sen, P. K., & Davis, C. E. (1985). Asymptotic equivalence of the Harrell-Davis median estimator and the sample median. *Communications in Statistics–Theory & Methods, 14*, 2129–2136.

Young, S. G., & Bowman, A. W. (1995). Nonparametric analysis of covariance. *Biometrics, 51*, 920–931.

Yuen, K. K. (1974). The two sample trimmed t for unequal population variances. *Biometrika, 61*, 165–170.

Zani, S., Riani, M., & Corbellini, A. (1998). Robust bivariate boxplots and multiple outlier detection. *Computational Statistics & Data Analysis, 28*, 257–270.

Zhao, J., & Wang, J. (2009). Robust testing procedures in heteroscedastic linear models. *Communications in Statistics – Simulation and Computation, 38*, 244–256.

Zhao, X., Lynch, J. G., Jr., & Chen, Q. (2010). Reconsidering Baron and Kenny: Myths and truths about mediation analysis. *Journal of Consumer Research, 37*, 197–206.

Zhou, W. (2008). Statistical inference for p(x<y). *Statistics in Medicine, 27*, 257–279.

Zimmerman, D. W. (2004). A note on preliminary tests of equality of variances. *British Journal of Mathematical and Statistical Psychology, 57*, 173–182.

Zou, C., Liu, Y., Wang, Z., & Zhang, R. (2010). Adaptive nonparametric comparison of regression curves. *Communications in Statistics – Theory and Methods, 39*, 1299–1320.

Zou, G. Y. (2007). Toward using confidence intervals to compare correlations. *Psychological Methods, 12*, 399–413.

Zu, J., & Yuan, K. H. (2010). Local influence and robust procedures for mediation analysis. *Multivariate Behavioral Research, 45*, 1–44.

Zuo, Y. (2003). Projection-based depth functions and associated medians. *Annals of Statistics, 31*, 1460–1490.

Zuo, Y. (2010). Is the t confidence interval $\bar{X} \pm t_\alpha (n-1)s/\sqrt{n}$ optimal? *American Statistician, 64*, 170–173.

Zuo, Y., Cui, H., & He, X. (2004). On the Stahel-Donoho estimator and depth-weighted means of multivariate data. *Annals of Statistics, 32*, 167–188.

Zuo, Y., Cui, H. & Young, D. (2004). Influence function and maximum bias of projection depth based estimators. *Annals of Statistics, 32*, 189–218.

Zuo, Y., & He, X. (2006). On the limiting distributions of multivariate depth-based rank sum statistics and related tests. *Annals of Statistics, 34*, 2879–2896.

Zuo, Y., Hengjian, C., & He, X. (2004). On the Stahel-Donoho estimator and depth-weighted means of multivariate data. *Annals of Statistics, 32*, 167–188.

Zuo, Y., Hengjian, C., & Young, D. (2004). Influence function and maximum bias of projection depth based estimators. *Annals of Statistics, 32*, 189–218.

Zuo, Y., & Serfling, R. (2000a). General notions of statistical depth functions. *Annals of Statistics, 28*, 461–482.

Zuo, Y., & Serfling, R. (2000b). Structural properties and convergence results for contours of sample statistical depth functions. *Annals of Statistics, 28*, 483–499.

Zuur, A. F., Ieno, E. N., & Meesters, E. (2009). *A Beginner's guide to R*. New York, NY: Springer.

Index